D0913990

Advanced Calculus

Advanced Calculus

Pure and Applied

Peter V. O'Neil
Department of Mathematics
College of William and Mary in Virginia

Macmillan Publishing Co., Inc.
NEW YORK

Collier Macmillan Publishers
LONDON

Printed in the United States of America

Macmillan Publishing Co., Inc.
866 Third Avenue, New York, New York 10022

Collier-Macmillan Canada, Ltd.

Library of Congress Cataloging in Publication Data

O'Neil, Peter V.
Advanced calculus, pure and applied.

Includes index.
1. Calculus. I. Title.
QA303.05 515 74-4646
ISBN 0-02-389320-6

Printing: 12345678 Year: 567890

This book is dedicated to my wife, Leslie, and our daughter, Alison Ann.

Preface

THIS BOOK was written partly with a view toward opposing what I regard as an unhealthy trend toward excessive abstraction in advanced calculus courses. By advanced calculus here I mean the two or three semesters of calculus following the standard first year. This course has a number of important functions. It must lay the foundations for abstract analysis at a more advanced level while computational skills are developed that will later be needed but taken for granted; it should contain enough applications from the real world, other disciplines, and mathematics itself to afford some perspective and appreciation of the scope and power of the methods being developed; and it should contain material both useful and interesting to students of varying backgrounds, not simply the small percentage who will go on to graduate school in mathematics.

These factors largely determined both the style of presentation and the choice of topics. Besides the usual material covered in courses at this level, I have included chapters on the calculus of variations, complex analysis, and Fourier series, integrals, and transforms. These provide a good opportunity to use many of the mathematical tools previously developed, and also greatly widen the scope of mathematically and physically interesting problems that can be treated. In addition, they provide some instructive contrasts and comparisons: for example, solution of the Dirichlet problem for the upper half-plane by conformal mapping, Fourier transform, and Fourier sine transform.

Topics are arranged for flexibility with regard to length of the course (two or three semesters) and makeup of the class (mathematics, science, or engineering concentrators, or a mixture of these). The last three chapters are largely independent, and may be included individually or omitted entirely, as the instructor wishes.

Concerning style of presentation, I feel that the theorem-proof method is suitable only for certain advanced or graduate courses, and I have attempted a conversational style here. This has resulted in statements of some definitions and theorems appearing in the body of a paragraph. Key words in such statements are italicized to make them more easily identifiable. In all cases, hypotheses of theorems are given in full. It is impossible (and, I think, undesirable) to rigorously prove every theorem in this type of course, but generally I included proofs that seemed to me likely to enhance the student's understanding of the theory being presented, or which contained interesting ideas or techniques, and omitted, or sketched briefly, proofs which seemed to me to be of such a technical nature as to contribute little to the student's mastery of the material at this level. A number of difficult theorems are proved, including the boundedness of a continuous function on a compact set, an implicit function theorem, and theorems on the pointwise convergence of Fourier series and integrals. Such topics as Banach and Hilbert space, measure, and L^p spaces are left to more advanced courses in analysis, and some topics (for example, the intermediate value theorem in R^n, and many of the proofs of theorems on integration in R^n) are left to sequences of exercises, with detailed hints for their solution, so that they may be easily included or omitted as the instructor chooses. Consistent with this philosophy, I have not attempted to place Fourier series (by which I mean here trigonometric series) in the wider and more elegant setting of general orthogonal expansions and mean-square convergence. However, careful attention is paid to the question of pointwise convergence of Fourier series, and uniform convergence of Fourier series is treated in the exercises.

I believe that problem solving plays a major role in any calculus course and so I have included problem sets following almost every section. These vary greatly in degree of difficulty and range from routine exercises in computation to exercises designed to augment or extend the theory of the section. I have avoided the temptation to star problems according to difficulty, as I think that this sometimes puts too heavy a psychological burden on the student (and, besides, I have never been able to consistently agree with myself as to how many stars to place before certain problems). Many of the more theoretical exercises come supplied with hints, some of which constitute detailed, step-by-step procedures for answering the exercise. Some exercises are worked out in full, or provided with additional hints, at the end of the book. I have not, however, provided answers to all, or even half, of the exercises, as I feel that answers without comment are of little value.

As a convenience for the reader, I have included an index of symbols, giving with each symbol an indication of its use throughout the book. There is also a chart of major topics and applications designed to indicate the flow and interplay of many of the concepts developed.

Finally, the only prerequisite for the course is the usual first year of calculus. I have included short appendixes treating two-by-two and three-by-three determinants, Cramer's rule for "small" systems, and general solutions to two differential equations which arise frequently in applications, in order to be able to make use of this material without requiring previous courses in linear algebra or differential equations.

PETER V. O'NEIL

Contents

Greek Letters Used

α	alpha
β	beta
Γ, γ	gamma
Δ, δ	delta
ε or ϵ	epsilon
ζ	zeta
η	eta
θ	theta
λ	lambda
μ	mu
ν	nu
ξ	xi
π	pi
ρ	rho
Σ, σ	sigma
τ	tau
Φ, φ	phi
ψ	psi

List of Symbols

$\{u_n\}_{n=1}^{\infty}$, $\{u_n\}$ — infinite sequence with nth term u_n.

$\lim_{n\to\infty} u_n$ — limit of a sequence $\{u_n\}$.

$\text{lub}\{u_n\}$ — least upper bound of a sequence $\{u_n\}$.

$\text{glb}\{u_n\}$ — greatest lower bound of a sequence $\{u_n\}$.

$\lim\inf\{u_n\}$ — limit infimum of $\{u_n\}$.

$\lim\sup\{u_n\}$ — limit supremum of $\{u_n\}$.

$\lim_{n\to\infty} f_n = g$ (pointwise on D) — g is the pointwise limit of $\{f_n\}$ on D.

$\lim_{n\to\infty} f_n = g$ (uniformly on D) — g is the uniform limit of $\{f_n\}$ on D.

$\sum_{n=1}^{\infty} a_n$ — infinite series with general term a_n.

S_n, or s_n — usually denotes the nth partial sum of an infinite series.

$\int_a^{\infty} f(x)\,dx$ — improper integral, defined as $\lim_{b\to\infty} \int_a^b f(x)\,dx$.

$\sum_{n=1}^{\infty} f_n = g$ (pointwise on D) — $\sum_{n=1}^{\infty} f_n$ converges pointwise to g on D.

$\sum_{n=1}^{\infty} f_n = g$ (uniformly on D) — $\sum_{n=1}^{\infty} f_n$ converges uniformly to g on D.

$\int_{-\infty}^{b} f(x)\,dx$ — improper integral, defined as $\lim_{a\to-\infty} \int_a^b f(x)\,dx$.

R^n — n-dimensional space, consisting of all n-tuples $(x_1, \ldots, x_n)$ with each x_i real.

$\|(a_1, \ldots, a_n)\|$ — magnitude or norm of $(a_1, \ldots, a_n)$, defined as $\left(\sum_{j=1}^{n} a_j^2\right)^{1/2}$

$(x_1, \ldots, x_n) \cdot (y_1, \ldots, y_n)$ — dot product of $(x_1, \ldots, x_n)$ with $(y_1, \ldots, y_n)$, defined by $\sum_{i=1}^{n} x_i y_i$.

$A \times B$ — cross product of three-vectors A and B.

$\{x \mid x \text{ satisfies property } P\}$ — set notation designating the set of objects x satisfying a given property P.

∂D — the set of boundary points of a set D.

mesh (P) — mesh of a partition or grid P.

$\underline{S}(P)$ — lower Darboux sum corresponding to a partition P.

$\overline{S}(P)$ — upper Darboux sum corresponding to a partition P.

$\underline{I}(f)$ — lower integral of f.

$\overline{I}(f)$ — upper integral of f.

$\int_a^b f$, or $\int_a^b f(x)\,dx$ — single Riemann integral of f over $[a, b]$.

$\mathscr{I}(D, G)$ — inner area of D relative to a grid G.

$\mathscr{O}(D, G)$ — outer area of D relative to a grid G.

$\underline{A}(D)$ — inner area of D.

$\overline{A}(D)$ — outer area of D.

$A(D)$ — area of D.

$\iint_D f$, or $\iint_D f(x, y)\,dx\,dy$ — double Riemann integral of f over D.

$\int_a^b \left(\int_{g_1(x)}^{g_2(x)} f(x, y)\,dy\right) dx$, $\int_a^b \int_{g_1(x)}^{g_2(x)} f(x, y)\,dy\,dx$, or $\int_a^b dx \int_{g_1(x)}^{g_2(x)} f(x, y)\,dy$ — two-fold iterated integrals.

$\underline{V}(D)$ — inner volume of D.

$\overline{V}(D)$ — outer volume of D.

$V(D)$ — volume of D.

$\iiint_D f$, $\iiint_D f(x, y, z)\,dx\,dy\,dz$ — triple Riemann integral of f over D.

$\int_a^b \int_{\alpha(x)}^{\beta(x)} \int_{\varphi(x,y)}^{(\psi x,y)} f(x, y, z)\,dz\,dy\,dx$, $\int_a^b \left(\int_{\alpha(x)}^{\beta(x)} \left(\int_{\varphi(x,y)}^{\psi(x,y)} f(x, y, z)\,dz\right) dy\right) dx$ or

$\int_a^b dx \int_{\alpha(x)}^{\beta(x)} dy \int_{\varphi(x,y)}^{\psi(x,y)} f(x, y, z)\,dz$ — three-fold iterated integrals.

$\frac{\partial f}{\partial x}(a, b), \left.\frac{\partial f}{\partial x}\right|_{(a,b)}, f_1(a, b), f_x(a, b)$ — equivalent notation for the partial derivative of f with respect to its first variable x, evaluated at (a, b).

$\frac{\partial f}{\partial y}(a, b), \left.\frac{\partial f}{\partial y}\right|_{(a,b)}, f_2(a, b), f_y(a, b)$ — equivalent notation for the partial derivative of f with respect to its second variable y, evaluated at (a, b).

$\frac{\partial^2 f}{\partial x_j\, \partial x_i}, \frac{\partial^3 f}{\partial x_j\, \partial x_i\, \partial x_k}$ — higher order partial derivatives with respect to the indicated variables.

$f_{ij}, f_{kij}, f_{xy}, f_{zxy}$ — subscript notation for higher order partials.

Δz or Δf — usually denotes function increment of z or f.

dz or df — usually denotes differential of z or f.

$(D_L f)(a_1, \ldots, a_n)$ — directional derivative of f at $(a_1, \ldots, a_n)$ in the direction specified by L.

$\nabla f(a_1, \ldots, a_n)$ — gradient of f at $(a_1, \ldots, a_n)$.

$\frac{\partial(F, G)}{\partial(x, y)}$ — denotes Jacobian of F and G with respect to variables x and y, given by

$$\begin{vmatrix} F_x & F_y \\ G_x & G_y \end{vmatrix}.$$

$\frac{\partial(F, G, H)}{\partial(x, y, z)}$ — denotes Jacobian of F, G, and H with respect to variables x, y, and z, given by:

$$\begin{vmatrix} F_x & F_y & F_z \\ G_x & G_y & G_z \\ H_x & H_y & H_z \end{vmatrix}.$$

λ — often denotes a Lagrange multiplier.

$C\colon [a, b] \to R^3$ — denotes a curve C in R^3 with parameter domain $[a, b]$.

$C = (\alpha, \beta, \gamma)$ — usually denotes a curve C with coordinate functions α, β, and γ.

$\int_C F$, or $\int_C f\,dx + g\,dy + h\,dz$ — line integral of vector-valued function $F = (f, g, h)$ over the curve C.

$s(t)$ — usually denotes arc-length function, over a given curve C.

curl F — curl of a vector F.

$\int_C p\,dx, \int_C p\,dy, \int_C p\,dz$ — line integrals over C of a scalar-valued function p.

$\int_C p\, ds$ — line integral over C of the scalar-valued function p, using arc length s as parameter.

$\oint_C F$ — line integral of F counterclockwise around closed curve C.

$\dfrac{dF}{d\eta}$ — normal derivative of F, given by $\nabla F \cdot \eta$.

$\nabla^2 F$ — Laplacian of F, given by $\dfrac{\partial^2 F}{\partial x^2} + \dfrac{\partial^2 F}{\partial y^2}$ in two dimensions, and by $\dfrac{\partial^2 F}{\partial x^2} + \dfrac{\partial^2 F}{\partial y^2} + \dfrac{\partial^2 F}{\partial z^2}$ in three dimensions.

$\Sigma: D \to R^3$ — denotes a surface Σ in R^3 with parameter domain D.

$\Sigma = (f, g, h)$ — usually denotes a surface Σ with coordinate functions f, g, and h.

$\eta(u, v)$ — usually denotes a normal vector to a surface, evaluated at the point (u, v) in the parameter domain.

$\iint_\Sigma F$ or $\iint_\Sigma f\, dy\, dz + g\, dz\, dx + h\, dy\, dz$ — surface integral of $F = (f, g, h)$ over Σ.

$\iint_\Sigma f\, dy\, dz$, $\iint_\Sigma f\, dz\, dx$, $\iint_\Sigma f\, dx\, dy$ — surface integrals of scalar-valued f over Σ.

$\iint_\Sigma f$ — surface integral of scalar-valued f over Σ.

$\partial\Sigma$ — boundary of Σ.

$V(\Sigma)$ — set of points bounded by a closed surface Σ, together with points on the graph of $\partial\Sigma$ itself.

T — often denotes kinetic energy.

V — often denotes potential energy.

L — often denotes a Lagrangian function, defined by $L = T - V$.

q_i — ith generalized coordinate of a mechanical system.

$\dot{q}_i$ — denotes time derivative of q_i, that is, dq_i/dt.

$a + ib$, or $a + bi$ — complex number.

$\operatorname{Re}(a + ib)$ — real part of $a + ib$, given by a.

$\operatorname{Im}(a + ib)$ — imaginary part of $a + ib$, given by b.

$\bar{z}$ — complex conjugate of z (if $z = a + ib$, then $\bar{z} = a - ib$).

$|z|$ — magnitude of z (if $z = a + ib$, then $|z| = \sqrt{a^2 + b^2}$).

$\bar{f}$ — conjugate function of f (given by $\bar{f}(z) = \overline{f(z)}$).

$\operatorname{Re} f$ — real part function formed from f ($(\operatorname{Re} f)(z) = \operatorname{Re}(f(z))$).

$\mathrm{Im}\, f$ — imaginary part function formed from f ($(\mathrm{Im}\, f)(z) = \mathrm{Im}(f(z))$).

$|f|$ — magnitude function formed from f ($|f|(z) = |f(z)|$).

$\arg(z)$ — any argument of z.

$\mathrm{Arg}(z)$ — principle argument of z, specified by $-\pi < \mathrm{Arg}(z) \leq \pi$.

$\log(z)$ — logarithm of z, denoting any number of the form $\ln(|z|) + i \arg(z)$.

$\mathrm{Log}(z)$ — principle logarithm of z, given by $\ln(|z|) + i\, \mathrm{Arg}(z)$.

$[z^\alpha]_P$ — principle value of z^α, given by $[z^\alpha]_P = \exp(\alpha\, \mathrm{Log}(z))$.

$\mathrm{Res}(f, a)$ — residue of f at a.

a_n, b_n — in the context of Fourier series, these denote Fourier coefficients of a given function.

$\hat{f}$ — Fourier transform of f.

$\hat{f}_C$ — Fourier cosine transform of f.

$\hat{f}_S$ — Fourier sine transform of f.

Guide to Applications

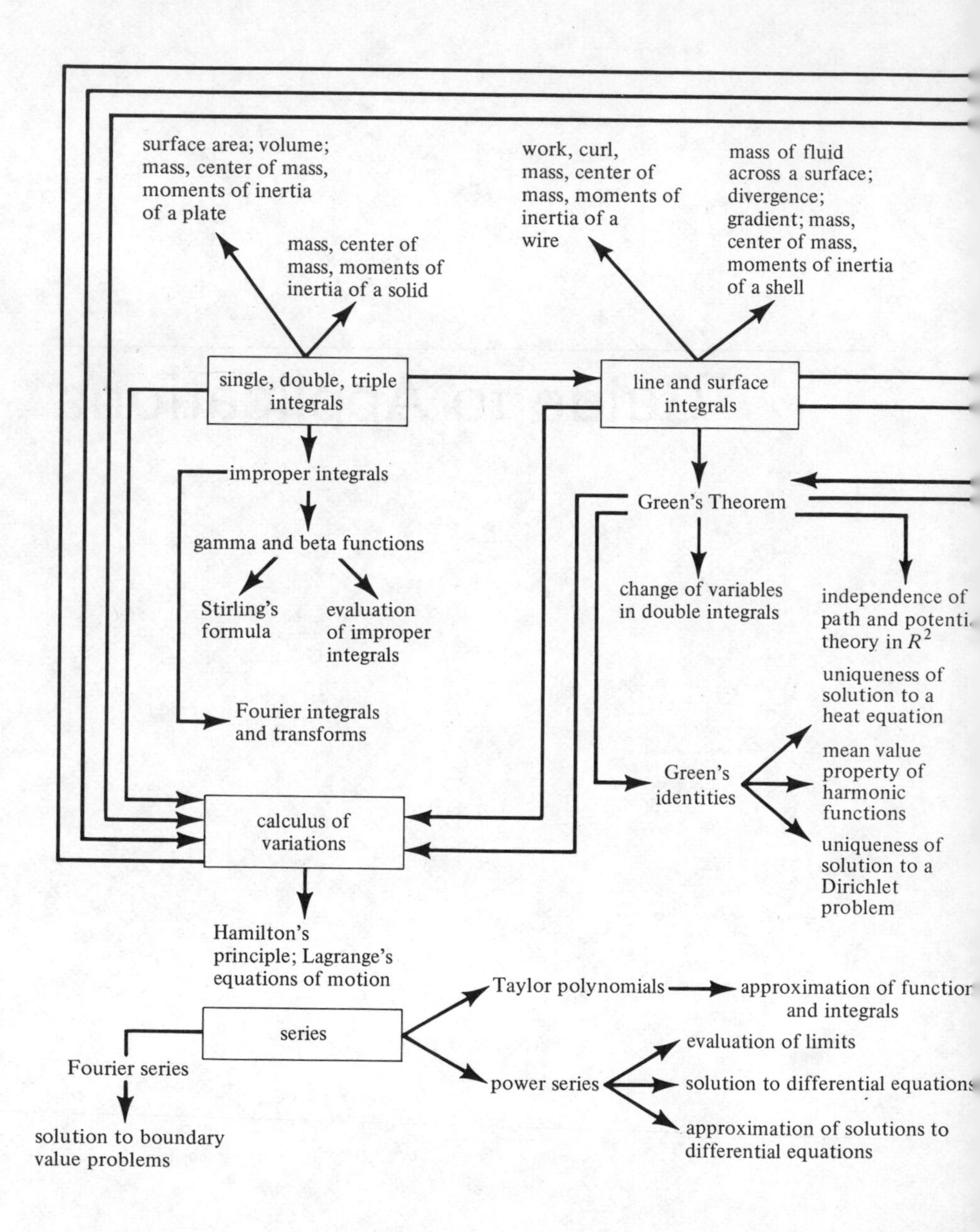
surface area; volume; mass, center of mass, moments of inertia of a plate
mass, center of mass, moments of inertia of a solid
work, curl, mass, center of mass, moments of inertia of a wire
mass of fluid across a surface; divergence; gradient; mass, center of mass, moments of inertia of a shell
single, double, triple integrals
line and surface integrals
improper integrals
gamma and beta functions
Stirling's formula
evaluation of improper integrals
Fourier integrals and transforms
Green's Theorem
change of variables in double integrals
independence of path and potenti
theory in R^2
uniqueness of solution to a heat equation
mean value property of harmonic functions
uniqueness of solution to a Dirichlet problem
Green's identities
calculus of variations
Hamilton's principle; Lagrange's equations of motion
series
Taylor polynomials
approximation of functior and integrals
Fourier series
solution to boundary value problems
power series
evaluation of limits
solution to differential equations
approximation of solutions to differential equations

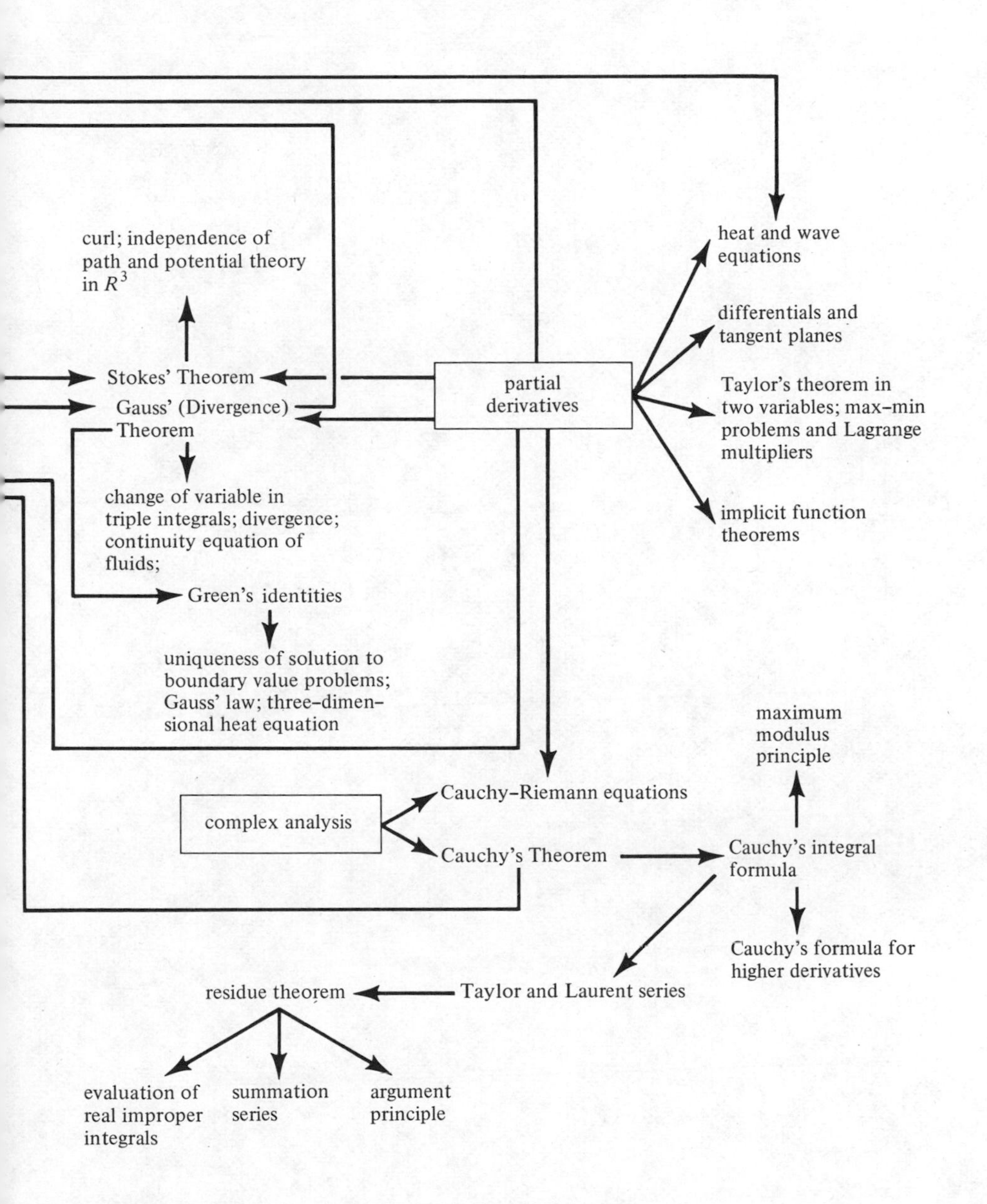
heat and wave equations
curl; independence of path and potential theory in R^3
differentials and tangent planes
Stokes' Theorem
partial derivatives
Taylor's theorem in two variables; max–min problems and Lagrange multipliers
Gauss' (Divergence) Theorem
implicit function theorems
change of variable in triple integrals; divergence; continuity equation of fluids;
Green's identities
uniqueness of solution to boundary value problems; Gauss' law; three-dimensional heat equation
maximum modulus principle
Cauchy–Riemann equations
complex analysis
Cauchy's Theorem
Cauchy's integral formula
Cauchy's formula for higher derivatives
residue theorem
Taylor and Laurent series
evaluation of real improper integrals
summation series
argument principle

Advanced Calculus

1

Sequences, Series, and Improper Integrals

1.1 Real Sequences

We begin by considering real sequences, that is, sequences $u_1, u_2, \ldots$, in which each u_i is a real number. Such a sequence may be denoted $\{u_n\}_{n=1}^{\infty}$, or, in some instances, just $\{u_n\}$. It is understood that in particular instances, different starting points or letters for the subscripts might be convenient, as with $\{u_n\}_{n=2}^{\infty}$, $\{b_r\}_{r=8}^{\infty}$, or $\{c_K\}_{K=-2}^{\infty}$. The number u_j is called the jth *term* of the sequence $\{u_n\}$.

The basic notion in the theory of sequences is that of convergence. We say that $\{u_n\}$ converges to the real number L, and write $\lim_{n \to \infty} u_n = L$ if all the terms u_n can be made arbitrarily close to L by choosing n sufficiently large. If we take the phrase "arbitrarily close" to mean within a specified tolerance ε, and the phrase "by choosing n sufficiently large" to mean that n is larger than a specified number N (which will in general depend upon ε), then we arrive at a suitable definition.

Definition

Let $\{u_n\}$ be a sequence and L a real number. Then $\lim_{n \to \infty} u_n = L$ if, given $\varepsilon > 0$, there is some N such that $|u_j - L| < \varepsilon$ for every positive integer j with $j \geq N$.

It is usually the case that N must be chosen larger the smaller one specifies ε. By way of illustration, we give some examples.

EXAMPLE

$$\lim_{n\to\infty} \frac{1}{n} = 0.$$

Of course, this is perfectly obvious intuitively. To prove the statement by using the definition, let $\varepsilon > 0$. We want N so that $|1/n - 0| < \varepsilon$ if $n \geq N$. In this case, simply observe that $1/n < \varepsilon$ when $n > 1/\varepsilon$. Thus choose $N = 1/\varepsilon$. If $n \geq N$, we have $1/n \leq 1/N = \varepsilon$. ∎

EXAMPLE

$$\lim_{n\to\infty} \frac{6n^2 + 2n + 1}{5n^2 + 3n} = \frac{6}{5}.$$

Again, the result is intuitively obvious. Simply rewrite

$$\frac{6n^2 + 2n + 1}{5n^2 + 3n} = \frac{6 + 2/n + 1/n^2}{5 + 3/n}$$

and note that the terms $2/n$, $1/n^2$, and $3/n$ all become negligible as n becomes larger.

To prove the limit assertion, let $\varepsilon > 0$. We seek N so that

$$\left| \frac{6n^2 + 2n + 1}{5n^2 + 3n} - \frac{6}{5} \right| < \varepsilon$$

whenever $n \geq N$.

To see how we should choose N, rewrite

$$\begin{aligned}\left| \frac{6n^2 + 2n + 1}{5n^2 + 3n} - \frac{6}{5} \right| &= \left| \frac{-8n + 5}{5(5n^2 + 3n)} \right| \\ &\leq \frac{8}{5}\frac{n}{5n^2 + 3n} + \frac{1}{5n^2 + 3n} = \frac{8}{5}\frac{1}{5n + 3} + \frac{1}{5n^2 + 3n} \\ &\leq \frac{2}{5n} + \frac{1}{5n} = \frac{3}{5n}.\end{aligned}$$

We therefore choose N so that $3/5n < \varepsilon$ if $n \geq N$. To do this it is good enough to put $3/5N = \varepsilon$, or $N = 3/5\varepsilon$. Then, if n is a positive integer and $n \geq N$, we have

$$\left| \frac{6n^2 + 2n + 1}{5n^2 + 3n} - \frac{6}{5} \right| < \frac{3}{5n} \leq \frac{3}{5N} = \varepsilon.$$

Note that here, as in the previous example, N becomes larger as ε is chosen smaller. ∎

Finding an appropriate choice of N was a little harder in this example than it was in the previous example, although some crude estimates along the way simplified things. An easier way to calculate many limits of this type is to observe that sequential limits behave nicely under the operations of addition, multiplication, division, and multiplication by a constant. That is, if $\lim_{n\to\infty} u_n = L$ and $\lim_{n\to\infty} v_n = M$, then

(1) $\lim_{n\to\infty} (u_n + v_n) = L + M$.

(2) $\lim_{n\to\infty} (au_n) = aL$ for any real number a.

(3) $\lim_{n\to\infty} (u_n \cdot v_n) = L \cdot M$.

And, if $M \neq 0$, and, at least for sufficiently large n, $v_n \neq 0$, then

(4) $\lim_{n\to\infty} (u_n/v_n) = L/M$.

The proofs of these are left to the exercises, with some hints as to procedure.

The last example can now be redone:

$$\lim_{n\to\infty} \frac{6n^2 + 2n + 1}{5n^2 + 3n} = \lim_{n\to\infty} \frac{6 + 2/n + 1/n^2}{5 + 3/n}$$
$$= \frac{\lim_{n\to\infty} 6 + \lim_{n\to\infty} (2/n) + \lim_{n\to\infty} (1/n^2)}{\lim_{n\to\infty} 5 + \lim_{n\to\infty} (3/n)} = \frac{6}{5}.$$

Of course, in practice one does not write the details as elaborately as this, the third step being an easy mental calculation.

It is important to realize at this stage the tremendous difference between proving a certain limit statement directly from the definition, and calculating a limit by some means or other. Generally the definition serves as a starting point and as a means of establishing theorems designed to help in calculating limits. Note, however, that one must guess a value for L before it is possible to show from the definition that $\lim_{n\to\infty} u_n = L$. This "guessing" part of the calculation usually requires some clever device or a more sophisticated theorem than we have yet developed.

As an example of a limit obtained by a trick, consider the following.

EXAMPLE

Let r be a given real number, and let $u_n = \sum_{j=0}^{n} r^j$ for $n = 1, 2, 3, \ldots$. What can we say about $\lim_{n\to\infty} u_n$? The trick here is to note that ru_n differs from u_n in only two terms. That is,

$$u_n = 1 + r + \cdots + r^n$$

and

$$ru_n = r + r^2 + \cdots + r^n + r^{n+1},$$

so

$$u_n - ru_n = 1 - r^{n+1}.$$

Then

$$u_n = \frac{1 - r^{n+1}}{1 - r},$$

assuming that $r \neq 1$. This gives us a closed-form expression for u_n, and we may replace, for $r \neq 1$, the problem of calculating $\lim_{n\to\infty} \sum_{j=0}^{n} r^j$ by that of calculating $\lim_{n\to\infty} \frac{1 - r^{n+1}}{1 - r}$, apparently a much easier limit. Consider some cases:

(1) If $-1 < r < 1$, then $r^{n+1} \to 0$ as $n \to \infty$, so

$$\lim_{n\to\infty} \frac{1 - r^{n+1}}{1 - r} = \frac{1}{1 - r}.$$

(2) If $r > 1$, then $r^{n+1} \to \infty$ as $n \to \infty$, so $(1 - r^{n+1})/(1 - r)$ approaches no finite limit.

(3) If $r < -1$, then r^{n+1} oscillates between positive and negative values, so $(1 - r^{n+1})/(1 - r)$ has no limit.

(4) If $r = -1$, then

$$\frac{1 - r^{n+1}}{1 - r} = \tfrac{1}{2}[1 - (-1)^{n+1}] = \begin{cases} 0 & \text{if } n \text{ is odd} \\ 1 & \text{if } n \text{ is even.} \end{cases}$$

Again, this oscillation precludes existence of a limit.

(5) If $r = 1$, then our closed-form expression for u_n is invalid. However, here $u_n = n + 1$, with no finite limit as $n \to \infty$.

In summary, then,

$$\lim_{n\to\infty} \sum_{j=0}^{n} r^j = \begin{cases} \dfrac{1}{1 - r} & \text{for } |r| < 1, \\ \text{is not finite or does not exist} & \text{for } r \leq -1 \text{ or } r \geq 1. \end{cases}$$ ▮

This example suggests the need for further terminology. The general situation that a sequence does not have a real number as a limit is described by saying that the sequence diverges. Divergence can happen in at least two ways. If the terms u_n become arbitrarily large as $n \to \infty$, we say that $\lim_{n\to\infty} u_n = \infty$. Similarly, we have a notion of $\lim_{n\to\infty} u_n = -\infty$. Finally, as with $u_n = (-1)^n$, the terms of the sequence may be bounded but not approach any one particular value. In this case the limit simply does not exist.

For the sake of completeness, we define infinite limits as follows.

Definition

(1) $\lim_{n\to\infty} u_n = \infty$ if, given $\delta > 0$, there is some N such that $u_n > \delta$ whenever n is a positive integer and $n \geq N$.

(2) $\lim_{n\to\infty} u_n = -\infty$ if, given $\delta > 0$, there is some N such that $u_n < -\delta$ whenever n is a positive integer and $n \geq N$.

We now consider the situation in which one is hard pressed to determine convergence in a straightforward way, or to guess a candidate for the limit by intuition or some other means. For example, let $E_n = (1 + 1/n)^n$ for $n = 1, 2, 3, \ldots$. In elementary calculus the base e for natural logarithms is usually defined as the limit of $\{E_n\}$, and the convergence question is deferred. Convergence of $\{E_n\}$ is far from obvious, however, and its limit, denoted e but numerically equal to 2.718281828459045235 36 . . ., is certainly not something one would be led to by guessing or some elementary device.

What we would like, in cases such as this, is to be able to establish convergence without having to produce a limit and go through an ε, N type of argument based directly on the definition. For $\{E_n\}$, the key is the monotonicity of the sequence.

In general, let $\{u_n\}$ be any real sequence, and suppose that either $u_{n+1} \leq u_n$ for all n, or $u_n \leq u_{n+1}$ for all n. In the first case we say that $\{u_n\}$ is *monotone nonincreasing*; in the second, *monotone nondecreasing*. Such sequences are called *monotone*.

To be specific for awhile, suppose that $\{u_n\}$ is a monotone nondecreasing sequence. Look at the numbers $u_1, u_2, \ldots$ on the real line (Fig. 1-1a). It may happen that the u_n's increase without bound. In this event, $\lim_{n\to\infty} u_n = \infty$. However, there may be a real number M such that $u_n \leq M$ for all n. Such a number is called an *upper bound* for $\{u_n\}$. In this case we look for the least upper bound,†

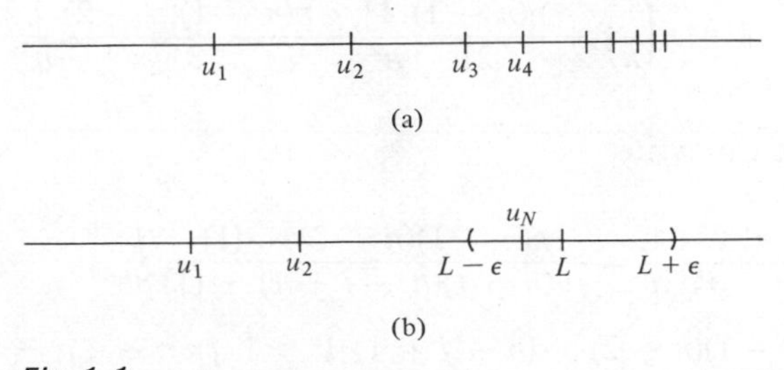

Fig. 1-1

† Existence of such a number is actually a fundamental property of the real number system, and is discussed in Appendix I of this chapter.

that is, an upper bound L with the property that every other upper bound is at least as large as L. Often L is denoted $\text{lub}\{u_n\}$. We now claim that $\lim_{n\to\infty} u_n = L$.

To prove this, choose any $\varepsilon > 0$, and look at the interval $(L - \varepsilon, L + \varepsilon)$, as in Fig. 1-1b. For some positive integer N, $L - \varepsilon < u_N$. If not, then we would have $u_n \leq L - \varepsilon$ for every n, and then $L - \varepsilon$ would be a smaller upper bound for $\{u_n\}$ than L, contradicting the choice of L. Having established the existence of such an N, we need only observe that $L - \varepsilon < u_N \leq u_n \leq L$ for each $n \geq N$. Put another way, $|u_n - L| < \varepsilon$ for $n \geq N$.

We state this result as a theorem.

Theorem

If $\{u_n\}$ is a monotone nondecreasing sequence that is bounded above, then $\{u_n\}$ converges, and $\lim_{n\to\infty}\{u_n\} = \text{lub}\{u_n\}$.

The importance of the theorem is not as a means of calculating a limit, as in general a least upper bound may be difficult to evaluate numerically, but of showing that a limit exists. Often it is fairly easy to show that a sequence is monotone nondecreasing and bounded above, even when the least upper bound is far from obvious.

EXAMPLE

We shall now prove that $(1 + 1/n)^n$ converges. As above, put $E_n = (1 + 1/n)^n$.

Step 1. $\{E_n\}$ is monotone nondecreasing. Here we use a binomial expansion. Recall that $(a + b)^n = \sum_{j=0}^{n}\binom{n}{j} a^{n-j}b^j$ for any real numbers a and b. Here $\binom{n}{j} = \frac{n!}{j!\,(n-j)!}$, $k!$ denotes the product of the integers from 1 to k if $k > 0$, and $0!$ is defined to be 1, as a matter of notational convenience.

Put $a = 1$ and $b = 1/n$, to obtain

$$\left(1 + \frac{1}{n}\right)^n = \sum_{j=0}^{n}\binom{n}{j}\left(\frac{1}{n}\right)^j$$

$$= 1 + n\left(\frac{1}{n}\right) + \frac{n(n-1)}{2!}\frac{1}{n^2} + \frac{n(n-1)(n-2)}{3!}\frac{1}{n^3} + \cdots + \left(\frac{1}{n}\right)^n.$$

In general, we can write

$$\binom{n}{j}\frac{1}{n^j} = \frac{n!}{j!\,(n-j)!}\frac{1}{n^j} = \frac{n(n-1)(n-2)\cdots(1)}{j!\,(n-j)(n-j+1)\cdots(1)}\frac{1}{n^j}$$

$$= \frac{n(n-1)(n-2)\cdots(n-j+1)}{j!}\frac{1}{n^j} = \frac{1}{j!}\left(\frac{n}{n}\frac{n-1}{n}\frac{n-2}{n}\cdots\frac{n-j+1}{n}\right)$$

$$= \frac{1}{j!}\left[\left(1 - \frac{1}{n}\right)\left(1 - \frac{2}{n}\right)\cdots 1 - \left(\frac{j-1}{n}\right)\right].$$

Thus

$$E_n = 2 + \frac{1}{2}\left(1 - \frac{1}{n}\right) + \frac{1}{3!}\left(1 - \frac{1}{n}\right)\left(1 - \frac{2}{n}\right) + \frac{1}{4!}\left(1 - \frac{1}{n}\right)\left(1 - \frac{2}{n}\right)\left(1 - \frac{3}{n}\right)$$
$$+ \cdots + \frac{1}{n!}\left(1 - \frac{1}{n}\right)\left(1 - \frac{2}{n}\right)\cdots\left(1 - \frac{n-1}{n}\right).$$

Similarly,

$$E_{n+1} = 2 + \frac{1}{2}\left(1 - \frac{1}{n+1}\right) + \frac{1}{3!}\left(1 - \frac{1}{n+1}\right)\left(1 - \frac{2}{n+1}\right)$$
$$+ \frac{1}{4!}\left(1 - \frac{1}{n+1}\right)\left(1 - \frac{2}{n+1}\right)\left(1 - \frac{3}{n+1}\right) + \cdots$$
$$+ \frac{1}{n!}\left(1 - \frac{1}{n+1}\right)\left(1 - \frac{2}{n+1}\right)\cdots\left(1 - \frac{n-1}{n+1}\right)$$
$$+ \frac{1}{(n+1)!}\left(1 - \frac{1}{n+1}\right)\cdots\left(1 - \frac{n}{n+1}\right).$$

Now, we note that

$$1 - \frac{1}{n} < 1 - \frac{1}{n+1},$$
$$1 - \frac{2}{n} < 1 - \frac{2}{n+1},$$
$$\vdots$$
$$1 - \frac{n-1}{n} < 1 - \frac{n-1}{n+1}.$$

Comparing successive terms in the expressions for E_n and E_{n+1}, we have

$$\frac{1}{2}\left(1 - \frac{1}{n}\right) < \frac{1}{2}\left(1 - \frac{1}{n+1}\right),$$
$$\frac{1}{3!}\left(1 - \frac{1}{n}\right)\left(1 - \frac{2}{n}\right) < \frac{1}{3!}\left(1 - \frac{1}{n+1}\right)\left(1 - \frac{2}{n+1}\right),$$
$$\vdots$$
$$\frac{1}{n!}\left(1 - \frac{1}{n}\right)\left(1 - \frac{2}{n}\right)\cdots\left(1 - \frac{n-1}{n}\right) < \frac{1}{n!}\left(1 - \frac{1}{n+1}\right)\left(1 - \frac{2}{n+1}\right)$$
$$\cdots\left(1 - \frac{n-1}{n+1}\right).$$

Further, E_{n+1} has the additional positive term

$$\frac{1}{(n+1)!}\left(1-\frac{1}{n+1}\right)\cdots\left(1-\frac{n}{n+1}\right).$$

Hence $E_n < E_{n+1}$.

Step 2. $\{E_n\}$ is bounded above. Here we make some estimates. Note that in general $1 - \dfrac{j}{n} < 1$ for $j = 1, 2, \ldots$. Replace each term of the form $1 - \dfrac{j}{n}$ in the binomial expansion of E_n by 1, to obtain $E_n < 2 + 1/2! + 1/3! + \cdots + 1/n!$. Further, for $j \geq 2$, $2^j \leq j!$, so $1/j! \leq 1/2^j$. Then

$$E_n < 2 + \frac{1}{2} + \frac{1}{2^2} + \cdots + \frac{1}{2^{n-1}} = 1 + \sum_{j=0}^{n-1}\left(\frac{1}{2}\right)^j.$$

The sequence $\{v_n\}$ defined by $v_n = \sum_{j=0}^{n-1} 1/2^j$ is monotone nondecreasing and converges to $1/(1-\frac{1}{2}) = 2$ by a previous example. Thus, for $n \geq 2$, $E_n \leq v_n < 3$. Since $E_1 = 2$, then $E_n < 3$ for all n. Hence $\{E_n\}$ converges.

In fact, since $\{E_n\}$ converges to its least upper bound, which is certainly larger than 2 and no greater than 3, we have the crude estimate $2 < e \leq 3$. ∎

For simplicity, we have restricted ourselves in all this discussion to monotone nondecreasing sequences. Clearly a dual situation holds for monotone nonincreasing sequences. If $\{u_n\}$ is monotone nonincreasing and $M \leq u_n$ for all n, then we call M a *lower bound* of $\{u_n\}$. If G is a lower bound, and $M \leq G$ for every lower bound M of $\{u_n\}$, then we call G a *greatest lower bound* of $\{u_n\}$, denoted $\text{glb}\{u_n\}$. The following is then easy to prove.

Theorem

If $\{u_n\}$ is monotone nonincreasing and bounded below, then $\{u_n\}$ converges and

$$\lim_{n\to\infty} u_n = \text{glb}\{u_n\}.$$

There are other theorems which are sometimes useful in showing that a sequence converges without having to know the limit. We say that $\{u_n\}$ satisfies the Cauchy condition, or is a Cauchy sequence, if the terms can be made arbitrarily close to one another by going out far enough in the sequence. More carefully, $\{u_n\}$ is a *Cauchy sequence* if, given $\varepsilon > 0$, there is some N such that $|u_n - u_m| < \varepsilon$ whenever $n, m \geq N$.

It is fairly obvious that every convergent sequence is Cauchy. For, if $u_n \to L$ as $n \to \infty$, then all the terms sufficiently far out are arbitrarily close to L, hence to each other. The converse is not so obvious, and constitutes the more important half of the theorem.

Theorem (Cauchy Convergence Criterion)

$\{u_n\}$ converges if and only if $\{u_n\}$ is a Cauchy sequence.

PROOF

First, the easy part. Suppose that $\{u_n\}$ converges to L. Let $\varepsilon > 0$. For some N, $|u_n - L| < \varepsilon/2$ if $n \geq N$. Then, for $n \geq N$ and $m \geq N$,

$$\begin{aligned} |u_n - u_m| &= |u_n - L - u_m + L| \\ &\leq |u_n - L| + |u_m - L| < \frac{\varepsilon}{2} + \frac{\varepsilon}{2} = \varepsilon. \end{aligned}$$

Hence $\{u_n\}$ is a Cauchy sequence.

Conversely, suppose that $\{u_n\}$ is a Cauchy sequence. We must produce a candidate for the limit. First, we dispose of a simple special case. Suppose that there are only finitely many distinct u_n's. That is, there are distinct numbers $\alpha_1, \ldots, \alpha_s$ such that, for any given n, u_n is either $\alpha_1, \alpha_2, \ldots,$ or α_s. If $s = 1$, then each $u_n = \alpha_1$, and $\lim_{n\to\infty} u_n = \alpha_1$. If $s > 1$, let ε be the smallest of the numbers $|\alpha_i - \alpha_j|$ for $i \neq j$. Since $\{u_n\}$ is a Cauchy sequence, and $\varepsilon > 0$, then, for some N, $|u_n - u_m| < \varepsilon$ for $m, n \geq N$. By choice of ε, this is impossible unless all u_n's are equal for $n \geq N$. Then, for some α_j, $u_n = \alpha_j$ for $n \geq N$, and $\{u_n\}$ converges to α_j.

Now consider the case that there are infinitely many distinct u_n's. To help clarify the argument, we proceed in steps.

Step 1. For some $A > 0$, $-A \leq u_n \leq A$ for every n. To show this, put $\varepsilon = 1$ in Cauchy's condition. For some N, which we may choose to be a positive integer, $|u_n - u_m| < 1$ for $n, m \geq N$. In particular, $|u_n - u_N| < 1$ for $n \geq N$. Then $1 - u_N \leq u_n \leq 1 + u_N$ for $n \geq N$. Now let α be the largest of the numbers $|u_1|, \ldots, |u_{N-1}|$. Then certainly $-1 - |u_N| - \alpha < u_n < 1 + |u_N| + \alpha$ for $n = 1, \ldots, N - 1$ and also for $n \geq N$, hence for all n. We may choose $A = 1 + |u_N| + \alpha$.

Step 2. There is a monotone nondecreasing, convergent sequence $\{a_n\}$ and a monotone nonincreasing, convergent sequence $\{b_n\}$ such that, for each n, $[a_n, b_n]$ contains infinitely many distinct u_j's. To show this, let $I_1 = [-A, A]$, with A as produced in step 1. At least one of the subintervals $[-A, 0]$, $[0, A]$ contains infinitely many distinct u_j's. Pick one such subinterval and call it I_2. Now bisect I_2 into two closed subintervals, and choose one of these containing infinitely many distinct u_j's. Call this subinterval I_3. Continuing in this way, produce a sequence of intervals $I_1, I_2, I_3, \ldots$, each a subinterval of the preceding one, and each containing infinitely many distinct u_j's. Denote $I_n = [a_n, b_n]$. Then $a_n \leq a_{n+1}$ and $b_{n+1} \leq b_n$. Further, $\{a_n\}$ is bounded above (by A), and $\{b_n\}$ is bounded below (by $-A$). Thus $\{a_n\}$ and $\{b_n\}$ both converge.

Step 3. $\lim_{n\to\infty} a_n = \lim_{n\to\infty} b_n$. Note that I_1 has length $2A$, I_2 has length A, I_3 has length $A/2$, I_4 has length $A/4$, and, in general I_n has length $A/2^{n-2}$. Then length of $I_n \to 0$ as $n \to \infty$. But the length of I_n is $b_n - a_n$, so

$$0 = \lim_{n\to\infty} (b_n - a_n) = \lim_{n\to\infty} b_n - \lim_{n\to\infty} a_n.$$

Step 4. For convenience, let L denote the common value of $\lim_{n\to\infty} a_n$ and $\lim_{n\to\infty} b_n$. We now claim that $\lim_{n\to\infty} u_n = L$.

To prove this, let $\varepsilon > 0$. We must produce N such that $|u_n - L| < \varepsilon$ if $n \geq N$.

First, choose K sufficiently large that $b_K - L < \varepsilon$ and $L - a_K < \varepsilon$. This can be done, as $a_n \leq L \leq b_n$ for all n and both $\{a_n\}$ and $\{b_n\}$ converge to L.

Let δ be the smaller of the two numbers $b_K - L$ and $L - a_K$, or, if one is zero, the nonzero one; then $0 < \delta < \varepsilon$. By the Cauchy condition, there is some N such that $|u_n - u_m| < \delta/2$ for $n, m \geq N$. Now, $[L - \delta/2, L + \delta/2]$ contains an interval I_M for some M. Since I_M contains infinitely many distinct u_j's, we can choose some u_s in I_M with $s \geq N$. Then, for any $n \geq N$,

$$|u_n - L| = |u_n - L + u_s - u_s| \leq |u_n - u_s| + |u_s - L|$$
$$< \frac{\delta}{2} + \frac{\delta}{2} = \delta < \varepsilon$$

(see Fig. 1-2). This completes the proof. ∎

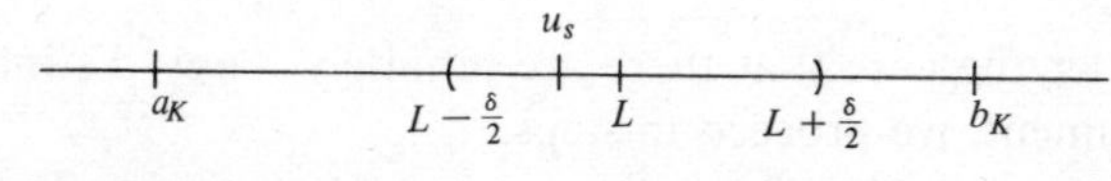

Fig. 1-2

We should remark that, in step 4, it is possible for $L - a_K$ or $b_K - L$ to be zero. If, say, we always chose an interval with left end point $-A$ in the bisection process of step 2, then each a_n would be $-A$, and we would have $L = -A$. But then we would have $b_n > L$ for each n, so $b_K - L$ would be positive. Similarly, we could have $b_n = A = L$ for each n, but then we would have $L - a_n > 0$ for each n.

We conclude this section with a simple example.

EXAMPLE

Let a and b be any two real numbers. Put $u_1 = a$, $u_2 = b$, $u_3 = \frac{1}{2}(u_1 + u_2)$, $u_4 = \frac{1}{2}(u_2 + u_3)$, and, in in general, $u_n = \frac{1}{2}(u_{n-1} + u_{n-2})$ for $n \geq 3$.

We claim that $\{u_n\}$ converges. We shall prove this by showing that $\{u_n\}$ is a Cauchy sequence.

First, put $d = |u_1 - u_2|$. Note that

$$|u_3 - u_2| = \left|\frac{1}{2}(u_1 + u_2) - u_2\right| = \frac{1}{2}|u_1 - u_2| = \frac{d}{2},$$

$$|u_4 - u_3| = \left|\frac{1}{2}(u_3 + u_2) - \frac{1}{2}(u_2 + u_1)\right| = \left|\frac{1}{2}(u_3 - u_1)\right|$$
$$= \left|\frac{1}{2}\left(\frac{1}{2}(u_2 + u_1) - u_1\right)\right| = \frac{1}{4}|u_2 - u_1| = \frac{d}{4},$$

and, in general,

$$|u_{n+1} - u_n| = \frac{d}{2^{n-1}}.$$

Next, for any positive integers n and p,

$$\begin{aligned}|u_{n+p} - u_n| &= \left|\sum_{j=1}^{p} (u_{n+j} - u_{n+j-1})\right| \le \sum_{j=1}^{p} |u_{n+j} - u_{n+j-1}| \\ &= \sum_{j=1}^{p} \frac{d}{2^{n+j-2}} = \frac{d}{2^{n-1}} \sum_{j=1}^{p} \frac{1}{2^{j-1}} = \frac{d}{2^{n-1}} \sum_{j=0}^{p-1} \frac{1}{2^j} = \frac{d}{2^{n-1}} \frac{1 - 1/2^p}{1 - 1/2}.\end{aligned}$$

The last step is by a previous example. Simplifying the last expression yields

$$|u_{n+p} - u_n| \le \frac{d}{2^{n-2}}\left(1 - \frac{1}{2^p}\right) < \frac{d}{2^{n-2}}.$$

Now, $d/2^{n-2} \to 0$ as $n \to \infty$. If $\varepsilon > 0$, we can find N such that $d/2^{n-2} < \varepsilon$ if $n \ge N$. Then, if $n \ge N$ and $m \ge N$, say $m = n + p$, we have

$$|u_m - u_n| = |u_{n+p} - u_n| < \frac{d}{2^{n-2}} < \varepsilon.$$

Thus $\{u_n\}$ is a Cauchy sequence, hence converges. ∎

The method used in the last example is fairly typical of applications of Cauchy's convergence criterion. From an estimate on the distance $|u_{n+1} - u_n|$ between successive terms, we obtained an estimate on $|u_{n+p} - u_n|$, which really is an estimate on $|u_m - u_n|$, as we can write m (or n) as $n + p$ (or $m + p$). Since the estimate we got depended only on n, and not on p, and went to zero as $n \to \infty$, we have only to choose N so that the estimate ($d/2^{n-2}$ in the above example) was as small as we wanted for $n \ge N$, in order to make $|u_n - u_m|$ as small as we wanted for n and m sufficiently large.

We conclude this section with a theorem that may at first sight appear somewhat uninteresting. However, we shall find it very useful in later work (e.g., look at the test for convergence of alternating series in Section 1.3).

Theorem

Let $\{a_n\}_{n=1}^{\infty}$ be a real sequence. Suppose that $\lim_{n\to\infty} a_{2n} = \lim_{n\to\infty} a_{2n-1} = L$. Then $\lim_{n\to\infty} a_n = L$.

PROOF

Let $\varepsilon > 0$. For some N_1, $|a_{2n} - L| < \varepsilon$ for $2n \ge N_1$. For some N_2, $|a_{2n-1} - L| < \varepsilon$ for $2n - 1 \ge N_2$. Choose N as the larger of the two numbers N_1 and N_2. If $k \ge N$ and k is even, then $k \ge N_1$ gives us $|a_k - L| < \varepsilon$. If $k \ge N$ and k is odd, then $k \geq N_2$ gives us $|a_k - L| < \varepsilon$. In any event, then, $|a_k - L| < \varepsilon$ whenever $k \geq N$, so $\lim_{n\to\infty} a_n = L$.

Exercises for Section 1.1

1. Let $\lim_{n\to\infty} u_n = L$ and $\lim_{n\to\infty} v_n = M$. Prove:

(a) $\lim(u_n + v_n) = L + M$. *Hint:* Let $\varepsilon > 0$. For some N_1, $|u_n - L| < \varepsilon/2$ if $n \geq N_1$. For some N_2, $|v_n - M| < \varepsilon/2$ if $n \geq N_2$. Now note that $|(u_n + v_n) - (L + M)| \leq |u_n - L| + |v_n - M|$.

(b) $\lim_{n\to\infty} (u_n \cdot v_n) = L \cdot M$. *Hint:* Write $|u_n \cdot v_n - LM| = |(u_n - L) \cdot v_n + L(v_n - M))|$.

(c) $\lim_{n\to\infty} (u_n/v_n) = L/M$, under the additional assumption that $v_n \neq 0$ for each n. *Hint:* Show that $\lim_{n\to\infty} 1/v_n = 1/M$, then use (b).

2. Let

$$v_n = \frac{a_0 + a_1 n + a_2 n^2 + \cdots + a_r n^r}{b_0 + b_1 n + b_2 n^2 + \cdots + b_s n^s},$$

where $a_0, \ldots, a_r, b_0, \ldots, b_s$ are real numbers, and the denominator is never zero. Discuss convergence or divergence of $\{v_n\}$. *Hint:* Consider separately the cases $r = s$, $r < s$, and $r > s$.

3. Determine convergence or divergence of $\{u_n\}$ in each of the following:

(a) $u_n = \sqrt{n}(\sqrt{n+1} - \sqrt{n})$ **(b)** $u_n = \left(1 + \dfrac{2}{n}\right)^n$

(c) $u_n = \dfrac{\cos(n)}{n}$ **(d)** $u_n = \dfrac{e^n}{n^2}$

(e) $u_n = \sin\left(\dfrac{n}{6}\right)$ **(f)** $u_n = \dfrac{\ln(n)}{n^b}$, b any positive number

(g) $u_n = \dfrac{[\ln(n)]^a}{n^b}$, where a and b are arbitrary positive numbers *Hint:* (g) follows from (f).

4. Let $\lim_{n\to\infty} u_n = 0$. Show that, for any positive number a, $\lim_{n\to\infty} u_n^a = 0$.

5. Let $a > 0$. Show that $\lim_{n\to\infty} a^{1/n} = 1$.

6. Show that $\lim_{n\to\infty} n^{1/n} = 1$.

7. Let $u_n \leq c_n \leq v_n$, and $\lim_{n\to\infty} u_n = \lim_{n\to\infty} v_n = L$. Show that $\lim_{n\to\infty} c_n = L$. If $u_n \leq c_n \leq v_n$, and $\lim_{n\to\infty} u_n = L$ and $\lim v_n = K \neq L$, does $\{c_n\}$ necessarily converge?

8. Suppose that $u_n \le v_n$ for all n, and both $\{u_n\}$ and $\{v_n\}$ converge. Is it true that $\lim_{n\to\infty} u_n \le \lim v_n$?

9. Suppose that $\lim_{n\to\infty} u_n = L$. For each positive integer n, let $v_n = (1/n) \sum_{j=1}^{n} u_j$. Show that $\lim_{n\to\infty} v_n = L$. ($\{v_n\}$ is called the *sequence of arithmetic means* of $\{u_n\}$.)

10. Show that $\{(1 + 1/n)^{n+1}\}_{n=1}^{\infty}$ converges. *Hint:* Show that the sequence is monotone nonincreasing.

11. Let a and b be any two real numbers. Put $u_1 = a$, $u_2 = b$, and, for $n \ge 3$, $u_n = \frac{1}{2}(u_{n-1} + u_{n-2})$. Show that $\lim_{n\to\infty} u_n = \frac{1}{3}(a + 2b)$.

12. The notion of continuity from elementary calculus can be characterized in terms of sequences. Suppose that $f(x)$ is defined in some interval $(a - r, a + r)$. Then f is continuous at a if $\lim_{x\to a} f(x) = f(a)$.

Prove the following: *f is continuous at a if and only if, given any sequence $\{u_n\}$, with $a - r < u_n < a + r$ for each n, and $\lim_{n\to\infty} u_n = a$, then $\lim_{n\to\infty} f(u_n) = f(a)$.* *Hint:* Sufficiency is easy—just use the definition of $\lim_{x\to a} f(x) = f(a)$. To prove necessity, suppose that $\lim_{n\to\infty} f(u_n) = f(a)$ whenever each u_n is in $(a - r, a + r)$ and $\lim_{n\to\infty} u_n = a$. We must show that f is continuous at a. Do this by contradiction. Show that if f is not continuous at a, then for some $\varepsilon > 0$, the following is true: *If n is a positive integer, there is some u_n with $a - r < u_n < a + r$ and $|u_n - a| < 1/n$ and $|f(u_n) - f(a)| \ge \varepsilon$.* Show that $\lim_{n\to\infty} u_n = a$, but $\lim_{n\to\infty} f(u_n) \ne f(a)$.

13. Let P be a positive number. Define a sequence $\{p_n\}$ by choosing p_1 as any number you like, except 0, and

$$p_n = \frac{1}{2}\left(p_{n-1} + \frac{P}{p_{n-1}}\right)$$

for $n \ge 2$. Show that $\lim_{n\to\infty} p_n = \sqrt{P}$. As illustrations, try estimating $\sqrt{2}$, $\sqrt{3}$, $\sqrt{4}$, and $\sqrt{5}$ by forming the first few terms of such sequences. *Hint:* Show first that $\{p_n\}$ is monotone nonincreasing by showing that $p_n - p_{n+1} > 0$. Next, show that $\sqrt{P}$ is a lower bound for $\{p_n\}$ by calculating $p_n^2 - P$. Hence conclude that $\{p_n\}$ converges, say to L. Show that $L = \frac{1}{2}(L + P/L)$, hence that $L = \sqrt{P}$.

14. Let $\{u_n\}$ be a sequence. A sequence $\{v_n\}$ is a subsequence of $\{u_n\}$ if there is an increasing function φ mapping positive integers to positive integers such that $v_n = u_{\varphi(n)}$ for $n = 1, 2, 3, \ldots$. For example, if $u_n = 1/n$, and $v_n = 1/2n$, then we can put $\varphi(n) = 2n$ to have $v_n = u_{\varphi(n)}$.

(a) Suppose that $\{u_n\}$ converges to L. Show that every subsequence of $\{u_n\}$ also converges to L.

(b) If $\{u_n\}$ diverges, what can be said about the subsequences of $\{u_n\}$?

15. A real number c is called a *cluster point* of $\{u_n\}$ if every open interval about c contains u_n for infinitely many values of n. For example, $\{\sin(n\pi/2)\}_{n=1}^{\infty}$ has 0, 1, and -1 as cluster points, while 1 and -1 are cluster points of $\{(-1)^n\}$.

(a) Prove that every bounded sequence has at least one cluster point. *Hint:* The idea is contained in steps 2 and 3 of the proof of the Cauchy convergence criterion.

(b) Prove that every bounded sequence has a convergent subsequence. *Hint:* Use (a).

(c) Prove that, if $\{u_n\}$ converges to a, then a is the cluster point of $\{u_n\}$.

16. Let $\{u_n\}$ be a bounded sequence (i.e., for some M, $-M \leq u_n \leq M$ for $n = 1, 2, 3, \ldots$).

(a) Prove that the set of cluster points of $\{u_n\}$ is bounded above and below.

(b) Let C be the set of cluster points of $\{u_n\}$. Prove that glb(C) and lub(C) are also cluster points of $\{u_n\}$.

17. Let $\{u_n\}$ be a bounded sequence, and define:

$$\liminf\{u_n\} = \text{smallest cluster point of } \{u_n\}$$

and $$\limsup\{u_n\} = \text{largest cluster point of } \{u_n\} \qquad \text{(see Exercise 16).}$$

For each of the following sequences, write down all the cluster points, and find lim sup and lim inf.

(a) $u_n = \dfrac{(-1)^n}{n}$ **(b)** $u_n = \sin\left(\dfrac{n\pi}{2}\right)$

(c) $u_n = 1 - (-1)^n$ **(d)** $u_n = \left(1 + \dfrac{1}{n}\right)\cos\left(\dfrac{n\pi}{2}\right)$

(e) $u_n = 200$ for $1 \leq n \leq 16$; $u_n = 1 - n$ for $17 \leq n \leq 206$; $u_n = \cos(\pi n)$ for $n \geq 207$.

18. Let $\{u_n\}$ be a bounded sequence. Prove that $\limsup\{u_n\}$ and $\liminf\{u_n\}$ are cluster points of $\{u_n\}$. *Hint:* This should be immediate from Exercises 16 and 17.

19. Establish the following alternative characterization of lim inf and lim sup for bounded sequences.

(a) $\liminf\{u_n\} = L$ if and only if, given $\varepsilon > 0$, then $u_n < L + \varepsilon$ for infinitely many values of n, and $u_n < L - \varepsilon$ for only finitely many values of n.

(b) $\limsup\{u_n\} = \bar{L}$ if and only if, given $\varepsilon > 0$, then $u_n > \bar{L} - \varepsilon$ for infinitely many values of n, and $u_n > \bar{L} + \varepsilon$ for only finitely many values of n.

20. Let $\{u_n\}$ be a bounded sequence. Prove that $\{u_n\}$ converges if and only if $\liminf\{u_n\} = \limsup\{u_n\}$. Also show that, in the case of convergence, $\limsup\{u_n\} = \lim_{n\to\infty} u_n$.

21. Let $\{u_n\}$ be a bounded sequence of positive numbers. Show that

$$\liminf\left\{\frac{u_{n+1}}{u_n}\right\} \le \liminf\{u_n^{1/n}\}$$
$$\leqslant \limsup\{u_n^{1/n}\} \le \limsup\left\{\frac{u_{n+1}}{u_n}\right\}.$$

22. Prove that existence of $\lim_{n\to\infty} u_{n+1}/u_n$ implies existence of $\lim_{n\to\infty} u_n^{1/n}$, and that, when both these limits exist, they are equal. *Hint:* Use Exercise 21. However, show by an example that existence of $\lim_{n\to\infty} u_n^{1/n}$ does not imply existence of $\lim_{n\to\infty} u_{n+1}/u_n$.

23. Prove that $\lim_{n\to\infty} n/(n!)^{1/n} = e$. *Hint:* Choose $u_n = n^n/n!$, compute the limit of u_{n+1}/u_n, and use Exercise 22.

24. Let $\{u_n\}$ be a sequence. Suppose that $\{u_{2n}\}$ converges to L and $\{u_{3n}\}$ converges to L. Then does $\lim_{n\to\infty} u_n = L$?

25. Let $u_n = \left(\sum_{j=1}^{n} 1/j\right) - \ln(n)$ for $n = 1, 2, 3, \ldots$. Show that $\{u_n\}$ converges. (The limit is called *Euler's constant* and is often denoted by γ. Numerically, $\gamma = 0.57721566\ldots$.)

26. This exercise offers an alternative proof that $\left\{\left(1 + \frac{1}{n}\right)^n\right\}$ converges. Show that the sequence is bounded above as in the text. Now argue as follows to show that the sequence is monotone nondecreasing.

(a) Show that it is sufficient to prove that

$$\left[\frac{(n+1)^2}{n(n+2)}\right]^n \le \frac{n+2}{n+1}.$$

(b) Show that $\binom{n}{j} \le n^j$ for $0 \le j \le n$ (by induction or by some other means).

(c) Write $(n+1)^2/n(n+2) = 1 + 1/a_n$, with $a_n = 1/n(n+2)$, and use the binomial formula to show that $(1 + 1/a_n)^n \le \sum_{j=0}^{n} (ja_j)^j$.

(d) Use (b) and (c) to conclude that

$$\left(1 + \frac{1}{a_n}\right)^n \le \sum_{n=0}^{\infty} (na_n)^n = \frac{n+2}{n+1},$$

using results from the text on geometric series.

27. Define a sequence $\{u_n\}$ recursively by $u_1 = u_2 = 1$, and $u_n = u_{n-1} + u_{n-2}$ for $n \ge 3$. Show that $u_n = (a^n - b^n)/\sqrt{5}$, where $a = (1 + \sqrt{5})/2$ and $b = (1 - \sqrt{5})/2$. (The sequence $\{u_n\}$ is called the *Fibonacci sequence*.) Show also that

$$\lim_{n\to\infty} \frac{u_n}{u_{n+1}} = \frac{\sqrt{5} - 1}{2}.$$

(This number is called the *golden mean*, thought by the ancient Greeks to be the ratio of the sides of the rectangle of most pleasing proportions.)

1.2 Sequences of Functions

In many instances we are concerned with sequences $\{f_n(x)\}_{n=1}^{\infty}$ whose terms are functions rather than real numbers. For example, let $f_n(x) = x^n$, for $n = 1, 2, \ldots$ and $0 \leq x \leq 1$. For any x in the permitted domain ([0, 1] in the example) we can ask for $\lim_{n\to\infty} f_n(x)$, if it exists. Further, we would naturally expect that this limit will itself be a function of x, in the sense that $\lim_{n\to\infty} f_n(x)$ may be different from $\lim_{n\to\infty} f_n(y)$ if we chose x different from y. In the example cited above, it is easy to see that

$$\lim_{n\to\infty} f_n(x) = \begin{cases} 0, & 0 \leq x < 1, \\ 1, & x = 1. \end{cases}$$

Thus the limit function has the graph shown in Fig. 1-3.

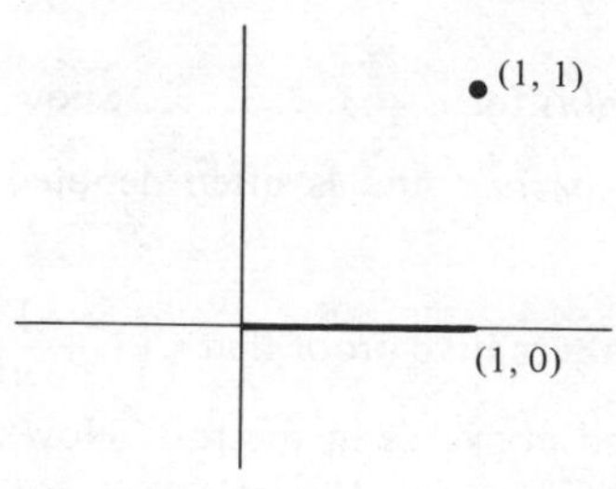

Fig. 1-3

In general, suppose that $f_n(x)$ is defined for each positive integer n, and each real number x in some specified set D. In many instances D will be an interval, but this is not necessary for first considerations. If $g(x)$ is defined for each x in D, we say that f_n *converges pointwise* to g on D if $\lim_{n\to\infty} f_n(x) = g(x)$ for each x in D. In this event we write $\lim_{n\to\infty} f_n = g$ (pointwise on D). In terms of the definition of sequential convergence, this may be rephrased as

$$\lim_{n\to\infty} f_n = g \text{ (pointwise on } D)$$

if given x in D and $\varepsilon > 0$, there exists N such that $|f_n(x) - g(x)| < \varepsilon$ for each positive integer $n \geq N$.

We have already seen one example of pointwise convergence, with $f_n(x) = x^n$, $D = [0, 1]$ and $g(x)$ as defined above. Here are some other examples.

EXAMPLE

Let $f_n(x) = \dfrac{1}{1 + nx^2}$, $-\infty < x < \infty$, $n = 1, 2, 3, \ldots$. Then $\lim_{n\to\infty} f_n(0) = 1$, but, if $x \neq 0$, then $\lim_{n\to\infty} f_n(x) = 0$. Then $\lim_{n\to\infty} f_n = g$ [pointwise on $(-\infty, \infty)$],

where

$$g(x) = \begin{cases} 0, & x \neq 0, \\ 1, & x = 0. \end{cases}$$

∎

EXAMPLE

Let $f_n(x) = \sum_{j=0}^{n} x^j$, $-1 < x < 1$. From the previous section we know that $\lim_{n\to\infty} f_n(x) = 1/(1 - x)$ for $-1 < x < 1$. Thus, $\lim_{n\to\infty} f_n = g$ [pointwise on $(-1, 1)$], where $g(x) = 1/(1 - x)$ for $-1 < x < 1$. ∎

EXAMPLE

Let $f_n(x) = [\sin(x)]^n$, $0 \le x \le 2\pi$. Then

$$\lim_{n\to\infty} f_n(x) = \begin{cases} 0, & 0 \le x < \frac{\pi}{2}, \frac{\pi}{2} < x < \frac{3\pi}{2}, \frac{3\pi}{2} < x < 2\pi, \\ 1, & x = \frac{\pi}{2}, \\ \text{does not exist}, & x = \frac{3\pi}{2}. \end{cases}$$

Here $\{f_n\}$ converges pointwise on $0 \le x < 3\pi/2$ and $3\pi/2 < x \le 2\pi$, but not on $[0, 2\pi]$. ∎

It is important to realize that, in the ε, N formulation of pointwise convergence, an N appropriate for a given ε will in general depend not only on ε, but on the point x chosen. To see this, look again at the example $f_n(x) = x^n$, $0 \le x \le 1$. At, say, $x = 0$, $f_n(x) = 0$ for all n, so $|f_n(x) - g(x)| = 0 < \varepsilon$ for all n and any $\varepsilon > 0$. Here we may take $N = 1$. Similarly, at $x = 1$, $|f_n(x) - g(x)| = 0 < \varepsilon$, for all n, and again $N = 1$ will work. However, at $x = \frac{1}{2}$, we have $|f_n(x) - g(x)| = 1/2^n$, and we may have to choose n very large to make $1/2^n < \varepsilon$ if ε is small. For example, if $\varepsilon = 1/2^{1,000,000}$, then a good choice for N would be 1,000,001. The point is that we are really looking at sequences $\{f_n(x)\}$, for each x, and convergence at each x is a whole new game. The sequences $\{f_n(x)\}$ and $\{f_n(y)\}$ for $x \neq y$ are handled completely independently.

Except in relatively simple cases, it is usually difficult to determine explicitly the pointwise limit g of a sequence $\{f_n\}$, even though we may know through some convergence test that g exists. As an example, take $f_n(x) = \sum_{j=1}^{n} jx^{j-1}$. We shall see later that $\{f_n(x)\}$ converges to $-1/(1 - x^2)$ for $-1 < x < 1$. At this point, however, this is hardly obvious. But even in those instances where $g(x)$ is not known explicitly, it may still be desirable to know certain things about g or to perform certain basic calculus operations on g. In particular, three natural questions arise:

(1) Is g continuous at x?
(2) How can we compute an integral of g?
(3) Does g have a derivative at x, and, if so, how can we calculate $g'(x)$?

It is not hard to see that all three questions involve an interchange of different kinds of limits or operations.

(1) *Continuity.* For g to be continuous at x, we need $\lim_{x\to a} g(x) = g(a)$. Now $g(x) = \lim_{n\to\infty} f_n(x)$, $g(a) = \lim_{n\to\infty} f_n(a)$. Further, if each f_n is continuous at a, then $f_n(a) = \lim_{x\to a} f_n(x)$. Continuity of g at a then reduces to the question: Does

$$\lim_{x\to a}\left[\lim_{n\to\infty} f_n(x)\right] = \lim_{n\to\infty}\left[\lim_{x\to a} f_n(x)\right]?$$

That is, can we interchange the order of $\lim_{x\to a}$ and $\lim_{n\to\infty}$?

(2) *Integration.* In the absence of an explicit knowledge of $g(x)$, our main hope of computing $\int_a^b g(x)\,dx$ is to compute $\int_a^b f_n(x)\,dx$ for each n, and ask whether $\lim_{n\to\infty}\int_a^b f_n(x)\,dx = \int_a^b g(x)\,dx$. That is: Does

$$\lim_{n\to\infty}\int_a^b f_n(x)\,dx = \int_a^b \left[\lim_{n\to\infty} f_n(x)\right] dx?$$

Here we are trying to interchange $\lim_{n\to\infty}$ and $\int_a^b$.

(3) *Differentiation.* As with integration, we must usually try to compute $g'(x)$ as $\lim_{n\to\infty} f_n'(x)$. The question here is: Does

$$\frac{d}{dx}\left[\lim_{n\to\infty} f_n(x)\right] = \lim_{n\to\infty}\left[\frac{d}{dx}(f_n(x))\right]?$$

The answer is yes only if d/dx and $\lim_{n\to\infty}$ can be taken in either order without changing the result.

Unfortunately, pointwise convergence is not sufficient to ensure positive answers to any of these questions, as the following examples show.

EXAMPLE

In the example $f_n(x) = x^n$, $0 \le x \le 1$, we have continuity of each f_n, but the pointwise limit of g is discontinuous at 1. ▮

EXAMPLE

Let $f_n(x) = nxe^{-nx^2}$, $0 \le x \le 1$. Here $\lim_{n\to\infty} f_n(x) = g(x) = 0$ for each x in $[0, 1]$. Now

$$\int_0^1 f_n(x)\,dx = -\tfrac{1}{2}e^{-nx^2}\Big|_0^1 = \tfrac{1}{2}(1 - e^{-n}).$$

Then

$$\lim_{n\to\infty}\int_0^1 f_n(x)\,dx = \lim_{n\to\infty}\tfrac{1}{2}(1 - e^{-n}) = \tfrac{1}{2}.$$

But $\int_0^1 \left[\lim_{n\to\infty} f_n(x)\right] dx = \int_0^1 g(x)\,dx = 0$. In this case, then,

$$\lim_{n\to\infty}\int_0^1 f_n(x)\,dx \neq \int_0^1 \left[\lim_{n\to\infty} f_n(x)\right] dx.$$

In addition, note that $f_n'(x) = ne^{-nx^2} - 2n^2x^2e^{-nx^2}$, for $0 \le x \le 1$. This sequence diverges at 0, hence certainly

$$\lim_{n\to\infty} f_n'(0) \neq \left[\lim_{n\to\infty} f_n(x)\right]'(0).$$ ∎

The natural step now is to look for conditions which will guarantee that things will work out better than in these last examples. It turns out that what is needed in general is a stronger kind of convergence called *uniform convergence*. Before becoming immersed in the details, we shall consider an example that contrasts pointwise convergence with what we shall later define to be uniform convergence.

EXAMPLE

Let $f_n(x) = nxe^{-nx}$. We shall consider convergence of this sequence for two different sets of values of x: $0 \le x < \infty$ and $\tau \le x < \infty$, where τ is an arbitrary positive number (imagine τ small to make things interesting).

In the first case, then, we are considering $f_n(x) = nxe^{-nx}$, $0 \le x < \infty$. Some of the f_n's are graphed in Fig. 1-4a. Note that f_n has its maximum at $1/n$, and that this maximum value is $f_n(1/n) = 1/e$. The important thing to observe is that, as $n \to \infty$, $1/n \to 0$, so f_n assumes its maximum closer to 0, then dips down more sharply toward the positive x-axis, which it approaches asymptotically as $x \to \infty$. In view of this behavior, examine the statement $\lim_{n\to\infty} f_n(x) = 0$, $0 \le x < \infty$, from a geometric point of view.

To say that $\lim_{n\to\infty} f_n(x) = 0$ means that, given $\varepsilon > 0$, there is some N such that $|f_n(x)| < \varepsilon$ if $n \ge N$. That is, $-\varepsilon < f_n(x) < \varepsilon$ if $n \ge N$. This means that,

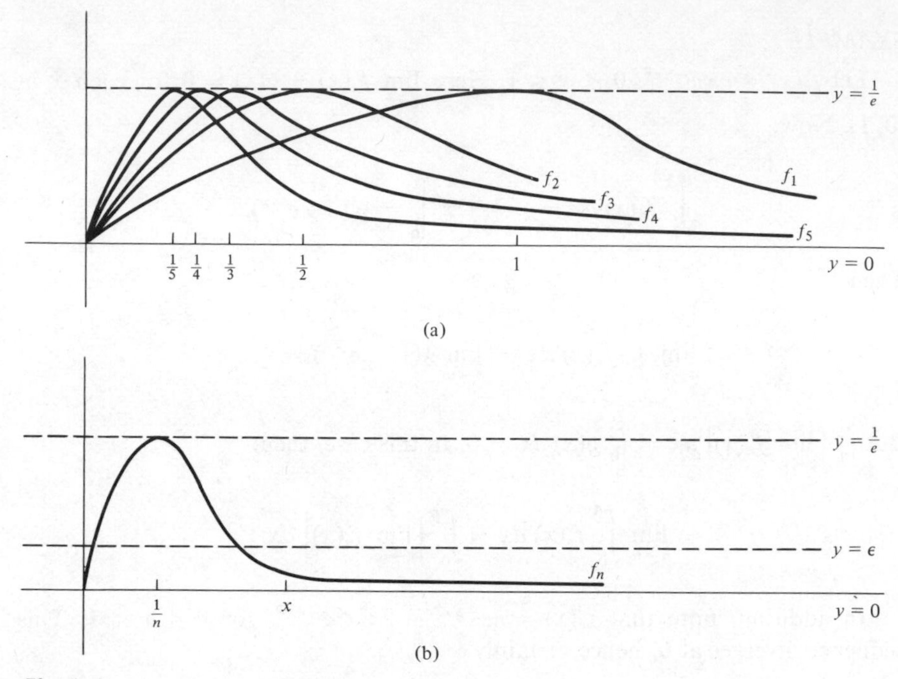

Fig. 1-4

for $n \geq N$, the graph of f_n at x is within the strip of width 2ε about the horizontal axis (Fig. 1-4b). If x is very large, then from the graphs it is apparent that $f_n(x)$ is within this strip for relatively small values of n. But if x is closer to 0, n must be taken very large to give f_n time to assume its maximum $1/e$ (outside the strip if $\varepsilon < 1/e$) at $1/n$ and then curve back down inside the strip. This suggests that we must choose N larger for x closer to 0 than for x farther out the axis, and that N depends upon x as well as upon ε.

Contrast this with the convergence on an interval $\tau \leq x < \infty$, where τ is positive but otherwise arbitrary. Of course, the limit is still the same: $\lim_{n \to \infty} f_n(x) = 0$, $\tau \leq x < \infty$. However, given $\varepsilon > 0$, we can now choose an N independent of x.

Geometrically, this is very plausible. If we choose N large enough so that $1/N$ is sufficiently close to 0 (how close we need depends on how small ε is), then for $n \geq N$, f_n will have time to take on its maximum at $1/n$ and drop back down inside the strip by the time x reaches τ.

Analytically, we argue as follows. Choose N sufficiently large that $N\tau e^{-N\tau} < \varepsilon$ and $1/N < \tau$. Then $f_N(x) = Nxe^{-Nx} < \varepsilon$ for all $x \geq \tau$, as $f_N(x)$ is a decreasing function for $x > 1/N$, and we chose N so that also $1/N < \tau$. If $n \geq N$, then $1/n \leq 1/N < \tau$, and $f_n(x) \leq f_N(x) < \varepsilon$ for every $x \geq \tau$. We may therefore choose N subject only to the conditions $N\tau e^{-N\tau} < \varepsilon$ and $1/N < \tau$. The choice, then, depends upon ε (and, of course, the number τ, which specifies our interval), but not upon x. Such an N will work for every x in $[\tau, \infty)$, and on this interval we call the convergence of $\{f_n\}$ *uniform*. ▮

We now give a definition of uniform convergence.

Definition

Let $f_n(x)$ be defined for $n = 1, 2, \ldots$ and x in some specified set D of real numbers. Let $g(x)$ be defined for each x in D. Then,

$$\lim_{n\to\infty} f_n = g \text{ (uniformly on } D)$$

if given $\varepsilon > 0$, there exists N such that $|f_n(x) - g(x)| < \varepsilon$ for each $n \geq N$ and x in D.

Compare this with the ε, N statement given before for $\lim_{n\to\infty} f_n = g$ (pointwise on D). There, we started with any positive ε, *and a point x in D*, and required that for some N, $|f_n(x) - g(x)| < \varepsilon$ if $n \geq N$. If a different point is chosen from D, probably a different N must be found, even for the same ε. When the convergence is uniform on D, we choose $\varepsilon > 0$, and then N can be found so that $|f_n(x) - g(x)| < \varepsilon$ whenever $n \geq N$, *for every x in D*.

The idea of uniform convergence is shown in Fig. 1.5. Specification of $\varepsilon > 0$ determines a strip of width 2ε about the graph of $y = g(x)$. For $\{f_n\}$ to converge to g uniformly on D, it is necessary that all the graphs $y = f_n(x)$, x in D, lie within the strip when n is at least as large as some number N. For pointwise convergence, the behavior of the entire graph does not concern us, just the sequence of numbers $\{f_n(x)\}_{n=1}^{\infty}$ for each x in D.

It should be obvious that uniform convergence implies pointwise convergence. That a sequence may converge pointwise but not uniformly is shown by the example $f_n(x) = nxe^{-nx}$, $0 \leq x < \infty$, with $g(x) = 0$.

We now return to the basic reason for considering uniform convergence.

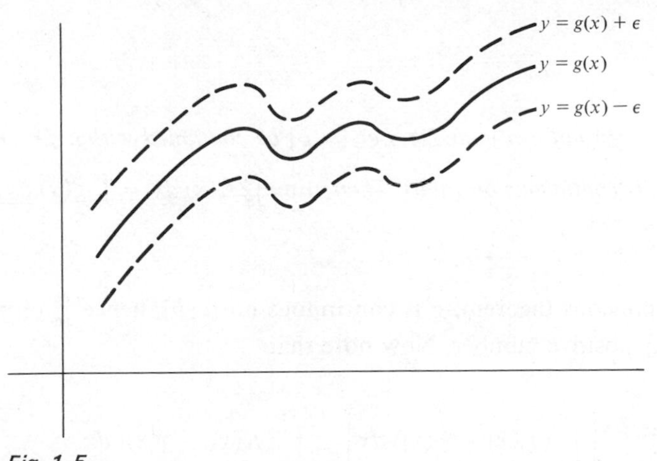

Fig. 1-5

We begin by showing that continuity of each of the f_n's at a point α in D is sufficient to guarantee continuity of the limit function g at α if the convergence is uniform.

Theorem

Let $\lim_{n \to \infty} f_n = g$ *(uniformly on D). Suppose that* α *is in D and each* f_n *is continuous at* α*. Then g is continuous at* α*.*

PROOF

Let $\varepsilon > 0$. We seek $\delta > 0$ such that, if $|x - \alpha| < \delta$ and $g(x)$ is defined, then $|g(x) - g(\alpha)| < \varepsilon$. First, note that, for any positive integer n, we can write

$$\begin{aligned} |g(x) - g(\alpha)| &= |g(x) - f_n(x) + f_n(x) - f_n(\alpha) + f_n(\alpha) - g(\alpha)| \\ &\leq |g(x) - f_n(x)| + |f_n(x) - f_n(\alpha)| + |f_n(\alpha) - g(\alpha)|. \end{aligned}$$

Since $\{f_n\}$ converges to g uniformly, we can choose N so that $|g(t) - f_n(t)| < \varepsilon/3$ for every t in D and $n \geq N$. In particular, choose a positive integer $K \geq N$ and we have $|f_K(\alpha) - g(\alpha)| < \varepsilon/3$. Now, f_K is continuous at α, so for some $\delta > 0$, $|f_K(x) - f_K(\alpha)| < \varepsilon/3$ whenever x is in D and $|x - \alpha| < \delta$. But also $|f_K(x) - g(x)| < \varepsilon/3$ for x in D, as $K \geq N$. Then

$$\begin{aligned} |g(x) - g(\alpha)| &\leq |g(x) - f_K(x)| + |f_K(x) - f_K(\alpha)| + |f_K(\alpha) - g(\alpha)| \\ &< \frac{\varepsilon}{3} + \frac{\varepsilon}{3} + \frac{\varepsilon}{3} = \varepsilon, \end{aligned}$$

whenever x is in D and $|x - \alpha| < \delta$. ∎

Next, we show that $\lim_{n \to \infty}$ and $\int_a^b$ can be interchanged when the convergence is uniform.

Theorem

Let $\lim_{n \to \infty} f_n = g$ *(uniformly on D). Let* $[a, b]$ *be contained within D, and suppose that each* f_n *is continuous on* $[a, b]$*. Then,* $\lim_{n \to \infty} \int_a^b f_n(x)\,dx = \int_a^b g(x)\,dx$*.*

PROOF

By the previous theorem, g is continuous on $[a, b]$, hence $\int_a^b g(x)\,dx$ exists. Let ε be any positive number. Now note that

$$\left| \int_a^b (f_n(x) - g(x))\,dx \right| \leq \int_a^b |f_n(x) - g(x)|\,dx.$$

By the uniform convergence of $\{f_n\}$, produce N such that $|f_n(x) - g(x)| < \varepsilon/(b - a)$ for $n \geq N$. Then, for $n \geq N$.

$$\int_a^b |f_n(x) - g(x)|\, dx \leq \int_a^b \frac{\varepsilon}{b - a}\, dx = \varepsilon.$$

But then $\left|\int_a^b f_n(x)\, dx - \int_a^b g(x)\, dx\right| < \varepsilon$ for $n \geq N$, implying that

$$\lim_{n\to\infty} \int_a^b f_n(x)\, dx = \int_a^b g(x)\, dx. \quad \blacksquare$$

The hypothesis that each f_n be continuous on $[a, b]$ is actually somewhat stronger than needed but is sufficient for our purposes without being too technical.

Finally, we consider the problem of interchanging $\lim_{n\to\infty}$ and d/dx. Here uniform convergence of the sequence of derivatives is the important hypothesis.

Theorem

Let $\lim_{n\to\infty} f_n = g$ *(pointwise on* $[a, b]$*). Suppose that* f_n' *is continuous on* $[a, b]$ *for each* n*, and that* $\{f_n'\}$ *converges uniformly on* $[a, b]$*. Let* $a < \alpha < b$*. Then* $\lim_{n\to\infty} f_n'(\alpha) = g'(\alpha)$.

PROOF

Put $h(x) = \lim_{n\to\infty} f_n'(x)$ for $a \leq x \leq b$. By the previous theorem,

$$\lim_{n\to\infty} \int_a^\alpha f_n'(t)\, dt = \int_a^\alpha h(t)\, dt.$$

But

$$\lim_{n\to\infty} \int_a^\alpha f_n'(t)\, dt = \lim_{n\to\infty} [f_n(\alpha) - f_n(a)] = g(\alpha) - g(a).$$

Then, $g(\alpha) = g(a) + \int_a^\alpha h(t)\, dt$. Now, h is continuous at α, as $\{f_n'\}$ converges uniformly and each f_n' is continuous. Then $\int_a^\alpha h(t)\, dt$ is differentiable at α. Then g is differentiable at α also, and

$$g'(\alpha) = \frac{d}{dx}\left[g(a) + \int_a^\alpha h(t)\, dt\right] = h(\alpha) = \lim_{n\to\infty} f_n'(\alpha). \quad \blacksquare$$

Exercises for Section 1.2

1. For each of the following sequences, determine uniform or pointwise convergence, on the given set.

(a) $f_n(x) = \sin^n(x)$, $0 \le x \le \pi$
(b) $f_n(x) = \sin^n(x)$, $0 \le x \le \pi/4$
(c) $f_n(x) = \dfrac{nx}{1 + nx}$, $0 \le x \le 1$
(d) $f_n(x) = \dfrac{x^n}{n}$, $0 \le x \le 1$
(e) $f_n(x) = \dfrac{xe^{-nx}}{n}$, $x \ge 0$
(f) $f_n(x) = xe^{-nx^2}$, $0 \le x \le 1$
Hint: It is often helpful to sketch graphs of the functions, noting in particular points where maxima and minima occur.

2. Prove the following Cauchy-type theorem: *Let $\{f_n\}$ be a sequence of functions defined on a set D. Then, $\{f_n\}$ converges uniformly on D if and only if, given $\varepsilon > 0$, there is some N such that $|f_n(x) - f_m(x)| < \varepsilon$ whenever $n, m \ge N$ and x is in D.* (This theorem enables us to consider the question of uniform convergence of a sequence without knowing the limit function.)

3. Let $f_n(x) = nx/(1 + n^2x^2)$, $-\infty < x < \infty$. Show that convergence of $\{f_n\}$ is uniform on any interval $[a, b]$ not containing 0, but is not uniform on any interval $[c, d]$ containing 0.

4. Let $\{f_n\}$ and $\{g_n\}$ converge uniformly on D.
(a) Prove that $\{f_n(x) + g_n(x)\}$ converges uniformly on D.
(b) Produce an example to show that $\{f_n(x) \cdot g_n(x)\}$ need not converge uniformly on D.

5. Let $f_n(x) = 1/x^{2n+1}$, $-1 \le x \le 1$. Investigate convergence of $\{f_n\}$.

6. Let $f_n(x) = x/(1 + nx^2)$, $-1 \le x \le 1$.
(a) Show that $\{f_n\}$ converges uniformly on $[-1, 1]$.
(b) Determine $\lim_{n\to\infty} f_n(x)$ for each x in $[-1, 1]$.
(c) Show that

$$\lim_{n\to\infty} f_n'(x) \ne \frac{d}{dx}\left[\lim_{n\to\infty} f_n(x)\right] \quad \text{on } [-1, 1].$$

(d) Explain why this example does not contradict the theorem on interchange of $\lim_{n\to\infty}$ and d/dx.

7. Let $\{f_n\}$ be a sequence of differentiable functions on $[a, b]$. Suppose that $\{f_n(\alpha)\}$ converges for some α, $a \le \alpha \le b$, and that $\{f_n'\}$ converges uniformly on $[a, b]$. Show that $\{f_n\}$ converges uniformly on $[a, b]$.

8. Evaluate $\lim_{n\to\infty} \int_1^3 e^{-nx^2}\, dx$. *Hint:* Let $g(x) = \lim_{n\to\infty} e^{-nx^2}$ for $1 \le x \le 3$ and show that the convergence is uniform.

9. Investigate convergence of $\{f_n\}$ on $[0, 1]$, with $f_n(x) = n(1 - x)x^n$.

10. Investigate convergence on $[-\pi/2, \pi/2]$ of $\{f_n\}$, with $f_n(x) = [\cos(x)]^{1/n}$.

1.3 Real Series

We now turn our attention to series, the study of which will depend heavily on our results on sequences. Suppose that $\{a_n\}$ is a sequence of real numbers. What significance, if any, can we attach to the symbol $\sum_{n=1}^{\infty} a_n$, by which we hope to mean a sum of all the terms of the sequence?

There is at least one obvious example in which a simple argument based on everyday experience and common sense provides a hint as to how to interpret $\sum_{n=1}^{\infty} a_n$. Consider the problem of walking 1 mile. Starting from point 0, we wish to reach point 1. Of course, this can be done. But think of the journey as broken up into steps as follows. First we have to walk half the distance, or $\frac{1}{2}$ mile, then half of the remaining distance, or $\frac{1}{4}$ mile, then half of the new remainder, or $\frac{1}{8}$ mile, and so on. Based on this argument, we should have to walk a total of

$$\frac{1}{2} + \frac{1}{4} + \frac{1}{8} + \cdots + \frac{1}{2^{100}} + \cdots,$$

or $\sum_{n=1}^{\infty} 1/2^n$ miles. Clearly, then, the symbol $\sum_{n=1}^{\infty} 1/2^n$ should mean the number 1. (Incidentally, part of this is *Zeno's paradox*. Zeno, however, was not much on common sense, and appeared to argue, from the "obvious impossibility" of adding infinitely many positive numbers and obtaining a finite number, that one cannot in fact walk 1 mile. We are arguing exactly the opposite. Since one can walk 1 mile, an infinite sum must make sense, at least in this instance.)

If you think about it for a while, it is easy to see that the problem of attaching a meaning to $\sum_{n=1}^{\infty} a_n$ can be thrown back into a problem of sequential convergence. Associate with the sequence $\{a_n\}$ a new sequence $\{S_n\}$ given by $S_n = \sum_{j=1}^{n} a_j$. We call $\{S_n\}$ the *sequence of partial sums* of $\{a_n\}$. Now, $\{S_n\}$ may or may not converge. If it does, it is natural to call its limit the *sum of the series*, and write $\lim_{n\to\infty} S_n$ as $\sum_{n=1}^{\infty} a_n$. We say then that the series $\sum_{n=1}^{\infty} a_n$ *converges*. If $\{S_n\}$ diverges, then we say that the series $\sum_{n=1}^{\infty} a_n$ *diverges*. In this case we do not attempt to attach a real number to the "sum of the terms" of $\{a_n\}$.

As an example of a convergent series, recall from an example in Section 1.1 that $\lim_{n\to\infty} \sum_{j=0}^{n} r^j = 1/(1 - r)$ if $|r| < 1$. That is, $\sum_{n=0}^{\infty} r^n = 1/(1 - r)$. This is called a *geometric series*. Note that this result is consistent with our physically motivated argument that $\sum_{n=1}^{\infty} 1/2^n = 1$, since

$$\sum_{n=1}^{\infty} \frac{1}{2^n} = \left(\sum_{n=0}^{\infty} \frac{1}{2^n}\right) - 1 = \frac{1}{1 - \frac{1}{2}} - 1 = 1.$$

Given a series $\sum_{n=1}^{\infty} a_n$ of real numbers, the first question one asks is whether or not the series converges. Actually calculating the sum of the series in the case of convergence is usually far more difficult, and we shall not worry about this problem until Section 1.5 and again in Chapter 9. So, for the remainder of this section, the object is to develop tests for convergence. The basic scheme is to first consider series whose terms are nonnegative, or, in some cases, strictly positive. The basic tests for such series are the comparison, ratio, root, and integral tests. We shall then consider arbitrary series, then series whose terms alternate in sign.

Before beginning this program, we shall make two preliminary observations, one frequently of use in calculations, the second sometimes useful in showing that a series diverges.

Suppose first that $\sum_{n=1}^{\infty} a_n$ and $\sum_{n=1}^{\infty} b_n$ converge. It is immediate from the corresponding statement on sequences that $\sum_{n=1}^{\infty} (a_n + b_n)$ converges, and that

$$\sum_{n=1}^{\infty} (a_n + b_n) = \sum_{n=1}^{\infty} a_n + \sum_{n=1}^{\infty} b_n.$$

Further, if t is a real number, then $\sum_{n=1}^{\infty} ta_n = t \sum_{n=1}^{\infty} a_n$.

For our second observation, suppose that $\sum_{n=1}^{\infty} a_n$ converges, say to s. Then $\lim_{n\to\infty} \sum_{j=1}^{n} a_j = s$. Given $\varepsilon > 0$, there exists N such that

$$\left| \sum_{j=1}^{n} a_j - s \right| < \frac{\varepsilon}{2}$$

if $n \geq N$. In particular,

$$\left| \sum_{j=1}^{n} a_j - s \right| < \frac{\varepsilon}{2} \quad \text{and} \quad \left| \sum_{j=1}^{n+1} a_j - s \right| < \frac{\varepsilon}{2},$$

if $n \geq N$. Then, for $n \geq N$,

$$\begin{aligned} |a_{n+1}| &= \left| \sum_{j=1}^{n+1} a_j - \sum_{j=1}^{n} a_j \right| = \left| \sum_{j=1}^{n+1} a_j - s + s - \sum_{j=1}^{n} a_j \right| \\ &\leq \left| \sum_{j=1}^{n+1} a_j - s \right| + \left| \sum_{j=1}^{n} a_j - s \right| < \frac{\varepsilon}{2} + \frac{\varepsilon}{2} = \varepsilon. \end{aligned}$$

This is the same as saying that $\lim_{n\to\infty} a_n = 0$ whenever $\sum_{n=1}^{\infty} a_n$ converges. Put another way:

Theorem

If $\lim_{n\to\infty} a_n \neq 0$, *then* $\sum_{n=1}^{\infty} a_n$ *diverges.*

It is important to realize that this theorem can never be used to establish convergence, only divergence. It is possible for $\lim_{n\to\infty} a_n = 0$ and $\sum_{n=1}^{\infty} a_n$ to diverge, as we shall soon see is the case with $\sum_{n=1}^{\infty} 1/n$.

EXAMPLES

$$\sum_{n=1}^{\infty} \frac{e^n}{n^2} \text{ diverges, as } \lim_{n\to\infty} \frac{e^n}{n^2} = \infty;$$

$$\sum_{n=1}^{\infty} (-1)^n \text{ diverges, as } \lim_{n\to\infty} (-1)^n \text{ does not exist.}$$

The last series is sometimes deceptive, as $-1 + 1 - 1 + 1 - 1 + 1 - 1 + \cdots$ might appear to have a lot of cancellations and so have a finite sum. But the partial sum of an even number of terms is 0, and of an odd number of terms, -1. The sequence of partial sums is then $\{(-1 + (-1)^n)/2\}_{n=1}^{\infty}$, which diverges. ∎

We now turn to nonnegative series, that is, series whose terms are all nonnegative. Suppose, then, until further notice, that $a_n \geq 0$ for $n = 1, 2, \ldots$. Note first that the partial sums of $\sum_{n=1}^{\infty} a_n$ form a monotone nondecreasing sequence, since $\sum_{j=1}^{n} a_j \leq \sum_{j=1}^{n+1} a_j$ if each $a_j \geq 0$. This observation suggests the following comparison test.

COMPARISON TEST

Suppose that $0 \leq a_n \leq b_n$ *for* $n = 1, 2, \ldots$. *Then,*

(1) *If* $\sum_{n=1}^{\infty} b_n$ *converges, so does* $\sum_{n=1}^{\infty} a_n$.

(2) *If* $\sum_{n=1}^{\infty} a_n$ *diverges, so does* $\sum_{n=1}^{\infty} b_n$.

PROOF

Let $A_n = \sum_{j=1}^{n} a_j$ and $B_n = \sum_{j=1}^{n} b_j$. Then $\{A_n\}$ and $\{B_n\}$ are monotone nondecreasing sequences, and $A_n \leq B_n$, for each n. Now, if $\sum_{n=1}^{\infty} a_n$ diverges, then so does $\{A_n\}$, hence $\{A_n\}$ is not bounded above. Then $\{B_n\}$ is also not bounded above, so $\lim_{n\to\infty} B_n = \infty$, and $\sum_{n=1}^{\infty} b_n$ diverges.

On the other hand, if $\sum_{n=1}^{\infty} b_n$ converges, then $\{B_n\}$ is bounded above, so $\{A_n\}$ is also bounded above. Then $\{A_n\}$ converges, so $\sum_{n=1}^{\infty} a_n$ converges. ∎

Be careful not to misapply this test. If $\sum_{n=1}^{\infty} a_n$ converges, possibly $\sum_{n=1}^{\infty} b_n$ diverges. And if $\sum_{n=1}^{\infty} b_n$ diverges, possibly $\sum_{n=1}^{\infty} a_n$ converges.

EXAMPLE

$\sum_{n=1}^{\infty} (1/n!)$ converges. There are easier ways to show this than by comparison, but we shall do it this way as an example. Note that

$$\frac{1}{n!} = \frac{1}{1 \cdot 2 \cdot 3 \cdots n} < \frac{1}{2^{n-1}} \qquad \text{if } n \geq 2.$$

Thus let $a_n = 1/n!$ and $b_n = 1/2^{n-1}$ for $n = 2, 3, \ldots$. Now $\sum_{n=2}^{\infty} b_n = \sum_{n=2}^{\infty} (1/2^{n-1}) = \sum_{n=1}^{\infty} 1/2^n$ converges. (This is a geometric series, already treated. In fact, it converges to 1.) Then $\sum_{n=2}^{\infty} (1/n!)$ converges. Then also $\sum_{n=1}^{\infty} (1/n!)$ converges. ∎

Note that, in this example, it turns out to be inconvenient to compare $\sum_{n=1}^{\infty} a_n$ with $\sum_{n=1}^{\infty} b_n$ but rather easy to compare $\sum_{n=2}^{\infty} a_n$ with $\sum_{n=2}^{\infty} b_n$. This is of no importance if all we want to do is establish convergence. Certainly $\sum_{n=1}^{\infty} a_n$ and $\sum_{n=2}^{\infty} a_n$ will have different sums if $a_1 \neq 0$, but if one converges, so does the other.

EXAMPLE

$\sum_{n=1}^{\infty} (1/n^n)$ converges. To prove this, show (by induction or otherwise) that $1/n^n < 1/n!$ for $n \geq 2$. Then use the fact that $\sum_{n=1}^{\infty} (1/n!)$ has been shown to converge. ∎

The limitations of the comparison test are that it applies only to nonnegative series, and also that it may be difficult to find a known convergent or divergent series to compare with. For example, $\sum_{n=1}^{\infty} n!/n^n$ converges, but this is most easily shown by the ratio test, which comes later. The reader might also try a comparison test on $\sum_{n=2}^{\infty} [\ln(\ln(n))/\ln(n)]$.

Our second test is applicable only to nonincreasing, positive series $(0 < a_{n+1} \leq a_n)$, and requires the notion of convergence of integrals of the form $\int_a^{\infty} f(x)\,dx$.

By analogy with the sequence of partial sums approach to the convergence of series, we say that $\int_a^\infty f(x)\,dx$ *converges* if $\lim_{b\to\infty} \int_a^b f(x)\,dx$ exists finite. If this limit does not exist, or is infinite, we say that $\int_a^\infty f(x)\,dx$ *diverges*.

As an example that will be useful later, consider $\int_1^\infty (1/x^p)\,dx$. If $p \neq 1$, then by direct integration we have

$$\int_1^b \frac{1}{x^p}\,dx = \frac{x^{1-p}}{1-p}\bigg|_1^b = \frac{b^{1-p}}{1-p} - \frac{1}{1-p}.$$

If $p > 1$, then $b^{1-p}/(1-p) \to 0$ as $b \to \infty$; if $p < 1$, then $b^{1-p}/(1-p) \to \infty$ as $b \to \infty$. Then

$$\int_1^\infty \frac{1}{x^p}\,dx \begin{cases} \text{converges to } \dfrac{1}{p-1} & \text{if } p > 1, \\ \text{diverges} & \text{if } p < 1. \end{cases}$$

For $p = 1$, $\int_1^b (1/x)\,dx = \ln(b) \to \infty$ as $b \to \infty$, so $\int_1^\infty (1/x)\,dx$ diverges.

The connection between "improper" integrals $\int_a^\infty f(x)\,dx$ and infinite series $\sum_{n=1}^{\infty} a_n$ can be seen graphically. Suppose that $0 < a_{n+1} \leq a_n$. Plot the points (n, a_n) on a graph, to get something like Fig. 1-6.

The partial sum $\sum_{j=1}^{n} a_j$ represents a sum of areas of rectangles of width 1 and height $a_1, a_2, \ldots, a_n$. Now invent in the obvious way a nonincreasing, continuous function f such that $f(n) = a_n$. For example, if $a_n = 1/n$, put $f(x) = 1/x$ for $x \geq 1$; if $a_n = e^n/n^n$, put $f(x) = e^x/x^x$. For any two successive positive integers j and $j + 1$, we then have

$$f(j+1) \leq f(x) \leq f(j) \qquad \text{for } j \leq x \leq j+1.$$

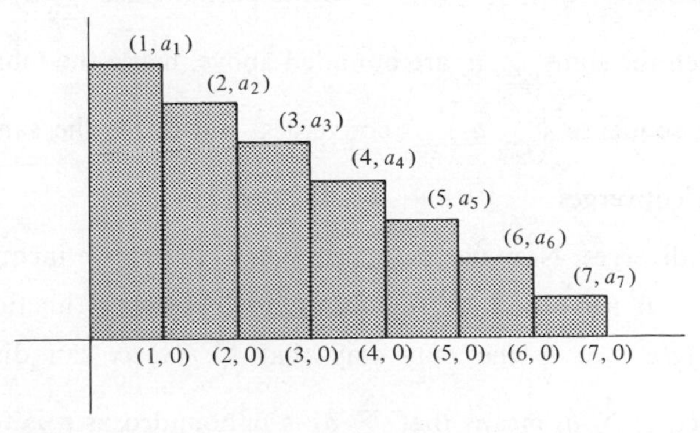

Fig. 1-6

Now $\int_j^{j+1} f(x)\,dx$ represents the area under the graph $y = f(x)$ for $j \le x \le j + 1$. This area is no less than that of the rectangle of height a_{j+1} and width 1, and no greater than that of the rectangle of height a_j, width 1 (Fig. 1-7).

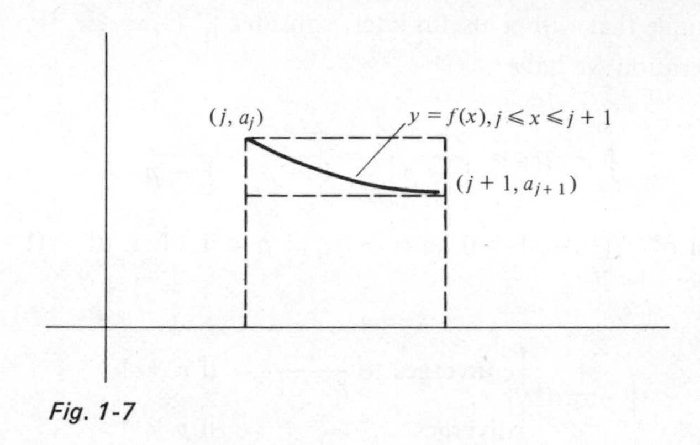

Fig. 1-7

For any j, then,

$$a_{j+1} \le \int_j^{j+1} f(x)\,dx \le a_j.$$

By summing, we have

$$\sum_{j=1}^{n} a_{j+1} = \sum_{j=2}^{n+1} a_j \le \sum_{j=1}^{n} \int_j^{j+1} f(x)\,dx = \int_1^{n+1} f(x)\,dx \le \sum_{j=1}^{n} a_j.$$

Now consider two cases:

(1) $\int_1^\infty f(x)\,dx$ converges. Since $f(x) > 0$ for $x \ge 1$, then $\int_1^{n+1} f(x)\,dx \le \int_1^\infty f(x)\,dx$, for all n. Then $\sum_{j=2}^{n+1} a_j \le \int_1^\infty f(x)\,dx$, so the partial sums $\sum_{j=2}^{n+1} a_j$ are bounded above. Then the sums $\sum_{j=1}^{n} a_j$ are bounded above, hence the monotone nondecreasing sequence $\left\{\sum_{j=1}^{n} a_j\right\}_{n=1}^{\infty}$ converges. But this is the same as saying that $\sum_{n=1}^{\infty} a_n$ converges.

(2) $\int_1^\infty f(x)\,dx$ diverges. Now use the other side of the above inequality. Here, since $f(x) > 0$ for $x \ge 1$, $\int_1^b f(x)\,dx$ is an increasing function of b, so $\lim_{b\to\infty} \int_1^b f(x)\,dx = \infty$ is the only way that $\int_1^\infty f(x)\,dx$ can diverge. Then $\int_1^{n+1} f(x)\,dx \le \sum_{j=1}^{n} a_j$ means that $\sum_{j=1}^{n} a_j$ is unbounded as n is taken larger.

The sequence $\left\{\sum_{j=1}^{n} a_j\right\}_{n=1}^{\infty}$ then diverges, hence so does $\sum_{j=1}^{\infty} a_j$.
We summarize this discussion:

INTEGRAL TEST

Let $f(x)$ be a positive, continuous, decreasing function for $x \geq 1$. Then $\sum_{n=1}^{\infty} f(n)$ and $\int_1^{\infty} f(x)\,dx$ either both converge or both diverge.

EXAMPLE

$$\sum_{n=1}^{\infty} \frac{1}{n^p} \begin{cases} \text{converges if } p > 1, \\ \text{diverges if } p \leq 1. \end{cases}$$

For, as already noted, $\int_1^{\infty} (1/x^p)\,dx$ converges if $p > 1$, diverges if $p \leq 1$. The divergent series $\sum_{n=1}^{\infty} (1/n)$ is usually called the *harmonic series.* ∎

EXAMPLE

Consider the series

$$-8 - 16 - 32 - \pi + e^{-7} + \frac{8}{e^8} + \frac{9}{e^9} + \frac{10}{e^{10}} + \cdots$$

$$= -8 - 32 - \pi + e^{-7} + \sum_{n=8}^{\infty} ne^{-n}.$$

An obvious modification of the integral test applies. The first five terms do not fit the general pattern of the other terms of the series, but these will not affect convergence anyway. Just look at $\sum_{n=8}^{\infty} ne^{-n}$. This will converge or diverge along with $\int_8^{\infty} xe^{-x}\,dx$. Now

$$\int_8^b xe^{-x}\,dx = -xe^{-x}\Big|_8^b + \int_8^b e^{-x}\,dx = -xe^{-x}\Big|_8^b - e^{-x}\Big|_8^b \to 9e^{-8} \quad \text{as } b \to \infty.$$

Then $\int_8^{\infty} xe^{-x}\,dx$ converges, so $\sum_{n=8}^{\infty} ne^{-n}$ converges. Then

$$-8 - 16 - 32 - \pi + e^{-7} + \sum_{n=8}^{\infty} ne^{-n}$$

converges also. ∎

EXAMPLE

$\sum_{n=1}^{\infty} [\ln(n)/n]$ diverges. We now have two ways of handling this:

(1) $1/n < \ln(n)/n$ for $n \geq 2$, and $\sum_{n=1}^{\infty} (1/n)$ diverges, so $\sum_{n=1}^{\infty} [\ln(n)/n]$ diverges by comparison.

(2) $\int_1^{\infty} [\ln(x)/x]\, dx = \lim_{b \to \infty} [\ln(b)]^2 = \infty$, so $\sum_{n=1}^{\infty} \ln(n)/n$ diverges. ∎

An important observation about the integral test is that so far its effectiveness depends upon how good the user is at integrating and taking limits. Some series just cannot be handled this way. An example is $\sum_{n=1}^{\infty} (1/n!)$, which we know converges by comparison with $\sum_{n=1}^{\infty} (1/2^n)$. To apply the integral test, we would be led to try $\int_1^{\infty} (1/x!)\, dx$, and $x!$ is not clear if x is not a nonnegative integer (in this context, note the gamma function in Chapter 7). Perhaps a better example is

$$\sum_{n=1}^{\infty} \frac{(n+1)(n+4)}{n^6(n+5)(n+6)}.$$

This would involve consideration of

$$\int_1^{\infty} \frac{(x+1)(x+4)}{x^6(x+5)(x+6)}\, dx,$$

a very unpleasant integral (which, however, can certainly be done by partial fractions). As we shall see in Section 1.6, there are ways of attacking convergence of improper integrals without directly integrating, so things will improve. However, the point is that the test does have its limitations.

The next test we shall consider is called the *ratio test*. Since it involves dividing successive terms of the series, we shall suppose here that all the a_n's are positive (although not necessarily monotone nonincreasing).

RATIO TEST

Let $a_n > 0$, $n = 1, 2, 3, \ldots$. *If* $\lim_{n \to \infty} (a_{n+1}/a_n) < 1$, *then* $\sum_{n=1}^{\infty} a_n$ *converges. If* $\lim_{n \to \infty} (a_{n+1}/a_n) > 1$, *then* $\sum_{n=1}^{\infty} a_n$ *diverges. If* $\lim_{n \to \infty} (a_{n+1}/a_n) = 1$, *no conclusion can be drawn.*

PROOF

Let $\lim_{n \to \infty} (a_{n+1}/a_n) = r$. Given $\varepsilon > 0$, there is some N such that $|(a_{n+1}/a_n) - r|$

$< \varepsilon$ for $n \geq N$. In particular, $r - \varepsilon < a_{N+1}/a_N < r + \varepsilon$. Then $(r - \varepsilon)a_N < a_{N+1} < (r + \varepsilon)a_N$. Similarly,

$$(r - \varepsilon)^2 a_N < (r - \varepsilon)a_{N+1} < a_{N+2} < (r + \varepsilon)a_{N+1} < (r + \varepsilon)^2 a_N,$$

and, in general,

$$(r - \varepsilon)^K a_N < a_{N+K} < (r + \varepsilon)^K a_N \qquad \text{for } K = 1, 2, \ldots.$$

Now consider cases.

(1) $r < 1$. Choose ε so small that $r + \varepsilon < 1$. Then the series $\sum_{K=1}^{\infty} (r + \varepsilon)^K$ converges. (It is a geometric series.) Then $\sum_{K=1}^{\infty} (r + \varepsilon)^K a_N = a_N \sum_{K=1}^{\infty} (r + \varepsilon)^K$ converges. By comparison, $\sum_{K=1}^{\infty} a_{N+K} = \sum_{j=N+1}^{\infty} a_j$ converges. Hence $\sum_{j=1}^{\infty} a_j$ converges.

(2) $r > 1$. Now choose ε sufficiently small that $r - \varepsilon > 1$. Then $\sum_{K=1}^{\infty} (r - \varepsilon)^K$ is a divergent geometric series, so $\sum_{K=1}^{\infty} (r - \varepsilon)^K a_N$ diverges. By comparison, $\sum_{K=1}^{\infty} a_{N+K} = \sum_{K=N+1}^{\infty} a_K$ diverges, hence $\sum_{j=1}^{\infty} a_j$ diverges.

(3) $r = 1$. Here no conclusion can be drawn. For example, $\sum_{n=1}^{\infty} 1/n$ diverges, and

$$\lim_{n \to \infty} \frac{1/(n+1)}{1/n} = 1.$$

But $\sum_{n=1}^{\infty} (1/n^2)$ converges, and

$$\lim_{n \to \infty} \frac{1/(n+1)^2}{1/n^2} = 1$$

also. ∎

EXAMPLE

$\sum_{n=1}^{\infty} (n!/n^n)$ converges. Here, $a_n = n!/n^n$, and

$$\frac{a_{n+1}}{a_n} = \frac{(n+1)!\, n^n}{(n+1)^{n+1} n!} = \frac{(n+1)n^n}{(n+1)^{n+1}} = \left(\frac{n}{n+1}\right)^n$$

$$= \frac{1}{(1 + 1/n)^n} \to \frac{1}{e} < 1 \qquad \text{as } n \to \infty. \quad ∎$$

As with the integral and comparison tests, the ratio test is no panacea. If the limit of the ratio a_{n+1}/a_n is 1, it tells us nothing.

It is even possible for the series to converge, with all $a_n > 0$, and a_{n+1}/a_n to have no limit as $n \to \infty$. This is the case with the series

$$\sum_{n=1}^{\infty} a_n = \frac{1}{2} + \frac{1}{3} + \frac{1}{2^2} + \frac{1}{3^2} + \frac{1}{2^3} + \frac{1}{3^3} + \cdots,$$

where $a_{2n} = 1/3^n$ and $a_{2n-1} = 1/2^n$, for $n = 1, 2, \ldots$. For

$$\frac{a_{2n+1}}{a_{2n}} = \frac{1/2^{n+1}}{1/3^n} = \frac{1}{2}\left(\frac{3}{2}\right)^n \to \infty \qquad \text{as } n \to \infty.$$

But

$$\frac{a_{2n+2}}{a_{2n+1}} = \left(\frac{1/3^{n+1}}{1/2^{n+1}}\right)\cdot\left(\frac{2}{3}\right)^{n+1} \to 0 \qquad \text{as } n \to \infty.$$

Thus the ratio of successive terms has no limit, as a_{n+1}/a_n becomes larger when n is even and large, and smaller when n is odd and large.

However, $\sum_{n=1}^{\infty} a_n$ does converge. Note that $a_{2n} = 1/3^n < 1/2^n$, so we may compare $\sum_{n=1}^{\infty} a_n$ with the series $\sum_{n=1}^{\infty} b_n$, where $b_{2n} = b_{2n-1} = 1/2^n$. To show that $\sum_{n=1}^{\infty} b_n$ converges, note that the partial sums are different for even and odd numbers of terms:

$$S_{2n} = \sum_{j=1}^{2n} b_j = \frac{1}{2} + \frac{1}{2} + \frac{1}{2^2} + \frac{1}{2^2} + \cdots + \frac{1}{2^n} + \frac{1}{2^n}$$

$$= 2\cdot\sum_{j=1}^{n}\frac{1}{2^j} = 2\left(1 - \frac{1}{2^n}\right) = 2 - \frac{1}{2^{n-1}},$$

$$S_{2n-1} = \sum_{j=1}^{2n+1} b_j = S_{2n} + b_{2n+1} = 2 - \frac{1}{2^{n-1}} + \frac{1}{2^{n+1}},$$

but both S_{2n} and S_{2n+1} go to 2 as $n \to \infty$. Hence $\sum_{j=1}^{\infty} b_j$ converges.

It should be noted that the ratio test yields a conclusion of divergence when $\lim_{n\to\infty} (a_{n+1}/a_n) = \infty$, although in the proof above we were assuming that the limit of a_{n+1}/a_n was a real number. The proof for this case is a simple extension of the idea of the proof of divergence when $\lim_{n\to\infty} (a_{n+1}/a_n) > 1$, and will be left to the reader. Actually, it is easy to show, in the case that $\lim_{n\to\infty} (a_{n+1}/a_n) = \infty$, that $\lim_{n\to\infty} a_n \neq 0$.

We now move on to our last test for positive series.

ROOT TEST

Let $a_n > 0$ *for* $n = 1, 2, 3, \ldots$. *If* $\lim_{n\to\infty} a_n^{1/n} < 1$, *then* $\sum_{n=1}^{\infty} a_n$ *converges. If* $\lim_{n\to\infty} a_n^{1/n} > 1$, *then* $\sum_{n=1}^{\infty} a_n$ *diverges. If* $\lim_{n\to\infty} a_n^{1/n} = 1$, *no conclusion can be drawn.*

PROOF

Put $\lim a_n^{1/n} = r$. Given $\varepsilon > 0$, there is some N such that $|a_n^{1/n} - r| < \varepsilon$, so $r - \varepsilon < a_n^{1/n} < r + \varepsilon$ for $n \geq N$; then, for $n \geq N$, $(r - \varepsilon)^n < a_n < (r + \varepsilon)^n$. If $r < 1$, choose ε so that $r + \varepsilon < 1$, and compare $\sum_{n=N}^{\infty} a_n$ with $\sum_{n=N}^{\infty} (r + \varepsilon)^n$, a convergent geometric series. If $r > 1$, choose ε so that $r - \varepsilon > 1$, and compare $\sum_{n=N}^{\infty} a_n$ with $\sum_{n=N}^{\infty} (r - \varepsilon)^n$, a divergent geometric series. For the case $r = 1$, look at these examples. First, $\sum_{n=1}^{\infty} [(n + 1)/n]^n$ diverges, as

$$\lim_{n\to\infty} \left(\frac{n+1}{n}\right)^n = \lim_{n\to\infty} \left(1 + \frac{1}{n}\right)^n = e \neq 0.$$

And

$$\left[\left(\frac{n+1}{n}\right)^n\right]^{1/n} = \frac{n+1}{n} \to 1 \qquad \text{as } n \to \infty.$$

On the other hand, $\sum_{n=1}^{\infty} (1/n^2)$ converges. But $1/n^{2/n} = n^{-2/n} = e^{-(2/n)\ln(n)} \to 1$ as $n \to \infty$, since $(2/n)\ln(n) \to 0$. Thus

$$\left(\frac{1}{n^2}\right)^{1/n} \to 1 \qquad \text{as } n \to \infty.$$

As with the ratio test, a series may converge, but $a_n^{1/n}$ have no limit as $n \to \infty$. This happens with $1/2 + 1/3 + 1/2^2 + 1/3^2 + \cdots$. Another difficulty (and this can happen with the ratio test as well) is that the limit one has to compute may not be so easy. For example, consider $\sum_{n=1}^{\infty} (n/2^{n+1})$. Here

$$\left(\frac{n}{2^{n+1}}\right)^{1/n} = \frac{n^{1/n}}{2^{(1+1/n)}}$$

and the limit as $n \to \infty$ is not obvious. In Appendix II of this chapter we offer some relief for this in the form of L'Hospital's Rule.

There are situations where the root test works and the ratio test fails. One such is a modification of a previous example. Consider the series $\sum_{n=1}^{\infty} a_n$, where $a_{2n} = 1/3^n$ and $a_{2n-1} = 1/3^{n+1}$. Then

$$\sum_{n=1}^{\infty} a_n = \frac{1}{9} + \frac{1}{3} + \frac{1}{27} + \frac{1}{9} + \frac{1}{81} + \frac{1}{27} + \cdots.$$

It is easy to check that

$$\frac{a_{n+1}}{a_n} = \begin{cases} \frac{1}{9} & \text{if } n \text{ is even,} \\ 3 & \text{if } n \text{ is odd.} \end{cases}$$

Then $\lim_{n \to \infty} (a_{n+1}/a_n)$ does not exist. However,

$$(a_{2n})^{1/2n} = \left(\frac{1}{3^n}\right)^{1/2n} = \frac{1}{\sqrt{3}}$$

and

$$(a_{2n-1})^{1/(2n-1)} = \left(\frac{1}{3^{n+1}}\right)^{1/(2n-1)} = \left(\frac{1}{3}\right)^{(n+1)/(2n-1)}$$

$$= \left(\frac{1}{3}\right)^{n/(2n-1)} \left(\frac{1}{3}\right)^{1/(2n-1)}.$$

As $n \to \infty$, clearly $(a_{2n})^{1/2n} \to 1/\sqrt{3}$. But, since $n/(2n-1) \to 1/2$ and $1/(2n-1) \to 0$, then $(a_{2n-1})^{1/(2n-1)} \to 1/\sqrt{3}$ also. Then $\lim_{n \to \infty} (a_n)^{1/n} = 1/\sqrt{3}$. Perhaps surprisingly, this cannot happen the other way around. If $\lim_{n \to \infty} (a_{n+1}/a_n) = r$, then $\lim_{n \to \infty} (a_n)^{1/n} = r$ also. We leave this to the exercises (note Exercise 15, Section 1.1).

The tests that we have developed so far are the most commonly used ones for nonnegative and positive series. Several more-delicate tests, and some refinements of the above ones, are given in the exercises.

We now turn briefly to arbitrary real series, dropping the assumption that $a_n \geq 0$. Such series are difficult to treat in general, and so far our only relevant result is that $\lim_{n \to \infty} a_n \neq 0$ implies that $\sum_{n=1}^{\infty} a_n$ diverges. However, often the question of convergence of an arbitrary series can be reduced to a question for a nonnegative series as follows. We claim that *convergence of* $\sum_{n=1}^{\infty} |a_n|$ *implies that of* $\sum_{n=1}^{\infty} a_n$. To prove this, we shall use the Cauchy convergence criterion. Before giving the argument, however, we note that such a device is needed because $\sum_{n=1}^{\infty} |a_n|$ and $\sum_{n=1}^{\infty} a_n$ have completely different partial sums, and even if both series converge, the sums will be different unless each a_n is nonnegative.

Now for the proof. Suppose that $\sum_{n=1}^{\infty} |a_n|$ converges. Given $\varepsilon > 0$, there is some N such that

$$\left| \sum_{j=1}^{n} |a_j| - \sum_{j=1}^{m} |a_j| \right| < \varepsilon$$

for $n, m \geq N$. This follows from applications of Cauchy's theorem to $\left\{\sum_{j=1}^{n} |a_j|\right\}_{n=1}^{\infty}$. Now, if $n > m \geq N$, then

$$\left|\sum_{j=1}^{n} a_j - \sum_{j=1}^{m} a_j\right| = \left|\sum_{j=m+1}^{n} a_j\right| \leq \sum_{j=m+1}^{n} |a_j|$$
$$= \left|\sum_{j=1}^{n} |a_j| - \sum_{j=1}^{m} |a_j|\right| < \varepsilon.$$

Hence $\left\{\sum_{j=1}^{n} a_j\right\}_{n=1}^{\infty}$ converges (by Cauchy's theorem), which means that $\sum_{j=1}^{\infty} a_j$ converges.

A series $\sum_{j=1}^{\infty} a_j$ is called *absolutely convergent* if $\sum_{j=1}^{\infty} |a_j|$ converges. As we have just seen, absolute convergence implies convergence. The value of this is that tests for nonnegative series can be applied to $\sum_{j=1}^{\infty} |a_j|$, and, if this converges, then so does $\sum_{j=1}^{\infty} a_j$. It may happen, however, that $\sum_{j=1}^{\infty} a_j$ converges, but $\sum_{j=1}^{\infty} |a_j|$ diverges. Such a series is called *conditionally convergent*. An example is $\sum_{n=1}^{\infty} [(-1)^n/n]$, which will be soon shown to converge. In general, then, unless the series is absolutely convergent, we have little in the way of a testing procedure for $\sum_{n=1}^{\infty} a_n$ in the absence of extra conditions on the a_n's. This is less serious than it might at first seem, as will become apparent when we treat power series in Section 1.5.

Another type of series for which a test is available is the alternating series, in which the terms alternate in sign.

TEST FOR ALTERNATING SERIES

Suppose that $0 \leq a_{n+1} \leq a_n$ *for* $n = 1, 2, \ldots,$ *and that* $a_n \to 0$ *as* $n \to \infty$. *Then* $\sum_{n=1}^{\infty} (-1)^{n+1} a_n$ *converges.*

The proof depends upon looking at the sequence of partial sums in the right way. Put $S_n = \sum_{j=1}^{n} (-1)^{j+1} a_j$. We want to show that $\{S_n\}$ converges. But S_n appears to behave differently depending upon whether n is even or odd. So we examine individually the sequences $\{S_{2n}\}_{n=1}^{\infty}$ and $\{S_{2n-1}\}_{n=1}^{\infty}$. It is easy to check (using the fact that $0 \leq a_{n+1} \leq a_n$) that $\{S_{2n}\}$ is monotone nondecreasing, and $\{S_{2n-1}\}$ monotone nonincreasing. But $\{S_{2n}\}$ is bounded above (by a_1), hence converges, say to L. And $\{S_{2n-1}\}$ is bounded below (e.g., by 0), so $\{S_{2n-1}\}$ converges, say to M. Finally,

$$\lim_{n \to \infty} (S_{2n} - S_{2n-1}) = L - M,$$

and, at the same time,

$$S_{2n} - S_{2n-1} = \sum_{j=1}^{2n} (-1)^{j+1} a_j - \sum_{j=1}^{2n-1} (-1)^{j+1} a_j = a_{2n} \to 0 \qquad \text{as } n \to \infty.$$

Hence $L = M$, and we conclude that $\sum_{j=1}^{\infty} (-1)^{j+1} a_j$ converges. ∎

EXAMPLE

$\sum_{n=1}^{\infty} [(-1)^{n+1}/n]$ converges, since $1/(n+1) \le 1/n$ and $1/n \to 0$ as $n \to \infty$. But

$$\sum_{n=1}^{\infty} \left| \frac{(-1)^{n+1}}{n} \right| = \sum_{n=1}^{\infty} \frac{1}{n}$$

diverges. Thus $\sum_{n=1}^{\infty} [(-1)^{n+1}/n]$ is conditionally convergent. ∎

A convergent series is generally difficult to sum. However, for convergent alternating series there is a way of estimating the sum to within an arbitrary tolerance. Suppose that $\sum_{j=1}^{\infty} (-1)^{j+1} a_j$ converges to L. In the notation of the proof of the alternating series test, $\{S_{2n}\}$ is monotone nondecreasing to L, and $\{S_{2n-1}\}$ monotone nonincreasing to L. Then $S_m \le L \le S_r$, whenever m is even and r is odd. In particular,

$$0 \le L - S_{2n} \le S_{2n+1} - S_{2n} = a_{2n+1}$$

and

$$0 \le S_{2n-1} - L \le S_{2n-1} - S_{2n} = a_{2n}.$$

Putting these together yields

$$|L - S_m| \le a_{m+1} \qquad \text{for } m = 1, 2, \ldots.$$

That is, the sum of the series differs from the sum of the first m terms by no more than the $(m + 1)$st term.

As an example, look again at $\sum_{n=1}^{\infty} [(-1)^{n+1}/n]$. The sum of this series is not obvious, but we can estimate the sum as closely as we like. If, say, we want $\sum_{n=1}^{\infty} [(-1)^{n+1}/n]$ to within $\frac{1}{10}$, we calculate

$$\left(1 - \frac{1}{2} + \frac{1}{3} - \frac{1}{4} + \frac{1}{5} - \frac{1}{6} + \frac{1}{7} - \frac{1}{8} + \frac{1}{9}\right),$$

and this differs from $\sum_{n=1}^{\infty} [(-1)^{n+1}/n]$ by no more than the absolute value of the next term, $\frac{1}{10}$.

In concluding this section, we make mention of the fact that the order of the terms in an infinite series is of extreme importance. That is, if $\sum_{n=1}^{\infty} a_n$ is a convergent series, and we rearrange terms to form a new series $\sum_{n=1}^{\infty} b_n$ (for example, $b_1 = a_{20}, b_{10} = a_{500}, b_{62} = a_{21}$, and so on), then $\sum_{n=1}^{\infty} b_n$ is an entirely new series; it may diverge, or converge, to a different sum than $\sum_{n=1}^{\infty} a_n$. This should not be too surprising if you recall that the sum of a series is the limit of the sequence of partial sums, and the partial sums of the rearranged series may bear no resemblance at all to those of the original series. There is a remarkable theorem due to Riemann, which says that the terms of any conditionally convergent series can be rearranged in such a way as to converge to any number we want. It is true, however, that any rearrangement of an absolutely convergent series will converge to the same sum. This is pursued in the exercises.

Insertion of parentheses into an infinite series can also cause problems. As an example, consider the series $\sum_{n=1}^{\infty} (-1)^n$, which diverges as $\lim_{n\to\infty} (-1)^n \neq 0$. If we write out the first few terms and insert parentheses,

$$-1 + (1 - 1) + (1 - 1) + (1 - 1) + (1 - 1) + \cdots,$$

we apparently get a series converging to -1. Written as

$$(-1 + 1) + (-1 + 1) + (-1 + 1) + (-1 + 1) + \cdots,$$

however, the series apparently converges to zero. Again, the key is the sequence of partial sums. The original series, $\sum_{n=1}^{\infty} (-1)^n$, has as its nth partial sum -1 if n is odd and 0 if n is even. The sequence of partial sums is then

$$\{[(-1) + (-1)^n]/2\}_{n=1}^{\infty},$$

which diverges. The series

$$-1 + (1 - 1) + (1 - 1) + (1 - 1) + \cdots$$

is really an entirely new series, $\sum_{n=1}^{\infty} b_n$, where $b_n = 0$ for $n \geq 2$ and $b_1 = -1$. This series converges to -1, the sequence of partial sums having every term equal to -1. Finally, $(-1 + 1) + (-1 + 1) + \cdots$ is a series with all partial sums zero. The problem here is that we started with a divergent series. We shall indicate in the exercises a proof that the sum of a convergent series is unaltered by insertion of parentheses (but leaving the order of the terms the same).

Exercises for Section 1.3

1. Test the following series for convergence. If the terms are not all nonnegative, test for absolute and conditional convergence.

(a) $\sum_{n=2}^{\infty} \frac{(-1)^n}{\ln(n)}$ **(b)** $\sum_{n=1}^{\infty} \frac{1}{n^2 + 2}$ **(c)** $\sum_{n=1}^{\infty} \frac{(-1)^n \ln(n)}{n}$

(d) $\sum_{n=1}^{\infty} \frac{n^2}{2^n}$ **(e)** $\sum_{n=1}^{\infty} n(\frac{2}{3})^{n+1}$ **(f)** $\sum_{n=1}^{\infty} \frac{n^3}{n^4 + 6}$

(g) $\sum_{n=1}^{\infty} \frac{n + 2}{n!}$ **(h)** $\sum_{n=2}^{\infty} \frac{\ln(n)}{2n^2 + 6}$ **(i)** $\sum_{n=1}^{\infty} \sin^2\left(\frac{1}{n}\right)$

(j) $\sum_{n=1}^{\infty} \frac{n^{256}}{e^{n^2}}$ **(k)** $\sum_{n=1}^{\infty} \frac{\arctan(n)}{n}$ **(l)** $\sum_{n=1}^{\infty} \frac{1}{n^{1+(1/n)}}$

2. Prove that $\sum_{n=1}^{\infty} a_n$ and $\sum_{n=1}^{\infty} a_{n+k}$ either both converge or both diverge (k is a positive integer). Also, if $\sum_{n=1}^{\infty} a_n = L$, then $\sum_{n=1}^{\infty} a_{n+k} = L - \sum_{j=1}^{k} a_j$.

3. Let $a_n > 0$ and $b_n > 0$ for $n = 1, 2, \ldots$. Suppose that $\frac{a_{n+1}}{a_n} \leq \frac{b_{n+1}}{b_n}$ for $n = 1, 2, \ldots$.

(a) Show that convergence of $\sum_{n=1}^{\infty} b_n$ implies convergence of $\sum_{n=1}^{\infty} a_n$.

(b) Show that divergence of $\sum_{n=1}^{\infty} a_n$ implies divergence of $\sum_{n=1}^{\infty} b_n$.

(c) Show by example that convergence of $\sum_{n=1}^{\infty} a_n$ yields no conclusion about $\sum_{n=1}^{\infty} b_n$, and divergence of $\sum_{n=1}^{\infty} b_n$, no conclusion about $\sum_{n=1}^{\infty} a_n$.

4. Let $a_n > 0$ for $n = 1, 2, \ldots$.

(a) Suppose that there is number r such that $a_{n+1}/a_n \leq r < 1$ for n sufficiently large (say $n \geq N$). Prove that $\sum_{n=1}^{\infty} a_n$ converges.

(b) Give an example of a series $\sum_{n=1}^{\infty} a_n$ in which the condition of (a) holds, but a_{n+1}/a_n has no limit as $n \to \infty$.

(c) Suppose that $a_{n+1}/a_n \geq 1$ for n sufficiently large, say $n \geq N$. Prove that $\sum_{n=1}^{\infty} a_n$ diverges.

(d) Give an example of a series $\sum_{n=1}^{\infty} a_n$ in which the condition of (c) holds, but $\lim_{n \to \infty} (a_{n+1}/a_n)$ does not exist finite.

5. Let $a_n > 0$ for $n = 1, 2, \ldots$. Suppose that $\limsup\{a_{n+1}/a_n\} < 1$. Show that $\sum_{n=1}^{\infty} a_n$ converges (see Exercise 17, Section 1.1). If $\limsup\{a_{n+1}/a_n\} > 1$, does $\sum_{n=1}^{\infty} a_n$ necessarily diverge? What conclusions do you draw if $\limsup\{a_{n+1}/a_n\} = 1$?

6. Let $a_n > 0$ and $b_n > 0$ for $n = 1, 2, \ldots$. Suppose that $a_n^{1/n} \le b_n^{1/n}$ for $n = 1, 2, \ldots$. What conclusions can you draw for each of the following?

(a) $\sum_{n=1}^{\infty} b_n$ converges

(b) $\sum_{n=1}^{\infty} a_n$ diverges

(c) $\sum_{n=1}^{\infty} b_n$ diverges

(d) $\sum_{n=1}^{\infty} a_n$ converges

7. Let $a_n > 0$ for $n = 1, 2, \ldots$.

(a) Suppose that, for some r, $(a_n)^{1/n} \le r < 1$ for $n \ge N$. Show that $\sum_{n=1}^{\infty} a_n$ converges.

(b) Give an example of a series $\sum_{n=1}^{\infty} a_n$ in which the condition of (a) is satisfied, but $a_n^{1/n}$ has no limit as $n \to \infty$.

(c) Suppose that $a_n^{1/n} \ge 1$ for $n \ge N$. Prove that $\sum_{n=1}^{\infty} a_n^{1/n}$ diverges.

(d) Give an example of a series $\sum_{n=1}^{\infty} a_n$ in which the condition of (c) holds, but $\lim_{n\to\infty} a_n^{1/n}$ does not exist finite.

8. Let $a_n > 0$ for $n = 1, 2, \ldots$. Suppose that $\limsup\{a_n^{1/n}\} < 1$. Show that $\sum_{n=1}^{\infty} a_n$ converges. If $\limsup\{a_n^{1/n}\} > 1$, does $\sum_{n=1}^{\infty} a_n$ necessarily diverge? What conclusion can be drawn if $\limsup\{a_n^{1/n}\} = 1$?

9. Suppose that $0 < a_{n+1} < a_n$ and $\sum_{n=1}^{\infty} a_n$ converges. Prove that $\lim_{n\to\infty} na_n = 0$. Is it also true that $\lim_{n\to\infty} n^2 a_n = 0$?

10. Prove the following Cauchy-type theorem: $\sum_{n=1}^{\infty} a_n$ *converges if and only if, given* $\varepsilon > 0$, *there is some* N *such that* $\left|\sum_{j=n}^{n+k} a_j\right| < \varepsilon$ *whenever* $k = 1, 2, \ldots$ *and* $n \ge N$.

11. Suppose that $\sum_{n=1}^{\infty} |a_n|$ converges, and, for $n = 1, 2, \ldots, -A \le b_n \le A$, where A is some real number. Prove that $\sum_{n=1}^{\infty} a_n b_n$ converges.

12. Suppose that $\lim_{n\to\infty} a_n = 0$, that $\sum_{n=1}^{\infty} |a_{n+1} - a_n|$ converges, and that, for some A, $\sum_{j=1}^{n} |b_j| \le A$ for $n = 1, 2, \ldots$. Prove that $\sum_{n=1}^{\infty} a_n b_n$ converges. (*Hint:* $\sum_{j=1}^{n} a_j b_j = a_n B_n - \sum_{j=1}^{n-1} (a_{j+1} - a_j) B_j$, where $B_j = \sum_{k=1}^{j} |b_k|$).

Use this result to prove the following: If $a_{n+1} < a_n$ and $\lim_{n\to\infty} a_n = 0$, and for some A, $\sum_{j=1}^{n} |b_j| \le A$ for $n = 1, 2, \ldots$, then $\sum_{n=1}^{\infty} a_n b_n$ converges.

13. Suppose that $a_n > 0$ and $b_n > 0$ for $n = 1, 2, \ldots$ and for some A and B, $A \le a_n/b_n \le B$. Show that

(a) If $\sum_{n=1}^{\infty} b_n$ converges, so does $\sum_{n=1}^{\infty} a_n$.

(b) If $\sum_{n=1}^{\infty} b_n$ diverges, so does $\sum_{n=1}^{\infty} a_n$.

14. Let $a_n > 0$ for $n = 1, 2, \ldots$. Suppose that $\sum_{n=1}^{\infty} 2^n a_{2^n}$ converges. Show that $\sum_{n=1}^{\infty} a_n$ converges.

15. Let $a_n > 0$ and $b_n > 0$ for $n = 1, 2, \ldots$. Suppose that $\sum_{n=1}^{\infty} b_n$ diverges and that

$$\lim_{n\to\infty} \left(\frac{1}{b_n} \frac{a_n}{a_{n+1}} - \frac{1}{b_{n-1}} \right) = r.$$

Then

(1) $r > 0$ implies that $\sum_{n=1}^{\infty} a_n$ converges.

(2) $r < 0$ implies that $\sum_{n=1}^{\infty} a_n$ diverges.

(When $b_n = 1/n$, this test is often called *Raabe's test*.)

16. Let $a_n > 0$ for $n = 1, 2, \ldots$. Suppose that for each $n \ge 2$, there is some b_n such that

$$\frac{a_n}{a_{n+1}} = 1 + \frac{1}{n} + \frac{b_n}{n \ln(n)}$$

and $\lim b_n > 1$. Then $\sum_{n=1}^{\infty} a_n$ converges. If, on the other hand, $\lim_{n\to\infty} b_n < 1$, then $\sum_{n=1}^{\infty} a_n$ diverges.

17. Make a list of all the convergence tests given in the section together with those in the preceding exercises (whether or not you proved their correctness). Then try them out on the following series:

(a) $\sum_{n=2}^{\infty} \frac{1}{n \ln(n)^p}$ (considers different values of p)

(b) $\sum_{n=3}^{\infty} \frac{1}{n \ln(n) \ln[\ln(n)]^p}$

(c) $\sum_{n=1}^{\infty} 2^{-n+(-1)^n}$

(d) $\sum_{n=1}^{\infty} a_n$, where $a_1 = \frac{1}{2}, a_2 = \frac{1 \cdot 3}{2 \cdot 4}, a_3 = \frac{1 \cdot 3 \cdot 5}{2 \cdot 4 \cdot 6}, \ldots, a_n = \frac{1 \cdot 3 \cdot 5 \cdots (2n-1)}{2 \cdot 4 \cdot 6 \cdots (2n)}$ for $n \geq 4$

(e) $\sum_{n=1}^{\infty} b_n$ with $b_n = \frac{1}{n} a_n$, a_n as in (d)

(f) $\sum_{n=1}^{\infty} a_n$, where $a_1 = 1$, $a_2 = \frac{\alpha\beta}{\gamma}$, $a_3 = \frac{\alpha(\alpha+1)(\beta)(\beta+1)}{\gamma(\gamma+1)}, \ldots,$

$$a_n = \frac{\alpha(\alpha+1)\cdots(\alpha+n-2)\beta(\beta+1)\cdots(\beta+n-2)}{\gamma(\gamma+1)\cdots(\gamma+n-2)} \text{ for } n \geq 4$$

Consider conditions on α, β, and γ to ensure convergence. *Hint:* Look at $\gamma > \alpha + \beta$. This series is called a *hypergeometric series.*

(g) $\sum_{n=1}^{\infty} \frac{\sin(n)}{\sqrt{n}}$

(h) $\sum_{n=1}^{\infty} \frac{n^2 + 2n + 1}{n^6 + 8n^2 + 4}$

(i) $\sum_{n=1}^{\infty} \frac{1}{(n^2+1)^{1/2}}$

(j) $\sum_{n=2}^{\infty} \frac{1}{\ln(n)^{\ln(n)}}$

(k) $\sum_{n=0}^{\infty} \frac{a_n}{10^n}$, where for each n, a_n is any integer from 0 to 9 inclusive

(l) $\sum_{n=1}^{\infty} \left[n \ln\left(\frac{2n+1}{2n-1}\right) - 1\right]$

(m) $\sum_{n=2}^{\infty} \left[1 + \frac{1}{n \ln(n)^p}\right]^{-n^q}$, p and q rcal

18. Suppose that $\sum_{n=1}^{\infty} a_n^2$ converges. Show that $\sum_{n=1}^{\infty} a_n/n$ converges.

19. Prove the following: *If* $\sum_{n=1}^{\infty} a_n$ *converges absolutely, and* $\sum_{n=1}^{\infty} b_n$ *is a rearrangement of* $\sum_{n=1}^{\infty} a_n$, *then* $\sum_{n=1}^{\infty} b_n$ *also converges absolutely, and* $\sum_{n=1}^{\infty} a_n = \sum_{n=1}^{\infty} b_n$. *Hint:* First, be clear as to what we mean by a rearrangement. Formulated precisely, we mean that there is a function φ assigning to each positive integer a positive integer (φ shifts the integers around) such that $\varphi(m) \neq \varphi(n)$ if $n \neq m$ and $b_n = a_{\varphi(n)}$.

Now to prove the theorem: consider two cases. First, say that each $a_n \geq 0$. Then each $b_n \geq 0$ also. If $S_n = \sum_{j=1}^{n} a_j$, then $\{S_n\}$ is a monotone nondecreasing sequence converging to $\sum_{n=1}^{\infty} a_n$. Show that the sequence of partial sums of $\sum_{n=1}^{\infty} b_n$ is bounded above by $\sum_{n=1}^{\infty} a_n$, hence $\sum_{n=1}^{\infty} b_n$ converges, and $\sum_{n=1}^{\infty} b_n \leq \sum_{n=1}^{\infty} a_n$. By reversing the argument $\left(\sum_{n=1}^{\infty} a_n \text{ is a rearrangement of } \sum_{n=1}^{\infty} b_n\right)$, conclude that also $\sum_{n=1}^{\infty} a_n \leq \sum_{n=1}^{\infty} b_n$.

Now consider the general case. The key is to show that if $\sum_{n=1}^{\infty} a_n$ converges absolutely, then $\sum_{n=1}^{\infty} a_n = \sum_{n=1}^{\infty} u_n - \sum_{n=1}^{\infty} v_n$, where $u_n \geq 0$ and $v_n \geq 0$. Then use the first case.

20. Prove the following: *If* $\sum_{n=1}^{\infty} a_n = L$, *then* $\sum_{n=1}^{\infty} b_n = L$, *where* $\sum_{n=1}^{\infty} b_n$ *is any series formed from* $\sum_{n=1}^{\infty} a_n$ *by inserting parentheses, leaving the order of the terms the same.*

21. For each n, put $c_n = \sum_{j=1}^{n} a_j b_{n-j}$. Then $\sum_{n=1}^{\infty} c_n$ is called the *Cauchy product series* of $\sum_{n=1}^{\infty} a_n$ and $\sum_{n=1}^{\infty} b_n$. Prove that $\sum_{n=1}^{\infty} c_n = \left(\sum_{n=1}^{\infty} a_n\right)\left(\sum_{n=1}^{\infty} b_n\right)$ when both $\sum_{n=1}^{\infty} a_n$ and $\sum_{n=1}^{\infty} b_n$ converge and, for each n, $a_n \geq 0$ and $b_n \geq 0$.

There is a theorem due to Mertens which draws the same conclusion under the weaker hypothesis that both series converge, and one converges absolutely. Try proving this as well, or look up a proof in the literature.

22. Let $a_n > 0$ for each n, and suppose that $\sum_{n=1}^{\infty} a_n$ diverges. Let $S_n = \sum_{j=1}^{n} a_j$. Show that $\sum_{n=1}^{\infty} [a_n/(S_n)^p]$ converges for $p > 1$ and diverges for $p \leq 1$.

23. Suppose that $a_n > 0$ for each n and that $\sum_{n=1}^{\infty} a_n^4$ converges. Prove that $\sum_{n=1}^{\infty} n^{-1/3} a_n^3$ converges.

24. Find the sum of the series

$$\sum_{n=1}^{\infty} \frac{1}{(4n-1)(4n+3)}.$$

Hint: Look at the sequence of partial sums.

25. Suppose that $a_n \neq 0$ and that $|a_{n+1}/a_n| \geq 1$ for $n \geq N$. Prove that $\sum_{n=1}^{\infty} a_n$ fails to converge even conditionally.

26. Let $u_n > 0$ for each positive integer n, and suppose that $\sum_{n=1}^{\infty} u_n$ diverges. Put $S_n = \sum_{j=1}^{n} u_j$ for $n = 1, 2, \ldots$.

(a) Prove that $\sum_{n=1}^{\infty} \frac{u_n}{1+u_n}$ diverges.

(b) Prove that $\sum_{n=1}^{\infty} \frac{u_n}{S_n}$ diverges.

(c) What can be said about the series $\sum_{n=1}^{\infty} \frac{u_n}{S_n^2}$?

27. Suppose that $\lim_{n\to\infty} nu_n = r$ and that r is nonzero (but may be infinite). Prove that $\sum_{n=1}^{\infty} u_n$ diverges.

1.4 Series of Functions

As with sequences, we can now consider series $\sum_{n=1}^{\infty} f_n$ in which the nth term is a function, defined over some set D of real numbers. Virtually all the machinery needed to handle such series was developed in Sections 1.2 and 1.3.

To begin, suppose that, for each x in a specified set D, the series $\sum_{n=1}^{\infty} f_n(x)$ converges, say to $g(x)$. We then say that $\sum_{n=1}^{\infty} f_n = g$ (pointwise on D). As examples,

$$\sum_{n=0}^{\infty} x^n = \frac{1}{1-x}, \qquad -1 < x < 1$$

(geometric series), and

$$\sum_{n=1}^{\infty} x^n = \frac{1}{1-x} - 1 = \frac{x}{1-x}, \qquad -1 < x < 1.$$

We say that $\sum_{n=1}^{\infty} f_n = g$ (uniformly on D) if the sequence $\left\{\sum_{j=1}^{n} f_j\right\}_{n=1}^{\infty}$ of partial sums converges uniformly to g on D.

As immediate consequences of the corresponding theorems for sequences, we have the following.

Theorem

Let $\sum_{n=1}^{\infty} f_n = g$ (uniformly on D). If α is in D, and each f_n is continuous at α, then g is continuous at α. That is,

$$\lim_{x\to\alpha} \sum_{n=1}^{\infty} f_n(x) = \sum_{n=1}^{\infty} \lim_{x\to\alpha} f_n(x).$$

Theorem

Let $\sum_{n=1}^{\infty} f_n = g$ *(uniformly on D). Suppose that* $[a, b]$ *is contained within D, and that each* f_n *is continuous on* $[a, b]$*. Then* $\sum_{n=1}^{\infty} \int_a^b f_n(x)\,dx = \int_a^b g(x)\,dx$*. That is,*

$$\sum_{n=1}^{\infty} \int_a^b f_n(x)\,dx = \int_a^b \left[\sum_{n=1}^{\infty} f_n(x)\right] dx.$$

Theorem

Let $\sum_{n=1}^{\infty} f_n = g$ *(pointwise on* $[a, b]$*). Suppose that each* f_n' *is continuous on* $[a, b]$*, and that* $\sum_{n=1}^{\infty} f_n'$ *converges uniformly on* $[a, b]$*. Then* $\sum_{n=1}^{\infty} f_n'(x) = g'(x)$ *for* $a < x < b$*. That is,*

$$\sum_{n=1}^{\infty} f_n'(x) = \left(\sum_{n=1}^{\infty} f_n\right)'(x).$$

It is often the case that interchange of various analytic processes ($\sum$, d/dx, $\int$) is extremely important in physical applications, as well as in purely mathematical considerations. We shall see many examples of this in later chapters, but here is a fairly simple one for immediate consumption.

Consider the problem of determining the shape at any time t of an elastic string that has been stretched between two points, set in an initial position, then released from rest with its end points held fixed, and allowed to vibrate freely. (Imagine a guitar string pulled out to some position and then released.)

To set up the problem in such a way as to be able to attack it mathematically, suppose that the string originally has length L, with its ends pegged down at $(0, 0)$ and $(0, L)$ on the x-axis. At time 0, the string is stretched to its initial position, given by the curve of $y = f(x)$. It is then released and allowed to vibrate in a vertical plane, with $u(x, t)$ denoting the y-coordinate at time $t > 0$ of the point originally at $(x, f(x))$. That is, at time t, the shape of the string is the graph of $y = u(x, t)$, $0 \leq x \leq L$ (see Fig. 1-8). We want to determine $u(x, t)$ for $0 \leq x \leq L$ and $t > 0$.

It is possible to derive a partial differential equation for u, but we will do this in Section 5.2. For now, argue on physical intuition (or take it on faith) that the motion should take the form of a superposition or sum of fundamental harmonics or waves. That is, $u(x, t)$ should have the general form

$$u(x, t) = \sum_{n=1}^{\infty} \alpha_n \sin\left(\frac{n\pi x}{L}\right) \cos(n\pi a t),$$

where a is a constant determined by the material the string is made from, and the α_n's are constants to be determined. A rigorous derivation will be given in Chapter 10. For now, concentrate on the problem of determining the α_n's.

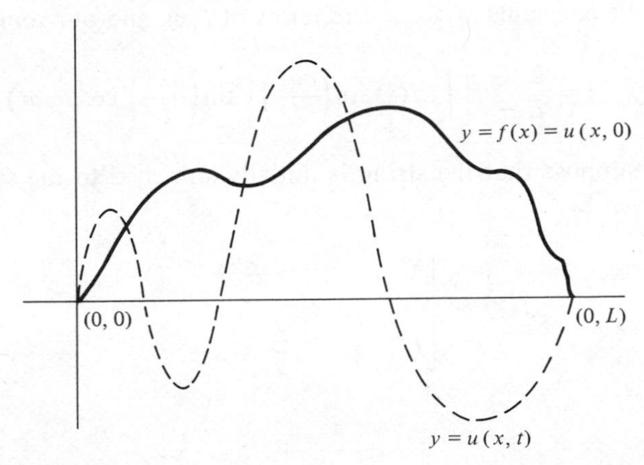

Fig. 1-8

Note that $u(0, t) = u(L, t) = 0$ for any time t, as required by the fixed end points of the string. The α_n's must be determined by our only other piece of data—the initial position of the string. We required that $u(x, 0) = f(x)$ for $0 \leq x \leq L$. That is,

$$u(x, 0) = f(x) = \sum_{n=1}^{\infty} \alpha_n \sin\left(\frac{n\pi x}{L}\right), \qquad 0 \leq x \leq L.$$

Here a clever trick involving interchange of $\sum$ and $\int$ comes to the rescue. Multiply by $\sin(k\pi x/L)$ to obtain

$$f(x) \sin\left(\frac{k\pi x}{L}\right) = \sum_{n=1}^{\infty} \alpha_n \sin\left(\frac{n\pi x}{L}\right) \sin\left(\frac{k\pi x}{L}\right).$$

Now integrate from 0 to L and interchange $\sum$ and $\int$:

$$\int_0^L f(x) \sin\left(\frac{k\pi x}{L}\right) dx = \sum_{n=1}^{\infty} \alpha_n \int_0^L \sin\left(\frac{n\pi x}{L}\right) \sin\left(\frac{k\pi x}{L}\right) dx.$$

But

$$\int_0^L \sin\left(\frac{n\pi x}{L}\right) \sin\left(\frac{k\pi x}{L}\right) dx = \begin{cases} 0, & \text{if } k \neq n, \\ \dfrac{L}{2}, & \text{if } k = n. \end{cases}$$

The infinite series on the right then collapses to the single term $\alpha_k \cdot L/2$, and we have, on the assumption that $\int_0^L \sum = \sum \int_0^L$,

$$\alpha_k = \frac{2}{L} \int_0^L f(x) \sin\left(\frac{k\pi x}{L}\right) dx, \qquad k = 1, 2, \ldots.$$

This gives us the constants $\alpha_1, \alpha_2, \ldots$ in terms of $f(x)$, and our solution is

$$u(x, t) = \frac{2}{L}\sum_{n=1}^{\infty}\left[\int_0^L f(\xi)\sin\left(\frac{n\pi\xi}{L}\right)\right]\sin\left(\frac{n\pi x}{L}\right)\cos(n\pi at)$$

For example, suppose that the string is initially stretched to the shape shown in Fig. 1-9, given by

$$f(x) = \begin{cases} x, & 0 \le x \le \dfrac{L}{2}, \\ L - x, & \dfrac{L}{2} \le x \le L. \end{cases}$$

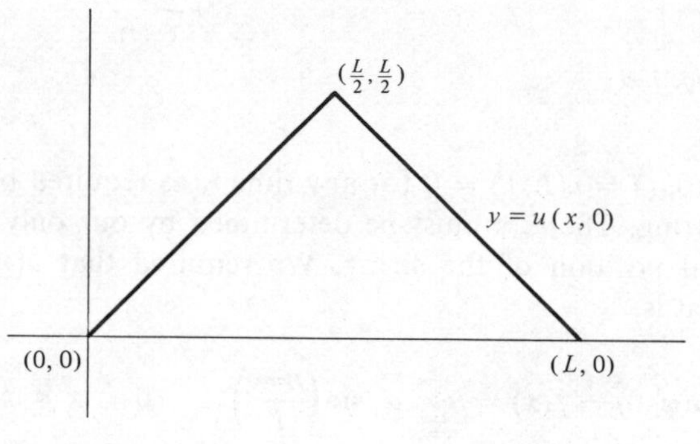

Fig. 1-9

This is what would happen if the string were picked up to a height of $L/2$ from its center point. We would then calculate

$$\begin{aligned}\int_0^L f(x)\sin\left(\frac{k\pi x}{L}\right)dx &= \int_0^{L/2} x\sin\left(\frac{k\pi x}{L}\right)dx + \int_{L/2}^{L}(L - x)\sin\left(\frac{k\pi x}{L}\right)dx \\ &= \frac{2L^2}{k^2\pi^2}\sin\left(\frac{k\pi}{2}\right) \\ &= \begin{cases} 0 & \text{if } k \text{ is even} \\ -\dfrac{2L^2}{(4n+3)^2\pi^2} & \text{if } k = 4n+3, n = 0, 1, \ldots, \\ \dfrac{2L^2}{(4n+1)^2\pi^2} & \text{if } k = 4n+1, n = 0, 1, \ldots. \end{cases}\end{aligned}$$

In this particular case, then, we would propose as the solution:

$$\begin{aligned}u(x, t) = {} & \frac{4L}{\pi^2}\sum_{n=1}^{\infty}\frac{1}{(4n+1)^2}\sin\left[\frac{(4n+1)\pi x}{L}\right]\cos[(4n+1)\pi at] \\ & -\frac{4L}{\pi^2}\sum_{n=1}^{\infty}\frac{1}{(4n+3)^2}\sin\left[\frac{(4n+3)\pi x}{L}\right]\cos[(4n+3)\pi at].\end{aligned}$$

So far, everything in this chapter (except the example above) has been a simple consequence of corresponding material on sequences. We can now give an important and useful test for uniform convergence of series which has no direct analogue for sequences.

WEIERSTRASS M-TEST

Suppose that $|f_n(x)| \leq M_n$ *for each* x *in* D, *and that* $\sum_{n=1}^{\infty} M_n$ *converges. Then* $\sum_{n=1}^{\infty} f_n$ *converges uniformly on* D.

PROOF

If x is in D, then $\sum_{n=1}^{\infty} |f_n(x)|$ converges by comparison with $\sum_{n=1}^{\infty} M_n$. Thus $\sum_{n=1}^{\infty} f_n(x)$ converges also. Put $g(x) = \sum_{n=1}^{\infty} f_n(x)$ for each x in D.

Now let $\varepsilon > 0$. We seek N such that $\left| g(x) - \sum_{j=1}^{n} f_j(x) \right| < \varepsilon$ if $n \geq N$ and x is in D. Now, for any x in D,

$$\left| g(x) - \sum_{j=1}^{n} f_j(x) \right| = \left| \sum_{j=1}^{\infty} f_j(x) - \sum_{j=1}^{n} f_j(x) \right|$$

$$= \left| \sum_{j=n+1}^{\infty} f_j(x) \right| \leq \sum_{j=n+1}^{\infty} |f_j(x)| \leq \sum_{j=n+1}^{\infty} M_j.$$

Choose N sufficiently large that $\sum_{j=n+1}^{\infty} M_j < \varepsilon$ if $n \geq N$. Then

$$\left| g(x) - \sum_{j=1}^{n} f_j(x) \right| < \varepsilon$$

for all $n \geq N$ and x in D, so $\left\{ \sum_{j=1}^{n} f_j \right\}_{n=1}^{\infty}$ converges uniformly to g on D. Then $\sum_{n=1}^{\infty} f_n$ converges uniformly to g on D, as was to be shown. ▮

EXAMPLE

$\sum_{n=1}^{\infty} [\sin(nx)/n^{3/2}]$ converges uniformly on $[0, \pi]$, for $|\sin(nx)/n^{3/2}| \leq 1/n^{3/2}$ and $\sum_{n=1}^{\infty} (1/n^{3/2})$ converges. In this example, the interval $[0, \pi]$ plays no role, and the series would converge uniformly on any set of real numbers.

Note that we have shown uniform convergence of the series without knowing the sum of the series, which is not obvious in this case. This is important, as finding the sum of a series is in general much more difficult than finding the limit of a sequence. ▮

EXAMPLE

$\sum_{n=1}^{\infty} e^{-nx}/(1 + nx)$ converges uniformly on $[\frac{1}{16}, 24]$. Here we have to be a little more clever than in the previous example. Note that, for any $n = 1, 2, \ldots,$

$$e^{-nx} \leq e^{-n/16}$$

and

$$\frac{1}{1 + nx} \leq \frac{1}{1 + n/16}$$

for $\frac{1}{16} \leq x \leq 24$. Then

$$\frac{e^{-nx}}{1 + nx} \leq \frac{e^{-n/16}}{1 + n/16} = \frac{16e^{-n/16}}{n + 16}.$$

Since $\sum_{n=1}^{\infty} 16e^{-n/16}/(n + 16)$ converges, then $\sum_{n=1}^{\infty} e^{-nx}/(1 + nx)$ converges uniformly on $[\frac{1}{16}, 24]$. As with the previous example, the sum of the series is not apparent. ∎

Exercises for Section 1.4

1. Test for pointwise and uniform convergence on the given set:

   (a) $\sum_{n=2}^{\infty} n(n - 1)x^{n-2}, \ -1 < x < 1$

   (b) $\sum_{n=2}^{\infty} n(n - 1)x^{n-2}, \ -\frac{1}{2} \leq x \leq \frac{1}{2}$

   (c) $\sum_{n=1}^{\infty} \frac{x}{n^2 + x^2}, \ -\infty < x < \infty$

   (d) $\sum_{n=1}^{\infty} ne^{-nx}, \ a \leq x \leq b$

   (e) $\sum_{n=1}^{\infty} e^{-nx} \sin(nx), \ 0 \leq x \leq \pi.$

2. Prove the following Cauchy-type theorem: $\sum_{n=1}^{\infty} f_n$ *converges uniformly on D if and only if, given* $\varepsilon > 0$, *there is some N such that*

$$\left| \sum_{j=n}^{n+k} f_j(x) \right| < \varepsilon$$

whenever $n \geq N$ *and x is in D, for each* $k = 1, 2, \ldots.$

3. Suppose that $\sum_{n=1}^{\infty} a_n$ converges. Prove that $\sum_{n=1}^{\infty} a_n x^n$ converges uniformly on [0, 1].

4. Prove the following theorem due to Abel: *Suppose that* $\{f_n\}$ *converges to* 0 *uniformly on D. Suppose that, for each n,* $\left|\sum_{j=1}^{n} h_j(x)\right| \leq M$ *for each x in D. Then,* $\sum_{n=1}^{\infty} f_n h_n$ *converges uniformly on D.*

5. Find the sum of the series

$$\sum_{n=1}^{\infty} \frac{x}{[1 + (n-1)x](1 + nx)}$$

and show that convergence is uniform on any interval not containing 0, nonuniform on any interval containing 0. *Hint:* Use a partial-fraction decomposition on

$$\frac{x}{[1 + (n-1)x](1 + nx)}.$$

6. Evaluate $\int_{\pi/2}^{\pi} \left[\sum_{n=1}^{\infty} e^{-nx} \cos(nx)\right] dx$. *Hint:* Your answer will be an infinite series.

7. (a) Show that $\sum_{n=0}^{\infty} (-1)^n x^n = 1/(1 + x)$ for $-1 < x < 1$. *Hint:* Look at the geometric series.
   (b) Show that the series converges uniformly in $[-\alpha, \alpha]$, where $0 < \alpha < 1$.
   (c) Show that

$$\ln(1 + x) = \sum_{n=0}^{\infty} \frac{(-1)^n x^{n+1}}{n+1}, \qquad -\alpha \leq x \leq \alpha.$$

8. Let $f(x) = \sum_{n=1}^{\infty} [\cos(nx)/n^3]$, $-\pi \leq x \leq \pi$.
   (a) Show that f is continuous on $[-\pi, \pi]$.
   (b) Evaluate $\int_0^{\pi} f(x)\, dx$.
   (c) Evaluate $f'(\pi/2)$. *Hint:* This answer will be an infinite series.

9. Let $f(x) = \sum_{n=1}^{\infty} [\sin(nx)/\sinh(nx)]$, $-\pi \leq x \leq \pi$.
   (a) Show that f is continuous on $[-\pi, \pi]$.
   (b) Evaluate $\int_{-\pi/2}^{\pi} f(x)\, dx$.
   (c) Show that f is twice differentiable on $(-\pi, \pi)$ and determine $f''(0)$.

10. Investigate convergence of $\sum_{n=1}^{\infty} [x^2/(1 + x^2)^n]$ on $[-1, 1]$.

11. Show that $\sum_{n=1}^{\infty} [\cos(nx)/n]$ converges uniformly on $[\pi/4, 3\pi/2]$. *Hint:* Use the condition of 4.

12. Show by an example that a uniformly convergent series of functions need not converge absolutely.

1.5 Power Series and Taylor's Theorem

We now particularize the discussion of Section 1.4 to an important special case, in which $f_n(x)$ is a polynomial of degree n of the form $a_n(x - a)^n$. A series $\sum_{n=0}^{\infty} a_n(x - a)^n$ is called a *power series*, with *center a* and *coefficient sequence* $\{a_n\}_{n=0}^{\infty}$. An example that we have already seen is the geometric series $\sum_{n=0}^{\infty} x^n$, with center 0 and coefficient sequence having each $a_n = 1$. For $|x| < 1$, this series converges to $1/(1 - x)$. Power series have a number of properties that make them particularly pleasant to work with, in addition to which they are also extremely useful in a wide variety of physical and mathematical problems.

Let us begin by considering the question of convergence of $\sum_{n=0}^{\infty} a_n(x - a)^n$. Clearly the series converges at a. Depending upon $\{a_n\}$, this may be the only point of convergence. But suppose, to avoid trivialities, that the series converges for some $b \neq a$. We claim, then, that the series converges absolutely at each point closer to a than b, that is, for each x with $|x - a| < |b - a|$.

To prove this, note that convergence of $\sum_{n=0}^{\infty} a_n(b - a)^n$ implies that $\lim_{n\to\infty} a_n(b - a)^n = 0$. Then, for some $M > 0$,

$$|a_n(b - a)^n| \leq M \qquad \text{for all } n.$$

Write

$$\begin{aligned} |a_n(x - a)^n| &= \left| a_n(b - a)^n \left(\frac{x - a}{b - a} \right)^n \right| \\ &= |a_n(b - a)^n| \left| \frac{x - a}{b - a} \right|^n \leq M \left| \frac{x - a}{b - a} \right|^n. \end{aligned}$$

We may now compare $\sum_{n=0}^{\infty} a_n(x - a)^n$ with $\sum_{n=0}^{\infty} M|(x - a)/(b - a)|^n$, the latter being M times a convergent geometric series, as $|(x - a)/(b - a)| < 1$ if $|x - a| < |b - a|$.

For convenience, we summarize this result.

Theorem

If $\sum_{n=0}^{\infty} a_n(b - a)^n$ converges, and $b \neq a$, then $\sum_{n=0}^{\infty} a_n(x - a)^n$ converges absolutely for all x with $|x - a| < |b - a|$.

Using this theorem, we may now delineate exactly three possibilities:

(1) Possibly $\sum_{n=0}^{\infty} a_n(x - a)^n$ converges only at $x = a$.

(2) At the other extreme, possibly $\sum_{n=0}^{\infty} a_n(x - a)^n$ converges for all real x.

(3) If not (1) and not (2), then $\sum_{n=0}^{\infty} a_n(x - a)^n$ converges for some $\alpha \neq a$ and diverges for some β. By the last theorem, $\sum_{n=0}^{\infty} a_n(x - a)^n$ diverges for any x with $|x - a| > |\beta - a|$ [for to converge at such an x would imply convergence of $\sum_{n=0}^{\infty} a_n(\beta - a)^n$, contradicting the choice of β]. The set of real numbers $|x - a|$, considered over values of x for which $\sum_{n=0}^{\infty} a_n(x - a)^n$ converges, is then bounded above, hence has a least upper bound, say r. Note now that, with this choice of r, $\sum_{n=0}^{\infty} a_n(x - a)^n$ converges whenever $|x - a| < r$ and diverges whenever $|x - a| > r$.

To sum up, then, a power series converges either only at its center, or on the whole real line, or on an interval about its center. The number r of case (3) is called the *radius of convergence* of the power series, and $(a - r, a + r)$ is called the *open interval of convergence*.

Behavior of $\sum_{n=0}^{\infty} a_n(x - a)^n$ at the points $a + r$ and $a - r$ in case (3) depends entirely upon the individual series. For example, $\sum_{n=0}^{\infty} x^n$ has $r = 1$ and $a = 0$, and diverges at 1 and -1; $\sum_{n=0}^{\infty} \frac{x^n}{n}$ has $a = 0$, $r = 1$, and diverges at 1, converges at -1.

The *complete interval of convergence* of a power series consists of all points at which the series converges. In case (1) this is a degenerate interval consisting of just the number a; in case (2) this is the whole real line; in case (3) it may be $(a - r, a + r)$, $[a - r, a + r)$, $(a - r, a + r]$, or $[a - r, a + r]$, depending upon convergence of the series at $a + r$ and $a - r$. As a means of consolidating the three cases, we will put $r = 0$ in case (1) and call the radius of convergence zero if the series converges only at its center, and $r = \infty$ in case (2), where the series converges on the whole real line. With this convention, a power series can always be said to have an interval domain of convergence. That this is not true of series in general is shown by $\sum_{n=1}^{\infty} 1/(x^2 - n^2)$, which converges for all real, noninterger x.

We now know a great deal about what to expect in the way of convergence of a power series. Ideally, we would like next some means of calculating the number r, which holds promise of being effective in a large number of cases one is likely to meet in practice. Here we encounter the second nice property of power series: generally r is not too difficult to compute from the coefficient sequence $\{a_n\}$. To see how this might be done, note that $\sum_{n=0}^{\infty} |a_n(x - a)^n|$ converges if $\lim_{n\to\infty} |a_n(x - a)^n|^{1/n} < 1$, and diverges if $\lim_{n\to\infty} |a_n(x - a)^n|^{1/n} > 1$

(assuming that this limit exists). But $|a_n(x - a)^n|^{1/n} = |a_n|^{1/n}|x - a|$. Assume that $\lim_{n\to\infty} |a_n|^{1/n}$ exists finite and is not zero; then

$$\sum_{n=0}^{\infty} |a_n(x - a)^n| \begin{cases} \text{converges} & \text{if } |x - a| < \dfrac{1}{\lim_{n\to\infty} |a_n|^{1/n}} \\ \text{diverges} & \text{if } |x - a| > \dfrac{1}{\lim_{n\to\infty} |a_n|^{1/n}} \end{cases}.$$

In this event the radius of convergence is given by

$$r = \frac{1}{\lim_{n\to\infty} |a_n|^{1/n}}.$$

If $\lim_{n\to\infty} |a_n|^{1/n} = 0$, then for any x, $\lim_{n\to\infty} |a_n|^{1/n}|x - a| = \lim_{n\to\infty} |a_n(x - a)^n|^{1/n} = 0$, and the series converges on the whole real line or has infinite radius of convergence. If $\lim_{n\to\infty} |a_n|^{1/n} = \infty$, then, for any $x \neq a$, $\lim_{n\to\infty} |a_n|^{1/n}|x - a| > 1$, so the series diverges for all $x \neq a$, hence has zero radius of convergence. In all cases, then, whenever $\lim_{n\to\infty} |a_n|^{1/n}$ exists, it tells us the radius of convergence of $\sum_{n=0}^{\infty} a_n(x - a)^n$.

A similar calculation, based upon the ratio rather than the root test, shows that, in the case that $\lim_{n\to\infty} |a_{n+1}/a_n|$ exists finite and is not zero, then the radius of convergence of $\sum_{n=0}^{\infty} a_n(x - a)^n$ is

$$\frac{1}{\lim_{n\to\infty} |a_{n+1}/a_n|}.$$

Further, $r = \infty$, if $\lim_{n\to\infty} |a_{n+1}/a_n| = 0$, and $r = 0$ if $\lim_{n\to\infty} |a_{n+1}/a_n| = \infty$.

For the eventuality that these limits do not exist, the Cauchy–Hadamard formula gives $1/r = \lim\sup\{|a_n|^{1/n}\}$ (see Exercise 29). This formula, although more widely applicable, is of course more difficult to compute with.

Note that we have only worked with absolute convergence in drawing our conclusion about the radius of convergence. This is justified by the fact that within its open interval of convergence a power series does converge absolutely.

EXAMPLES

$\sum_{n=0}^{\infty} x^n/n!$ converges for all real x. For $a_n = 1/n!$ here, and

$$\lim_{n\to\infty} \frac{a_{n+1}}{a_n} = \lim_{n\to\infty} \frac{n!}{(n + 1)!} = \lim_{n\to\infty} \frac{1}{n + 1} = 0,$$

implying an infinite radius of convergence.

$\sum_{n=0}^{\infty} (\frac{2}{3})^n(x - 2)^n$ converges for $|x - 2| < \frac{3}{2}$, since $\lim_{n\to\infty} [(\frac{2}{3})^n]^{1/n} = \lim_{n\to\infty} \frac{2}{3} = \frac{2}{3}$. The open interval of convergence of $\sum_{n=0}^{\infty} (\frac{2}{3})^n(x - 2)^n$ is then $(2 - \frac{3}{2}, 2 + \frac{3}{2})$, or $(\frac{1}{2}, \frac{7}{2})$. At the end points, we must test separately. At $\frac{7}{2}$, we have $\sum_{n=0}^{\infty} (\frac{2}{3})^n(\frac{7}{2} - 2)^n = \sum_{n=0}^{\infty} (\frac{2}{3})^n(\frac{3}{2})^n = \sum_{n=0}^{\infty} (1)^n$, a divergent geometric series. At $\frac{1}{2}$, we have

$$\sum_{n=0}^{\infty} \left(\frac{2}{3}\right)^n \left(\frac{1}{2} - 2\right)^n = \sum_{n=0}^{\infty} (-1)^n,$$

again a divergent series. The interval of convergence of $\sum_{n=0}^{\infty} (\frac{2}{3})^n(x - 2)^n$ is then $(\frac{1}{2}, \frac{7}{2})$. ∎

We now turn to the properties that make power series such a powerful tool in a variety of applications. We shall show first that convergence is uniform on any closed interval within the open interval of convergence.

Theorem

Suppose that the closed interval $[c, d]$ lies within the open interval of convergence of $\sum_{n=0}^{\infty} a_n(x - a)^n$. Then $\sum_{n=0}^{\infty} a_n(x - a)^n$ converges uniformly on $[c, d]$.

PROOF

Pick a point γ in $[c, d]$ at maximum distance from a. Since γ is in the open interval of convergence, then $\sum_{n=0}^{\infty} |a_n(\gamma - a)^n|$ converges. Further, if $c \leq x \leq d$, then

$$|a_n(x - a)^n| \leq |a_n(\gamma - a)^n|.$$

The conclusion is now immediate by the Weierstrass M-test with

$$M_n = |a_n(\gamma - a)^n|.$$ ∎

Since each term of a power series is a polynomial, hence continuous, we then have immediately the following:

Theorem

A power series converges to a continuous function on each closed interval $[c, d]$ within its open interval of convergence. Further,

$$\int_c^d \left[\sum_{n=0}^{\infty} a_n(x - a)^n\right] dx = \sum_{n=0}^{\infty} \int_c^d a_n(x - a)^n\, dx$$

$$= \sum_{n=0}^{\infty} \frac{a_n}{n + 1} [(d - a)^{n+1} - (c - a)^{n+1}].$$

Finally, we would like to know that we can differentiate a power series term by term. That is, if $f(x) = \sum_{n=0}^{\infty} a_n(x-a)^n$, we want to be sure that $f'(x) = \sum_{n=1}^{\infty} na_n(x-a)^{n-1}$ within the interval of convergence. The implications of this will be far reaching. Noting that $\sum_{n=1}^{\infty} na_n(x-a)^{n-1}$ is again a power series, we will naturally be led to try to write $f''(x) = \sum_{n=2}^{\infty} n(n-1)a_n(x-a)^{n-2}$, $f'''(x) = \sum_{n=3}^{\infty} n(n-1)(n-2)a_n(x-a)^{n-3}$, and so on indefinitely. The only hole in this line of thought is that conceivably the process of differentiation shrinks the interval of convergence. In fact, on the face of it, we may, after k differentiations, reach a series whose radius of convergence is zero. We shall now show that this does not happen.

Lemma

$\sum_{n=0}^{\infty} a_n(x-a)^n$ *and* $\sum_{n=1}^{\infty} na_n(x-a)^{n-1}$ *have the same open interval of convergence.*

PROOF

Let r be the radius of convergence of $\sum_{n=0}^{\infty} a_n(x-a)^n$ and $\hat{r}$ that of

$$\sum_{n=1}^{\infty} na_n(x-a)^{n-1}.$$

We shall first show that $r \le \hat{r}$. To do this, suppose that $\sum_{n=0}^{\infty} a_n(x-a)^n$ converges at $x_0 \ne a$. Let $|x-a| < |x_0-a|$. We shall prove that $\sum_{n=1}^{\infty} na_n(x-a)^{n-1}$ converges. Note that

$$|na_n(x-a)^{n-1}| = \frac{n}{|x_0-a|}|a_n||x_0-a|^n\left|\frac{x-a}{x_0-a}\right|^{n-1}.$$

Now, $\sum_{n=0}^{\infty} a_n(x_0-a)^n$ converges, so $\lim_{n\to\infty} a_n(x_0-a)^n = 0$. For some $M > 0$, $|a_n(x_0-a)^n| \le M$ for all n. Then

$$|na_n(x-a)^{n-1}| \le \frac{n}{|x_0-a|}M\left|\frac{x-a}{x_0-a}\right|^{n-1}.$$

A straightforward application of the ratio test shows that

$$\sum_{n=1}^{\infty} \frac{Mn}{|x_0-a|}\left|\frac{x-a}{x_0-a}\right|^{n-1}$$

converges. For

$$\lim_{n\to\infty} \frac{\dfrac{M(n+1)}{|x_0 - a|}\left|\dfrac{x-a}{x_0-a}\right|^n}{\dfrac{Mn}{|x_0 - a|}\left|\dfrac{x-a}{x_0-a}\right|^{n-1}} = \lim_{n\to\infty}\left(\frac{n+1}{n}\right)\left|\frac{x-a}{x_0-a}\right| = \left|\frac{x-a}{x_0-a}\right| < 1.$$

Hence, by comparison, $\sum_{n=1}^{\infty} |na_n(x-a)^{n-1}|$ converges. But then $\sum_{n=1}^{\infty} na_n(x-a)^{n-1}$ converges for all x with $|x - a| < |x_0 - a|$, so $\hat{r}$ is at least as large as r.

We now show that $\hat{r} = r$. If $\sum_{n=0}^{\infty} a_n(x-a)^n$ converges on the whole real line, so does $\sum_{n=1}^{\infty} na_n(x-a)^{n-1}$, by the above argument.

Suppose, then, that $0 < r < \infty$. Let $|y - a| > r$, and suppose that y is in the open interval of convergence of $\sum_{n=1}^{\infty} na_n(x-a)^{n-1}$. Then $\sum_{n=1}^{\infty} |na_n(y-a)^{n-1}|$ converges. But

$$\begin{aligned} |a_n(y-a)^n| &= |na_n(y-a)^{n-1}| \cdot \left|\frac{y-a}{n}\right| \\ &\leq |na_n(y-a)^{n-1}| \end{aligned}$$

for n sufficiently large that $|(y-a)/n| \leq 1$, say for $n \geq N$. Then $\sum_{n=N}^{\infty} a_n(y-a)^n$ converges by comparison, a contradiction, as $|y - a| > r$. Thus $r = \hat{r}$, and the proof is complete. ∎

It might appear that we have made the proof a good deal harder than necessary, since

$$\lim_{n\to\infty}\left|\frac{a_{n+1}}{a_n}\right| = \lim_{n\to\infty}\left|\frac{(n+1)a_{n+1}}{na_n}\right|$$

would seem to imply in one line that the two series have the same radius of convergence. This argument works when these limits exist. However, the proof given is valid even for those cases where a_{n+1}/a_n has no limit as $n \to \infty$ and so is to be preferred in general.

However, the hard work is now behind us and we can proceed to the result that we have been aiming for.

Theorem

Suppose that $[c, d]$ is contained within the open interval of convergence of $\sum_{n=0}^{\infty} a_n(x-a)^n$. Then

$$\frac{d}{dx}\left[\sum_{n=0}^{\infty} a_n(x-a)^n\right] = \sum_{n=1}^{\infty} na_n(x-a)^{n-1} \qquad \text{for } c \leq x \leq d.$$

The proof consists of simply observing that $\sum_{n=1}^{\infty} na_n(x-a)^{n-1}$ has the same open interval of convergence as the original series, hence converges uniformly on $[c, d]$, and that $d[a_n(x-a)^n]/dx = na_n(x-a)^{n-1}$. ∎

Note that $\sum_{n=1}^{\infty} na_n(x-a)^{n-1}$ is again a power series, and we can apply our theorems again on $[c, d]$ to obtain

$$\frac{d^2}{dx^2}\left[\sum_{n=0}^{\infty} a_n(x-a)^n\right] = \sum_{n=2}^{\infty} n(n-1)a_n(x-a)^{n-2}.$$

In fact, we have now shown that we can continue differentiating term by term indefinitely, as we conjectured before we should be able to do. At the kth step, we get

$$\frac{d^k}{dx^k}\left[\sum_{n=0}^{\infty} a_n(x-a)^n\right] = \sum_{n=k}^{\infty} n(n-1)\cdots(n-k+1)(x-a)^{n-k}.$$

EXAMPLE

If $0 < \alpha < 1$, then, on $[-\alpha, \alpha]$ we have

$$\frac{1}{1-x} = \sum_{n=0}^{\infty} x^n,$$

$$\frac{1}{(1-x)^2} = \sum_{n=1}^{\infty} nx^{n-1},$$

$$\frac{2}{(1-x)^3} = \sum_{n=2}^{\infty} n(n-1)x^{n-2},$$

$$\frac{6}{(1-x)^4} = \sum_{n=3}^{\infty} n(n-1)(n-2)x^{n-3}, \quad \text{and so on.}$$ ∎

The reader may sense here the germ of a method for suming series. Given, for example, the series $\sum_{n=3}^{\infty} n(n-1)(n-2)x^{n-3}$, a clever enough person might recognize it as the third derivative of $\sum_{n=0}^{\infty} x^n$, which he knows to be $1/(1-x)$, and thereby sum the given series to $6/(1-x)^4$. We shall develop this method in the applications at the end of the chapter. For now, we want to consider the reverse of the situation treated up to this point.

Within its interval of convergence, a power series represents a function f, defined by $f(x) = \sum_{n=0}^{\infty} a_n(x-a)^n$. We now ask: Given $f(x)$ in an interval containing a, can we find $a_0, a_1, \ldots$ so that $f(x) = \sum_{n=0}^{\infty} a_n(x-a)^n$? That is, can f be represented by a power series?

Clearly f will have to have some nice properties if it is to have such a representation. At the very least, f will have to have derivatives of all orders in some interval about a. What else is needed, if anything, is not obvious.

At this point, let us back off to a simpler problem which eventually will yield some insight into the expansion problem. Suppose that, instead of attempting to write $f(x)$ as a power series, we try to approximate $f(x)$ by a polynomial $P_m(x)$ of degree m, written in powers of $x - a$. By approximate we mean here that we want

$$P_m(a) = f(a),\ P_m'(a) = f'(a), \ldots, P_m^{(m)}(a) = f^{(m)}(a),$$

where superscripts in parentheses denote order of derivatives, and, by convention, $f^{(0)}(x) = f(x)$. We need assume here only that f be m times differentiable at a.

To find $P_m(x)$ explicitly, write

$$P_m(x) = \sum_{j=0}^{m} a_j(x - a)^j.$$

For $f(a) = P_m(a)$, we need $a_0 = f(a)$. Next

$$P_m'(x) = a_1 + 2a_2(x - a) + \cdots + ma_m(x - a)^{m-1},$$

so we have $P_m'(a) = f'(a) = a_1$.

Continuing in this way, we get

$$P_m''(a) = 2a_2 = f(a), \qquad \text{so } a_2 = \frac{f''(a)}{2},$$

and, in general,

$$a_k = \frac{f^{(k)}(a)}{k!}, \qquad k = 0, 1, \ldots, m.$$

Thus

$$P_m(x) = \sum_{j=0}^{m} \frac{f^{(j)}(a)}{j!}(x - a)^j.$$

The question now arises as to how good this approximation is near a. One suspects that it might be fairly good if m is large, as the graph of P_m has the same slope, acceleration, ..., as that of f at a, and $f(a) = P_m(a)$ ensures that the graphs both pass through $(a, f(a))$.

To obtain some estimate of the difference between $f(x)$ and $P_m(x)$ near a, we adopt a new point of view which will eventually give us $P_m(x)$ from $f(x)$, but with an error term as well.

Note that, if f' is continuous on $[c, d]$, and a and x are in $[c, d]$, then

$$\int_a^x f'(t)\,dt = f(x) - f(a),$$

so

$$f(x) = f(a) + \int_a^x f'(t)\,dt = P_0(x) + \int_a^x f'(t)\,dt.$$

If f'' is also continuous on $[c, d]$, we can now integrate by parts:

$$f(x) = P_0(x) + \int_a^x f'(t)\,dt \qquad \begin{aligned} u &= f'(t) \\ dv &= dt \\ v &= -(x - t) \end{aligned}$$

$$= P_0(x) + \left[-(x - t)\cdot f'(t)\Big|_a^x - \int_a^x -(x - t)f''(t)\,dt \right]$$

$$= P_0(x) + (x - a)f'(a) + \int_a^x (x - t)f''(t)\,dt$$

$$= P_1(x) + \int_a^x (x - t)f''(t)\,dt.$$

If f''' is continuous on $[c, d]$, we can integrate by parts again [with $u = f''(t)$, $dv = (x - t)\,dt$, $v = -(x - t)^2/2$] to obtain

$$f(x) = P_1(x) - \frac{f''(t)(x - t)^2}{2}\Big|_a^x - \int_a^x -\frac{(x - t)^2 f'''(t)}{2}\,dt$$

$$= P_1(x) + \frac{f''(a)(x - a)^2}{2} + \int_a^x \frac{f'''(t)(x - t)^2}{2}\,dt$$

$$= P_2(x) + \int_a^x \frac{f'''(t)(x - t)^2}{2}\,dt.$$

We can keep this up as long as f has continuous derivatives on $[c, d]$. *At step k, we obtain*

$$f(x) = P_k(x) + \int_a^x \frac{(x - t)^k f^{(k+1)}(t)}{k!}\,dt.$$

This is called *Taylor's theorem with remainder*. What we have gained by these calculations is that we always obtain $f(x) = P_k(x) +$ an error term. That is, the expression

$$\int_a^x \frac{(x - t)^k f^{(k+1)}(t)}{k!}\,dt$$

gives us a means of estimating the difference between $f(x)$ and the approximating polynomial $P_k(x)$, hence a measure of how good the approximation is. To illustrate one crude estimate we might make without any further assumptions on f, note that, if $|f^{(k+1)}(t)| \leq M$ for $c \leq t \leq d$, then

$$\left| \int_a^x \frac{f^{(k+1)}(t)(x-t)^k}{k!}\, dt \right| \leq \frac{M}{k!} \int_a^x |x-t|^k\, dt = \frac{M}{(k+1)!} |x-a|^{k+1},$$

so that

$$|f(x) - P_k(x)| \leq \frac{M|x-a|^{k+1}}{(k+1)!},$$

which is small if x is close to a.

There are many different expressions for the remainder which one can derive. For example, by a mean value theorem for integrals, there is some θ between a and x such that

$$\int_a^x \frac{f^{(k+1)}(t)(x-t)^k}{k!}\, dt = \frac{f^{(k+1)}(\theta)}{k!} \int_a^x (x-t)^k\, dt = \frac{f^{(k+1)}(\theta)(x-a)^{k+1}}{(k+1)!}.$$

With this remainder, Taylor's theorem says that, for some θ between a and x,

$$f(x) = P_k(x) + \frac{f^{(k+1)}(\theta)(x-a)^{k+1}}{(k+1)!}.$$

Note that when $k = 0$, this reduces to a mean value theorem: $f(x) = f(a) + f'(\theta)(x-a)$ *for some θ between x and a.* For this reason, Taylor's theorem with remainder is sometimes called an *extended* mean value theorem.

This ends our digression into polynomial approximations. Returning to the problem of expanding f in a power series about a, suppose that f has derivatives of all orders in some interval $[c, d]$, with $c < a < d$. Write, for x near a,

$$\begin{aligned} f(x) &= P_m(x) + R_m(x) \\ &= \sum_{j=0}^{m} \frac{f^{(j)}(a)(x-a)^j}{j!} + R_m(x), \end{aligned}$$

where m is any positive integer and $R_m(x)$ is the remainder term. Then

$$\left| f(x) - \sum_{j=0}^{m} \frac{f^{(j)}(a)(x-a)^j}{j!} \right| \leq |R_m(x)|.$$

If $|R_m(x)| \to 0$ as $m \to \infty$, we will then have convergence of

$$\sum_{j=0}^{\infty} \frac{f^{(j)}(a)(x-a)^j}{j!}$$

to $f(x)$, giving us exactly the representation sought. We call

$$\sum_{j=0}^{\infty} \frac{f^{(j)}(a)(x-a)^j}{j!}$$

the *Taylor series* (or, if $a = 0$, the *MacLaurin series*) of f about a.

In summary, then, we can expand $f(x)$ in a Taylor series

$$\sum_{n=0}^{\infty} \frac{f^{(n)}(a)(x-a)^n}{n!},$$

valid in an interval $|x - a| < r$ if f has all orders of derivatives in this interval and if, for each x in the interval, $R_n(x) \to 0$ as $n \to \infty$.

A function f that can be represented by a Taylor series about a is said to be analytic at a. Existence of derivatives of all orders, with no conditions on the remainder term, does not ensure analyticity. For example, let

$$f(x) = \begin{cases} e^{-1/x^2} & \text{for } x \neq 0, \\ 0 & \text{for } x = 0. \end{cases}$$

The reader can show (with some effort) that $f^{(j)}(0) = 0$ for $j = 0, 1, 2, \ldots$. The Taylor series then converges to 0 for all x, and to $f(x)$ only at $x = 0$. Thus the series does not represent $f(x)$ in any interval about 0.

It is natural at this point to ask how many different power-series representations a given $f(x)$ might have about a. The answer is that the Taylor series is the only possibility. For suppose that $f(x) = \sum_{n=0}^{\infty} a_n(x-a)^n$ in $(a - r, a + r)$. Then $f(a) = a_0$ by substitution. We can differentiate the series term by term within $(a - r, a + r)$ to obtain $f'(x) = \sum_{n=1}^{\infty} na_n(x-a)^{n-1}$. Then $a_1 = f'(a)$. Similarly, $f''(x) = \sum_{n=2}^{\infty} n(n-1)a_n(x-a)^{n-2}$, so $a_2 = f''(a)/2$. Continuing in this way yields $a_n = f^{(n)}(a)/n!$, and the series is a Taylor series.

Some examples of Taylor series expansions about zero of familiar functions are

$$e^x = \sum_{n=0}^{\infty} \frac{x^n}{n!}, \qquad -\infty < x < \infty,$$

$$\sin(x) = \sum_{n=0}^{\infty} \frac{(-1)^n x^{2n+1}}{(2n+1)!}, \qquad -\infty < x < \infty,$$

$$\cos(x) = \sum_{n=0}^{\infty} \frac{(-1)^n x^{2n}}{(2n)!}, \qquad -\infty < x < \infty,$$

$$\ln(1 + x) = \sum_{n=1}^{\infty} \frac{(-1)^n x^n}{n}, \qquad -1 < x < 1$$

[this result can be computed directly, with

$$\frac{1}{n!}\frac{d^n}{dx^n}\ln(1 + x)\bigg|_{x=0} = (-1)^n$$

or by integrating $1/(1 + x) = \sum_{n=0}^{\infty} (-1)^n x^n$ term by term, as in Exercise 7(c), Section 1.4]; and

$$\arctan(x) = \sum_{n=1}^{\infty} \frac{(-1)^{n-1}x^{2n-1}}{2n - 1}, \qquad |x| < 1.$$

We conclude this section with some examples that show a variety of applications of power-series expansions and Taylor polynomial approximations.

EXAMPLE

Approximate $\sin(x)$ near 0 by a polynomial to within a tolerance of 10^{-3}. To do this, write

$$\sin(x) = P_m(x) + \frac{1}{m!}\int_0^x (x - t)^m \frac{d^{m+1}}{dt^{m+1}} [\sin(t)]\, dt.$$

The phrase "near 0" is a little inexact, so let us say that we want the approximation for $-\varepsilon \le x \le \varepsilon$, with $\varepsilon < 1$. Thus we must choose m so that

$$|\sin(x) - P_m(x)| \le \frac{1}{10^3} \qquad \text{for } -\varepsilon \le x \le \varepsilon.$$

Now

$$\frac{d^{m+1}}{dt^{m+1}}\sin(t) = \begin{cases} \pm\sin(t) \\ \pm\cos(t) \end{cases} \qquad \text{depending upon } m.$$

In any case,

$$\left|\frac{d^{m+1}}{dt^{m+1}}\sin(t)\right| \le 1.$$

Further,

$$|x - t|^m \le \varepsilon^m \qquad \text{if } -\varepsilon \le x \le \varepsilon \text{ and } 0 \le t \le x, \text{ or } -x \le t \le 0$$

(see Fig. 1-10). Then

$$\left|\frac{1}{m!}\int_0^x (x - t)^m \frac{d^{m+1}}{dt^{m+1}}\sin(t)\, dt\right| \le \frac{1}{m!}|x|\varepsilon^m \le \frac{\varepsilon^{m+1}}{m!}.$$

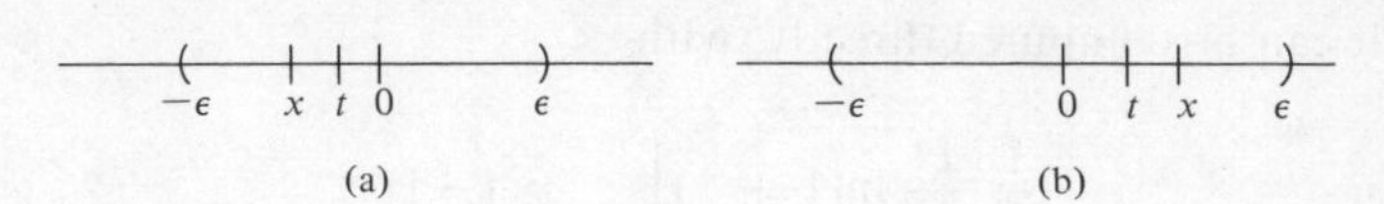

Fig. 1-10

So choose m sufficiently large that $\varepsilon^{m+1}/m! \leq 10^{-3}$. For example, if $\varepsilon = \frac{1}{2}$, take $m = 5$, as then

$$\frac{\varepsilon^{m+1}}{m!} = \frac{(\frac{1}{2})^6}{5!} = \frac{1}{7680} < \frac{1}{10^3}.$$

With $m = 5$, our approximating polynomial would be

$$P_5(x) = \sum_{j=0}^{5} \frac{d^j}{dx^j} \sin(x)\Big|_{x=0} \left(\frac{1}{j!}\right) x^j$$

$$= x - \frac{x^3}{3!} + \frac{x^5}{5!}.$$

In this particular example, $P_6(x) = P_5(x)$, as

$$\frac{d^6}{dx^6} \sin(x)\Big|_{x=0} = 0.$$

We can then take P_6 as our approximation to $\sin(x)$ on $[-\varepsilon, \varepsilon]$, with an error not expeeding $\varepsilon^7/6!$. With $\varepsilon = \frac{1}{2}$ as before, this gives us

$$\left|\sin(x) - \left(x - \frac{x^3}{3!} + \frac{x^5}{5!}\right)\right| < \frac{1}{2^7 6!}$$

for $-\frac{1}{2} < x < \frac{1}{2}$, a pretty good approximation considering that we need only a fifth-degree polynomial. ∎

EXAMPLE

We shall now show how the result of the previous example might be useful in a numerical calculation. Suppose that we want to compute

$$\int_{1/4}^{1/2} \frac{\sin(x)}{x}\, dx.$$

Since $\sin(x)/x$ has no elementary antiderivative, the fundamental theorem of calculus is of no use here. So, we approximate. On $[\frac{1}{4}, \frac{1}{2}]$,

$$\left|\frac{\sin(x)}{x} - \frac{P_6(x)}{x}\right| = \frac{1}{x}\,|\sin(x) - P_6(x)|$$

$$\leq \frac{1}{\frac{1}{4}} \frac{1}{2^7 6!} = \frac{1}{2^5 6!}$$

using the estimate obtained above. Then

$$\left| \int_{1/4}^{1/2} \frac{\sin(x)}{x}\, dx - \int_{1/4}^{1/2} \frac{P_6(x)}{x}\, dx \right| \le \int_{1/4}^{1/2} \left| \frac{\sin(x)}{x} - \frac{P_6(x)}{x} \right| dx$$

$$\le \int_{1/4}^{1/2} \frac{1}{2^5 6!}\, dx = \frac{1}{2^7 6!}.$$

Then

$$\int_{1/4}^{1/2} \frac{1}{x} P_6(x)\, dx - \frac{1}{2^7 6!} \le \int_{1/4}^{1/2} \frac{\sin(x)}{x}\, dx \le \int_{1/4}^{1/2} \frac{1}{x} P_6(x)\, dx + \frac{1}{2^7 6!}.$$

Since

$$\int_{1/4}^{1/2} \frac{1}{x} P_6(x)\, dx = \int_{1/4}^{1/2} \left(1 - \frac{x^2}{6} + \frac{x^4}{120}\right) dx,$$

we can easily calculate this integral, obtaining an estimate to within $1/2^7 6!$ of the desired integral.

By taking m larger, we can increase the accuracy to any tolerance we want, at the expense of more work in computing. ∎

EXAMPLE

Series expansions are often useful in taking limits. For example, consider

$$\lim_{x \to \infty} x[\ln(x + 1) - \ln(x)].$$

Note that $\ln(1 + x) - \ln(x) = \ln(1 + 1/x)$. This suggests that we begin with the expansion

$$\ln(1 + t) = \sum_{n=1}^{\infty} \frac{(-1)^{n+1} t^n}{n}, \qquad |t| < 1,$$

obtained above. Putting $t = 1/x$, then for $|x| > 1$ we have

$$\ln\left(1 + \frac{1}{x}\right) = \sum_{n=1}^{\infty} \frac{(-1)^{n+1}}{n x^n}.$$

Then

$$x \ln\left(1 + \frac{1}{x}\right) = \sum_{n=1}^{\infty} \frac{(-1)^{n+1}(1/x)^{n-1}}{n} \qquad \text{for } |x| > 1.$$

This is not a power series, but it is a valid series expansion of $x \ln(1 + 1/x)$ for $x > 1$ or $x < -1$. We are interested in $x > 1$, as we want to let $x \to \infty$.

By the theorem of Section 1.3 on alternating series, $\sum_{n=1}^{\infty} [(-1)^{n+1}/nx^{n-1}]$ differs from its first term, 1, by no more than the absolute value of its second term, $1/2x$. Then

$$\left| \sum_{n=1}^{\infty} \frac{(-1)^{n+1}}{nx^{n-1}} - 1 \right| \leq \left| \frac{1}{2x} \right|.$$

Then

$$\left| x \ln\left(1 + \frac{1}{x}\right) - 1 \right| \leq \left| \frac{1}{2x} \right|, \qquad |x| > 1.$$

As $x \to \infty$, $1/2x \to 0$, and we conclude that

$$\lim_{x \to \infty} x \ln\left(1 + \frac{1}{x}\right) = 1.$$

The reader should also do this problem by L'Hospital's Rule (see Appendix II at the end of this chapter). ∎

EXAMPLE

It is not always convenient or necessary to calculate all the numbers $f^{(n)}(a)/n!$ to obtain a Taylor series for $f(x)$. Often it is possible to work from known series and manipulate to obtain the series we want. For example, to expand $1/(1 - x)$ about $\frac{1}{2}$, write

$$\frac{1}{1-x} = \sum_{n=0}^{\infty} x^n,$$

an expansion about 0 for $-1 < x < 1$. Now translate terms

$$\frac{1}{1-x} = \frac{1}{\frac{1}{2} - (x - \frac{1}{2})} = \frac{2}{2(\frac{1}{2} - [x - \frac{1}{2}])} = \frac{2}{1 - 2(x - \frac{1}{2})}$$

$$= 2\left(\frac{1}{1 - 2(x - \frac{1}{2})}\right) = 2 \sum_{n=0}^{\infty} [2(x - \tfrac{1}{2})]^n = \sum_{n=0}^{\infty} 2^{n+1}(x - \tfrac{1}{2})^n,$$

valid for $|2(x - \frac{1}{2})| < 1$, or $|x - \frac{1}{2}| < \frac{1}{2}$.

Note that the open interval of convergence, $(0, 1)$, extends from the center of the series, $\frac{1}{2}$, to the "nearest" point of divergence, 1, so that the radius of convergence is $\frac{1}{2}$. It is only coincidental here that the center and the radius of convergence are both $\frac{1}{2}$.

As a second illustration of this method, suppose that we want to expand $1/(x^2 + 2x + 2)$ about -1. Write

$$x^2 + 2x + 2 = (x + 1)^2 + 1 = 1 - (-(x + 1)^2).$$

Then

$$\frac{1}{x^2 + 2x + 2} = \frac{1}{1 - [-(x+1)^2]} = \sum_{n=0}^{\infty} [-(x+1)^2]^n = \sum_{n=0}^{\infty} (-1)^n (x+1)^{2n},$$

a valid representation about -1 for $|x - (-1)| < 1$, or $-2 < x < 0$. ▌

EXAMPLE

Sum the series $\sum_{n=2}^{\infty} n(n-1)(-1)^n x^n$. To attack a problem of this type, we try to exploit the fact that a power series can be differentiated and integrated term by term within its open interval of convergence.

Note first that

$$\lim_{n\to\infty} \left| \frac{(n+1)n(-1)^{n+1}}{(n-1)n(-1)^n} \right| = 1,$$

so the radius of convergence here is 1. Put

$$f(x) = \sum_{n=2}^{\infty} (-1)^n n(n-1)x^n, \qquad -1 < x < 1.$$

The factors n and $n - 1$ suggest derivatives, but the power of x is wrong for that. Rewrite

$$\begin{aligned} f(x) &= x^2 \sum_{n=2}^{\infty} (-1)^n n(n-1)x^{n-2} \\ &= x^2 g(x), \qquad -1 < x < 1, \end{aligned}$$

where $g(x) = \sum_{n=2}^{\infty} (-1)^n n(n-1)x^{n-2}$. Integrate once, using 0 as a lower limit, just for convenience:

$$\int_0^x g(t)\,dt = \sum_{n=2}^{\infty} (-1)^n n(n-1) \int_0^x t^{n-2}\,dt = \sum_{n=2}^{\infty} (-1)^n n x^{n-1} = h(x).$$

Integrate once more:

$$\int_0^x h(t)\,dt = \sum_{n=2}^{\infty} (-1)^n n \int_0^x t^{n-1}\,dt = \sum_{n=2}^{\infty} (-1)^n x^n = p(x).$$

Now, $p(x)$ is recognizable, since $\sum_{n=0}^{\infty} (-1)^n x^n = 1/(1+x)$ means that

$$p(x) = \frac{1}{1+x} - 1 + x = \frac{x^2}{1+x}.$$

Then

$$h(x) = p'(x) = \frac{2x + x^2}{(1 + x)^2}$$

and

$$g(x) = h'(x) = \frac{2}{(1 + x)^3}.$$

Finally, $f(x) = x^2g(x) = 2x^2/(1 + x)^3$, and we have summed the series. ▌

EXAMPLE

In this example we shall indicate how power series can be used to solve differential equations. Consider the equation

$$y'' - 2xy' + \lambda y = 0 \qquad (\lambda \text{ constant}).$$

This is *Hermite's equation* and arises in the quantum mechanical treatment of the harmonic oscillator. Try a power-series solution $y(x) = \sum_{n=0}^{\infty} a_n x^n$. Assuming that there is such a solution in some interval about 0, we have

$$y'(x) = \sum_{n=1}^{\infty} na_n x^{n-1} \quad \text{and} \quad y'' = \sum_{n=2}^{\infty} n(n-1)a_n x^{n-2}.$$

Substitute into the equation

$$0 = \sum_{n=2}^{\infty} n(n-1)a_n x^{n-2} + \sum_{n=1}^{\infty} -2na_n x^n + \sum_{n=0}^{\infty} \lambda a_n x^n.$$

In order to be able to find the coefficient of each power of x, write

$$\sum_{n=2}^{\infty} n(n-1)a_n x^{n-2} = \sum_{n=0}^{\infty} (n+2)(n+1)a_{n+2}x^n.$$

Then

$$0 = \sum_{n=0}^{\infty} (n+2)(n+1)a_{n+2}x^n + \sum_{n=1}^{\infty} -2na_n x^n + \sum_{n=0}^{\infty} \lambda a_n x^n$$

$$= \sum_{n=1}^{\infty} [(n+2)(n+1)a_{n+2} - 2na_n + \lambda a_n]x^n + 2a_2 + \lambda a_0.$$

The only way this series can be zero for all x in some interval is for every power of x to have coefficient zero. Then $2a_2 + \lambda a_0 = 0$ and, for $n \geq 1$,

$$(n+2)(n+1)a_{n+2} + (\lambda - 2n)a_n = 0.$$

Then $a_2 = -\lambda a_0/2$, and, for $n \geq 1$,

$$a_{n+2} = \frac{(2n - \lambda)a_n}{(n + 1)(n + 2)}.$$

This allows us to solve for all even-indexed terms as multiples of a_0, and all odd-indexed terms as multiples of a_1. In particular,

$$a_2 = \frac{-\lambda a_0}{2},$$

$$a_4 = \frac{(4 - \lambda)a_2}{3 \cdot 4} = \frac{-(4 - \lambda)\lambda a_0}{1 \cdot 2 \cdot 3 \cdot 4},$$

$$a_6 = \frac{(8 - \lambda)a_4}{5 \cdot 6} = \frac{-(8 - \lambda)(4 - \lambda)\lambda a_0}{1 \cdot 2 \cdot 3 \cdot 4 \cdot 5 \cdot 6},$$

$$a_8 = \frac{(12 - \lambda)a_6}{7 \cdot 8} = \frac{-(12 - \lambda)(8 - \lambda)(4 - \lambda)\lambda a_0}{8!},$$

and, in general,

$$a_{2n} = \frac{-(4n - 4 - \lambda)(4n - 8 - \lambda)\cdots(4 - \lambda)\lambda a_0}{(2n)!} \qquad \text{for } n = 1, 2, 3, \ldots.$$

Next, since a_1 does not appear in our equations for the a_i's, we conclude that a_1 is arbitrary, and

$$a_3 = \frac{(2 - \lambda)a_1}{2 \cdot 3}$$

$$a_5 = \frac{(6 - \lambda)a_3}{4 \cdot 5} = \frac{(6 - \lambda)(2 - \lambda)a_1}{5!}$$

$$a_7 = \frac{(10 - \lambda)a_5}{6 \cdot 7} = \frac{(10 - \lambda)(6 - \lambda)(2 - \lambda)a_1}{7!}.$$

In general,

$$a_{2n+1} = \frac{(4n - 2 - \lambda)(4n - 6 - \lambda)\cdots(2 - \lambda)a_1}{(2n + 1)!}$$

for $n = 1, 2, \ldots$.

The solution can then be split into two series, one containing all the even powers, one all the odd powers of x. We obtain

$$y(x) = a_0\left[1 - \frac{\lambda x^2}{2!} - \frac{\lambda(4 - \lambda)x^4}{4!} - \frac{(8 - \lambda)(4 - \lambda)\lambda x^6}{6!} - \cdots\right]$$
$$+ a_1\left[x + \frac{(2 - \lambda)x^3}{3!} + \frac{(6 - \lambda)(2 - \lambda)x^5}{5!} + \frac{(10 - \lambda)(6 - \lambda)(2 - \lambda)x^7}{7!} + \cdots\right],$$

with a_0 and a_1 arbitrary. It is easy to check that both the series of even powers and the series of odd powers converge for all real x.

It is a curious fact that for this particular equation we obtain finite series, or polynomial, solutions for certain choices of λ, a_0, and a_1. For example:

a_0	a_1	λ	$y(x)$
1	0	0	1
0	1	2	x
1	0	4	$1 - 2x^2$
0	1	6	$x - \frac{2}{3}x^3$
1	0	8	$1 - 4x^2 + \frac{4}{3}x^4$
0	1	10	$1 - \frac{4}{3}x^3 + \frac{4}{15}x^5$

In general, when $\lambda = 2n$, we can obtain as solution a polynomial of degree n. Except for a constant factor customarily introduced because of the physics of the oscillator problem, these polynomials are called *Hermite polynomials.* ▮

EXAMPLE

In this example we illustrate the use of a polynomial approximation to approximate the solution to the simple pendulum problem. Consider the differential equation

$$\ddot{\theta} + \frac{g}{l}\sin(\theta) = 0.$$

Here a dot denotes d/dt, t is time, l is the length of the pendulum, g is the magnitude of acceleration due to gravity, and $\theta(t)$ is the angle of displacement from the vertical at time t. We have for convenience put a unit mass at the end of the pendulum (see Fig. 1-11).

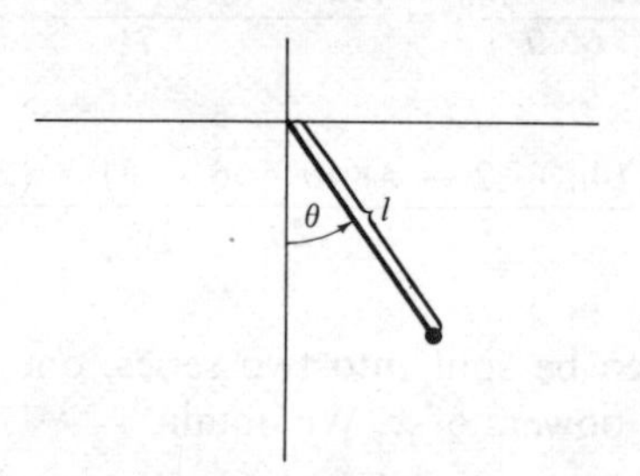

Fig. 1-11

Considering the simplicity of the physical situation, the equation is surprisingly difficult to solve exactly, because of the sin(θ) term. So we try an approximation. For small oscillations, θ is near 0, so

$$\sin(\theta) \approx \theta.$$

Since $\sin(\theta) = \theta - \theta^3/3! + \theta^5/5! + \cdots$, this approximation is good to within an error not exceeding $|\theta^3/3!|$, which is small if θ is small.

Instead of $0 = \ddot{\theta} + (g/l)\sin(\theta)$, then, we consider

$$0 = \ddot{\theta} + \frac{g}{l}\theta.$$

This is easy to solve, and we obtain (see Appendix II of Chapter 10)

$$\theta(t) = A\sin\left(\sqrt{\frac{g}{l}}\,t\right) + B\cos\left(\sqrt{\frac{g}{l}}\,t\right),$$

where A and B are constants whose determination requires more information, such as initial position and initial velocity. If, say, $\theta(0) = \pi/64$ and $\theta'(0) = 0$ (so that the pendulum is released from rest from an initial position having $\theta = \pi/64$), then we obtain, by substitution:

$$\theta(0) = B = \frac{\pi}{64},$$

$$\theta'(0) = A = 0.$$

The solution to our approximate differential equation is then

$$\theta(t) = \frac{\pi}{64}\cos\left(\sqrt{\frac{g}{l}}\,t\right),$$

and this is taken as an approximate solution to $\ddot{\theta} + (g/l)\sin(\theta) = 0$ good for small θ. (How small you must take θ depends upon how good an approximation you want.)

There is more to this example than is at first apparent. We have assumed that, in approximating a differential equation by approximating one of its terms, the solution to the new equation approximates the solution to the original. There is a rather technical sense in which one can justify such a procedure in some (but not all) instances. In the case of the simple pendulum, laboratory experiment shows the above approximation to be very good. ▌

Exercises for Section 1.5

1. Write a Taylor polynomial for each of the following functions, about the given point a, and approximate the function on the given interval to within the error indicated.

(a) $\arctan(x)$, $a = 0$, on $[-\frac{1}{2}, \frac{1}{2}]$ to within $\frac{1}{20}$

(b) $\cos(x)$, $a = \pi/2$, on $[\pi/4, 3\pi/4]$ to within $\frac{1}{100}$

(c) e^{-x^2}, $a = 0$, on $[-1, 1]$ to within $\frac{1}{200}$

(d) $\ln(1 + x)$, $a = 0$, on $[-\frac{1}{8}, \frac{1}{8}]$ to within $\frac{1}{100}$

2. Approximate $\int_0^1 [\sin(x)/x]\,dx$ to within $\frac{1}{50}$. *Hint:* The integrand is really a function $f(x)$ given by

$$f(x) = \begin{cases} \dfrac{\sin(x)}{x} & \text{for } x \neq 0, \\ 1 & \text{for } x = 0. \end{cases}$$

This makes f continuous at 0, as $\lim_{x \to 0} \sin(x)/x = 1$.

3. Approximate $\int_0^1 e^{-x^2}\,dx$ to within $\frac{1}{100}$.

4. Approximate

$$\int_0^{\pi/2} \frac{1 - \cos(x)}{x}\,dx$$

to within $\frac{1}{50}$.

5. Write out the first six terms of the Taylor series for each of the following:
(a) $\cos(3x + 2)$ about $\pi/4$
(b) $\sin(x^2)$ about 0
(c) $3x^5 + 4x^3 + 8x^2 + 2x + 1$ about 2, with error 0
(d) $1/(1 + x^{3/2})$ about 0
(e) $\sqrt{1 + x^2}$ about 1

6. Sum the series $\sum_{n=5}^{\infty} n(n-1)x^{n-2}$.

7. Sum the series

$$\sum_{n=0}^{\infty} \frac{x^{n+2}}{(n+1)(n+2)}.$$

8. Sum the series

$$\sum_{n=2}^{\infty} \frac{(-1)^n x^{n+1}}{n+1}.$$

9. Use the Cauchy product of two series to multiply $\sum_{n=0}^{\infty} (x^n/n!)$ by itself to obtain $\sum_{n=0}^{\infty} (2^n x^n/n!)$, thus showing that $(e^x)^2 = e^{2x}$ (see Exercise 21, Section 1.3).

10. Use the Cauchy product series to multiply together the MacLaurin series for $\sin(x)$ and $\cos(x)$ and verify directly that $\sin(2x) = 2\sin(x)\cos(x)$.

11. Find the radius of convergence, open interval of convergence, and interval of convergence of each of the following:

(a) $\sum_{n=0}^{\infty} \dfrac{e^n(x-2)^2}{n!}$ **(b)** $\sum_{n=0}^{\infty} (\tfrac{2}{3})^n(x-\pi)^n$

(c) $\sum_{n=1}^{\infty} \dfrac{(-1)^n x^n}{n}$ **(d)** $\sum_{n=4}^{\infty} \dfrac{\ln(n)(x-2)^{n+1}}{n}$

(e) $\sum_{n=1}^{\infty} \frac{(-1)^{n+1}x^{2n}}{2n+1}$ (f) $\sum_{n=1}^{\infty} \frac{n!\,(x-3)^n}{n^n}$

(g) $\sum_{n=1}^{\infty} \frac{n^n(x-3)^n}{n!}$

12. Find the sum of the series $\sum_{n=0}^{\infty} (n+4)x^n$.

13. Find the first eight terms in the Taylor series about 0 of $e^{\cos(x)}$.

14. Use a Taylor polynomial about 0 for e^x to compute e to within $\frac{1}{2500}$.

15. Use a Taylor polynomial approximation about 0 for $\arctan(x)$ to approximate $\pi/4$ (hence π) to within an error not exceeding $\frac{1}{100}$.

16. Sum the series $\sum_{n=0}^{\infty} x^{3n+2}$.

17. Solve the Legendre differential equation

$$(1-x^2)y'' - 2xy' + \lambda(\lambda+1)y = 0,$$

where λ is constant, by putting $y(x) = \sum_{j=0}^{\infty} a_j x^j$. Show that when λ is chosen as a positive integer n, one can obtain as solution a polynomial of degree n. Except for a constant factor, this is called a *Legendre polynomial.*

Show that, in general, the radius of convergence of the solution of the form $\sum_{j=0}^{\infty} a_j x^j$ is 1.

18. Obtain a solution to Bessel's equation:

$$y'' + \frac{1}{x}y' + \left(1 - \frac{n^2}{x^2}\right)y = 0 \qquad (n \text{ an integer})$$

by putting $y(x) = \sum_{j=0}^{\infty} a_j x^{n+j}$.

A solution of this form is called a *Bessel function of order n.* Bessel functions occur in solutions to problems in astronomy, elasticity, wave propagation, and in a variety of other contexts.

19. Sum the series

$$\sum_{n=0}^{\infty} \frac{(-1)^n x^{2n+1}}{2n+1}.$$

20. Sum the series $\sum_{n=1}^{\infty} n^2 x^{n-1}$.

21. Approximate $\int_{1/2}^{1/3} \ln(1+t^2)\,dt$ to within $\frac{1}{100}$.

22. Find the first six terms in the Taylor expansion of $e^{-x^2}\ln(x^2+3x+1)$ about 0:

(a) By finding the series for e^{-x^2} and $\ln(x^2+3x+1)$ separately, and multiplying together enough of the terms to obtain the coefficients of $1, x, \ldots, x^5$.

(b) By computing the coefficients directly from the derivative formula.

23. Find the first five terms in the Taylor series for $\sin^2(x)\arctan(x)$ about 0, using two different methods. *Hint:* See (a) and (b) of Exercise 22.

24. Approximate $\int_0^1 e^{x^2+x}\,dx$ to within $\frac{1}{1000}$.

25. Find a series solution about 0 for the differential equation $y'' + y = 0$. Compare your solution with the Taylor series for $\sin(x)$ and $\cos(x)$ about 0.

26. Find the first six terms in the Taylor series for $\arctan(e^x)$ about $\pi/2$. Give an estimate for the difference between the complete Taylor series and the sum of the first six terms in the interval $[0, \pi]$.

27. Find a series solution about 0 for the differential equation $y'' - ay = 0$, with a some positive constant. Compare this solution with the Taylor expansions of $\sinh(\sqrt{a}x)$ and $\cosh(\sqrt{a}x)$ about 0.

28. Find a series solution about 0 for the differential equation $y' = ay$, with a some constant. Compare the solution with the Taylor series about 0 for e^{ax}.

29. Prove the Cauchy–Hadamard formula for the radius of convergence of $\sum_{n=0}^{\infty} a_n(x-a)^n$: $1/r = \limsup\{|a_n|^{1/n}\}$.

30. Find the radius of convergence of $\sum_{n=0}^{\infty} a_n x^n$, where $a_{2n} = 3^n$ and $a_{2n+1} = 2^n$ for $n = 0, 1, 2, \ldots$. *Hint:* Try the formulas in the text, and then use Exercise 29.

31. Give an example to show that the radius of convergence of $\sum_{n=0}^{\infty} a_n(x-a)^n$ is not necessarily given by

$$\frac{1}{r} = \limsup\left\{\frac{a_{n+1}}{a_n}\right\}.$$

Hint: Look at Exercise 30.

32. For each of the following functions, prove that $R_n(x) \to 0$ as $n \to \infty$ for x in the given interval.

(a) e^{-x} and **(b)** $\sin(x)$: $-L \leq x \leq L$, where L is any positive number

(c) $\ln(1+x)$, $-\frac{1}{2} \leq x \leq \frac{1}{2}$

(d) $\arctan(x)$, $-\pi/4 \leq x \leq \pi/4$

(e) $\cosh(3x)$, $-1 \leq x \leq 1$

(f) 2^x, $-1 \leq x \leq \frac{1}{2}$

33. Suppose that $\sum_{n=0}^{\infty} a_n(x-a)^n$ has radius of convergence $r\,(0 < r < \infty)$ and that the series converges at $a + r$. Prove that the series converges uniformly on $[a, a+r]$.

34. Suppose that $\sum_{n=0}^{\infty} a_n(x - a)^n$ has radius of convergence $r(0 < r < \infty)$, and that the series converges at $a + r$ to L. Prove that

$$\lim_{x \to a+r} \sum_{n=0}^{\infty} a_n(x - a)^n = L.$$

35. Use the Cauchy–Hadamard formula (Exercise 29) to give a short proof that $\sum_{n=0}^{\infty} a_n(x - a)^n$ and $\sum_{n=1}^{\infty} na_n(x - a)^{n-1}$ have the same radius of convergence.

36. Let m be any real number. Derive the binomial series for $(1 + x)^m$ by writing $(1 + x)^m$ equal to a Taylor polynomial plus remainder, and then showing that, for $|x| < 1$, the remainder goes to zero as $n \to \infty$. *Hint:* Use the derivative form of the remainder given in the text.

1.6 Improper Integrals

In elementary calculus the symbol $\int_a^b f(x)\,dx$ is usually reserved for situations in which f is a continuous function over a closed interval $[a, b]$. In developing an integral test for series, however, we saw one instance in which it was useful to consider an integral over an infinite interval. While that particular setting was mathematical, improper integrals also arise in physical situations. For example, in quantum mechanics, the Schrödinger equation for a harmonic oscillator is

$$\frac{d^2\psi}{dx^2} + \frac{8\pi^2}{m}(W - 2\pi^2 m\nu^2 x^2)\psi = 0,$$

with h = Planck's constant, m = mass, W = total energy, and ν = vibrational frequency. The wave function $\psi(x)$ is defined for $-\infty < x < \infty$ and has the interpretation that $\int_a^b |\psi(x)|^2\,dx$ is the probability that the particle is in the interval $[a, b]$. This necessitates that $\int_{-\infty}^{\infty} |\psi(x)|^2\,dx = 1$, which is to say that the particle is certainly somewhere.

We now consider improper integrals in more detail than in Section 1.3. If $\int_a^b f(x)\,dx$ is defined for every $b > a$, and if $\lim_{b\to\infty} \int_a^b f(x)\,dx$ exists finite, we say that $\int_a^{\infty} f(x)\,dx$ converges, and we assign to $\int_a^{\infty} f(x)\,dx$ the value of this limit. Similarly, $\int_{-\infty}^{d} f(x)\,dx = \lim_{c\to-\infty} \int_c^d f(x)\,dx$ and

$$\int_{-\infty}^{\infty} f(x)\,dx = \lim_{\varepsilon\to\infty} \int_{-\varepsilon}^{\varepsilon} f(x)\,dx,$$

whenever these limits exist finite. In the last case, there is a fine point which we shall mention only briefly. Strictly speaking, we should take

$$\int_{-\infty}^{\infty} f(x)\,dx = \lim_{\substack{d\to\infty \\ c\to-\infty}} \int_c^d f(x)\,dx,$$

with the understanding that d and c may go to $+\infty$ and $-\infty$, respectively, at different rates. This may give a different result than $\lim_{\varepsilon \to \infty} \int_{-\varepsilon}^{\varepsilon} f(x)\,dx$, which is called the *Cauchy principal value* of $\int_{-\infty}^{\infty} f(x)\,dx$. For example, consider $\int_{-\infty}^{\infty} x^3\,dx$. Since

$$\int_{-\varepsilon}^{\varepsilon} x^3\,dx = \frac{x^4}{4}\bigg|_{-\varepsilon}^{\varepsilon} = \frac{\varepsilon^4}{4} - \frac{\varepsilon^4}{4} = 0,$$

then $\int_{-\infty}^{\infty} x^3\,dx = 0$ in the sense of a Cauchy principal value. But if we take, say, $\lim_{d \to \infty} \int_{-d}^{2d} x^3\,dx$, we get $\lim_{d \to \infty} \dfrac{15d^4}{4} = \infty$, which is another matter entirely. However, it turns out to be convenient in a number of applications to use the Cauchy principal value in speaking of $\int_{-\infty}^{\infty} f(x)\,dx$, and this is what we shall do.

As with series, it is convenient to use the word "divergence" to cover all cases in which an integral does not converge in the sense defined above. Note that we can always trade in an integral $\int_{-\infty}^{b} f(x)\,dx$ for one of the form $\int_{c}^{\infty} g(x)\,dx$ by a change of variable. Put $t = -x$. Then

$$\int_{a}^{b} f(x)\,dx = \int_{-a}^{-b} -f(-t)\,dt = \int_{-b}^{-a} f(-t)\,dt.$$

As $a \to -\infty$, $-a \to +\infty$, so $\int_{-\infty}^{b} f(x)\,dx$ will converge or diverge the same as $\int_{-b}^{\infty} f(-x)\,dx$.

It is obvious that $\int_{a}^{\infty} f(x)\,dx + \int_{a}^{\infty} g(x)\,dx = \int_{a}^{\infty} [f(x) + g(x)]\,dx$ when both integrals on the left converge. Note, however, that the formula may fail if both integrals on the left diverge. For example,

$$\int_{1}^{\infty} \frac{1}{x}\,dx \qquad \text{and} \qquad \int_{1}^{\infty} \frac{-1}{1+x}\,dx$$

both diverge, but

$$\int_{1}^{\infty} \left(\frac{1}{x} - \frac{1}{1+x}\right) dx = \int_{1}^{\infty} \frac{dx}{x(1+x)}$$

converges (by direct integration, or by the comparison test a little further on). It should also be obvious that $\int_{a}^{\infty} \alpha f(x)\,dx = \alpha \int_{a}^{\infty} f(x)\,dx$ if α is real and $\int_{a}^{\infty} f(x)\,dx$ converges.

For the remainder of this section we want to develop some ways of determining convergence or divergence of improper integrals without having to carry out the integration and take a limit each time, since both these processes may involve practical difficulties (e.g., try $\int_0^\infty e^{-x^2}\,dx$ this way). In view of the analogy between $\int_a^\infty$ and $\sum_{n=K}^{\infty}$, it is not surprising that some of the tests we shall develop have direct analogues in tests for series, although this analogy must not be stretched too far. [For example, convergence of $\sum_{n=1}^{\infty} a_n$ implies that $\lim_{n\to\infty} a_n = 0$; convergence of $\int_a^\infty f(x)\,dx$ does not imply that $\lim_{x\to\infty} f(x) = 0$. This is Exercise 5.]

As with series, we begin by considering $f(x) \ge 0$ on $x \ge a$. Then $\int_a^b f(x)\,dx$ is a nondecreasing function of b. If $\int_a^b f(x)\,dx$ increases without bound as $b \to \infty$, then $\lim_{b\to\infty} \int_a^b f(x)\,dx = \infty$ and $\int_a^\infty f(x)\,dx$ diverges. If, however, $\int_a^b f(x)\,dx$ is bounded for $b > a$, then $\lim_{b\to\infty} \int_a^b f(x)\,dx$ exists finite, and $\int_a^\infty f(x)\,dx$ converges. This is the integral version of the theorem on monotone nondecreasing sequences. From this, we obtain a comparison test.

COMPARISON TEST

Let $0 \le f(x) \le g(x)$ for $x \ge a$. Then

(1) *If $\int_a^\infty f(x)\,dx$ diverges, so does $\int_a^\infty g(x)\,dx$.*
(2) *If $\int_a^\infty g(x)\,dx$ converges, so does $\int_a^\infty f(x)\,dx$.*

To prove (1), note that $\int_a^b f(x)\,dx \le \int_a^b g(x)\,dx$ whenever $b \ge a$. Then $\int_a^b g(x)\,dx \to \infty$ as $b \to \infty$ if $\int_a^\infty f(x)\,dx$ diverges.

To prove (2), note that $\int_a^b f(x)\,dx$ is bounded above for all $b \ge a$ by $\int_a^\infty g(x)\,dx$, hence $\int_a^\infty f(x)\,dx$ converges, by the remarks preceding the statement of the theorem. ∎

As with series, do not mix (1) and (2). If $\int_a^\infty g(x)\,dx$ diverges, $\int_a^\infty f(x)\,dx$ may converge or diverge. And if $\int_a^\infty f(x)\,dx$ converges, we can draw no conclusion about $\int_a^\infty g(x)\,dx$. We leave it to the exercises to construct an example of this.

EXAMPLE

$$\int_1^\infty \frac{dx}{x^2 + 2x + 4}$$

converges. This can be done by integrating, but it is easier to observe that

$$\frac{1}{x^2 + 2x + 4} < \frac{1}{x^2} \qquad \text{for } x \ge 1,$$

and recall that $\int_1^\infty (1/x^2)\,dx$ converges. ∎

As with series, we say that $\int_a^\infty f(x)\,dx$ *converges absolutely* if $\int_a^\infty |f(x)|\,dx$ converges. If $\int_a^\infty f(x)$ converges, but $\int_a^\infty |f(x)|\,dx$ diverges, we say that $\int_a^\infty f(x)\,dx$ *converges conditionally*. We shall see shortly that $\int_1^\infty [\sin(x)/x]\,dx$ converges conditionally, but first we show that absolute convergence implies convergence. Note that

$$-f(x) \leq f(x) \leq |f(x)|,$$

so that

$$0 \leq |f(x)| + f(x) \leq 2|f(x)|.$$

If $\int_a^\infty |f(x)|\,dx$ converges, then so does $\int_a^\infty (|f(x)| + f(x))\,dx$ by comparison, hence so does $\int_a^\infty f(x)\,dx$.

In situations where the integrand is a product of two functions satisfying certain properties, the following test is often useful.

DIRICHLET TEST

Let f, g, and g′ be continuous for $x \geq a$. Suppose that

(1) $\lim_{x \to \infty} g(x) = 0.$
(2) $\int_a^\infty |g'(x)|\,dx$ *converges.*
(3) *For some M,* $\left|\int_a^x f(t)\,dt\right| \leq M$ *for all* $x \geq a$.

Then

$$\int_a^\infty f(x)g(x)\,dx \text{ converges.}$$

PROOF

Integrate by parts, putting $F(x) = \int_a^x f(t)\,dt$ and $F'(t) = f(t)$. Then

$$\begin{aligned}\int_a^x f(t)g(t)\,dt = \int_a^x g(t)F'(t)\,dt &= g(t)F(t)\Big|_a^x - \int_a^x F(t)g'(t)\,dt \\ &= g(x)F(x) - g(a)F(a) - \int_a^x F(t)g'(t)\,dt \\ &= g(x)F(x) - \int_a^x F(t)g'(t)\,dt.\end{aligned}$$

Now, by (3), $|F(x)| \leq M$ for all $x \geq a$. Then $0 \leq |g(x)F(x)| \leq M|g(x)| \to 0$ as $x \to \infty$, hence $g(x)F(x) \to 0$ as $x \to \infty$. Further, $|F(x)g'(x)| \leq M|g'(x)|$, and

$\int_a^\infty |g'(x)|\,dx$ converges, so $\int_a^\infty |F(x)g'(x)|\,dx$ converges. Then $\int_a^\infty F(t)g'(t)\,dt$ converges. Then $\lim_{x\to\infty} \int_a^x F(t)g'(t)\,dt = \lim_{x\to\infty} \int_a^x f(t)g(t)\,dt$ exists finite, so $\int_a^\infty f(t)g(t)\,dt$ converges. ∎

Note the similarity between this test and the test for series given in Exercise 12, Section 1.3.

EXAMPLE

$\int_1^\infty x^{-1/8}\cos(x)\,dx$ converges. Put $g(x) = x^{-1/8}$, $f(x) = \cos(x)$ in Dirichlet's test. It is easy to verify that hypotheses (1) through (3) are satisfied. ∎

EXAMPLE

$\int_1^\infty [\sin(x)/x]\,dx$ converges conditionally. For convergence, put $g(x) = 1/x$, $f(x) = \sin(x)$ in Dirichlet's test. We now show that $\int_1^\infty |\sin(x)/x|\,dx$ diverges. For $x \geq 1$, $|\sin(x)/x|$ looks something like the graph in Fig. 1-12. As $x \to \infty$, $|\sin(x)/x| \to 0$. Further, $|\sin(x)/x| = 0$ when x is an integer multiple of π. The graph suggests that we look at

$$\int_{n\pi}^{(n+1)\pi} \left|\frac{\sin(x)}{x}\right| dx \qquad \text{for } n = 1, 2, 3, \ldots.$$

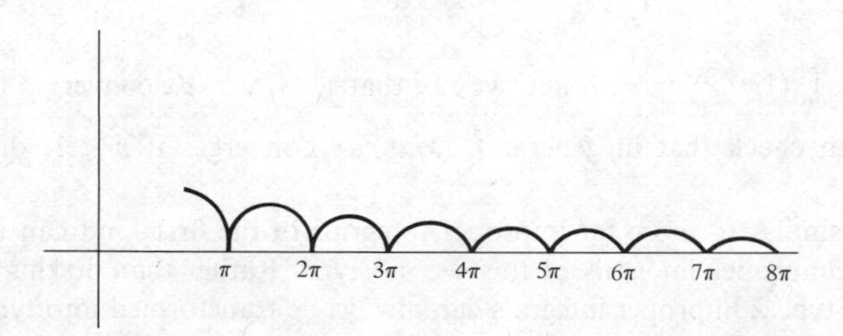

Fig. 1-12

Make an estimate:

$$\int_{n\pi}^{(n+1)\pi} \left|\frac{\sin(x)}{x}\right| dx \geq \frac{1}{(n+1)\pi} \int_{n\pi}^{(n+1)\pi} |\sin(x)|\,dx,$$

since

$$\frac{1}{(n+1)\pi} \leq \frac{1}{x} \qquad \text{for } n\pi \leq x \leq (n+1)\pi.$$

But

$$\int_{n\pi}^{(n+1)\pi} |\sin(x)|\,dx = |-\cos((n+1)\pi) + \cos(n\pi)| = 2.$$

Then

$$\int_{n\pi}^{(n+1)\pi} \left| \frac{\sin(x)}{x} \right| dx \geq \frac{2}{(2n+1)\pi}.$$

Now $\sum_{n=1}^{\infty} 2/(2n+1)\pi$ is a divergent series [it behaves like $\sum_{n=1}^{\infty} (1/n)$]. Then we can make $\sum_{n=1}^{\infty} 2/(2n+1)\pi$ as large as we like by taking N large enough. This means that we can make

$$\int_{\pi}^{N\pi} \left| \frac{\sin(x)}{x} \right| dx = \sum_{n=1}^{N-1} \int_{n\pi}^{(n+1)\pi} \left| \frac{\sin(x)}{x} \right| dx$$

arbitrarily large; hence $\int_1^{\infty} |\sin(x)/x| \, dx$ diverges. ∎

In concluding this section, we mention briefly a second kind of improper integral, in which the interval is bounded, but the integrand is unbounded, as, for example, with $\int_0^1 (1/\sqrt{x}) \, dx$. As with the first kind of improper integral such a situation can be handled by taking a limit, in this case $\lim_{\varepsilon \to 0+} \int_{\varepsilon}^1 (1/\sqrt{x}) \, dx$. Since for $0 < \varepsilon < 1$,

$$\int_{\varepsilon}^{1} \frac{1}{\sqrt{x}} \, dx = 2\sqrt{x} \Big|_{\varepsilon}^{1} = 2 - 2\sqrt{\varepsilon},$$

then $\lim_{\varepsilon \to 0+} \int_0^1 (1/\sqrt{x}) \, dx = 2$, and we say that $\int_0^1 (1/\sqrt{x}) \, dx$ converges to 2. The reader can check that in general $\int_0^1 (1/x^p) \, dx$ converges if $p < 1$, diverges if $p \geq 1$.

Tests similar to those for improper integrals of the first kind can be developed for improper integrals of this second type. Rather than do this here, we note that type 2 improper integrals can always be transformed into type 1 by a change of variables.

In particular, suppose that we want to consider $\int_a^b f(x) \, dx$, where f is continuous but unbounded on $(a, b]$. We want to consider $\lim_{\varepsilon \to 0+} \int_{a+\varepsilon}^b f(x) \, dx$. Put $x = a + 1/t$. Then $t = 1/(x - a)$, for $a < x \leq b$. When $x = a + \varepsilon$, $t = 1/\varepsilon$; when $x = b$, $t = 1/(b - a)$. Then, for $0 < \varepsilon < b - a$,

$$\int_{a+\varepsilon}^{b} f(x) \, dx = \int_{1/\varepsilon}^{1/(b-a)} f(a + 1/t)(-1/t^2) \, dt = \int_{1/(b-a)}^{1/\epsilon} (1/t^2) f(a + 1/t) \, dt.$$

As $\varepsilon \to 0+$, $1/t \to +\infty$, so $\lim_{\varepsilon \to 0+} \int_{a+\varepsilon}^b f(x) \, dx = \lim_{c \to \infty} \int_{1/(b-a)}^c (1/t^2) f(a + 1/t) \, dt$. Then $\int_a^b f(x) \, dx$, type 2, converges exactly when $\int_{1/(b-a)}^c (1/t^2) f(a + 1/t) \, dt$, type 1, converges. As an example, application of the transformation $x = 1/t$ changes $\int_2^1 (dx/\sqrt{x})$ into $\int_1^{\infty} (1/t^{3/2}) \, dt$.

In similar fashion, if f is continuous but unbounded on $[a, b)$, we can handle the type 2 improper integral $\int_a^b f(x)\,dx$ by putting $x = b - 1/t$ and considering the type 1 integral $\int_{1/(b-a)}^{\infty} (1/t^2) f(b - 1/t)\,dt$.

Exercises for Section 1.6

1. Test the following for convergence. In the case that the function is not strictly nonnegative in the relevant interval, test for conditional convergence.

(a) $\displaystyle\int_1^{\infty} \sin(x^2)\,dx$

(b) $\displaystyle\int_1^{\infty} \frac{x}{x^2 + 4}\,dx$

(c) $\displaystyle\int_0^{\infty} e^{-x^2}\,dx$

(d) $\displaystyle\int_1^{\infty} \frac{\cos(x)}{x^p}\,dx$ (determine values of p giving convergence and values giving divergence)

(e) $\displaystyle\int_1^{\infty} \frac{\ln(x)}{x}\,dx$

(f) $\displaystyle\int_0^{1} \ln(x)\,dx$

(g) $\displaystyle\int_a^{\infty} \frac{p(x)}{q(x)}\,dx$, where $p(x)$ is a polynomial of degree n, $q(x)$ is a polynomial of degree m, $q(x) \neq 0$ for $x \geq a$, and $m \geq n + 2$

(h) $\displaystyle\int_{-\infty}^{\infty} \frac{dx}{1 + x^2}$ (Cauchy principal value)

(i) $\displaystyle\int_1^{\infty} \frac{1 - \sin(x)}{x}\,dx$

2. Suppose that $g(x)$ is continuous for $x \geq a$ and g is a strictly decreasing function with limit 0 as $x \to \infty$. What can you conclude about $\int_a^{\infty} g(x)\cos(x)\,dx$ and $\int_a^{\infty} g(x)\sin(x)\,dx$?

3. Prove the ratio test for improper integrals: If $f(x)$ is continuous and positive for $x \geq a$, and

$$\lim_{x\to\infty} \frac{f(x + 1)}{f(x)} < 1,$$

then $\int_a^{\infty} f(x)\,dx$ converges.

What happens if $\displaystyle\lim_{x\to\infty} \frac{f(x + 1)}{f(x)} \geq 1$?

4. Prove the root test for improper integrals: If $f(x)$ is continuous and positive for $x \geq a$, and $\lim_{x\to\infty} [f(x)]^{1/x} < 1$, then $\int_a^\infty f(x)\,dx$ converges.

What conclusion do you draw if $\lim_{x\to\infty} [f(x)]^{1/x} \geq 1$?

5. Give an example of a function f for which $\int_1^\infty f(x)\,dx$ converges, but $\lim_{x\to\infty} f(x) \neq 0$.

6. It sometimes happens that an improper integral becomes proper with a change of variables. As an example, consider

$$\int_0^1 \frac{x}{\sqrt{1-x^2}}\,dx.$$

(a) Evaluate by direct integration.
(b) Make the substitution $x = \cos(\theta)$ to obtain a proper integral.

7. Test the following for convergence (absolute and conditional):

(a) $\int_1^\infty (x^3 + 1)^{-3/2}$

(b) $\int_4^\infty \frac{dx}{(x+1)^{1/4}\ln(x)}$

(c) $\int_0^{\pi/2} \frac{x\,dx}{\cos(x)}$

(d) $\int_3^\infty \frac{dx}{\ln[\ln(x)]}$

(e) $\int_3^\infty \frac{dx}{[\ln(\ln(x))]^2}$

(f) $\int_0^\infty \sin[\cos(x)]\,dx.$

Appendix I: Least Upper Bounds

Suppose that A is a set of real numbers [for example, the interval $(0, 1)$ consisting of all x with $0 < x < 1$; or the set of rationals less than π; or the set of numbers e^n, for $n = 1, 2, \ldots$]. If there is some number M such that $x \leq M$ for each x in A, then A is said to be bounded above, and M is an upper bound. An axiom of the real number system is that every nonempty set A of real numbers that is bounded above has a least upper bound. That is, there is some number L such that:

(1) $x \leq L$ for each x in A.
(2) $L - \varepsilon$ is not an upper bound of A for any $\varepsilon > 0$.

Another way of putting (2) is to say that, given $\varepsilon > 0$, there is some x in A with $L - \varepsilon < x \leq L$.

We denote L by lub(A). As examples, lub((0, 1)) = 1, lub((0, 1]) = 1, and lub $A = \pi$ if A consists of all numbers $\pi - 1/n$, for $n = 1, 2, 3, \ldots$.

If we are considering sequences, as in the chapter, we can think of A as consisting of all the u_i's and thus speak of the least upper bound of $\{u_n\}$ as the least upper bound of the set A containing all the terms of the sequence.

Turning things around, we say that m is a lower bound of A if $m \leq x$ for each x in A. If A is bounded below, in the sense that a lower bound exists, then A has a greatest lower bound, denoted glb(A) and characterized by the properties that glb(A) $\leq x$ for each x in A, but, given any $\varepsilon > 0$, there is some x in A with glb(A) $\leq x <$ glb(A) $+ \varepsilon$.

Once we assume that every set bounded above has a least upper bound, it is easy to prove that every set bounded below has a greatest lower bound. For suppose that A is bounded below. Let B consist of all $-x$ with x in A. [So, for example, if $A = (1, \frac{3}{2})$, then $B = (-\frac{3}{2}, -1)$]. B is then bounded above, and the reader can check that glb(A) = $-$lub(B).

In closing this appendix, we caution the reader not to be misled by the phrasing "A has a least upper bound" into thinking that lub(A) must belong to A. That this is not the case is shown by the example lub((0, 1)) = 1, noted above. Similarly, glb(A) need not actually be a member of A.

Appendix II: L'Hospital's Rule

In applying the ratio and root tests for series, we often encounter limits in which the usual rules, such as limit of quotient = quotient of limits, break down. We shall give here a method, called *L'Hospital's Rule*, which is often effective in such circumstances.

To be specifix, we want to consider the problem of evaluating

$$\lim_{x \to \alpha} \frac{f(x)}{g(x)},$$

where α is a real number, $+\infty$ or $-\infty$, and we are interested in the cases

(1) $f(x) \to 0$ and $g(x) \to 0$ as $x \to \alpha$, or
(2) $f(x) \to \pm\infty$ and $g(x) \to \pm\infty$ as $x \to \alpha$.

We shall assume that f and g are differentiable in the relevant interval. That is, if $\alpha = +\infty$, we need f and g differentiable for x sufficiently large, say $x \geq M$; if α is real, and we want a limit as $x \to \alpha$ from both sides, we will want $f'(x)$ and $g'(x)$ to exist for $\alpha - \varepsilon < x < \alpha + \varepsilon$ for some ε; and if α is real and we want a one-sided limit, say $x \to \alpha$ from the right, we want $f'(x)$ and $g'(x)$ to exist in some interval $\alpha < x < \alpha + \varepsilon$.

L'Hospital's Rule then says: If

$$\lim_{x \to \alpha} \frac{f'(x)}{g'(x)} = L \qquad \text{(where } L \text{ is real, } +\infty \text{ or } -\infty\text{)},$$

then also

$$\lim_{x\to\alpha} \frac{f(x)}{g(x)} = L.$$

We shall indicate a proof shortly. But first we shall look at some examples.

EXAMPLE

$$\lim_{x\to\infty} x[\ln(x+1) - \ln(x)] = 1.$$

We derived this result using power series in Section 1.5. To use L'Hospital's Rule, write

$$x[\ln(1+x) - \ln(x)] = \frac{\ln(1+x) - \ln(x)}{1/x}.$$

As $x \to \infty$, note that $1/x \to 0$ and $\ln(x+1) - \ln(x) \to 0$. Further,

$$\frac{\dfrac{d[\ln(1+x) - \ln(x)]}{dx}}{\dfrac{d(1/x)}{dx}} = \frac{\dfrac{1}{x+1} - \dfrac{1}{x}}{-1/x^2} = \frac{x}{x+1} \to 1 \quad \text{as } x \to \infty.$$

Hence $\lim_{x\to\infty} x[\ln(x+1) - \ln(x)] = 1.$ ▮

EXAMPLE

$\lim_{x\to 0+} x^x = 1$. Here, write $x^x = e^{x\ln(x)}$. If we can determine $\lim_{x\to 0+} x\ln(x)$, then we can find $\lim_{x\to 0+} e^{x\ln(x)}$. Now $x\ln(x) = \ln(x)/(1/x)$. As $x \to 0+$, $\ln(x) \to -\infty$ and $1/x \to \infty$. Further,

$$\frac{\dfrac{d[\ln(x)]}{dx}}{\dfrac{d[1/x]}{dx}} = \frac{1/x}{-1/x^2} = -x \to 0 \quad \text{as } x \to 0+.$$

So $\lim_{x\to 0+} x\ln(x) = 0$, and $\lim_{x\to 0+} e^{x\ln(x)} = e^{\lim_{x\to\infty} x\ln(x)} = e^0 = 1$. Note that in this example we rewrote $x\ln(x)$ as $\ln(x)/(1/x)$ to put it into a form accessible by L'Hospital's Rule. ▮

EXAMPLE

$$\lim_{x\to 0+} \left[\frac{\sin(x)}{x}\right]^{1/x^2} = e^{-1/6}.$$

Write

$$\left[\frac{\sin(x)}{x}\right]^{1/x^2} = \exp\left[\frac{1}{x^2}\ln\left(\frac{\sin(x)}{x}\right)\right],$$

and concentrate on $\lim_{x\to0+} (1/x^2)\ln[\sin(x)/x]$. As $x \to 0$, $x^2 \to 0$. Further, $\sin(x)/x \to 1$ as $x \to 0+$, so $\ln[\sin(x)/x] \to 0$. Now

$$\frac{\dfrac{d[\ln(\sin(x)/x)]}{dx}}{\dfrac{d(x^2)}{dx}} = \frac{\dfrac{d[\ln(\sin(x)) - \ln(x)]}{dx}}{2x}$$

$$= \frac{\dfrac{\cos(x)}{\sin(x)} - \dfrac{1}{x}}{2x} = \frac{x\cos(x) - \sin(x)}{2x^2\sin(x)}.$$

As $x \to 0+$, $x\cos(x) - \sin(x) \to 0$, and $2x^2\sin(x) \to 0$, so we try L'Hospital's Rule again:

$$\frac{\dfrac{d[x\cos(x) - \sin(x)]}{dx}}{\dfrac{d[2x^2\sin(x)]}{dx}} = \frac{\cos(x) - x\sin(x) - \cos(x)}{4x\sin(x) + 2x^2\cos(x)}$$

$$= \frac{-\sin(x)}{4\sin(x) + 2x\cos(x)}.$$

Again, numerator and denominator both go to zero as $x \to 0+$, so we try once more:

$$\frac{\dfrac{d[-\sin(x)]}{dx}}{\dfrac{d[4\sin(x) + 2x\cos(x)]}{dx}} = \frac{-\cos(x)}{4\cos(x) + 2\cos(x) - 2x\sin(x)}$$

$$\to -\tfrac{1}{6} \qquad \text{as } x \to 0.$$

Thus, working back, $\lim_{x\to0+} [\ln(\sin(x)/x)/x^2] = -\frac{1}{6}$. Then $\lim_{x\to0+} [\sin(x)/x]^{1/x^2} = e^{-1/6}$. ▌

We shall now sketch the idea of a proof of L'Hospital's Rule. The key is a generalization due to Cauchy of the mean value theorem.

Theorem

Let f and g be continuous on $[a, b]$ and differentiable on (a, b). Suppose that $g(b) \neq g(a)$ and that $f'(x)$ and $g'(x)$ are never both zero on (a, b). Then, for some ξ, $a < \xi < b$ and

$$\frac{f(b) - f(a)}{g(b) - g(a)} = \frac{f'(\xi)}{g'(\xi)}.$$

PROOF

Let $h(x) = f(x) - f(a) - \dfrac{f(b) - f(a)}{g(b) - g(a)}[g(x) - g(a)]$. Then $h(a) = h(b) = 0$,

and h is continuous on $[a, b]$ and differentiable on (a, b). By Rolle's theorem there is a ξ in (a, b) such that $h'(\xi) = 0$. Then

$$f'(\xi) - \frac{f(b) - f(a)}{g(b) - g(a)} g'(\xi) = 0,$$

proving the generalized mean value theorem. ∎

We shall now show how one case of L'Hospital's Rule follows from this, the other cases being left to the exercises. Suppose, then, that $\alpha = \infty$ and $f(x) \to 0$ and $g(x) \to 0$ as $x \to \infty$. Suppose that $\lim_{x \to \infty} [f'(x)/g'(x)] = L$, and that L is a real number. We want to show that $\lim_{x \to \infty} [f(x)/g(x)] = L$ also.

We shall assume that $g'(x) \neq 0$ for, say, $x \geq K$. Let $\varepsilon > 0$. For some $M \geq K$,

$$\left| \frac{f'(x)}{g'(x)} - L \right| < \varepsilon$$

if $x \geq M$. Then, for $x \geq M$,

$$L - \varepsilon < \frac{f'(x)}{g'(x)} < L + \varepsilon.$$

Now, for any choice of distinct x and y greater than M, we can find ξ between x and y such that

$$\frac{f(x) - f(y)}{g(x) - g(y)} = \frac{f'(\xi)}{g'(\xi)}.$$

Rewrite

$$\frac{f'(\xi)}{g'(\xi)} = \frac{\dfrac{f(x) - f(y)}{g(x) - g(y)}}{1 - \dfrac{g(y)}{g(x)}}.$$

Since $\xi \geq M$, then

$$L - \varepsilon < \frac{\dfrac{f(x) - f(y)}{g(x) - g(x)}}{1 - \dfrac{g(y)}{g(x)}} < L + \varepsilon.$$

Letting $y \to \infty$, we have $f(y) \to 0$ and $g(y) \to 0$, so that

$$L - \varepsilon < \frac{f(x)}{g(x)} < L + \varepsilon$$

for $x \geq M$. Then, $\lim_{x \to \infty} [f(x)/g(x)] = L$ also.

The other cases [$L = \infty$; L real and $f(x), g(x) \to \pm\infty$ as $x \to \alpha$; and so on] are handled similarly.

It should be remarked in closing that the method has been discussed in the context of limits of continuous functions, but applies easily to limits of sequences under appropriate conditions. For example, the sequence $\{n^{1/n}\}_{n=1}^{\infty}$ has limit 1 as $n \to \infty$, since $\lim_{x \to \infty} x^{1/x} = 1$.

Exercises

1. Evaluate the following or conclude that the limit does not exist:

(a) $\displaystyle\lim_{x \to -\infty} \frac{\ln(x + 1/x)}{\sin(1/x)}$

(b) $\displaystyle\lim_{x \to \infty} e^{\sin(x)}$

(c) $\displaystyle\lim_{x \to 0+} [\ln(1/x)]^x$

(d) $\displaystyle\lim_{x \to \infty} \frac{\arctan(x)}{\operatorname{arccot}(x)}$

(e) $\displaystyle\lim_{x \to 0} \frac{\sin(\pi x)}{x}$

(f) $\displaystyle\lim_{x \to \infty} \frac{e^{-x^2}}{x} \int_0^x t^2 e^{t^2}\, dt$

(g) $\displaystyle\lim_{x \to 0} \left[\frac{\tan(x)}{x}\right]^{1/x^2}$

(h) $\displaystyle\lim_{x \to \infty} x^{\ln(1/x)}$

2. Prove L'Hospital's Rule for the cases:

(a) $f(x) \to \infty$, $g(x) \to \infty$ as $x \to \infty$, α real
(b) $f(x) \to \infty$, $g(x) \to -\infty$ as $x \to \alpha$, α real
(c) $f(x) \to 0$, $g(x) \to 0$ as $x \to \alpha+$, α real
(d) $f(x) \to \infty$, $g(x) \to \infty$ as $x \to -\infty$
(e) $f(x) \to -\infty$, $g(x) \to \infty$ as $x \to \infty$

3. Construct an example in which $\lim_{x \to \infty} [f(x)/g(x)]$ exists finite, but $\lim_{x \to \infty} [f'(x)/g'(x)]$ does not exist finite. Why does this not contradict L'Hospital's Rule?

4. Evaluate each of the following or conclude that the limit does not exist:

(a) $\lim_{x \to \infty} \left(\frac{x+1}{x}\right)^{1/x}$

(b) $\lim_{x \to \infty} \frac{\ln(x)}{x^p}$ (consider different possibilities for p)

(c) $\lim_{x \to \infty} \left(1 + \frac{3x}{2}\right)^{1/x}$

(d) $\lim_{x \to 0} [\cos(3x)]^{2/x^2}$

(e) $\lim_{x \to \infty} \frac{\sin(x)}{x}$

5. What happens if you try to apply L'Hospital's rule to $\lim_{x \to \infty} [\cos(x)/x]$?

2

Vectors in R^n

2.1 Addition and Scalar Multiplication in R^n

The major distinction between advanced and elementary calculus is in the consideration of functions of n variables in the former, and the restriction to functions of one variable in the latter. Before we begin the analysis of functions of n variables, we must say something about the geometry and algebra of n-tuples.

First, by an n-vector we mean an n-tuple $(x_1, \ldots, x_n)$ of real numbers. The number x_j is the jth coordinate of $(x_1, \ldots, x_n)$. If we denote n-dimensional space by R^n, then R^n consists of all n-tuples, or n-vectors. We can visualize R^1 geometrically as the real line by associating with each real number a point on the line; R^2 as the plane, consisting of points (x, y); and R^3 as 3-space, consisting of points (x, y, z).

The two basic algebraic operations with n-vectors are addition and scalar multiplication, defined by

$$(x_1, \ldots, x_n) + (y_1, \ldots, y_n) = (x_1 + y_1, \ldots, x_n + y_n)$$

and

$$\alpha(x_1, \ldots, x_n) = (\alpha x_1, \ldots, \alpha x_n)$$

for n-vectors $(x_1, \ldots, x_n)$, $(y_1, \ldots, y_n)$, and α real.

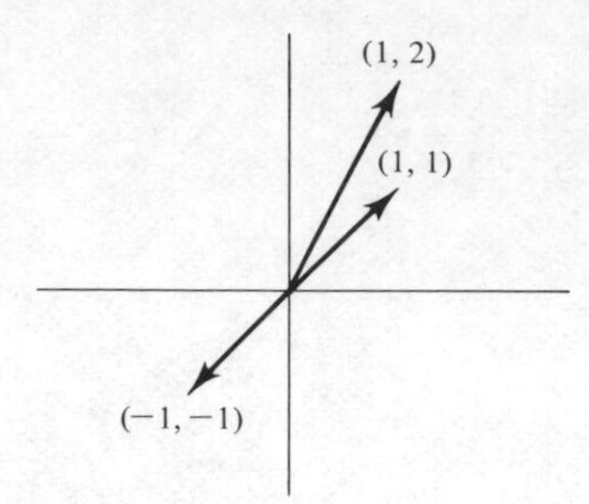

Fig. 2-1

It is sometimes useful to think of $(x_1, \ldots, x_n)$ not only as a point in R^n, but also as a directed line segment from the origin $(0, \ldots, 0)$ to the geometric point $(x_1, \ldots, x_n)$. This directed line segment is usually written as an arrow, with the blunt end at $(0, \ldots, 0)$ and the point at $(x_1, \ldots, x_n)$. Figure 2-1 shows some vectors in R^2. This way of thinking is particularly common in applications, where a directed line segment is often taken as a schematic representation of a force. Then the length represents the magnitude of the force, and the direction, the direction of the force. For example, the arrow from (0, 0) to (2, 3) represents a force of magnitude $\sqrt{13}$ with horizontal component 2 units and vertical component 3 units.

This interpretation suggests that the important things about a vector are its direction and length, not the particular point chosen for the blunt end of the arrow. For example, the arrow from (1, 1) to (3, 4) has the same magnitude and direction as that from (0, 0) to (2, 3), and so represents the same vector (Fig. 2-2).

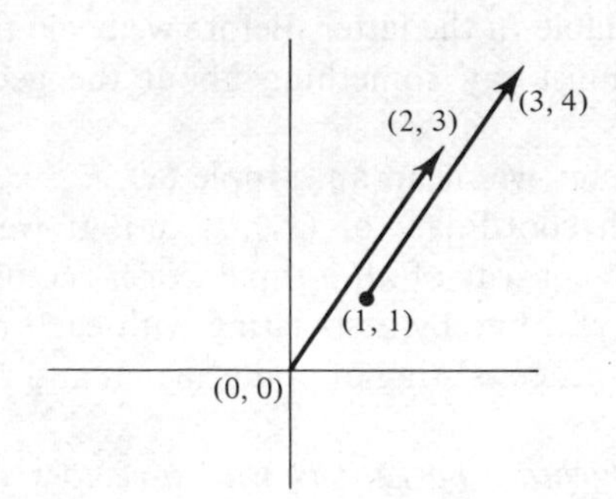

Fig. 2-2

Addition and scalar multiplication have most of the properties we would expect from our everyday experience in R^1. We list some below, with capital letters denoting n-vectors, and Greek letters denoting scalars (real numbers):

(1) $A + B = B + A$ (commutativity)
(2) $(A + B) + C = A + (B + C)$ (associativity)

(3) $(\alpha\beta)A = \alpha(\beta A)$
(4) $(\alpha + \beta)A = \alpha A + \beta A$
(5) $\alpha(A + B) = \alpha A + \alpha B$

The proofs are simple. For example, in (5), put $A = (a_1, \ldots, a_n)$ and $B = (b_1, \ldots, b_n)$. Then

$$\begin{aligned}
\alpha(A + B) &= \alpha((a_1, \ldots, a_n) + (b_1, \ldots, b_n)) \\
&= \alpha(a_1 + b_1, \ldots, a_n + b_n) \\
&= (\alpha(a_1 + b_1), \ldots, \alpha(a_n + b_n)) \\
&= (\alpha a_1 + \alpha b_1, \ldots, \alpha a_n + \alpha b_n) \\
&= (\alpha a_1, \ldots, \alpha a_n) + (\alpha b_1, \ldots, \alpha b_n) \\
&= \alpha(a_1, \ldots, a_n) + \alpha(b_1, \ldots, b_n) \\
&= \alpha A + \alpha B.
\end{aligned}$$

Geometrically, addition is usually thought of as obeying a parallelogram law, which we illustrate for 2-vectors in Fig. 2-3. Suppose that we want to add (a_1, a_2) to (b_1, b_2). Represent each vector as a directed line segment from the origin. Draw the other two sides of the indicated parallelogram. A simple geometric argument (note the shaded similar triangles) now shows that $(a_1 + b_1, a_2 + b_2)$ is represented by the arrow from the origin to the opposite corner of the parallelogram. Note that the side parallel to the arrow from (0, 0) to (b_1, b_2) also represents the vector (b_1, b_2), and is $(a_1 + b_1, a_2 + b_2) - (a_1, a_2)$. This

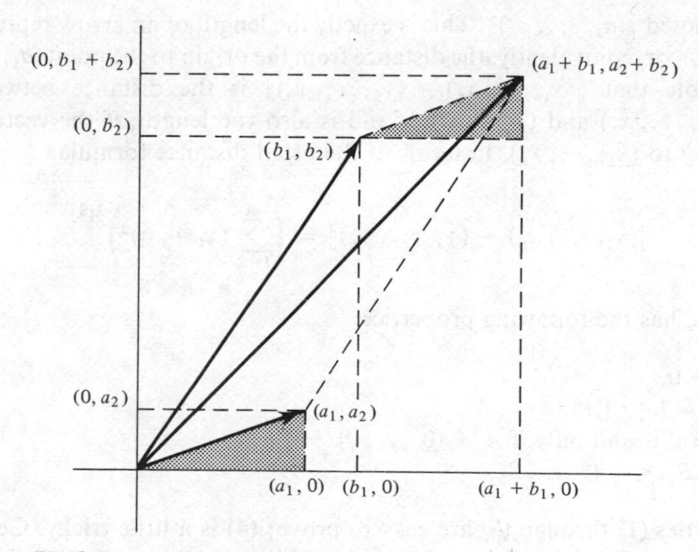

Fig. 2-3

means that an arrow with initial point (p, q) and terminal point (u, v) represents the vector $(u - p, v - q)$, with immediate generalization to n-vectors for $n > 2$.

Scalar multiplication also has a simple geometric interpretation. If $\alpha > 0$, αA is a vector in the same direction as A, but with length multiplied by a factor of α. If $\alpha < 0$, αA is in the opposite direction from A, with length scaled by a factor $|\alpha|$ (see Fig. 2-4).

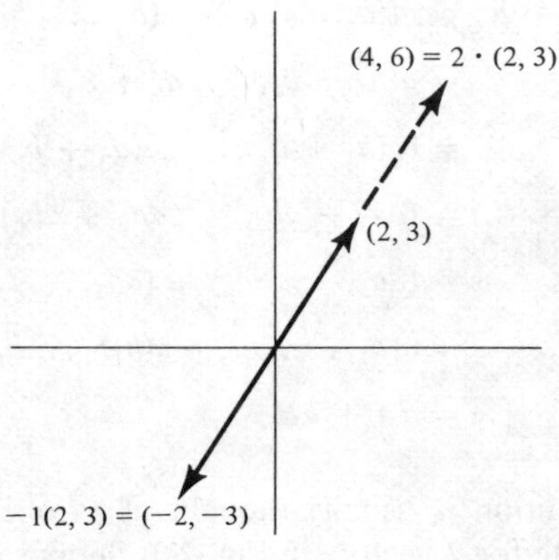

Fig. 2-4

The *magnitude of a vector* $(a_1, \ldots, a_n)$ is defined as

$$\left(\sum_{i=1}^{n} a_i^2\right)^{1/2}$$

and is denoted $\|(a_1, \ldots, a_n)\|$. This is exactly the length of an arrow representing $(a_1, \ldots, a_n)$, or, equivalently, the distance from the origin to the point $(a_1, \ldots, a_n)$ in R^n. Note that $\|(x_1, \ldots, x_n) - (y_1, \ldots, y_n)\|$ is the distance between the points $(x_1, \ldots, x_n)$ and $(y_1, \ldots, y_n)$, and is also the length of the vector from $(y_1, \ldots, y_n)$ to $(x_1, \ldots, x_n)$. In terms of the usual distance formula,

$$\|(x_1, \ldots, x_n) - (y_1, \ldots, y_n)\| = \left(\sum_{i=1}^{n} (x_i - y_i)^2\right)^{1/2}.$$

Magnitude has the following properties:

(1) $\|A\| \geq 0$.
(2) $\|\alpha A\| = |\alpha| \cdot \|A\|$.
(3) $\|A\| = 0$ if and only if $A = (0, \ldots, 0)$.
(4) $\|A + B\| \leq \|A\| + \|B\|$.

Properties (1) through (3) are easy to prove; (4) is a little tricky. Geometrically, in view of the parallelogram law for addition, (4) says that the length of one side of a triangle cannot exceed the sum of the lengths of the other two sides. We shall defer a proof until later, when it will be easier.

Exercises for Section 2.1

1. Write out proofs of properties (1) through (4) of vector addition and scalar multiplication.

2. Write out proofs of properties (1) through (3) of magnitude.

2.2 Dot Product

The *dot product* of two n-vectors is defined by

$$(x_1, \ldots, x_n) \cdot (y_1, \ldots, y_n) = \sum_{i=1}^{n} x_i y_i.$$

Note that this operation produces a real number, not another vector.

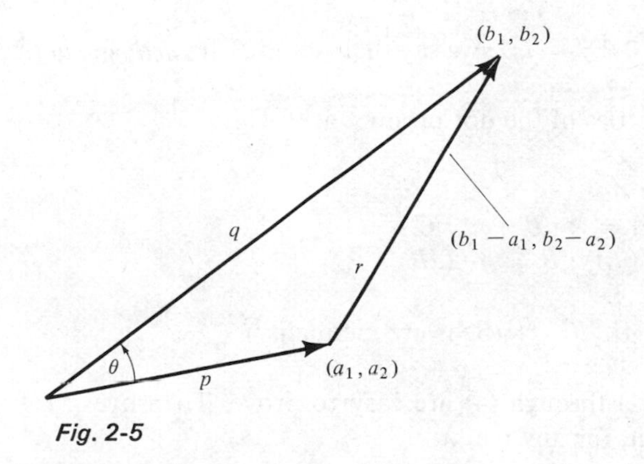

Fig. 2-5

This may seem a rather artificial way of producing a real number from two vectors. It is geometrically motivated by the following consideration. For convenience, we restrict our attention temporarily to the plane, although the same argument can be applied in R^3. Consider (a_1, a_2) and (b_1, b_2) as shown in Fig. 2-5. Complete the triangle by drawing the vector $(b_1 - a_1, b_2 - a_2)$ from (a_1, a_2) to (b_1, b_2). With θ as the angle between (a_1, a_2) and (b_1, b_2), and the side lengths labeled p, q, and r as shown, the law of cosines says:

$$q^2 = p^2 + r^2 - 2pr \cos(\theta)$$

Now

$$p = \|(b_1, b_2)\| = \sqrt{b_1^2 + b_2^2},$$
$$q = \|(b_1 - a_1, b_2 - a_2)\| = \sqrt{(b_1 - a_1)^2 + (b_2 - a_2)^2},$$
$$r = \|(a_1, a_2)\| = \sqrt{a_1^2 + a_2^2}.$$

Then

$$(b_1 - a_1)^2 + (b_2 - a_2)^2 = b_1^2 + b_2^2 + a_1^2 + a_2^2 - 2\sqrt{b_1^2 + b_2^2}\sqrt{a_1^2 + a_2^2}\cos(\theta).$$

After some cancellations, we get

$$\cos(\theta) = \frac{a_1 b_1 + a_2 b_2}{\sqrt{a_1^2 + a_2^2}\sqrt{b_1^2 + b_2^2}} = \frac{(a_1, a_2) \cdot (b_1, b_2)}{\|(a_1, a_2)\| \|(b_1, b_2)\|}.$$

This extends to three dimensions and suggests that we define, for any n, the angle between two nonzero n-vectors A and B to be given by

$$\cos(\theta) = \frac{A \cdot B}{\|A\| \|B\|}.$$

Since $\cos(\theta) = 0$ if $\theta = \pi/2$, we say that A and B are *orthogonal* or *perpendicular* if $A \cdot B = 0$.

Some properties of the dot product are:

(1) $A \cdot B = B \cdot A$.
(2) $A \cdot (B + C) = A \cdot B + A \cdot C$
(3) $\alpha(A \cdot B) = (\alpha A) \cdot B = A \cdot (\alpha B)$.
(4) $\|A\|^2 = A \cdot A$.
(5) $|A \cdot B| \leq \|A\| \|B\|$ (Schwarz inequality).

Properties (1) through (4) are easy to prove. To prove (5), begin with the observation that, for any real λ:

$$\begin{aligned} 0 \leq \|A + \lambda B\|^2 &= (A + \lambda B) \cdot (A + \lambda B) \\ &= A \cdot A + \lambda^2 B \cdot B + 2\lambda(A \cdot B) \\ &= \|A\|^2 + \lambda^2 \|B\|^2 + 2\lambda A \cdot B. \end{aligned}$$

Assuming that $\|B\| \neq 0$, put $\lambda = A \cdot B/\|B\|^2$ and we obtain by direct substitution:

$$0 \leq \frac{1}{\|B\|^2} [\|A\|^2 \|B\|^4 - (A \cdot B)^2 \|B\|^2]$$

or, equivalently,

$$0 \leq \|A\|^2\|B\|^2 - (A \cdot B)^2.$$

Then $|A \cdot B| \leq \|A\|\,\|B\|$.

There remains the case that $\|B\| = 0$. Then $B = (0, \ldots, 0)$, and $A \cdot B = 0 \leq \|A\|\,\|B\|$.

The Schwarz inequality gives us an easy proof of the triangle inequality:

$$\begin{aligned}\|A + B\|^2 = (A + B)\cdot(A + B) &= \|A\|^2 + \|B\|^2 + 2A \cdot B \\ &\leq \|A\|^2 + \|B\|^2 + 2\|A\|\,\|B\| \\ &= (\|A\| + \|B\|)^2.\end{aligned}$$

Hence $\|A + B\| \leq \|A\| + \|B\|$.

Exercises for Section 2.2

1. Write out proofs of properties (1) through (4) of the dot product.
2. Find the angle (in terms of arccos) between the following vectors. Which pairs, if any, are orthogonal?
   (a) (2, 1, 6), (−3, 1, 4)
   (b) (0, 0, 1), (1, 1, 0)
   (c) (1, 1), (2, 4)
   (d) (1, 1, 3, 0, 4), (0, 0, 1, 1, 2)
3. What conditions on A and B will give us $A \cdot B = \|A\|\,\|B\|$?
4. What conditions on A and B will give us $\|A + B\| = \|A\| + \|B\|$?

2.3 Cross Product

In this section we restrict ourselves to 3-vectors (which may be thought of as lying in the xy-plane by fixing the third coordinate at zero).

The *cross product* of (a_1, a_2, a_3) and (b_1, b_2, b_3) is given by

$$(a_1, a_2, a_3) \times (b_1, b_2, b_3) = (a_2b_3 - a_3b_2, a_3b_1 - a_1b_3, a_1b_2 - a_2b_1).$$

In determinant notation (see the Appendix of this chapter) we can write

$$(a_1, a_2, a_3) \times (b_1, b_2, b_3) = \left(\begin{vmatrix} a_2 & a_3 \\ b_2 & b_3 \end{vmatrix}, \begin{vmatrix} a_3 & a_1 \\ b_3 & b_1 \end{vmatrix}, \begin{vmatrix} a_1 & a_2 \\ b_1 & b_2 \end{vmatrix}\right),$$

in which the subscripts display a cyclic pattern.

Before discussing the geometry of cross products, we state some identities which are easy (but sometimes tedious) to prove.

(1) $A \times B = -(B \times A)$
(2) $A \times A = (0, \ldots, 0)$
(3) $(\alpha A) \times B = \alpha(A \times B) = A \times (\alpha B)$
(4) $A \cdot (B \times C) = B \cdot (C \times A) = C \cdot (A \times B)$
(5) $(A + B) \times C = A \times C + B \times C$
(6) $A \times (B + C) = A \times B + A \times C$
(7) $\|A + B\|^2 = \|A\|^2 \cdot \|B\|^2 - (A \cdot B)^2$

We would now like to know something about the magnitude and direction of $A \times B$. From (7), $A \times B$ has length

$$\sqrt{\|A\|^2\|B\|^2 - (A \cdot B)^2},$$

which can be simplified somewhat by writing:

$$\begin{aligned} \|A\|^2\|B\|^2 - (A \cdot B)^2 &= \|A\|^2\|B\|^2 - \|A\|^2\|B\|^2 \cos^2(\theta) \\ &= \|A\|^2\|B\|^2 \sin^2(\theta). \end{aligned}$$

Then

$$\|A \times B\| = \|A\| \, \|B\| \sin(\theta),$$

where θ is the angle between A and B. This is, incidentally, the area of a parallelogram with sides A and B having an angle θ between them.

Next, what is the direction of $A \times B$? Put $C = A$ in (4) to obtain

$$B \cdot (A \times A) = A \cdot (A \times B).$$

But, $A \times A = (0, \ldots, 0)$, so $A \cdot (A \times B) = 0$. Then A is orthogonal to $A \times B$.

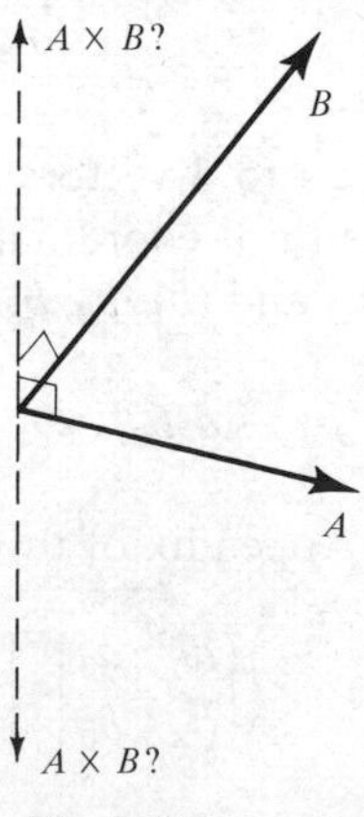

Fig. 2-6

Similarly, B is orthogonal to $A \times B$. This leaves two possibilities for $A \times B$, as shown in Fig. 2-6. To get some feeling for which it might be, note that

$$(1, 0, 0) \times (0, 1, 0) = (0, 0, 1).$$

In general, if you place your right hand so that the fingers are pointing in the sense of rotation from A to B, your thumb should point along $A \times B$.

Exercises for Section 2.3

1. Prove properties (1) through (7) for the cross product.
2. A parallelogram has vertices at (0, 0), (1, 1), (4, 2), and (3, 1). Use vector methods to find its area.
3. Let $A = (a_1, a_2, a_3)$, $B = (b_1, b_2, b_3)$, and $C = (c_1, c_2, c_3)$. Show that

$$A \cdot (B \times C) = \begin{vmatrix} a_1 & a_2 & a_3 \\ b_1 & b_2 & b_3 \\ c_1 & c_2 & c_3 \end{vmatrix}.$$

4. Prove Lagrange's identity:

$$(A \times B) \cdot (C \times D) = (A \cdot C)(B \cdot D) - (A \cdot D)(B \cdot C).$$

5. Suppose that $A \times B = (0, 0, 0)$ for every vector B. Prove that $A = (0, 0, 0)$.

2.4 Applications of Vector Algebra

Vector algebra and operations can often be used to greatly simplify geometric calculations in 3-space. We give several illustrations.

EXAMPLE

Find the equation of the line through (1, 2, 3) and $(\pi, e, 17)$. Argue as follows. The vector $(\pi - 1, e - 2, 14)$ has (1, 2, 3) as initial point and $(\pi, e, 17)$ as terminal point. If (x, y, z) is any point on the line in question, then the vector $(x - 1, y - 2, z - 3)$, being in the same direction as $(\pi - 1, e - 2, 14)$, must be some multiple of $(\pi - 1, e - 2, 14)$. Then, for some t,

$$(x - 1, y - 2, z - 3) = t(\pi - 1, e - 2, 14).$$

Then

$$x = 1 + t(\pi - 1),$$

$$y = 2 + t(e - 2),$$

and

$$z = 3 + 14t.$$

As t takes on all real values, the point (x, y, z) moves along the line. When $t = 0$, (x, y, z) is at $(1, 2, 3)$, and when $t = 1$, (x, y, z) is at $(\pi, e, 17)$. For $0 < t < 1$, (x, y, z) is between these two points on the line, and for $t > 1$ and $t < 0$, beyond these two points.

The above equations specify points (x, y, z) on the line in parametric form, with t as parameter. If we eliminate t, we get

$$\frac{x-1}{\pi-1} = \frac{y-2}{e-2} = \frac{z-3}{14}$$

as the equation of the same line. ∎

EXAMPLE

We shall find it necessary in the sequel to be able to find the equation of a plane through a given point and perpendicular to a given vector (i.e., every vector in the plane should be perpendicular to the given vector).

Let (a_1, a_2, a_3) be the given point and (n_1, n_2, n_3) the vector normal to the plane we seek. Let (x, y, z) denote the coordinates of an arbitrary point on the plane. Then the vector $(x - a_1, y - a_2, z - a_3)$ lies in the plane, hence is orthogonal to (n_1, n_2, n_3). Then

$$(n_1, n_2, n_3) \cdot (x - a_1, y - a_2, z - a_3) = 0.$$

The plane is then the locus of points (x, y, z), satisfying

$$n_1(x - a_1) + n_2(y - a_2) + n_3(z - a_3) = 0.$$ ∎

EXAMPLE

The law of sines says that, for any triangle with angles and side lengths as shown in Fig. 2-7,

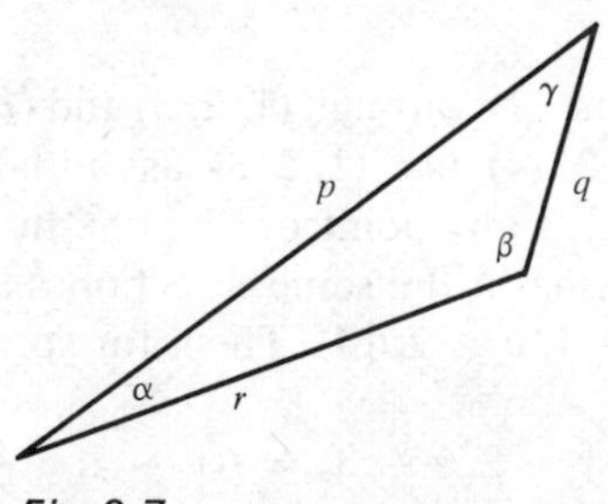

Fig. 2-7

$$\frac{p}{\sin(\beta)} = \frac{q}{\sin(\alpha)} = \frac{r}{\sin(\gamma)}.$$

To prove this, think of the triangle as being composed of two vectors R and Q and their sum $P = R + Q$, as shown in Fig. 2-8. Now, in general,

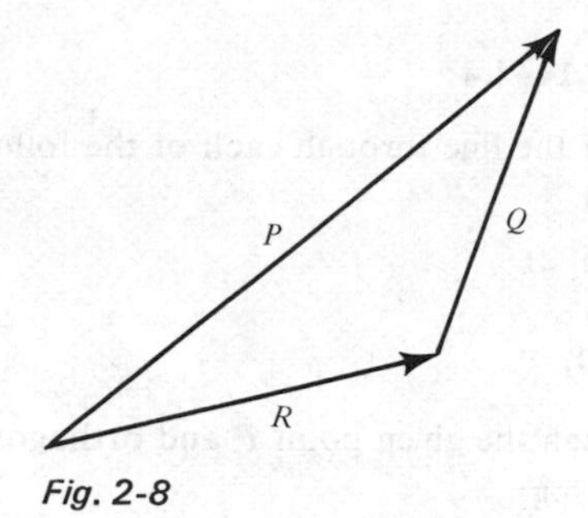

Fig. 2-8

$$\|A \times B\| = \|A\| \, \|B\| \sin(\theta),$$

where θ is the angle between A and B. Then

$$\sin(\alpha) = \frac{\|P \times R\|}{\|P\| \, \|R\|},$$

$$\sin(\beta) = \frac{\|R \times Q\|}{\|R\| \, \|Q\|},$$

and

$$\sin(\gamma) = \frac{\|P \times Q\|}{\|P\| \, \|Q\|}.$$

Now $p = \|P\|$, $q = \|Q\|$, and $r = \|R\|$. Then

$$\frac{p}{\sin(\beta)} = \frac{\|P\| \, \|R\| \, \|Q\|}{\|R \times Q\|}, \qquad \frac{q}{\sin(\alpha)} = \frac{\|Q\| \, \|P\| \, \|R\|}{\|P \times R\|},$$

and

$$\frac{r}{\sin(\gamma)} = \frac{\|R\| \, \|Q\| \, \|P\|}{\|P \times Q\|}.$$

The numerators on the right are all the same, so we need only show that the denominators are equal. Now $P = Q + R$, so

$$\|P \times Q\| = \|(Q + R) \times Q\| = \|Q \times Q + R \times Q\| = \|R \times Q\|.$$

Similarly,

$$\|P \times R\| = \|(Q + R) \times R\| = \|Q \times R + R \times R\| = \|Q \times R\| = \|R \times Q\|,$$

completing the proof. ∎

Exercises for Sections 2.1—2.4

1. Find the equation of the line through each of the following pairs of points:
   (a) $(0, 0, 0), (1, 2, 3)$
   (b) $(\pi, e, \sqrt{2}), (1, 16, 2)$
   (c) $(2, 2, 3), (4, 4, 6)$
   (d) $(1, 6, 0), (2, 8, 10)$

2. Find the plane through the given point P and orthogonal to the given vector N in each of the following:
   (a) $P = (1, 1, 6), N = (2, 2, 4)$
   (b) $P = (1, 2, 4), N = (2, 8, 3)$
   (c) $P = (2, 16, \pi), N = (8, 1, 7)$
   (d) $P = (8, 4, 4), N = (0, 0, 1)$

3. Find the angle between the following pairs of vectors (answers are in terms of arccos):
   (a) $(2, 1, 4), (1, 0, 0)$
   (b) $(6, 1, 2), (2, 1, 3)$
   (c) $(2, 2, 3), (2, 2, 4)$
   (d) $(0, 0, 1), (2, 0, 0)$

4. Find a vector perpendicular to the given plane in each of the following:
   (a) $2x + 37 + 4z = 8$
   (b) $8x - 3y + 16z = 12$
   (c) $x + y = 0$
   (d) $x + 16y + 2z = 0$
   (e) $ax + by + cz = d$, a, b, c, d real numbers with $a^2 + b^2 + c^2 \neq 0$

5. Find a vector through $(2, 1, 6)$ and orthogonal to the plane $6x + 3y + 2z = 8$.

6. Show that the distance from a point $P = (x_0, y_0, z_0)$ to the plane $ax + by + cz = d$ is

$$\frac{|ax_0 + by_0 + cz_0 + d|}{\sqrt{a^2 + b^2 + c^2}}$$

   provided that $a^2 + b^2 + c^2 \neq 0$. (The distance from a point to a plane is the length of a vector from the point to a point in the plane and perpendicular to the plane.)

7. Prove that every line is the intersection of two planes.

8. In each of the following, determine whether the planes are parallel, coincide, or intersect. If the planes intersect, find the line of intersection.
   (a) $8x + 3y + 2z = 16$, $8x + 3y + 2z = 15$
   (b) $2x + y + z = 8$, $6x + 3y + 3z = 24$
   (c) $8x + 3y + 2z = 4$, $8x - 3y + 4z = 8$
   (d) $x - y = 0$, $x + y = 0$

9. In R^n, produce n vectors $v_1, \ldots, v_n$ such that, for any P in R^n, there are real numbers $\alpha_1, \ldots, \alpha_n$ such that $P = \sum_{i=1}^{n} \alpha_i v_i$.

10. Let $v_1, \ldots, v_n$ be in R^n, with the property that $v_i \cdot v_j = 0$ if $i \neq j$. Let $\alpha_1, \ldots, \alpha_n$, $\beta_1, \ldots, \beta_n$ be real numbers. Show that

(a) $\left(\sum_{i=1}^{n} \alpha_i v_i\right) \cdot \left(\sum_{i=1}^{n} \beta_i v_i\right) = \sum_{i=1}^{n} \alpha_i \beta_i \|v_i\|^2$

(b) $\left\| \sum_{i=1}^{n} \alpha_i v_i \right\| = \left(\sum_{i=1}^{n} \alpha_i^2\right)^{1/2}$ if $\|v_i\| = 1$ for $i = 1, \ldots, n$.

11. Find the volume of the parallelopiped in R^3 having as three sides the vectors $(\frac{1}{2}, \frac{1}{2}, 0)$, $(\frac{2}{3}, \frac{1}{2}, 0)$, $(0, 0, \sqrt{3})$.

12. Let $F(t) = (f_1(t), \ldots, f_n(t))$, where $f_1, \ldots, f_n$ are functions defined for $a \leq t \leq b$.

(a) Define $\lim_{t \to c} F(t) = (\lim_{t \to c} f_1(t), \ldots, \lim_{t \to c} f_n(t))$, assuming that these limits exist finite. Prove that:

$$\lim_{t \to c} F(t) = (a_1, \ldots, a_n) \text{ if and only if } \lim_{t \to c} \|F(t) - (a_1, \ldots, a_n)\| = 0.$$

(b) Define $F'(c) = (f_1'(c), \ldots, f_n'(c))$, assuming that these derivatives exist. Show that

$$F'(c) = (a_1, \ldots, a_n)$$

$$\text{if and only if } \lim_{\Delta t \to 0} \left\| \frac{F(c + \Delta t) - F(c)}{\Delta t} - (a_1, \ldots, a_n) \right\| = 0.$$

(c) Let $G(t) = (g_1(t), \ldots, g_n(t))$. Show that

$$\frac{d[F(t) \cdot G(t)]}{dt} = F(t) \cdot \frac{dG(t)}{dt} + G(t) \cdot \frac{dF(t)}{dt}$$

and

$$\frac{d[F(t) \times G(t)]}{dt} = F(t) \times \frac{dG(t)}{dt} + \left[\frac{dF(t)}{dt}\right] \times G(t).$$

13. Suppose that vectors A, B, and C form a parallelopiped in 3-space. Show that the volume of the parallelopiped is $A \cdot (B \times C)$.

14. Prove that points P, Q, and S are on a straight line in R^3 if and only if there is some real number a such that $S = (1 - a)P + aQ$.

15. Show that the midpoints of the sides of any quadrilateral are vertices of a parallelogram.

16. Prove that the sum of the squares of the lengths of the diagonals of any quadrilateral is twice the sum of the squares of the lengths of the line segments connecting midpoints of opposite sides.

2.5 Sequences in R^n

We conclude this chapter with a brief consideration of sequences of n-tuples. This material will be of use when we treat functions of n variables in Chapter 3. Suppose that $(x_{j1},\ldots, x_{jn})$ is in R^n for $j = 1, 2,\ldots$. Then we call $\{(x_{j1},\ldots, x_{jn})\}_{j=1}^{\infty}$ a *sequence of n-vectors*. Intuitively, it is natural to say that such a sequence converges to $(a_1,\ldots, a_n)$ if the points $(x_{j1},\ldots, x_{jn})$ can be made arbitrarily close to $(a_1,\ldots, a_n)$ by choosing j large enough. Put more carefully: $\{(x_{j1},\ldots, x_{jn})\}_{j=1}^{\infty}$ *converges to* $(a_1,\ldots, a_n)$ *if, given* $\varepsilon > 0$, *there is some* N *such that* $\|(x_{j1},\ldots, x_{jn}) - (a_1,\ldots, a_n)\| < \varepsilon$ *whenever* $j \geq N$.

Note that this is the same as saying that

$$\lim_{j\to\infty} \|(x_{j1},\ldots, x_{jn}) - (a_1,\ldots, a_n)\| = 0.$$

Fortunately, a sequence in R^n can be treated by looking at n real sequences, making all the material of Section 1.1 available to us. To see how this can be done, start with $\{(x_{j1},\ldots, x_{jn})\}_{j=1}^{\infty}$, and make a separate sequence out of the terms occurring in each coordinate place. This gives us n real sequences: $\{x_{j1}\}_{j=1}^{\infty}, \{x_{j2}\}_{j=1}^{\infty},\ldots, \{x_{jn}\}_{j=1}^{\infty}$. These are the *coordinate sequences* of

$$\{(x_{j1},\ldots, x_{jn})\}_{j=1}^{\infty}.$$

For example, the coordinate sequences of $\{(1/j, \sin(j)/j, (1 + 1/j)^j)\}_{j=1}^{\infty}$ are $\{1/j\}_{j=1}^{\infty}$, $\{\sin(j)/j\}_{j=1}^{\infty}$, and $\{(1 + 1/j)^j\}_{j=1}^{\infty}$. It seems reasonable that

$$\{(x_{j1},\ldots, x_{jn})\}_{j=1}^{\infty}$$

should approach $(a_1,\ldots, a_n)$ as a limit exactly with the kth coordinate sequence converges to a_k. We shall now prove that this is the case.

Theorem

$\lim_{j\to\infty} (x_{j1},\ldots, x_{jn}) = (a_1,\ldots, a_n)$ *if and only if* $\lim_{j\to\infty} x_{jk} = a_k$ *for* $k = 1,\ldots, n$.

PROOF

Suppose first that $\lim_{j\to\infty} (x_{j1},\ldots, x_{jn}) = (a_1,\ldots, a_n)$. Let $\varepsilon > 0$. For some N, $\|(x_{j1} - a_1,\ldots, x_{jn} - a_n)\| < \varepsilon$ for $j \geq N$. Then, for $k = 1,\ldots, n$, we have

$$|x_{jk} - a_k| \leq \|(x_{j1} - a_1,\ldots, x_{jn} - a_n)\| < \varepsilon$$

for $j \geq N$, hence

$$\lim_{j\to\infty} x_{jk} = a_k.$$

Conversely, suppose that $x_{jk} \to a_k$ for $k = 1,\ldots, n$, as $j \to \infty$. Given $\varepsilon > 0$, and given k, $1 \leq k \leq n$, there is some N_k such that $|x_{jk} - a_k| < \varepsilon/\sqrt{n}$ for

$j \geq N_k$. Choose N as the largest of the numbers $N_1, \ldots, N_n$. Then, for $j \geq N$, we have also $j \geq N_k$ for $k = 1, \ldots, n$, hence

$$\|(x_{j1}, \ldots, x_{jn}) - (a_1, \ldots, a_n)\| = \left[\sum_{k=1}^{n} (x_{jk} - a_k)^2\right]^{1/2}$$
$$< \left[\sum_{k=1}^{n} \left(\frac{\varepsilon}{\sqrt{n}}\right)^2\right]^{1/2} = \varepsilon. \quad ▮$$

EXAMPLE

$$\lim_{j\to\infty} \left(\frac{1}{j}, \frac{\sin(j)}{j}, \left(1 + \frac{1}{j}\right)^j\right) = (0, 0, e).$$

For, $1/j \to 0$, $\sin(j)/j \to 0$, and $(1 + 1/j)^j \to e$ as $j \to \infty$. ▮

In closing this section, we note that

$$\lim_{j\to\infty} [(x_{j1}, \ldots, x_{jn}) + (y_{j1}, \ldots, y_{jn})] = \lim_{j\to\infty} (x_{j1}, \ldots, x_{jn}) + \lim_{j\to\infty} (y_{j1}, \ldots, y_{jn})$$

whenever the two last limits exist, and, for any real number α,

$$\lim_{j\to\infty} [\alpha(x_{j1}, \ldots, x_{jn})] = \alpha \lim_{j\to\infty} (x_{j1}, \ldots, x_{jn}).$$

The proofs of these are easy, either directly from the definition of convergence, or by the above theorem.

Exercises for Section 2.5

1. Determine whether or not the following converge:

(a) $\{[(1 + 2/n)^n, 1/2^n, (6n + 1)/(4n + 2)]\}_{n=1}^{\infty}$

(b) $\{[\ln(n + 1)/\ln(n), e^{-n^2}]\}_{n=1}^{\infty}$

(c) $\{[1/n, n^3/n!, e^n/\ln(n + 1)]\}_{n=1}^{\infty}$

2. Let $\{(x_{j1}, \ldots, x_{jn})\}_{j=1}^{\infty}$ and $\{(y_{j1}, \ldots, y_{jn})\}_{j=1}^{\infty}$ be sequences of n-vectors, converging respectively to A and B.

(a) Is it true that $\lim_{j\to\infty} [(x_{j1}, \ldots, x_{jn}) \cdot (y_{j1}, \ldots, y_{jn})] = A \cdot B$?

(b) Suppose now that $n = 3$. Is it true that

$$\lim_{j\to\infty} [(x_{j1}, x_{j2}, x_{j3}) \times (y_{j1}, y_{j2}, y_{j3})] = A \times B?$$

3. Prove the following Cauchy-type theorem for R^n: *Let V_k be in R^n for $k =$ 1, 2, ... Then $\{V_k\}_{k=1}^{\infty}$ converges if and only if, given $\varepsilon > 0$, there is some N such that $\|V_p - V_q\| < \varepsilon$ whenever $p, q \geq N$. Hint:* Use Cauchy's theorem as given in Chapter 1, concentrating on the coordinate sequences of $\{V_k\}_{k=1}^{\infty}$.

Appendix: Determinants and Cramer's Rule

In this chapter the notion of a determinant came up as a useful notational device, and it will again as well, so we include here a brief discussion of 2 by 2 and 3 by 3 determinants.

The symbol

$$\begin{vmatrix} a & b \\ c & d \end{vmatrix}$$

denotes a determinant with two rows and two columns (hence a 2 by 2 determinant), and stands for the product $ad - bc$. Here the letters a, b, c, and d can be numbers (real or complex) or functions. For example,

$$\begin{vmatrix} e^x & \sin(x) \\ 1 & x^2 \end{vmatrix}$$

stands for $x^2e^x - \sin(x)$, and

$$\begin{vmatrix} 1 & 16 \\ \pi & 26 \end{vmatrix} = 26 - 16\pi.$$

Three by three determinants (three rows and three columns) are given by

$$\begin{vmatrix} a & b & c \\ d & e & f \\ g & h & i \end{vmatrix} = a\begin{vmatrix} e & f \\ h & i \end{vmatrix} - b\begin{vmatrix} d & f \\ g & i \end{vmatrix} + c\begin{vmatrix} d & e \\ g & h \end{vmatrix}$$

$$= a(ei - fh) - b(di - fg) + c(dh - eg).$$

Again, the entries may be numbers or functions. As examples,

$$\begin{vmatrix} 1 & 1 & 2 \\ \pi & 1 & 3 \\ 2 & 1 & 4 \end{vmatrix} = 5 - 3\pi,$$

and

$$\begin{vmatrix} e^x & 1 & \sin(x) \\ x & x^2 & 2x \\ \cos(x) & \cosh(x) & 14 \end{vmatrix} = e^x[14x^2 - 2x\cosh(x)] - [14x - 2x\cos(x)] + \sin(x)[x\cosh(x) - x^2\cos(x)].$$

It is possible to continue on and consider n by n determinants for $n = 4, 5, \ldots$, but as n increases in size, the labor involved in evaluating the determinant becomes prohibitive, and we shall have use only for the 2 by 2 and 3 by 3 cases.

Determinants arise in a number of different contexts. One is Cramer's Rule, which allows us to write down solutions to certain systems of linear equations in terms of determinants. In the 2 by 2 case, Cramer's Rule says that the solution to the system

$$\begin{aligned} ax + by &= A \\ cx + dy &= B, \end{aligned}$$

is

$$x = \frac{\begin{vmatrix} A & b \\ B & d \end{vmatrix}}{\begin{vmatrix} a & b \\ c & d \end{vmatrix}}, \qquad y = \frac{\begin{vmatrix} a & A \\ c & B \end{vmatrix}}{\begin{vmatrix} a & b \\ c & d \end{vmatrix}},$$

provided that $ad - bc \neq 0$. We call $\begin{vmatrix} a & b \\ c & d \end{vmatrix}$ the determinant of the system of equations. Observe that it consists of the coefficients of x (first column) and y (second column), and that the solution for x (y) is obtained by replacing the first (second) column with the column of quantities appearing on the right side of the equations.

In the 3 by 3 case, Cramer's rule says that the solution to

$$\begin{aligned} ax + by + cz &= A \\ dx + ey + fz &= B \\ gx + hy + kz &= C \end{aligned}$$

is

$$x = \frac{\begin{vmatrix} A & b & c \\ B & e & f \\ C & h & k \end{vmatrix}}{D}, \qquad y = \frac{\begin{vmatrix} a & A & c \\ d & B & f \\ g & C & k \end{vmatrix}}{D}, \qquad z = \frac{\begin{vmatrix} a & b & A \\ d & e & B \\ g & h & C \end{vmatrix}}{D},$$

where

$$D = \begin{vmatrix} a & b & c \\ d & e & f \\ g & h & k \end{vmatrix}$$

is the determinant of the system and is assumed to be nonzero. Again, we obtain the "first" unknown, x, by replacing the first column of D with the column on the right; the "second" unknown y by replacing the second column of D with the column on the right; and the "third" unknown z by replacing the third column of D with the column on the right. The efficiency of this method is in having a rule by which the solutions can be written down by inspection from the original system of equations.

The reader should satisfy himself or herself that the above expressions actually do provide solutions to the given systems of equations.

Determinants will also be useful later when we treat partial derivatives and develop the Jacobian notation. In particular, implicit partial differentiation and use of differentials in taking partial derivatives will often be rendered computationally easier by use of determinant notation and Cramer's Rule.

3

Functions of n Variables

3.1 Continuity

Very often one independent and one dependent variable are insufficient to construct an accurate mathematical model of a situation. For example, in describing the motion of a vibrating elastic membrane stretched across a fixed frame in the plane, we shall see that we need three space variables and a time variable. In this chapter we lay the groundwork for the integration and differentiation of functions of n variables by developing the notions of limit and continuity, along with some of their ramifications.

By a real-valued function of n real variables we mean a rule that assigns a real number to each n-vector in some specified set. For example, we might have

$$f(x, y) = x^2 + y^2 + \sin(xy) \qquad \text{for all real } x \text{ and } y,$$

or

$$g(x, y, z) = \sin(x) + z\ln(y) \qquad \text{for } 0 \le x \le \pi \text{ and all } z,$$

or

$$h(x_1, \ldots, x_n) = \sum_{i=1}^{n} x_i^2, \qquad \text{where } 0 \le x_i \le 1 \text{ for } i = 1, \ldots, n.$$

An important fact about functions of one variable is that continuity on a closed interval implies boundedness. Recall that this fails if the interval is not closed. [For example, if $f(x) = 1/x$, for $0 < x \le 1$, then f is continuous on $(0, 1]$, but unbounded.] We would like to develop an analogous theorem for functions of n variables. This requires not only an extension of the notion of continuity, but also some n-dimensional analogue of the closed interval.

Continuity itself is a simple concept to extend to n dimensions. To say that f is continuous at $(a_1, \ldots, a_n)$ means that $f(x_1, \ldots, x_n)$ can be made arbitrarily close to $f(a_1, \ldots, a_n)$ by choosing $(x_1, \ldots, x_n)$ sufficiently close to $(a_1, \ldots, a_n)$.

Definition

f is *continuous* at $(a_1, \ldots, a_n)$ if, given $\varepsilon > 0$, there is some $\delta > 0$ such that $|f(x_1, \ldots, x_n) - f(a_1, \ldots, a_n)| < \varepsilon$ whenever $f(x_1, \ldots, x_n)$ is defined and $\|(x_1, \ldots, x_n) - (a_1, \ldots, a_n)\| < \delta$.

Put in terms of scalar quantities, this says that $|f(x_1, \ldots, x_n) - f(a_1, \ldots, a_n)| \to 0$ as $\|(x_1, \ldots, x_n) - (a_1, \ldots, a_n)\| \to 0$. *We say that f is continuous on a set D if f is continuous at each point of D.*

If it is clear from context that f is a function of n variables, then we need not emphasize this by writing $f(x_1, \ldots, x_n)$, but can simply write, say, $f(x)$, with the understanding that $x = (x_1, \ldots, x_n)$ is an n-vector. The definition of continuity then becomes: f is continuous at a if, given $\varepsilon > 0$, there exists some $\delta > 0$ such that $|f(x) - f(a)| < \varepsilon$ whenever $f(x)$ is defined and $\|x - a\| < \delta$. This reduces immediately to the usual one-variable definition of continuity if we replace the R^n-norm $\| \quad \|$ by absolute values $| \quad |$ in the case that x is a real number and not an n-tuple. In the remainder of the book, we shall write out vectors in full coordinate form whenever this seems to clarify or simplify things, but shall often use single letters for n-vectors where there appears to be no danger of ambiguity. The savings in space effected by this notation should be obvious.

As with functions of one variable, sums, products, quotients, and compositions of continuous functions are continuous wherever they are well defined. For example, suppose that

$$f(x, y, z) = \frac{x^3 e^{xy} + 3x \sin(yz) + yxz^3}{x^2 z + y + 2}.$$

In the absence of additional information, we understand f to be defined for all 3-vectors (x, y, z) such that $x^2 z + y + 2$ does not vanish. In view of the continuity of the individual functions pieced together by arithmetic operations to form f (powers of the variables, sine, and an exponential function), we should expect that f is continuous wherever it is defined. A straightforward ε, δ proof for such a function would be very messy, but a proof based on the fact that f is a quotient of functions which in turn are sums of products of elementary functions would be fairly simple. Exercises 6, 21, 22, and 23 are designed to help in formulating such proofs.

Given a function f of n variables, it is sometimes useful to invent a function of k variables ($k < n$) by fixing $n - k$ of the variables at specific values. For example, if $f(x_1, \ldots, x_{10})$ is a function of 10 variables, then $f(x_1, a_2, a_3, x_4, x_5, a_6, a_7, a_8, a_9, a_{10})$ is a function of the three variables x_1, x_4, x_5 for given $a_2, a_3, a_6, a_7, a_8, a_9$, and a_{10}. It is not difficult to establish from the definition that such a function will be continuous at (a_1, a_4, a_5) whenever f is continuous at $(a_1, \ldots, a_{10})$. Exercise 5 addresses this type of situation. In particular, *a continuous function of n variables is continuous in any one of the variables separately if the other $n - 1$ variables are fixed at given values.*

The converse of this last statement is not in general true. For example, let

$$f(x, y) = \begin{cases} \dfrac{xy}{x^2 + y^2} & \text{for } x^2 + y^2 \neq 0, \\ 0 & \text{for } x = y = 0. \end{cases}$$

Then the function $g(x)$ defined by $g(x) = f(x, 0)$ is continuous at 0 [in fact, $g(x) = 0/x^2 = 0$ for $x \neq 0$, and $g(0) = f(0, 0) = 0$]. Similarly, $h(y) = f(0, y)$ is continuous at 0, as $h(y) = 0/y^2 = 0$ if $y \neq 0$, and $h(0) = 0$. But f is not continuous at $(0, 0)$. One way to see this is to note that $f(x, x) = \frac{1}{2}$, so it is always possible to find points arbitrarily close to $(0, 0)$ at which the function values differ from $f(0, 0)$ by at least $\frac{1}{2}$.

There is a sequential characterization of continuity which, though not of great practical value in establishing continuity of specific functions, is nevertheless of importance in deriving general properties of continuous functions.

Theorem

f is continuous at a point a in R^n if and only if, given any sequence $\{x_j\}$ converging to a, for which each $f(x_j)$ is defined, then $\{f(x_j)\}$ converges to $f(a)$.

PROOF

First we do the easy part. Suppose that f is continuous at a. Let $x_j \to a$ as $j \to \infty$, and suppose that each $f(x_j)$ is defined. To show that $f(x_j) \to f(a)$, let $\varepsilon > 0$. For some $\delta > 0$, $|f(y) - f(a)| < \varepsilon$ whenever $f(y)$ is defined and $\|y - a\| < \delta$. Now choose N so that $\|x_j - a\| < \delta$ for $j \geq N$. This can be done by convergence of $\{x_j\}$ to a. Then, for $j \geq N$, we have $\|f(x_j) - f(a)\| < \varepsilon$, so that $f(x_j) \to f(a)$ as $j \to \infty$.

The converse is more subtle. Suppose now that $f(x_j) \to f(a)$ whenever $x_j \to a$ and each $f(x_j)$ is defined. We must show that f is continuous at a.

We proceed by contradiction. Suppose that f is not continuous at a. Then, for some $\varepsilon > 0$, there exists, for each positive integer j, some n-vector z_j such that $\|z_j - a\| < 1/j$ and $|f(z_j) - f(a)| \geq \varepsilon$. Now $z_j \to a$ as $j \to \infty$, since $1/j \to 0$. But then, by hypothesis, $f(z_j) \to f(a)$, which is impossible if $|f(z_j) - f(a)| \geq \varepsilon$ for every j. Hence f must be continuous at a. ∎

EXAMPLE

We illustrate here the importance in the theorem of having $f(x_j) \to f(a)$ for *every* sequence $\{x_j\}$ on which f is defined and converging to a. Put

$$f(x, y) = \begin{cases} \sin\left(\dfrac{1}{xy}\right) & \text{for } xy \neq 0, \\ 0 & \text{for } x = 0 \text{ or } y = 0. \end{cases}$$

The sequence $\{(1/\sqrt{j\pi}, 1/\sqrt{j\pi})\}_{j=1}^{\infty}$ converges to $(0, 0)$, and $f(1/\sqrt{j\pi}, 1/\sqrt{j\pi}) = \sin(j\pi) = 0$ for $j = 1, 2, \ldots$. Thus $f(1/\sqrt{j\pi}, 1/\sqrt{j\pi}) \to f(0, 0)$ as $j \to \infty$.

However, f is not continuous at $(0, 0)$. For the sequence

$$\left\{\left(1\Big/\sqrt{(2j + 1)\frac{\pi}{2}}, 1\Big/\sqrt{(2j + 1)\frac{\pi}{2}}\right)\right\}_{j=1}^{\infty}$$

also converges to $(0, 0)$, but $\sin((2j + 1)\pi/2)$ oscillates between $+1$ and -1, and so has no limit as $j \to \infty$. The point is that we can find 2-vectors (a, b) arbitrarily close to $(0, 0)$ at which the function values maintain a distance of 1 from $f(0, 0)$. ∎

We now turn to the problem of generalizing the notion of "closed interval" to n dimensions. It turns out that the important word is not "interval" but "closed." We need some preliminary definitions and notation.

By an *r-neighborhood* about a point a in R^n we mean the set of points in R^n at a distance less than r from a, that is, the set of all points x in R^n with $\|x - a\| < r$. Such a set can be conveniently described by developing the notation $\{x \mid x \text{ is in } R^n \text{ and } \|x - a\| < r\}$, read "the set of all x such that x is in R^n and $\|x - a\| < r$."

On the real line, an r-neighborhood about a (or of a) is the open interval $(a - r, a + r)$; in the plane, an r-neighborhood of (a_1, a_2) consists of all points within a circle of radius r about (a_1, a_2); in 3-space, an r-neighborhood of (a_1, a_2, a_3) consists of all points inside a sphere of radius r centered at (a_1, a_2, a_3), and so on.

Now let D be any set of points in R^n. By a *boundary point* of D we mean a point a in R^n with the property that every r-neighborhood (i.e., for every $r > 0$) of a contains a point in D and a point not in D. Denote by ∂D the set of boundary points of D. Then,

$$\partial D = \{P \mid P \text{ is a boundary point of } D\}.$$

We say that D is *closed* if D contains all its boundary points. Finally, a point P is *interior* to D if some r-neighborhood of P is wholly contained in (i.e., consists only of points of) D.

These notions should be clarified by some examples.

EXAMPLE

In R^2, let $D = \{(x, y) \mid x^2 + y^2 < 1\}$. Then D consists of all points in the plane inside the unit circle about the origin. Here the set of boundary points is given by $\partial D = \{(x, y) \mid x^2 + y^2 = 1\}$. Only for points P on the unit circle is it true that *every* r-neighborhood of P contains both points inside and points outside the unit circle. Here, D is not closed, as there are boundary points of D not in D (in fact, D contains none of its boundary points in this example). Each point of D has some neighborhood lying wholly in D (e.g., Q in Fig. 3-1), and so is an interior point of D. Each point outside the unit circle has some neighborhood lying wholly outside of D (e.g., S of Fig. 3-1), so that each point of $\{(x, y) | x^2 + y^2 > 1\}$ is interior to this set.

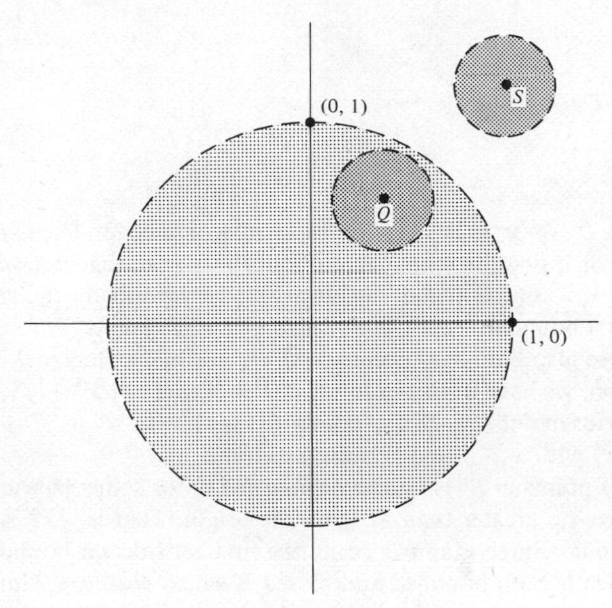

Fig. 3-1

If we put $M = \{(x, y) \mid x^2 + y^2 \leq 1\}$, then $\partial M = \{(x, y) \mid x^2 + y^2 = 1\}$, and M contains all its boundary points, hence is closed. ▮

EXAMPLE

In R^2, let $D = \{(x, y) \mid x \geq 2 \text{ and } y \geq 4\}$ (see Fig. 3-2). Then ∂D consists of all points $(2, y)$ with $y \geq 4$ and all points $(x, 4)$ with $x \geq 2$. Since all such points are in D, then D is closed. Interior points of D are exactly the points (x, y) with $x > 2$ and $y > 4$. ▮

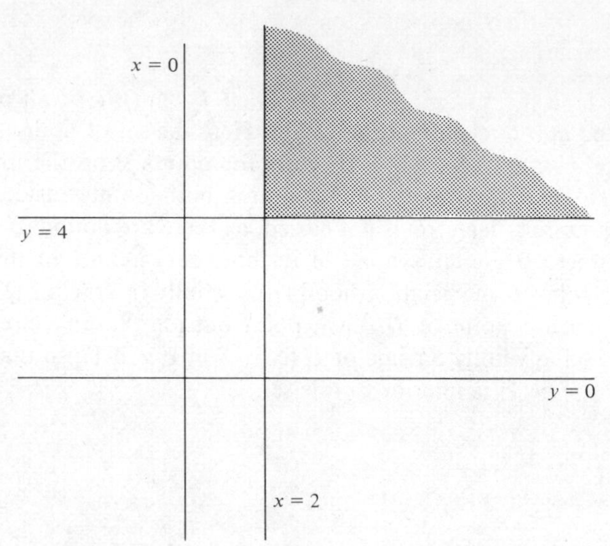

Fig. 3-2

EXAMPLE

In R^3, let $D = \{(x, y, z) \mid x^2 + y^2 + z^2 \leq 1 \text{ or } x > 3\}$. Then, D consists of all points on or inside the unit sphere about the origin, together with all points (x, y, z) with $x > 3$ (and y and z arbitrary). Every point on the surface of the unit sphere is a boundary point, and is also in D. But points $(3, y, z)$, with y and z arbitrary, are also boundary points, and are not in D. Hence D is not closed. In set notation, we have in this example $\partial D = \{(x, y, z) \mid x^2 + y^2 + z^2 = 1 \text{ or } x = 3\}$. Interior points of D are points (x, y, z) with $x^2 + y^2 + z^2 < 1$ and points (x, y, z) with $x > 3$ and y and z arbitrary. ∎

A set D of points in R^n is said to be *bounded* if, for some M, each point of D is at a distance no greater than M from the origin. That is, $\|x\| \leq M$ for each x in D. In the last three examples, only the first considers a bounded set.

A set which is both bounded and closed is called *compact*. Thus,

$$\{(x, y) \mid 0 \leq x \leq 1 \text{ and } 0 \leq y \leq 1\}$$

is compact. (This is a solid square in the plane).

Finally, a function f defined at each point of D is said to be *bounded* on D if, for some M, $|f(x)| \leq M$ for each x in D.

We are now in a position to state the main result of this section.

Theorem

Let f be continuous on a set D in R^n.
If D is compact, then f is bounded.

A proof of this theorem hinges on several extremely delicate properties of sequences and compact sets. Before becoming lost in the details, let us provide an overview by outlining the basic strategy of the proof.

OUTLINE OF THE PROOF

Suppose that the conclusion of the theorem is false. We shall attempt to derive a contradiction based on the sequential characterization of continuity given above.

If f is not bounded on D, then, for each positive integer j, there is some point z_j in D such that $|f(z_j)| \geq j$. Now, the sequence $\{z_j\}$ may or may not converge. We therefore produce a nondecreasing sequence of positive integers, $J_1 < J_2 < \cdots$ such that $\{z_{J_i}\}_{i=1}^{\infty}$ does converge. That is, we pick out certain terms of the original sequence in such a way as to form a new, convergent sequence. Our ability to do this will be seen shortly (I below) to depend upon the compactness of D.

Denote by c the limit of the sequence just formed. The assumption that D is closed will enable us to show (II below) that c is in D. Hence f is continuous at c, so that $f(z_{J_i}) \rightarrow f(c)$ as $i \rightarrow \infty$. But this is clearly impossible, as $|f(z_{J_i})| \geq J_i \rightarrow \infty$ as $i \rightarrow \infty$, by choice of the original sequence $\{z_j\}$. This contradiction will prove the theorem.

An examination of this outline reveals that we have two crucial steps to justify in order to make the argument complete. Both have to do with sequences on compact sets, and are in fact independent of the context in which they are applied above. We take them one at a time, using full coordinate notation in I to clarify the ideas involved.

I. If D is compact, and $(z_{j1}, \ldots, z_{jn})$ is in D for each j, then we can find a sequence $\{J_j\}_{j=1}^{\infty}$ of positive integers, with $J_1 < J_2 < \cdots$, such that the sequence $\{(z_{J_k1}, \ldots, z_{J_kn})\}_{k=1}^{\infty}$ converges. To show this, we begin by looking at the individual coordinate sequences $\{z_{jk}\}_{j=1}^{\infty}$ for $k = 1, \ldots, n$. First, consider $\{z_{j1}\}$. If, for some real number α, $z_{j1} = \alpha$ for infinitely many values of j, then simply choose $H_1 < H_2 < \cdots$ such that $z_{H_k1} = \alpha$ for $k = 1, \ldots$.

If there is no such α, proceed as follows. Since D is bounded, then each coordinate sequence of $\{(z_{j1}, \ldots, z_{jn})\}$ is bounded. For some M, $-M \leq z_{j1} \leq M$ for all j. Since there are infinitely many distinct z_{j1}'s, at least one of the intervals $[-M, 0]$, $[0, M]$ contains infinitely many z_{j1}'s. Choose one such interval, calling it I_1, and choose some z_{H_11} in I_1. Bisect I_1, retaining one of the subintervals, call it I_2, which contains infinitely many z_{j1}'s. Choose z_{H_21} in I_2, with $H_1 < H_2$. Continuing in this way, produce intervals $I_1, I_2, \ldots$ with z_{H_k1} in I_k and $H_k < H_{k+1}$. Now, the length of I_1 is M, that of I_2, $M/2$, and, in general, that of I_k, $M/2^{k-1}$. Since $M/2^{k-1} \rightarrow 0$ as $k \rightarrow \infty$, then $\{z_{H_j1}\}_{j=1}^{\infty}$ forms a Cauchy sequence, hence converges.

Now move on the second coordinate sequence, $\{z_{j2}\}_{j=1}^{\infty}$, but keep only the terms with j taking on the values $H_1, \ldots$ produced above. Since $\{z_{H_k2}\}_{k=1}^{\infty}$ is a bounded sequence, we can play the same game with it that we just did with $\{z_{j1}\}$ above to produce integers $L_1 < L_2 < \cdots$ such that $\{z_{L_k2}\}_{k=1}^{\infty}$ converges.

Since each L_j is some H_i, then $\{z_{L_k 1}\}_{k=1}^{\infty}$ converges to the same limit as $\{z_{H_k 1}\}_{k=1}^{\infty}$. We therefore have $L_1 < L_2 < \cdots$ such that $\{z_{L_k 1}\}$ and $\{z_{L_k 2}\}$ both converge.

Continue this process until after n steps each coordinate sequence has been treated. This yields positive integers $J_1 < J_2 < \cdots$ such that $\{z_{J_k 1}\}, \{z_{J_k 2}\}, \ldots, \{z_{J_k n}\}$ all converge. Then $\{(z_{J_k 1}, \ldots, z_{J_k n})\}_{k=1}^{\infty}$ converges to some n-tuple in R^n. Note that so far we have used only the boundedness of D.

II. The second crucial step is to show that the limit of the sequence produced in step I is in D. Here we use the fact that D is closed. The coordinate notation needed in step I to describe the construction of the desired sequence can be dispensed with in this step.

Let c be the limit of $\{(z_{J_k 1}, \ldots, z_{J_k n})\}$, and suppose that c is not in D. For each $r > 0$, the r-neighborhood of c has a point not in D in it (namely c itself). But each r-neighborhood of c also has a point of D in it. For $(z_{J_k 1}, \ldots, z_{J_k n}) \to c$, so that by choosing J_k sufficiently large we can make $(z_{J_k 1}, \ldots, z_{J_k n})$ as close to c as we like. Then c is a boundary point of D. But, D is closed, hence contains all its boundary points. This gives us a contradiction to the assumption that c is not in D. Thus c is in D, and the theorem is proved. ∎

The importance of this theorem will soon justify the considerable effort we have put into it. As a first consequence, we prove a theorem on existence of maxima and minima.

Theorem

Let f be continuous on a compact set D. Then f achieves a maximum and a minimum on D. That is, there are points P and Q in D such that

$$f(P) \leq f(x) \leq f(Q)$$

for every x in D.

Before proving the theorem, consider an example of how a continuous function might fail to achieve a maximum and minimum on a given set. Suppose that

$$f(x, y) = \frac{1}{x^2 + y^2}$$

for $0 \neq x^2 + y^2 < 1$. Here we are thinking of D as the set $\{(x, y) \mid 0 < x^2 + y^2 < 1\}$, which consists of all points inside the unit circle about the origin, except for the origin itself. Clearly f has no maximum value on D; we can make $f(x, y)$ as large as we like by choosing (x, y) close to $(0, 0)$. Similarly, f has no minimum on D. Since $x^2 + y^2 < 1$, then $f(x, y) > 1$ for every (x, y) in D. But for no point (x, y) does $f(x, y) = 1$, although we can come arbitrarily close by choosing (x, y) closer to the circle $x^2 + y^2 = 1$. The problem is that $(0, 0)$ and all points on the unit circle are boundary points, and none of these is in D.

We now prove the theorem. We know that f is bounded on D. Let M be the least upper bound and m the greatest lower bound of the numbers $f(x)$, x in D.

That is, $m \leq f(x) \leq M$ for each x in D, and, if $A \leq f(x) \leq B$ for each x in D, then $A \leq m$ and $M \leq B$.

We first show that $f(Q) = M$ for some Q in D. Suppose not. Then $M - f(x) > 0$ for each x in D. Put

$$g(x) = \frac{1}{M - f(x)}$$

for each x in D. Then g is continuous on D, and compactness of D ensures that g is bounded on D. For some S, then, $0 < g(x) \leq S$ for each x in D. Then

$$\frac{1}{M - f(x)} \leq S$$

for each x in D. Solving for $f(x)$ yields

$$f(x) \leq M - \frac{1}{S}$$

for each x in D. Since $S > 0$, $M - 1/S < M$, contradicting the way we chose M. Hence g is not continuous on D, implying that $M - f(Q) = 0$ for some Q in D.

By considering $h(x) = 1/[f(x) - m]$ for each x in D, and using a similar argument, the reader can check that $f(P) - m = 0$ for some P in D, completing the proof. ∎

We conclude this section with a statement of an n-dimensional version of the intermediate value theorem. Recall the theorem for the one-variable case. If f is continuous on $[a, b]$, and t is any number between $f(a)$ and $f(b)$, then there is some z between a and b with $t = f(z)$. That is, a function continuous on an interval $[a, b]$ takes on all intermediate values between $f(a)$ and $f(b)$.

The n-dimensional analogue of an interval is a connected set. In R^n we say that a set D is *arcwise connected*, or, more briefly, *connected*, if we can connect any pair of points in D by a continuous arc lying entirely in D. This means that, if P and Q are points in D, then there are continuous functions $\alpha_1, \alpha_2, \ldots, \alpha_n$ defined on some interval $[a, b]$ such that $(\alpha_1(a), \ldots, \alpha_n(a)) = Q$, $(\alpha_1(b), \ldots, \alpha_n(b)) = P$, and $(\alpha_1(t), \ldots, \alpha_n(t))$ is in D for each t in $[a, b]$. We call the locus of the points $(\alpha_1(t), \ldots, \alpha_n(t))$ an *arc* from Q to P in D, and the arc is called *continuous* because each coordinate function α_i is assumed to be continuous.

We can now state the following:

Theorem

If f is continuous on a compact, connected set D of points in R^n, Q and P are in D, and L is a real number with $f(Q) < L < f(P)$, then there is a point S in D such that $f(S) = L$.

Actually, our hypotheses are somewhat more restrictive than necessary, but will make the proof outlined in Exercise 25 somewhat easier for the reader. A somewhat shorter way of stating the theorem is that $f(D)$ is an interval when D is connected [here $f(D) = \{f(x) \mid x \text{ is in } D\}$]. This is illustrated in Fig. 3-3.

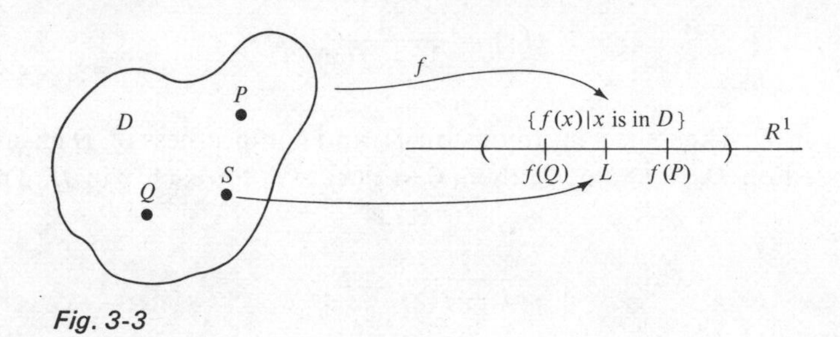

Fig. 3-3

Exercises for Section 3.1

1. In each of the following, determine ∂D and whether or not the set is closed or connected:

(a) $D = \left\{\left(\frac{1}{n}, \frac{1}{m}\right) \;\middle|\; m, n \text{ are positive integers}\right\}$

(b) $D = \{(x, y, z) \mid \|(x, y, z) - (2, 1, 3)\| \geq 4\}$

(c) $D = \{(x, y) \mid x = y \text{ or } y \geq 8\}$

(d) $D = \{(x, y) \mid 3x + 2y = 6\}$

2. Determine whether or not each of the following is continuous at the given point:

(a) $f(x, y = \cos(x + y)$; $(1, 2)$

(b) $g(x, y) = \begin{cases} \dfrac{1}{\cos(2xy)} & \text{for } xy \neq 0; \\ 1 & \text{for } xy = 0 \end{cases}$ $(0,0)$

(c) $h(x, y) = \begin{cases} \dfrac{x^2y}{x^2 + y^4} & \text{for } x^2 + y^4 \neq 0; \\ 0 & x = y = 0. \end{cases}$ $(0, 0)$

(d) $g(x, y, z) = \begin{cases} x \sin\left(\dfrac{1}{z}\right) + y \cos\left(\dfrac{1}{z}\right) + z \sin(x + y) & zx \neq 0 \\ 0 & zx = 0 \end{cases}$; $(0, 0, 0)$

(e) $\varphi(x, y, z, t) = \begin{cases} \dfrac{x + y + z + t}{\sqrt{x + y + z}}, & x + y + z \neq 0 \\ \frac{1}{2}, & x + y + z = 0 \end{cases}$; $(0, 0, 0, 0)$

(f) $p(x, y) = e^x \sin(y)$; at an arbitrary point (a, b)

3. Let $f(x, y) = \sin(xy)/xy$ for $xy \neq 0$. Is it possible to define a value for $f(0, 0)$ in such a way that the new f is continuous at the origin?

4. Give examples of sets in R^2, R^3, and R^4 which are
   (a) Compact but not connected
   (b) Connected but not compact
   (c) Closed but not compact or connected
   (d) Bounded but not connected or closed

5. Write out a rigorous proof of the following. Suppose that f is continuous at $(a_1, a_2, a_3, a_4, a_5)$. Let $g(x, y) = f(a_1, x, a_3, a_4, y)$ for all x and y such that $f(a_1, x, a_3, a_4, y)$ is defined. Then g is continuous at (a_2, a_5).

6. Let a and b be continuous in some neighborhood N_1 of (u_0, v_0). Let f be continuous in some neighborhood N_2 of (x_0, y_0), with $x_0 = a(u_0, v_0)$ and $y_0 = b(u_0, v_0)$. Suppose that $(a(u, v), b(u, v))$ is in N_2 for any (u, v) in N_1. Let $F(u, v) = f(a(u, v), b(u, v))$ for (u, v) in N_1. Show that F is continuous at (u_0, v_0).

7. Let $D = \{(x, y, z) \mid x, y$ and z are irrational and $0 \leq x \leq 1,\ 0 \leq y \leq 1,\ 0 \leq z \leq 1\}$. Determine ∂D and whether or not D is closed.

8. For any set D of points in R^n, the complement D^c of D consists of all points of R^n not in D. That is,

$$D^c = \{P \mid P \text{ is in } R^n \text{ and } P \text{ is not in } D\}.$$

   Define D to be open exactly when D^c is closed. Prove the following:
   (a) D is closed exactly when D^c is open.
   (b) D is open if and only if ∂D has no points in common with D.
   (c) Every r-neighborhood of $(x_1, \ldots, x_n)$ is open.

9. Give examples of sets in R^2, R^3, and R^4 which are
   (a) Closed and not open
   (b) Open and not closed
   (c) Neither open nor closed
   Hint: In (c) show that, for any n, R^n is the only nonempty set that is both open and closed.

10. Let D consist of all points lying on some plane in R^3. Is D open, closed, or neither? What are the boundary points of D?

11. Investigate continuity of $f(x, y) = (x + y)\ln(x + y)$, defined for $x > 0$ and $y > 0$.

12. Give an ϵ, δ argument for continuity of $f(x + y) = x^2 + 2xy + x^2y$ at $(2, 3)$.

13. Give an ε, δ argument for continuity of $f(x, y, z) = z^2 + x^2 + y^2 + 2xyz$ at $(1, 1, 3)$.

14. Prove the following: f is continuous at (x_0, y_0) if and only if, given $\varepsilon > 0$, there is some $\delta > 0$ such that $|f(x, y) - f(x_0, y_0)| < \varepsilon$ whenever $f(x, y)$ is defined and $|x - x_0| < \delta$ and $|y - y_0| < \delta$.

15. Let x be a boundary point of a set D in R^n. Prove that there is a sequence $\{u_n\}$ with each u_n in D and $\lim_{n\to\infty} u_n = x$. *Hint:* For any positive integer n, the $1/n$-neighborhood of x has a point u_n of D in it. Prove that the sequence thus formed converges to x.

16. Let D be a compact set in R^n, and $\{u_n\}$ a sequence with each u_n in D. Prove that there is a subsequence $\{v_n\}$ of $\{u_n\}$ which converges, and that $\lim_{n\to\infty} v_n$ is in D. *Hint:* Go back and look at I in the proof that a function continuous on a compact set is bounded.

17. Let f be a real-valued function continuous on a compact set D in R^n. Denote by $f(D)$ the set of all values $f(x)$ for x in D. Prove that $f(D)$ is compact in R^1. *Hint:* By the main theorem of this section, $f(D)$ is a bounded set in R^1, hence it remains to show that $f(D)$ is closed. Let x be a boundary point of $f(D)$. Use Exercise 15 to produce a sequence $\{u_n\}$ with each u_n in $f(D)$ and $\lim_{n\to\infty} u_n = x$. Produce, by the definition of $f(D)$, an element U_n in D such that $f(U_n) = u_n$ for each positive integer n. Use Exercise 16 to produce a subsequence $\{V_n\}$ of $\{U_n\}$ such that $\{V_n\}$ converges to some point, say X, in D. Now show that $f(X) = x$, hence conclude that x is in $f(D)$ and that $f(D)$ is closed.

18. Let f be a real-valued function defined on a set D of points in R^n. Define the graph of f by $G(f) = \{(x_1, \ldots, x_n, f(x_1, \ldots, x_n)) \mid (x_1, \ldots, x_n) \text{ is in } D\}$. Thus $G(f)$ is a set of points in R^{n+1}. Suppose now that D is compact. Prove that f is continuous if and only if $G(f)$ is compact in R^{n+1}.

19. Let f be continuous on a compact set D in R^n. Suppose also that f is one to one [i.e., $f(x) \neq f(y)$ if $x \neq y$]. If x is in $f(D)$ (see Exercise 17), define $g(x) = y$ in D if and only if $f(y) = x$. Then g is called the *inverse* of f. Prove that g is also continuous, in the sense that, given x in $f(D)$ and $\varepsilon > 0$, there exists $\delta > 0$ such that $\|g(x) - g(y)\| < \varepsilon$ whenever y is in $f(D)$ and $|x - y| < \delta$.

20. Let f be a real-valued function defined on some set D of points in R^n. We say that f is uniformly continuous on D if, given $\varepsilon > 0$, there is some $\delta > 0$ such that $|f(x) - f(y)| < \varepsilon$ whenever x and y are in D and $\|x - y\| < \delta$.

(a) Show that uniform continuity of f on D implies continuity of f on D.

(b) For each $n = 1, 2, \ldots$, give an example of D and f such that f is continuous on the set D of points in R^n, but not uniformly continuous on D. *Hint:* For $n = 1$, look at $f(x) = 1/x$, $0 < x \leq 1$.

(c) Suppose now that D is compact. Prove that continuity of f on D implies uniform continuity of f on D. *Hint:* Suppose that f is not uniformly continuous on D. Then, for some $\varepsilon > 0$, it is true that, given any $\delta > 0$, there are points x and y in D such that $\|x - y\| < \delta$ but $|f(x) - f(y)| \geq \varepsilon$. By choosing $\delta = 1/n$ (n any positive integer), produce x_n and y_n in D such that $\|x_n - y_n\| < 1/n$ and $\|f(x_n) - f(y_n)\| \geq \varepsilon$. Use the compactness of D (see Exercise 16) to produce a subsequence $\{X_n\}$ of $\{x_n\}$ such that $\{X_n\}$ converges, say to A, and show that A is in D. For each n, let $Y_n = y_{\varphi(n)}$, where φ is an increasing function on the positive integers to the positive integers such that $X_n = x_{\varphi(n)}$ (see Exercise 14, Section 1.1). Prove that $\{Y_n\}$ also converges to A. Now use continuity of f to conclude that $\lim_{n\to\infty} [f(X_n) - f(Y_n)] = 0$, contradicting the fact that $|f(X_n) - f(Y_n)| \geq \varepsilon$ for all n.

(d) Suppose that, for some constant $K > 0$, $|f(x) - f(y)| \leq K\|x - y\|$ for each x, y in D. Prove that f is uniformly continuous on D.

(e) Use (d) to show that $\cos(x)$ is uniformly continuous on $(0, 1)$. *Hint:* If $0 < x < y < 1$, then $\cos(y) - \cos(x) = [-\sin(\xi)](y - x)$ for some ξ between x and y, by the mean value theorem.

21. Let f and g be continuous at the n-vector a. Prove that $f + g$ is continuous at a. [Here $(f + g)(x) = f(x) + g(x)$ wherever $f(x)$ and $g(x)$ are defined.]

22. Let f and g be continuous at the n-tuple a. Prove that fg is continuous at a. [Here $(fg)(x) = f(x)g(x)$ wherever $f(x)$ and $g(x)$ are defined.]

23. Let f and g be continuous at the n-tuple a. Prove that f/g is continuous at a, if $g(a) \neq 0$. [Here $(f/g)(x) = f(x)/g(x)$ wherever $f(x)$ and $g(x)$ are defined and $g(x) \neq 0$.]

24. Let D be a compact, connected set in R^n.

(a) Prove that there do not exist nonempty, closed sets A and B such that (1) each point of D is either in A or in B; (2) A and B have no points in common; and (3) each point in A or B is also in D. *Hint:* Suppose that such A and B exist. Let P be in B and Q in A. Let C be an arc from Q to P in D. Now proceed in steps.

Step 1. Show that some r-neighborhood of Q has no points of B in it. [If Q is the only point of A in some r-neighborhood of Q, then show that Q is in ∂B.]

Step 2. Let $W = \{\|Q - x\| \mid x \text{ is in } A \text{ and on } C\}$. Show that W is bounded above.

Step 3. Let $r = \text{lub}\{\|Q - x\| \mid x \text{ is in } A \text{ and on } C\}$. Show that $r > 0$.

Step 4. Produce S in D such that $r = \|Q - S\|$. To do this, note that, for each positive integer j, there is some z_j on C and in A with $\|z_j - Q\| > r - 1/j$. Show that $\{z_j\}$ converges, say to S, that S is in D, and that $r = \|S - Q\|$.

Step 5. Derive a contradiction by showing that S is in A, and also in B.

(b) In the proof outlined in (a), was it necessary to use the hypothesis that D be compact?

25. Prove the n-dimensional intermediate value theorem. *Hint:* Suppose that there is no S in D with $f(S) = L$. Let $A = \{x \mid x \text{ is in } D \text{ and } f(x) < L\}$, and $B = \{x \mid x \text{ is in } D \text{ and } f(x) > L\}$. Show that A and B are closed, nonempty, have no points in common, and that every point of D is either in A or in B. Then use the result of Exercise 24.

3.2 Limits

Just as the notion of continuity generalizes easily to n dimensions, so does the notion of limit. We say that a real-valued function f has the real number L as a limit as x approaches the n-tuple a if we can make $f(x)$ arbitrarily close to L by choosing x close to a.

Definition

$\lim_{x \to \infty} f(x) = L$ if, given $\varepsilon > 0$, there is some $\delta > 0$ such that $|f(x) - L| < \varepsilon$ whenever $f(x)$ is defined and $0 < \|x - a\| < \delta$.

It is important to understand in this definition that f need not be defined at a, and, even if $f(a)$ is defined, that it need have no relation to L. For example, if

$$g(x, y) = \begin{cases} xy & \text{for } (x, y) \neq (0, 0), \\ \alpha & \text{for } x = y = 0, \end{cases}$$

where α is any real number, then $g(x, y) \to 0$ as (x, y) approaches the origin. The value of α is irrelevant. If, however, we want to know whether or not g is continuous at $(0, 0)$, then $g(0, 0)$ becomes important. In this example, g is continuous at $(0, 0)$ exactly when $\alpha = 0$. In general, f is continuous at a exactly when $\lim_{x \to a} f(x) = f(a)$.

Note that the definition of limit in n variables reduces to the usual one-variable definition when x and a are real numbers. There is, however, an important difference between limits in one variable and limits in two or more variables. A single variable can approach a number a on the line in at most two ways, from the right and from the left. By contrast, an n-tuple x can approach an n-tuple a along infinitely many different paths, and the function values must approach L along all possible paths if the limit is to be L.

As an example, let

$$f(x, y) = \begin{cases} \dfrac{xy}{x^2 + y^2} & \text{if } x^2 + y^2 \neq 0, \\ 0 & \text{if } x = y = 0. \end{cases}$$

We claim that $\lim_{(x,y) \to (0,0)} f(x, y)$ does not exist. As $(x, y) \to (0, 0)$ along the line $y = x$, we have $f(x, x) = \frac{1}{2}$ if $x \neq 0$. As $(x, x) \to (0, 0)$, then $f(x, x) \to \frac{1}{2}$. But along, say the x-axis, $(x, y) = (x, 0)$, and $f(x, y) = f(x, 0) = 0$, so $f(x, 0) \to 0$ as $(x, 0) \to (0, 0)$. Since we can make $f(x, y)$ arbitrarily close to $\frac{1}{2}$ and also to 0 by coming arbitrarily close to $(0, 0)$ along different paths, then f can have no limit at $(0, 0)$ (see Fig. 3-4).

These considerations suggest a sequential characterization of limits very like that given in the last section for continuity.

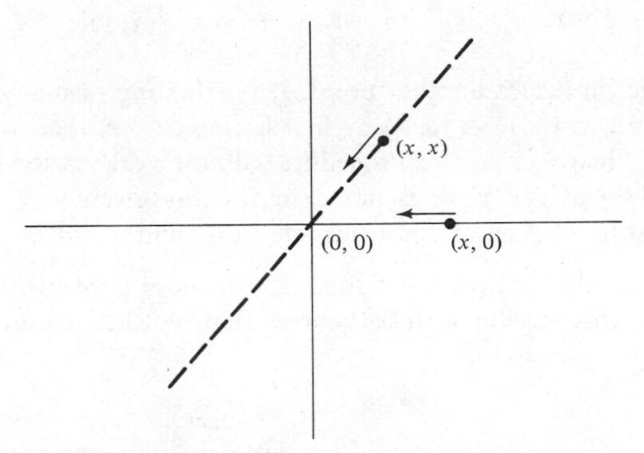

Fig. 3-4

Theorem

$\lim_{x \to a} f(x) = L$ *if and only if given any sequence* $\{x_j\}$ *of n-vectors converging to a, and for which each* $f(x_j)$ *is defined, then the sequence* $\{f(x_j)\}$ *converges to L.*

We leave it as an instructive exercise for the student to write down a complete proof. Follow the argument of the proof in Section 3.1, with L in place of $f(a)$.

As an illustration, we use the theorem on the last example. The sequence $\{(1/n, 1/n)\}_{n=1}^{\infty}$ converges to (0, 0), and $f(1/n, 1/n) \to \frac{1}{2}$ as $n \to \infty$. But $\{(1/n, 0)\}_{n=1}^{\infty}$ also converges to (0, 0), and $f(1/n, 0) = 0 \to 0 \neq \frac{1}{2}$, so f has no limit at (0, 0). Generally, showing that a limit exists can be a very delicate matter. The method of using different paths of approach, and the last theorem, can often be used to show that a limit does not exist, or to find the limit, knowing that it exists, but is rarely of use in showing existence. To illustrate an existence proof by direct use of the definition, consider

$$g(x, y) = \begin{cases} x^2 \sin\left(\dfrac{1}{y}\right) + y^2 \sin\left(\dfrac{1}{x}\right) & \text{for } x \neq 0 \text{ and } y \neq 0, \\ 0 & \text{if } x = 0 \text{ or } y = 0. \end{cases}$$

Then $\lim_{(x,y) \to (0,0)} g(x, y) = 0$. To prove this, let $\varepsilon > 0$. We seek $\delta > 0$ such that $|g(x, y)| < \varepsilon$ if $0 < \|(x, y)\| < \delta$ or, equivalently, $0 < x^2 + y^2 < \delta^2$.

Now, if $x \neq 0$ and $y \neq 0$, then $|g(x, y)| \leq |x^2||\sin(1/y)| + |y^2||\sin(1/x)| \leq x^2 + y^2$. Putting $\delta = \sqrt{\varepsilon}$, then

$$|g(x, y)| < \varepsilon \qquad \text{whenever } 0 < x^2 + y^2 < \delta^2,$$

holds for $x \neq 0$ and $y \neq 0$. If $x = 0$ or $y = 0$, then certainly

$$g(x, y) = 0 < \varepsilon$$

regardless of δ. Thus $|g(x, y)| < \varepsilon$ whenever $0 < \|(x, y)\| < \delta$, and we are done.

This example illustrates another point. It is tempting in many instances to try to take a limit as $(x, y) \to (a, b)$ by first letting $x \to a$, then $y \to b$, or perhaps first $y \to b$, then $x \to a$. This procedure will not work. In the last example, the limit as $x \to 0$ of $g(x, y)$ does not exist for any given $y \neq 0$, as $\sin(1/x)$ oscillates too badly. In general, $\lim_{x\to a}(\lim_{y\to b})$, $\lim_{y\to b}(\lim_{x\to a})$, and $\lim_{(x,y)\to(a,b)}$ have nothing to do with one another, with similar remarks for more than two variables.

We conclude this section with a theorem that will be of use a number of times later.

Theorem

Let $\lim_{x\to a} f(x) = L \neq 0$. Then, for some $\delta > 0$, $f(x)$ has the same sign as L whenever $0 < \|x - a\| < \delta$ and $f(x)$ is defined.

PROOF

For some $\delta > 0$, $|f(x) - L| < |L|/2$ whenever $0 < \|x - a\| < \delta$ and $f(x)$ is defined. For such x, we then have:

$$L - \frac{|L|}{2} < f(x) < L + \frac{|L|}{2}.$$

If $L > 0$, then $0 < L/2 < f(x) < 3L/2$. If $L < 0$, then $3L/2 < f(x) < L/2 < 0$. ∎

If f is continuous at a, then $L = f(a)$, and the theorem reads: *If $f(a) \neq 0$, then $f(x)$ assumes the same sign as $f(a)$ in some r-neighborhood of a.*

Exercises for Section 3.2

1. In each of the following, prove that the limit exists, or show that the limit does not exist:

(a) $\displaystyle\lim_{(x,y)\to(0,0)} \frac{x^4 y}{y^2 + x^4}$ **(b)** $\displaystyle\lim_{(x,y)\to(a,b)} \frac{y + x}{x^2 + e^{1/y}}$

(c) $\displaystyle\lim_{(x,y,z)\to(0,0,0)} \frac{2x^2yz}{x + y + z}$ **(d)** $\displaystyle\lim_{(x,y)\to(1,2)} \frac{(x - 1)(y - 2)}{2x + 3y}$

(e) $\displaystyle\lim_{(x,y)\to(0,0)} \left(x \sin\left(\frac{1}{x}\right) + y \sin\left(\frac{1}{y}\right) + e^{xy}\right)$ **(f)** $\displaystyle\lim_{(x,y)\to(0,0)} xy \ln(xy)$

2. In each part of Exercise 1, investigate the iterated limits. [*Note:* In (c) there are six iterated limits to examine.]

3. Prove the following: If

$$\lim_{(x,y)\to(a,b)} f(x, y) = L \quad \text{and} \quad \lim_{x\to a} \lim_{y\to b} f(x, y)$$

exists, then $\lim_{x\to a} \lim_{y\to b} f(x, y) = L$.

Similarly, if $\lim\limits_{(x,y)\to(a,b)} f(x, y) = L$ and $\lim\limits_{y\to b}\lim\limits_{x\to a} f(x, y)$ exists, then

$$\lim_{y\to b}\lim_{x\to a} f(x, y) = L.$$

4. Suppose that $\lim\limits_{(x,y)\to(a,b)} f(x, y)$, $\lim\limits_{x\to a}\lim\limits_{y\to b} f(x, y)$, and $\lim\limits_{y\to b}\lim\limits_{x\to a} f(x, y)$ all exist. Show that all three are equal.

5. Give an example in which $\lim\limits_{y\to b}\lim\limits_{x\to a} f(x, y) = \lim\limits_{x\to a}\lim\limits_{y\to b} f(x, y)$, but

$$\lim_{(x,y)\to(a,b)} f(x, y)$$

does not exist.

6. Give an example in which $\lim\limits_{y\to b}\lim\limits_{x\to a} f(x, y)$ and $\lim\limits_{(x,y)\to(a,b)} f(x, y)$ fail to exist, but $\lim\limits_{x\to a}\lim\limits_{y\to b} f(x, y)$ exists.

7. Prove the following: *Let f be continuous at the n-tuple a. Let $\alpha_1, \ldots, \alpha_n$ be n functions of one variable such that $\lim\limits_{t\to c} \alpha_i(t) = a_i$ for $i = 1, \ldots, n$. Then*

$$\lim_{t\to c} f(\alpha_1(t), \ldots, \alpha_n(t)) = (a_1, \ldots, a_n).$$

8. Give an example of a function of two variables such that $f(x, y) \to L$ as $(x, y) \to (0, 0)$ along any straight-line path, but $\lim\limits_{(x,y)\to(0,0)} f(x, y)$ does not exist.

4

Multiple Integrals

4.1 Introduction

We would like to develop a concept of integration of functions of n variables over sets in R^n, in such a way as to retain the usual properties we associate with ordinary Riemann integrals $\int_a^b f(x)\,dx$. In order to get some insight into how this might be done, it will be helpful to review the development of the integral of a function of one variable over an interval on the real line. We shall also need to develop a fairly precise notion of area for sets of points in the plane.

4.2 Riemann Integral for Functions of One Variable

Suppose that f is defined and bounded on an interval $[a, b]$. There are several approaches which lead to the usual Riemann integral $\int_a^b f(x)\,dx$. We shall outline two of these.

I. Form a partition P of $[a, b]$ by inserting points

$$a = x_0 < x_1 < \cdots < x_{n-1} < x_n = b.$$

The end points a and b are often relabeled as shown as a matter of notational convenience. The mesh of P, denoted mesh(P), is the largest distance between successive partition points:

$$\text{mesh}(P) = \max_{1 \le i \le n} (x_i - x_{i-1}).$$

We now say that f is *Riemann integrable* on $[a, b]$ if there is a real number L with the following property: *Given any* $\varepsilon > 0$, *there is some* $\delta > 0$ *such that, for any partition* P *of* $[a, b]$ *with* $\text{mesh}(P) < \delta$, *and any choice* of ξ_i *in* $[x_{i-1}, x_i]$ *for* $i = 1, \ldots, n$, *we have*

$$\left| \sum_{i=1}^{n} f(\xi_i)(x_i - x_{i-1}) - L \right| < \varepsilon.$$

We then call L the *Riemann integral* of f over $[a, b]$, and denote it $\int_a^b f(x)\,dx$, or, more concisely, $\int_a^b f$, if we wish. The sum $\sum_{i=1}^{n} f(\xi_i)(x_i - x_{i-1})$ occurring in the definition is called a *Riemann sum* corresponding to the partition P and to the choice of points ξ_i in the intervals $[x_{i-1}, x_i]$.

II. Form a partition $P: a = x_0 < x_1 \cdots < x_n = b$ of $[a, b]$, and form the sums

$$\underline{S}(P) = \sum_{i=1}^{n} m_i(x_i - x_{i-1})$$

and

$$\bar{S}(P) = \sum_{i=1}^{n} M_i(x_i - x_{i-1}),$$

where

$$m_i = \text{glb}\{f(x) \mid x_{i-1} \le x \le x_i\}$$

and

$$M_i = \text{lub}\{f(x) \mid x_{i-1} \le x \le x_i\}.$$

These are, respectively, the lower and upper Darboux sums for f corresponding to the partition P of $[a, b]$.

It is not too difficult to show (see Exercise 1) that, for any partitions P and W of $[a, b]$,

$$\underline{S}(P) \le \bar{S}(W).$$

Thus the numbers $\underline{S}(P)$, considered over all possible partitions of $[a, b]$, are bounded above, and the numbers $\bar{S}(P)$ are bounded below. Put

$$\underline{I}(f) = \text{lub}\{\underline{S}(P) \mid P \text{ is a partition of } [a, b]\}$$

and

$$\bar{I}(f) = \text{glb}\{\bar{S}(P) \mid P \text{ is a partition of } [a, b]\}.$$

These are the upper and lower integrals, respectively, of f on $[a, b]$. These numbers always exist (see Appendix I of Chapter 1), and from a previous remark, $\underline{I}(f) \leq \bar{I}(f)$. It may happen that $\underline{I}(f) < \bar{I}(f)$ for the function f under consideration (see Exercise 3). If $\underline{I}(f) = \bar{I}(f)$, *then we say that f is integrable on* $[a, b]$, *and denote* $\underline{I}(f)$ [*or* $\bar{I}(f)$] *as* $\int_a^b f(x)\,dx$.

We now have two different definitions of integrability of a function, and of the resulting number, $\int_a^b f(x)\,dx$. Obviously this situation makes sense only if functions integrable in the sense of I are also integrable in the sense of II, and conversely, and also if then the numbers L of I, and $\underline{I}(f)$ [or $\bar{I}(f)$] of II, agree. A proof that this is in fact the case is sketched in Exercise 5, with Exercise 4 devoted to necessary and sufficient conditions that a function be integrable in the sense of II.

Offhand, it is not obvious what kinds of functions turn out to be integrable. In Exercises 6 and 7 we outline arguments showing that continuous functions, and monotone nondecreasing (or nonincreasing) functions defined on $[a, b]$, are Riemann integrable on $[a, b]$.

We list below some of the more commonly used properties of the Riemann integral. Proofs are sketched in the remaining exercises at the end of this section.

(1) $\int_a^b (cf(x) + dg(x))\,dx = c\int_a^b f(x)\,dx + d\int_a^b g(x)\,dx$ for real numbers c and d, whenever the integrals on the right exist. This is called linearity of the integral.
(2) $\int_a^b f(x)\,dx \leq \int_a^b g(x)\,dx$ if $f(x) \leq g(x)$ for $a \leq x \leq b$.
(3) $\int_a^b f(x)\,dx = \int_a^c f(x)\,dx + \int_c^b f(x)\,dx$.
(4) $\int_a^b f(x)\,dx = -\int_b^a f(x)\,dx$. (This is simply a convenient definition.)
(5) $\left|\int_a^b f(x)\,dx\right| \leq \int_a^b |f(x)|\,dx$.
(6) If $m \leq f(x) \leq M$ for $a \leq x \leq b$, then

$$m(b - a) \leq \int_a^b f(x)\,dx \leq M(b - a).$$

(7) If f and g are continuous on $[a, b]$ and $g(x) \geq 0$ for $a \leq x \leq b$, then, for some ξ in $[a, b]$,

$$\int_a^b f(x)g(x)\,dx = f(\xi)\int_a^b g(x)\,dx.$$

In particular, when $g(x) = 1$ for $a \leq x \leq b$, then for some ξ in $[a, b]$,

$$\int_a^b f(x)\,dx = f(\xi)(b - a).$$

(8) If $g'(x) = f(x)$ on $[a, b]$, then $\int_a^b f(x)\,dx = g(b) - g(a)$. This is called the *fundamental theorem of integral calculus.*

(9) If f is continuous on $[a, b]$, then

$$\frac{d}{dx}\int_a^x f(t)\,dt = f(x) \qquad \text{for } a \le x \le b.$$

This is another form of the fundamental theorem of integral calculus.

Exercises for Section 4.2

1. Let f be defined and bounded on $[a, b]$. Let P and W be partitions of $[a, b]$. Prove that $\underline{S}(P) \le \bar{S}(W)$. *Hint:* Break the proof into two steps:

Step 1. If Q is a partition obtained from a partition P by inserting additional partition points, show that $\underline{S}(P) \le \underline{S}(Q) \le \bar{S}(Q) \le \bar{S}(P)$. To do this, first concentrate on showing that $\underline{S}(P) \le \underline{S}(Q)$. Let $P\colon a = x_0 < x_1 < \cdots < x_n = b$. Look at a typical subinterval $[x_{i-1}, x_i]$ which has been subdivided in forming Q, say $x_{i-1} = y_0 < y_1 < \cdots < y_m = x_i$. Now

$$m_i = \text{glb}\{f(x) \mid x_{i-1} \le x \le x_i\} \le m_j^* = \text{glb}\{f(x) \mid y_{j-1} \le x \le y_j\}$$

for $j = 1, \ldots, m$. Then $m_i(x_i - x_{i-1}) = m_i \sum_{j=1}^{m} (y_j - y_{j-1}) = \sum_{j=1}^{m} m_i(y_j - y_{j-1})$ $\le \sum_{j=1}^{m} m_j^*(y_j - y_{j-1})$. Thus each term in the sum for $\underline{S}(P)$ is bounded above by a sum of terms in the sum for $\underline{S}(Q)$. Conclude that $\underline{S}(P) \le \underline{S}(Q)$. Next, show that $\underline{S}(Q) \le \bar{S}(Q)$. Finally, show that $\bar{S}(Q) \le \bar{S}(P)$ by adapting the argument used for the lower sums.

Step 2. Now let P and W be any partitions of $[a, b]$. Form a new partition Q by using all the partition points of both P and W. Then from step 1 conclude that $\underline{S}(P) \le \underline{S}(Q) \le \bar{S}(Q) \le \bar{S}(W)$.

2. Prove that $\underline{I}(f) \le \bar{I}(f)$ for f bounded on $[a, b]$. *Hint:* Make good use of Exercise 1.

3. Define f on $[0, 1]$ by putting

$$f(x) = \begin{cases} 0 & \text{if } 0 \le x \le 1 \text{ and } x \text{ is rational,} \\ 1 & \text{if } 0 \le x \le 1 \text{ and } x \text{ is irrational.} \end{cases}$$

Show that f is not integrable in the sense of II on $[0, 1]$. *Hint:* It is not difficult to show directly that $\underline{I}(f) = 0$ and $\bar{I}(f) = 1$, by using the fact that every non-degenerate interval on the real line contains at least one rational and at least one irrational number.

4. Prove the following theorem: *Let f be bounded on $[a, b]$. Then f is Riemann integrable in the sense of II if and only if, given $\varepsilon > 0$, there is a partition P of $[a, b]$ such that $\bar{S}(P) - \underline{S}(P) < \varepsilon$.* *Hint:* Suppose first that f is Riemann integrable in the sense of II, and let $\varepsilon > 0$. Show first, using the definitions of $\underline{I}(f)$ and $\bar{I}(f)$ that there exist partitions W and Q such that $\bar{S}(W) - \bar{I}(f) < \varepsilon/2$ and $\underline{I}(f) - \underline{S}(Q) < \varepsilon/2$. Form a new partition P by using all the partition points of both W and Q. Then use Exercise 1 and the hint to its solution (step 1) to show that $0 \le \bar{S}(P) - \bar{I}(f) \le \bar{S}(W) - \bar{I}(f) < \varepsilon/2$, and $0 \le \underline{I}(f) - \underline{S}(P) \le \underline{I}(f) - \underline{S}(Q) < \varepsilon/2$. Now use the fact that $\underline{I}(f) = \bar{I}(f)$ to conclude that $\bar{S}(P) - \underline{S}(P) < \varepsilon$.

The converse is easier and is left to the reader.

5. Let f be bounded in $[a, b]$. Show that f is integrable in the sense of I exactly when f is integrable in the sense of II. Further, if f is integrable in either sense, then the value of the integral obtained from I agrees with that obtained from II. *Hint:* Suppose first that f is integrable in the sense of I. Let $\varepsilon > 0$. For some δ, $\left|\sum_{i=1}^{n} f(\xi_i)(x_i - x_{i-1}) - L\right| < \varepsilon$ for any partition $P: a = x_0 < x_1 < \cdots < x_n = b$ with mesh$(P) < \delta$ and any choice of ξ_i in $[x_{i-1}, x_i]$. Here L is the value of the integral obtained from I.

Now choose any partition P with mesh$(P) < \delta$.Find ξ_i and η_i in $[x_{i-1}, x_i]$ with $0 \le M_i - f(\xi_i) < \varepsilon/2(b - a)$ and $0 \le f(\eta_i) - m_i < \varepsilon/2(b - a)$. Show that $\left|L - \sum_{i=1}^{n} M_i(x_i - x_{i-1})\right| < \varepsilon/2$ and $\left|L - \sum_{i=1}^{n} m_i(x_i - x_{i-1})\right| < \varepsilon/2$. Hence show that $\bar{S}(P) - \underline{S}(P) < \varepsilon$, and conclude from Exercise 4 that f is Riemann integrable in the sense of II, and also that the values obtained for $\int_a^b f$ from I and II agree.

Conversely: Suppose that f is integrable in the sense of II. Let $\varepsilon > 0$. Choose a partition $P: a = x_0 < x_1 < \cdots < x_n = b$ with $\bar{S}(P) < \bar{I}(f) + \varepsilon/2$, and let $|f(x)| \le R$ for $a \le x \le b$. Put $\delta = \min_{1 \le j \le n} (x_j - x_{j-1})$. Now choose a partition Q with mesh$(Q) < \delta$ and also mesh$(Q) < \varepsilon/4nR$. Form a partition W by using all the partition points of both P and Q. Show that $0 \le \bar{S}(Q) - \bar{I}(f) = (\bar{S}(Q) - \bar{S}(W)) + (\bar{S}(W) - \bar{I}(f))$, hence conclude that $0 \le \bar{S}(W) - \bar{I}(f) < \varepsilon/2$ Next show that $0 \le \bar{S}(Q) - \bar{S}(W) < \varepsilon/2$. Then $\bar{S}(Q) - \bar{I}(f) < \varepsilon$. Similarly, show that $\underline{I}(f) - \underline{S}(Q) < \varepsilon$. If we now choose ξ_i in the ith subinterval $[y_{i-1}, y_i]$ of Q, then $\underline{S}(Q) = \sum_{i=1}^{n} m_i(y_i - y_{i-1}) \le \sum_{i=1}^{n} f(\xi_i)(y_i - y_{i-1}) \le \sum_{i=1}^{n} M_i(y_i - y_{i-1}) \le \bar{S}(Q)$. Now conclude that $\left|\sum_{i=1}^{n} f(\xi_i)(y_i - y_{i-1}) - \bar{I}(f)\right| < \varepsilon$, hence f is Riemann integrable in the sense of I. The above argument also shows that the same value for $\int_a^b f$ is obtained from I and II.

6. Let f be continuous on $[a, b]$. Prove that f is Riemann integrable on $[a, b]$. *Hint:* Let $\varepsilon > 0$. By Exercise 20(b), Section 3.1, there is some $\delta > 0$ such that $|f(x) - f(y)| < \varepsilon/(b - a)$ whenever x and y are in $[a, b]$ and $|x - y| < \delta$. Let $P: a = x_0 < x_1 < \cdots < x_n = b$ be a partition of $[a, b]$ with mesh$(P) < \delta$. Show that there are ξ_i, η_i in $[x_{i-1}, x_i]$ with $f(\xi_i) = m_i$ and $f(\eta_i) = M_i$. Now show that $\bar{S}(P) - \underline{S}(P) < \varepsilon$.

7. Let f be monotone nondecreasing (or nonincreasing) on $[a, b]$. Prove that f is Riemann integrable on $[a, b]$. *Note:* You cannot assume continuity of f on $[a, b]$.

8. Prove that $\int_a^b f + \int_a^b g = \int_a^b (f + g)$ whenever two integrals on the left exist. [Here the function $f + g$ is defined by $(f + g)(x) = f(x) + g(x)$, whenever $f(x)$ and $g(x)$ are defined.] *Hint:* This is most easily proved using formulation I of the Riemann integral.

9. Prove that $\int_a^b cf = c \int_a^b f$ when c is a real number and f is integrable on $[a, b]$. [Here cf is the function defined by $(cf)(x) = cf(x)$ whenever $f(x)$ is defined.]

10. Give proofs for properties (2), (3), (5), (6), and (7) from the list of properties given in the section for $\int_a^b f$.

11. Prove that $\dfrac{d}{dx} \displaystyle\int_a^t f(t)\, dt = f(x)$ if f is continuous on $[a, b]$. *Hint:* Proceed directly from the limit definition of derivative applied to the function $g(x) = \int_a^x f(t)\, dt$, and make use of the uniform continuity of f on $[a, b]$.

4.3 Area

Before attempting to develop a concept of a Riemann integral for functions of two or more variables, it will be useful to first get some feeling for a notion of area applied to sets of points in the plane. Obviously we have to have some initial information to get us started. We shall assume that the area of a rectangle is given by the usual formula.

Now let D be any bounded set of points in the plane. Then, for some numbers a, b, c, and d, we have $a \le x \le b$ and $c \le y \le d$ for each (x, y) in D. Partition the intervals $[a, b]$ and $[c, d]$:

$$a = x_0 < x_1 < \cdots < x_n = b$$

and

$$c = y_0 < y_1 < \cdots < y_m = d,$$

and form a rectangular grid G over D in the plane by drawing vertical lines $x = x_i$, $i = 0, 1, \ldots, n$, and horizontal lines $y = y_i$, $i = 0, 1, \ldots, m$. The mesh of the grid, denoted mesh(G), is the length of the diameter of the largest rectangle in the grid. The rectangles in G may be separated into three classes:

(1) Those having no points in common with D.
(2) Those containing only interior points of D.
(3) Those containing at least one boundary point of D (see Fig. 4-1).

Grid rectangles in (1) can be ignored. The sum of the areas of rectangles in (2) is called the *inner area* of D relative to G, and is denoted $\mathscr{I}(D, G)$. Finally, $\mathscr{I}(D, G)$ plus the sum of the areas of rectangles in (3) is called the *outer area* of D relative to G, and is denoted $\mathscr{O}(D, G)$. Clearly

$$0 \le \mathscr{I}(D, G) \le \mathscr{O}(D, G).$$

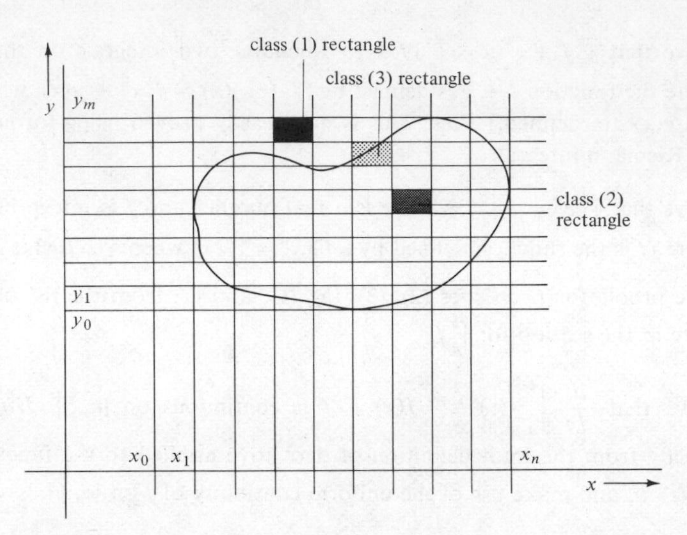

Fig. 4-1

The numbers $\mathscr{I}(D, G)$, taken over all grids G over D, are bounded above [e.g., by $(b - a)(d - c)$], and the numbers $\mathscr{O}(D, G)$ are bounded below (e.g., by 0). We call the least upper bound of the numbers $\mathscr{I}(D, G)$, taken over all grids G over D, the *inner area* $\underline{A}(D)$ of D. The greatest lower bound of the numbers $\mathscr{O}(D, G)$, taken over all grids G over D, is the *outer area* $\overline{A}(D)$ of D.

Note that $0 \leq \underline{A}(D) \leq \overline{A}(D)$. If $\underline{A}(D) = \overline{A}(D)$, we say that D *has area*, and we call the common value of the inner area and the outer area, the area $A(D)$ of D. If $\underline{A}(D) < \overline{A}(D)$, we say that D *does not have area*. In the exercises, we give examples of sets that do not have area and leave it to the reader to ponder some of the details (with a few hints).

We now make an observation that will be very important for the integral we want to develop in the next section. Note that, for any grid G over D, $\mathscr{O}(D, G) - \mathscr{I}(D, G)$ is the sum of the areas of rectangles of G in class (3), that is, a sum of rectangles covering the boundary ∂D of D. Hence $\overline{A}(D) - \underline{A}(D)$ should be $\overline{A}(\partial D)$, and we conclude that D has area $[\overline{A}(D) = \underline{A}(D)]$ exactly when $\overline{A}(\partial D) = 0$. Since $0 \leq \underline{A}(\partial D) \leq \overline{A}(\partial D)$, then $\overline{A}(\partial D) = 0$ forces $\underline{A}(\partial D) = 0$ also, and we have the fundamental result that D has area exactly when ∂D has zero area. (Note here that to say that ∂D has zero area is very different from saying that ∂D does not have area.)

Obviously these remarks on area are of a rather technical nature; calculating $\underline{A}(D)$ and $\overline{A}(D)$ for a given set D is usually extremely difficult, even for fairly simple examples. The student can convince himself of this by attempting it for $D = \{(x, y) \mid x^2 + y^2 \leq 1\}$. However, in developing the double integral it is important to have at least some solid basis for a notion of area, and, in particular, for the connection between D having area and ∂D having zero area.

Exercises for Section 4.3

1. Prove that any finite set of points in the plane has area zero.

2. Let D consist of all points (x, y) with $0 \leq x \leq 1$, $0 \leq y \leq 1$, and x and y both rational. Show that D does not have area by directly calculating $\underline{A}(D)$ and $\bar{A}(D)$.

3. Let D_1 and D_2 be bounded sets in the plane, having only boundary points of both sets in common. Suppose that ∂D_1 and ∂D_2 both have zero area. Show that the set D consisting of points in D_1 together with points in D_2 has area (this set is called the union of D_1 and D_2, denoted $D_1 \cup D_2$). Show also that $A(D_1 \cup D_2) = A(D_1) + A(D_2)$.

4. Let f be differentiable on the interval $[a, b]$. Show that the graph of f [that is, the set of points $(x, f(x))$ for $a \leq x \leq b$] has zero area.

5. Let $D = \{(x, 1/n) \mid 0 \leq x \leq 1 \text{ and } n = 1, 2, 3, \ldots\}$. Determine whether or not D has area, and compute the area of D if D has area.

6. Let $D = \{(n, 1/n) \mid n = 1, 2, 3, \ldots\}$. Determine whether or not D has area, and, if D has area, compute it.

7. Let D consist of all points (x, y) with $1/(n+1) \leq x \leq 1/n$, $1/(n+1) \leq y \leq 1/n$, for $n = 1, 2, 3, \ldots$. Determine whether or not D has area. If D has area, compute it.

4.4 Definition of the Double Integral Over a Rectangle

A convenient way of approaching the notion of a double integral of a function of two variables over a set in the plane is to first consider the special case that D is a rectangle, say

$$D = \{(x, y) \mid a \leq x \leq b \text{ and } c \leq y \leq d\},$$

where $a < b$ and $c < d$. Suppose, then, that f is defined and bounded on D. As in Section 4.1, we have two equivalent means of approach.

I. Form a grid G over D by partitioning $a = x_0 < x_1 < \cdots < x_n = b$ and $c = y_0 < y_1 < \cdots < y_m = d$. (Here, because D is a rectangle, G exactly covers D.) Note that there are mn rectangles in the grid. Let L be a real number. We say that the integral of f over D *exists*, and *equals* L, if, given $\varepsilon > 0$, there is some $\delta > 0$ such that $\left|\sum_{i=1}^{mn} f(\xi_i, \eta_i)A(R_i) - L\right| < \varepsilon$ whenever G is chosen with mesh less than δ, and (ξ_i, η_i) is any point of the ith rectangle R_i, $i = 1, \ldots, mn$. We then denote L by $\iint_D f$, or, if you prefer, $\iint_D f(x, y)\, dx\, dy$.

II. Analogous to II in Section 4.1, we can associate with each grid G over D an upper and a lower Darboux sum, given respectively by

$$\bar{S}(G) = \sum_{i=1}^{mn} M_i A(R_i) \quad \text{and} \quad \underline{S}(G) = \sum_{i=1}^{mn} m_i A(R_i),$$

where $M_i = \text{lub}\{f(x, y) \mid (x, y) \text{ is in } R_i\}$ and $m_i = \text{glb}\{f(x, y) \mid (x, y) \text{ is in } R_i\}$, for $i = 1, \ldots, mn$.

As in Section 4.1, we can show that for any grids G_1 and G_2 over D,

$$\underline{S}(G_1) \leq \bar{S}(G_2).$$

The proof mirrors the argument outlined in Exercise 1 of Section 4.2. Using this inequality, it is immediate that the numbers $\underline{S}(G)$, taken over all grids G over D, are bounded above, and that the numbers $\bar{S}(G)$ are bounded below. Put $\underline{I}(f) = \text{lub}\{\underline{S}(G) \mid G \text{ is a grid over } D\}$, and $\bar{I}(f) = \text{glb}\{\bar{S}(G) \mid G \text{ is a grid over } D\}$. These numbers exist by the previous remark, and the above inequality yields $\underline{I}(f) \leq \bar{I}(f)$. If $\underline{I}(f) = \bar{I}(f)$, we say that f is *Riemann integrable* over D, or on D, and denote $\underline{I}(f)$ [or $\bar{I}(f)$] as $\iint_D f$, or $\iint_D f(x, y)\, dx\, dy$.

It is not too difficult to adapt the argument outlined in Exercise 5, Section 4.2, to show that I and II above lead to equivalent notions of integrability, and to the same value for the integral. Exercise 3 of Section 4.2 can also be adapted to the two-variable case to produce an example of a bounded function that fails to be integrable over a bounded set.

We shall now prove our fundamental theorem on integrability. Remember that D in this section is restricted to be a rectangular region.

Theorem

If f is continuous on D except possibly at points that make up a set of zero area, then f is integrable on D.

PROOF

Let $\varepsilon > 0$. We can partition D by a grid G consisting of rectangles $R_1, \ldots, R_k, R_{k+1}, \ldots, R_n$ such that $R_1, \ldots, R_k$ cover the points of discontinuity of f in D, and $\sum_{i=1}^{k} A(R_i) < \varepsilon$.

Now, $R_{k+1}, \ldots, R_n$ cover a compact subset K of D on which f is continuous, hence uniformly continuous (see Exercise 20, Section 3.1). Then, for some $\delta_1 > 0$, $|f(x) - f(y)| < \varepsilon$ whenever x and y are in K and $\|x - y\| < \delta_1$. Now let G' be any grid over D, containing rectangles $r_1, \ldots, r_s, \ldots, r_t$, with $r_{s+1}, \ldots, r_t$ all those rectangles contained within K. Now

$$\bar{S}(G') - \underline{S}(G') = \sum_{i=1}^{s} (M_i - m_i)A(r_i) + \sum_{i=s+1}^{t} (M_i - m_i)A(r_i).$$

Suppose that mesh$(G') < \delta_1$. Then $M_i - m_i < \varepsilon$ for $i = s + 1, \ldots, t$. Then $\sum_{i=s+1}^{t} (M_i - m_i)A(r_i) \le \sum_{i=s+1}^{t} \varepsilon A(r_i) \le \varepsilon A(K) \le \varepsilon A(D)$.

If now $|f(x)| \le C$ for all x in D, then $\sum_{i=1}^{s} (M_i - m_i)A(r_i) \le 2C \sum_{i=1}^{s} A(r_i)$. Since the set of points of discontinuity of f has zero area, and $r_1, \ldots, r_s$ cover the points of discontinuity of f, then we may choose δ_2 so that $\sum_{i=1}^{s} A(r_i) \le \sum_{i=1}^{k} A(R_i) + \varepsilon < 2\varepsilon$ if mesh$(G') < \delta_2$. Thus, if we choose mesh$(G') < \min(\delta_1, \delta_2)$, we have

$$\bar{S}(G') - \underline{S}(G') < \varepsilon A(D) + 4C\varepsilon = [4C + A(D)]\varepsilon,$$

which can be made arbitrarily small. Since $\bar{I}(f) - \underline{I}(f) \le \bar{S}(G') - \underline{S}(G')$, we conclude that $\underline{I}(f) = \bar{I}(f)$, so that f is integrable over D. ∎

It is also the case that, if the set of discontinuities of f on D has positive area, then f is not Riemann integrable over D. However, we shall not attempt to prove this here.

Exercises for Section 4.4

In these exercises D is a rectangle, as in the section, and f is bounded on D.

1. Prove that $\underline{S}(P) \le \bar{S}(W)$ for any grids P and W over D.

2. Prove that $\underline{I}(f) \le \bar{I}(f)$.

3. Produce an example of f and D such that $\underline{I}(f) < \bar{I}(f)$.

4. Prove the following theorem: *f is integrable in the same sense of* II *over D if and only if, given $\varepsilon > 0$, there exists a grid G over D such that $\bar{S}(G) - \underline{S}(G) < \varepsilon$.*

5. Prove that the formulations I and II are equivalent.

6. Define f over the unit square $0 \le x \le 1$, $0 \le y \le 1$ by

$$f(x, y) = \begin{cases} x + y & \text{if } 0 \le x \le 1,\ 0 \le y \le 1, \text{ and } x \ne y, \\ 0 & \text{if } x = y. \end{cases}$$

Determine whether or not f is integrable over D.

7. Let f be defined over the unit square $0 \le x \le 1$, $0 \le y \le 1$ by

$$f(x, y) = \begin{cases} 2 & \text{if } 0 \le x \le 1,\ 0 \le y \le 1, \text{ but } y \ne 1/n \text{ for integer } n, \\ 1 & \text{if } (x, y) = (0, 1/n) \text{ for } n = 1, 2, \ldots, \\ 3 & \text{if } (x, y) = (x, 1/n) \text{ for } x \ne 0, \text{ and } n = 1, 2, \ldots. \end{cases}$$

Determine whether or not f is integrable over D. *Hint:* For Exercises 6 and 7, see if the sets of points on which f is discontinuous have zero area.

4.5 Extension of the Double Integral to Bounded Sets in the Plane

We can now very easily extend the notion of integrability to bounded functions defined over bounded sets which need not be rectangles. Let D be any bounded set in the plane, and suppose that f is a bounded function defined over D. Choose any rectangle $\mathscr{D}$ containing D, and put

$$g(x) = \begin{cases} f(x) & \text{for } x \text{ in } D, \\ 0 & \text{for } x \text{ in } \mathscr{D} \text{ but not in } D. \end{cases}$$

Then g is defined and bounded on a rectangle $\mathscr{D}$. If g is integrable on $\mathscr{D}$, we say that f is *integrable* over D, and we put

$$\iint_D f = \iint_{\mathscr{D}} g.$$

It is important to realize that, in order for this definition to be an effective one, we must show that it does not depend upon the particular choice of rectangular region containing D. This is a simple matter. If $\mathscr{D}$ and $\mathscr{D}'$ are both rectangles containing D, then the rectangle $\mathscr{D}''$ formed by taking points common to both $\mathscr{D}$ and $\mathscr{D}'$ is also a rectangle containing D. Putting $h(x) = f(x)$ for x in D, and $h(x) = 0$ for x in $\mathscr{D}''$ but not in D, we then have

$$\iint_{\mathscr{D}} g = \iint_{\mathscr{D}''} g = \iint_{\mathscr{D}''} h = \iint_{\mathscr{D}'} h,$$

as we wanted to show.

We now prove a theorem giving sufficient conditions for integrability.

Theorem

Let D be a set of points in the plane, and suppose that D has area. Let f be defined and bounded on D, and continuous at all interior points of D. Then f is integrable over D.

PROOF

Choose any rectangle $\mathscr{D}$ containing D, and put $g(x) = f(x)$ if x is in D, and $g(x) = 0$ if x is in $\mathscr{D}$ but not in D. By assumption, any discontinuities of g in $\mathscr{D}$ must appear in ∂D, which is a set of zero area in $\mathscr{D}$. By the theorem of Section 4.4, g is integrable over $\mathscr{D}$, hence by definition f is integrable over D. ∎

We now list some of the more useful properties of the double integral. Proofs will be left to the exercises.

(1) $\iint_D cf + dg = c \iint_D f + d \iint_D g$ whenever the two integrals on the right exist.

(2) $\iint_D f \leq \iint_D g$ if $f(x) \leq g(x)$ for each x in D.

(3) $\iint_D f = \iint_{D_1} f + \iint_{D_2} f$ if both the integrals on the right exist, and D_1 and D_2 have only boundary points of both sets in common.

(4) $\left|\iint_D f\right| \leq \iint_D |f|$.

(5) If $m \leq f(x) \leq M$ for each x in D, then $mA(D) \leq \iint_D f \leq MA(D)$. Hence, for some number B in $[m, M]$, $\iint_D f = BA(D)$.

(6) If f is continuous on D, then for some ξ in D,

$$\iint_D f = f(\xi)A(D).$$

The last part of (5), and (6), are referred to as *mean value theorems.*

Exercises for Section 4.5

1. Let $f(x, y) = 2$ for $x^2 + y^2 < 1$, but $f(x, y) = 3$ if $x^2 + y^2 = 1$. Determine whether or not f is integrable over $\{(x, y) \mid x^2 + y^2 \leq 1\}$.

2. Let $f(x, y) = 1$ for $x^2 + y^2 \leq 1$ but $(x, y) \neq (0, 1/n)$ for $n = 1, 2, \ldots$. Let $f(0, 1/n) = 1/n$ for $n = 1, 2, \ldots$. Determine whether or not f is integrable over $\{(x, y) \mid x^2 + y^2 \leq 1\}$. *Hint:* Here the theorem of this section does not apply, so enclose the unit disc in a rectangle and work with theorems from Section 4.4.

3. Let $f(x, y) = x + y$ for $0 < x < 1$, $x^2 < y < x$, and $f(x, x) = 2$ for $0 \leq x \leq 1$, and $f(x, x^2) = 3$ for $0 \leq x \leq 1$. Determine whether or not f is integrable over the set $\{(x, y) \mid 0 \leq x \leq 1 \text{ and } x^2 \leq y \leq x\}$.

4. Write out proofs of properties (1) through (6) of the double integral listed at the end of the section.

4.6 Physical Interpretations of Double Integrals

Thus far we have not attempted to develop any techniques for convenient evaluation of double integrals. Before doing this in the next two sections, we shall describe several physical interpretations which might help the student get some feeling for the concept of the double integral.

A. Volume As an Integral

Under certain conditions ($f(x) > 0$, f continuous on $[a, b]$), $\int_a^b f(x)\,dx$ is the area between the graph of $y = f(x)$, $a \leq x \leq b$, and the x-axis. If f is a function of two variables, the graph of $z = f(x, y)$, for (x, y) in D, may be thought of as a surface in 3-space consisting of the locus of points $(x, y, f(x, y))$. Assuming that $f(x, y) > 0$ and that f is continuous on the compact set D (which we assume has

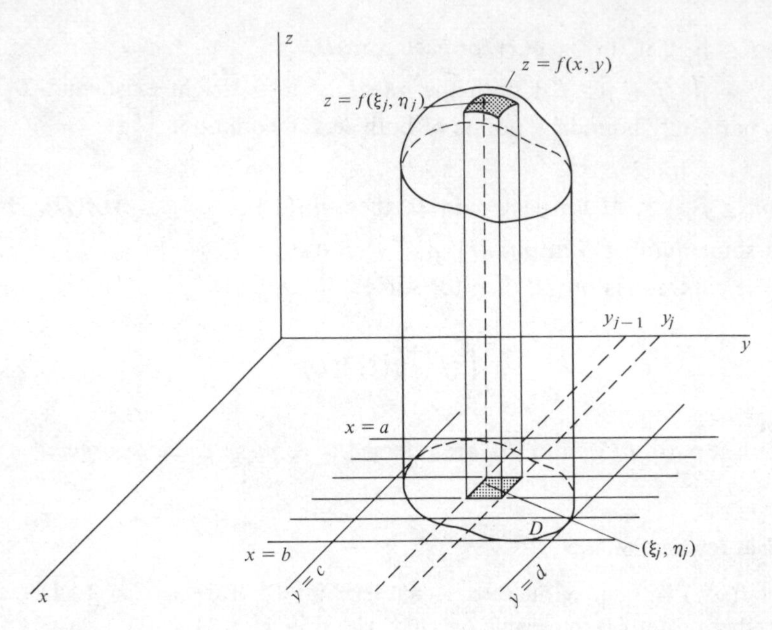

Fig. 4-2

area), we claim that $\iint_D f$ gives the volume V between the surface and the xy-plane. (In most such applications, D is connected as well as compact, although this extra condition is not really needed.)

Picture D and f as in Fig. 4-2, and consider the problem of finding V. We might begin by attempting an approximation. Partition D by a grid $a = x_0 < x_1 < \cdots < x_n = b$, $c = y_0 < y_1 < \cdots < y_m = d$, with $R_1, \ldots, R_k$ the rectangles of the grid which contain only interior points of D. Form the Riemann sum $\sum_{j=1}^{k} f(\xi_j, \eta_j)A(R_j)$, where (ξ_j, η_j) is any point of R_j. Now observe that the number $f(\xi_j, \eta_j)A(R_j)$ represents the volume of a rectangular box with height $f(\xi_j, \eta_j)$ and base rectangle R_j. The Riemann sum is then a sum of volumes of boxes, each having one of the R_j's as its base, and as its height, the distance from a point (ξ_j, η_j) in R_j to a point $(\xi_j, \eta_j, f(\xi_j, \eta_j))$ on the surface above. The Riemann sum approximates the volume under the surface by replacing the surface with rectangular boxes with heights adjusted to fit the surface at particular points.

As we take finer partitions, with more and smaller rectangles, the boxes more closely approximate the shape of the surface itself. In the "limit" used in forming the double integral, the mesh of the grids goes to zero, and the error incurred by approximating a surface by flat-topped boxes becomes smaller, tending to zero, so that the exact volume is given by $\iint_D f$.

This argument exactly parallels the interpretation of $\int_a^b f(x)\,dx$ as an area when $f(x) > 0$ on $[a, b]$. As in the case of a single integral, not every double integral can be interpreted as a volume. If, for example, $f(x, y)$ changes sign, then some cancellations can occur. This is analogous to what happens with $\int_0^{2\pi} \sin(x)\,dx = 0$.

We note in passing that $\iint_D f$ is the area of D when $f(x, y) = 1$ for every (x, y) in D. This is immediate by looking at the Riemann sums, which for this f are $\sum_{j=1}^{k} A(R_j)$, or $\mathscr{I}(D, G)$. Assuming that D has area, these Riemann sums then approach $\underline{A}(D)$, which will then be the area of D.

Another way of looking at this is that $\iint_D 1$ gives the volume of a surface of constant height 1 and base D, and that this volume should be $A(D)(1) = A(D)$ (see Fig. 4-3).

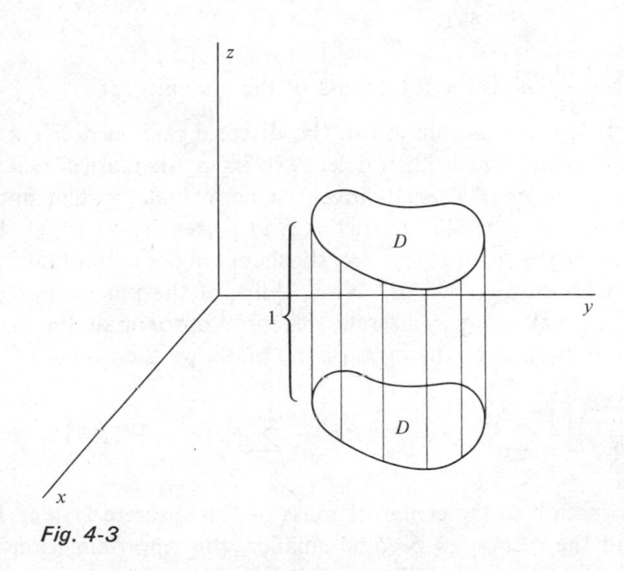

Fig. 4-3

B. Mass of a Flat Plate

Suppose that we have a thin flat sheet of some material, and we know its density (mass/unit area) at each point. We want to determine the mass. Set up a coordinate system so that the sheet can be represented as a region D in the xy-plane. Let $\rho(x, y)$ be the density at (x, y) in D. Since density = mass per unit area, it is natural to think of $\iint_D \rho$ as the mass of the sheet.

In more detail, partition D as usual, forming rectangles $R_1, \ldots, R_k$ which contain only interior points of D. If (ξ_j, η_j) is in R_j, then $\rho(\xi_j, \eta_j)A(R_j)$ approximates the mass of the part of the plate covered by R_j. Then $\sum_{j=1}^{k} \rho(\xi_j, \eta_j)A(R_j)$ approximates the mass of the plate, with the approximation improving as the mesh of the grid is taken smaller. In this "limit" we obtain $\iint_D \rho$ as the exact value of the mass of the plate.

C. Center of Mass of a Flat Plate

As in Section 4.6B, suppose that we have a flat plate of known density and that we want to determine the center of mass. Here we have to assume something. We shall suppose that we know how to find the center of mass of a discrete (finite) system of particles. By way of review, suppose that we have particles $P_1, \ldots, P_s$ at locations $(x_1, y_1), \ldots, (x_s, y_s)$ in the plane, and that the mass of P_j is m_j. The center of mass of the system is then at $(\bar{x}, \bar{y})$, where

$$\bar{x} = \frac{\sum_{j=1}^{s} x_j m_j}{\sum_{j=1}^{s} m_j} \quad \text{and} \quad \bar{y} = \frac{\sum_{j=1}^{s} y_j m_j}{\sum_{i=1}^{s} m_j}.$$

Here, of course, $\sum_{j=1}^{s} m_j$ is the total mass of the system.

We want to apply this solution in the discrete case to derive a solution in the case of the plate, which has (at least in its mathematical description) infinitely many particles. As usual, invent a coordinate system and place the plate down to cover a region D. Let $\rho(x, y) =$ density at (x, y). Partition D by a grid of rectangles, with $R_1, \ldots, R_k$ those rectangles containing only interior points of D. Choose (ξ_j, η_j) in R_j. Now, think of the particles of plate at the points (ξ_j, η_j) as making up a discrete system. Approximate the mass of R_j by $\rho(\xi_j, \eta_j)A(R_j)$. Letting m be the mass of the plate, we then think of

$$\left(\frac{1}{m}\sum_{j=1}^{k} \xi_j \rho(\xi_j, \eta_j)A(R_j), \frac{1}{m}\sum_{j=1}^{k} \xi_j \rho(\xi_j, \eta_j)A(R_j)\right)$$

as an approximation to the center of mass of this discrete system. In the limit as $k \to \infty$ and the rectangles become smaller, this approximation approaches the center of mass of the plate. But in this limiting process,

$$\frac{1}{m}\sum_{j=1}^{k} \xi_j \rho(\xi_j, \eta_j)A(R_j) \to \frac{1}{m}\iint_D x\rho(x, y)\,dx\,dy,$$

and

$$\frac{1}{m}\sum_{j=1}^{k} \xi_j \rho(\xi_j, \eta_j)A(R_j) \to \frac{1}{m}\iint_D y\rho(x, y)\,dx\,dy.$$

Using the result from Section 4.6B, that $m = \iint_D \rho(x, y)\,dx\,dy$, we now have the coordinates of the center of mass:

$$\bar{x} = \frac{\iint_D x\rho(x, y)\,dx\,dy}{\iint_D \rho(x, y)\,dx\,dy} \quad \text{and} \quad \bar{y} = \frac{\iint_D y\rho(x, y)\,dx\,dy}{\iint_D \rho(x, y)\,dx\,dy},$$

Exercises for Section 4.6

Rework the motivating arguments that lead to double-integral formulas for volume, mass, and center of mass given above, using formulation II of Section 4.4 of the Riemann integral, instead of formulation I as used in this section.

4.7 Evaluation of Double Integrals by Iterated Integrals—A Geometric Approach

Thus far we know in theory what a symbol like $\iint_D f$ means, and several physical applications, but we have not as yet discussed any practical means of turning the symbol into a number. We now give a geometric argument which suggests a method for doing this. A more analytic treatment will be deferred until the next section. The basic strategy is to try to reduce the problem of evaluating $\iint_D f$ to two problems of evaluating single integrals, for which the fundamental theorem of calculus is available.

To get some insight, let us begin with the case that $f(x, y) > 0$ on D, so that we can interpret $\iint_D f$ as a volume. We shall outline another way of finding this volume, then compare results. We need an assumption about D. Suppose that the boundary of D consists of two pieces, a lower and an upper piece, which can be written respectively as the graphs of $y = g_1(x)$ and $y = g_2(x)$, $a \leq x \leq b$, with g_1 and g_2 continuous (see Fig. 4-4).

Now partition $[a, b]$ with subintervals by inserting points: $a = x_0 < x_1 < x_2 < \cdots < x_n = b$. To be specific, look at a typical subinterval $[x_{k-1}, x_k]$. The

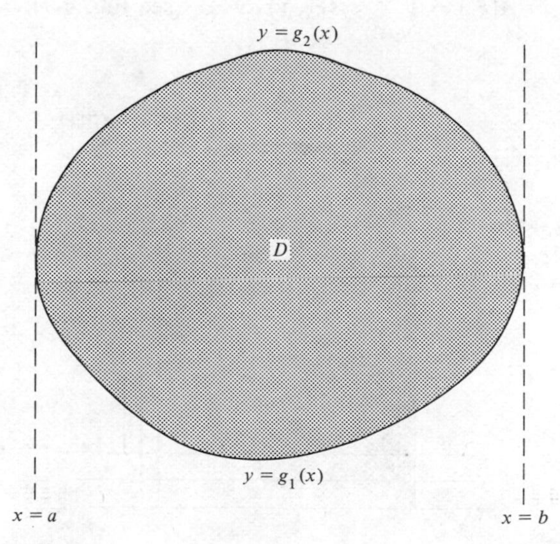

Fig. 4-4

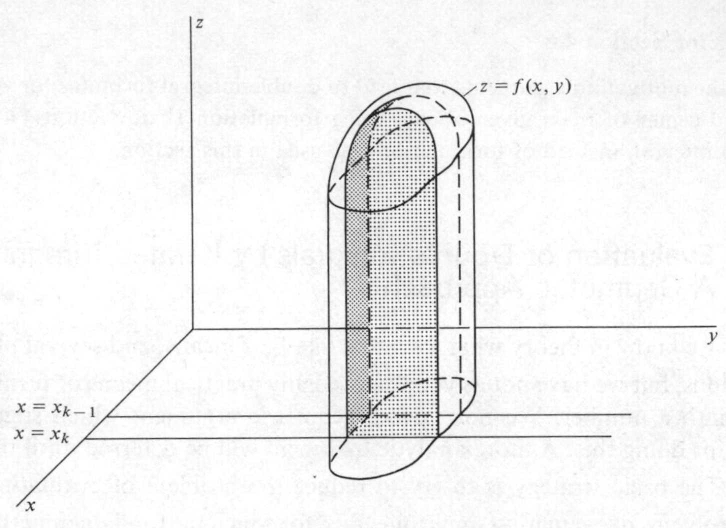

Fig. 4-5

parallel planes $x = x_k$, $x = x_{k-1}$ bound a slice of the volume determined by $z = f(x, y)$ (see Figs. 4-5 and 4-6). Let V_k denote the volume of this slice. To approximate V_k, choose any ξ_k in $[x_{k-1}, x_k]$. The plane $x = \xi_k$ cuts C in two points, at $(\xi_k, g_1(\xi_k), 0)$ and $(\xi_k, g_2(\xi_k), 0)$ (Fig. 4-6). The curve $z = f(\xi_k, y)$, $g_1(\xi_k) \le y \le g_2(\xi_k)$ is the intersection of the surface $z = f(x, y)$ with the plane $x = \xi_k$. The area under this curve is

$$A(\xi_k) = \int_{g_1(\xi_k)}^{g_2(\xi_k)} f(\xi_k, y)\, dy \qquad \text{(see Fig. 4-7).}$$

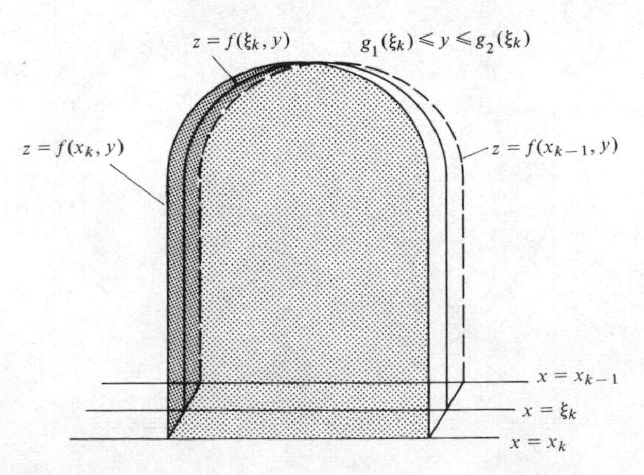

Fig. 4-6

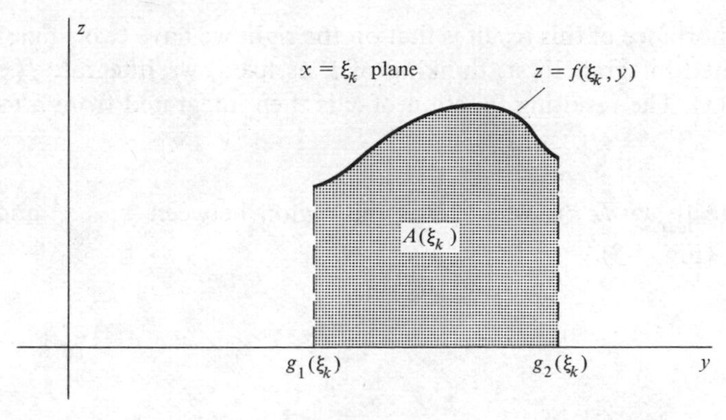

Fig. 4-7

Now, $A(\xi_k)$ is the area of a slice through V_k. We then approximate V_k as $A(\xi_k)$ (width of V_k), which is $A(\xi_k)(x_k - x_{k-1})$. Then

$$\sum_{k=1}^{n} V_k = \sum_{k=1}^{n} A(\xi_k)(x_k - x_{k-1})$$

approximates V. Now recognize the sum on the right as the Riemann sum for $\int_a^b A(x)\,dx$, where, for any x in $[a, b]$,

$$A(x) = \int_{g_1(x)}^{g_2(x)} f(x, y)\,dy.$$

Letting $n \to \infty$ and each $x_k - x_{k-1} \to 0$, we then have

$$\sum_{k=1}^{n} V_k \to V.$$

But also

$$\sum_{k=1}^{n} V_k = \sum_{k=1}^{n} A(\xi_k)(x_k - x_{k-1}) \to \int_a^b A(x)\,dx = \int_a^b \left[\int_{g_1(x)}^{g_2(x)} f(x, y)\,dy\right] dx.$$

We conclude that

$$V = \int_a^b \left[\int_{g_1(x)}^{g_2(x)} f(x, y)\,dy\right] dx.$$

But, from Section 4.6, we also have $V = \iint_D f$. Therefore, on geometrical grounds, we conclude that

$$\iint_D f(x, y)\,dx\,dy = \int_a^b \left[\int_{g_1(x)}^{g_1(x)} f(x, y)\,dy\right] dx.$$

The importance of this result is that on the right we have two single integrals, or an iterated integral. First, thinking of x as fixed, we integrate $f(x, y)$ from $g_1(x)$ to $g_2(x)$. The resulting function of x is then integrated from a to b.

EXAMPLE

Evaluate $\iint_D xy\,dx\,dy$, where D is the region between $y = x^2$ and $y^2 = x$, $0 \le x \le 1$ (Fig. 4-8).

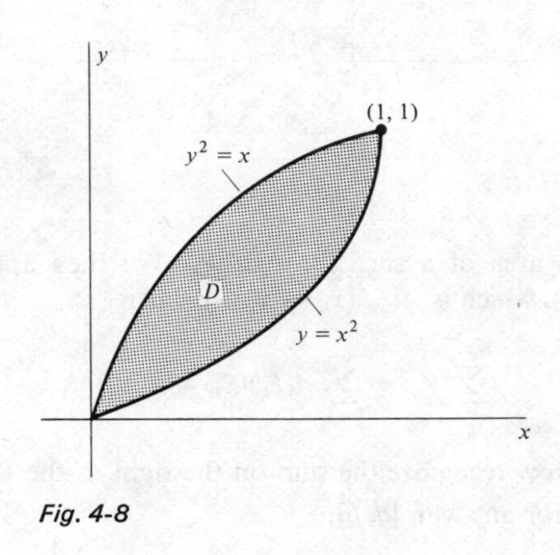

Fig. 4-8

Here $g_1(x) = x^2$ and $g_2(x) = \sqrt{x}$. Write

$$\iint_D xy = \int_0^1 \left(\int_{x^2}^{\sqrt{x}} xy\,dy \right) dx.$$

First, the inner integral:

$$\int_{x^2}^{\sqrt{x}} xy\,dy = \frac{xy^2}{2}\bigg|_{x^2}^{\sqrt{x}} = \frac{x}{2}(x - x^4) = \frac{x^2}{2} - \frac{x^5}{2}.$$

Then

$$\iint_D xy = \int_0^1 \left(\frac{x^2}{2} - \frac{x^5}{2} \right) dx = \frac{x^3}{6}\bigg|_0^1 - \frac{x^6}{12}\bigg|_0^1 = \frac{1}{6} - \frac{1}{12} = \frac{1}{12}.$$ ∎

EXAMPLE

Evaluate $\iint_D \sin(xy)$, where D is the region bounded by the x-axis and $y = 2x$, $\pi/4 \le x \le \pi/2$ (see Fig. 4-9). Write $\iint_D \sin(xy) = \int_{\pi/4}^{\pi/2} \int_0^{2x} \sin(xy)\,dy\,dx$. Now

$$\int_0^{2x} \sin(xy)\,dy = -\frac{1}{x}\cos(xy)\bigg|_0^{2x} = -\frac{1}{x}\cos(2x^2) + \frac{1}{x},$$

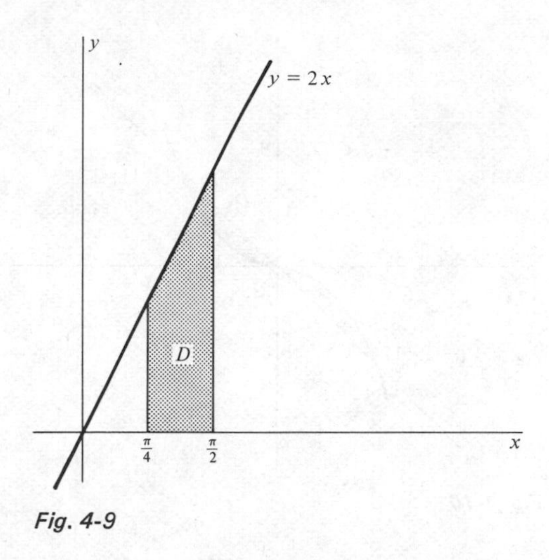

Fig. 4-9

Then

$$\iint_D \sin(xy) = \int_{\pi/4}^{\pi/2} \left[-\frac{1}{x}\cos(2x^2) + \frac{1}{x} \right] dx.$$

Now $\int_{\pi/4}^{\pi/2} (1/x)\, dx = \ln(\pi/2) - \ln(\pi/4) = \ln(2)$. But

$$\int_{\pi/4}^{\pi/2} -\frac{1}{x}\cos(2x^2)\, dx$$

cannot be done so easily, as $\cos(2x^2)/x$ has no elementary antiderivative. We could, however, expand $\cos(2x^2)/x$ in a Taylor series, say about $3\pi/4$ (which is midway between $\pi/4$ and $\pi/2$), and approximate the integral to any desired degree of accuracy. ∎

The above derivation of an iterated integral expression for $\iint_D f$ was based upon interpretation of the double integral as a volume. This was strictly a matter of convenience in presenting the idea from a physically plausible point of view. In general, it is not necessary to assume that $f(x, y) \geq 0$ on D in order to use this method. Further, the assumption that the boundary of D consists of two continuous pieces, $y = g_1(x)$ and $y = g_2(x)$, can often be modified by breaking D up into smaller regions $D_1, \ldots, D_s$ and writing $\iint_D f$ as $\sum_{i=1}^{s} \iint_{D_i} f$, provided that D_i overlaps D_j on a set of zero area for $i \neq j$. We illustrate below.

EXAMPLE

Evaluate $\iint_D x^2 y\, dx\, dy$, where D is the region bounded by $y = x^3$ and $y = x$, $-1 \leq x \leq 1$ (Fig. 4-10). Here we must break D up into two regions, D_1 and D_2, as shown in Fig. 4-10. Note that D_1 can be described as the set of points

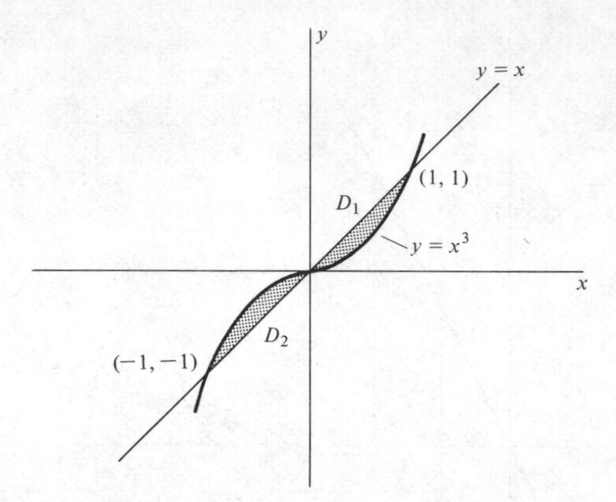

Fig. 4-10

(x, y) with $x^3 \le y \le x$, $0 \le x \le 1$, and D_2 as the set of (x, y) with $x \le y \le x^3$, $-1 \le x \le 0$. Then

$$\iint_D x^2y = \iint_{D_1} x^2y + \iint_{D_2} x^2y = \int_0^1 \left(\int_{x^3}^{x} x^2y\,dy\right) dx + \int_{-1}^{0} \left(\int_{x}^{x^3} x^2y\,dy\right) dx$$

$$= \int_0^1 \left(\frac{x^2y^2}{2}\bigg|_{x^3}^{x}\right) dx + \int_{-1}^{0} \left(\frac{x^2y^2}{2}\bigg|_{x}^{x^3}\right) dx$$

$$= \int_0^1 \left(\frac{x^4}{2} - \frac{x^8}{2}\right) dx + \int_{-1}^{0} \left(\frac{x^8}{2} - \frac{x^4}{2}\right) dx = 0. \qquad \blacksquare$$

Some limitations inherent in this method of evaluating double integrals should be apparent by this time. For one thing, either of the single integrals may be very difficult. For another, the region D may have to be broken up into many pieces to get the kind of description we need, and this can involve considerable ingenuity.

A partial remedy for both difficulties can sometimes be found in reversing the order of integration in the iterated integral. In arguing that $\iint_D f = \int_a^b \left[\int_{g_1(x)}^{g_2(x)} f(x, y)\,dy\right] dx$, we started out on the x-axis, and split the boundary of D into two pieces, each representable with y as a function of x. If instead we started out on the y-axis, wrote the boundary of D as two curves $x = h_1(y)$ and $x = h_2(y)$, $c \le y \le d$, and then repeated essentially the same argument as above, with the point of view rotated by 90°, then we would obtain

$$\iint_D f = \int_c^d \left[\int_{h_1(y)}^{h_2(y)} f(x, y)\,dx\right] dy.$$

This is simply another iterated integral expression for $\iint_D f$, with the reverse order of integration from the first one we derived.

As a matter of notation, sometimes the parentheses are omitted in writing iterated integrals, leaving just

$$\int_a^b \int_{g_1(x)}^{g_2(x)} f(x, y)\, dy\, dx \qquad \text{and} \qquad \int_c^d \int_{h_1(y)}^{h_2(y)} f(x, y)\, dx\, dy.$$

These are also sometimes seen written as

$$\int_a^b dx \int_{g_1(x)}^{g_2(x)} f(x, y)\, dy \qquad \text{and} \qquad \int_c^d dy \int_{h_1(y)}^{h_2(y)} f(x, y)\, dx,$$

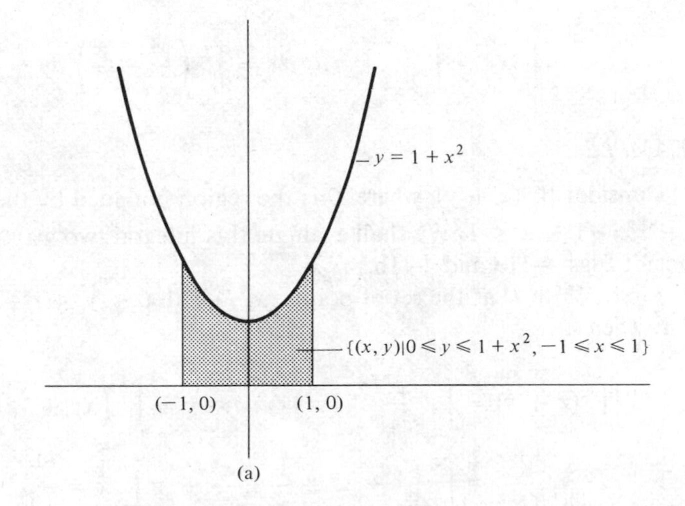

(a)

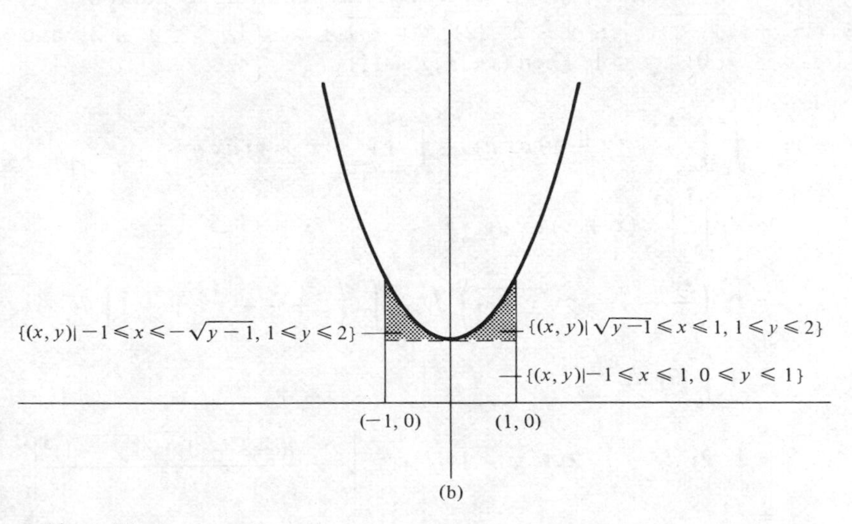

(b)

Fig. 4-11

respectively. It is usually clear from the limits of integration which variable is to be integrated first. Note, however, that the symbol $\iint_D f(x, y)\,dx\,dy$ simply stands for the double integral, with no significance to the dx preceding the dy part of the symbol.

EXAMPLE

Look again at $\iint_D xy$, with D bounded by $y = x^2$ and $y = \sqrt{x}$, $0 \le x \le 1$. This time we write D as the set of points (x, y) with $y^2 \le x \le \sqrt{y}$, $0 \le y \le 1$. Then

$$\iint_D xy = \int_0^1 \int_{y^2}^{\sqrt{y}} xy\,dy\,dx = \int_0^1 \left(\frac{y^2}{2} - \frac{y^5}{2}\right) dy = \tfrac{1}{12}. \qquad \blacksquare$$

EXAMPLE

Consider $\iint_D (x + y)$, where D is the region bounded by the x-axis and $y = 1 + x^2$, $-1 \le x \le 1$. We shall evaluate this integral two ways. (It may help to consult Figs. 4-11a and 4-11b.)

First: Write D as the set of points (x, y) with $0 \le y \le 1 + x^2$ and $-1 \le x \le 1$. Then

$$\iint_D (x + y) = \int_{-1}^1 \int_0^{1+x^2} (x + y)\,dy\,dx = \int_0^1 \left(xy + \frac{y^2}{2}\right)\Bigg|_0^{1+x^2} dx$$

$$= \int_0^1 \left(x + x^3 + \frac{1}{2} + \frac{x^4}{2} + x^2\right) dx = \frac{28}{15}.$$

To reverse the order of integration, we have to think of D as being made up of three pieces, consisting of (x, y) satisfying the following conditions: (1) $-1 \le x \le -\sqrt{y - 1}$, $1 \le y \le 2$; (2) $\sqrt{y - 1} \le x \le 1$, $1 \le y \le 2$; and (3) $-1 \le x \le 1$, $0 \le y \le 1$. Then (see Fig. 4-11)

$$\iint_D (x + y) = \int_1^2 \int_{-1}^{-\sqrt{y-1}} (x + y)\,dx\,dy + \int_1^2 \int_{\sqrt{y-1}}^2 (x + y)\,dx\,dy$$

$$+ \int_0^1 \int_{-1}^1 (x + y)\,dx\,dy$$

$$= \int_1^2 \left(\frac{3y}{2} - 1 - y\sqrt{y - 1}\right) dy + \int_1^2 \left(1 + \frac{y}{2} - y\sqrt{y - 1}\right) dy$$

$$+ \int_0^1 2y\,dy$$

$$= \int_0^2 2y\,dy - \int_1^2 2y\sqrt{y - 1}\,dy = y^2\Big|_0^2 + \frac{4(-2 - 3y)\sqrt{(y - 1)^3}}{15}\Bigg|_1^2$$

$$= \tfrac{28}{15}. \qquad \blacksquare$$

EXAMPLE

Evaluate $\iint_D e^{y/x}\,dx\,dy$, where D is bounded by $y = x$, $y = 2x$, $x = 1$, and $x = 2$ (see Fig. 4-12). In one order of integration, the problem is simple:

$$\iint_D e^{y/x}\,dy\,dx = \int_1^2 \int_x^{2x} e^{y/x}\,dy\,dx = \int_1^2 \left\{ xe^{y/x} \Big|_x^{2x} \right\} dx$$

$$= \int_1^2 (xe^2 - xe)\,dx = (e^2 - e)\int_1^2 x\,dx = \tfrac{3}{2}(e^2 - e).$$

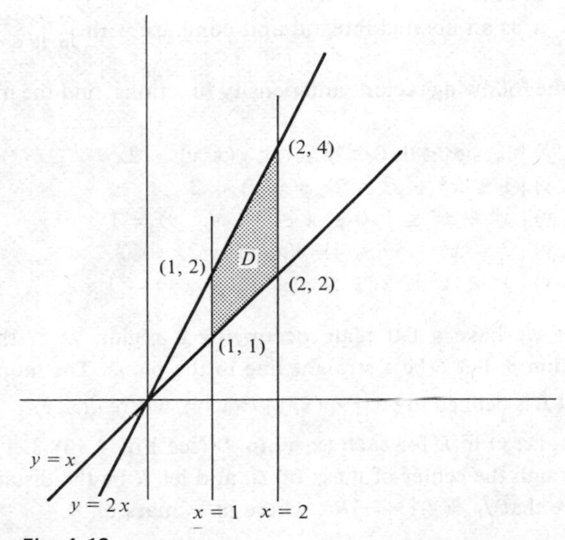

Fig. 4-12

But try reversing the order of integration, and we encounter two problems. First, we must break D up into more than one region:

$$\iint_D e^{y/x}\,dx\,dy = \int_1^2 \int_1^y e^{y/x}\,dx\,dy + \int_2^4 \int_{y/2}^2 e^{y/x}\,dx\,dy.$$

This is not a serious drawback. A greater problem is that $e^{y/x}$ has no elementary antiderivative with respect to x, so $\int_1^y e^{y/x}\,dx$ and $\int_{y/2}^2 e^{y/x}\,dx$ are very difficult to evaluate in a useful form. In this example, then, there is a considerable advantage in one order of integration over the other. ∎

Exercises for Section 4.7

1. Evaluate each of the following:

(a) $\iint_D xy$, $D = \{(x, y) \mid 0 \le x \le 1, 0 \le y \le x\}$

(b) $\iint_D y \sin(x)$, $D = \{(x, y) \mid 0 \le x \le 2y, 0 \le y \le \pi/4\}$
(c) $\iint_D y^2 \cos(yx)$, $D = \{(x, y) \mid x^2 \le y \le x, 0 \le y \le 1\}$
(d) $\iint_D (x + y)$, $D = \{(x, y) \mid 0 \le x \le 1, 1 \le y \le 2x\}$
(e) $\iint_D \cos(xy)$, $D = \{(x, y) \mid 0 \le x \le 1, 0 \le y \le \pi/4\}$

2. Redo the problems of Exercise 1 with all possible orders of integration.

3. Show that $\int_0^1 \int_0^{x-1/2} x\, dy\, dx$ is not a double integral of x over a region D. *Hint:* Suppose that $\int_0^1 \int_0^{x-1/2} x\, dy\, dx = \iint_D x$. What would D have to be? Find D, then write $\iint_D x$ as an iterated integral and compare with $\int_0^1 \int_0^{x-1/2} x\, dy\, dx$.

4. For each of the following regions and density functions, find the mass and center of mass.
(a) $D = \{(x, y) \mid 0 \le x \le 1, 0 \le y \le 1\}$, $\rho(x, y) = 2xy$
(b) $D = \{(x, y) \mid 1 \le x^2 + y^2 \le 2\}$, $\rho(x, y) = 3$
(c) $D = \{(x, y) \mid x^2 + y^2 \le 1, 0 \le x \le 1\}$, $\rho(x, y) = 1$
(d) $D = \{(x, y) \mid 2 \le x^2 + y^2 \le 4\}$, $\rho(x, y) = 2xy$
(e) $D = \{(x, y) \mid x^2 + y^2 \le x, \frac{1}{4} \le x \le \frac{3}{4}\}$, $\rho(x, y) = 1$

5. Suppose that we have a flat plate occupying a region D of the plane, with density function ρ. Let L be a straight line in the plane. The moment of inertia I_L of D about L is defined to be $\iint_D \rho(x, y) d^2(x, y)$, where $d(x, y)$ = perpendicular distance from (x, y) to L for each (x, y) in D (see Fig. 4-13). Let L_0 be parallel to L and through the center of mass of D, and let R be the distance between L and L_0. Prove that $I_L = I_{L_0} + mR^2$, where m = mass of D.

6. Verify the result of Exercise 5 in each of the following cases by computing I_L, I_{L_0}, and mR^2.

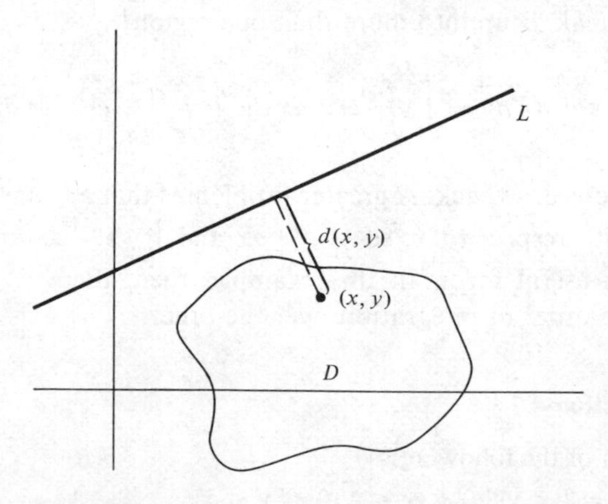

Fig. 4-13

(a) D is the triangle with vertices $(0, 0)$, $(1, 1)$, $(0, 1)$; L is the line $x = 1$; $\rho(x, y) = 1$.
(b) D is the square with vertices $(0, 0)$, $(0, 1)$, $(1, 0)$, $(1, 1)$; L is the line $y = 4$; $\rho(x, y) = 1$.
(c) D as in (b), L is the line $x + y = 8$, $\rho(x, y) = 2$.

7. Find the volume of each of the following regions by integration:
(a) $D = \{(x, y, z) \mid 0 \leq x^2 + y^2 + z^2 \leq 1, z \geq 0\}$ (hemisphere)
(b) $D = \{(x, y, z) \mid z^2 \geq x^2 + y^2, 0 \leq z \leq 1\}$(cone)
(c) $D = \{(x, y, z) \mid z \geq x^2 + y^2, 0 \leq z \leq 1\}$(parabolic bowl)
(d) $D = \{(x, y, z) \mid x^2 + y^2 \leq 1, 0 \leq z \leq 1\}$(cylinder)

8. Find the center of mass of the region bounded by $x^{2n} + y^{2n} = 1$, $x \geq 0$ and $y \geq 0$, if $\rho(x, y) = 1$. What happens to the region and to the center of mass if you let $n \to \infty$?

4.8 Evaluation of Double Integrals by Iterated Integrals—An Analytic Approach

The purpose of this short section is to state and prove a theorem that gives some mathematical backing to the geometrically based argument of the last section, which said that certain double integrals can be evaluated in terms of iterated single integrals.

Theorem

Let α and β be continuous on the interval $[a, b]$, and let D be the set

$$\{(x, y) \mid a \leq x \leq b \text{ and } \alpha(x) \leq y \leq \beta(x)\}.$$

Let f be continuous on D. Then

$$\iint_D f = \int_a^b \left(\int_{\alpha(x)}^{\beta(x)} f(x, y)\, dy \right) dx.$$

PROOF

Enclose D in a rectangle R, say $R = \{(x, y) \mid a \leq x \leq b \text{ and } c \leq y \leq d\}$. Put $F(p) = f(p)$ if p is in D, and $F(p) = 0$ if p is in R but not in D. Put, for each x in $[a, b]$, $g(x) = \int_{\alpha(x)}^{\beta(x)} f(x, y)\, dy$, and observe that also $g(x) = \int_{\alpha(x)}^{\beta(x)} F(x, y)\, dy$. The idea now is to compare Riemann sums for the integral of g over $[a, b]$ with Riemann sums for $\iint_R F$.

Form a Riemann sum $\sum_{i=1}^{n} g(\xi_i)(x_i - x_{i-1})$ for g by partitioning: $a = x_0 < x_1 < \cdots < x_n = b$ and choosing ξ_i in $[x_{i-1}, x_i]$, $i = 1, \ldots, n$. Next, partition $[c, d]: c = y_0 < y_1 < \cdots < y_m = d$. These two partitions determine a grid G over R. Now, F is discontinuous in R only possibly at points $(x, \alpha(x))$ and $(x, \beta(x))$, $a \leq x \leq b$, and these make up a set of zero area in R.

Now use a mean value theorem for single integrals to write

$$g(\xi_i) = \int_c^d F(\xi_i, y)\, dy = \sum_{j=1}^{m} \int_{y_{j-1}}^{y_j} F(\xi_i, y)\, dy = \sum_{j=1}^{m} F(\xi_i, \eta_j)(y_j - y_{j-1})$$

for some η_j in $[y_{j-1}, y_j]$, $j = 1, \ldots, m$. Then the Riemann sum for g becomes

$$\sum_{i=1}^{n} \sum_{j=1}^{m} F(\xi_i, \eta_j)(y_j - y_{j-1})(x_i - x_{i-1}),$$

which is a Riemann sum for F on R. We conclude that $\iint_R F = \int_a^b g(x)\, dx$. Since $\iint_R F = \iint_D f$, we then have $\iint_D f = \int_a^b \left(\int_{\alpha(x)}^{\beta(x)} f(x, y)\, dy \right) dx$, as we wanted to show. ∎

Exercises for Section 4.8

State and prove a theorem similar to that given in this section, but providing for evaluation of $\iint_D f$ in terms of an iterated integral $\int_c^d \int_{a(y)}^{b(y)} f(x, y)\, dx\, dy$.

4.9 Multiple Integrals in *n* Dimensions

The development of the Riemann integral in higher dimensions exactly parallels that for $n = 2$. We shall sketch the ideas for $n = 3$, and leave higher values of n to the student's imagination. First we need a notion of volume which is more solidly based than that used in Section 4.7. We parallel the development of Section 4.3. Suppose that D is a bounded set in R^3, so that, for each (x, y, z) in D,

$$a \le x \le b,$$

$$c \le y \le d,$$

and

$$t \le z \le r.$$

Partition these intervals to form a grid G over D consisting of cubical boxes. The mesh of G is the length of the diagonal of the box of maximum volume (the volume of such a box being the product of the lengths of its sides). These boxes fall into three classes:

(1) Those having no points in common with D.
(2) Those containing only interior points of D.
(3) Those containing at least one boundary point of D.

Ignore grid boxes in (1). The sum of the volumes of the boxes in (2) is called the inner volume of D, relative to G, and is denoted $\mathscr{I}(D, G)$. Finally, $\mathscr{I}(D, G)$

plus the sum of the volumes of cubicals in (3) is the outer volume of D relative to G, denoted $\mathcal{O}(D, G)$. The inner volume $\underline{V}(D)$ is the least upper bound, taken over all grids G over D, of the numbers $\mathcal{I}(D, G)$. The outer volume $\overline{V}(D)$ is the greatest lower bound, taken over all grids G over D, of the numbers $\mathcal{O}(D, G)$. If $\underline{V}(D) = \overline{V}(D)$, we say that D *has volume*, and we define this volume to be $V(D) = \underline{V}(D)$. If $\underline{V}(D) < \overline{V}(D)$, we say that D does not have volume. Note that D has volume if and only if ∂D has zero volume.

We can now develop the notion of a triple integral of a bounded function over a cube D in R^3. Associate with each grid G over D an upper and lower Darboux sum, given respectively by

$$\overline{S}(G) = \sum_{i=1}^{k} M_i V(R_i) \qquad \text{and} \qquad \underline{S}(G) = \sum_{i=1}^{k} m_i V(R_i),$$

where $M_i = \text{lub}\{f(x) \mid x \text{ is in } R_i\}$ and $m_i = \text{glb}\{f(x) \mid x \text{ is in } R_i\}$, and $R_1, \ldots, R_k$ are the boxes in the grid. The lower integral of a bounded function f over D is given by $\underline{I}(f) = \text{lub}\{\underline{S}(G) \mid G \text{ is a grid over } D\}$, and the upper integral by $\overline{I}(f) = \text{glb}\{\overline{S}(G) \mid G \text{ is a grid over } D\}$. If $\underline{I}(f) = \overline{I}(f)$, then we say that f is *integrable* over D, and denote $\underline{I}(f)$ by the symbol $\iiint_D f$, or $\iiint_D f(x, y, z)\, dx\, dy\, dz$.

If now D is any bounded set of points in R^3, and D has volume, then we enclose D in a cube R and put $g(x) = f(x)$ for x in D, and $g(x) = 0$ for x in R but not in D. We then define $\iiint_D f = \iiint_R g$ when the latter exists.

As the reader can see, the last few paragraphs are virtually identical, with an extension to one more dimension, of the development of area and double integrals given before. We have left out a number of details included in the previous sections, but these should be easy enough for the reader to fill in.

In a similar vein, it does not take too much imagination to expect that, under reasonable conditions on f and D, $\iiint_D f$ should be accessible as a three-fold iterated integral, and this is exactly the case. The three-dimensional analogue of the theorem of the last section is this: If $D = \{(x, y, z) \mid a \le x \le b, \alpha(x) \le y \le \beta(x), \text{ and } \varphi(x, y) \le z \le \psi(x, y)\}$, with α and β continuous on $[a, b]$, and φ and ψ continuous for $a \le x \le b$, $\alpha(x) \le y \le \beta(x)$, and if f is continuous on D, then

$$\iiint_D f = \int_a^b \left(\int_{\alpha(x)}^{\beta(x)} \left[\int_{\varphi(x,y)}^{\psi(x,y)} f(x, y, z)\, dz \right] dy \right) dx.$$

In all, there may be six different orders of integration available for $\iiint_D f$, depending upon possible descriptions of D. For example, if D consists of (x, y, z) with $c \le z \le d$, $p_1(z) \le x \le p_2(z)$, and $q_1(x, z) \le y \le q_2(x, z)$, then

$$\iiint_D f = \int_c^d \left(\int_{p_1(z)}^{p_2(z)} \left[\int_{q_1(x,z)}^{q_2(x,z)} f(x, y, z)\, dy \right] dx \right) dz,$$

assuming, of course, certain continuity conditions on the functions involved. As with double integrals, the parentheses are often omitted, leaving

$$\int_a^b \int_{\alpha(x)}^{\beta(x)} \int_{\varphi(x,y)}^{\psi(x,y)} f(x, y, z)\, dx\, dy\, dz.$$

One also sometimes encounters such notation as

$$\int_a^b dx \int_{\alpha(x)}^{\beta(x)} dy \int_{\varphi(x,y)}^{\psi(x,y)} f(x, y, z)\, dz$$

for the same iterated integral.

EXAMPLE

Evaluate $\iiint_D xyz\, dx\, dy\, dz$, where D is the region bounded by the hemisphere $x^2 + y^2 + z^2 = 1$, $z \geq 0$. We can write D as the set of points (x, y, z) with $-1 \leq x \leq 1$, $-\sqrt{1 - x^2} \leq y \leq \sqrt{1 - x^2}$, $0 \leq z \leq \sqrt{1 - x^2 - y^2}$. Then

$$\iiint_D xyz = \int_{-1}^{1} \int_{-\sqrt{1-x^2}}^{\sqrt{1-x^2}} \int_0^{\sqrt{1-x^2-y^2}} xyz\, dz\, dy\, dx.$$

The inner integral is

$$\int_0^{\sqrt{1-x^2-y^2}} xyz\, dz = \frac{xyz^2}{2}\bigg|_0^{\sqrt{1-x^2-y^2}} = \frac{xy}{2} - \frac{x^3y}{2} - \frac{xy^3}{2}.$$

Next we compute

$$\int_{-\sqrt{1-x^2}}^{\sqrt{1-x^2}} \int_0^{\sqrt{1-x^2-y^2}} xyz\, dz\, dy = \int_{-\sqrt{1-x^2}}^{\sqrt{1-x^2}} \left(\frac{xy}{2} - \frac{x^3y}{2} - \frac{xy^3}{2}\right) dy$$
$$= \left(\frac{xy^2}{4} - \frac{x^3y^2}{4} - \frac{xy^4}{8}\right)\bigg|_{-\sqrt{1-x^2}}^{\sqrt{1-x^2}} = 0.$$

Hence

$$\iiint_D xyz = \int_{-1}^{1} 0\, dx = 0. \qquad \blacksquare$$

EXAMPLE

Evaluate $\iiint_D xy \sin(z)$ over the cube $0 \leq x \leq 1$, $0 \leq y \leq 1$, $0 \leq z \leq 1$. Here

$$\iiint_D xy \sin(z) = \int_0^1 \int_0^1 \int_0^1 xy \sin(z)\, dz\, dy\, dx$$
$$= \int_0^1 x\, dx \int_0^1 y\, dy \int_0^1 \sin(z)\, dz = \tfrac{1}{2} \cdot \tfrac{1}{2}[1 - \cos(1)].$$

Note that the iterated integrals split into a product of three integrals. This is because the limits of integration do not contain any of the variables involved in the integration, and the function being integrated is a product of functions of the individual variables. ▮

EXAMPLE

Evaluate $\iiint_D x\cos(z)$, where D consists of points on and inside the cylinder $x^2 + y^2 = 1$, $0 \le z \le 1$. Here D consists of points (x, y, z) with $0 \le z \le 1$ and $x^2 + y^2 \le 1$, or, put another way,

$$-1 \le x \le 1, \qquad -\sqrt{1 - x^2} \le y \le \sqrt{1 - x^2}, \qquad 0 \le z \le 1.$$

Then

$$\iiint_D x\cos(z) = \int_{-1}^{1}\int_{-\sqrt{1-x^2}}^{\sqrt{1-x^2}}\int_0^1 x\cos(z)\,dz\,dy\,dx$$

$$= \int_0^1 \cos(z)\,dz \int_{-1}^{1}\int_{-\sqrt{1-x^2}}^{\sqrt{1-x^2}} x\,dy\,dx = \sin(1)\left[\int_{-1}^{1} (xy)\Big|_{-\sqrt{1-x^2}}^{\sqrt{1-x^2}}\,dx\right]$$

$$= \sin(1)\left(\int_{-1}^{1} 2x\sqrt{1 - x^2}\,dx\right)$$

$$= 2\sin(1)\left(\frac{-1}{3}\right)(1 - x^2)^{3/2}\Big|_{-1}^{1} = 0. \qquad ▮$$

This is all that we shall say for now about evaluation of multiple integrals. Later we shall develop an extremely powerful change-of-variables technique. However, this must be put off until after we have developed partial derivatives.

Exercises for Section 4.9

1. Give an example of a set D in R^3 which is bounded but has no volume.

2. Let D be a bounded set in the plane. Suppose that D has area, and let $D^* = \{(x, y, 0) \mid (x, y) \text{ is in } D\}$. Prove that $V(D^*) = 0$.

3. Let D be a bounded set in R^3.
   (a) Prove that $\underline{V}(D) < \overline{V}(D)$.
   (b) Prove that D has volume if and only if $V(\partial D) = 0$.

4. Let D be a bounded set having volume in R^3, and f a bounded function defined on D.
   (a) Prove that, for any grids G and G' over D, $\underline{S}(G) \le \overline{S}(G')$.
   (b) Prove that $\underline{I}(f) \le \overline{I}(f)$.
   (c) Give an example of f and D such that $\underline{I}(f) < \overline{I}(f)$.

(d) Prove that f is Riemann integrable over D if and only if, given $\varepsilon > 0$, there exists a grid G over D such that $\bar{S}(G) - \underline{S}(G) < \varepsilon$.

5. Develop the triple integral from the point of view of Riemann sums, and prove that this formulation is equivalent to that given in terms of Darboux sums.

6. Prove the theorem stated in this section, equating a triple integral of f over D to a threefold iterated integral.

7. Evaluate the following triple integrals:

(a) $\iiint_D e^x y \sin(z)$, D the unit cube $0 \leq x \leq 1, 0 \leq y \leq 1, 0 \leq z \leq 1$.

(b) $\iiint_D (x + y + z)$, D the region bounded by the coordinate planes and the plane $x + y + z = 1$.

(c) $\iiint_D 1$, D the region given by $x^2 + y^2 + z^2 \leq 1, x \geq 0, y \geq 0$.

(d) $\iiint_D x^2 + y$, D the region given by $x^2 + y^2 \leq 2, 0 \leq z \leq 1$.

4.10 Improper Multiple Integrals

Just as integrals of the form $\int_a^\infty f(x)\,dx$ arise in the one-variable case, it sometimes happens that multiple integrals of continuous functions must be attempted over unbounded regions. As an example from quantum mechanics, consider the problem of determining the motion of an electron about a nucleus. In spherical coordinates, the wave function $\psi(r, \theta, \varphi)$ has the interpretation that $\iiint_V |\psi|^2\,dr\,d\theta\,d\varphi$ is the probability that the electron is in V. Since the probability is 1 that the particle is somewhere in R^3, then we need $\iiint_{R^3} |\psi|^2 = 1$. This condition plays a vital role in determining certain physical constants associated with the motion.

For simplicity, we shall begin with improper double integrals. The difficulties encountered, and the ways around them, are similar for higher-dimension integrals. Note first that a connected region in the plane can be unbounded in many more ways than an interval on the line. In R^1 the only unbounded intervals are of the form $[a, \infty)$, (a, ∞), $(-\infty, a]$, $(-\infty, a)$, and R^2 itself. In R^2, however, things are not as predictable. For example, the regions

$$\{(x, y) \mid x \geq 0, y \geq 0\}$$

$$\{(x, y) \mid -1 \leq x \leq 1, y \geq 0\}$$

$$\{(x, y) \mid -x < y < x, x \geq 0\}$$

and

$$\{(x, y) \mid x^2 + y^2 \geq 2\}$$

are all unbounded, but have no apparent other features in common. This means that we cannot expect to handle $\iint_D f$ for unbounded D with as simple a limit as $\lim_{b\to\infty} \int_a^b f(x)\,dx = \int_a^\infty f(x)\,dx$.

One means of approach is to define C_r as the circular region $x^2 + y^2 \leq r^2$. As $r \to \infty$, the discs C_r spread out over the whole plane in the sense that eventually any given point will lie in some C_r. Given an unbounded region D, eventually some C_R will touch the region D. For each $r \geq R$, C_r will then have points of D in it. Let K_r consist of the points common to C_r and D for $r \geq R$. As $r \to \infty$, K_r will in the limit cover all of D, but for any given r, K_r is a bounded region. If $\lim_{r \to \infty} \iint_{K_r} f$ exists finite, we then define this limit to be $\iint_D f$, and say that $\iint_D f$ *converges*.

A subtle difficulty in this procedure is that conceivably a different value for $\iint_D f$ could be obtained by choosing regions other than discs. If, say, we let S_r be the region bounded by the square with corner vertices at (r, r), $(-r, r)$, $(r, -r)$, $(-r, -r)$, and T_r the set of points common to S_r and D (for values of r such that S_r and D intersect), then it is not entirely obvious that $\lim_{r \to \infty} \iint_{T_r} f = \lim_{r \to \infty} \iint_{K_r} f$. This is analogous to the problem of the Cauchy principal value for single improper integrals.

In fact, it is not always true that the value we get for $\iint_D f$ is the same whether we use discs or squares in the limit process described above. However, in the exercises we indicate a proof that the same value is achieved if $f(x, y) \geq 0$ on D. With this assumption, the same value will also be obtained using certain other shapes of regions in the limit, but this is a very involved theorem to develop in thoroughly acceptable mathematical form, and we shall not attempt to do so here.

We pause to illustrate the ideas developed so far.

EXAMPLE

Evaluate $\iint_D e^{-x-y}\, dx\, dy$, where D consists of all points in the first quadrant. Put another way,

$$D = \{(x, y) \mid x \geq 0 \text{ and } y \geq 0\}.$$

For any $r > 0$, let S_r consist of all points bounded by the square with vertices at (r, r), $(r, -r)$, $(-r, r)$, $(-r, -r)$. If T_r consists of all points common to S_r and D, then (Fig. 4-14)

$$T_r = \{(x, y) \mid 0 \leq x \leq r \text{ and } 0 \leq y \leq r\}.$$

Now

$$\begin{aligned} \iint_{T_r} e^{-x}e^{-y}\, dx\, dy &= \int_0^r \int_0^r e^{-x}e^{-y}\, dx\, dy \\ &= \int_0^r e^{-x}\, dx \int_0^r e^{-y}\, dy \\ &= (-e^{-r} + 1)(-e^{-r} + 1). \end{aligned}$$

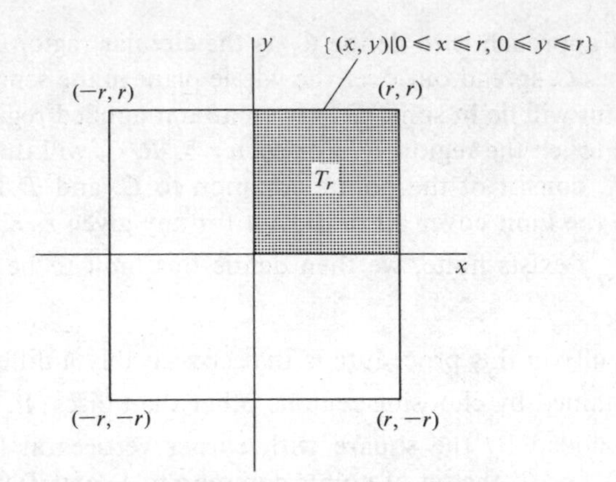

Fig. 4-14

As $r \to \infty$, $1 - e^{-r} \to 1$, so $\iint_{T_r} e^{-x-y} \to 1$. Then $\iint_D e^{-x-y} = 1$. We could use discs here, but not as conveniently. If

$$S_r = \{(x, y) \mid x^2 + y^2 \leq r^2\}$$

and K_r consists of the points common to S_r and D (see Fig. 4-15), then

$$K_r = \{(x, y) \mid x^2 + y^2 \leq r^2 \text{ and } x \geq 0 \text{ and } y \geq 0\}.$$

Compute

$$\iint_{K_r} e^{-x-y} = \int_0^r \int_0^{\sqrt{r^2-x^2}} e^{-x-y}\, dy\, dx.$$

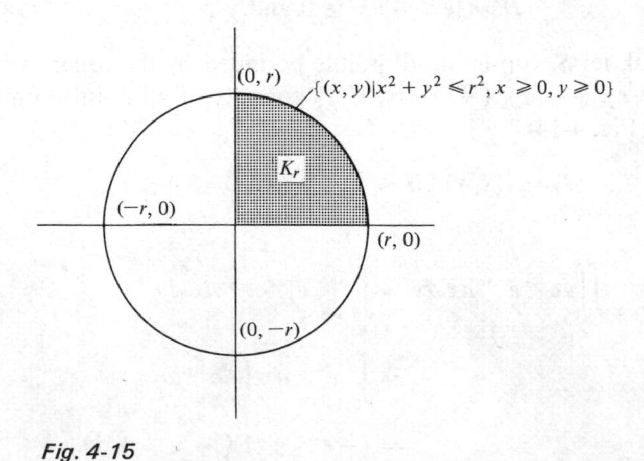

Fig. 4-15

Now

$$\int_0^{\sqrt{r^2-x^2}} e^{-x}e^{-y}\,dy = e^{-x}(-e^{-y})\Big|_0^{\sqrt{r^2-x^2}} = e^{-x}(1 - e^{-\sqrt{r^2-x^2}}),$$

so

$$\iint_{K_r} e^{-x-y} = \int_0^r e^{-x}(1 - e^{-\sqrt{r^2-x^2}})\,dx = \int_0^r e^{-x}\,dx - \int_0^r e^{-x-\sqrt{r^2-x^2}}\,dx.$$

This integration is far less pleasant than the one countered using squares. Immediately $\int_0^r e^{-x}\,dx = 1 - e^{-r} \to 1$ as $r \to \infty$. For the other integral, put $x = rt$ to obtain

$$\int_0^r e^{-x-\sqrt{r^2-x^2}}\,dx = \int_0^1 e^{-rt-\sqrt{r^2-r^2t^2}}r\,dt = r\int_0^1 e^{-r(t+\sqrt{1-t^2})}\,dt.$$

Now the minimum value of $t + \sqrt{1-t^2}$ on $[0, 1]$ is $\sqrt{2}$. Then

$$e^{-r(t+\sqrt{1-t^2})} \leq e^{-\sqrt{2}r}$$

for $0 \leq t \leq 1$. Then

$$0 \leq \int_0^r e^{-x-\sqrt{r^2-x^2}}\,dx = r\int_0^1 e^{-r(t+\sqrt{1-t^2})}\,dt \leq r\int_0^1 e^{-\sqrt{2}r}\,dt = re^{-\sqrt{2}r}.$$

But $re^{-\sqrt{2}r} \to 0$ as $r \to \infty$, and we conclude that

$$\int_0^r e^{-x-\sqrt{r^2-x^2}}\,dx \to 0$$

as $r \to \infty$. Note that we have evaluated the limit without actually performing the integration. At any rate, this gives us $\lim_{r\to\infty} \iint_{K_r} e^{-x-y} = 1$, the same result derived above (more easily) using squares. ∎

EXAMPLE

Consider $\iint_D e^{-x}\sin(y)\,dx\,dy$, where

$$D = \{(x, y) \mid x \geq 0 \text{ and } 0 \leq y \leq \pi\}.$$

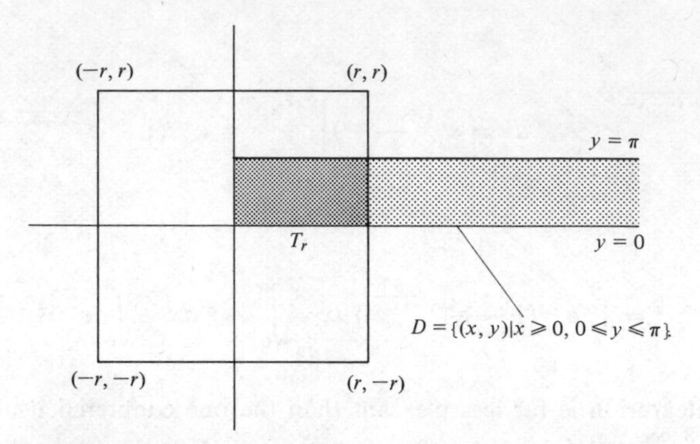

Fig. 4-16

If we use squares, with S_r and T_r as in the above discussion (Fig. 4-16), then

$$\iint_{T_r} e^{-x}\sin(y) = \int_0^r \int_0^\pi e^{-x}\sin(y)\,dy\,dx = 2\int_0^r e^{-x}\,dx = 2 - 2e^{-r}.$$

As $r \to \infty$, $2 - 2e^{-r} \to 2$, and we are led to put $\iint_D e^{-x}\sin(y) = 2$.

Use of circles here incurs a nasty integration, as in the last example. Here K_r is the crosshatched region of Fig. 4-17, and, for $r > \pi$,

$$\iint_{K_r} e^{-x}\sin(y)\,dy\,dx = \int_0^\pi \int_0^{\sqrt{r^2-y^2}} e^{-x}\sin(y)\,dx\,dy.$$

Now

$$\int_0^{\sqrt{r^2-y^2}} e^{-x}\sin(y)\,dx = -e^{-x}\sin(y)\Big|_0^{\sqrt{r^2-y^2}} = \sin(y) - e^{-\sqrt{r^2-y^2}}\sin(y),$$

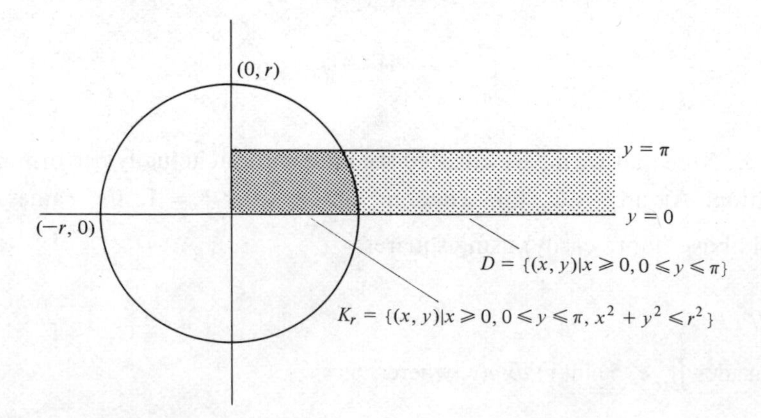

Fig. 4-17

and $\int_0^\pi e^{-\sqrt{r^2-y^2}} \sin(y)\,dy$ is not obvious. Again, however, we do not really want the value of the integral but the limit of the integral as $r \to \infty$. Make an estimate:

$$|e^{-\sqrt{r^2-y^2}} \sin(y)| \leq e^{-\sqrt{r^2-\pi^2}}$$

for $0 \leq y \leq \pi$, so

$$0 \leq \int_0^\pi e^{-\sqrt{r^2-y^2}} \sin(y)\,dy \leq \int_0^\pi e^{-\sqrt{r^2-\pi^2}}\,dy = \pi e^{-\sqrt{r^2-\pi^2}} \to 0 \qquad \text{as } r \to \infty.$$

Hence $\int_0^\pi e^{-\sqrt{r^2-y^2}} \sin(y)\,dy \to 0$ as $r \to \infty$, and we have

$$\iint_D e^{-x} \sin(y) = \lim_{r \to \infty} \int_0^\pi \sin(y)\,dy = 2,$$

as above. ∎

At this point we should remark that life is not as difficult as we have made it seem. In the last two examples, we encountered, in turn,

$$\lim_{r \to \infty} \int_0^1 re^{-r(t+\sqrt{1-t^2})}\,dt$$

and

$$\lim_{r \to \infty} \int_0^\pi e^{-\sqrt{r^2-y^2}} \sin(y)\,dy,$$

which we evaluated with some effort. If, however, we could have interchanged $\int$ and $\lim_{r \to \infty}$, things would have been much easier, since

$$\lim_{r \to \infty} re^{-r(t+\sqrt{1-t^2})} = \lim_{r \to \infty} e^{-\sqrt{r^2-y^2}} \sin(y) = 0,$$

and we would then have had $\int_0^1 0\,dt$ and $\int_0^\pi 0\,dy$ in the above. Interchange of a limit and an integral is analogous to interchange of a limit and an infinite sum, and will be considered in some detail in Chapter 7 when we discuss functions defined in terms of parameters that appear in integrals.

We conclude this section with a brief mention of type 2 improper double integrals and improper integrals in higher dimensions. First, the latter. Suppose that we want to consider $\iiint_D f$, where D is an unbounded region in R^3.

The basic idea is to proceed as in the two-dimensional case, with spheres replacing circles and cubes replacing squares. In dimensions 4, 5, or higher, these methods extend immediately.

Finally, we consider the analogue in higher dimensions of the type 2 improper integral of Chapter 1. Above, we were looking at $\iint_D f$, with D unbounded but f

continuous on D. Now we want to consider the case that D is bounded, but f is not. We shall simply illustrate the idea with an example and leave it to the reader to draw the obvious conclusions about how to proceed in general.

EXAMPLE

Consider

$$\iint_D \frac{1}{\sqrt{x+y}}. \qquad \text{where } D = \{(x, y) \mid 0 \le x \le 1 \text{ and } 0 \le y \le 1\}.$$

The problem is that $1/\sqrt{x+y}$ is unbounded in D. The idea is to enclose the trouble point, $(0, 0)$, in a square S_r, as shown, with $r < 1$. Denote by A_r the region inside D and outside S_r (shaded in the diagram, Fig. 4-18), evaluate $\iint_{A_r} 1/\sqrt{x+y}$, and then let $r \to 0$. If this limit exists, we call this limit the value of $\iint_D 1/\sqrt{x+y}$. If not, we say that $\iint_D 1/\sqrt{x+y}$ diverges. Now

$$\begin{aligned}
\iint_{A_r} \frac{1}{\sqrt{x+y}} &= \int_0^1 \int_r^1 \frac{1}{\sqrt{x+y}}\, dy\, dx + \int_r^1 \int_0^r \frac{1}{\sqrt{x+y}}\, dy\, dx \\
&= \int_0^1 2\sqrt{x+y}\,\Big|_r^1\, dx + \int_r^1 2\sqrt{x+y}\,\Big|_0^r\, dx \\
&= \int_0^1 (2\sqrt{x+1} - 2\sqrt{x+r})\, dx + \int_r^1 (2\sqrt{x+r} - 2\sqrt{x})\, dx \\
&= \tfrac{4}{3}(x+1)^{3/2}\big|_0^1 - \tfrac{4}{3}(x+r)^{3/2}\big|_0^1 + \tfrac{4}{3}(x+r)^{3/2}\big|_r^1 - \tfrac{4}{3}x^{3/2}\big|_r^1 \\
&= \tfrac{4}{3}(2^{3/2}) - \tfrac{8}{3} + \tfrac{8}{3}r^3 - \tfrac{4}{3}(2^{3/2})r^{3/2} \to \tfrac{4}{3}(2^{3/2} - 2) \qquad \text{as } r \to 0.
\end{aligned}$$

Thus $\iint_D 1/\sqrt{x+y}$ converges, and we assign to it the value $\frac{4}{3}(2^{3/2} - 2)$, or, equivalently, $\frac{8}{3}(\sqrt{2} - 1)$. ∎

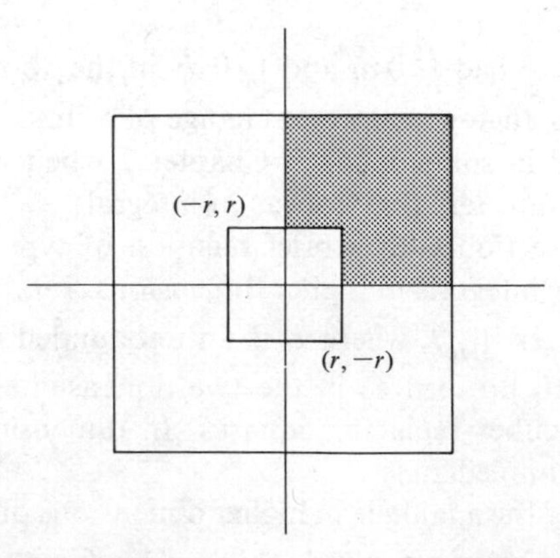

Fig. 4-18

Exercises for Section 4.10

1. Determine convergence or divergence of each of the following:

(a) $\iint_D xy \sin(y/x)$, $D = \{(x, y) \mid x^2 + y^2 \leq 1\}$

(b) $\iint_D e^{-x^2-y^2}$, $D = \{(x, y) \mid x \geq 0, y \geq 0\}$

(c) $\iiint_D e^{-x^2-y^2-z^2}$, $D = \{(x, y, z) \mid 1 \leq x \leq 3, 1 \leq y \leq 3, z \geq 0\}$

(d) $\iint_V \cos(x)e^{-xy}$, $V = \{(x, y) \mid y \geq 0, 0 \leq x \leq 2\pi\}$

2. Prove the assertion made in the text that the limit of $\iint_{K_r} f$ as $r \to \infty$, using discs, is the same as the limit of $\iint_{T_r} f$ as $r \to \infty$, using squares, if $f(x, y) \geq 0$ in D and if both limits exist finite. *Hint:* Let $A_r = \iint_{K_r} f$ and $B_r = \iint_{T_r} f$. Show first that $\{A_r\}_{r=1}^{\infty}$ and $\{B_r\}_{r=1}^{\infty}$ are monotone nondecreasing. Let $A = \lim_{r\to\infty} A_r$ and $B = \lim_{r\to\infty} B_r$.

Next, show that, given n, then for some N, $A_n \leq B_N$. Hence show that $A \leq B$. Finally, given m, for some M, $B_m \leq A_M$. Hence $B \leq A$.

3. Suppose that f and g are continuous on D, an unbounded region in the plane, and that $0 \leq f(x, y) \leq g(x, y)$ for (x, y) in D.

(a) Does convergence of $\iint_D g$ imply convergence of $\iint_D f$?

(b) Does divergence of $\iint_D f$ imply divergence of $\iint_D g$?

5

Partial Differentiation

5.1 Partial Derivatives

Imagine that we are attempting to describe changes in a system defined by specifying one variable in terms of n others, say $w = f(x_1, \ldots, x_n)$. It might be of interest to know how w changes if just one variable, say x_i, is changed. If we want an instantaneous rate of change of w with x_i, understanding that all the other variables are held fixed, then it would seem natural to just differentiate with respect to x_i and treat the other x_j's as constants at the fixed values. This is called partial differentiation of w with respect to x_i.

In order to be more precise, let us first consider the two-variable case, say $z = f(x, y)$. The *partial derivative with respect to* x at (a, b) is defined to be

$$\lim_{\Delta x \to 0} \frac{f(a + \Delta x) - f(a, b)}{\Delta x}$$

if this limit exists finite, and is denoted

$$\frac{\partial f}{\partial x}(a, b), \left.\frac{\partial f}{\partial x}\right|_{(a,b)}, f_1(a, b), f_x(a, b), z_1(a, b), \text{ or } z_x(a, b).$$

There are other notations as well, but these should suffice.

Similarly, the *partial derivative with respect to* y *at* (a, b) is

$$\lim_{\Delta y \to 0} \frac{f(a, b + \Delta y) - f(a, b)}{\Delta y}$$

if this limit exists. Notations for this derivative include

$$\frac{\partial f}{\partial y}(a, b), \left.\frac{\partial f}{\partial y}\right|_{(a,b)}, f_2(a, b), f_y(a, b), z_2(a, b), \text{ or } z_y(a, b).$$

From the limit definitions, it is obvious that we compute $\partial f(a, b)/\partial x$ by differentiating with respect to x according to the usual rules, carrying y along as a constant, and then substituting $x = a$, $y = b$. A similar remark holds for $\partial f(a, b)/\partial y$. If we are thinking of these derivatives as new functions of x and y, without reference to a specific point (a, b), then we write just $\partial f/\partial x$, $\partial f/\partial y$, f_1, f_2, f_x, and so on.

EXAMPLE

If $f(x, y) = e^x \sin(y^2x)$, then

$$\frac{\partial f}{\partial x} = e^x \sin(y^2x) + y^2e^x \cos(y^2x)$$

and

$$\frac{\partial f}{\partial y} = 2yxe^x \cos(y^2x).$$

At $(0, \pi)$ we have

$$\frac{\partial f(0, \pi)}{\partial x} = \pi^2 \quad \text{and} \quad \frac{\partial f(0, \pi)}{\partial y} = 0. \quad \blacksquare$$

This idea of a partial derivative extends to as many variables as we need. For example, if $w = f(x, y, z)$, then

$$\frac{\partial w}{\partial x} = \lim_{\Delta x \to 0} \frac{f(x + \Delta x, y, z) - f(x, y, z)}{\Delta x}.$$

$$\frac{\partial w}{\partial y} = \lim_{\Delta y \to 0} \frac{f(x, y + \Delta y, z) - f(x, y, z)}{\Delta y}.$$

and

$$\frac{\partial w}{\partial z} = \lim_{\Delta z \to 0} \frac{f(x, y, z + \Delta z) - f(x, y, z)}{\Delta z}$$

whenever the limits exist finite.

In the one-variable case, a derivative may be thought of geometrically as a slope of a tangent to a curve. A similar interpretation applies to partial derivatives of functions of two variables. Suppose that $z = f(x, y)$. Think of the graph of z as a surface or locus of points $(x, y, f(x, y))$ in R^3, and suppose that $\partial f(a, b)/\partial x$ exists.

If we fix y at b, then $z = g(x) = f(x, b)$ may be thought of as a curve on the surface in the plane $y = b$ [i.e., the curve of intersection of the plane $y = b$ with the surface $z = f(x, y)$]. The slope of the tangent to this curve at $(a, b, f(a, b))$ is

$$g'(a) = \lim_{\Delta x \to 0} \frac{g(a + \Delta x) - g(a)}{\Delta x} = \lim_{\Delta x \to 0} \frac{f(a + \Delta x, b) - f(a, b)}{\Delta x} = \frac{\partial f}{\partial x}(a, b)$$

(see Fig. 5-1).

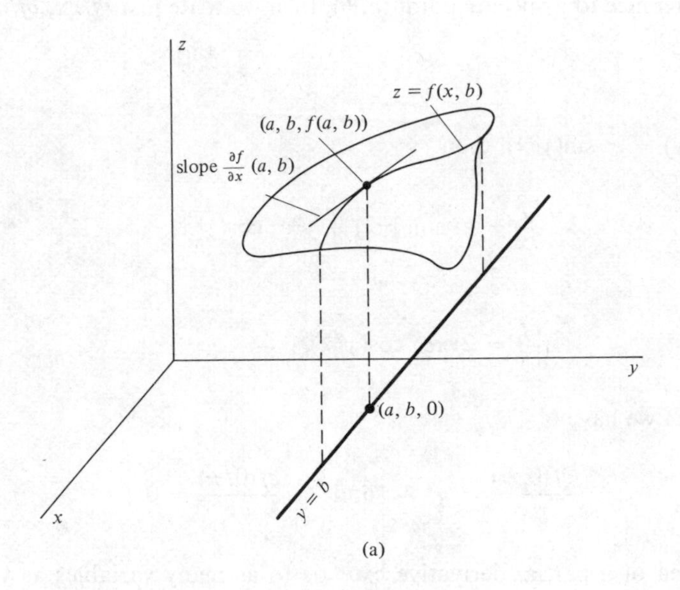

(a)

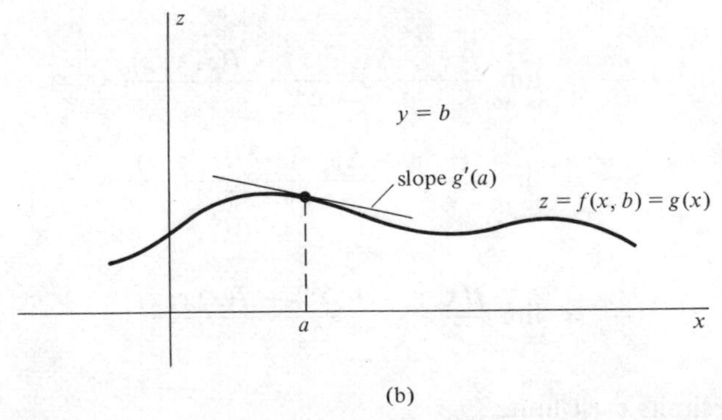

(b)

Fig. 5-1

Similarly, if we fix x at a, then $z = f(a, y)$ is a curve on the surface in the plane $x = a$, and $\partial f(a, b)/\partial y$ is the slope of the tangent at the point $(a, b, f(a, b))$.

Basically, then, what we are calculating with $\partial f/\partial x$ and $\partial f/\partial y$ (in this two-variable case) are slopes of curves lying on a surface in planes parallel to the xz- and yz-planes, respectively.

In general, if f is a function of n-variables, and $\partial f/\partial x_i$ is defined, then $\partial f/\partial x_i$ is again a function of n variables, and we can attempt to partially differentiate it, say with respect to x_j. This second partial $\partial(\partial f/\partial x_i)/\partial x_j$ is usually written $\partial^2 f/\partial x_j\, \partial x_i$, or sometimes f_{ij}. In many instances, we can continue indefinitely, obtaining $\partial^3 f/\partial x_k\, \partial x_j\, \partial x_i$, and so on. If $i = j$, we usually write $\partial^2 f/\partial x_i\, \partial x_i$ as $\partial^2 f/\partial x_i^2$ and $\partial^3 f/\partial x_k\, \partial x_i\, \partial x_i$ as $\partial^3 f/\partial x_k\, \partial x_i^2$. Similarly, we have $\partial^3 f/\partial x_i^2\, \partial x_k$, $\partial^4 f/\partial x_i^2\, \partial x_j^2$, $\partial^5 f/\partial x_i^3\, \partial x_j\, \partial x_k$, and so on. In subscript notation for partials, we might write these as $f_{x_i x_i}$, $f_{x_i x_i x_k}$, $f_{x_k x_i x_i}$, $f_{x_j x_j x_i x_i}$, $f_{x_k x_j x_i x_i x_i}$, respectively.

Exercises for Section 5.1

1. Compute the indicated partial derivatives.

(a) $f(x, y) = e^x \sin(y);\ \dfrac{\partial^2 f}{\partial x^2}, \dfrac{\partial^2 f}{\partial y^2}, \dfrac{\partial^2 f}{\partial x\, \partial y}, \dfrac{\partial^2 f}{\partial y\, \partial x}$

(b) $f(x, y, z) = \dfrac{xy}{x^2 + y^2 + z^2};\ \dfrac{\partial^3 f}{\partial x\, \partial y\, \partial z}(1, 1, 1)$

(c) $g(x, y) = \arccos(xy);\ \dfrac{\partial^2 g}{\partial x\, \partial y}$

(d) $h(x, y, z, w) = e^{xyzw};\ \dfrac{\partial^5 h}{\partial x\, \partial y\, \partial w\, \partial z^2}$

2. Let

$$f(x, y) = \begin{cases} \dfrac{xy}{x^2 + y^2} & \text{for } (x, y) \neq (0, 0), \\ 0 & \text{for } x = y = 0. \end{cases}$$

Does $f_x(0, 0)$ or $f_y(0, 0)$ exist?

3. Let $f(x, y) = (x^2 + y^2)^{1/2}$.

(a) Compute $f_x(0, 0)$ and $f_y(0, 0)$.

(b) Investigate the existence of $\partial^2 f(0, 0)/\partial x^2$, $\partial^2 f(0, 0)/\partial y^2$, $\partial^2 f(0, 0)/\partial x\, \partial y$, and $\partial^2 f(0, 0)/\partial y\, \partial x$.

4. Let $f(x, y) = \begin{cases} \dfrac{xy(x^2 - y^2)}{x^2 + y^2} & \text{for } x^2 + y^2 \neq 0, \\ 0 & \text{for } x = y = 0. \end{cases}$

(a) Compute $f_x(0, 0)$.

(b) Compute $f_y(0, 0)$.

(c) Compute $f_x(a, b)$ and $f_y(a, b)$ for $(a, b) \neq (0, 0)$.

(d) Compute $(f_x)_y(0, 0)$, and $(f_y)_x(0, 0)$.

(e) Compute $(f_x)_y(a, b)$ and $(f_y)_x(a, b)$ for $(a, b) \neq (0, 0)$.

5. Let $z = 3x^2 + 2y^2 + x + y = f(x, y)$ for x, y real.

(a) Find the equation of the line tangent to the curve $z = f(x, 3)$ at the point $(1, 3, 25)$.

(b) Find the equation of the line tangent to the curve $z = f(1, y)$ at (1, 3, 25).

(c) Find the equation of the tangent plane to $z = f(x, y)$ at (1, 3, 25). That is, find the equation of a plane containing the straight lines requested in (a) and (b).

6. Show that $u(x, y, z) = 1/\sqrt{x^2 + y^2 + z^2}$ is a solution to Laplace's equation: $\partial^2 u/\partial x^2 + \partial^2 u/\partial y^2 + \partial^2 u/\partial z^2 = 0$ for $(x, y, z) \neq (0, 0, 0)$.

7. Show that $A_x + B_y + C_z = 0$ for $(x, y, z) \neq (0, 0, 0)$, if

$$A(x, y, z) = \frac{x}{(x^2 + y^2 + z^2)^{3/2}},$$

$$B(x, y, z) = \frac{y}{(x^2 + y^2 + z^2)^{3/2}},$$

and

$$C(x, y, z) = \frac{z}{(x^2 + y^2 + z^2)^{3/2}}.$$

8. Let $f(x, y) = x^2 + xy^3$. Use the definition of partial derivative to determine $f_x(x_0, y_0)$, $f_y(x_0, y_0)$, $\partial^2 f(x_0, y_0)/\partial x^2$ and $\partial^3 f(x_0, y_0)/\partial x\, \partial y\, \partial x$.

9. Let $f(x, y) = e^x \sin(y)$. Compute $f_x(1, 1)$ and $f_y(1, 1)$ directly from the definition.

10. Show that $u(x, y) = \ln[(x - a)^2 + (y - b)^2]$ satisfies the two-dimensional Laplace equation: $\partial^2 u/\partial x^2 + \partial^2 u/\partial y^2 = 0$.

5.2 Partial Differential Equations of Physics

A large number of phenomena in science and engineering are studied by means of mathematical models in which functional relationships between the various quantities involved are constructed on the basis of observation. Since systems are generally observed in motion, the resulting equations generally involve derivatives. And, since more than one independent and dependent variable are often needed, these derivatives are often partial. It is for this reason that partial differential equations (i.e., equations involving partial derivatives) play such a vital role in modern technology.

In this section we shall attempt to give two elementary illustrations of the use of partial derivatives in setting up mathematical descriptions of physical phenomena. The processes we shall look at are simple cases of wave propagation and heat flow.

A. One-Dimensional Wave Equation

Suppose an elastic string is pegged down at both ends, then set in some initial configuration and released with a given initial velocity and allowed to vibrate freely in a vertical plane. The problem is to accurately describe the ensuing motion of the string.

To set up a mathematical model, we first suppose that the string is thin and perfectly elastic. Draw a set of axes and place the fixed ends of the string at

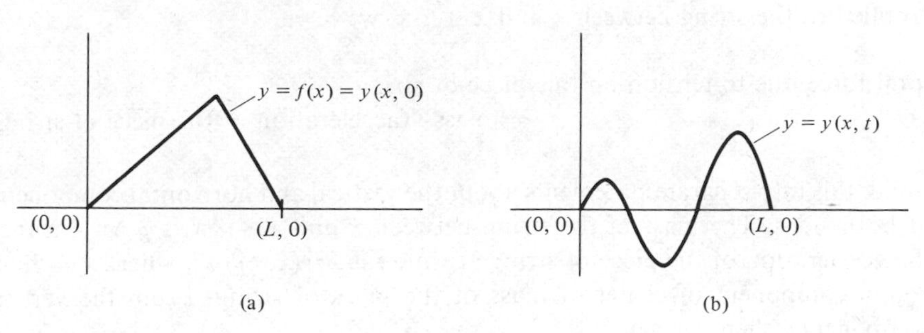

Fig. 5-2

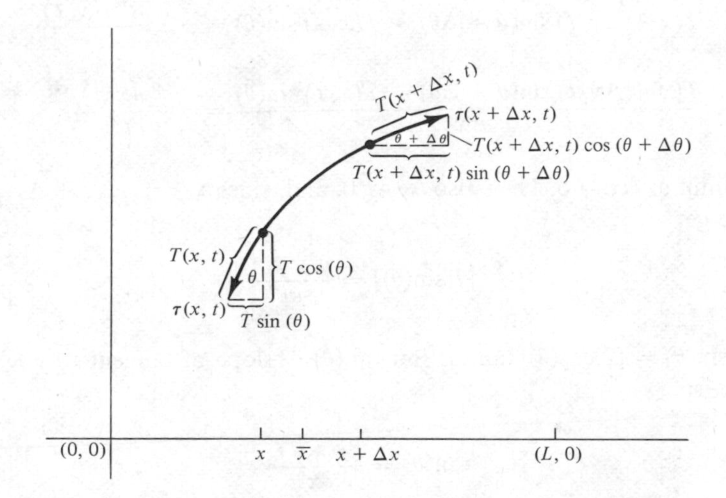

Fig. 5-3

(0, 0) and (0, L), with the string stretched tightly between. Assume that damping effects such as air resistance are negligible, and that, once stretched to an initial position and set in motion, the string is confined to a vertical plane.

We begin the mathematical part by supposing that some function $y = f(x)$ gives the initial stretched position of the string. At time t after the release, suppose that the particle originally at $(x, f(x))$ is at some point $(x, y(x, t))$ (see Figs. 5-2a and 5-2b). Let $T(x, t)$ denote the magnitude of the tension $\tau(x, t)$ in the string at time t and point $(x, y(x, t))$, and suppose that $\tau(x, t)$ always acts tangent to the string at $(x, y(x, t))$. Finally, let ρ be the mass per unit length of the string, and suppose that ρ is constant. What we hope to derive is an equation for y, the function that will tell us the shape of the string at any time t. To do this, look at a piece of string, say between x and $x + \Delta x$, at time t (see Fig. 5-3).

Newton's law of motion says that

$$\text{force} = \text{mass} \cdot \text{acceleration}.$$

Applied to the string between x and $x + \Delta x$, we have

total force due to tension on this piece of string
= mass $\cdot$ (acceleration of the piece of string).

Break this into separate statements about the vertical and horizontal components of both sides. The mass of the string between x and $x + \Delta x$ is $\rho\,\Delta x$. Further, the acceleration of the piece of string at time t is $\partial^2 y(\bar{x}, t)/\partial t^2$, where $\bar{x}$ = horizontal component of center of mass of the piece of string. From the vertical components, then, we get

$$T(x + \Delta x, t)\sin(\theta + \Delta\theta) - T(x, t)\sin(\theta) = \rho\,\Delta x\,\frac{\partial^2 y(\bar{x}, t)}{\partial t^2},$$

or

$$\frac{T(x + \Delta x, t)\sin(\theta + \Delta\theta) - T(x, t)\sin(\theta)}{\Delta x} = \rho\,\frac{\partial^2 y(\bar{x}, t)}{\partial t^2}.$$

Take a limit as $\Delta x \to 0$. Then also $\Delta\theta \to 0$, and, since $x < \bar{x} < x + \Delta x$, $\bar{x} \to x$. Then we get

$$\frac{\partial}{\partial x}[T\sin(\theta)] = \frac{\partial^2 y(x, t)}{\partial t^2}.$$

Now $T\sin(\theta) = [T\cos(\theta)]\tan(\theta)$. But $\tan(\theta)$ = slope of tangent to $y = y(x, t)$ at time t, so

$$\tan(\theta) = \frac{\partial y(x, t)}{\partial x}.$$

Then

$$\frac{\partial}{\partial x}\left[T\cos(\theta)\frac{\partial y}{\partial x}\right] = \rho\,\frac{\partial^2 y}{\partial t^2}.$$

Now look at horizontal components in Newton's law. We have no acceleration in the horizontal direction (by assumption in the model), so

$$T(x + \Delta x, t)\cos(\theta + \Delta\theta) - T(x, t)\cos(\theta) = 0.$$

Then

$$T(x + \Delta x, t)\cos(\theta + \Delta\theta) = T(x, t)\cos(\theta),$$

which is to say that $T(x, t)\cos(\theta)$ is independent of x, and does not change as x is varied. Then

$$\frac{\partial}{\partial x}\left[T\cos(\theta)\frac{\partial y}{\partial x}\right] = T\cos(\theta)\frac{\partial^2 y}{\partial x^2} = \rho\,\frac{\partial^2 y}{\partial t^2}.$$

Putting the constant $T\cos(\theta)$ equal to α, we have

$$\frac{\partial^2 y}{\partial x^2} = \frac{\rho}{\alpha}\frac{\partial^2 y}{\partial t^2},$$

which is the one-dimensional wave equation.

The partial differential equation above does not convey all the information we have about y, nor does it specify y uniquely. We also have the boundary conditions

$$y(0, t) = y(L, t) = 0,$$

which tell us that the ends of the string are held fixed. In addition, we have the initial shape of the string: $y(x, 0) = f(x)$. Finally, we must know the initial velocity:

$$\frac{\partial y(x, 0)}{\partial t} = g(x).$$

The differential equation, together with the boundary and initial conditions, completely determines $y(x, t)$ for $0 \leq x \leq L$, $t \geq 0$, hence completely describes the motion of the string. We shall obtain the solution by Fourier methods in Chapter 10.

B. Heat Conduction in a Thin Solid Rod

Suppose that we have a solid, straight rod of length L, made of some homogeneous material, and having uniform cross section (e.g., circular). Heat energy is introduced into the rod, but we assume that there is no heat loss through the sides.

For convenience in setting up the model, align the rod along the x-axis between $(0, 0)$ and $(L, 0)$. We shall assume that the temperature is the same throughout any given cross section at any given time t, and so depends only on x and t. (Experimentally, this assumption works well when the rod is much longer than it is wide.) We denote the temperature at time t in the cross section at x as $u(x, t)$.

We now derive a differential equation for u. Begin with an observation. If we place two flat plates of the same shape and area A, at a distance d apart, and if the temperatures on the plates are T_1 and T_2, then the amount of heat per unit time passing from the warmer to the cooler plate is directly proportional to A and the temperature difference $|T_1 - T_2|$, and inversely proportional to d. Letting K be the constant of proportionality, this says that

$$\frac{\text{amount of heat}}{\text{unit time}} = \frac{KA|T_1 - T_2|}{d}.$$

The constant K depends on the material in the rod.

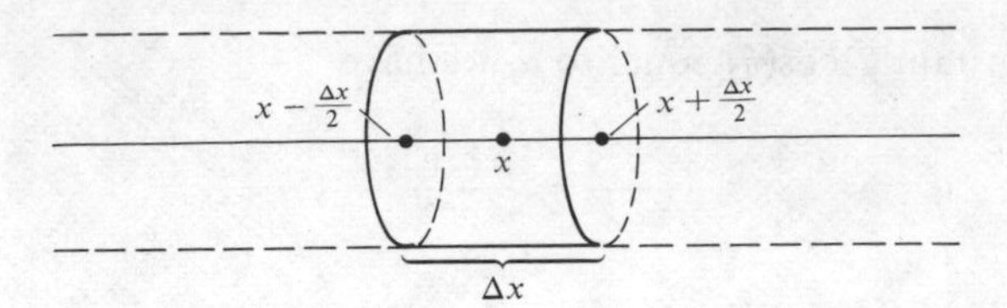

Fig. 5-4

To apply this observation to our problem, consider as parallel plates the cross sections of rod at points $x - \Delta x/2$ and $x + \Delta x/2$ (see Fig. 5-4). The amount of heat per unit time at time t flowing from left to right between these cross sections is

$$\frac{-KA(u(x + \Delta x/2, t) - u(x - \Delta x/2, t))}{\Delta x}.$$

The minus sign in front acknowledges that heat will flow from left to right when $u(x - \Delta x/2, t)$ is greater than $u(x + \Delta x/2, t)$.

Take a limit as Δx goes to 0 to obtain the instantaneous rate of heat flow $H(x, t)$ across the cross section at x at time t. Here we encounter a practical difficulty, however. This limit is not so easy, because it does not look like the limit in the definition of partial derivative. Recall that

$$\frac{\partial u(x, t)}{\partial x} = \lim_{h \to 0} \frac{u(x + h, t) - u(x, t)}{h}.$$

But note that the increment h here can approach zero through both positive and negative values. So write

$$\begin{aligned}\frac{u(x + \Delta x/2, t) - u(x - \Delta x/2, t)}{\Delta x} &= \frac{u(x + \Delta x/2, t) - u(x, t)}{\Delta x} \\ &\quad - \frac{u(x - \Delta x/2, t) - u(x, t)}{\Delta x} \\ &= \frac{\frac{1}{2}[u(x + \Delta x/2, t) - u(x, t)]}{\Delta x/2} \\ &\quad - \frac{\frac{1}{2}[u(x - \Delta x/2, t) - u(x, t)]}{\Delta x/2}.\end{aligned}$$

In the limit, as $\Delta x \to 0$,

$$\frac{\frac{1}{2}[u(x + \Delta x/2, t) - u(x, t)]}{\Delta x/2} \to \frac{1}{2}\frac{\partial u(x, t)}{\partial x}$$

(simply take $h = \Delta x/2$), and also

$$-\frac{\frac{1}{2}[u(x - \Delta x/2, t) - u(x, t)]}{\Delta x/2} \to \frac{1}{2}\frac{\partial u(x, t)}{\partial x}$$

(take $h = -\Delta x/2$). Thus

$$\begin{aligned} H(x, t) &= \lim_{\Delta x \to 0} \frac{-KA[u(x + \Delta x/2, t) - u(x - \Delta x/2, t)]}{\Delta x} \\ &= -KA\left(\frac{1}{2}\frac{\partial u}{\partial x} + \frac{1}{2}\frac{\partial u}{\partial x}\right) = -KA\frac{\partial u(x, t)}{\partial x}. \end{aligned}$$

Now look at increments in $H(x, t)$ due to increments in x:

$H(x, t) - H(x + \Delta x, t) =$ instantaneous rate of heat flow from left to right across the section at $(x + \Delta x)$ − instantaneous rate of heat flow from left to right across the section at x

= net rate of flow of heat into the segment of rod between x and $x + \Delta x$

$$= -KA\frac{\partial u(x, t)}{\partial x} + KA\frac{\partial u(x + \Delta x, t)}{\partial x}$$

$$= KA\left(\frac{\partial u(x + \Delta x, t)}{\partial x} - \frac{\partial u(x, t)}{\partial x}\right).$$

Next, use the fact that the average change in temperature Δu in a time interval Δt in the segment of rod from x to $x + \Delta x$ is directly proportional to the amount of heat $(H(x, t) - H(x + \Delta x, t)) \cdot \Delta t$ and inversely proportional to the mass Δm of the segment. Let Q be the constant of proportionality. Then

$$\Delta u = \frac{Q(H(x, t) - H(x + \Delta x, t))\,\Delta t}{\Delta m}.$$

Now $\Delta m = \rho A\,\Delta x$, where A = area of cross section and ρ = density. (Both A and ρ are constant by assumption.) Then substituting from above, we have

$$\begin{aligned} \Delta u &= \frac{QKA\left(\frac{\partial u(x + \Delta x, t)}{\partial x} - \frac{\partial u(x, t)}{\partial x}\right)\Delta t}{\rho A\,\Delta x} \\ &= \frac{QK\left(\frac{\partial u(x + \Delta x, t)}{\partial x} - \frac{\partial u(x, t)}{\partial x}\right)\Delta t}{\Delta x}. \end{aligned}$$

It is physically plausible that the average change Δu in u over the time interval Δt of the segment from x to $x + \Delta x$ should be the actual temperature change at a point ξ between x and $x + \Delta x$. Then for some ξ,

$$u(\xi, t + \Delta t) - u(\xi, t) = \frac{\frac{QK}{\rho}\left(\frac{\partial u(x + \Delta x, t)}{\partial x} - \frac{\partial u(x, t)}{\partial x}\right)\Delta t}{\Delta x}$$

or

$$\frac{u(\xi, t + \Delta t) - u(\xi, t)}{\Delta t} = \frac{\frac{QK}{\rho}\left(\frac{\partial u(x + \Delta x, t)}{\partial x} - \frac{\partial u(x, t)}{\partial x}\right)}{\Delta x}$$

Take a limit as $\Delta t \to 0$ and as $\Delta x \to 0$. Since ξ is between x and $x + \Delta x$, then $\xi \to x$ and we get

$$\frac{\partial u}{\partial t} = \frac{QK}{\rho}\frac{\partial^2 u}{\partial x^2}.$$

This is the heat equation for an insulated homogeneous rod of uniform cross section.

As with the wave equation, the heat equation alone does not completely determine $u(x, t)$. We need the initial temperature at each point: $u(x, 0) = f(x)$, together with a mathematical statement of the condition that there is no heat transfer across the ends of the rod, owing to the insulation hypothesis. This condition takes the form

$$0 = \frac{\partial u(0, t)}{\partial x} = \frac{\partial u(L, t)}{\partial x}.$$

We shall see other examples of partial differential equations of mathematical physics later. These two were given here to illustrate the use of partial derivatives in setting up a mathematical description of a physical problem, and to provide some experience in working with partial derivatives as instantaneous rates of change with respect to specified variables.

5.3 Differentiability, Differentials, and Tangent Planes

A function f of one variable is said to be differentiable at a if $f'(a)$ exists. In this event, $f'(a)$ is the slope of the tangent to $y = f(x)$ at $x = a$. For functions of two or more variables, we do not have a derivative in the usual sense, so we rely on geometric considerations to arrive at a suitable notion of differentiability. By analogy with the one-variable case, we say, on a temporary informal basis, that a function of two variables is differentiable at (a, b) if the surface $z = f(a, b)$ has a tangent plane at $(a, b, f(a, b))$.

This raises the question as to what is meant by a tangent plane. In one variable, a line L is tangent to $y = f(x)$ at P if, for any other point Q on the graph, the angle α between the line PQ and L approaches zero as $Q \to P$ along

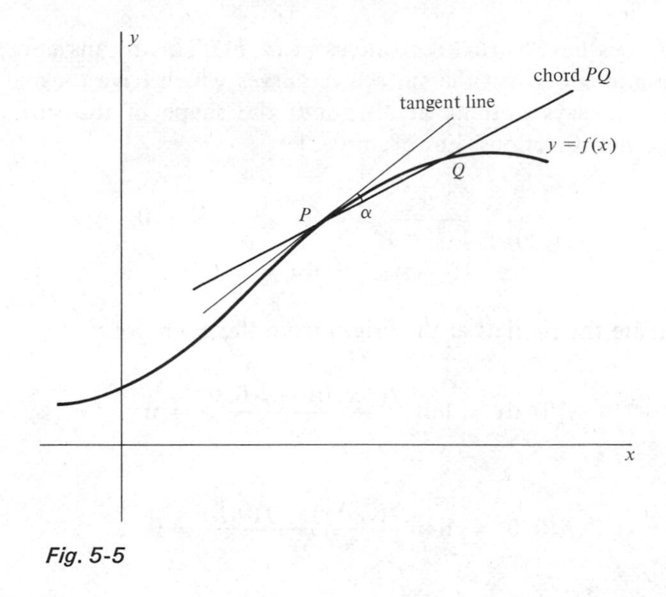

Fig. 5-5

the graph (Fig. 5-5). For a surface S, we say that a plane T intersecting S at P is tangent to S at P if, for any point Q on S, the angle between the line PQ and T goes to zero as $Q \to P$ from any direction on S. Intuitively, this says that T fits closer to S as we near P from any direction.

One implication of this is that we would expect $z = f(x, y)$ to be continuous at every point where its graph has a tangent plane. Continuity, however, is not sufficient for a tangent plane to exist. For example, $z = \sqrt{x^2 + y^2}$ has no tangent plane at the origin but is continuous there (Fig. 5-6). In this example, z has no partial derivatives at (0, 0).

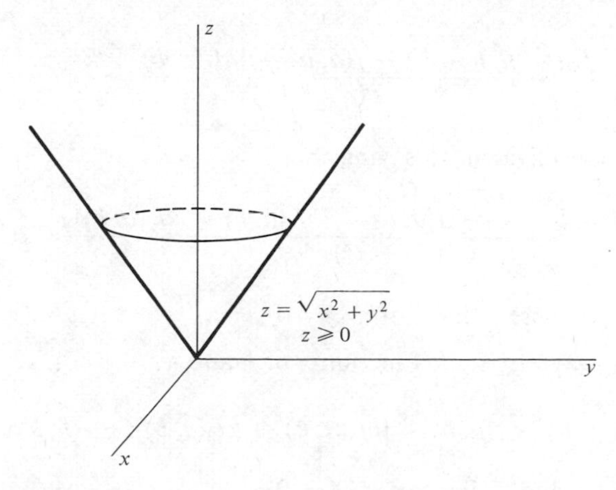

Fig. 5-6

What if f does have partial derivatives at (a, b)? This means only that the planes $x = a$ and $y = b$ cut the surface in curves which have tangent lines at $(a, b, f(a, b))$, and says nothing at all about the shape of the surface when viewed from other directions. For example, let

$$f(x, y) = \begin{cases} \dfrac{xy}{x^2 + y^2} & \text{for } x^2 + y^2 \neq 0, \\ 0 & \text{for } x = y = 0. \end{cases}$$

We can calculate the partials at the origin from the definition:

$$f_1(0, 0) = \lim_{\Delta x \to 0} \frac{f(\Delta x, 0) - f(0, 0)}{\Delta x} = 0$$

and

$$f_2(0, 0) = \lim_{\Delta y \to 0} \frac{f(0, \Delta y) - f(0, 0)}{\Delta y} = 0.$$

Thus f has partials everywhere, even though, from Section 3.1, f is not continuous at $(0, 0)$. To show that $z = f(x, y)$ has no tangent plane at $(0, 0, 0)$, note that the x- and y-axes both lie on the graph of $z = f(x, y)$, so the tangent plane at the origin would have to be the xy-plane. But the xy-plane does not fit the surface very well near the origin. For example, the points $(x, x, \frac{1}{2})$ are all on the surface for $x \neq 0$, and are all $\frac{1}{2}$ unit away from the xy-plane, no matter how close x is to zero. The upshot of all this is that existence of partials at a point in no way guarantees either continuity or existence of a tangent plane at the point.

An analytic condition equivalent to the existence of a tangent plane, and which we shall take as our formal definition of differentiability, is this: Suppose that $f(x, y)$ is defined for all (x, y) in some r-neighborhood of (a, b). Suppose that $\partial f(a, b)/\partial x$ and $\partial f(a, b)/\partial y$ exist. Then f is *differentiable* at (a, b) if and only if

$$\lim_{(h,k) \to (0,0)} \frac{f(a + h, b + k) - f(a, b) - [hf_1(a, b) + kf_2(a, b)]}{(h^2 + k^2)^{1/2}} = 0.$$

This condition is equivalent to saying that

$$\frac{f(a + h, b + k) - f(a, b) - [hf_1(a, b) + kf_2(a, b)]}{(h^2 + k^2)^{1/2}} = \varepsilon(h, k)$$

for $h^2 + k^2 \neq 0$, where $\lim_{(h,k) \to (0,0)} \varepsilon(h, k) = 0$.

Put another way, for some function ε of h and k,

$$f(a + h, b + k) - f(a, b) - [hf_1(a, b) + kf_2(a, b)] = \varepsilon(h, k)\sqrt{h^2 + k^2}$$

for $(h, k) \neq (0, 0)$, and $\lim_{(h,k) \to (0,0)} \varepsilon(h, k) = 0$.

There are various other ways of writing the same condition. First, the term $\sqrt{h^2 + k^2}$ can be replaced by $|h| + |k|$ without changing the mathematical content. Another equivalent formulation is to assert the existence of $\varphi_1(h, k)$ and $\varphi_2(h, k)$ such that $\varphi_1(h, k) \to 0$ and $\varphi_2(h, k) \to 0$ as $(h, k) \to (0, 0)$, and

$$f(a + h, b + k) - f(a, b) = hf_1(a, b) + kf_2(a, b) + h\varphi_1(h, k) + k\varphi_2(h, k).$$

Finally, and this is purely a notational matter, often Δx and Δy are used to denote the increments h and k, respectively. When this is done, Δz, or Δf, is often used to denote the resulting increment $f(a + \Delta x, b + \Delta y) - f(a, b)$ in f. Then our original condition above becomes

$$\lim_{(\Delta x, \Delta y) \to (0,0)} \frac{\Delta z - f_1(a, b)\,\Delta x - f_2(a, b)\,\Delta y}{[(\Delta x)^2 + (\Delta y)^2]^{1/2}} = 0.$$

We shall defer to the exercises a rigorous proof of the equivalence of this analytic condition to the existence of a tangent plane. However, it might be helpful in understanding what the analytic condition says, to go back to the one-variable case, where $y = f(x)$ has a tangent line at a exactly when $f'(a)$ exists. The analytic expression for this is the existence of a number $f'(a)$ such that

$$\lim_{h \to 0} \frac{f(a + h) - f(a)}{h} = f'(a).$$

This says that

$$\lim_{h \to 0} \left(\frac{f(a + h) - f(a)}{h} - f'(a) \right) = 0.$$

Write

$$\frac{f(a + h) - f(a)}{h} - f'(a) = \frac{f(a + h) - f(a) - hf'(a)}{h}.$$

We then have a tangent line at $(a, f(a))$ exactly when

$$\lim_{h \to 0} \frac{f(a + h) - f(a) - hf'(a)}{h} = 0.$$

This is exactly the one-dimensional analogue of

$$\lim_{(h,k) \to (0,0)} \frac{f(a + h, b + k) - f(a, b) - (hf_1(a, b) + kf_2(a, b))}{(h^2 + k^2)^{1/2}} = 0.$$

In one variable, $f'(a) \cdot h$ is called the *differential* of f at a due to increment h, and is often denoted dy or df. It seems natural then to call $hf_1(a, b) + kf_2(a, b)$

the differential of f at (a, b) due to increments h in the x-direction and k in the y-direction. This is usually denoted dz or df. Note that, given (a, b), df is a function of the two variables h and k. In terms of increments and differentials, the differentiability condition can be written

$$\lim_{(h,k)\to(0,0)} \frac{\Delta f - df}{\sqrt{h^2 + k^2}} = 0.$$

This says that the differential df is a better approximation to Δf as h and k are chosen smaller, and that the difference $\Delta f - df$ goes to zero faster than $(h^2 + k^2)^{1/2}$. We shall see a geometric way of looking at this at the end of the section.

Very often in writing a differential, h is written as dx and k as dy. If $z = f(x, y)$, we then often write

$$dz = \frac{\partial f}{\partial x}\,dx + \frac{\partial f}{\partial y}\,dy.$$

This extends immediately to n variables. If $z = f(x_1, \ldots, x_n)$, then

$$dz = \frac{\partial f}{\partial x_1}\,dx_1 + \frac{\partial f}{\partial x_2}\,dx_2 + \cdots + \frac{\partial f}{\partial x_n}\,dx_n,$$

where $dx_1, \ldots, dx_n$ are arbitrary numbers and the partials $\partial f/\partial x_i$ are evaluated at some point $(a_1, \ldots, a_n)$ under consideration.

Likewise, the limit formulation of differentiability extends to higher dimensions: f is *differentiable* at $(a_1, \ldots, a_n)$ if

$$\lim_{(\Delta x_1,\ldots,\Delta x_n)\to(0,\ldots,0)} \frac{\Delta f - df}{((\Delta x_1)^2 + \cdots + (\Delta x_n)^2)^{1/2}} = 0,$$

where

$$\Delta f = f(a_1 + \Delta x_1, \ldots, a_n + \Delta x_n) - f(a_1, \ldots, a_n)$$

and

$$df = f_1(a_1, \ldots, a_n)\,\Delta x_1 + \cdots + f_n(a_1, \ldots, a_n)\,\Delta x_n.$$

We pause for an example.

Let $f(x, y) = 2x^2y$. Then f is differentiable at $(2, 3)$. To prove this, first calculate:

$$f_1(2, 3) = 4xy|_{(2,3)} = 24$$

and

$$f_2(2, 3) = 2x^2|_{(2,3)} = 8.$$

Then

$$df = 24\,\Delta x + 8\,\Delta y.$$

Next

$$\begin{aligned}\Delta f &= f(2 + \Delta x, 3 + \Delta y) - f(2, 3)\\ &= 2(2 + \Delta x)^2(3 + \Delta y) - 24\\ &= 6(\Delta x)^2 + 24\,\Delta x + 8\,\Delta y + 2(\Delta y)(\Delta x)^2 + 8(\Delta x)(\Delta y).\end{aligned}$$

Then

$$\Delta f - df = 6(\Delta x)^2 + 2(\Delta x)^2(\Delta y) + 8(\Delta x)(\Delta y).$$

Then we have to show that

$$\lim_{(\Delta x, \Delta y)\to(0,0)} \frac{6(\Delta x)^2 + 2(\Delta x)^2(\Delta y) + 8(\Delta x)(\Delta y)}{[(\Delta x)^2 + (\Delta y)^2]^{1/2}} = 0.$$

To do this, note first that, for $(\Delta x)^2 + (\Delta y)^2 \neq 0$,

$$\left|\frac{6(\Delta x)^2 + 2(\Delta x)^2(\Delta y) + 8(\Delta x)(\Delta y)}{\sqrt{(\Delta x)^2 + (\Delta y)^2}}\right|$$

$$\leq \frac{6(\Delta x)^2}{\sqrt{(\Delta x)^2 + (\Delta y)^2}} + \frac{2(\Delta x)^2|\Delta y|}{\sqrt{(\Delta x)^2 + (\Delta y)^2}} + \frac{8|\Delta x|\,|\Delta y|}{\sqrt{(\Delta x)^2 + (\Delta y)^2}}.$$

Now examine the last three terms individually:

$$\frac{6(\Delta x)^2}{\sqrt{(\Delta x)^2 + (\Delta y)^2}}\begin{cases}\leq 6|\Delta x| & \text{if } \Delta x \neq 0,\\ = 0 & \text{if } \Delta x = 0;\end{cases}$$

$$\frac{2(\Delta x)^2|\Delta y|}{\sqrt{(\Delta x)^2 + (\Delta y)^2}}\begin{cases}\leq 2|\Delta x|\,|\Delta y| & \text{if } \Delta x \neq 0,\\ = 0 & \text{if } \Delta x = 0;\end{cases}$$

$$\frac{8|\Delta x|\,|\Delta y|}{\sqrt{(\Delta x)^2 + (\Delta y)^2}} \leq \begin{cases}8|\Delta x| & \text{if } \Delta x \neq 0,\\ 8|\Delta y| & \text{if } \Delta y \neq 0.\end{cases}$$

In sum, then, certainly

$$\left|\frac{6(\Delta x)^2 + 2(\Delta x)^2(\Delta y) + 8(\Delta x)(\Delta y)}{\sqrt{(\Delta x)^2 + (\Delta y)^2}}\right| \leq 6|\Delta x| + 2|\Delta x|\,|\Delta y| + 8|\Delta x| + 8|\Delta y|.$$

Since $14|\Delta x| + 2|\Delta x|\,|\Delta y| + 8|\Delta y|$ can be made arbitrarily close to zero by choosing Δx and Δy small, we conclude that

$$\lim_{(\Delta x, \Delta y)\to(0,0)} \frac{\Delta f - df}{[(\Delta x)^2 + (\Delta y)^2]^{1/2}} = 0. \qquad \blacksquare$$

It is interesting to ask whether any other numbers might be substituted for $\partial f(a, b)/\partial x$ and $\partial f(a, b)/\partial y$ in the limit formulation of differentiability at (a, b). Suppose, for example, that

$$\lim_{(\Delta x, \Delta y) \to (0,0)} \left(\frac{f(a + \Delta x, b + \Delta y) - f(a, b) - A\,\Delta x - B\,\Delta y}{\sqrt{(\Delta x)^2 + (\Delta y)^2}} \right) = 0.$$

Choose $\Delta y = 0$ to obtain

$$\lim_{\Delta x \to 0} \left(\frac{f(a + \Delta x, b) - f(a, b) - A\,\Delta x}{|\Delta x|} \right) = 0.$$

If $\Delta x > 0$, this becomes

$$\lim_{\Delta x \to 0} \left(\frac{f(a + \Delta x, b) - f(a, b)}{\Delta x} - A \right) = 0.$$

If $\Delta x < 0$, we have

$$\lim_{\Delta x \to 0} \left(\frac{f(a + \Delta x, b) - f(a, b)}{-\Delta x} + A \right) = 0.$$

In either case, we have the limit defining $\partial f(a, b)/\partial x$, and we conclude that $A = \partial f(a, b)/\partial x$. Similar reasoning, with $\Delta y \neq 0$ and $\Delta x = 0$, yields $B = \partial f(a, b)/\partial y$.

In practice, combinations of elementary functions, such as the two-variable polynomial of the above example, are differentiable at points where they are well defined. For example, one would expect that $e^{x+y} \cos(x) \sin(y)$ is differentiable everywhere. A proof of this based directly on the definition is messy, and, in fact, even treatment of simple functions such as polynomials is cumbersome if we always have to resort to the limit argument. There is a criterion for differentiability which is much easier to apply in practice, but we need more machinery to prove it and so will hold off on this until Section 5.5.

We have already noted that we would expect a differentiable function to be continuous. We now give a proof of this for the n-variable case.

Theorem

If f is differentiable at $(a_1, \ldots, a_n)$, then f is continuous at $(a_1, \ldots, a_n)$.

PROOF

We must show that $f(a_1 + \Delta x_1, \ldots, a_n + \Delta x_n) - f(a_1, \ldots, a_n) \to 0$ as $(\Delta x_1, \ldots, \Delta x_n) \to (0, \ldots, 0)$. Since f is differentiable at $(a_1, \ldots, a_n)$, there is some $\varepsilon(\Delta x_1, \ldots, \Delta x_n)$ such that $\varepsilon(\Delta x_1, \ldots, \Delta x_n) \to 0$ as $(\Delta x_1, \ldots, \Delta x_n) \to (0, \ldots, 0)$ and

$$\Delta f - \sum_{i=1}^{n} \frac{\partial f(a_1, \ldots, a_n)}{\partial x_i} \Delta x_i = \varepsilon(\Delta x_1, \ldots, \Delta x_n) \left[\sum_{i=1}^{n} (\Delta x_i)^2 \right]^{1/2}.$$

As $(\Delta x_1, \ldots, \Delta x_n) \to (0, \ldots, 0)$, the right side tends to zero, and so does

$$\sum_{i=1}^{n} \frac{\partial f(a_1, \ldots, a_n)}{\partial x_i} \Delta x_i.$$

Hence also $\Delta f \to 0$, and f is continuous at $(a_1, \ldots, a_n)$. ∎

That the converse is not true has already been shown by the example $f(x, y) = \sqrt{x^2 + y^2}$ at $(0, 0)$.

We conclude this section with a practical problem. Assuming that the graph of $z = f(x, y)$ has a tangent plane at $(a, b, f(a, b))$, how can we find it?

Here vector methods come in handy. Let (x, y, z) be any point on the tangent plane T. Then $(x - a, y - b, z - f(a, b))$ may be thought of as the vector from $(a, b, f(a, b))$ to (x, y, z), hence a vector that lies in T.

Put this aside for a moment and recall the geometric meanings of $\partial f(a, b)/\partial x$ and $\partial f(a, b)/\partial y$. We saw that $\partial f(a, b)/\partial x$ is the slope of the tangent line to the curve $z = f(x, b)$ at $(a, b, f(a, b))$. The vector $(1, 0, \partial f(a, b)/\partial x)$ is in the direction of this tangent line, which must lie in T. Hence $(1, 0, \partial f(a, b)/\partial x)$ is parallel to T (see Fig. 5-7).

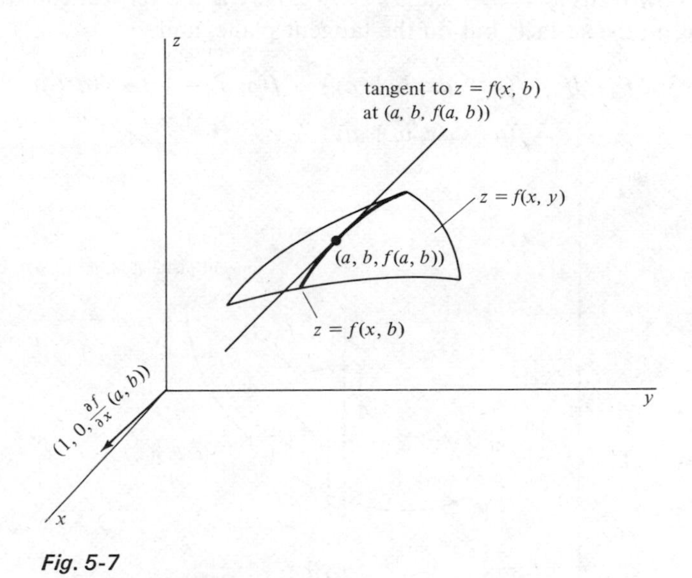

Fig. 5-7

By similar reasoning, the vector $(0, 1, \partial f(a, b)/\partial y)$ is parallel to the tangent line to $z = f(a, y)$ at $(a, b, f(a, b))$, hence also is parallel to T.

We now have two vectors parallel to T. Their cross product is

$$(1, 0, f_x(a, b)) \times (0, 1, f_y(a, b)) = (-f_x(a, b), -f_y(a, b), 1),$$

and this is normal to both $(1, 0, f_x(a, b))$ and $(0, 1, f_y(a, b))$, hence also to T, and, in particular, to $(x - a, y - b, z - f(a, b))$, which lies in T. Then

$$(-f_x(a, b), -f_y(a, b), 1) \cdot (x - a, y - b, z - f(a, b)) = 0.$$

This works out to

$$f_x(a, b)(x - a) + f_y(a, b)(y - b) = z - f(a, b).$$

Every point (x, y, z) on T satisfies this equation. Conversely, every point (x, y, z) satisfying this equation is on T. In this sense the above equation completely specifies T. The numbers $(-f_x(a, b), -f_y(a, b), 1)$ are often called *direction numbers* of T.

We can use the equation of the tangent plane to clarify the geometric sense in which df approximates Δf at a point. If we choose $dx = x - a$ and $dy = y - b$, then the differential df of f at (a, b) due to increments dx and dy is

$$\frac{\partial f(a, b)}{\partial x} \cdot (x - a) + \frac{\partial f(a, b)}{\partial y} \cdot (y - b),$$

which is exactly the left side of the tangent-plane equation. The tangent-plane equation now reads $df = z - f(a, b)$. Now z here is the vertical coordinate of a point, not on the surface, but on the tangent plane, and

$$\begin{aligned}\Delta f - df &= f(a + dx, b + dy) - f(a, b) - (z - f(a, b)) \\ &= f(a + dx, b + dy) - z\end{aligned}$$

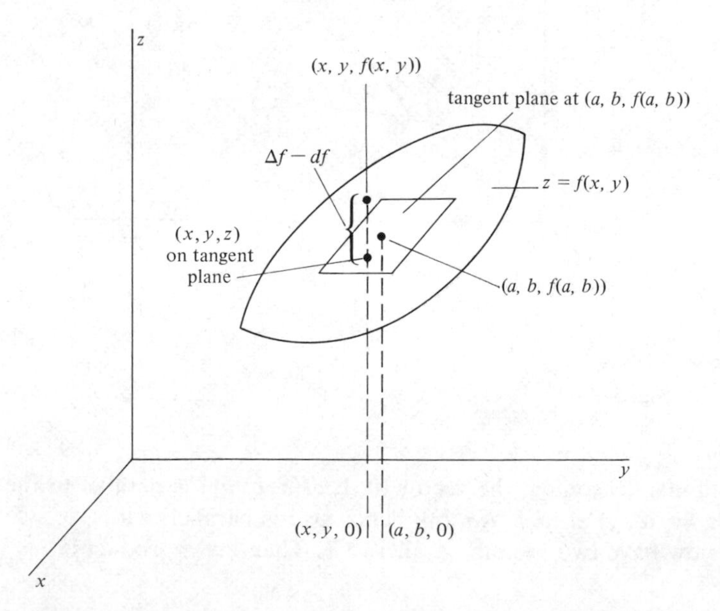

Fig. 5-8

is the vertical distance from the surface to the tangent plane above the point (x, y) in the xy-plane (Fig. 5-8). As might be expected from the definition of tangent plane, this distance becomes small fast as $(x, y) \to (a, b)$.

Exercises for Section 5.3

1. Find the tangent plane to each of the following surfaces at the indicated point. Also find a normal to the surface at the point.

(a) $z = e^x \cos(y)$, $(0, \pi, -1)$
(b) $z = x^2 + y^2$, $(1, 1, 2)$
(c) $z = \sqrt{x^2 + y^2}$, $(1, 1, \sqrt{2})$
(d) $z = \sqrt{4 - x^2 - y^2}$, $(1, 1, \sqrt{2})$
(e) $z = x + y + 2$, $(0, 0, 2)$

2. Prove that the following are equivalent:

(1) For some $\varepsilon(h, k)$, $\lim_{(h,k)\to(0,0)} \varepsilon(h, k) = 0$ and

$$f(a + h, b + k) - f(a, b) = hf_x(a, b) + kf_y(a, b) + \varepsilon(h, k)\sqrt{h^2 + k^2}.$$

(2) $$\lim_{(h,k)\to(0,0)} \frac{f(a + h, b + k) - f(a, b) - hf_x(a, b) - kf_y(a, b)}{\sqrt{h^2 + k^2}} = 0.$$

(3) $$\lim_{(h,k)\to(0,0)} \frac{f(a + h, b + k) - f(a, b) - hf_x(a, b) - kf_y(a, b)}{|h| + |k|} = 0.$$

(4) For some φ_1 and φ_2,

$$\lim_{(h,k)\to(0,0)} \varphi_1(h, k) = \lim_{(h,k)\to(0,0)} \varphi_2(h, k) = 0,$$

and

$$f(a + h, b + k) - f(a, b) = hf_x(a, b) + kf_y(a, b) + h\varphi_1(h, k) + k\varphi_2(h, k).$$

3. Prove that each of the following is differentiable at the given point:

(a) $f(x, y) = x^2 + y^2$, $(1, 2)$
(b) $f(x, y) = 2x + 3y + 6$, $(2, 4)$
(c) $f(x, y) = e^x y^2$, $(0, 1)$
(d) $f(x, y) = \sqrt{xy}$, $(2, 2)$

4. Let $f(x, y) = e^{xy} \sin(x)$. Show that

$$\lim_{(\Delta x,\Delta y)\to(0,0)} \frac{\Delta f - df}{\sqrt{(\Delta x)^2 + (\Delta y)^2}} = 0,$$

where Δf and df are calculated at the point $(1, 1)$.

5. Let $f(x, y, z) = x^2 + y^2 + z^2 + e^{xyz}$. Show that

$$\lim_{(\Delta x,\Delta y,\Delta z)\to(0,0,0)} \left(\frac{\Delta f - df}{\sqrt{(\Delta x)^2 + (\Delta y)^2 + (\Delta z)^2}}\right) = 0,$$

with Δf and df calculated at $(0, 2, 1)$.

6. Show that $f(x, y, z, w) = xyzw - 2xy^2 - z^2w$ is differentiable at $(0, 1, 1, \pi)$.

7. Show that $g(x_1, \ldots, x_n) = \sum_{i=1}^{n} x_i^2$ is differentiable at the origin $(0, \ldots, 0)$ for each positive integer n.

8. Let f and g be differentiable functions of two variables. Show that

(a) $d(fg) = f \cdot dg + g \cdot df$.

(b) $d\left(\frac{f}{g}\right) = \frac{g \cdot df - f \cdot dg}{g^2}$, provided that $g(x, y) \neq 0$.

9. Is $f(x, y) = |x + y|$ differentiable at $(0, 0)$?

10. Let $z = \sqrt{x^2 + y^2}$. We have already observed that z is not differentiable at the origin due to the fact that there is no tangent plane to the surface at $(0, 0, 0)$. Show that z is not differentiable at $(0, 0)$ analytically by showing that $\frac{\partial z}{\partial x}(0, 0)$ and $\frac{\partial z}{\partial y}(0, 0)$ do not exist.

11. In this exercise we outline a proof that the graph of $z = f(x, y)$ has a tangent plane at $(a, b, f(a, b))$ exactly when

$$\lim_{(\Delta x, \Delta y) \to (0,0)} \frac{\Delta f - \frac{\partial f}{\partial x}(a, b)\, \Delta x - \frac{\partial f}{\partial y}(a, b)\, \Delta y}{[(\Delta x)^2 + (\Delta y)^2]^{1/2}} = 0,$$

under the assumptions that f is continuous at (a, b) and that (a, b) is an interior point of the set over which f is defined. Denote $P_0 = (a, b, f(a, b))$. If $\mathscr{P}$ is a plane through P_0 and not parallel to the z-axis, then we can write $\mathscr{P}$ in the form $z - f(a, b) = A(x - a) + B(y - b)$, for some constants A and B. Let $P = (a + \Delta x, b + \Delta y, f(a + \Delta x, b + \Delta y))$, and show that the angle between P_0P and the normal to $\mathscr{P}$ is given by

$$\cos(\theta) = \frac{A\, \Delta x + B\, \Delta y - \Delta z}{[A^2 + B^2 + 1]^{1/2}[(\Delta x)^2 + (\Delta y)^2 + (\Delta z)^2]^{1/2}},$$

where $\Delta z = f(a + \Delta x, b + \Delta y) - f(a, b)$. Let ψ be the angle between P_0P and $\mathscr{P}$. Show that $\cos(\theta) = \sin(\psi)$. Now note that $\mathscr{P}$ is tangent to the surface at $(a, b, f(a, b))$ if and only if $\psi \to 0$ as $P \to P_0$. Show from the formula for $\cos(\theta)$ that this is equivalent to

$$\lim_{(\Delta x, \Delta y) \to (0,0)} \frac{\Delta z - (A\, \Delta x + B\, \Delta y)}{\sqrt{(\Delta x)^2 + (\Delta y)^2 + (\Delta z)^2}} = 0.$$

Now show that this is equivalent to

$$\lim_{(\Delta x, \Delta y)\to(0,0)} \frac{\Delta z - (A\,\Delta x + B\,\Delta y)}{[(\Delta x)^2 + (\Delta y)^2]^{1/2}} = 0.$$

5.4 Partial Derivatives of Composite Functions

In this section we consider a question of both theoretical and practical importance. Suppose that f and g are functions of two variables, defined on some set of points $\mathscr{D}$ of the (s, t)-plane. Suppose that F is a function of two variables defined on some set D of points in the (x, y)-plane, and that $(f(s, t), g(s, t))$ is in D whenever (s, t) is in $\mathscr{D}$. We can then consider the composite function H defined by $H(s, t) = F(f(s, t), g(s, t))$ for (s, t) in $\mathscr{D}$. The question is: What can be said about H, knowing certain things about f, g, and F? (See Fig. 5-9.)

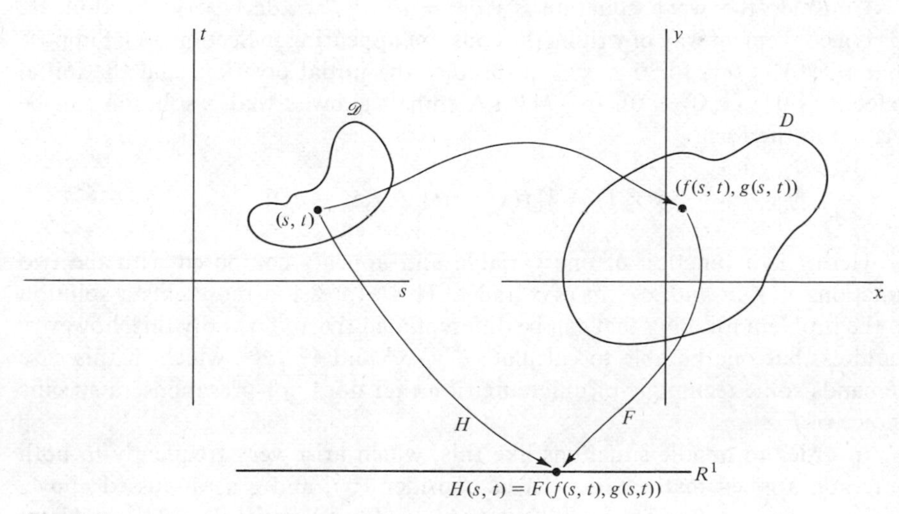

Fig. 5-9

It is fairly obvious that H is continuous at (s_0, t_0) if f and g are continuous at (s_0, t_0) and F is continuous at $(f(s_0, t_0), g(s_0, t_0))$. For, if $(s, t) \to (s_0, t_0)$ in $\mathscr{D}$, then $f(s, t) \to f(s_0, t_0)$ and $g(s, t) \to g(s_0, t_0)$ by continuity of f and g. Then $(f(s, t), g(s, t)) \to (f(s_0, t_0), g(s_0, t_0))$. Then, by continuity of F, $F(f(s, t), g(s, t)) \to F(f(s_0, t_0), g(s_0, t_0))$.

The question of differentiability of H is more difficult, and also involves a practical problem: How do we calculate partial derivatives $\partial H/\partial s$ and $\partial H/\partial t$ of H? Ideally we would like to be able to express these in terms of the partials of the functions f, g, and F whose composition determines H.

In order to clarify the ideas involved, let us contrast two examples.

EXAMPLE

Let $F(x, y) = e^{xy}\sin(xy^2) + 2x\cos(y)$ for all (x, y). Let $f(s, t) = \sqrt{s}t^2$ for $s \geq 0$, all real t, and $g(s, t) = s^3\sqrt{t}$ for all real s and $t \geq 0$. Then $H(s, t) = F(f(s, t), g(s, t))$ is defined for all $s \geq 0$, $t \geq 0$. Explicitly,

$$\begin{aligned} H(s, t) &= e^{f(s,t)g(s,t)}\sin(f(s, t)g(s, t)^2) + 2f(s, t)\cos(g(s, t)) \\ &= e^{s^{7/2}t^{5/2}}\sin(s^{13/2}t^3) + 2\sqrt{s}t^2\cos(s^3\sqrt{t}) \qquad \text{for } s \geq 0 \text{ and } t \geq 0. \end{aligned}$$

In this case there is no real need for any theorems about composite functions, as we have $H(s, t)$ explicitly in terms of s and t and can work with $H(s, t)$ directly. For example, we can easily compute $\partial H/\partial s$ directly in terms of s and t. ▌

EXAMPLE

Consider the wave equation $\partial^2 y/\partial t^2 = a^2\,\partial^2 y/\partial x^2$ derived in Section 5.2 (a^2 is a convenient way of writing the constant appearing in Section 5.2). Suppose that $y(x, 0) = f(x)$ for $0 \leq x \leq L$ specifies the initial position, and the initial velocity is 0 [$y_t(x, 0) = 0$]. In 1747 d'Alembert showed that a solution can be written in the form

$$y(x, t) = \tfrac{1}{2}[f(x + at) + f(x - at)].$$

Here f is a function of one variable and appears composed with the two functions $x + at$ and $x - at$ of x and t. This expression represents a solution to the problem for any f that can be differentiated twice. To verify this, however, requires that one be able to calculate $\partial^2 y/\partial x^2$ and $\partial^2 y/\partial t^2$, which in this case demands some technique of differentiation that does not presuppose a specific choice of f. ▌

In order to handle situations like this, which arise very frequently in both pure and applied mathematics, let us consider F, f, and g as discussed above, and *suppose that f and g are differentiable at (s_0, t_0) and F is differentiable at (x_0, y_0), where $x_0 = f(s_0, t_0)$ and $y_0 = g(s_0, t_0)$. With $H(s, t) = F(f(s, t), g(s, t))$, we shall show that:*

(1) H is differentiable at (s_0, t_0).

(2)
$$\frac{\partial H(s_0, t_0)}{\partial s} = \frac{\partial F(x_0, y_0)}{\partial x}\frac{\partial f(s_0, t_0)}{\partial s} + \frac{\partial F(x_0, y_0)}{\partial y}\frac{\partial g(s_0, t_0)}{\partial s}$$

and

$$\frac{\partial H(s_0, t_0)}{\partial t} = \frac{\partial F(x_0, y_0)}{\partial x}\frac{\partial f(s_0, t_0)}{\partial t} + \frac{\partial F(x_0, y_0)}{\partial y}\frac{\partial g(s_0, t_0)}{\partial t}.$$

(3) The differential of H at (s_0, t_0) due to increments dx and dy is given by

$$dH = \frac{\partial F(x_0, y_0)}{\partial x}\, dx + \frac{\partial F(x_0, y_0)}{\partial y}\, dy,$$

where

$$dx = \frac{\partial f(s_0, t_0)}{\partial s}\, ds + \frac{\partial f(s_0, t_0)}{\partial t}\, dt \quad \text{and} \quad dy = \frac{\partial g(s_0, t_0)}{\partial s}\, ds + \frac{\partial g(s_0, t_0)}{\partial t}\, dt.$$

To prove these, put, for convenience, $z = H(s, t)$. Since F is differentiable at (x_0, y_0), we can write, for Δx and Δy sufficiently small,

$$F(x_0 + \Delta x, y_0 + \Delta y) - F(x_0, y_0)$$
$$= \frac{\partial F(x_0, y_0)}{\partial x}\,\Delta x + \frac{\partial F(x_0, y_0)}{\partial y}\,\Delta y + \varepsilon_1\sqrt{(\Delta x)^2 + (\Delta y)^2},$$

where $\varepsilon_1 \to 0$ as $(\Delta x, \Delta y) \to (0, 0)$.

Similarly, f and g are differentiable at (s_0, t_0), so we can write

$$f(s_0 + \Delta s, t_0 + \Delta t) - f(s_0, t_0)$$
$$= \frac{\partial f(s_0, t_0)}{\partial s}\,\Delta s + \frac{\partial f(s_0, t_0)}{\partial t}\,\Delta t + \varepsilon_2\sqrt{(\Delta s)^2 + (\Delta t)^2},$$

and

$$g(s_0 + \Delta s, t_0 + \Delta t) - g(s_0, t_0)$$
$$= \frac{\partial g(s_0, t_0)}{\partial s}\,\Delta s + \frac{\partial g(s_0, t_0)}{\partial t}\,\Delta t + \varepsilon_3\sqrt{(\Delta s)^2 + (\Delta t)^2},$$

where

$$\varepsilon_2, \varepsilon_3 \to 0 \text{ as } (\Delta s, \Delta t) \to (0, 0).$$

Now

$$H(s_0 + \Delta s, t_0 + \Delta t) - H(s_0, t_0)$$
$$= F(f(s_0 + \Delta s, t_0 + \Delta t), g(s_0 + \Delta s, t_0 + \Delta t)) - F(f(s_0, t_0), g(s_0, t_0)).$$

Think of $x = f(s, t)$ and $y = g(s, t)$, and choose

$$\Delta x = f(s_0 + \Delta s, t_0 + \Delta t) - f(s_0, t_0) = f(s_0 + \Delta s, t_0 + \Delta t) - x_0$$

and

$$\Delta y = g(s_0 + \Delta s, t_0 + \Delta t) - g(s_0, t_0) = g(s_0 + \Delta s, t_0 + \Delta t) - y_0.$$

That is, we choose increments in x and y dependent upon the increments in s and t. Then

$$
\begin{aligned}
H(s_0 + \Delta s, t_0 + \Delta t) &- H(s_0, t_0) \\
&= F(x_0 + \Delta x, y_0 + \Delta y) - F(x_0, y_0) \\
&= \frac{\partial F(x_0, y_0)}{\partial x} \Delta x + \frac{\partial F(x_0, y_0)}{\partial y} \Delta y + \varepsilon_1 \sqrt{(\Delta x)^2 + (\Delta y)^2} \\
&= \frac{\partial F(x_0, y_0)}{\partial x} \left[\frac{\partial f(s_0, t_0)}{\partial s} \Delta s + \frac{\partial f(s_0, t_0)}{\partial t} \Delta t + \varepsilon_2 \sqrt{(\Delta s)^2 + (\Delta t)^2} \right] \\
&\quad + \frac{\partial F(x_0, y_0)}{\partial y} \left[\frac{\partial g(s_0, t_0)}{\partial s} \Delta s + \frac{\partial g(s_0, t_0)}{\partial t} \Delta t + \varepsilon_3 \sqrt{(\Delta s)^2 + (\Delta t)^2} \right] \\
&\quad + \varepsilon_1 \sqrt{(\Delta x)^2 + (\Delta y)^2} \\
&= \left[\frac{\partial F(x_0, y_0)}{\partial x} \frac{\partial f(s_0, t_0)}{\partial s} + \frac{\partial F(x_0, y_0)}{\partial y} \frac{\partial g(s_0, t_0)}{\partial s} \right] \Delta s \\
&\quad + \left[\frac{\partial F(x_0, y_0)}{\partial x} \frac{\partial f(s_0, t_0)}{\partial t} + \frac{\partial F(x_0, y_0)}{\partial y} \frac{\partial g(s_0, t_0)}{\partial t} \right] \Delta t \\
&\quad + \frac{\partial F(x_0, y_0)}{\partial x} \varepsilon_2 \sqrt{(\Delta s)^2 + (\Delta t)^2} + \frac{\partial F(x_0, y_0)}{\partial y} \varepsilon_3 \sqrt{(\Delta s)^2 + (\Delta t)^2} \\
&\quad + \varepsilon_1 \sqrt{(\Delta x)^2 + (\Delta y)^2} \\
&= \left[\frac{\partial F(x_0, y_0)}{\partial x} \frac{\partial f(s_0, t_0)}{\partial s} + \frac{\partial F(x_0, y_0)}{\partial y} \frac{\partial g(s_0, t_0)}{\partial s} \right] \Delta s \\
&\quad + \left[\frac{\partial F(x_0, y_0)}{\partial x} \frac{\partial f(s_0, t_0)}{\partial t} + \frac{\partial F(x_0, y_0)}{\partial y} \frac{\partial g(s_0, t_0)}{\partial t} \right] \Delta t \\
&\quad + \left[\varepsilon_1 \sqrt{\frac{(\Delta x)^2 + (\Delta y)^2}{(\Delta s)^2 + (\Delta t)^2}} + \varepsilon_2 \frac{\partial F(x_0, y_0)}{\partial x} + \varepsilon_3 \frac{\partial F(x_0, y_0)}{\partial y} \right] \\
&\quad \times \sqrt{(\Delta s)^2 + (\Delta t)^2}. \\
&= A \, \Delta s + B \, \Delta t + \varepsilon_4 \sqrt{(\Delta s)^2 + (\Delta t)^2}.
\end{aligned}
$$

We claim now that $\varepsilon_4 \to 0$ as $(\Delta s, \Delta t) \to (0, 0)$. Since $\varepsilon_2 \to 0$ and $\varepsilon_3 \to 0$ as $(\Delta s, \Delta t) \to (0, 0)$, all we have to worry about is the term

$$\varepsilon_1 \left[\frac{(\Delta x)^2 + (\Delta y)^2}{(\Delta s)^2 + (\Delta t)^2} \right]^{1/2}.$$

A simple calculation yields

$$
\begin{aligned}
\sqrt{\frac{(\Delta x)^2 + (\Delta y)^2}{(\Delta s)^2 + (\Delta t)^2}} = \frac{1}{\sqrt{(\Delta s)^2 + (\Delta t)^2}} \Big[& (f_s(s_0, t_0)^2 + \varepsilon_2^2 + \varepsilon_3^2 + g_s(s_0, t_0)^2)(\Delta s)^2 \\
& + (f_t(s_0, t_0)^2 + \varepsilon_2^2 + \varepsilon_3^2 + g_t(s_0, t_0)^2)(\Delta t)^2 \\
& + 2\left(\frac{\partial f(s_0, t_0)}{\partial s} \frac{\partial f(s_0, t_0)}{\partial t} + \frac{\partial g(s_0, t_0)}{\partial s} \frac{\partial g(s_0, t_0)}{\partial t} \right)(\Delta s)(\Delta t) \Big].
\end{aligned}
$$

But

$$\frac{(\Delta s)^2}{\sqrt{(\Delta s)^2 + (\Delta t)^2}} \begin{cases} \le \dfrac{(\Delta s)^2}{\sqrt{(\Delta s)^2}} = |\Delta s| \to 0 \text{ as } (\Delta s, \Delta t) \to (0, 0) & \text{if } \Delta s \ne 0, \\ = 0 & \text{if } \Delta s = 0. \end{cases}$$

Similarly,

$$\frac{(\Delta t)^2}{\sqrt{(\Delta s)^2 + (\Delta t)^2}} \to 0 \qquad \text{as } (\Delta s, \Delta t) \to (0, 0).$$

Finally,

$$\frac{(\Delta s)(\Delta t)}{\sqrt{(\Delta s)^2 + (\Delta t)^2}} \begin{cases} = 0 & \text{if } \Delta s \text{ or } \Delta t \text{ is zero}, \\ \le \dfrac{(\Delta s)(\Delta t)}{\sqrt{(\Delta s)^2}} = \pm\Delta t \to 0 \text{ as } (\Delta s, \Delta t) \to (0, 0) & \\ & \text{if } \Delta s \ne 0 \text{ and } \Delta t \ne 0. \end{cases}$$

In sum, then, $\varepsilon_4 \to 0$ as $(\Delta s, \Delta t) \to (0, 0)$. Then H is differentiable at (s_0, t_0), and we know from Section 5.3 that

$$\frac{\partial H(s_0, t_0)}{\partial s} = A = \frac{\partial F(x_0, y_0)}{\partial x}\frac{\partial f(s_0, t_0)}{\partial s} + \frac{\partial F(x_0, y_0)}{\partial y}\frac{\partial g(s_0, t_0)}{\partial s}$$

and

$$\frac{\partial H(s_0, t_0)}{\partial t} = B = \frac{\partial F(x_0, y_0)}{\partial x}\frac{\partial f(s_0, t_0)}{\partial t} + \frac{\partial F(x_0, y_0)}{\partial y}\frac{\partial g(s_0, t_0)}{\partial t}.$$

So far we have proved (1) and (2). For (3), the differential of H at (s_0, t_0) due to the increments ds and dt is

$$\begin{aligned} dH &= \frac{\partial H(s_0, t_0)}{\partial s}\,ds + \frac{\partial H(s_0, t_0)}{\partial t}\,dt \\ &= \left[\frac{\partial F(x_0, y_0)}{\partial x}\frac{\partial f(s_0, t_0)}{\partial s} + \frac{\partial F(x_0, y_0)}{\partial y}\frac{\partial g(s_0, t_0)}{\partial s}\right] ds \\ &\quad + \left[\frac{\partial F(x_0, y_0)}{\partial x}\frac{\partial f(s_0, t_0)}{\partial t} + \frac{\partial F(x_0, y_0)}{\partial y}\frac{\partial g(s_0, t_0)}{\partial t}\right] dt \\ &= \frac{\partial F(x_0, y_0)}{\partial x}\left[\frac{\partial f(s_0, t_0)}{\partial s}\,ds + \frac{\partial f(s_0, t_0)}{\partial t}\,dt\right] \\ &\quad + \frac{\partial F(x_0, y_0)}{\partial x}\left[\frac{\partial g(s_0, t_0)}{\partial s}\,ds + \frac{\partial g(s_0, t_0)}{\partial t}\,dt\right] \\ &= \frac{\partial F(x_0, y_0)}{\partial x}\,dx + \frac{\partial F(x_0, y_0)}{\partial y}\,dy, \end{aligned}$$

and the proof is complete. ∎

We have been purposely elaborate in our notation on p. 187 because it is important for the reader to understand the interrelationships between the variables ($x = f(s, t)$, $y = g(s, t)$, $H(s, t) = F(f(s, t), g(s, t))$), and where the partial derivatives are being evaluated [$\partial f/\partial t$, $\partial g/\partial t$, $\partial f/\partial s$, $\partial g/\partial s$ at (s_0, t_0), $\partial F/\partial x$, $\partial F/\partial y$ at (x_0, y_0), $x_0 = f(s_0, t_0)$, $y_0 = g(s_0, t_0)$]. However, if all this is understood, then we usually just write

$$\frac{\partial H}{\partial s} = \frac{\partial F}{\partial x}\frac{\partial f}{\partial s} + \frac{\partial F}{\partial y}\frac{\partial g}{\partial s} = \frac{\partial F}{\partial x}\frac{\partial x}{\partial s} + \frac{\partial F}{\partial y}\frac{\partial y}{\partial s}$$

and

$$\frac{\partial H}{\partial t} = \frac{\partial F}{\partial x}\frac{\partial f}{\partial t} + \frac{\partial F}{\partial y}\frac{\partial g}{\partial t} = \frac{\partial F}{\partial x}\frac{\partial x}{\partial t} + \frac{\partial F}{\partial y}\frac{\partial y}{\partial t}$$

This is called the *chain rule for differentiating a composition*

$$H(s, t) = F(f(s, t), g(s, t)).$$

The same kind of reasoning that led to this chain rule can be used to derive chain-rule formulas for other compositions as well. We state two typical examples, assuming suitable differentiability conditions on the functions involved as we go along.

EXAMPLE

If $y = f(x)$ and $x = g(s, t)$, then $y = f(g(s, t))$, and

$$\frac{\partial y}{\partial s} = f'(x)\frac{\partial g}{\partial s}$$

and

$$\frac{\partial y}{\partial t} = f'(x)\frac{\partial g}{\partial t}.$$

Here $\partial y/\partial s$, $\partial y/\partial t$, $\partial g/\partial s$, and $\partial g/\partial t$ are evaluated at some (s_0, t_0), and $f'(x)$ at x_0, where $x_0 = g(s_0, t_0)$.

As an application, consider again d'Alembert's solution to the wave equation. We want to show that

$$y(x, t) = \tfrac{1}{2}[f(x + at) + f(x - at)]$$

satisfies $y_{tt} = a^2 y_{xx}$ with initial conditions $y(x, 0) = f(x)$ and $y_t(x, 0) = 0$. So we start calculating partials by the chain rule:

$$\begin{aligned}\frac{\partial y}{\partial x} &= \frac{1}{2}\left[f'(x + at)\frac{\partial(x + at)}{\partial x} + f'(x - at)\frac{\partial(x - at)}{\partial x}\right]\\ &= \tfrac{1}{2}[f'(x + at) + f'(x - at)];\end{aligned}$$

$$\frac{\partial^2 y}{\partial x^2} = \frac{1}{2}\left[f''(x+at)\frac{\partial(x+at)}{\partial x} + f''(x-at)\frac{\partial(x-at)}{\partial x}\right]$$

$$= \tfrac{1}{2}[f''(x+at) + f''(x-at)];$$

$$\frac{\partial y}{\partial t} = \frac{1}{2}\left[f'(x+at)\frac{\partial(x+at)}{\partial t} + f'(x-at)\frac{\partial(x-at)}{\partial t}\right]$$

$$= \tfrac{1}{2}[af'(x+at) - af'(x-at)];$$

$$\frac{\partial^2 y}{\partial t^2} = \frac{a}{2}\left[f''(x+at)\frac{\partial(x+at)}{\partial t} - f''(x-at)\frac{\partial(x-at)}{\partial t}\right]$$

$$= \frac{a}{2}[af''(x+at) + af''(x-at)] = a^2\frac{\partial^2 y}{\partial x^2}.$$

Further, $y(x, 0) = f(x)$ and

$$\frac{\partial y(x,0)}{\partial t} = \tfrac{1}{2}[af'(x) - af'(x)] = 0.$$ ∎

EXAMPLE

Suppose that $w = \varphi(x, y, z)$ and $x = f(u, v)$, $y = g(u, v)$, and $z = h(u, v)$. Then $w = \varphi(f(u, v), g(u, v), h(u, v)) = W(u, v)$. The first partials (assuming the proper differentiability conditions) are

$$\frac{\partial W}{\partial u} = \frac{\partial \varphi}{\partial x}\frac{\partial f}{\partial u} + \frac{\partial \varphi}{\partial y}\frac{\partial g}{\partial u} + \frac{\partial \varphi}{\partial z}\frac{\partial h}{\partial u}$$

and

$$\frac{\partial W}{\partial v} = \frac{\partial \varphi}{\partial x}\frac{\partial f}{\partial v} + \frac{\partial \varphi}{\partial y}\frac{\partial g}{\partial v} + \frac{\partial \varphi}{\partial z}\frac{\partial h}{\partial v}.$$

Here $\partial W/\partial u$, $\partial W/\partial v$, $\partial f/\partial u, \ldots, \partial h/\partial v$ are evaluated at some (u_0, v_0), and $\partial\varphi/\partial x$, $\partial\varphi/\partial y$, and $\partial\varphi/\partial z$ at (x_0, y_0, z_0), where $x_0 = f(u_0, v_0)$, $y_0 = g(u_0, v_0)$, and $z_0 = h(u_0, v_0)$.

Assuming that we can differentiate again, we have

$$\frac{\partial^2 W}{\partial u^2} = \frac{\partial}{\partial u}\left(\frac{\partial \varphi}{\partial x}\frac{\partial f}{\partial u} + \frac{\partial \varphi}{\partial y}\frac{\partial g}{\partial u} + \frac{\partial \varphi}{\partial z}\frac{\partial h}{\partial u}\right)$$

$$= \frac{\partial \varphi}{\partial x}\frac{\partial^2 f}{\partial u^2} + \frac{\partial \varphi}{\partial y}\frac{\partial^2 g}{\partial u^2} + \frac{\partial \varphi}{\partial z}\frac{\partial^2 h}{\partial u^2}$$

$$+ \frac{\partial f}{\partial u}\left(\frac{\partial}{\partial x}\left(\frac{\partial \varphi}{\partial x}\right)\frac{\partial f}{\partial u} + \frac{\partial}{\partial y}\left(\frac{\partial \varphi}{\partial x}\right)\frac{\partial g}{\partial u} + \frac{\partial}{\partial z}\left(\frac{\partial \varphi}{\partial x}\right)\frac{\partial h}{\partial u}\right)$$

$$+ \frac{\partial g}{\partial u}\left(\frac{\partial}{\partial x}\left(\frac{\partial \varphi}{\partial y}\right)\frac{\partial f}{\partial u} + \frac{\partial}{\partial y}\left(\frac{\partial \varphi}{\partial y}\right)\frac{\partial g}{\partial u} + \frac{\partial}{\partial z}\left(\frac{\partial \varphi}{\partial y}\right)\frac{\partial h}{\partial u}\right)$$

$$+ \frac{\partial h}{\partial u}\left(\frac{\partial}{\partial x}\left(\frac{\partial \varphi}{\partial z}\right)\frac{\partial f}{\partial u} + \frac{\partial}{\partial y}\left(\frac{\partial \varphi}{\partial z}\right)\frac{\partial g}{\partial u} + \frac{\partial}{\partial z}\left(\frac{\partial \varphi}{\partial z}\right)\frac{\partial h}{\partial u}\right),$$

with similar expressions for $\partial^2 W/\partial v^2$, $\partial^2 W/\partial u\,\partial v$, and $\partial^2 W/\partial v\,\partial u$.

Note that the complication in the last three terms arises from chain-rule differentiation of $\partial\varphi/\partial x$, $\partial\varphi/\partial y$, and $\partial\varphi/\partial z$. Each of these is in general a function of x, y, and z again, and must be differentiated by chain rule in the same way we differentiated W originally. ▮

Exercises for Section 5.4

1. Let $F(x, y) = 2xy + e^x \sin(y)$. Let $x = e^s t$, $y = 2st$, and $H(s, t) = F(x(s, t), y(s, t))$.
   (a) Determine $H(s, t)$ explicitly by substitution.
   (b) Calculate $\partial H/\partial s$, $\partial H/\partial t$, $\partial^2 H/\partial s\, \partial t$, $\partial^2 H/\partial t\, \partial s$ from (a).
   (c) Compute the partials in (b) by the chain rule, and compare with the results obtained in (b).

2. Let $w = F(x, y, z, t)$. Let $x = f(u, v, p)$, $y = g(u, v, p)$, $z = h(u, v, p)$, and $t = a(u, v, p)$. Give chain-rule formulas for $\partial w/\partial u$, $\partial w/\partial v$, $\partial w/\partial p$, $\partial^2 w/\partial u^2$, $\partial^2 w/\partial v^2$, $\partial^2 w/\partial u\, \partial v$, $\partial^3 w/\partial u\, \partial v\, \partial p$ in terms of partials of F with respect to x, y, z, and t, and partials of f, g, h, and a with respect to u, v, and p.

3. The equation

$$\frac{\partial^2 u}{\partial x^2} + \frac{\partial^2 u}{\partial y^2} = 0$$

is Laplace's equation and plays a role in such phenomena as heat flow and vibration of a membrane. Put $x = r\cos(\theta)$ and $y = r\sin(\theta)$, and $w(r, \theta) = u(r\cos(\theta), r\sin(\theta))$. Show that, in these polar coordinates, Laplace's equation becomes

$$\frac{\partial^2 w}{\partial r^2} + \frac{1}{r^2}\frac{\partial^2 w}{\partial \theta^2} + \frac{1}{r}\frac{\partial w}{\partial r} = 0,$$

assuming that $u_{yx} = u_{xy}$.

4. Laplace's equation in three dimensions is

$$\frac{\partial^2 u}{\partial x^2} + \frac{\partial^2 u}{\partial y^2} + \frac{\partial^2 u}{\partial z^2} = 0.$$

   (a) Use cylindrical coordinates: $x = r\cos(\theta)$, $y = r\sin(\theta)$, $z = z$, and put $w(r, \theta, z) = u(r\cos(\theta), r\sin(\theta), z)$. Show that Laplace's equation becomes

$$\frac{\partial^2 w}{\partial r^2} + \frac{1}{r^2}\frac{\partial^2 w}{\partial \theta^2} + \frac{1}{r}\frac{\partial w}{\partial r} + \frac{\partial^2 w}{\partial z^2} = 0,$$

   assuming that $u_{xy} = u_{yx}$.
   (b) In spherical coordinates: $x = \rho\cos(\theta)\sin(\varphi)$, $y = \rho\sin(\theta)\sin(\varphi)$, $z = \rho\cos(\varphi)$. Let $S(\rho, \varphi, \theta) = u(\rho\cos(\theta)\sin(\varphi), \rho\sin(\theta)\sin(\varphi), \rho\cos(\varphi))$. Show that

$$\frac{\partial^2 S}{\partial \rho^2} + \frac{1}{\rho^2}\frac{\partial^2 S}{\partial \theta^2} + \frac{1}{\rho^2\sin^2(\varphi)}\frac{\partial^2 S}{\partial \varphi^2} + \frac{2}{\rho}\frac{\partial S}{\partial \rho} + \frac{\cot(\varphi)}{\rho^2}\frac{\partial S}{\partial \varphi} = 0.$$

5. Let $z = f(x + y)$, where $x = f(u, v)$ and $y = g(u, v)$. Find $\partial z/\partial u$, $\partial z/\partial v$, $\partial^2 z/\partial u^2$, $\partial^2 z/\partial v^2$, $\partial^2 z/\partial u\, \partial v$, and $\partial^2 z/\partial v\, \partial u$.

6. Let $w = f(g(u, v), h(u), q(v))$. Find $\partial w/\partial u$, $\partial w/\partial v$, $\partial^2 w/\partial u^2$, $\partial^2 w/\partial v^2$, $\partial^2 w/\partial u\, \partial v$, and $\partial^2 w/\partial v\, \partial u$.

7. Let $w = f(x)$, $x = g(u, v)$, and $u = \varphi(s)$, $v = \xi(t)$. Find $\partial w/\partial s$, $\partial w/\partial t$, $\partial^2 w/\partial s^2$, and $\partial^2 w/\partial t^2$.

8. Let $u = u(w)$ and $w = \sqrt{x^2 + y^2 + z^2}$. Show that

$$\left(\frac{du}{dw}\right)^2 = \left(\frac{\partial u}{\partial x}\right)^2 + \left(\frac{\partial u}{\partial y}\right)^2 + \left(\frac{\partial u}{\partial z}\right)^2$$

and

$$\frac{\partial^2 u}{\partial x^2} + \frac{\partial^2 u}{\partial y^2} + \frac{\partial^2 u}{\partial z^2} = \frac{d^2 u}{dw^2} + \frac{2}{w}\frac{du}{dw}.$$

9. A function f of two variables is said to be positively homogeneous of degree r (r is some real number, possibly zero) if $f(tx, ty) = t^r f(x, y)$ for $t > 0$, provided that $f(tx, ty)$ and $f(x, y)$ are defined.

(a) For each positive integer n, given an example of a function positively homogeneous of degree n.

(b) Prove Euler's theorem: *If f is positively homogeneous of degree r, then* $x(\partial f/\partial x) + y(\partial f/\partial y) = rf(x, y)$. *Hint:* Compute

$$\frac{\partial(f(tx, ty))}{\partial t} = \frac{\partial(t^r f(x, y))}{\partial t}$$

using the chain rule on the left side, and then put $t = 1$.

10. Let f be positively homogeneous of degree r (see Exercise 9). Show that

$$x^2 \frac{\partial^2 f}{\partial x^2} + 2xy \frac{\partial^2 f}{\partial x\, \partial y} + y^2 \frac{\partial^2 f}{\partial y^2} = r(r - 1)f(x, y).$$

Hint: Begin by computing both sides of

$$\frac{\partial f(tx, ty)}{\partial t} = \frac{\partial(t^r f(x, y))}{\partial t},$$

then differentiate both sides with respect to t.

11. Let f be positively homogeneous of degree r. Let $g(x, y) = (x^2 + y^2)^{K/2} f(x, y)$. Show that

$$\frac{\partial^2 g}{\partial x^2} + \frac{\partial^2 g}{\partial y^2} = (x^2 + y^2)^{K/2}\left(\frac{\partial^2 f}{\partial x^2} + \frac{\partial^2 f}{\partial y^2}\right) + K(2r + K)(x^2 + y^2)^{-1+K/2} f(x, y).$$

5.5 Applications of Chain-Rule Differentiation

In an example in Section 5.4 we saw the use of chain-rule differentiation to verify that

$$y(x, t) = \tfrac{1}{2}[f(x + at) + f(x - at)]$$

satisfied

$$\frac{\partial^2 y}{\partial t^2} = a^2 \frac{\partial^2 y}{\partial x^2}, \qquad y(x, 0) = f(x), \qquad \frac{\partial y(x, 0)}{\partial t} = 0.$$

As applications go, this was not too impressive, as one might naturally (and rightly) suspect that d'Alembert derived the solution by some other means, and knew it to be a solution without having to substitute into the differential equation. We shall now give a number of more substantial applications.

A. Mean Value Theorem

For functions of one variable, the mean value theorem says: If f is continuous on $[a, b]$ and differentiable on (a, b), then for some ξ between a and b,

$$f'(\xi) = \frac{f(b) - f(a)}{b - a}.$$

Geometrically, this says that, at some point between a and b, the tangent to $y = f(x)$ is parallel to the line between $(a, f(a))$ and $(b, f(b))$ (see Fig. 5-10).

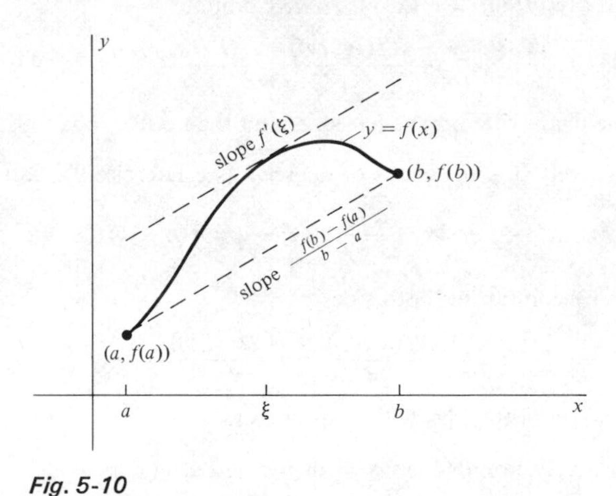

Fig. 5-10

If we write $b = a + h$, then ξ can be expressed as $a + \theta h$ for some θ between 0 and 1. This gives us

$$f'(a + \theta h) = \frac{f(a + h) - f(a)}{h}.$$

or, equivalently,

$$hf'(a + \theta h) = f(a + h) - f(a),$$

for some θ with $0 < \theta < 1$. This is a frequently seen alternative version of the mean value theorem.

Observe that, in this form, the theorem says that the increment in function values when x changes from a to $a + h$ is equal to the differential due to h at some point between a and $a + h$.

This generalizes directly to two or more variables. In two variables, *suppose that f is continuous on the line segment from (a, b) to $(a + h, b + k)$, and differentiable there except possibly at the end points (a, b) and $(a + h, b + k)$. We claim that, for some θ between* 0 *and* 1,

$$f(a + h, b + k) - f(a, b) = h\frac{\partial f(a + \theta h, b + \theta k)}{\partial x} + k\frac{\partial f(a + \theta h, b + \theta k)}{\partial y},$$

Note that the left side is an increment Δf due to incrementing a by h and b by k; the right side is the differential due to h and k at some intermediate point between (a, b) and $(a + h, b + k)$.

To prove our assertion, we appeal directly to the mean value theorem for functions of one variable by putting

$$g(t) = f(a + th, b + tk), \qquad 0 \le t \le 1.$$

Then g is continuous on $[0, 1]$ and differentiable on $(0, 1)$ by the results of Section 5.4. Further, for $0 < t < 1$, we have, putting $x = a + th$ and $y = b + tk$:

$$g'(t) = \frac{\partial f}{\partial x}\frac{\partial x}{\partial t} + \frac{\partial f}{\partial y}\frac{\partial y}{\partial t} = h\frac{\partial f}{\partial x} + k\frac{\partial f}{\partial y}.$$

By the mean value theorem, applied to g, there is some θ, $0 < \theta < 1$, such that

$$g(1) - g(0) = g'(\theta).$$

Now

$$g(1) - g(0) = f(a + h, b + k) - f(a, b)$$

and

$$g'(\theta) = h\frac{\partial f(a + \theta h, b + \theta k)}{\partial x} + k\frac{\partial f(a + \theta h, b + \theta k)}{\partial y},$$

giving us exactly what we wanted to prove.

This argument can be carried over almost verbatim to the case of three, four, or more variables. We shall leave it to the reader to do this in the exercises.

As an application of the mean value theorem for functions of two variables, we shall prove the following theorem: If f is differentiable on a set D, having the property that there is a polygonal path in D between any two of its points, and if $\partial f/\partial x = \partial f/\partial y = 0$ in D, then f is a constant function on D.

If you think of partial derivatives as rates of change, this is a very plausible result. It says that, if f does not change with x ($\partial f/\partial x = 0$), or with $y(\partial f/\partial y = 0)$, then $f(x, y)$ is constant on D.

Analytically, we argue as follows. Let (a_1, b_1) and (a_2, b_2) be any two distinct points of D. Join (a_1, b_1) to (a_2, b_2) by a series of straight-line segments in D, say from (a_1, b_1) to P_1, to $P_2, \ldots,$ to P_n, to (a_2, b_2). This can be done because of the assumption on D. Look at any one of these line segments, say from $P_j = (x_j, y_j)$ to $P_{j+1} = (x_{j+1}, y_{j+1})$. Write $x_{j+1} = x_j + h$, $y_{j+1} = y_j + k$, and use the mean value theorem to obtain:

$$f(x_{j+1}, y_{j+1}) - f(x_j, y_j) = h\,\frac{\partial f(x_j + \theta h, y_j + \theta k)}{\partial x} + k\,\frac{\partial f(x_j + \theta h, y_j + \theta k)}{\partial y}.$$

for some θ, $0 < \theta < 1$. By assumption, the partials on the right are zero, hence $f(x_{j+1}, y_{j+1}) = f(x_j, y_j)$. It is now easy to follow along the polygonal path from (a_1, b_1) to (a_2, b_2) and conclude that $f(a_1, b_1) = f(a_2, b_2)$, hence that $f(x, y)$ is the same number for each (x, y) in D.

B. Sufficient Conditions for Differentiability

We remarked in Section 5.4 that, in most instances, a proof of differentiability by direct appeal to the definition is very cumbersome. Fortunately, it is also generally unnecessary. We shall now give, for the two-variable case, a set of fairly convenient conditions sufficient for differentiability at a point.

Theorem

Let f be continuous in some r-neighborhood of (a, b). Let $\partial f/\partial x$ and $\partial f/\partial y$ be defined in this neighborhood and continuous at (a, b). Then f is differentiable at (a, b).

PROOF

In the following argument, h and k are always assumed sufficiently small that $(a + h, b + k)$ is in the given r-neighborhood of (a, b). That is, $h^2 + k^2 < r^2$. Begin by writing

$$\begin{aligned} &f(a + h, b + k) - f(a, b) \\ &\qquad = f(a + h, b + k) - f(a + h, b) + f(a + h, b) - f(a, b). \end{aligned}$$

Put $g(t) = f(a + h, b + tk)$, $0 \le t \le 1$, and apply the mean value theorem to g on $[0, 1]$, as in Section 5.5A, to write, for some θ with $0 < \theta < 1$,

$$\begin{aligned} f(a + h, b + k) - f(a + h, b) &= g(1) - g(0) \\ &= g'(\theta) = k \cdot \frac{\partial f(a + h, b + \theta k)}{\partial y}. \end{aligned}$$

Similarly, apply the mean value theorem on $[0, 1]$ to $h(t) = f(a + th, b) - f(a, b)$ to obtain, for some δ with $0 < \delta < 1$,

$$f(a + h, b + k) - f(a, b) = h(1) - h(0) = h'(\delta).$$

Then, collecting our terms, we have

$$f(a + h, b + k) - f(a, b) = h\frac{\partial f(a + \delta h, b)}{\partial x} + k\frac{\partial f(a + h, b + \theta k)}{\partial y}.$$

Then

$$\begin{aligned} f(a + h, b + k) - f(a, b) - h\frac{\partial f(a, b)}{\partial x} - k\frac{\partial f(a, b)}{\partial y} \\ = h\left(\frac{\partial f(a + \delta h, b)}{\partial x} - \frac{\partial f(a, b)}{\partial x}\right) + k\left(\frac{\partial f(a + h, b + \theta k)}{\partial y} - \frac{\partial f(a, b)}{\partial y}\right). \end{aligned}$$

Then

$$\frac{\Delta f - df}{\sqrt{h^2 + k^2}} = \frac{h\left(\dfrac{\partial f(a + \delta h, b)}{\partial x} - \dfrac{\partial f(a, b)}{\partial x}\right)}{\sqrt{h^2 + k^2}} + \frac{k\left(\dfrac{\partial f(a + h, b + \theta k)}{\partial y} - \dfrac{\partial f(a, b)}{\partial y}\right)}{\sqrt{h^2 + k^2}}.$$

Our objective is to show that

$$\frac{\Delta f - df}{\sqrt{h^2 + k^2}} \to 0 \qquad \text{as } (h, k) \to (0, 0).$$

But note that

$$\left|\frac{h\left(\dfrac{\partial f(a + \delta h, b)}{\partial x} - \dfrac{\partial f(a, b)}{\partial x}\right)}{\sqrt{h^2 + k^2}}\right| \le \left|\frac{\partial f(a + \delta h, b)}{\partial x} - \frac{\partial f(a, b)}{\partial x}\right| \to 0$$

as $(h, k) \to (0, 0)$, due to the continuity of $\partial f/\partial x$ at (a, b).

Similarly, by continuity of $\partial f/\partial y$ at (a, b),

$$\left|\frac{k\left(\dfrac{\partial f(a + h, b + \theta k)}{\partial y} - \dfrac{\partial f(a, b)}{\partial y}\right)}{\sqrt{h^2 + k^2}}\right| \le \left|\frac{\partial f(a + h, b + \theta k)}{\partial y} - \frac{\partial f(a, b)}{\partial y}\right| \to 0$$

as $(h, k) \to (0, 0)$. Hence

$$\lim_{(h,k)\to(0,0)} \frac{\Delta f - df}{\sqrt{h^2 + k^2}} = 0,$$

and f is differentiable at (a, b). ∎

C. Interchange of Order in Mixed Partials

If you write down a "reasonable" function f of two variables, you will probably find that the mixed partials $\partial^2 f/\partial x\, \partial y$ and $\partial^2 f/\partial y\, \partial x$ are equal. This is not always the case [see Exercise 4(c), Section 5.1]. However, interchange of the order of differentiation is valid under rather general conditions. We shall now show (using subscripts to denote partials for convenience) that $f_{12}(a, b) = f_{21}(a, b)$ if f, f_1, f_2, f_{12}, and f_{21} are all defined in some r-neighborhood of (a, b) and f_{12} and f_{21} are continuous at each point of the neighborhood.

The proof begins with a trick and proceeds through four applications of the mean value theorem. To begin, put

$$F(h) = f(a + h, b + h) - f(a + h, b) - f(a, b + h) + f(a, b),$$

where h is sufficiently small that $(a + h, b + h)$ is within the given neighborhood of (a, b) (i.e., $\sqrt{2}|h| < r$.)

Let

$$g(x) = f(x, b + h) - f(x, b);$$

then

$$F(h) = g(a + h) - g(a).$$

Now, for x between a and $a + h$, $g'(x) = f_1(x, b + h) - f_1(x, b)$. By the mean value theorem applied to g, there is some θ_1, with $0 < \theta_1 < 1$ such that

$$\begin{aligned} g(a + h) - g(a) &= g'(a + \theta_1 h) \cdot h \\ &= [f_1(a + \theta_1 h, b + h) - f_1(a + \theta_1 h, b)] \cdot h. \end{aligned}$$

Then

$$F(h) = h(f_1(a + \theta_1 h, b + h) - f_1(a + \theta_1 h, b)).$$

Now put $s(y) = f_1(a + \theta_1 h, y)$. Then $s'(y) = f_{12}(a + \theta_1 h, y)$ for y between b and $b + h$. Applying the mean value theorem to s yields the existence of some number θ_2, $0 < \theta_2 < 1$, such that

$$s(b + h) - s(b) = s'(b + \theta_2 h) \cdot h.$$

Then, piecing everything together gives us

$$\begin{aligned} h(s(b + h) - s(b)) &= hf_1(a + \theta_1 h, b + h) - hf_1(a + \theta_1 h, b) \\ &= s'(b + \theta_2 h) \cdot h^2 = h^2 f_{12}(a + \theta_1 h, b + \theta_2 h). \\ &= F(h). \end{aligned}$$

The part we are interested in is

$$F(h) = h^2 f_{12}(a + \theta_1 h, b + \theta_2 h).$$

Now put this aside for the moment and start all over again with

$$G(h) = f(a + h, y) - f(a, y).$$

Note that $F(h) = G(b + h) - G(b)$. By the mean value theorem, there is some θ_3, $0 < \theta_3 < 1$, such that

$$\begin{aligned} F(h) &= G(b + h) - G(b) = G'(b + \theta_3 h) \cdot h \\ &= [f_2(a + h, b + \theta_3 h) - f_2(a, b + \theta_3 h)] \cdot h. \end{aligned}$$

Now put $S(x) = f_2(x, b + \theta_3 h)$. Then $S'(x) = f_{21}(x, b + \theta_3 h)$ for x between a and $a + h$, and a fourth application of the mean value theorem gives us some θ_4 between 0 and 1 such that

$$S(a + h) - S(a) = S'(a + \theta_4 h) \cdot h.$$

Then

$$f_2(a + h, b + \theta_3 h) - f_2(a, b + \theta_3 h) = f_{21}(a + \theta_4 h, b + \theta_3 h) \cdot h.$$

Then

$$\begin{aligned} F(h) &= h(f_2(a + h, b + \theta_3 h) - f_2(a, b + \theta_3 h)) \\ &= h^2 f_{21}(a + \theta_4 h, b + \theta_3 h). \end{aligned}$$

Now total everything up:

$$F(h) = h^2 f_{12}(a + \theta_1 h, b + \theta_2 h) = h^2 f_{21}(a + \theta_4 h, b + \theta_3 h).$$

Then, finally, we get

$$f_{12}(a + \theta_1 h, b + \theta_2 h) = f_{21}(a + \theta_4 h, b + \theta_3 h).$$

Now let $h \to 0$. By continuity of f_{12} and f_{21} at (a, b), we have

$$f_{12}(a + \theta_1 h, b + \theta_2 h) \to f_{12}(a, b)$$

and

$$f_{21}(a + \theta_4 h, b + \theta_3 h) \to f_{21}(a, b).$$

Hence $f_{12}(a, b) = f_{21}(a, b)$, and we are through. ∎

The reader should note that this is an example of a strictly nonintuitive proof. It starts with a definition of a function $F(h)$ right out of thin air, and proceeds through a number of mathematically correct, but apparently unmotivated, steps which finally lead to the desired result. The reader should go over the steps very carefully, keeping the final objective in mind, and construct in his own mind an outline of the main idea of the proof and the way the various steps piece together to contribute to the end result.

The theorem we have proved extends to higher-order derivatives; but things become progressively messier very quickly, and the above result will suffice for our needs.

D. Taylor's Theorem in Two Variables

An argument very similar to that used in Section 5.5A to extend the mean value theorem can also be used to develop a Taylor's theorem for functions of two variables. Suppose that $f(x, y)$ is defined for all (x, y) in some r-neighborhood of (a, b). We would like to write $f(x, y)$, for (x, y) close to (a, b), as a polynomial in $x - a$ and $y - b$, plus an error term. To do this, put

$$g(t) = f(a + th, b + tk)$$

for $0 \leq t \leq 1$, with h and k arbitrary except that $(a + th, b + tk)$ must be in the r-neighborhood of (a, b) for $0 \leq t \leq 1$. For this, we need $\sqrt{h^2 + k^2} < r$. Assuming that g has $m + 1$ derivatives on $[0, 1]$, we can apply Taylor's theorem to g and write, for some θ between 0 and 1:

$$g(t) = g(0) + g'(0) + \frac{g''(0)t^2}{2!} + \cdots + \frac{g^{(m)}(0)t^m}{m!} + \frac{g^{(m+1)}(\theta)t^{m+1}}{(m+1)!}.$$

In particular, for some θ, $0 < \theta < 1$ and

$$g(1) = g(0) + g'(0) + \frac{g''(0)}{2!} + \cdots + \frac{g^{(m)}(0)}{m!} + \frac{g^{(m+1)}(\theta)}{(m+1)!}.$$

Since $g(1) = f(a + h, b + k)$ and $g(0) = f(a, b)$, then we have

$$f(a + h, b + k) = f(a, b) + g'(0) + \frac{g''(0)}{2!} + \cdots + \frac{g^{(m)}(0)}{m!} + \frac{g^{(m+1)}(\theta)}{(m+1)!}.$$

There remains to calculate $g'(0), g''(0), \ldots, g^{(m)}(0)$. By the chain rule,

$$\begin{aligned} g'(t) &= f_1(a + th, b + tk)\frac{\partial(a + th)}{\partial t} + f_2(a + th, b + tk)\frac{\partial(b + tk)}{\partial t} \\ &= hf_1(a + th, b + tk) + kf_2(a + th, b + tk), \end{aligned}$$

with subscripts denoting, as usual, differentiation with respect to the first or second variable. As a space saver, we shall write this equation as just

$$g'(t) = hf_1 + kf_2.$$

Differentiate again by the chain rule:

$$\begin{aligned} g''(t) &= h(f_{11}h + f_{12}k) + k(f_{21}h + f_{22}k) \\ &= h^2 f_{11} + hk(f_{12} + f_{21}) + k^2 f_{22}. \end{aligned}$$

Assuming that $f_{12} = f_{21}$, then

$$g''(t) = h^2 f_{11} + 2hkf_{12} + k^2 f_{22}.$$

Next,

$$\begin{aligned} g^{(3)}(t) &= h^2(f_{111}h + f_{112}k) + 2hk(f_{121}h + f_{122}k) + k^2(f_{221}h + f_{222}k) \\ &= h^3 f_{111} + h^2 k f_{112} + 2h^2 k f_{121} + 2hk^2 f_{122} + k^2 h f_{221} + k^3 f_{222}. \end{aligned}$$

Note that $f_{112} = (f_1)_{12}$ and $f_{121} = (f_1)_{21}$. Assume that the mixed partials of f_1 are equal; then $f_{121} = f_{112}$. Similarly, we shall suppose that $f_{122} = f_{221}$. [Note that $f_{122} = (f_{12})_2 = (f_{21})_2 = (f_2)_{12} = (f_2)_{21} = f_{221}$ if $f_{12} = f_{21}$ and $(f_2)_{12} = (f_2)_{21}$.] Then

$$g^{(3)}(t) = h^3 f_{111} + 3h^2 k f_{112} + 3hk^2 f_{122} + k^3 f_{222}.$$

We can go on for as long as the chain rule can be successfully applied to the partials resulting from the previous step. However, we have gone far enough to point out a pattern that is beginning to emerge, which will help us write down $g^{(n)}(t)$ for any n. Think of $h(\partial/\partial x) + k(\partial/\partial y)$ as an operator, with a product

$$\left(h\frac{\partial}{\partial x} + k\frac{\partial}{\partial y}\right)f\bigg|_{(a,b)}$$

defined to mean

$$h\frac{\partial f(a, b)}{\partial x} + k\frac{\partial f(a, b)}{\partial y}.$$

We can now define powers of this operator, which behave just like powers of a sum $A + B$. For the second power, we have, if $\partial^2/\partial x\,\partial y = \partial^2/\partial y\,\partial x$,

$$\left(h\frac{\partial}{\partial x} + k\frac{\partial}{\partial y}\right)^2 = h^2\frac{\partial}{\partial x^2} + 2hk\frac{\partial^2}{\partial x\,\partial y} + k^2\frac{\partial^2}{\partial y^2}.$$

That is, for any f such that $\partial^2 f/\partial x\,\partial y = \partial^2 f/\partial y\,\partial x$ at (α, β),

$$\left(h\frac{\partial}{\partial x} + k\frac{\partial}{\partial y}\right)^2 f\bigg|_{(\alpha,\beta)} = h^2\frac{\partial^2 f(\alpha,\beta)}{\partial x^2} + 2hk\frac{\partial^2 f(\alpha,\beta)}{\partial x\,\partial y} + k^2\frac{\partial^2 f(\alpha,\beta)}{\partial y^2}.$$

Similarly, if

$$\frac{\partial^3 f}{\partial x\,\partial y\,\partial x} = \frac{\partial^3 f}{\partial x\,\partial x\,\partial y} \quad \text{and} \quad \frac{\partial^3 f}{\partial y\,\partial y\,\partial x} = \frac{\partial^3 f}{\partial x\,\partial y\,\partial y},$$

then

$$\left(h\frac{\partial}{\partial x} + k\frac{\partial}{\partial y}\right)^3 f\bigg|_{(\alpha,\beta)} = h^3\frac{\partial^3 f}{\partial x^3}\bigg|_{(\alpha,\beta)} + 2h^2k\frac{\partial^3 f}{\partial x^2\,\partial y}\bigg|_{(\alpha,\beta)}$$
$$+ 2hk^2\frac{\partial^3 f}{\partial x\,\partial y^2}\bigg|_{(\alpha,\beta)} + k^3\frac{\partial^3 f}{\partial y^3}\bigg|_{(\alpha,\beta)}.$$

In general, we can expand the operator $(h(\partial/\partial x) + k(\partial/\partial y))^n$ by the binomial formula:

$$\left(h\frac{\partial}{\partial x} + k\frac{\partial}{\partial y}\right)^n = \sum_{j=0}^{n}\binom{n}{j}h^{n-j}k^j\frac{\partial^n}{\partial x^{n-j}\partial y^j}.$$

This gives us a systematic way of computing $g^{(n)}(t)$ for $n = 1, 2, \ldots$. Specifically,

$$g^{(n)}(t) = \left(h\frac{\partial}{\partial x} + k\frac{\partial}{\partial y}\right)^n f\bigg|_{(a+th,b+tk)} = \sum_{j=0}^{n}\binom{n}{j}h^{n-j}k^j\frac{\partial^n f}{\partial x^{n-j}\partial y^j}\bigg|_{(a+th,b+tk)}.$$

In particular,

$$g^{(n)}(0) = \sum_{j=0}^{n}\binom{n}{j}h^{n-j}k^j\frac{\partial^n f}{\partial x^{n-j}\partial y^j}\bigg|_{(a,b)}$$

for $n = 0, 1, \ldots, m$, and

$$g^{(m+1)}(\theta) = \sum_{j=0}^{m+1}\binom{m+1}{j}h^{m+1-j}k^j\frac{\partial^{m+1} f}{\partial x^{m+1-j}\partial y^j}\bigg|_{(a+\theta h,b+\theta k)}$$

These values for $g(0)$, $g'(0), \ldots, g^{(m)}(0)$, and $g^{(m+1)}(\theta)$, when substituted into the above Taylor expansion of $g(1)$, give a two-variable polynomial expansion of $f(a + h, b + k)$ in powers of h and k, with coefficients expressed in terms of partials of f at (a, b), plus an error term given by the expression in terms of f for $g^{(m+1)}(\theta)$.

All of this is a lot to digest in one dose, but an example will indicate the ideas involved and also show that the mechanics are not in general difficult to carry out.

EXAMPLE

Let $f(x, y) = e^x \sin(y)$, for all real x and y. Here all partials are continuous everywhere, and

$$\frac{\partial^2 f}{\partial x\,\partial y} = \frac{\partial^2 f}{\partial y\,\partial x}, \frac{\partial^3 f}{\partial x^2\,\partial y} = \frac{\partial^3 f}{\partial x\,\partial y\,\partial x} = \frac{\partial^3 f}{\partial y\,\partial x^2},$$

and so on.

We shall attempt to approximate $f(x, y)$ near the origin by a fourth-degree polynomial in x and y. Write $g(t) = f(0 + th, 0 + tk) = f(th, tk) = e^{th}\sin(tk)$ for $0 \le t \le 1$. Then, from the above remarks,

$$g(1) = f(h, k) = e^h \sin(k) = g(0) + g'(0) + \frac{g''(0)}{2!} + \frac{g^{(3)}(0)}{3!} + \frac{g^{(4)}(0)}{4!} + R_5,$$

where R_5 is the remainder term. Now

$$g(0) = 0;$$

$$g'(0) = \left(h\frac{\partial}{\partial x} + k\frac{\partial}{\partial y}\right)f\Big|_{(0,0)} = he^x\sin(y)|_{(0,0)} + ke^x\cos(y)|_{(0,0)} = k,$$

$$g''(0) = \left(h\frac{\partial}{\partial x} + k\frac{\partial}{\partial y}\right)^2 f\Big|_{(0,0)} = h^2\frac{\partial^2 f(0,0)}{\partial x^2} + 2hk\frac{\partial^2 f(0,0)}{\partial x\,\partial y} + k^2\frac{\partial^2 f(0,0)}{\partial y^2}$$

$$= h^2e^x\sin(y)|_{(0,0)} + 2hke^x\cos(y)|_{(0,0)} - k^2e^x\sin(y)|_{(0,0)} = 2hk,$$

$$g^{(3)}(0) = \left(h\frac{\partial}{\partial x} + k\frac{\partial}{\partial y}\right)^3 f\Big|_{(0,0)} = h^3\frac{\partial^3 f(0,0)}{\partial x^3} + 3h^2k\frac{\partial^3 f(0,0)}{\partial x^2\,\partial y}$$

$$+ 3hk^2\frac{\partial^3 f(0,0)}{\partial x\,\partial y^2} + k^3\frac{\partial^3 f(0,0)}{\partial y^3} = 3h^2k - k^3,$$

$$g^{(4)}(0) = \left(h\frac{\partial}{\partial x} + k\frac{\partial}{\partial y}\right)^4 f\Big|_{(0,0)} = h^4\frac{\partial^4 f(0,0)}{\partial x^4} + 4h^3k\frac{\partial^4 f(0,0)}{\partial x^3\,\partial y}$$

$$+ 6h^2k^2\frac{\partial^4 f(0,0)}{\partial x^2\,\partial y^2} + 4hk^3\frac{\partial^4 f(0,0)}{\partial x\,\partial y^3} + k^4\frac{\partial^4 f(0,0)}{\partial y^4}$$

$$= 4h^3k - 4hk^3.$$

Then

$$e^h\sin(k) = k + \frac{2hk}{2!} + \frac{3h^2k - k^3}{3!} + \frac{4h^3k - 4hk^3}{4!} + R_5$$

$$= k + hk + \frac{h^2k}{2} - \frac{k^3}{6} + \frac{h^3k}{6} - \frac{hk^3}{6} + R_5.$$

This fourth-degree polynomial in h and k approximates $e^h \sin(k)$ near (0,0), with an error given by R_5. For some θ between 0 and 1,

$$R_5 = \frac{g^{(5)}(\theta)}{5!} = \frac{1}{5!}\left(h\frac{\partial}{\partial x} + k\frac{\partial}{\partial y}\right)^5 f\bigg|_{(\theta h, \theta k)}.$$ ∎

This two-variable form of Taylor's theorem will be useful in the next section, when we develop a test for finding maxima and minima of functions of two variables.

E. Directional Derivatives

It is often the case that we are interested in the instantaneous rate of change of $f(x_1, \ldots, x_n)$ as $(x_1, \ldots, x_n)$ varies in a given direction. To begin with the two-variable case, suppose that f is a function of two variables, defined in some region of the xy-plane, containing a particular point (a, b) as well as all the points on some directed line segment L from (a, b).

If θ is the angle of inclination of L to the x-axis, then (Fig. 5-11) any (x, y) on L can be written as $(a + r\cos(\theta), b + r\sin(\theta))$, where $r \geq 0$. For (x, y) constrained to lie on L, then, $f(x, y)$ becomes $f(a + r\cos(\theta),\ b + r\sin(\theta))$, which is a function of r alone. The *directional derivative* of f at (a, b) in the direction L from (a, b) is somewhat naturally defined as the derivative of $f(a + r\cos(\theta), b + r\sin(\theta))$, with respect to r, evaluated at (a, b) (i.e., at $r = 0$). Assuming that we can apply the chain rule to f,

$$\frac{df(a + r\cos(\theta), b + r\sin(\theta))}{dr}\bigg|_{r=0} = \left(\frac{\partial f}{\partial x}\frac{\partial x}{\partial r} + \frac{\partial f}{\partial y}\frac{\partial y}{\partial r}\right)\bigg|_{r=0}$$

$$= \frac{\partial f(a, b)}{\partial x}\cos(\theta) + \frac{\partial f(a, b)}{\partial y}\sin(\theta).$$

We shall denote this derivative as $(D_L f)(a, b)$.

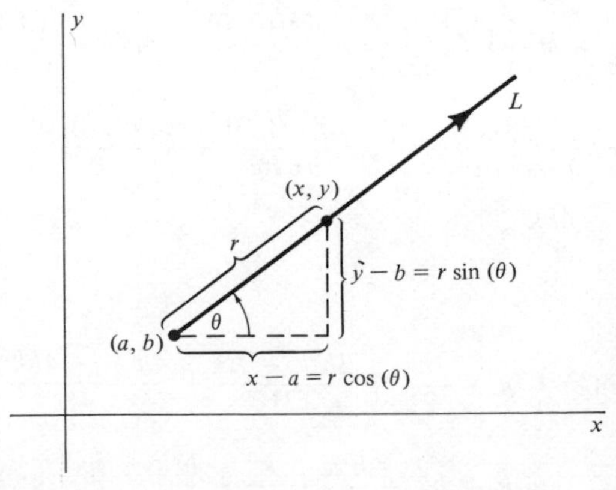

Fig. 5-11

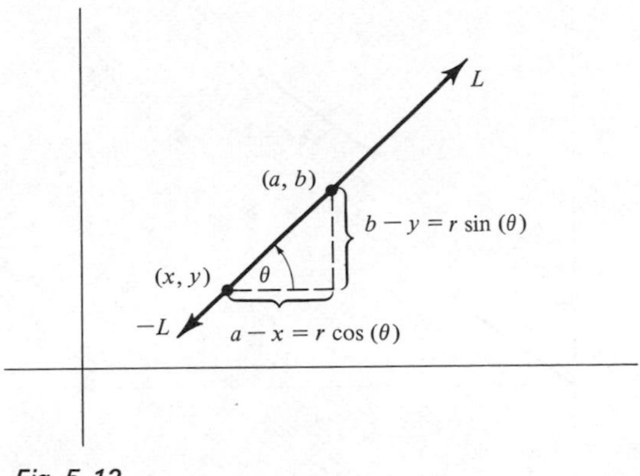

Fig. 5-12

Note what happens if we reverse the direction of L, or, in vector terminology, take the derivative along $-L$ instead of along L. If (x, y) is on $-L$, then $x = a - r\cos(\theta)$, and $y = b - r\sin(\theta)$, with θ as in Fig. 5-12. Then $f(x, y) = f(a - r\cos(\theta), b - r\sin(\theta))$, and

$$\begin{aligned}(D_{-L}f)(a, b) &= \frac{\partial f}{\partial x}\frac{\partial x}{\partial r} + \frac{\partial f}{\partial y}\frac{\partial y}{\partial r}\bigg|_{r=0} \\ &= \frac{\partial f(a, b)}{\partial x}(-\cos(\theta)) + \frac{\partial f(a, b)}{\partial y}(-\sin(\theta)) \\ &= -(D_L f)(a, b).\end{aligned}$$

As we might expect, reversing the direction changes the sign of the directional derivative.

There are various ways of looking at directional derivatives. Note that $(\cos(\theta), \sin(\theta))$ is a unit vector in the direction L of Fig. 5-11, and that

$$(D_L f)(a, b) = \left(\frac{\partial f(a, b)}{\partial x}, \frac{\partial f(a, b)}{\partial y}\right) \cdot (\cos(\theta), \sin(\theta)),$$

a dot product of 2-vectors. The vector $(f_x(a, b), f_y(a, b))$ is called the gradient of f at (a, b) and is denoted $\nabla f(a, b)$. This is read "del f at (a, b)." Then

$$(D_L f)(a, b) = \nabla f(a, b) \cdot u,$$

where u is a unit vector in the direction of L.

Another approach to the directional derivative can be made in terms of limits. We can think of $(D_L f)(a, b)$ as $\lim_{\Delta s \to 0} \Delta f/\Delta s$, where Δf is a function increment $f(x, y) - f(a, b)$, with (x, y) in the given direction, and Δs is the distance from (a, b) to (x, y). To see this, write L as a vector (h, k). Then (x, y) is in the

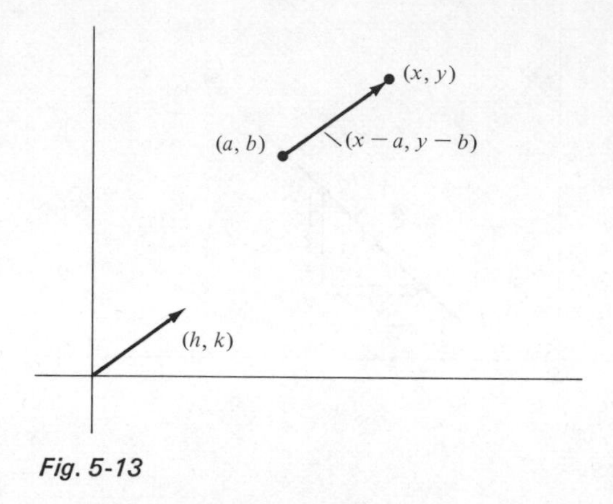

Fig. 5-13

direction L from (a, b) exactly when $(x - a, y - b)$ is parallel to (h, k) (see Fig. 5-13). Then, for some $t > 0$, $x - a = th$ and $y - b = tk$. Further, $\Delta s = \sqrt{t^2h^2 + t^2k^2} = t\|L\|$. Then, as a limit,

$$\lim_{\Delta s \to 0} \frac{\Delta f}{\Delta s} = \lim_{t \to 0+} \frac{f(a + th, b + tk) - f(a, b)}{t\|L\|}.$$

Now suppose that f is differentiable at (a, b). Then we can write

$$f(a + th, b + tk) - f(a, b) = f_x(a, b)th + f_y(a, b)tk + \varepsilon\sqrt{(th)^2 + (tk)^2},$$

where $\varepsilon \to 0$ as $t \to 0$. This gives us

$$\frac{\Delta f}{\Delta s} = \frac{f(a + th, b + tk) - f(a, b)}{t\|L\|} = f_x(a, b)\frac{h}{\|L\|} + f_y(a, b)\frac{k}{\|L\|} + \varepsilon.$$

As $t \to 0+$, the right side approaches the limit

$$\nabla f(a, b) \cdot \left(\frac{h}{\|L\|}, \frac{k}{\|L\|}\right).$$

Since $(h/\|L\|, k/\|L\|)$ is a unit vector in the direction of L, then the last expression is exactly $(D_L f)(a, b)$, showing that

$$\lim_{\Delta s \to 0} \frac{\Delta f}{\Delta s} = (D_L f)(a, b),$$

as given by the previous approach, when f is differentiable at (a, b).

Thus far we have concentrated on the two-variable case because the diagrams and vectors are easier to draw and visualize. However, the dot product and limit formulations just given generalize easily to higher dimensions. If f is a function of n-variables, then the *gradient* of f at $(a_1, \ldots, a_n)$ is defined to be the vector

$$\left(\frac{\partial f}{\partial x_1}(a_1, \ldots, a_n), \ldots, \frac{\partial f}{\partial x_n}(a_1, \ldots, a_n)\right)$$

denoted $\nabla f(a_1, \ldots, a_n)$. If L is a unit vector, then the *directional derivative* of f at $(a_1, \ldots, a_n)$ in the direction of L is given by

$$(D_L f)(a_1, \ldots, a_n) = \nabla f(a_1, \ldots, a_n) \cdot L.$$

As above, when f is differentiable at $(a_1, \ldots, a_n)$, it is easy to show that this is equivalent to

$$\lim_{\Delta s \to 0} \frac{\Delta f}{\Delta s},$$

where $\Delta f = f(x_1, \ldots, x_n) - f(a_1, \ldots, a_n)$ with $(x_1, \ldots, x_n)$ in the direction of L from $(a_1, \ldots, a_n)$ and $\Delta s = \|(x_1, \ldots, x_n) - (a_1, \ldots, a_n)\|$.

EXAMPLE

Find the directional derivative of $e^{xyz}\sin(xz)$ at $(1, \pi, 2)$ in the direction $(3, 1, 4)$. First,

$$\nabla f(x, y, z)$$
$$= (yze^{xyz}\sin(xz) + e^{xyz}z\cos(xz), xze^{xyz}\sin(xz), xye^{xyz}\sin(xz) + e^{xyz}x\cos(xz)).$$

Then

$$\nabla f(1, \pi, 2) = (2\pi e^{2\pi}\sin(2) + e^{2\pi}2\cos(2), 2e^{2\pi}\sin(2), e^{2\pi}\sin(2) + e^{2\pi}\cos(2)).$$

Now, a unit vector in the direction of $(3, 1, 4)$ is $(1/\sqrt{26})(3, 1, 4)$. Then

$$(D_{(3,1,4)})e^{xyz}\sin(xz)(1, \pi, 2)$$
$$= \frac{1}{\sqrt{26}}[6\pi e^{2\pi}\sin(2) + 6e^{2\pi}\cos(2) + 2e^{2\pi}\sin(2) + 4\pi e^{2\pi}\sin(2) + 4e^{2\pi}\cos(2)]$$
$$= \frac{e^{2\pi}}{\sqrt{26}}[(10\pi + 2)\sin(2) + 10\cos(2)].$$

As an exercise, we shall derive the same result as a limit. We want to compute

$$\lim_{\Delta s \to 0} \frac{\Delta f}{\Delta s},$$

where Δf and Δs are as defined above.

To compute Δf explicitly, suppose that (x, y, z) is in the direction determined by L from $(1, \pi, 2)$. Then $(x - 1, y - \pi, z - 2)$ is parallel to L, so, for some $t > 0$, $x - 1 = 36t$, $y - \pi = t$, and $z - 2 = 4t$. Then $(x, y, z) = (1 + 3t, \pi + t, 2 + 4t)$, and

$$f(x, y, z) = f(1 + 3t, \pi + t, 2 + 4t) = e^{(1+3t)(\pi+t)(2+4t)} \sin[(1 + 3t)(2 + 4t)].$$

Then

$$\begin{aligned}\Delta f &= f(x, y, z) - f(1, \pi, 2)\\ &= e^{(1+3t)(2+4t)} \sin[(1 + 3t)(2 + 4t)] - e^{2\pi} \sin(2)\\ &= \exp[12t^3 + (12\pi + 6)t^2 + (10\pi + 2)t + 2\pi] \sin(2 + 10t + 12t^2) - e^{2\pi} \sin(2).\end{aligned}$$

Next, $\Delta s = \|(x, y, z) - (3, 1, 4)\| = t\sqrt{26}$. Then

$$\frac{\Delta f}{\Delta s} = \frac{1}{\sqrt{26}} \frac{\begin{array}{r}\exp[12t^3 + (12\pi + 6)t^2 + (10\pi + 2)t + 2\pi]\\ \times \sin(2 + 10t + 12t^2) - e^{2\pi} \sin(2)\end{array}}{t}.$$

As $t \to 0+$, both numerator and denominator in the right-hand term go to zero. In order to invoke L'Hospital's Rule, consider the quotient of the derivatives:

$$\begin{aligned}&[36t^2 + 2t(12\pi + 6) + (10\pi + 2)] \exp[12t^3 + (12\pi + 6)t^2 + (10\pi + 2)t + 2\pi]\\ &\qquad \times \sin(2 + 10t + 12t^2) + \exp[12t^3 + (12\pi + 6)t^2 + (10\pi + 2)t + 2\pi]\\ &\qquad \times (24t + 10) \cos(2 + 10t + 12t^2).\end{aligned}$$

This approaches $(10\pi + 2)e^{2\pi} \sin(2) + e^{2\pi}(10) \cos(2)$ as $t \to 0$. Then

$$\lim_{\Delta s \to 0} \frac{\Delta f}{\Delta s} = \frac{e^{2\pi}}{\sqrt{26}} [(10\pi + 2) \sin(2) + 10 \cos(2)],$$

in agreement with the result obtained first by the other method. ∎

Exercises for Section 5.5

1. Use the method of Section 5.5A to prove the following: If f is continuous on the line segment from (a, b, c) to $(a + h, b + k, c + p)$, and differentiable at all points on this segment, then, for some θ, $0 < \theta < 1$ and

$$\begin{aligned}&f(a + h, b + k, c + p) - f(a, b, c)\\ &\quad = h \frac{\partial f(a + \theta h, b + \theta k, c + \theta p)}{\partial x} + k \frac{\partial f(a + \theta h, b + \theta k, c + \theta p)}{\partial y}\\ &\qquad + p \frac{\partial f(a + \theta h, b + \theta k, c + \theta p)}{\partial z}\end{aligned}$$

2. Give an example of a region D in the plane, and a function f defined on D such that $\partial f/\partial x = \partial f/\partial y = 0$ at each point of D but $f(x, y)$ is not constant on D. *Hint:* The reason that this does not contradict the theorem proved in the text is that here we have made no assumption to the effect that any two points of D can be connected by a polygonal line in D.

3. State and prove an n-dimensional analogue of the mean value theorems of this section and Exercise 1.

4. Prove the following: If f is differentiable at each point of a set D of points in R^3, and if any pair of points of D can be joined by a polygonal path in D, and if $\partial f/\partial x = \partial f/\partial y = \partial f/\partial z = 0$ in D, then f is a constant function on D.

5. Use the theorem of Section 5.5B to show that each of the following is differentiable at the indicated point. Also prove differentiability directly from the definition.

(a) $x^2 + 2xy + y^2$, $(1, 2)$
(b) e^{x+y}, $(1, 1)$
(c) $e^x \ln(y)$, $(0, \pi/2)$
(d) $xy + \ln(xy)$, $(1, 1)$

6. Prove the following: If f is continuous in some r-neighborhood of (a, b, c), and if $\partial f/\partial x$, $\partial f/\partial y$, and $\partial f/\partial z$ are defined in this neighborhood and continuous at (a, b, c), then f is differentiable at (a, b, c).

7. Let f be differentiable at $(a_1, \ldots, a_n)$. Show that

$$\nabla f(a_1, \ldots, a_n) \cdot u = \lim_{\Delta s \to 0} \frac{\Delta f}{\Delta s},$$

where u is a unit vector in R^n, $\Delta f = f(x_1, \ldots, x_n) - f(a_1, \ldots, a_n)$ for $(x_1, \ldots, x_n)$ in the direction from $(a_1, \ldots, a_n)$ given by u, and $\Delta s = \|(x_1, \ldots, x_n) - (a_1, \ldots, a_n)\|$.

8. Compute each of the following directional derivatives, first by using the gradient, then as a limit.

(a) $(D_{(1,2,1)}x^2yz)(1, 1, 1)$
(b) $(D_{(2,2,1,4)}e^{xyzw})(0, 1, 1, \pi)$
(c) $(D_{(0,1)} \ln(x) \cos(y))(2, \pi)$
(d) $(D_{(1,0)}e^x \sin(y))(0, \pi)$

9. Prove that $(D_L f)(a, b) + (D_L g)(a, b) = (D_L(f + g))(a, b)$.

10. Obtain Taylor polynomial approximations in two variables for each of the following, up to the indicated degree terms, and about the given point. In each case, attempt to estimate the error in the approximation.

(a) $e^x \ln(y)$, about $(0, 1)$, up to and including fourth-degree terms.
(b) $\sin(x) \cos(y)$, about $(0, \pi/2)$, up to and including fifth-degree terms.
(c) $\dfrac{1}{(1 - x)(1 - y)}$ about $(0, 0)$, up to and including fourth-degree terms.
(d) x^3y^2 about $(1, 2)$, up to and including fifth-degree terms.

11. In each of the problems in Exercise 10, the given function $f(x, y)$ can be written as $\alpha(x) \cdot \beta(y)$, for appropriate choices of α and β. For (a) through (d), expand $\alpha(x)$ and $\beta(y)$ about the relevant points in one-variable Taylor polynomials and multiply these together. Then compare with the solutions obtained by the methods of D [e.g., in (a), put $\alpha(x) = e^x$, $\beta(y) = \ln(y)$. Expand $\alpha(x)$ about 0, $\beta(y)$ about 1].

12. Let f be differentiable at (a, b). Show that the tangent plane to $z = f(x, y)$ at (a, b) is given by

$$z - f(a, b) = \nabla f(a, b) \cdot (x - a, y - b).$$

5.6 Applications of Partial Derivatives to Max–Min Problems

It is often of interest to know where a function of n variables, defined on some set D of points in R^n, achieves maximum and minimum values. At the very least, if f is continuous and D is compact, we know that maxima and minima do occur.

In many instances we have restricted ourselves to functions of two variables with the understanding that the generalization to higher dimensions was immediate. This is not the case here. We shall treat max–min problems for functions of two variables in some detail, but there will be no obvious means of going to three, four, or more variables. It is simply the case that the tests for maxima and minima become very complicated very fast as the number of variables increases.

Suppose then that f is a function of two variables, defined over some set D of points in the plane, and that (ξ, η) is an interior point of D. We say that f has a *relative, or local, maximum* (*minimum*) at (ξ, η) if, for some $r > 0, f(x, y) \leq f(\xi, \eta)$ $(f(x, y) \geq f(\xi, \eta))$ for every (x, y) in the r-neighborhood of (ξ, η); that is, f has a local max (min) at (ξ, η) if $f(\xi, \eta)$ is at least as large (no larger than) $f(x, y)$ for all (x, y) sufficiently close about (ξ, η) in D.

In the same spirit, f has an *absolute maximum* (*minimum*) at (a, b) in D if $f(x, y) \leq f(a, b)$ $(f(x, y) \geq f(a, b))$ for all (x, y) in D. Here we explicitly remove the condition that (a, b) be an interior point of D. It may be in ∂D.

As an example, let $f(x, y) = \sqrt{x^2 + y^2}$ for all (x, y). Then f has an absolute and local min at $(0, 0)$ but no relative or absolute max, as we can make $\sqrt{x^2 + y^2}$ as we like by choosing x and/or y large.

By contrast, consider $g(x, y) = \sqrt{x^2 + y^2}$ for $x^2 + y^2 \leq 4$. Then g has an absolute (and relative) min at $(0, 0)$, and an absolute (but not relative) max at any point (x, y) with $x^2 + y^2 = 4$.

In general, it is not so easy to spot points at which a function of two variables has a max or min. We therefore look for some systematic procedure to follow in searching them out. As usual, a review of the one-variable case provides some clues. Consider a function g defined and continuous on some closed interval $[a, b]$. To find points at which g has a max or min (either relative or absolute), we would normally proceed as follows:

(1) Examine the critical points α, $a < \alpha < b$, at which $g'(\alpha) = 0$. Then g has a relative max at α if $g''(\alpha) < 0$, and a relative min if $g''(\alpha) > 0$. If $g''(\alpha) = 0$, g may have either a max, min, or inflection at α. [As an example, $g(x) = x^3$ has an inflection point at 0, where the graph $y = g(x)$ crosses its horizontal tangent.]
(2) Examine points $a < \alpha < b$ at which $g'(\alpha)$ does not exist [e.g., $g(x) = |x|$ has a min at 0, but no derivative there].
(3) Test the end points separately. At a and b, the derivative test yields no information. For example, $g(x) = 1/x$ has a min at 4 and a max at 3 on $[3, 4]$. At neither point does $g'(x)$ vanish.

In attempting to generalize this procedure to the two-variable case, let us assume that f is continuous on some compact, connected set D of points in the plane. The connectedness is not really so important, as if D is disconnected we can examine each connected piece of D separately for local max and min points of f. By assuming that D is compact, however, we are assured of the existence of max and min points of f in D.

First, what kind of points shall we call critical? A simple argument suggests that critical points should be points interior (i.e., not on the boundary) to D at which $\partial f/\partial x$ and $\partial f/\partial y$ vanish. For suppose, say, that (ξ, η) is interior to D, and that f has a local max at (ξ, η). For some r, then, all points in the r-neighborhood about (ξ, η) are also in D. Examine the partial derivatives at (ξ, η). First,

$$\frac{\partial f(\xi, \eta)}{\partial x} = \lim_{\Delta x \to 0} \frac{f(\xi + \Delta x, \eta) - f(\xi, \eta)}{\Delta x}.$$

If we choose Δx small and positive, then $(\xi + \Delta x, \eta)$ is within a distance r of (ξ, η), and $f(\xi + \Delta x, \eta) - f(\xi, \eta) \leq 0$, as f has a relative max at (ξ, η). Then (Fig. 5-14)

$$\frac{f(\xi + \Delta x, \eta) - f(\xi, \eta)}{\Delta x} \leq 0,$$

and as we let $\Delta x \to 0$ through positive values, we have

$$\frac{\partial f(\xi, \eta)}{\partial x} \leq 0.$$

If we choose Δx small and negative, however, then still $f(\xi + \Delta x, \eta) - f(\xi, \eta) \leq 0$, but now

$$\frac{f(\xi + \Delta x, \eta) - f(\xi, \eta)}{\Delta x} \geq 0,$$

so that in the limit as $\Delta x \to 0$ from the left, $\partial f(\xi, \eta)/\partial x \geq 0$. Combining these two results yields $\partial f(\xi, \eta)/\partial x = 0$. Similar reasoning leads to $\partial f(\xi, \eta)/\partial y = 0$.

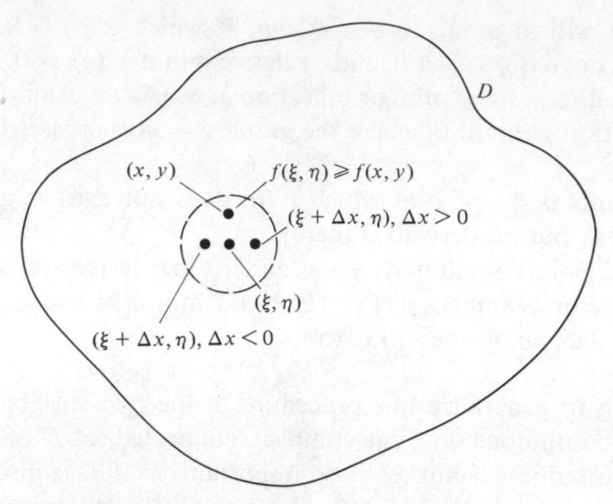

Fig. 5-14

If f has a relative min at (ξ, η), then essentially the same reasoning leads again to

$$\frac{\partial f(\xi, \eta)}{\partial x} = \frac{\partial f(\xi, \eta)}{\partial y} = 0.$$

We therefore have proved the following.

Theorem

If f has a local max or local min at an interior point (ξ, η) of D, and if the partials of f exist at (ξ, η), then

$$\frac{\partial f(\xi, \eta)}{\partial x} = \frac{\partial f(\xi, \eta)}{\partial y} = 0.$$

The converse of this is not true. If

$$\frac{\partial f(\xi, \eta)}{\partial x} = \frac{\partial f(\xi, \eta)}{\partial y} = 0,$$

then f need not have a local max or min at (ξ, η), even if (ξ, η) is interior to D. Nevertheless, interior points at which the partials vanish are certainly the first places to look for local max and min points.

Places where the partials vanish, but the function has neither a max nor a min, are the two-dimensional analogues of inflection points for functions of one variable. As an example, let $f(x, y) = x^2 - y^2$ and $D = \{(x, y) \mid x^2 + y^2 \leq 4\}$. Then

$$\frac{\partial f(0, 0)}{\partial x} = \frac{\partial f(0, 0)}{\partial y} = 0.$$

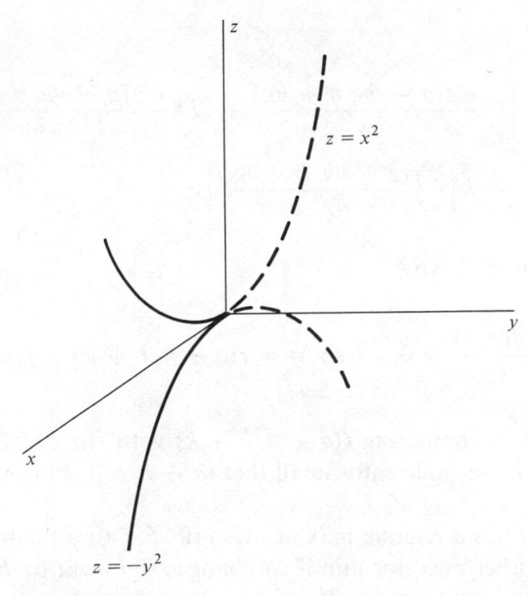

Fig. 5-15

But f has neither a local max nor a local min at (0, 0). For $f(0, 0) = 0$, but $f(x, 0) > 0$ for x arbitrarily close to 0 while $f(0, y) < 0$ for y arbitrarily small. In this example it is easy to see what is happening graphically. When viewed along the x-axis (Fig. 5-15, in the plane $y = 0$), the graph looks like $z = x^2$, which has a min at $x = 0$. But when viewed along the y-axis (in the plane $x = 0$), the graph looks like $z = -y^2$, which has a max at $y = 0$. The surface resembles a saddle at the origin, motivating the term *saddle point* for an interior point where both partials vanish but the function has neither a local max nor a local min.

We can now outline a procedure for attempting to determine max and min points of f on D, paralleling exactly the one-variable case.

(1) Examine the critical points.
(2) Examine interior points at which one or both partials fail to exist.
(3) Test the points on the boundary of D.

It must be admitted that we have no satisfactory procedure for handling (2) or (3) in general. For (1) the difficulty is that we lack something like a second derivative test which tells us what the first derivative test has found. We now derive such a test, basing our argument on Taylor's theorem for two variables.

Suppose that (a, b) is a critical point of f and that f and its first and second partials are continuous in some neighborhood about (a, b). By Taylor's theorem, we may expand f about (a, b) in a two-variable polynomial of degree 1 plus remainder:

$$f(a + h, b + k) = f(a, b) + h\frac{\partial f(a, b)}{\partial x} + k\frac{\partial f(a, b)}{\partial y} + R_2,$$

where

$$R_2 = \frac{1}{2}\left[h^2 \frac{\partial^2 f(a + \theta h, b + \theta k)}{\partial x^2} + 2hk \frac{\partial^2 f(a + \theta h, b + \theta k)}{\partial x\, \partial y} + k^2 \frac{\partial^2 f(a + \theta h, b + \theta k)}{\partial y^2}\right],$$

for some θ, $0 < \theta < 1$. Now

$$\frac{\partial f(a, b)}{\partial x} = \frac{\partial f(a, b)}{\partial y} = 0, \qquad \text{so } \Delta f = f(a + h, b + k) - f(a, b) = R_2.$$

We are interested in comparing $f(a + h, b + k)$ with $f(a, b)$ when h and k are small [say, h and k are sufficiently small that $(a + h, b + k)$ is within a distance r of (a, b)].

Observe that f has a relative max at (a, b) if $\Delta f \leq 0$; a relative min at (a, b) if $\Delta f \geq 0$, and neither max nor min if Δf changes sign near (a, b). We therefore look for sufficient conditions for $R_2 \geq 0$ or $R_2 \leq 0$, or for R_2 to change sign. For convenience, write

$$\alpha = \frac{\partial^2 f(a + \theta h, b + \theta k)}{\partial x^2}, \qquad \beta = \frac{\partial^2 f(a + \theta h, b + \theta k)}{\partial y^2},$$

and

$$\gamma = \frac{\partial^2 f(a + \theta h, b + \theta k)}{\partial x\, \partial y}.$$

Then $\Delta f = \frac{1}{2}[h^2\alpha + 2hk\gamma + k^2\beta]$. We will also write

$$A = \frac{\partial^2 f(a, b)}{\partial x^2}, \qquad B = \frac{\partial^2 f(a, b)}{\partial y^2}, \qquad C = \frac{\partial^2 f(a, b)}{\partial x\, \partial y}.$$

Note that, as $(h, k) \to (0, 0)$, we have $\alpha \to A$, $\beta \to B$, and $\gamma \to C$ by continuity of the second partials.

Now consider the following cases.

(1) $AB - C^2 > 0$. Then necessarily A and B are both nonzero and of the same sign. If $A < 0$, choose δ sufficiently small that $\alpha < 0$ and $\alpha\beta - \gamma^2 > 0$ for $0 < \theta < 1$, $h^2 + k^2 < \delta^2$. This can be done by continuity of the second partials of f. Then

$$\Delta f = \frac{1}{2\alpha}\,[(\alpha h + \gamma k)^2 + (\alpha\beta - \gamma^2)k^2] < 0,$$

for $h^2 + k^2 < \delta^2$, hence f has a local max at (a, b). Similarly, if $A > 0$, choose δ so that $\alpha > 0$ and $\alpha\beta - \gamma^2 > 0$ for $0 < \theta < 1$, $h^2 + k^2 < \delta^2$. Then

$$\Delta f = \frac{1}{2\alpha}[(\alpha h + \gamma k)^2 + (\alpha\beta - \gamma^2)k^2] > 0,$$

so f has a relative min at (a, b).

(2) $AB - C^2 < 0$. We shall show that f has neither a relative max nor a relative min at (a, b). Suppose first that $A \neq 0$. Put $k = 0$ in the increment formula and calculate

$$\lim_{h\to 0}\frac{\Delta f}{h^2} = \lim_{h\to 0}\frac{1}{h^2}\frac{1}{2}\left(h^2\frac{\partial^2 f(a + \theta h, b)}{\partial x^2}\right) = \frac{1}{2}\frac{\partial^2 f(a, b)}{\partial x^2} = \frac{A}{2}.$$

Next, put $h = -tC$ and $k = tA$ to calculate

$$\begin{aligned}\lim_{t\to 0}\frac{\Delta f}{t^2} &= \lim_{t\to 0}\frac{1}{2t^2}[t^2C^2 f_{xx}(a - \theta tC, b + \theta tA) \\ &\quad - 2t^2ACf_{yx}(a - \theta tC, b + \theta tA) + t^2A^2 f_{yy}(a - \theta tC, b + \theta tA)] \\ &= \tfrac{1}{2}(C^2A - 2AC^2 + A^2B) = \frac{A}{2}(AB - C^2).\end{aligned}$$

Since $AB - C^2 < 0$, then $A/2$ and $A(AB - C^2)/2$ differ in sign. This shows that Δf takes on positive and negative values arbitrarily close to (a, b) for different choices of h and k, hence f has neither a relative max nor a relative min at (a, b). A similar argument holds if $B \neq 0$.

(3) $A = B = 0$. Since $AB - C^2 < 0$, then $C \neq 0$. The argument now proceeds as above. Putting $h = k$, calculate

$$\begin{aligned}\lim_{h\to 0}\frac{\Delta f}{h^2} &= \lim_{h\to 0}\frac{1}{2h^2}[h^2 f_{xx}(a + \theta h, b + \theta k) + 2h^2 f_{yx}(a + \theta h, b + \theta k) \\ &\quad + h^2 f_{yy}(a + \theta h, b + \theta k)] = C,\end{aligned}$$

and, putting $h = -k$, obtain $\lim_{h\to 0}(\Delta f/h^2) = -C$. As above, f can have neither a relative max nor a relative min at (a, b).

By way of summary:

If $AB - C^2 > 0$: $\begin{cases}\text{if } A, B < 0\text{—local max at } (a, b) \\ \text{if } A, B > 0\text{—local min at } (a, b).\end{cases}$

If $AB - C^2 < 0$: neither a relative max nor a relative min at (a, b).

If $AB - C^2 = 0$, no conclusion may be drawn. We leave it to the reader to show by example that in this case f may have a relative max, a relative min, or neither at (a, b).

It is important to realize that this test for functions of two variables has the same limitations as the analogous derivative test for functions of one variable. Further, the test provides only sufficient conditions, not necessary ones. Analogous tests can be developed for functions of more than two variables, using Taylor's theorem, but we shall not pursue such tests here.

By way of illustration, we shall consider a problem of finding absolute and relative maxima and minima for a function on a compact set. This will give some idea of the difficulties inherent in the consideration of boundary points.

EXAMPLE

Let $f(x, y) = (x - 2)^2 y + y^2 - y$, for $x \geq 0$, $y \geq 0$, and $x + y \leq 4$. Thus (x, y) is confined to the triangular region bounded by the x-axis, the y-axis, and the line $x + y = 4$.

We first look for critical points by examining the partial derivatives of interior points. We have

$$\frac{\partial f}{\partial x} = 2y(x - 2) \qquad \text{and} \qquad \frac{\partial f}{\partial y} = (x - 2)^2 + 2y - 1.$$

Setting these equal to zero and solving yields only one critical point, $(2, \frac{1}{2})$. The second partials are given by

$$\frac{\partial^2 f}{\partial x^2} = 2y, \qquad \frac{\partial^2 f}{\partial y^2} = 2, \qquad \frac{\partial^2 f}{\partial x\, \partial y} = 2(x - 2).$$

Using the notation of the preceding discussion, we then have, at the point $(2, \frac{1}{2})$,

$$AB - C^2 = (1)(2) - 0^2 = 2 > 0.$$

Thus we have a relative max or min at $(2, \frac{1}{2})$. Since A (and also B) are positive, then we have a local min at $(2, \frac{1}{2})$.

We now have to see what happens at the boundary points. Split the boundary into three pieces, consisting of boundary points on the x-axis, those on the y-axis, and those on the line $x + y = 4$. Treat these pieces one at a time.

On the x-axis part of the boundary, we have $y = 0$, $0 \leq x \leq 4$. Put $g(x) = f(x, 0) = 0$ for $0 \leq x \leq 4$. This is constant, having a max and a min at each value of x in $[0, 4]$.

On the y-axis part of the boundary, we have $x = 0$, $0 \leq y \leq 4$. Put $h(y) = f(0, y) = y^2 + 3y$. This is a strictly increasing function of y in $[0, 4]$, with a min at $y = 0$ and a max at $y = 4$.

Finally, on $x + y = 4$, put $x = y - 4$ and write $h(y) = f(y - 4, y) = y^3 - 11y^2 + 35y$. Now $h'(y) = 3y^2 - 22y + 35$, so $h'(y) = 0$ for $y = 5$ and

$y = \frac{7}{3}$. Since 5 is outside the interval $[0, 4]$, the only relevant value is $y = \frac{7}{3}$. Since $h''(\frac{7}{3}) = -8 < 0$, then h has a relative max at $\frac{7}{3}$.

Totaling up our information from analysis of the function on the boundary, we have to examine values of f at $(0, 0)$, $(0, 4)$, $(4, 0)$, and $(\frac{5}{3}, \frac{7}{3})$. We have $f(0, 0) = 0$, $f(0, 4) = 28$, $f(4, 0) = 0$, and $f(\frac{5}{3}, \frac{7}{3}) = \frac{91}{27}$. By direct comparison, then, $f(x, y)$ has a maximum value of 28 and a minimum value of 0 on the boundary.

Finally, note that $f(2, \frac{1}{2}) = -\frac{1}{4}$ [recall that $(2, \frac{1}{2})$ was the critical point we found at the beginning of this example]. Thus f has an absolute maximum at $(0, 4)$ on the boundary, and an absolute minimum at the interior point $(2, \frac{1}{2})$. ▮

Note that in this example our success in finding what we wanted on the boundary hinged on being able to break the boundary up into pieces for which equations could easily be found and substituted into the original function, giving us a one-variable max–min problem on each piece of boundary. If the boundary is very complicated, this is not always practical, so that boundary points are generally difficult to treat in max–min problems in two variables.

Exercises for Section 5.6

1. Suppose that f has a local min at the point (ξ, η) interior to D. Prove that $f_x(\xi, \eta) = f_y(\xi, \eta) = 0$, assuming of course that these partials exist.

2. For each of (a) and (b), give an example of a function f defined on a set D such that

   (a) $AB - C^2 = 0$, and f has neither local max or min at (a, b).
   (b) $AB - C^2 = 0$, and f has a local max at (a, b).

3. Determine points in the given sets where the following have local maxima or minima.

   (a) $x^2 + 2xy + y^2$, on $\{(x, y) \mid x^2 + y^2 \leq 4\}$
   (b) $e^{xy} \sin(x)$, on $\{(x, y) \mid |x| \leq 1, |y| \leq 1\}$
   (c) $\ln(x + y) \cdot (x^2 + y^2)$, on $\{(x, y) \mid 2 \leq x^2 + y^2 \leq 4\}$
   (d) $\arctan(xy)$, on $\{(x, y) \mid 0 \leq x \leq 1 \text{ and } 0 \leq y \leq 1\}$

4. Find the point on the surface $z = x^2 + y^2$ that is nearest the point $(1, 2, 3)$.

5. Suppose that a function f of three variables has a local max or min at a point (ξ, η, ζ) interior to D. Show that

$$f_x(\xi, \eta, \zeta) = f_y(\xi, \eta, \zeta) = f_z(\xi, \eta, \zeta) = 0.$$

6. Extend the result of Exercise 5 to functions of n variables.

7. Examine $ax^2 + bxy + cy^2$ for relative maxima and minima in the plane, with a, b, and c real numbers. (Consider possibilities for a, b, and c which will give critical points and allow application of the second derivative test.)

8. For each of the problems of Exercise 3, look for maxima and minima on the boundary of the given region, and determine in each case the absolute maximum and absolute minimum on the given region.

5.7 Implicit Function Theorems

In many instances a relationship between variables does not specify one explicitly in terms of the others, as with $y = x^2$ or $z = e^x \sin(y)$, but is given by one or more equations which the variables must satisfy, as with, say $x^2 + y^2 = 4$, or $e^x \sin(y) - zxy + 2 = 0$. In such instances it may not be obvious (or even true) that the equation can be solved (at least in theory) for one variable in terms of the others, determining, say, $y = f(x)$ or $z = g(x, y)$.

The idea behind the implicit function theorems is to give sufficient conditions for systems of one or more equations in two or more variables to determine a functional relationship between the variables. Perhaps of equal importance, the theorems also tell us how to calculate the derivatives of the functions whose existence they assert.

We begin with the case of one equation in two variables, which the reader is probably familiar with, at least from a computational point of view, from elementary calculus. The problem is this. Suppose that we have an equation $F(x, y) = 0$, for (x, y) in a specified set D. Can we solve for, say, $y = f(x)$, for x in some interval I? That is, is there a function f and an interval I such that $(x, f(x))$ is in D when x is in I, and also $F(x, f(x)) = 0$ for every x in I?

The answer is, probably not. One problem is that we have given very little information about F. This can be rectified by placing suitable restrictions as needed. More serious is that the problem as stated is a global one—it asks for a solution to $F(x, y) = 0$ in all of D. As we shall see, it is more reasonable just to seek a solution near some point (a, b), where $F(a, b) = 0$. That is, we shall seek local solutions.

To clarify this distinction, consider a simple example. Let $F(x, y) = x^2 + y^2 - 4$. The equation $F(x, y) = 0$ is satisfied by certain pairs (x, y) which geometrically determine a circle of radius 2 about the origin in the xy-plane. Obviously we can solve for, say, y, by writing $y = \pm\sqrt{4 - x^2}$ for $-2 \leq x \leq 2$. The trouble with this is the $\pm$ part. For $-2 < x < 2$, we have two possibilities for y, hence no single functional relationship between x and y. However, choose a point, say $(\sqrt{2}, \sqrt{2})$ on the circle, different from $(2, 0)$ or $(-2, 0)$. We can solve for $y = f(x)$ satisfying $F(x, y) = 0$ locally, around $(\sqrt{2}, \sqrt{2})$, by putting $y = +\sqrt{4 - x^2}$ for x in an interval about $\sqrt{2}$, for example, $\sqrt{2} - \frac{1}{10} < x < \sqrt{2} + \frac{1}{10}$. The size of the interval is somewhat arbitrary, as long as it is not too big. In this example we cannot have $|x| > 2$. Similarly, about $(-\sqrt{3}, -1)$, we can solve for $y = -\sqrt{4 - x^2}$, say with $-\sqrt{3} - \frac{1}{20} < x < \sqrt{3} + \frac{1}{20}$ and we have $-1 = -\sqrt{4 - 3}$ and $F(x, -\sqrt{4 - x^2}) = 0$ for each x in $(-\sqrt{3} - \frac{1}{20}, -\sqrt{3} + \frac{1}{20})$ (see Fig. 5-16).

Note that, even with the very simple F of this example, it is not always possible to obtain even a local solution, say $y = f(x)$ for $a - \varepsilon < x < a + \varepsilon$, about each (a, b) with $F(a, b) = 0$. For example, $F(2, 0) = 0$. But there is no interval $2 - \varepsilon < x < 2 + \varepsilon$ in which we can specify $y = f(x)$ with $f(2) = 0$ and

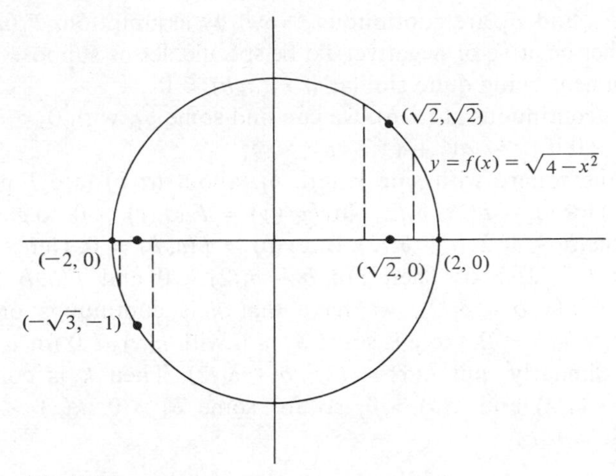

Fig. 5-16

$F(x, f(x)) = 0$. In fact, if $2 < x < 2 + \varepsilon$, no matter how small the positive number ε is, then there is no real y with $F(x, y) = 0$.

This example was designed to emphasize the distinction between local and global solutions, and why one must generally be content to look for the former. The example $F(x, y) = 2xe^y + 3x^3y + \sin(xy)$ will illustrate a second point. The equation $F(x, y) = 0$ is satisfied at $(0, 0)$. Assuming that there is a solution $y = g(x)$ in some interval $-\varepsilon < x < \varepsilon$, what can we say about g? Is g continuous, is g differentiable, and, if so, what is $g'(x)$? Note also that the ability in theory to assert the existence of g' does not help to calculate $g'(x)$ for any x. Such matters are the point to the following theorem.

Implicit Function Theorem (for one function of two variables)

Suppose that F, $\partial F/\partial x$, and $\partial F/\partial y$ are continuous in some δ-neighborhood about the point (a, b), and that $F(a, b) = 0$, but that $\partial F(a, b)/\partial y \neq 0$. Then there is a positive number ε and a function f defined for $a - \varepsilon < x < a + \varepsilon$ such that $f(a) = b$ and

(1) *$F(x, f(x)) = 0$ for $a - \varepsilon < x < a + \varepsilon$.*
(2) *f is continuous on $(a - \varepsilon, a + \varepsilon)$.*
(3) *f is differentiable on $(a - \varepsilon, a + \varepsilon)$, and, for $a - \varepsilon < x < a + \varepsilon$,*

$$f'(x) = \frac{\dfrac{-\partial F(x, f(x))}{\partial x}}{\dfrac{\partial F(x, f(x))}{\partial y}}.$$

We shall not prove the theorem in complete detail but discuss the idea of the proof to see how the argument can be adapted to more complicated situations. To begin with, we have a circular region, of radius δ, about (a, b), at every point of which F, F_x, and F_y are continuous. Now, by assumption, $F_y(a, b) \neq 0$, so $F_y(a, b)$ is either positive or negative. To be specific, let us suppose that $F_y(a, b) > 0$, the argument being quite similar if $F_y(a, b) < 0$.

Since F_y is continuous at (a, b), we can find some δ_1, with $0 < \delta_1 \leq \delta$, such that $F_y(x, y) > 0$ if $(x - a)^2 + (y - b)^2 < \delta_1^2$.

Look at the square with side length δ_1, about (a, b) (see Fig. 5-17). Put $g(y) = F(a, y)$ for $|y - b| < \delta_1/2$. Now $g'(y) = F_y(a, y) > 0$, so g is an increasing function on $(b - \delta_1/2, b + \delta_1/2)$. But $g(b) = F(a, b) = 0$. Hence $g(b - \delta_1/2) < 0$ and $g(b + \delta_1/2) > 0$. Then $F(a, b - \delta_1/2) < 0$ and $F(a, b + \delta_1/2) > 0$. Putting $h(x) = F(x, b - \delta_1/2)$, we have that h is continuous on $(a - \delta_1/2, a + \delta_1/2)$. Since $h(a) < 0$, there is some $\delta_2 > 0$ with $h(x) < 0$ for $a - \delta_2/2 < x < a + \delta_2/2$. Similarly, put $k(x) = F(x, b + \delta_1/2)$. Then k is continuous on $(a - \delta_1/2, a + \delta_1/2)$ and $k(a) > 0$, so for some $\delta_3 > 0$, $k(x) > 0$ whenever $a - \delta_3 < x < a + \delta_3$.

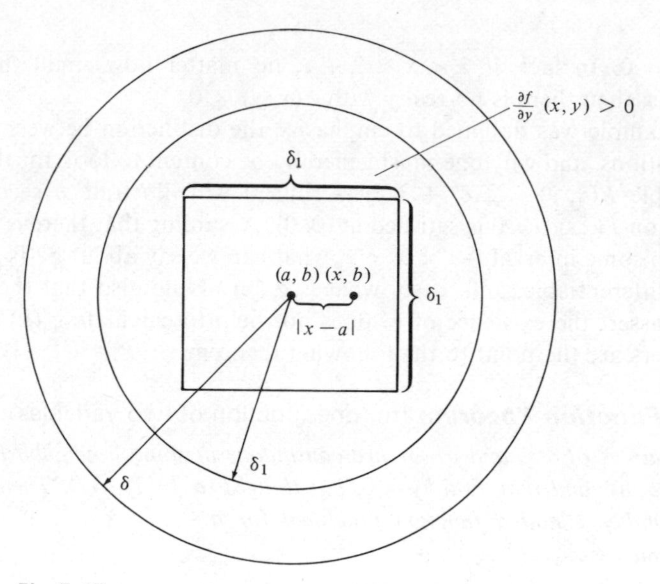

Fig. 5-17

We can now pick the ε of the theorem. Let 2ε be the smallest of the numbers δ_1, δ_2, δ_3. Note that, whenever $|x - a| < \varepsilon$, then $h(x) = F(x, b - \delta_1/2) < 0$ and $k(x) = F(x, b + \delta_1/2) > 0$.

Now choose any x in $(a - \varepsilon, a + \varepsilon)$, and put $r(y) = F(x, y)$, $b - \delta_1/2 < y < b + \delta_1/2$. Then r is continuous. Also, $r(b - \delta_1/2) < 0$ and $r(b + \delta_1/2) > 0$. By the intermediate value theorem, there is some y with $b - \delta_1/2 < y < b + \delta_1/2$ and $r(y) = 0$. Further, as r is an increasing function, there is only one such y. Denote $y = f(x)$.

We then have, for each x in $(a - \varepsilon, a + \varepsilon)$, a unique number $f(x)$ in

$(b - \delta_1/2, b + \delta_1/2)$ with $F(x, f(x)) = 0$. Existence of f as called for in conclusion (1) of the theorem has now been established.

Continuity of f on $(a - \varepsilon, a + \varepsilon)$ can be established by a routine ε, δ argument. Once this has been done, $f'(x)$ is obtained as follows. Since F, $\partial F/\partial x$, and $\partial F/\partial y$ are continuous about (a, b), F is differentiable in some neighborhood about (a, b). Then we can write, for Δx and Δy sufficiently small, and (x, y) in this neighborhood.

$$F(x + \Delta x, y + \Delta y) - F(x, y) = \frac{\partial F(x, y)}{\partial x}\Delta x + \frac{\partial F(x, y)}{\partial y}\Delta y + \varphi_1 \Delta x + \varphi_2 \Delta y,$$

where $\varphi_1, \varphi_2 \to 0$ as $(\Delta x, \Delta y) \to (0, 0)$. Now write $y = f(x)$, and, given Δx, choose $\Delta y = f(x + \Delta x) - f(x)$. Then we have

$$\begin{aligned} &F(x + \Delta x, f(x + \Delta x)) - F(x, f(x)) \\ &\qquad = \frac{\partial F(x, f(x))}{\partial x}\Delta x + \frac{\partial F(x, f(x))}{\partial y}\Delta y + \varphi_1 \Delta x + \varphi_2 \Delta y. \end{aligned}$$

Since $y = f(x)$ is a solution to $F(x, y) = 0$ in the neighborhood obtained above, then $F(x + \Delta x, f(x + \Delta x)) = 0 = F(x, y)$. Then

$$\frac{\Delta y}{\Delta x} = \frac{F_x(x, f(x)) + \varphi_1}{-F_y(x, f(x)) - \varphi_2}.$$

Since f is continuous, then $\Delta y \to 0$ as $\Delta x \to 0$. Hence $\varphi_1, \varphi_2 \to 0$ as $\Delta x \to 0$. Then

$$f'(x) = \lim_{\Delta x \to 0} \frac{\Delta f}{\Delta x} = \frac{-F_x(x, f(x))}{F_y(x, f(x))},$$

where $y = f(x)$ in the evaluation of the partials.

This not only establishes the existence of $f'(x)$, for x sufficiently close to a, but gives a way of calculating $f'(x)$ in terms of partials of F, which we can figure out without knowing $f(x)$ explicitly.

Note that the derivative formula we have just obtained is consistent with what we should expect from the chain rule. If $F(x, f(x)) = g(x) = 0$ for $a - \varepsilon < x < a + \varepsilon$, then $g'(x) = 0 = F_x + F_y y_x = F_x + F_y f'(x)$, so $f'(x) = -F_x/F_y$ whenever $F_y \neq 0$. Note also that the theorem could be turned around to give sufficient conditions for a solution $x = h(y)$ to exist. In this case we would have $F(h(y), y) = 0$ for $b - \varepsilon < y < b + \varepsilon$, and the assumption that $\partial F(a, b)/\partial y \neq 0$ would be replaced by the assumption that $\partial F(a, b)/\partial x \neq 0$.

We now generalize to the case of one function of three variables. The proof will very much parallel that just sketched for the two-variable case.

Implicit Function Theorem (for one function of three variables)

Suppose that F, $\partial F/\partial x$, $\partial F/\partial y$, and $\partial F/\partial z$ are continuous in some δ-neighborhood of (a, b, c), and $F(a, b, c) = 0$, but $\partial F(a, b, c)/\partial z \neq 0$. Then there is a positive *number* ε *and a function* g *defined for* $|x - a| < \varepsilon$, $|y - b| < \varepsilon$, *such that* $g(a, b) = c$ *and*

(1) $F(x, y, g(x, y)) = 0$ *for* $|x - a| < \varepsilon$, $|y - b| < \varepsilon$.
(2) g *is continuous on* $\{(x, y) \mid |x - a| < \varepsilon, |y - b| < \varepsilon\}$.
(3) g *has partials in* $|x - a| < \varepsilon$, $|y - b| < \varepsilon$, *given by*

$$\frac{\partial g(x, y)}{\partial x} = \frac{\dfrac{-\partial F(x, y, g(x, y))}{\partial x}}{\dfrac{\partial F(x, y, g(x, y))}{\partial z}}$$

and

$$\frac{\partial g(x, y)}{\partial y} = \frac{\dfrac{-\partial F(x, y, g(x, y))}{\partial y}}{\dfrac{\partial F(x, y, g(x, y))}{\partial z}},$$

PROOF

Suppose for definiteness that $\partial F(a, b, c)/\partial z > 0$. Then there is some δ_1, with $\partial F(x, y, z)/\partial z > 0$ for $(x - a)^2 + (y - b)^2 + (z - c)^2 < \delta_1^2$. For any given (x, y) with $|x - a| \leq \delta_1/2$ and $|y - b| \leq \delta_1/2$, $F(x, y, z)$ is an increasing function of z for $|z - c| \leq \delta_1/2$, as $\partial F(x, y, z)/\partial z > 0$. Since $F(a, b, c) = 0$, then $F(a, b, c - \delta_1/2) < 0$ and $F(a, b, c + \delta_1/2) > 0$. Hence by continuity we may choose $\delta_2 > 0$ so that $\delta_2 < \delta_1/2$, $F(x, y, c - \delta_1/2) < 0$, and $F(x, y, c + \delta_1/2) > 0$ for $|x - a| < \delta_2/2$, $|y - b| < \delta_2/2$ (see Fig. 5-18).

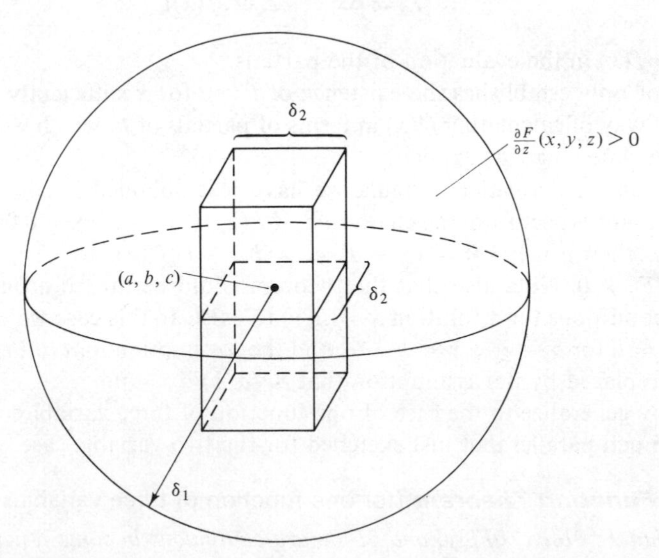

Fig. 5-18

Now, for a given (x, y), with $|x - a| \leq \delta_2/2$ and $|y - b| \leq \delta_2/2$, $F(x, y, z)$ is a continuous function of z for $|z - c| \leq \delta_1/2$. By the intermediate value theorem, there is some z with $|z - c| \leq \delta_1/2$ and $F(x, y, z) = 0$, as $F(x, y, c - \delta_1/2) < 0$ and $F(x, y, c + \delta_1/2) > 0$. Such z is unique, as $F(x, y, z)$ is increasing as a function of z. Thus we denote such z as $g(x, y)$, and the existence of g is established.

We now have, for each (x, y) with $|x - a| \leq \delta_2/2$ and $|y - b| \leq \delta_2/2$ a unique number $g(x, y)$ such that $F(x, y, g(x, y)) = 0$. To show that g is continuous on $|x - a| \leq \delta_1/2$, $|y - b| \leq \delta_1/2$, we first prove continuity at (a, b).

Let $\varepsilon > 0$. For convenience, we may suppose that $\varepsilon \leq \delta_2/2$. Now, $F(a, b, c - \varepsilon) < 0$ and $F(a, b, c + \varepsilon) > 0$. Choose $\delta > 0$ so that $F(x, y, c + \varepsilon) > 0$ and $F(x, y, c - \varepsilon) < 0$ if $|x - a| < \delta$, $|y - b| < \delta$. For such (x, y), there is some unique z with $|z - c| < \varepsilon$ and $F(x, y, z) = 0$. Since then $z = g(x, y)$, and we already have $c = g(a, b)$, then $|g(x, y) - g(a, b)| < \varepsilon$ for $|x - a| < \delta$, $|y - b| < \delta$, and g is continuous at (a, b).

A similar argument applies at any (x, y) with $|x - a| < \delta_2/2$, $|y - b| < \delta_2/2$. For, given $|x_0 - a| < \delta_2/2$ and $|y_0 - b| < \delta_2/2$, then put $z_0 = g(x_0, y_0)$ and observe that $F(x_0, y_0, z_0 + \delta_2/2) > 0$ and $F(x_0, y_0, z_0 - \delta_2/2) < 0$; hence choose $\delta > 0$ so that $\delta < \delta_2/2$ and $F(x, y, z_0 + \varepsilon) > 0$ and $F(x, y, z_0 - \varepsilon) < 0$ if $|x - x_0| < \delta$ and $|y - y_0| < \delta$. Since $\delta < \delta_2/2$, then for each such (x, y), we have $z = g(x, y)$ satisfying $|z - z_0| < \varepsilon$, that is, $|g(x, y) - g(x_0, y_0)| < \varepsilon$.

We must now calculate $\partial g/\partial x$ and $\partial g/\partial y$. First, we consider $\partial g(x, y)/\partial x$, for $|x - a| < \delta_2/2$ and $|y - b| < \delta_2/2$. For Δx and Δz sufficiently small, we havc, by differentiability of F,

$$F(x + \Delta x, y, z + \Delta z) - F(x, y, z) \\ = F_x(x, y, z)\,\Delta x + F_z(x, y, z)\,\Delta z + \varphi_1\,\Delta x + \varphi_2\,\Delta z,$$

where $\varphi_1, \varphi_2 \to 0$ as $\Delta x, \Delta z \to 0$. Choose $z = g(x, y)$ and $\Delta z = g(x + \Delta x, y) - g(x, y)$. By continuity of g, note that $\Delta z \to 0$ as $\Delta x \to 0$. Then we have

$$F(x + \Delta x, y, g(x + \Delta x, y)) - F(x, y, g(x, y)) \\ = F_x(x, y, z)\,\Delta x + F_z(x, y, z)\,\Delta z + \varphi_1\,\Delta x + \varphi_2\,\Delta z.$$

Now $F(x + \Delta x, y, g(x + \Delta x, y)) = 0 = F(x, y, g(x, y))$, so

$$\frac{\Delta z}{\Delta x} = \frac{\dfrac{-\partial F(x, y, g(x, y))}{\partial x} - \varphi_1}{\dfrac{\partial F(x, y, g(x, y))}{\partial z} + \varphi_2}.$$

In the limit as $\Delta x \to 0$, $\Delta z/\Delta x \to \partial g(x, y)/\partial x$, and we have

$$\frac{\partial g(x, y)}{\partial x} = \frac{\dfrac{-\partial F(x, y, g(x, y))}{\partial x}}{\dfrac{\partial F(x, y, g(x, y))}{\partial z}}.$$

Similarly,

$$\frac{\partial g(x, y)}{\partial y} = \frac{\dfrac{-\partial F(x, y, g(x, y))}{\partial y}}{\dfrac{\partial F(x, y, g(x, y))}{\partial z}},$$

completing the proof. ∎

Again, we note that we can with ease adapt the theorem to the problem of solving for y in terms of x and z, or x in terms of y and z. In the latter case, replace the condition $\partial F(a, b, c)/\partial z \neq 0$ by $\partial F(a, b, c)/\partial x \neq 0$, and in the former, $\partial F(a, b, c)/\partial y \neq 0$.

We shall now state an implicit function theorem for one function of n variables. From the above considerations the reader should have no trouble in understanding the ideas involved.

Implicit Function Theorem (for one function of n variables)

Suppose that F, $\partial F/\partial x_1$, $\partial F/\partial x_2, \ldots, \partial F/\partial x_n$ are continuous in some δ-neighborhood of $(a_1, \ldots, a_n)$. Suppose also that $F(a_1, \ldots, a_n) = 0$, but $\partial F(a_1, \ldots, a_n)/\partial x_n \neq 0$. Then there is a positive number ε and a function g defined for $|x_1 - a_1| < \varepsilon, \ldots, |x_{n-1} - a_{n-1}| < \varepsilon$ such that $a_n = g(a_1, \ldots, a_{n-1})$ and

(1) *$F(x_1, \ldots, x_{n-1}, g(x_1, \ldots, x_{n-1})) = 0$ for $|x_1 - a_1| < \varepsilon, \ldots, |x_{n-1} - a_{n-1}| < \varepsilon$.*
(2) *g is continuous on $|x_1 - a_1| < \varepsilon, \ldots, |x_{n-1} - a_{n-1}| < \varepsilon$.*
(3) *For $|x_1 - a_1| < \varepsilon, \ldots, |x_{n-1} - a_{n-1}| < \varepsilon$, and $1 \leq j \leq n - 1$,*

$$\frac{\partial g(x_1, \ldots, x_{n-1})}{\partial x_j} = \frac{\dfrac{-\partial F(x_1, \ldots, x_{n-1}, g(x_1, \ldots, x_{n-1}))}{\partial x_j}}{\dfrac{\partial F(x_1, \ldots, x_{n-1}, g(x_1, \ldots, x_{n-1}))}{\partial x_n}}.$$

We now turn to the problem of implicit functions defined by two or more functions of three or more variables. Here we shall be less rigorous and simply expand on the ideas suggested by the above situations.

Suppose, to begin, that we have two equations $F(x, y, z) = 0$ and $G(x, y, z) = 0$. We might expect, under reasonable conditions, to be able to solve for two of the variables in terms of the third.

To be concrete, suppose that $F(a, b, c) = G(a, b, c) = 0$, and that F, $\partial F/\partial x$, $\partial F/\partial y$, and $\partial F/\partial z$ are continuous in some neighborhood about (a, b, c). If $\partial F(a, b, c)/\partial z \neq 0$, we can then solve $F(x, y, z) = 0$ for $z = f(x, y)$ in some square region $|x - a| < \delta$, $|y - b| < \delta$ of the xy-plane. Substitute this into G to obtain $G(x, y, z) = G(x, y, f(x, y)) = H(x, y)$. Note that $c = f(a, b)$, so $H(a, b) = 0$. We can then attempt to eliminate, say, y from H. This can be done if H, $\partial H/\partial x$, $\partial H/\partial y$ are continuous about (a, b) and $\partial H(a, b)/\partial y \neq 0$. We can then solve for $y = g(x)$ such that $G(x, g(x), f(x, g(x)) = 0$ in some interval $|x - a| < \delta_1$, so that $z = f(x, g(x))$, and $y = g(x)$ are in terms of x. Next, we want to obtain $g'(x)$ and $\partial f/\partial x$ and $\partial f/\partial y$. First, $F(x, y, f(x, y)) = 0$ for $|x - a| < \delta$, $|y - b| < \delta$, so we have

$$\frac{\partial f}{\partial x} = \frac{\dfrac{-\partial F(x, y, f(x, y))}{\partial x}}{\dfrac{\partial F(x, y, f(x, y))}{\partial z}}$$

and

$$\frac{\partial f}{\partial y} = \frac{\dfrac{-\partial F(x, y, f(x, y))}{\partial y}}{\dfrac{\partial F(x, y, f(x, y))}{\partial z}},$$

as we might expect from previous implicit function theorems.

To obtain $g'(x)$, note that $G(x, g(x), f(x, g(x))) = 0$ for $|x - a| < \delta_1$. Then, for $|x - a| < \delta_1$,

$$\frac{\partial G}{\partial x} + \frac{\partial G}{\partial y}\frac{\partial g}{\partial x} + \frac{\partial G}{\partial z}\frac{\partial f}{\partial x} = 0.$$

Then

$$\frac{\partial g}{\partial x} = g'(x) = \frac{\dfrac{-\partial G(x, g(x), f(x, g(x)))}{\partial x} - \dfrac{\partial G(x, g(x), f(x, g(x)))}{\partial z}\dfrac{\partial f(x, g(x))}{\partial x}}{\dfrac{\partial G(x, g(x), f(x, g(x)))}{\partial y}}.$$

As a second example, suppose we have, say, two equations in four variables:

$$F(x, y, u, w) = 0$$

$$G(x, y, u, w) = 0.$$

Suppose that $F(a, b, c, d) = G(a, b, c, d) = 0$. Under reasonable conditions, we can solve $F(x, y, u, w) = 0$ for $w = f(x, y, u)$ in some region $|x - a| < \delta$, $|y - b| < \delta$, $|u - c| < \delta$. Then, putting $H(x, y, u) = G(x, y, u, f(x, y, u))$, we solve for $u = g(x, y)$ in some square $|x - a| < \delta_1$, $|y - b| < \delta_1$. Then

$$w = f(x, y, g(x, y)) = h(x, y)$$

and

$$u = g(x, y).$$

Again, using the chain rule, we can calculate $\partial w/\partial x$, $\partial w/\partial y$, $\partial w/\partial u$, $\partial u/\partial x$, and $\partial u/\partial y$. We shall do this in detail for illustration.

Note first that

$$F(x, y, g(x, y), f(x, y, g(x, y))) = 0$$

for $|x - a| < \delta_1$, $|y - b| < \delta_1$. By the chain rule (here x and y are both independent, so $\partial y/\partial x = \partial x/\partial y = 0$):

$$F_x + F_u u_x + F_w w_x = 0$$

and

$$F_y + F_u u_y + F_w w_y = 0.$$

Similarly,

$$G_x + G_u u_x + G_w w_x = 0$$

and

$$G_y + G_u u_y + G_w w_y = 0.$$

Pair these off as follows:

$$F_u u_x + F_w w_x = -F_x,$$
$$G_u u_x + G_w w_x = -G_x,$$

and

$$F_u u_y + F_w w_y = -F_y,$$
$$G_u u_y + G_w w_y = -G_y.$$

From the first two, solve for

$$\frac{\partial u}{\partial x} = \frac{\begin{vmatrix} -F_x & F_w \\ -G_x & G_w \end{vmatrix}}{\begin{vmatrix} F_u & F_w \\ G_u & G_w \end{vmatrix}} = \frac{-F_x G_w + F_w G_x}{F_u G_w - F_w G_u},$$

$$\frac{\partial w}{\partial x} = \frac{\begin{vmatrix} F_u & -F_x \\ G_u & -G_x \end{vmatrix}}{\begin{vmatrix} F_u & F_w \\ G_u & G_w \end{vmatrix}} = \frac{-F_u G_x + F_x G_u}{F_u G_w + G_u F_w}.$$

For the last two, obtain

$$\frac{\partial u}{\partial y} = \frac{\begin{vmatrix} -F_y & F_w \\ -G_y & G_w \end{vmatrix}}{\begin{vmatrix} F_u & F_w \\ G_u & G_w \end{vmatrix}} = \frac{-F_y G_w + F_w G_y}{F_u G_w - F_w G_u},$$

and

$$\frac{\partial w}{\partial y} = \frac{\begin{vmatrix} F_u & -F_y \\ G_u & -G_y \end{vmatrix}}{\begin{vmatrix} F_u & F_w \\ G_u & G_w \end{vmatrix}} = \frac{-F_uG_y + F_yG_u}{F_uG_w - F_wG_u}.$$

These give $\partial u(x, y)/\partial x, \ldots, \partial w(x, y, g(x, y))/\partial y$ in terms of partials of F and G.

There is a useful notational device here. The Jacobian $\partial(A, B)/\partial(\alpha, \beta)$ is defined as $\begin{vmatrix} A_\alpha & A_\beta \\ B_\alpha & B_\beta \end{vmatrix}$. Then

$$\frac{\partial u}{\partial x} = \frac{-\dfrac{\partial(F, G)}{\partial(x, w)}}{\dfrac{\partial(F, G)}{\partial(u, w)}}, \qquad \frac{\partial w}{\partial x} = \frac{-\dfrac{\partial(F, G)}{\partial(u, x)}}{\dfrac{\partial(F, G)}{\partial(u, w)}},$$

$$\frac{\partial u}{\partial y} = \frac{-\dfrac{\partial(F, G)}{\partial(y, w)}}{\dfrac{\partial(F, G)}{\partial(u, w)}}, \qquad \frac{\partial w}{\partial y} = \frac{-\dfrac{\partial(F, G)}{\partial(u, y)}}{\dfrac{\partial(F, G)}{\partial(u, w)}}.$$

"Cancellation" of symbols (neglecting the minus signs) then gives a check on the calculations. For example,

$$\frac{\dfrac{\not\partial(\not F, \not G)}{\partial(x, \not w)}}{\dfrac{\not\partial(\not F, \not G)}{\partial(u, \not w)}} \text{ “}=\text{” } \frac{\partial u}{\partial x}.$$

Finally, we calculate $\partial w/\partial u$. Note that

$$F(x, y, u, f(x, y, u)) = 0 \qquad \text{for } |x - a| < \delta, |y - b| < \delta, |u - c| < \delta.$$

Then

$$\frac{\partial F}{\partial u} + \frac{\partial F}{\partial w}\frac{\partial w}{\partial u} = 0,$$

so

$$\frac{\partial w}{\partial u} = \frac{-\dfrac{\partial F}{\partial u}}{\dfrac{\partial F}{\partial w}}.$$

Obviously we can consider new implicit function theorems ad infinitum. We shall stop here, with the hope that the student has gotten some feeling for sufficient conditions for existence in implicit-function situations, and, equally important, the techniques for calculating derivatives of the solution functions in terms of derivatives of the original data functions.

As an application, suppose that a surface S in 3-space is described as a locus of points satisfying $F(x, y, z) = 0$, and that we want the equation of the tangent plane (which we suppose exists) to S at (a, b, c). In practice we may not be able to solve for $z = f(x, y)$ and use the previous formula. However, if in theory such a solution is possible around (a, b), then we have the tangent plane given by $f_x(a, b)(x - a) + f_y(a, b)(y - b) = z - c$. Now

$$f_x(a, b) = \frac{-F_x(a, b, f(a, b))}{F_z(a, b, f(a, b))} \quad \text{and} \quad f_y(a, b) = \frac{-F_y(a, b, f(a, b))}{F_z(a, b, f(a, b))}.$$

so our tangent plane is

$$F_1(a, b, f(a, b))(x - a) + F_2(a, b, f(a, b))(y - b) + F_3(a, b, f(a, b))(z - c) = 0.$$

The numbers $F_1(a, b, f(a, b))$, $F_2(a, b, f(a, b))$, $F_3(a, b, f(a, b))$ are called *direction numbers* of the plane, and the vector $(F_1(a, b, f(a, b)),\ F_2(a, b, f(a, b)),\ F_3(a, b, f(a, b)))$ is *normal* to the plane.

Exercises for Section 5.7

1. Let $F(x, y, z) = x^3y + e^x \sin(yz) + \ln(x + y + z)$ for $x > 0, y > 0, z > 0$.

(a) Show that the equation $F(x, y, z) = 0$ can be solved in some neighborhood of $(1, 0)$ for $z = f(x, y)$ such that $f(1, 0) = 0$.

(b) Compute $\partial z(1, 0)/\partial x$ and $\partial z(1, 0)/\partial y$.

(c) Assume that these partials exist, and compute

$$\frac{\partial^2 z(1, 0)}{\partial x^2}, \quad \frac{\partial^2 z(1, 0)}{\partial y^2}, \quad \frac{\partial^2 z(1, 0)}{\partial x\, \partial y}, \quad \frac{\partial^2 z(1, 0)}{\partial y\, \partial x}.$$

2. Let $F(x, y, z) = 2^x yz$, $G(x, y, z) = y/x + \sin(xyz)$.

(a) Show that these equations can be solved for $y = g(x)$, $z = h(x)$ in some neighborhood of 1 such that $g(1) = 0$ and $h(1) = 1$.

(b) Evaluate $h'(1)$ and $g'(1)$.

(c) Evaluate $h''(1)$ and $g''(1)$.

3. Let F, G, and H be functions of (x, y, z, u, v), continuous with continuous partials in some neighborhood of (a, b, c, d, p). Assume that $F(a, b, c, d, p) = G(a, b, c, d, p) = H(a, b, c, d, p) = 0$. Suppose that the equations can be solved for $z = f(x, y)$, $u = g(x, y)$, $v = h(x, y)$ in some neighborhood of (a, b).

(a) Compute $\partial z/\partial x$, $\partial z/\partial y$, $\partial g/\partial x$, $\partial g/\partial y$, $\partial h/\partial x$, $\partial h/\partial y$, stating as you go along the assumptions you are required to make to justify the steps.

(b) Compute $\partial^2 z/\partial x\, \partial y$, $\partial^2 g/\partial x^2$, and $\partial^2 h/\partial y^2$.

4. Let $F(x, y, z, w) = 0$, $G(x, y, z, w) = 0$, $H(x, y, z, w) = 0$ implicitly define $x = f(y)$, $z = g(y)$, and $w = h(y)$.
(**a**) Compute $f'(y)$, $g'(y)$, and $h'(y)$.
(**b**) Compute $f''(y)$, $g''(y)$, and $h''(y)$.

5. Suppose that $F(x, y, z, u, v, w) = 0$, $G(x, y, z, u, v, w) = 0$, and $H(x, y, z, u, v, w) = 0$ define x, u, and w as functions of y, z, and v.
(**a**) Compute $\partial x/\partial y$, $\partial x/\partial z$, $\partial x/\partial v$.
(**b**) Compute $\partial^2 u/\partial y^2$, $\partial^2 u/\partial z^2$, $\partial^2 u/\partial y\, \partial z$.
(**c**) Compute $\partial^3 w/\partial y\, \partial z\, \partial w$.

6. Suppose that the equations $F(x, y, z) = 0$, $G(x, y, z) = 0$ define surfaces in R^3, and that these surfaces intersect at (a, b, c). Find an expression for the angle between the normals to these surfaces at (a, b, c).

7. Two functions $f(x, y)$ and $g(x, y)$ are said to be functionally dependent over a set D in the plane if there is some function $H(t)$ such that $H(f(x, y)) = g(x, y)$ for (x, y) in D. Prove the following test for functional dependence: Suppose that f and g are continuous with continuous first partials in a set D, that the Jacobian $\partial(f, g)/\partial(x, y) = 0$ at each (x, y) in D, and that $\partial f(x_0, y_0)/\partial x \neq 0$. Then, there is some $H(t)$ such that $H(f(x, y)) = g(x, y)$ for each (x, y) in some neighborhood of (x_0, y_0) in D. *Hint:* Show that the equation $z = f(x, y)$ can be solved for $y = A(x, z)$ in some neighborhood of (x_0, z_0), where $z_0 = f(x_0, y_0)$. Next, show that $\partial g(x, A(x, z))/\partial x = 0$ around (x_0, z_0). Hence conclude that $g(x, A(x, z))$ is independent of x. Put $H(z) = g(x, A(x, z))$.

8. Show that $\sin(x + y)$ and $\tan(x + y)$ are functionally dependent in the region $|x| \leq \pi/4$, $|y| \leq \pi/4$.

9. Show that $\ln(x^2 - y^2)$ and $\exp(x^4 - 2x^2y^2 + y^4)$ are functionally dependent in the region $x^2 - y^2 > 0$.

10. Show that $\frac{1}{3}(x^2 + 4y^2 + 4xy)$ and $\sin[\pi(x + 2y)]$ are functionally dependent in the entire plane.

11. In each of Exercise 8, 9, and 10, produce H such that $H(f(x, y)) = g(x, y)$.

5.8 Use of Differentials in Differentiating Implicitly Defined Functions

The technique described in Section 5.7 for computing derivatives of implicitly defined functions is called *implicit differentiation*. Differentials provide a different mechanism for computing these derivatives. We shall illustrate the method with an example.

Suppose that $u = f(x, y)$, $v = g(x, y)$ are defined implicitly in $|x - a| < \delta$, $|y - b| < \delta$ by equations $F(x, y, u, v) = 0 = G(x, y, u, v)$. Then

$$F(x, y, f(x, y), g(x, y)) = 0$$

and

$$G(x, y, f(x, y), g(x, y)) = 0$$

for all (x, y) with $|x - a| < \delta$ and $|y - b| < \delta$. Taking differentials, we have

$$dF = F_x\,dx + F_y\,dy + F_u\,du + F_v\,dv = 0$$

and

$$dG = G_x\,dx + G_y\,dy + G_u\,du + G_v\,dv = 0.$$

Now

$$du = f_x\,dx + f_y\,dy \qquad \text{and} \qquad dv = g_x\,dx + g_y\,dy.$$

Then

$$F_x\,dx + F_y\,dy + F_u(f_x\,dx + f_y\,dy) + F_v(g_x\,dx + g_y\,dy) = 0$$

and

$$G_x\,dx + G_y\,dy + G_u(f_x\,dx + f_y\,dy) + G_v(g_x\,dx + g_y\,dy) = 0.$$

Collecting terms, we have

$$(F_x + F_u f_x + F_v g_x)\,dx + (F_y + F_u f_y + F_v g_y)\,dy = 0$$

and

$$(G_x + G_u f_x + G_v g_x)\,dx + (G_y + G_u f_y + G_v g_y)\,dy = 0.$$

Recall now that in the definition of differential, the symbols dx and dy are independent variables. Choose $dx = 1$ and $dy = 0$ to obtain

$$\frac{-\partial F}{\partial x} = F_u f_x + F_v g_x$$

and

$$\frac{-\partial G}{\partial x} = G_u f_x + G_v g_x.$$

These can be solved for f_x and g_x, yielding

$$f_x = \frac{\dfrac{-\partial(F, G)}{\partial(x, v)}}{\dfrac{\partial(F, G)}{\partial(u, v)}} \qquad \text{and} \qquad g_x = \frac{\dfrac{-\partial(F, G)}{\partial(u, x)}}{\dfrac{\partial(F, G)}{\partial(u, v)}}.$$

Similarly, putting $dx = 0$ and $dy = 1$ yields

$$\frac{-\partial F}{\partial y} = F_u f_y + F_v g_y$$

and

$$\frac{-\partial G}{\partial y} = G_u f_y + G_v g_y.$$

Solving these for f_y and g_y, we obtain

$$f_y = \frac{-\dfrac{\partial(F, G)}{\partial(y, v)}}{\dfrac{\partial(F, G)}{\partial(u, v)}} \quad \text{and} \quad g_y = \frac{-\dfrac{\partial(F, G)}{\partial(u, y)}}{\dfrac{\partial(F, G)}{\partial(u, v)}}.$$

The reader can check that these agree with the results derived by the chain rule at the end of Section 5.7.

There is a slightly different way of looking at this method. From the first equations ($dF = 0$ and $dG = 0$), write

$$F_u\,du + F_v\,dv = -F_x\,dx - F_y\,dy$$

and

$$G_u\,du + G_v\,dv = -G_x\,dx - G_y\,dy.$$

Solve these for du and dv:

$$du = \frac{\begin{vmatrix} -F_x\,dx - F_y\,dy & F_v \\ -G_x\,dx - G_y\,dy & G_v \end{vmatrix}}{\begin{vmatrix} F_u & F_v \\ G_u & G_v \end{vmatrix}} = \frac{-\dfrac{\partial(F, G)}{\partial(x, v)}}{\dfrac{\partial(F, G)}{\partial(u, v)}}\,dx + \frac{-\dfrac{\partial(F, G)}{\partial(y, v)}}{\dfrac{\partial(F, G)}{\partial(u, v)}}\,dy$$

and

$$dv = \frac{\begin{vmatrix} F_u & -F_x\,dx - F_y\,dy \\ G_u & -G_x\,dx - G_y\,dy \end{vmatrix}}{\begin{vmatrix} F_u & F_v \\ G_u & G_v \end{vmatrix}} = \frac{-\dfrac{\partial(F, G)}{\partial(u, x)}}{\dfrac{\partial(F, G)}{\partial(u, v)}}\,dx + \frac{-\dfrac{\partial(F, G)}{\partial(u, y)}}{\dfrac{\partial(F, G)}{\partial(u, v)}}\,dy.$$

Then

$$du = f_x\,dx + f_y\,dy = \frac{-\dfrac{\partial(F, G)}{\partial(x, v)}}{\dfrac{\partial(F, G)}{\partial(u, v)}}\,dx + \frac{-\dfrac{\partial(F, G)}{\partial(y, v)}}{\dfrac{\partial(F, G)}{\partial(u, v)}}\,dy$$

and

$$dv = g_x\,dx + g_y\,dy = \frac{\dfrac{-\partial(F, G)}{\partial(u, x)}}{\dfrac{\partial(F, G)}{\partial(u, v)}}\,dx + \frac{\dfrac{-\partial(F, G)}{\partial(u, y)}}{\dfrac{\partial(F, G)}{\partial(u, v)}}\,dy.$$

Equating coefficients of dx and dy in these equations yields the previously derived formulas for f_x, f_y, g_x, and g_y. Note that it is the independence of dx and dy which ensures that the coefficient of dx on the left equals that of dx on the right, and so on.

The method we have just illustrated is quite generally applicable to any situation in which implicit differentiation is called for to compute partials. Sometimes the method of differentials seems a bit easier, although this is largely a function of personal taste.

Exercises for Section 5.8

1. Suppose that $F(x, y, z) = 0$ defines $z = f(x, y)$ implicitly. Use the method of differentials to compute $\partial f/\partial x$ and $\partial f/\partial y$, and compare with the results of Section 5.7.

2. Suppose that $F(x, y, z) = 0 = G(x, y, z)$ define $z = f(x)$ and $y = g(x)$. Use the method of differentials to compute $f'(x)$ and $g'(x)$.

3. Suppose that $F(x, y, z, u, v) = 0 = G(x, y, z, u, v)$ define $u = f(x, y, z)$, $v = g(x, y, z)$. Use the method of differentials to give expressions for $\partial u/\partial x$, $\partial u/\partial y$, $\partial u/\partial z$, $\partial v/\partial x$, $\partial v/\partial y$, and $\partial v/\partial z$.

4. Use the method of differentials to obtain $\partial u/\partial x$, $\partial u/\partial y$, $\partial u/\partial v$, $\partial u/\partial w$, $\partial z/\partial x$, $\partial z/\partial y$, $\partial z/\partial v$, and $\partial z/\partial w$ if $A(x, y, z, u, v, w) = B(x, y, z, u, v, w) = 0$, implicitly define $u = u(x, y, v, w)$ and $z = z(x, y, v, w)$.

5. Use the method of differentials to compute dw/dx, dz/dx, and dy/dx if $A(x, y, z, w) = B(x, y, z, w) = C(x, y, z, w) = 0$, define $w = w(x)$, $z = z(x)$, and $y = y(x)$.

6. Let f, g, and h be functions of x and y. Show that

$$\frac{\partial f}{\partial y}\frac{\partial(g, h)}{\partial(x, y)} + \frac{\partial g}{\partial y}\frac{\partial(h, f)}{\partial(x, y)} + \frac{\partial h}{\partial y}\frac{\partial(f, g)}{\partial(x, y)} = 0,$$

assuming that all the partials involved exist.

7. Let f, g, and h be functions of x, y, and z. Let $x = F(u, v)$, $y = G(u, v)$, and $z = H(u, v)$. Show that

$$\frac{\partial(f, g)}{\partial(u, v)} = \frac{\partial(f, g)}{\partial(y, z)}\frac{\partial(y, z)}{\partial(u, v)} + \frac{\partial(f, g)}{\partial(z, x)}\frac{\partial(z, x)}{\partial(u, v)} + \frac{\partial(f, g)}{\partial(x, y)}\frac{\partial(x, y)}{\partial(u, v)}.$$

5.9 Lagrange Multipliers

We have already considered max–min (or extremal) problems in some detail for functions of two independent variables. We want now to consider the problem of determining the max or min of a function, subject to one or more constraints. Such problems arise quite naturally in both mathematics and the physical sciences. As one example, suppose that we have a particle moving with a certain known velocity $v(x, y, z, t)$, but constrained to remain on the surface of a sphere $x^2 + y^2 + z^2 = r^2$. We may then wish to determine an extremum of the velocity, subject to the condition that $x^2 + y^2 + z^2 - r^2 = 0$ for all t.

To be specific, suppose that we want to find extrema of $f(x, y, z)$ on a set D, subject to a side condition given by $g(x, y, z) = 0$. One method worth trying is to solve $g(x, y, z) = 0$ for $z = h(x, y)$, and substitute into $f(x, y, z)$ to obtain $G(x, y) = f(x, y, h(x, y))$. Previously developed methods for functions of two variables may then be applied to G. This method has serious drawbacks, however, as h may be hard to find, or may exist only locally.

A second approach, due to Lagrange, is the method of Lagrange multipliers. Think of $g(x, y, z) = 0$ as describing a surface S in 3-space. In seeking extrema for $f(x, y, z)$ subject to $g(x, y, z) = 0$, we are looking for relative maxima and minima of $f(x, y, z)$ with (x, y, z) constrained to lie on the given surface. We shall now examine this problem geometrically. Suppose that $f(x, y, z)$ has a local extremum at (a, b, c), and $f(a, b, c) = k$. Think of $f(x, y, z) = k$ as a surface Σ. Assuming that Σ has a tangent plane at (a, b, c), this plane has normal vector $(f_x(a, b, c), f_y(a, b, c), f_z(a, b, c))$. Similarly, the tangent plane to S at (a, b, c) has normal vector $(g_x(a, b, c), g_y(a, b, c), g_z(a, b, c))$. We now claim that these two tangent planes coincide. To show this, it is sufficient to show that one of these normal vectors is a constant multiple of the other.

A proof of this can be achieved by using an implicit function theorem. We assume that not all of $g_x(a, b, c), g_y(a, b, c), g_z(a, b, c)$ are zero. Say, for example, that $g_z(a, b, c) \neq 0$. Then we can solve $g(x, y, z) = 0$ (in theory) for $z = h(x, y)$ around (a, b), and substitute to obtain $f(x, y, z) = f(x, y, h(x, y))$. For convenience, call $w = f(x, y, h(x, y))$. For a critical point of w, we want $w_x = w_y = 0$. Now we are supposing that f has a relative max or min at (a, b, c), and $c = h(a, b)$, so we want $w_x(a, b) = w_y(a, b) = 0$. We now must calculate these derivatives. From $g(x, y, h(x, y)) = 0$ around (a, b), we have

$$h_x = \frac{-g_x}{g_z} \quad \text{and} \quad h_y = \frac{-g_y}{g_z}.$$

From $w = f(x, y, h(x, y))$, we then have by the chain rule,

$$w_x = f_x + f_z h_x = f_x - \frac{f_z g_x}{g_z}$$

and

$$w_y = f_y + f_z h_y = f_y - \frac{f_z g_y}{g_z}.$$

Then

$$w_x(a, b) = 0 = f_x(a, b, c) - \frac{f_z(a, b, c)g_x(a, b, c)}{g_z(a, b, c)}$$

and

$$w_y(a, b) = 0 = f_y(a, b, c) - \frac{f_z(a, b, c)g_y(a, b, c)}{g_z(a, b, c)}.$$

Then

$$f_x(a, b, c)g_z(a, b, c) = f_z(a, b, c)g_x(a, b, c)$$

and

$$f_y(a, b, c)g_z(a, b, c) = f_z(a, b, c)g_y(a, b, c).$$

This says that, for some constant of proportionality λ [here, in fact, $\lambda = f_z(a, b, c)/g_z(a, b, c)$], we have

$$f_x(a, b, c) = \lambda g_x(a, b, c),$$
$$f_y(a, b, c) = \lambda g_y(a, b, c),$$

and

$$f_z(a, b, c) = \lambda g_z(a, b, c).$$

But this is exactly what we wanted to show: that $f(x, y, z) = k$ and $g(x, y, z) = 0$ have the same tangent plane at the extremal point (a, b, c). But the last equations actually suggest something more. Rewritten, they say that

$$\frac{\partial(f - \lambda g)}{\partial x} = \frac{\partial(f - \lambda g)}{\partial y} = \frac{\partial(f - \lambda g)}{\partial z} = 0$$

at (a, b, c).

In other words, the function $f - \lambda g$, with λ a constant (called a *Lagrange multiplier*) to be determined, has its critical points where f has relative extrema subject to the constraint $g(x, y, z) = 0$.

A procedure to use in solving our constrained max–min problem, then, is to put $p(x, y, z) = f(x, y, z) - \lambda g(x, y, z)$, and solve for the critical points of p and for λ from the four equations $p_x = p_y = p_z = 0$, $g(x, y, z) = 0$. Note that the last is really $p_\lambda = 0$, so we can also adopt the point of view that p is a function of x, y, z, and λ, and we are looking for its critical points by setting the four partials equal to 0.

The method extends to functions of more than three variables, and to problems with more than one constraint. For example, if we added to the above problem another constraint $q(x, y, z) = 0$, then we would take two Lagrange multipliers λ_1 and λ_2, set

$$p(x, y, z, \lambda_1, \lambda_2) = f(x, y, z) - \lambda_1 g(x, y, z) - \lambda_2 q(x, y, z),$$

and solve the five equations

$$p_x = p_y = p_z = 0,$$

$$p_{\lambda_1} = g = 0 \qquad \text{and} \qquad p_{\lambda_2} = q = 0.$$

EXAMPLE

Here is an example from mathematics. What is the minimum of $\sum_{i=1}^{n} x_i^2$ subject to the constraint $\sum_{i=1}^{n} a_i x_i = 1$, where $a_1, \ldots, a_n$ are given and not all zero. Put $f(x_1, \ldots, x_n) = \sum_{i=1}^{n} x_i^2$ for any real $x_1, \ldots, x_n$, and $g(x_1, \ldots, x_n) = \sum_{i=1}^{n} a_i x_i$. Now, for $i = 1, \ldots, n$,

$$\frac{\partial(f - \lambda g)}{\partial x_i} = f_{x_i} - \lambda g_{x_i} = 2x_i - \lambda a_i.$$

Then

$$(f - \lambda g)_{x_i} = 0 \text{ exactly when } x_i = \frac{\lambda a_i}{2}.$$

To solve for λ, we have

$$\sum_{i=1}^{n} a_i x_i - 1 = 0 = \sum_{i=1}^{n} a_i \left(\frac{\lambda a_i}{2}\right) - 1 = \lambda \sum_{i=1}^{n} \frac{a_i^2}{2} - 1.$$

Then

$$\lambda = \frac{2}{\sum_{i=1}^{n} a_i^2},$$

so we must take

$$x_i = \frac{\lambda a_i}{2} = \frac{a_i}{\sum_{i=1}^{n} a_i^2}.$$

For these values of $x_1, \ldots, x_n$ we obtain

$$f(x_1, \ldots, x_n) = \sum_{i=1}^{n} \left(\frac{a_i}{\sum_{j=1}^{n} a_j^2}\right)^2 = \frac{\sum_{i=1}^{n} a_i^2}{\left(\sum_{j=1}^{n} a_j^2\right)^2} = \frac{1}{\sum_{j=1}^{n} a_j^2}.$$

It is not immediate that we have a minimum. Perhaps we have found a maximum value for f. To show that this is not the case, the reader can demonstrate by clever choice of $x_1, \ldots, x_n$ that $\sum_{i=1}^{n} x_i^2$ can be made arbitrarily large and still have $\sum_{i=1}^{n} a_i x_i = 1$. ∎

EXAMPLE

We now give an example from physics. Suppose that we have a particle of mass m in a rectangular box of sides a, b, and c. In quantum mechanics one shows that the ground-state energy of the particle is $(h^2/8m)(1/a^2 + 1/b^2 + 1/c^2)$, where h is Planck's constant. Problem: Determine the shape of the box that will minimize the energy, subject to the constraint that the volume remain unchanged.

Put $E(a, b, c) = 1/a^2 + 1/b^2 + 1/c^2$ (the constant $h^2/8m$ appearing in the energy expression will not affect the values of a, b, and c we seek). The constraint is: volume $= abc =$ constant, so, for some given k, $abc = k$. Put $g(a, b, c) = E(a, b, c) - \lambda(abc - k)$. We now attempt to solve the equations

$$g_a = \frac{-2}{a^3} - \lambda bc = 0, \qquad g_b = \frac{-2}{b^3} - \lambda ac = 0, \qquad g_c = \frac{-2}{c^3} - \lambda ab = 0.$$

Before bringing in the equation $g_\lambda = 0 = abc - k$, note that so far we have

$$\lambda abc = \frac{-2}{a^3} = \frac{-2}{b^3} = \frac{-2}{c^3}.$$

But $abc = k$ is the constraint equation, so we have, by direct substitution,

$$\lambda k = \frac{-2}{a^3} = \frac{-2}{b^3} = \frac{-2}{c^3}.$$

Then, without even solving explicitly for λ, we conclude that $a = b = c$, and we should choose a cube for the box containing the particle.

As is the case in so many physical applications, we obviously have a minimum here—no maximum exists. ∎

Exercises for Section 5.9

1. Redo Exercise 5, Section 5.6, using Lagrange multipliers.
2. Find the dimensions of the box of maximum volume having edges parallel to the coordinate axes and corners on the ellipsoid $(x^2/a^2) + (y^2/b^2) + (z^2/c^2) = 1$.
3. Find the minimum distance between the curves $x^2/4 + y^2 = 1$ and $y = x^3 + 27$.

4. Find the minimum distance between the surfaces $x^2 + y^2 + z^2 = 1$ and $x + y + z = 10$.

5. Suppose that a, b, and c are positive constants, and $x \geq 0$, $y \geq 0$, and $z \geq 0$ such that $ayz + bzx + cxy = 3abc$. Show that $xyz \leq abc$.

6. Let $x_1, \ldots, x_n$ be positive. Prove that

$$(x_1 \cdots x_n)^{1/n} \leq \frac{\sum_{i=1}^{n} x_i}{n}.$$

7. Find the maximum area of a rectangle that can be inscribed in the ellipse $(x^2/a^2) + (y^2/b^2) = 1$, with edges parallel to the coordinate axes and corners on the ellipse.

8. Suppose that x, y, and z are positive, and $x + y + z = k$, a given constant. What extra condition must be placed on x, y, and z if we require xyz to be a maximum?

9. Let f be a differentiable function, defined on the whole real line, and let (a, b) be a point in the plane not on the graph $y = f(x)$. Let L be the line segment of shortest length between (a, b) and a point on $y = f(x)$. Show that L is perpendicular to $y = f(x)$ at this point.

10. In the example that involves minimizing $\sum_{i=1}^{n} x_i^2$ subject to $\sum_{i=1}^{n} a_i x_i = 1$, show that, given $a_1, \ldots, a_n$ not all zero, and any $K > 0$, it is possible to choose $x_1, \ldots, x_n$ so as to make $\sum_{i=1}^{n} a_i x_i = 1$ and $\sum_{i=1}^{n} x_i^2 > K$. Thus conclude that we really did find a minimum in that example.

11. Let S_1 be the surface given by $x^2 + y^2 + z^2 = 1$, and S_2 the surface given by $(x - 8)^2 + (y - 10)^2 + (z - 11)^2 = 1$. Find points P_1 on S_1 and P_2 on S_2 so that the distance between P_1 and P_2 is as small as possible. Then find points Q_1 on S_1 and Q_2 on S_2 so that the distance between Q_1 and Q_2 is as large as possible.

12. Find the point on the parabolic bowl $z = x^2 + y^2$, $x^2 + y^2 \leq 1$, at minimum distance from the plane $x + z = 81$.

13. Obtain a formula for the shortest distance from the point (x_0, y_0, z_0) to the plane $ax + by + cz = d$.

14. Find the triangle of maximum area that can be inscribed in a circle of radius R.

6

Line and Surface Integrals

6.1 Line Integrals of Vector-Valued Functions

Rather than rush into the mathematics, let us begin with a simple physical example. Suppose that, at each point (x, y, z) of some region D of R^3, we are given a vector (to be thought of as a force) $F(x, y, z) = (f(x, y, z), g(x, y, z), h(x, y, z))$. Suppose that C is a curve in D from P_0 to P_1, and we want to calculate the work W done by F in moving a particle from P_0 to P_1 along C.

In order to derive some expression for W, we shall begin with the elementary physics notion of work. If $T = (\alpha, \beta, \gamma)$ is a constant force, the work done by T in moving a particle from (a, b, c) to $(a + \Delta x, b + \Delta y, c + \Delta z)$ along a straight-line path is defined to be $T \cdot (\Delta x, \Delta y, \Delta z)$, the dot product of the force with the displacement. Our problems are that $F(x, y, z)$ may vary from point to point, and C need not be a straight line.

As usual, we go from the elementary to the more sophisticated case by taking a limit of approximations. Choose points $Q_0 = P_0, Q_1, \ldots, Q_{n-1}, Q_n = P_1$ on C from P_0 to P_1, say $Q_i = (\alpha_i, \beta_i, \gamma_i)$. By joining each Q_{i-1} to Q_i by a line segment, we form a polygonal approximation to C (Fig. 6-1). Think of F as constant on each line segment, say by choosing $(\bar{x}_i, \bar{y}_i, \bar{z}_i)$ on the segment from Q_{i-1} to Q_i and replacing $F(x, y, z)$ by $F(\bar{x}_i, \bar{y}_i, \bar{z}_i)$ on this segment. We then take $F(\bar{x}_i, \bar{y}_i, \bar{z}_i) \cdot (\alpha_i - \alpha_{i-1}, \beta_i - \beta_{i-1}, \gamma_i - \gamma_{i-1})$ as an approximation to the work done by F from Q_{i-1} to Q_i along C, leading to

$$\sum_{i=1}^{n} F(\bar{x}_i, \bar{y}_i, \bar{z}_i) \cdot (\alpha_i - \alpha_{i-1}, \beta_i - \beta_{i-1}, \gamma_i - \gamma_{i-1})$$

as an approximation to W.

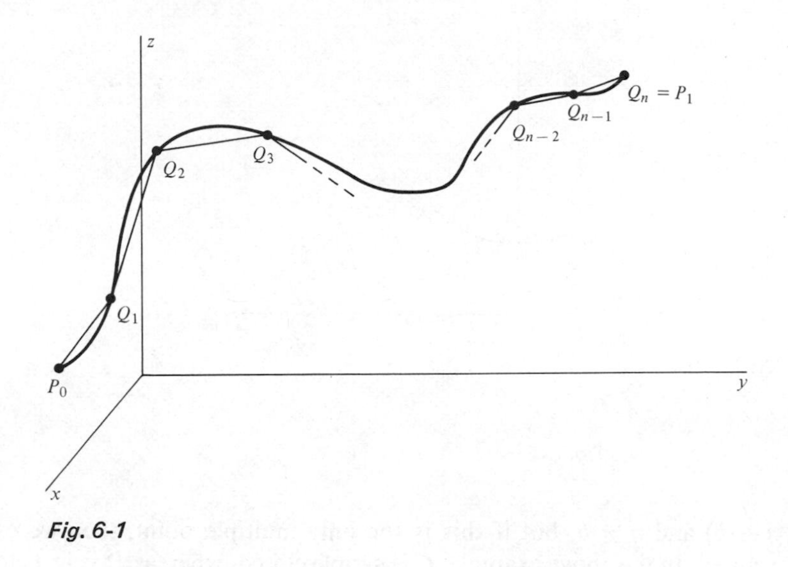

Fig. 6-1

Under reasonable conditions (about which more later) we would expect this approximation to improve as n is taken larger and the points Q_i closer together on C, so that we obtain an exact expression for W as a kind of limit similar to those used to define single and multiple integrals. This limit is called the *line integral* of F over C.

In order to clarify the kind of limit we are talking about, and to develop effective means of computing it, we must say a little more about what we mean by a curve in R^3. For our purposes here, it is convenient to define a *curve* in R^3 as a function, say C, which associates with each t in a closed interval $[a, b]$, a point $C(t)$ in R^3. Often all this is abbreviated to $C: [a, b] \to R^3$. If we write the coordinates of $C(t)$ as $(\alpha(t), \beta(t), \gamma(t))$, then α, β, and γ are called the *coordinate functions* of C, and we often write $C = (\alpha, \beta, \gamma)$. The graph of C is the locus of points $C(t)$, for $a \leq t \leq b$.

Note the distinction between C and the graph of C. This can be important in both theory and applications. For example, let $C(t) = (\cos(t), \sin(t), 1)$, for $0 \leq t \leq 2\pi$, and $D(t) = (\cos(t), \sin(t), 1)$ for $0 \leq t \leq 4\pi$. Then the graphs of C and D both consist of the points on the unit circle $x^2 + y^2 = 1$ in the plane $z = 1$ (see Fig. 6-2). But, as $t: 0 \to 2\pi$, $C(t)$ goes from $(1, 0, 1)$ to $(1, 0, 1)$ once around this circle, while, as $t: 0 \to 4\pi$, $D(t)$ goes around twice. This can make a difference in calculating something like work around C or D. As a convenience, we often speak of a "point on C." By this we of course always mean a point on the graph of C.

We will call a curve $C: [a, b] \to R^3$ *closed* if $C(a) = C(b)$, and *simple* if C has no multiple points [that is, $C(t) \neq C(\bar{t})$ if $t \neq \bar{t}$]. Thinking of t as time, C is simple if at different times, $C(t)$ is always some place where it has not been before. Finally, C is called *simple closed* if $C(a) = C(b)$, but $C(t) \neq C(\bar{t})$ for $a \leq t < \bar{t} \leq b$. In other words, a closed curve cannot be simple, as

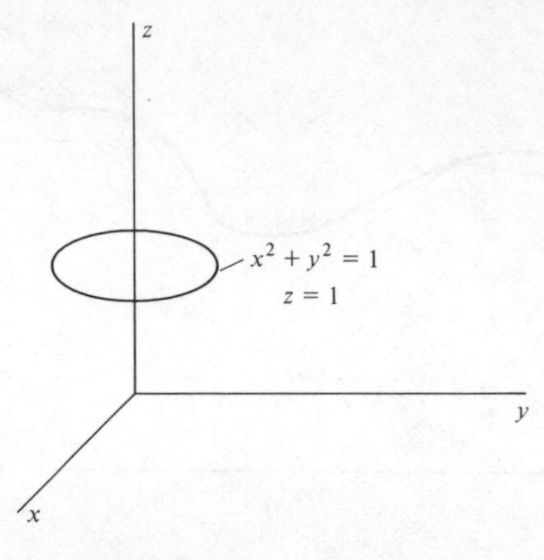

Fig. 6-2

$C(a) = C(b)$ and $a \neq b$, but if this is the only multiple point, then we call C simple closed. In the above example, C is simple closed, whereas D is just closed.

We say that C is *continuous* at t_0 if each coordinate function is continuous at t_0. This means that $C(t)$ can be made arbitrarily close to $C(t_0)$ by taking t close to t_0. If C is continuous at each point of $[a, b]$, then we say that C is continuous.

Similarly, C is differentiable at t_0 if α, β, and γ are differentiable at t_0. We then write $C'(t_0) = (\alpha'(t_0), \beta'(t_0), \gamma'(t_0))$.

We can now formulate a definition of line integral and a method of evaluation.

Definition

Let $F(x, y, z)$ be a 3-vector for each (x, y, z) in D. Let $C: [a, b] \to R^3$ be a curve with graph contained in D, and let L be a real number.

Suppose, given $\varepsilon > 0$, that there is some $\delta > 0$ such that, for any partition $a = t_0 < t_1 < \cdots < t_n = b$ with $|t_i - t_{i-1}| < \delta$, and any choice of ξ_i, $t_{i-1} \leq \xi_i \leq t_i$ for $i = 1, \ldots, n$, we have

$$\left| \sum_{i=1}^{n} F(C(\xi_i)) \cdot (C(t_i) - C(t_{i-1})) - L \right| < \varepsilon.$$

Then we call L the *line integral* of F over C, and write

$$L = \int_C F.$$

A close examination of the sums $\sum_{i=1}^{n} F(C(\xi_i)) \cdot (C(t_i) - C(t_{i-1}))$ suggests a means of computing $\int_C F$ under certain conditions on F and C. Write

$$F(x, y, z) = (f(x, y, z), g(x, y, z), h(x, y, z))$$

and

$$C(t) = (\alpha(t), \beta(t), \gamma(t)), \qquad a \le t \le b.$$

Now expand out the sum:

$$F(C(\xi_i)) \cdot (C(t_i) - C(t_{i-1})) = \sum_{i=1}^{n} [f(\alpha(\xi_i), \beta(\xi_i), \gamma(\xi_i)) \cdot (\alpha(t_i) - \alpha(t_{i-1}))$$
$$+ g(\alpha(\xi_i), \beta(\xi_i), \gamma(\xi_i)) \cdot (\beta(t_i) - \beta(t_{i-1}))$$
$$+ h(\alpha(\xi_i), \beta(\xi_i), \gamma(\xi_i)) \cdot (\gamma(t_i) - \gamma(t_{i-1}))].$$

For convenience, let us center our attention on one part of this sum, say $\sum_{i=1}^{n} f(\alpha(\xi_i), \beta(\xi_i), \gamma(\xi_i)) \cdot (\alpha(t_i) - \alpha(t_{i-1}))$. If α' exists in each $[t_{i-1}, t_i]$, then by the mean value theorem there is some ζ_i in (t_{i-1}, t_i) such that

$$\alpha(t_i) - \alpha(t_{i-1}) = \alpha'(\zeta_i)(t_i - t_{i-1}).$$

Then our sum becomes

$$\sum_{i=1}^{n} f(\alpha(\xi_i), \beta(\xi_i), \gamma(\xi_i)) \cdot \alpha'(\zeta_i)(t_i - t_{i-1}).$$

This looks like a Riemann sum for $\int_a^b f(\alpha(t), \beta(t), \gamma(t)) \cdot \alpha'(t)\, dt$, except that for an honest Riemann sum we would have $\xi_i = \zeta_i$ for $i = 1, \ldots, n$. Intuitively, it should not make much difference if $\xi_i \neq \zeta_i$, as both ξ_i and ζ_i are in $[t_{i-1}, t_i]$, and $t_i - t_{i-1} \to 0$ in the limit defining the Riemann integral anyway. To make a long story short, there is a delicate theorem due to Duhamel, which says that our intuition is correct, and that the above Riemann-like sums do approach $\int_a^b f(\alpha(t), \beta(t), \gamma(t))\alpha'(t)\, dt$ if $f(\alpha(t), \beta(t), \gamma(t))$, and $\alpha'(t)$ are continuous on $[a, b]$ except perhaps at finitely many points.

We now go back and pick up the other two sums and summarize the whole situation: *Suppose that* $F(x, y, z) = (f(x, y, z), g(x, y, z), h(x, y, z))$ *is defined for* (x, y, z) *in a region* D *of* R^3 *with* f, g, *and* h *continuous on* D, *and* $C: [a, b] \to R^3$ *is a curve with graph in* D *and having* α', β', *and* γ' *continuous on* $[a, b]$. *Then*

$$\int_C F = \int_a^b [f(\alpha(t), \beta(t), \gamma(t))\alpha'(t) + g(\alpha(t), \beta(t), \gamma(t))\beta'(t)$$
$$+ h(\alpha(t), \beta(t), \gamma(t))\gamma'(t)]\, dt.$$

In vector notation, we have the equivalent, but more compact, expression:

$$\int_C F = \int_a^b F(C(t)) \cdot C'(t)\, dt.$$

EXAMPLE

Let $F(x, y, z) = (x, y^2, xyz)$ for all (x, y, z). Let $C(t) = (t^2, \sin(t), 2 - t)$ for $0 \le t \le 3$. Then

$$\int_C F = \int_0^3 F(t^2, \sin(t), 2 - t) \cdot C'(t)\,dt$$

$$= \int_0^3 (t^2, \sin^2(t), t^2 \sin(t)(2 - t)) \cdot (2t, \cos(t), -1)\,dt$$

$$= \int_0^3 \{2t^3 + \sin^2(t)\cos(t) - t^2(2 - t)\sin(t)\}\,dt.$$

This is an ordinary Riemann integral, which the student can evaluate if he wishes. ∎

We note in passing here that the vector $C'(t)$ appearing in $\int_C F(C(t) \cdot C'(t))\,dt$ has a useful geometric interpretation: it is the tangent to C at $C(t)$. To see this, note that $C(t + \Delta t) - C(t)$ is the vector from $C(t)$ to $C(t + \Delta t)$, and that $(1/\Delta t)[C(t + \Delta t) - C(t)]$ has the same direction as $C(t + \Delta t) - C(t)$. As $\Delta t \to 0$, the vector $(1/\Delta t)(C(t + \Delta t) - C(t))$ approaches (at least intuitively, see Fig. 6-3) the tangent to C at $C(t)$. But, as $\Delta t \to 0$,

$$\frac{1}{\Delta t}[C(t + t) - C(t)]$$

$$= \left(\frac{\alpha(t + \Delta t) - \alpha(t)}{\Delta t}, \frac{\beta(t + \Delta t) - \beta(t)}{\Delta t}, \frac{\gamma(t + \Delta t) - \gamma(t)}{\Delta t}\right)$$

$$\to (\alpha'(t), \beta'(t), \gamma'(t)).$$

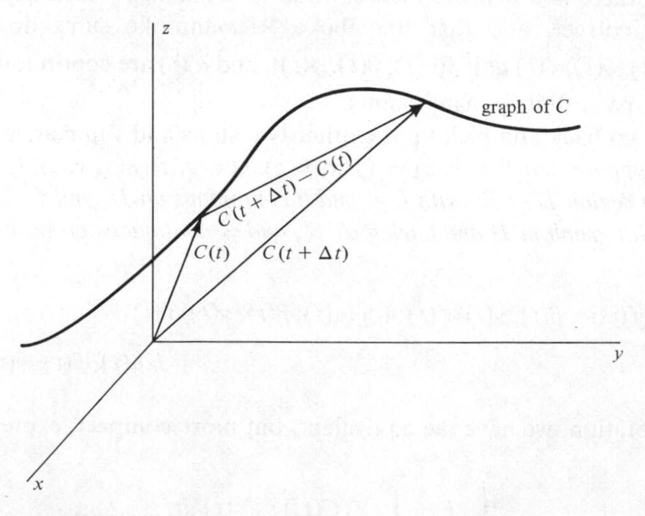

Fig. 6-3

There is a differential notation for line integrals which is often encountered. Note that $\alpha'(t)\,dt = d\alpha$, $\beta'(t)\,dt = d\beta$, and $\gamma'(t)\,dt = d\gamma$. Then $\int_C F$ is often written as $\int_C f\,d\alpha + g\,d\beta + h\,d\gamma$ or $\int_C f\,dx + g\,dy + h\,dz$. This is purely a matter of notation, and does not affect the method of evaluation.

In many instances, particularly in physical applications, a curve is described in words [e.g., the straight line from (1, 0, 1) to (2, 1, 3)] and the reader has to construct a parametric representation. Here a difficulty arises: What if there are several different parametric representations? For example, let the curve be the quarter circle from (1, 0, 1) to (0, 1, 1) in the plane $z = 1$. One person might write $C(t) = (\cos(t), \sin(t), 1)$, $t: 0 \to \pi/2$, and another $D(r) = (\sqrt{1 - r^2}, r, 1)$, $r: 0 \to 1$. Then C and D are different functions. Nevertheless, the graph of each is a quarter circle from (1, 0, 1) to (0, 1, 1), traversed once. The important feature in this example is that a change of variables $r = \sin(t)$ transforms D to C in the sense that $D(\sin(t)) = C(t)$.

In general, suppose that $C: [a, b] \to R^3$ and $D: [c, d] \to R^3$ are curves in R^3. *We say that C and D are equivalent if there is some* φ, *defined on* $[a, b]$, *such that* $\varphi'(t) > 0$ *on* (a, b) *and* $D(\varphi(t)) = C(t)$ *for* $a \le t \le b$.

We shall now show that *line integrals of F over equivalent curves are equal.* Consider

$$\int_C F = \int_a^b F(C(t)) \cdot C'(t)\,dt$$

and

$$\int_D F = \int_c^d F(D(r)) \cdot D'(r)\,dr.$$

Write $F(x, y, z) = (f(x, y, z), g(x, y, z), h(x, y, z))$,

$$C(t) = (\alpha(t), \beta(t), \gamma(t))$$

and

$$D(r) = (u(r), v(r), w(r)).$$

Then

$$\int_a^b F(C(t)) \cdot C'(t)\,dt = \int_a^b [f(C(t))\alpha'(t) + g(C(t))\beta'(t) + h(C(t))\gamma'(t)]\,dt$$

and

$$\int_c^d F(D(r)) \cdot D'(r)\,dr = \int_c^d [f(D(r))u'(r) + g(D(r))v'(r) + h(D(r))w'(r)\,dr].$$

Examine the integrals on the right in both of these equations. First, make a change of variables $r = \varphi(t)$ in $\int_c^d f(D(r))u'(r)\,dr$ to obtain

$$\int_c^d f(D(r))u'(r)\,dr = \int_a^b f(D(\varphi(t)))u'(\varphi(t))\varphi'(t)\,dt.$$

But $D(\varphi(t)) = C(t)$ implies that $u(\varphi(t)) = \alpha(t)$, so

$$\alpha'(t) = \frac{du(\varphi(t))}{dt} = u'(\varphi(t))\varphi'(t).$$

Then

$$\int_a^b f(D(\varphi(t))u'(\varphi(t))\varphi'(t)\,dt = \int_a^b f(C(t))\alpha'(t)\,dt.$$

Similarly,

$$\int_c^d g(D(r))v'(r)\,dr = \int_a^b g(C(t))\beta'(t)\,dt$$

and

$$\int_c^d h(D(r))w'(r)\,dr = \int_a^b h(C(t))\gamma'(t)\,dt.$$

Then $\int_C F = \int_D F$, as we wanted to show.

EXAMPLE

Let $C(t) = (\cos(t), \sin(t), 1)$, $0 \le t \le \pi/2$. Let $D(r) = (\sqrt{1 - r^2}, r, 1)$, $0 \le r \le 1$. Let $F(x, y, z) = (x, y^2, z)$. Then

$$\begin{aligned}\int_C F &= \int_0^{\pi/2} (\cos(t), \sin^2(t), 1)\cdot(-\sin(t), \cos(t), 0)\,dt \\ &= \int_0^{\pi/2} (-\cos(t)\sin(t) + \cos(t)\sin^2(t))\,dt \\ &= \frac{\cos^2(t)}{2}\Big|_0^{\pi/2} + \frac{\sin^3(t)}{3}\Big|_0^{\pi/2} = -\frac{1}{2} + \frac{1}{3} = -\frac{1}{6}\end{aligned}$$

and

$$\begin{aligned}\int_D F &= \int_0^1 (\sqrt{1 - r^2}, r^2, 1)\cdot(-r(1 - r^2)^{-1/2}, 1, 0)\,dr \\ &= \int_0^1 (-r + r^2)\,dr = \frac{-r^2}{2}\Big|_0^1 + \frac{r^3}{3}\Big|_0^1 = -\frac{1}{2} + \frac{1}{3} = -\frac{1}{6}.\end{aligned}$$

This freedom (within certain reasonable bounds) to choose different parametric representations for a given curve can often be useful. For example, we shall on occasion find it convenient to parametrize C so that we are working with a unit tangent. This can be done as follows. If $a \leq t \leq b$, let $s(t) =$ length of the graph of C from $C(a)$ to $C(t)$. Assuming that α', β', and γ' are continuous, then

$$s(t) = \int_a^t (\alpha'(\xi)^2 + \beta'(\xi)^2 + \gamma'(\xi)^2)^{1/2}\, d\xi.$$

Now s is a strictly increasing function of t, as $s'(t) = \sqrt{\alpha'(\xi)^2 + \beta'(\xi)^2 + \gamma'(\xi)^2} > 0$, and so at least in theory can be inverted to write $t = t(s)$, $0 \leq s \leq L$, where $L =$ length of C. Putting $D(s) = C(t(s))$, $0 \leq s \leq L$, we obtain a curve D equivalent to C. Now the tangent $C'(t)$ to C at $C(t)$ might have any nonzero length, but $D'(s)$ is always of unit length. For

$$\begin{aligned} |D'(s)| &= \left|\frac{d(C(t(s))}{ds}\right| = |C'(t)t'(s)| \\ &= |(\alpha'(t), \beta'(t), \gamma'(t))|\,|t'(s)| = \sqrt{\alpha'(t)^2 + \beta'(t)^2 + \gamma'(t)^2}\,|t'(s)|. \end{aligned}$$

But

$$s'(t) = \sqrt{\alpha'(t)^2 + \beta'(t)^2 + \gamma'(t)^2},$$

so

$$t'(s) = \frac{1}{\sqrt{\alpha'(t)^2 + \beta'(t)^2 + \gamma'(t)^2}},$$

giving us $|D'(s)| = 1$.

In practice, solution for t in terms of s, as required in the above consideration, can be very difficult. However, often the nature of a problem suggests arc length as the natural parameter to use. We shall see examples of this in the next section.

When dealing with line integrals, the letter s will always be reserved for arc length.

We shall pause for an example to illustrate the last few paragraphs.

EXAMPLE

Compute $\int_C (x, 2y, 2xz)$, where C is the quarter circle $x^2 + y^2 = 1$, $0 \leq x \leq 1$, $y \geq 0$.

Here we may parametrize C by θ: $C(\theta) = (\cos(\theta), \sin(\theta), 0)$, $0 \leq \theta \leq \pi/2$. Then

$$\begin{aligned} \int_C (x, 2y, 2xz) &= \int_0^{\pi/2} (\cos(\theta), 2\sin(\theta), 2\cos(\theta)\sin(\theta)) \cdot (-\sin(\theta), \cos(\theta), 0)\, d\theta \\ &= 0. \end{aligned}$$

Here we are actually using arc length as parameter, as the distance from $C(0)$ to $C(\theta)$ along C is

$$s(\theta) = \int_0^\theta \sqrt{\sin^2(\xi) + \cos^2(\xi)}\, d\xi = \theta$$

(see Fig. 6-4). Note that, in using arc length as parameter, we automatically have a unit tangent. The tangent to C at $C(\theta)$ is $C'(\theta) = (-\sin(\theta), \cos(\theta), 0)$, which has unit length for every θ. ∎

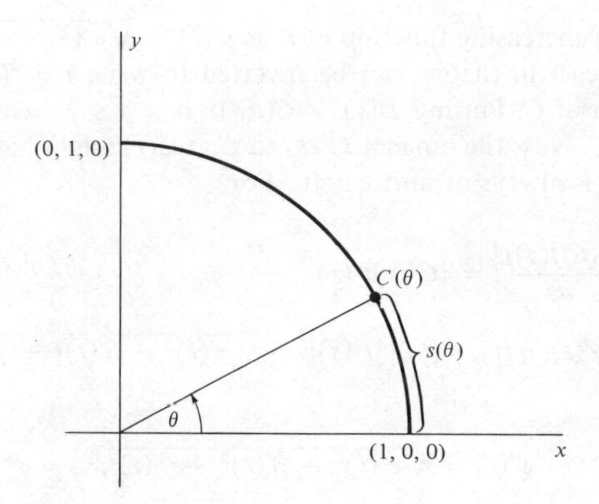

Fig. 6-4

There are several facts about line integrals of vector-valued functions which have not yet been stated but which are important to understand. We shall approach these intuitively, without going through rigorous proofs.

First, line integrals have the linearity property that we usually associate with any kind of integration:

$$\int_C (\alpha F + \beta G) = \alpha \int_C F + \beta \int_C G$$

whenever $\int_C F$ and $\int_C G$ exist, and α and β are real numbers.

There is another kind of additivity for line integrals. It may happen that the curve C we are interested in integrating over has been formed by stringing together two or more curves. For example, if $C_1: [a, b] \to R^3$ and $C_2: [b, d] \to R^3$, and $C_1(b) = C_2(b)$, then we can form $C: [a, d] \to R^3$ by

$$C(t) = \begin{cases} C_1(t), & a \leq t \leq b, \\ C_2(t), & b \leq t \leq d. \end{cases}$$

(see Fig. 6-5).

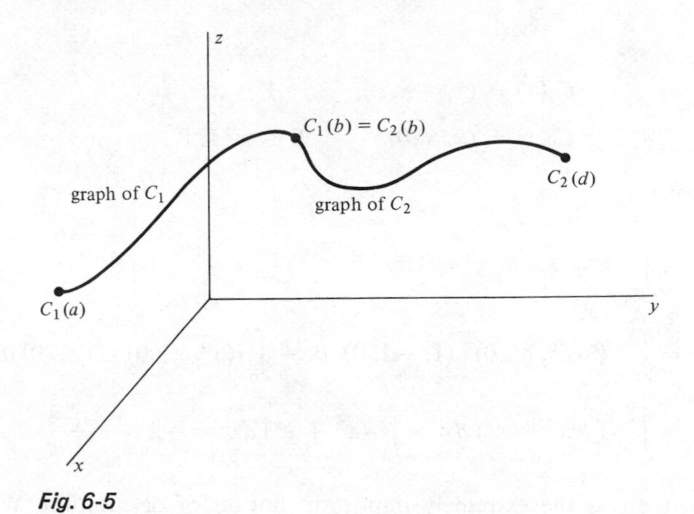

Fig. 6-5

Assuming that $\int_{C_1} F$ and $\int_{C_2} F$ exist, then so does $\int_C F$ and

$$\int_C F = \int_{C_1} F + \int_{C_2} F.$$

EXAMPLE

As an example of how this might be useful, consider $\int_C (xy, e^x, 0)$, where $C(x) = (x, |x|, 0)$ for $-1 \leq x \leq 1$ (see Fig. 6-6). Here we cannot use $\int_C F = \int_C F(C(t)) \cdot C'(t)\, dt$, as C has no tangent at the origin. However, we can write C as a "sum" of two pieces:

$$C_1(x) = (x, |x|, 0), \qquad -1 \leq x \leq 0,$$
$$C_2(x) = (x, |x|, 0), \qquad 0 \leq x \leq 1.$$

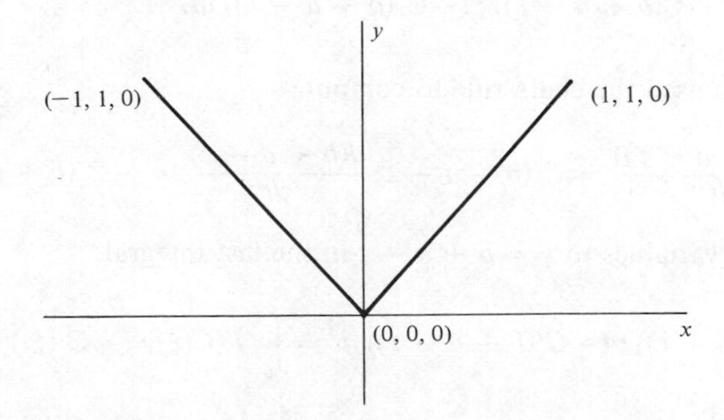

Fig. 6-6

Then

$$C_1(x) = (x, -x, 0), \qquad -1 \le x \le 0$$

and

$$C_2(x) = (x, x, 0), \qquad 0 \le x \le 1.$$

Then we have

$$\begin{aligned}\int F &= \int_{C_1} F + \int_{C_2} F \\ &= \int_{-1}^{0} (-x^2, e^x, 0) \cdot (1, -1, 0)\, dx + \int_0^1 (x^2, e^x, 0) \cdot (1, 1, 0)\, dx \\ &= \int_{-1}^{0} (-x^2 - e^x)\, dx + \int_0^1 (x^2 + e^x)\, dx = -2 + e + \frac{1}{e}. \end{aligned}$$ ∎

Finally, there is the extremely important notion of orientation. When we evaluate $\int_C F$ as $\int_a^b F(C(t)) \cdot C'(t)\, dt$, we are tacitly thinking of t as going from a to b, and $C(t)$ from $C(a)$ to $C(b)$ on the graph of C. This orients C by giving it an initial and terminal point. Since $\int_a^b = -\int_b^a$, it is reasonable to suppose that a reversal in orientation should change the sign of the line integral.

To put this on firmer ground, put

$$D(t) = C(b + a - t), \qquad a \le t \le b.$$

Note that, when $t\colon a \to b$, $b + a - t\colon b \to a$; thus as $t\colon a \to b$, $D(t)$ moves from $C(b)$ to $C(a)$. Then graph C = graph D, but the initial point of C is the terminal point of D, and vice versa. Observe the relationship between $\int_C F$ and $\int_D F$:

$$\begin{aligned}\int_D F = \int_a^b F(D(t)) \cdot D'(t)\, dt &= \int_a^b F(C(b + a - t)) \cdot \frac{d}{dt}(C(b + a - t))\, dt \\ &= \int_a^b F(C(b + a - t)) \cdot (-C'(b + a - t))\, dt.\end{aligned}$$

Here we have used the chain rule to compute

$$\frac{d(C(b + a - t))}{dt} = C'(b + a - t)\frac{d(b + a - t)}{dt} = -C'(b + a - t).$$

Now change variables to $\xi = b + a - t$ in the last integral:

$$\begin{aligned}\int_a^b F(C(b + a - t)) \cdot (-C'(b + a - t))\, dt &= \int_b^a F(C(\xi) \cdot (-C'(\xi))(-1)\, d\xi \\ &= -\int_a^b F(C(\xi)) \cdot C'(\xi)\, d\xi = -\int_C F.\end{aligned}$$

Thus $\int_C F = -\int_D F$, as we predicted should happen. Very often, D, as defined above, is denoted $-C$. Then we can write

$$\int_{C} F = -\int_{C} F.$$

In practice, we never actually go to the trouble of putting $D(t) = C(b + a - t)$ if we want to reverse orientation in a computation. To compute $\int_{-C} F$, just compute $\int_C F$ and take its negative.

We conclude this section with two more examples, using differential notation for our line integrals.

EXAMPLE

Compute $\int_C x\,dx + 3yz\,dy + xyz\,dz$, where $C(t) = (2t^2, t, t/2)$, $0 \le t \le 1$. Note that the question may be rephrased; evaluate $\int_C (x, 3yz, xyz)$ along the given C. We write the line integral as

$$\int_0^1 \left[2t^2(4t) + 3t\left(\frac{t}{2}\right)(1) + (2t^2)t\left(\frac{t}{2}\right)\left(\frac{1}{2}\right)\right] dt$$

$$= \int_0^1 \left[8t^3 + \left(\frac{3}{2}\right)t^2 + 2t^4\right] dt = \frac{29}{10}. \quad \blacksquare$$

EXAMPLE

Compute $\int_C -y\,dx + x^2\,dy$, where C is the parabola $y = x^2$ from $(0, 0)$ to $(1, 1)$, in the xy-plane. Again, note that we are integrating $\int_C (-y, x^2, 0)$. Parametrizing C by x, we can write $C(x) = (x, x^2, 0)$, $0 \le x \le 1$. Then

$$\int_C -y\,dx + x^2\,dy = \int_0^1 [-x^2 + x(2x)]\,dx = \tfrac{1}{6}.$$

If we use y as parameter, then we would write C as $C(y) = (\sqrt{y}, y, 0)$, $0 \le y \le 1$. Then

$$\int_C -y\,dx + x^2\,dy = \int_0^1 [-y(\tfrac{1}{2})y^{-1/2} + y]\,dy = \tfrac{1}{6}. \quad \blacksquare$$

Exercises for Section 6.1

1. Compute the following line integrals:

(a) $\int_C (x, y, z)$, $C(t) = (t^2, 2t, 0)$ for $2 \le t \le 3$

(b) $\int_C (x^2y, y^2x, z^2)$, $C(t) = (t, t^2, t^3)$, $0 \le t \le 4$

(c) $\int_C (3xy, e^z, 2)$, $C(t) = (\sin(t), \cos(t), t)$, $0 \le t \le \pi/2$

2. Compute the following:

(a) $\int_C x\,dx + y\,dy + z^2\,dz$, $C(t) = (1/t, t^2, 2)$, $1 \le t \le 3$

(b) $\int_C y^2x\,dx + 2z\,dy + e^x\,dz$, $C(t) = (0, 2t, t)$, $0 \le t \le 1$

(c) $\int_C a\,dx + b\,dy + c\,dz$, C the straight line from $(0, 1, 0)$ to $(6, 8, 4)$

(d) $\int_C x^2\,dx + y^2\,dy + 3\,dz$, C the quarter-circle from $(3, 0, 1)$ to $(0, 3, 1)$ in the plane $z = 1$

(e) $\int_C z\,dx + x\,dy + y\,dz$, C the parabola $y = x^2$ from $(0, 0)$ to $(2, 4)$ in the xy-plane

(f) $\int_C \sin(x)\,dx + yz\,dy + x^2yz\,dz$, C the path consisting of straight lines from $(0, 0, 0)$ to $(1, 3, 2)$, then to $(2, 4, 8)$, then to $(16, 1, 5)$, and back to $(0, 0, 0)$

3. Compute the work done by a force $F = (8x, 6y, 2z)$ in moving a particle along the straight line from $(0, 0, 0)$ to $(8, 1, 6)$.

4. Compute the line integral of $(-y/(x^2 + y^2), x/(x^2 + y^2), 0)$ on any circle about the origin in the xy-plane, oriented counterclockwise.

5. Show that the concept of a Riemann integral is a special case of the notion of line integral. *Hint:* Compare $\int_a^b f(x)\,dx$ with $\int_C (f(x), 0, 0)$, where $C(t) = (t, 0, 0)$ for $a \le t \le b$.

6. Let $C: [a, b] \to R^3$ be a curve with coordinate functions α, β, γ which are differentiable at t_0. In the text we put $C'(t_0) = (\alpha'(t_0), \beta'(t_0), \gamma'(t_0))$. Show that this is a reasonable thing to do by proving that

$$\lim_{\Delta t \to 0} \left\| \frac{1}{\Delta t}(C(t_0 + \Delta t) - C(t_0)) - (\alpha'(t_0), \beta'(t_0), \gamma'(t_0)) \right\| = 0.$$

7. Let $C(t) = (\alpha(t), \beta(t), \gamma(t))$, for $a \le t \le b$. Suppose that α, β, and γ have continuous derivatives on $[a, b]$. Give a heuristic argument to justify defining the length of C as

$$\int_a^b \sqrt{\alpha'(t)^2 + \beta'(t)^2 + \gamma'(t)^2}\,dt$$

Hint: Inscribe a polygonal path in C (as in Fig. 6-1) by partitioning $a = t_0 < t_1 < \cdots < t_n = b$, and put $Q_i = C(t_i)$, $i = 0, \ldots, n$. Approximate the length of C as the sum of the lengths of these line segments and use a mean value theorem argument (similar to that used in deriving a Riemann integral formula for evaluating line integrals) to reduce this approximation to a Riemann sum.

8. Let $C(x) = \begin{cases} \left(x, \sin\left(\dfrac{1}{x}\right), 0\right) & \text{for } 0 < x \le \dfrac{1}{\pi}, \\ (0, 0, 0) & \text{for } x = 0. \end{cases}$

Show that the polygonal approximations prescribed in Exercise 7 to arrive at a formula for length fails for this curve C, in the sense that such approximations can be made arbitrarily large by appropriate choice of the partition points of $[0, 1/\pi]$.

6.2 Application of Line Integrals to the Curl of a Vector

If $F(x, y, z) = (f(x, y, z), g(x, y, z), h(x, y, z))$ for each (x, y, z) in some region D of R^3, and the partials of F exist in D, then the curl of F is the vector defined by

$$\text{curl } F = \left(\frac{\partial h}{\partial y} - \frac{\partial g}{\partial z}, \frac{\partial f}{\partial z} - \frac{\partial h}{\partial x}, \frac{\partial g}{\partial x} - \frac{\partial f}{\partial y}\right).$$

An easy way to remember this is to think of a "symbolic determinant,"

$$\begin{vmatrix} ① & ② & ③ \\ \dfrac{\partial}{\partial x} & \dfrac{\partial}{\partial y} & \dfrac{\partial}{\partial z} \\ f & g & h \end{vmatrix},$$

where the numbers along the top row refer to the component of curl F obtained by taking the cofactor of that number.

Curl plays an important role in a large variety of physical problems. In this section we shall use line integrals to try to develop some feeling for the physical significance of curl.

Begin by recalling that $\int_C F$ for a closed curve C may be thought of as the work done in moving a particle once around C. If we think of $F(x, y, z)$ as the velocity of a fluid at point (x, y, z), measured with reference to some fixed frame, then $F(C(t)) \cdot C'(t)$ is the tangential component of the velocity around C, and $\int_C F = \int_a^b F(C(t)) \cdot C'(t)\, dt$ gives a measure of the extent to which the fluid is rotating around C.

Suppose now that C encloses a plane area A containing a point (x_0, y_0, z_0). It is natural to think of $\lim_{A \to 0} \frac{1}{A} \int_C F$ as a measure of the rotation of the fluid at (x_0, y_0, z_0) in the plane of C, where in the limit we understand that C is continuously shrunk to the point (x_0, y_0, z_0) [imagine C as being made out of string and being drawn through a hole at (x_0, y_0, z_0)].

We shall now give a plausibility argument that this limit is, in fact, the component of curl F at (x_0, y_0, z_0) in the direction of the normal to the plane of C (Fig. 6-7). That is, if η is a unit normal to the plane of C, then

$$(\text{curl } F|_{(x_0, y_0, z_0)}) \cdot \eta = \lim_{A \to 0} \frac{1}{A} \int_C F.$$

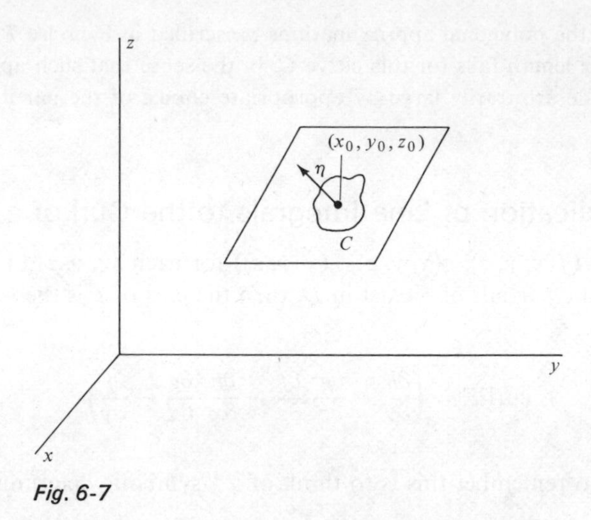

Fig. 6-7

The complete details in the general case somewhat obscure the point to be made, so we shall consider a special case. Suppose that $z_0 = 0$, putting our point in the xy-plane, and C is a rectangle about (x_0, y_0, z_0) as shown in Fig. 6-8.

Here $\eta = (0, 0, 1)$, $A = (\Delta x)(\Delta y)$, and C consists of the four sides of the rectangle, which we shall orient counterclockwise in our line integral. We want to consider

$$\lim_{(\Delta x, \Delta y)\to(0,0)} \frac{1}{(\Delta x)(\Delta y)} \int_C F.$$

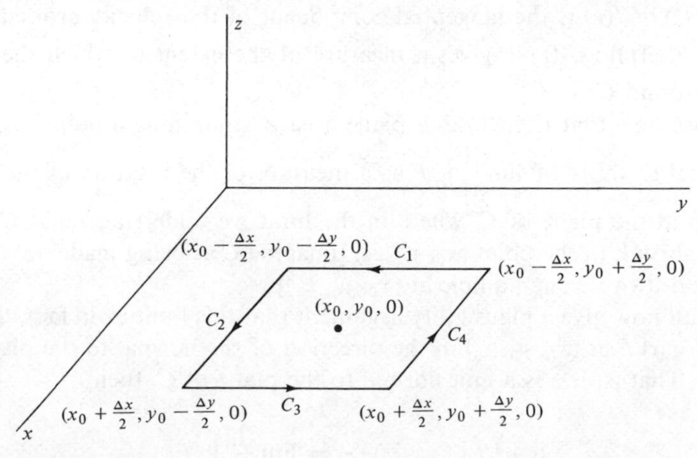

Fig. 6-8

In order to carry out the line integration, we break C up into four straight-line pieces, with orientation as indicated:

$$C_1(y) = \left(x_0 - \frac{\Delta x}{2}, y, 0\right), \; y: y_0 + \frac{\Delta y}{2} \to y_0 - \frac{\Delta y}{2},$$

$$C_2(x) = \left(x, y_0 - \frac{\Delta y}{2}, 0\right), \; x: x_0 - \frac{\Delta x}{2} \to x_0 + \frac{\Delta x}{2},$$

$$C_3(y) = \left(x_0 + \frac{\Delta x}{2}, y, 0\right), \; y: y_0 - \frac{\Delta y}{2} \to y_0 + \frac{\Delta y}{2},$$

$$C_4(x) = \left(x, y_0 + \frac{\Delta y}{2}, 0\right), \; x: x_0 + \frac{\Delta x}{2} \to x_0 - \frac{\Delta x}{2}.$$

These have tangents, respectively, $C_1'(y) = (0, 1, 0)$, $C_2'(x) = (1, 0, 0)$, $C_3'(y) = (0, 1, 0)$, $C_4'(x) = (1, 0, 0)$. Then

$$\int_C F = \int_{y_0 + \Delta y/2}^{y_0 - \Delta y/2} (f, g, h) \cdot (0, 1, 0)\, dy + \int_{x_0 - \Delta x/2}^{x_0 + \Delta x/2} (f, g, h) \cdot (1, 0, 0)\, dx$$
$$+ \int_{y_0 - \Delta y/2}^{y_0 + \Delta y/2} (f, g, h) \cdot (0, 1, 0)\, dy + \int_{x_0 + \Delta x/2}^{x_0 - \Delta x/2} (f, g, h) \cdot (1, 0, 0)\, dx.$$

Then, combining integrals where possible, we have

$$\frac{1}{(\Delta x)(\Delta y)} \int_C F$$
$$= \frac{1}{(\Delta x)(\Delta y)} \int_{y_0 - \Delta y/2}^{y_0 + \Delta y/2} \left[g\left(x_0 + \frac{\Delta x}{2}, y, 0\right) - g\left(x_0 - \frac{\Delta x}{2}, y, 0\right) \right] dy$$
$$- \frac{1}{(\Delta x)(\Delta y)} \int_{x_0 - \Delta x/2}^{x_0 + \Delta x/2} \left[f\left(x, y_0 + \frac{\Delta y}{2}, 0\right) - f\left(x, y_0 - \frac{\Delta y}{2}, 0\right) \right] dx.$$

Use a mean value theorem for integrals to write

$$\int_{y_0 - \Delta y/2}^{y_0 + \Delta y/2} \left[g\left(x_0 + \frac{\Delta x}{2}, y, 0\right) - g\left(x_0 - \frac{\Delta x}{2}, y, 0\right) \right] dy$$
$$= \left[g\left(x_0 + \frac{\Delta x}{2}, \xi, 0\right) - g\left(x_0 - \frac{\Delta x}{2}, \xi, 0\right) \right] \Delta y$$

for some ξ between $y_0 - \Delta y/2$ and $y_0 + \Delta y/2$.

Similarly, for some ζ between $x_0 - \Delta x/2$ and $x_0 + \Delta x/2$, we have

$$\int_{x_0-\Delta x/2}^{x_0+\Delta x/2} \left[f\left(x, y_0 + \frac{\Delta y}{2}, 0\right) - f\left(x, y_0 - \frac{\Delta y}{2}, 0\right) \right] dx$$

$$= \left[f\left(\zeta, y_0 + \frac{\Delta y}{2}, 0\right) - f\left(\zeta, y_0 - \frac{\Delta y}{2}, 0\right) \right] \Delta x.$$

Then

$$\lim_{(\Delta x, \Delta y)\to(0,0)} \frac{1}{(\Delta x)(\Delta y)} \int_C F$$

$$= \lim_{\Delta x\to 0} \frac{g(x_0 + \Delta x/2, \xi, 0) - g(x_0 - \Delta x/2, \xi, 0)}{\Delta x}$$

$$- \lim_{\Delta y\to 0} \frac{f(\zeta, y_0 + \Delta y/2, 0) - f(\zeta, y_0 - \Delta y/2, 0)}{\Delta y}$$

$$= g_x(x_0, y_0, 0) - f_y(x_0, y_0, 0) = (\text{curl } F|_{(x_0,y_0,0)}) \cdot (0, 0, 1).$$

Note that the above limits do not look exactly like the limits defining g_x and f_y. However, they are in the form we saw for these partials in the derivation of the heat equation in Section 5.2. We have also used the fact that $\xi \to y_0$ and $\zeta \to x_0$ as $\Delta y \to 0$ and $\Delta x \to 0$, respectively.

By taking rectangles in the xz- and yz-planes, an argument like that just given shows that

$$\lim_{(\Delta x, \Delta z)\to(0,0)} \frac{1}{(\Delta x)(\Delta z)} \int_C F = (\text{curl } F|_{(x_0,0,z_0)}) \cdot (0, 1, 0)$$

and

$$\lim_{(\Delta y, \Delta z)\to(0,0)} \frac{1}{(\Delta z)(\Delta y)} \int_C F = (\text{curl } F|_{(0,y_0,z_0)}) \cdot (1, 0, 0).$$

Exercises for Section 6.2

1. Adapt the argument outlined in this section to show that

$$\lim_{(\Delta x, \Delta z)\to(0,0)} \frac{1}{(\Delta x)(\Delta z)} \int_C F = (\text{curl } F|_{(x_0,0,z_0)}) \cdot (0, 1, 0)$$

when C is a rectangle about $(x_0, 0, z_0)$ in the xz-plane, and

$$\lim_{(\Delta y, \Delta z)\to(0,0)} \frac{1}{(\Delta y)(\Delta z)} \int_C F = (\text{curl } F|_{(0,y_0,z_0)}) \cdot (1, 0, 0)$$

when C is a rectangle about $(0, y_0, z_0)$ in the yz-plane.

2. Let C be a rectangle about (x_0, y_0, z_0) in the plane P. Show that

$$\lim_{A \to 0} \frac{1}{A} \int_C F = (\text{curl } F|_{(x_0, y_0, z_0)}) \cdot \eta,$$

where η is normal to P and has unit length, and A is the area enclosed by C.

3. For each of the following, compute $(\text{curl } F|_{(x_0, y_0, z_0)}) \cdot \eta$ directly, then by using a limit as developed in this section.

(a) $F(x, y, z) = (z, x, y)$, $\eta = (1/\sqrt{2}, 1/\sqrt{2}, 0)$, $(x_0, y_0, z_0) = (2, 1, 3)$.

(b) $F(x, y, z) = (xyz, 3x^2, yz)$, $\eta = (1/\sqrt{3}, 1/\sqrt{3}, 1/\sqrt{3})$, $(x_0, y_0, z_0) = (0, 1, 1)$.

(c) $F(x, y, z) = (axy, byz, czx)$, with a, b, c constant, $\eta = (1, 0, 0)$, $(x_0, y_0, z_0) = (1, 1, 1)$.

6.3 Line Integrals of Scalar-Valued Functions and Applications

In Section 6.1 we saw $\int_C f\,dx + g\,dy + h\,dz$ as an alternative notation for $\int_C (f, g, h)$. In view of the linearity of line integrals, we can write

$$\int_C f\,dx + g\,dy + h\,dz = \int_C f\,dx + \int_C g\,dy + \int_C h\,dz.$$

This is like writing $\int_C (f, g, h) = \int_C (f, 0, 0) + \int_C (0, g, 0) + \int_C (0, 0, h)$. However, we gain something by doing this, namely, a notion of the line integral of a scalar-valued function, since f, g, and h are scalar-valued. In particular, if $p(x, y, z)$ is a real number for each (x, y, z) on the graph of C, then we define

$$\int_C p\,dx = \int_C (p, 0, 0),$$

$$\int_C p\,dy = \int_C (0, p, 0),$$

and

$$\int_C p\,dz = \int_C (0, 0, p).$$

EXAMPLE

Let $C(t) = (\cos(t), \sin(t), 2t^2)$, $0 \le t \le \pi$, and $p(x, y, z) = x + 2y + z$. Then

$$\int_C p\,dx = \int_0^\pi (\cos(t) + 2\sin(t) + 2t^2) \cdot (-\sin(t))\,dt$$

$$= \int_0^\pi (-\sin(t)\cos(t) - 2\sin^2(t) - 2t^2 \sin(t))\,dt$$

$$= 8 - 2\pi^2.$$

Next

$$\int_C p\,dy = \int_0^\pi [\cos(t) + 2\sin(t) + 2t^2]\cos(t)\,dt$$

$$= \int_0^\pi [\cos^2(t) + 2\sin(t)\cos(t) + 2t^2\cos(t)]\,dt = -1 - 4\pi.$$

Finally,

$$\int_C p\,dz = \int_0^\pi [\cos(t) + 2\sin(t) + 2t^2]4t\,dt = 8\pi - 6. \quad ▮$$

If we parametrize C by arc length s, then the symbol $\int_C p\,ds$ denotes $\int_0^L p(x(s), y(s), z(s))\,ds$, with L the length of C. Evaluation of $\int_C p\,ds$ does not require that we actually write C in terms of s. If $C: [a, b] \to R^3$, and $C(t) = (\alpha(t), \beta(t), \gamma(t))$, then $ds = (\alpha'(t)^2 + \beta'(t)^2 + \gamma'(t)^2)^{1/2}$ and

$$\int_C p\,ds = \int_a^b p(\alpha(t), \beta(t), \gamma(t))\sqrt{\alpha'(t)^2 + \beta'(t)^2 + \gamma'(t)^2}\,dt.$$

EXAMPLE

Let $p(x, y, z) = x^2yz$, and $C(t) = (t, t/2, 2t)$, $0 \leq t \leq 1$. Then

$$\int_C x^2yz\,dx = \int_0^1 t^2\left(\frac{t}{2}\right)(2t)\,dt = \frac{1}{5};$$

$$\int_C x^2yz\,dy = \int_0^1 t^2\left(\frac{t}{2}\right)(2t)\left(\frac{1}{2}\right)\,dt = \frac{1}{10};$$

$$\int_C x^2yz\,dz = \int_0^1 t^2\left(\frac{t}{2}\right)(2t)2\,dt = \frac{2}{5};$$

and

$$\int_C x^2yz\,ds = \int_0^1 t^2\left(\frac{t}{2}\right)(2t)\sqrt{1 + \frac{1}{4} + 4}\,dt$$

$$= \int_0^1 \sqrt{\frac{21}{4}}\,t^4\,dt = \frac{\sqrt{21}}{10}. \quad ▮$$

We conclude this section with some applications of line integrals of scalar-valued functions.

EXAMPLE

Suppose that we want to compute the mass of a thin wire. Imagine that the curve $C: [0, L] \to R^3$ describes the shape of the wire, using arc length s as parameter for convenience, and let $\rho(x, y, z)$ be the mass per unit length (density).

Subdivide $[0, L]$ by inserting points $0 = t_0 < t_1 < \cdots < t_n = L$, and choose $t_{i-1} \le \xi_i \le t_i$, $i = 1, \ldots, n$. Then

$$\sum_{i=1}^{n} \rho(\alpha(\xi_i), \beta(\xi_i), \gamma(\xi_i))\, \Delta s_i$$

approximates the mass, with Δs_i = length of the graph of C from $C(t_{i-1})$ to $C(t_i)$. In the limit as $n \to \infty$ and $t_i - t_{i-1} \to 0$, this gives us

$$\int_0^L \rho(\alpha(s), \beta(s), \gamma(s))\, ds,$$

or $\int_C \rho\, ds$, as the mass of the wire. ▮

Suppose that we want the center of mass of the wire of the previous example. By reasoning similar to that used for the center of mass of two- and three-dimensional solids (Section 4.1), the reader can check that the center of mass is at $(\bar{x}, \bar{y}, \bar{z})$, where

$$\bar{x} = \frac{\int_C x\rho\, ds}{\int_C \rho\, ds}, \qquad \bar{y} = \frac{\int_C y\rho\, ds}{\int_C \rho\, ds}, \qquad \bar{z} = \frac{\int_C z\rho\, ds}{\int_C \rho\, ds}.$$ ▮

EXAMPLE

The student who has had some experience with moments of inertia can check that the moments I_x, I_y, and I_z about the x-, y-, and z-axes, respectively, of the wire of the last two examples, are given by

$$I_x = \int_C \rho(y^2 + z^2)\, ds,$$

$$I_y = \int_C \rho(x^2 + z^2)\, ds,$$

and

$$I_z = \int_C \rho(x^2 + y^2)\, ds.$$ ▮

Exercises for Section 6.3

1. In each of the following, compute $\int_C p\, dx$, $\int_C p\, dy$, $\int_C p\, dz$, and $\int_C p\, ds$.
   (a) $p(x, y, z) = x^2 y \sin(z)$, $C(t) = (t, 2t, t/2)$, $0 \le t \le 1$
   (b) $p(x, y, z) = z^2 + x + y$, $C(t) = (e^t, \cos(t), 2t)$, $0 \le t \le \pi$
   (c) $p(x, y, z) = \sqrt{x^2 + y^2}$, $C(t) = (\cos(t), \sin(t), t^2)$, $0 \le t \le \pi/2$
   (d) $p(x, y, z) = |\sin(x)|yz$, $C(t) = (2t, 1, t^2)$, $\pi \le t \le 3\pi/2$

2. For the following curves and density functions, compute the mass, center of mass, and moments of inertia:

(a) The semicircle $x^2 + y^2 = 1$, $-1 \le x \le 1$, $y \ge 0$, $z = 0$, with $\rho(x, y, z) = x + y$.

(b) The parabola $y = x^2$, $z = 1$, $0 \le x \le 1$, with $\rho(x, y, z) = 2$.

(c) The square with vertices $(1, 1, 0)$, $(1, -1, 0)$, $(-1, 1, 0)$, $(-1, -1, 0)$ with
$$\rho(x, y, z) = \begin{cases} 2 & \text{from } (1, 1) \text{ to } (-1, 1), \\ 1 & \text{on the other three sides.} \end{cases}$$

(d) The triangle with vertices $(1, 0, 0)$, $(0, 1, 0)$, $(0, 0, 2)$, with $\rho(x, y, z) = 8$.

(e) The polygon with sides the straight lines joining $(0, 0, 0)$ to $(1, 0, 0)$ to $(0, 1, 0)$ to $(0, 0, 1)$ to $(-1, -1, -1)$ and back to $(0, 0, 0)$, $\rho(x, y, z) = xyz$.

6.4 Green's Theorem in the Plane

We are now in a position to describe an extremely important relationship between double integrals and line integrals in the plane. Throughout this section, we shall consider two-vector-valued functions, $F(x, y) = (P(x, y), Q(x, y))$, integrated over curves in the plane, say $C(t) = (\alpha(t), \beta(t))$. We integrate F over C by taking $\int_C P\,dx + Q\,dy$, or, if you like, think of F as

$$F(x, y) = (P(x, y), Q(x, y), 0).$$

It will be convenient later to settle on some terminology and notation. A curve C (in 2- or 3-space) is called *smooth* if its coordinate functions have continuous derivatives. Geometrically this means not only that C has a tangent at each point, but also that the tangent varies continuously as we move along the curve. A curve is *piecewise smooth* if it is made up of a finite number of smooth pieces and the coordinate functions are continuous on their interval of definition. A simple piecewise smooth curve will be called a *path.*

We mentioned in Section 6.1 the notion of orientation on $C: [a, b] \to R^3$, the point of view being that we travel *from* $C(a)$ *to* $C(b)$ as t varies from a to b. If C is closed, then $C(a) = C(b)$, but we can still think of C as oriented in the sense that $C(t)$ travels around C in one direction as t varies from a to b, and in the opposite direction as t varies from b to a. By convention, for a closed curve C in the xy-plane, we think of C as positively oriented if $C(t)$ travels counterclockwise as t goes from a to b, and negatively oriented if $C(t)$ travels clockwise as $t: a \to b$. This choice of orientation is motivated by the right-hand rule, which has the thumb pointing along the positive z-axis of the usual xyz-frame if the right hand is placed with fingers turning from the positive x-axis to the positive y-axis (Fig. 6-9).

When C is a closed curve, we often write $\oint_C$ instead of just $\int_C$. An arrow placed in the symbol can be used to indicate the sense of the orientation. Thus $\oint_C$ indicates a line integral around C, taken counterclockwise. Note that $\oint_C = -\oint_C$.

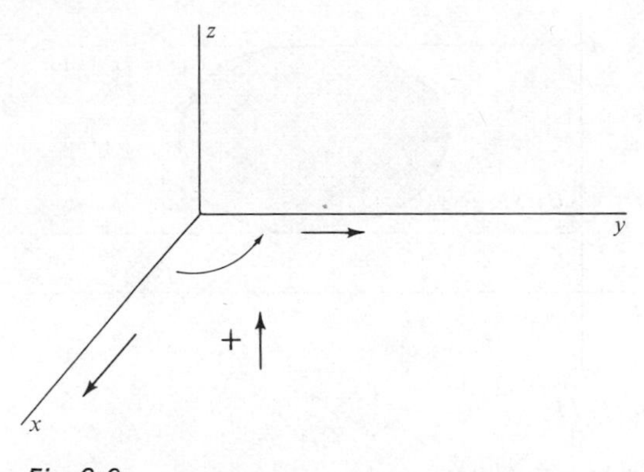

Fig. 6-9

We now state the main result of this section.

Green's Theorem

Let C be a closed path, enclosing a set D of points in the xy-plane. Let P, Q, P_y, and Q_x be continuous on C and in D. Then

$$\oint_C P\,dx + Q\,dy = \iint_D (Q_x - P_y).$$

A rigorous proof involves some very subtle difficulties. Rather than get in over everybody's head, we shall give a proof for a special case, and follow this with a plausibility argument for more general situations.

The special case is concerned with the description of D and C. Suppose

(1) There is an interval $[a, b]$ on the x-axis, and there are functions f_1 and f_2 such that C consists of two pieces, an upper piece $C_2: y = f_2(x)$, and a lower piece, $C_1: y = f_1(x)$, $a \leq x \leq b$ (see Fig. 6-10). Then

$$D = \{(x, y) \mid a \leq x \leq b \text{ and } f_1(x) \leq y \leq f_2(x)\}.$$

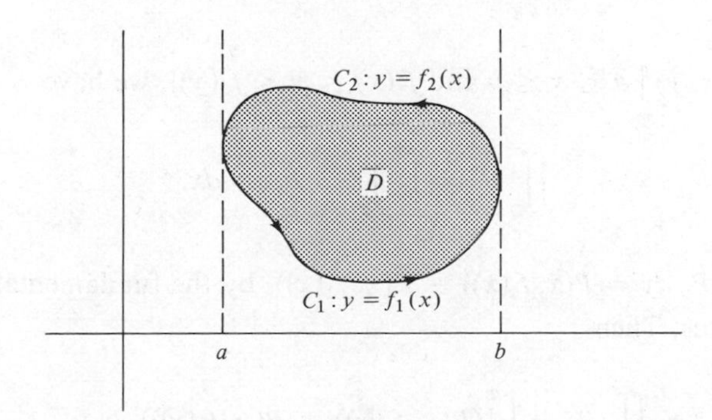

Fig. 6-10

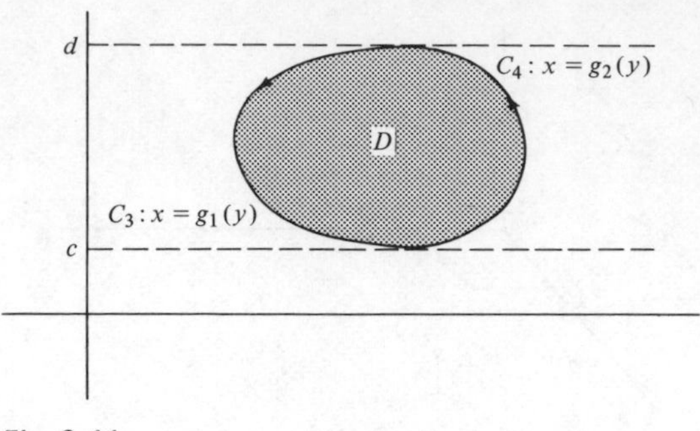

Fig. 6-11

(2) There are, at the same time, an interval $[p, q]$ on the y-axis, and functions g_1 and g_2 such that C also consists of a left piece $C_3: x = g_1(y)$ and a right piece $C_4: x = g_2(y)$, $p \le y \le q$ (see Fig. 6-11). Now

$$D = \{(x, y) \mid p \le y \le q \text{ and } g_1(y) \le x \le g_2(y)\}.$$

Under these circumstances, Green's Theorem can be verified directly by integrating both sides and comparing results. First, look at $\oint_C P(x, y)\, dx$, which we write as $\int_{C_1} P\, dx + \int_{C_2} P\, dx$, using description (1) of C assumed above. To keep the counterclockwise orientation, we parametrize:

$$C_1(x) = (x, f_1(x)),\ x: a \to b$$
$$C_2(x) = (x, f_2(x)),\ x: b \to a.$$

Thus

$$\begin{aligned}\oint_C P\, dx &= \int_a^b P(x, f_1(x))\, dx + \int_b^a P(x, f_2(x))\, dx \\ &= \int_a^b (-P(x, f_2(x)) + P(x, f_1(x)))\, dx.\end{aligned}$$

Using $D = \{(x, y) \mid a \le x \le b \text{ and } f_1(x) \le y \le f_2(y)\}$, we have

$$\iint_D P_y = \int_a^b \int_{f_1(x)}^{f_2(x)} P_y\, dy\, dx.$$

Now $\int_{f_1(x)}^{f_2(x)} P_y\, dy = P(x, f_2(x)) - P(x, f_1(x))$, by the fundamental theorem of integral calculus. Then

$$\iint_D P_y = \int_a^b (P(x, f_2(x)) - P(x, f_1(x)))\, dx.$$

By comparison, $\oint_C P\,dx = -\iint_D P_y$.

We now use the description (2) of C to evaluate $\oint_C Q\,dx$. Write C in two pieces:

$$C_3(y) = (g_1(y), y),\ y\colon q \to p,$$
$$C_4(y) = (g_2(y), y),\ y\colon p \to q.$$

Note that we must vary y from q to p on C_3 to keep a counterclockwise sense of rotation. Then

$$\oint_C Q\,dx = \int_q^p Q(g_1(y), y)\,dy + \int_p^q Q(g_2(y), y)\,dy$$
$$= \int_p^q [Q(g_2(y), y) - Q(g_1(y), y)]\,dy.$$

Finally, using $D = \{(x, y) \mid p \le y \le q \text{ and } g_1(y) \le x \le g_2(y)\}$, we have

$$\iint_D Q_x = \int_p^q \int_{g_1(y)}^{g_2(y)} Q_x\,dx\,dy = \int_p^q [Q(g_2(y), y) - Q(g_1(y), y)]\,dy = \oint_C Q\,dx.$$

Adding these results now gives us

$$\oint_C P\,dx + Q\,dy = \oint_C P\,dx + \oint_C Q\,dy$$
$$= \iint_D -P_y + \iint_D Q_x = \iint_D (Q_x - P_y).$$

This argument is not perfectly general because not every plane closed path C enclosing a region D admits both description (1) and (2) simultaneously. This is the case with the region shown in Fig. 6-12a. Often, however, a complicated region D can be subdivided into subregions $D_1, D_2, \ldots, D_n$ by inserting

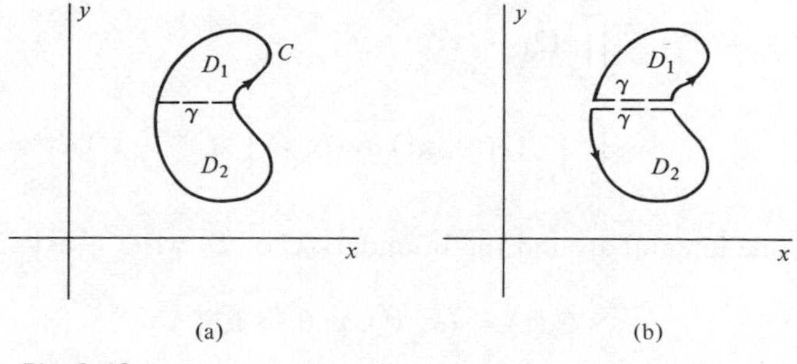

Fig. 6-12

interior curves (γ in Fig. 6-12a) in such a way that each D_i and its boundary curve has the kind of description we need. We can then apply Green's theorem to each D_i and its boundary. Now

$$\iint_{D_1}\left(\frac{\partial Q}{\partial x}-\frac{\partial P}{\partial y}\right)+\cdots+\iint_{D_n}\left(\frac{\partial Q}{\partial x}-\frac{\partial P}{\partial y}\right)=\iint_{D}\left(\frac{\partial Q}{\partial x}-\frac{\partial P}{\partial y}\right)$$

on the one side. On the other side, the boundary L_i of D_i consists of (1) a part of C not common to any other D_j, and (2) an inserted interior curve, common to exactly one other D_j. As we form $\oint_{L_1} P\,dx + Q\,dy + \cdots + \oint_{L_n} P\,dx + Q\,dy$, we eventually pick up all the pieces of C to form $\oint_C P\,dx + Q\,dy$, and we also integrate over each inserted interior piece twice, once in each direction, canceling the line integrals over the interior curves and leaving just $\oint_C P\,dx + Q\,dy$ (see Fig. 6-12b, where the boundary of D_1 consists of C_1 and γ, and of D_2, C_2, and γ. In applying Green's theorem to D_1 and to D_2, we form

$$\iint_{D_1}(Q_x - P_y) + \iint_{D_2}(Q_x - P_y) = \iint_{D}(Q_x - P_y)$$

and we also form

$$\begin{aligned}\int_{C_1} P\,dx + Q\,dy + \int_{-\gamma} P\,dx + Q\,dy + \int_{\gamma} P\,dx + Q\,dy + \int_{C_2} P\,dx + Q\,dy \\ = \oint_C P\,dx + Q\,dy.\Big)\end{aligned}$$

We conclude this section with a computational example, and devote the next section to some applications.

EXAMPLE

Let D be bounded by $y = x^2$ and $y = x$, $0 \le x \le 1$. Let $P(x, y) = x^2y$ and $Q(x, y) = 2xy$ (see Fig. 6-13).
Then

$$\begin{aligned}\iint_D (Q_x - P_y) &= \iint_D (2y - x^2) \\ &= \int_0^1 \int_{x^2}^{x} (2y - x^2)\,dy\,dx = \int_0^1 (x^2 - x^3)\,dx = \tfrac{1}{12}.\end{aligned}$$

For the line integral around the boundary C of D, write C as

$$\begin{aligned}C_1(x) &= (x, x^2),\ x\colon 0 \to 1,\\ C_2(x) &= (x, x),\ x\colon 1 \to 0.\end{aligned}$$

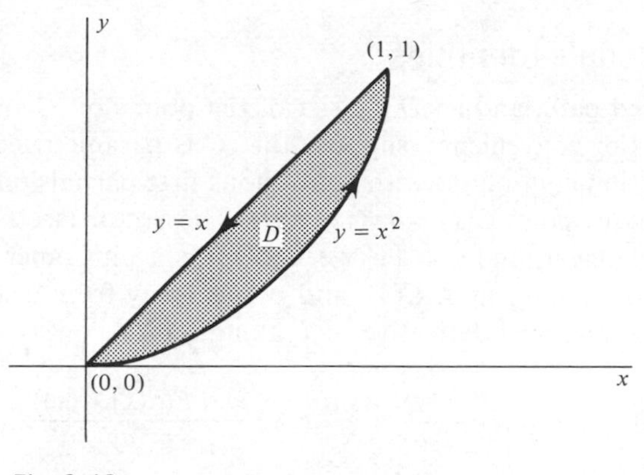

Fig. 6-13

Then

$$\oint_C x^2y\,dx + 2xy\,dy = \int_{C_1} x^2y\,dx + 2xy\,dy + \int_{C_2} x^2y\,dx + 2xy\,dy$$

$$= \int_0^1 x^2(x^2)\,dx + \int_0^1 2x(x^2)(2x)\,dx + \int_1^0 x^2x\,dx + \int_1^0 2xx\,dx$$

$$= \int_0^1 (x^4 + 4x^4 - x^3 - 2x^2)\,dx = \tfrac{1}{12}. \quad \blacksquare$$

Exercises for Section 6.4

1. Verify Green's theorem for each of the following:
   (a) $P(x, y) = 2xy$, $Q(x, y) = x + y$, C is the square with vertices $(1, 1)$, $(1, -1)$, $(-1, 1)$, $(-1, -1)$.
   (b) $P = e^x \sin(y)$, $Q = xy$, C is the triangle with vertices $(0, 0)$, $(2, 0)$, $(1, 1)$.
   (c) $P = x^4$, $Q - y^2$, C is the unit circle about the origin.
   (d) $P = x + y$, $Q = x^2 - y^2$, D is the region bounded by $y = x^3$ and $y = x$.

2. Let D be the region enclosed by the closed path C in the xy-plane. Show that

$$\text{area of } D = -\oint_C y\,dx = \oint_C x\,dy = \tfrac{1}{2}\oint_C (-y\,dx + x\,dy).$$

6.5 Applications of Green's Theorem

The purpose of this section is to illustrate the use of Green's theorem in a variety of different situations. We begin with Green's identities, which we shall use a number of times and which also have applications in various branches of physics (classical mechanics, to name just one).

A. Green's Identities

Let C be a closed path, and let D consist of the points on C and in the region enclosed by C. For convenience, suppose that C is parametrized by arc length. Let F be a function continuous with continuous first partials on D.

Now look at any point $C(s) = (x(s), y(s))$. The vector (see Fig. 6-14) $(x'(s), y'(s))$ is a unit tangent, and $\eta = (y'(s), -x'(s))$ is a unit outer normal [i.e., is perpendicular to the tangent at $C(s)$, and points away from instead of into the region D]. The directional derivative of F along η is

$$\nabla F|_{C(s)} \cdot \eta = \frac{\partial F(x(s), y(s))}{\partial x} y'(s) - \frac{\partial F(x(s), y(s))}{\partial y} x'(s).$$

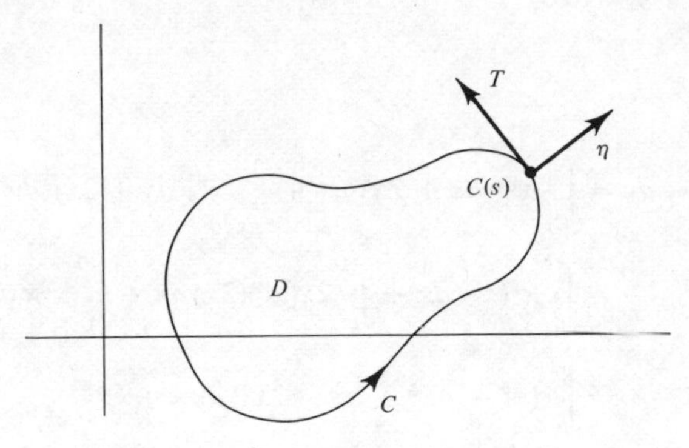

Fig. 6-14

We shall denote this as

$$\frac{dF}{d\eta}.$$

By Green's theorem,

$$\begin{aligned}\oint_C \frac{dF}{d\eta}\, ds &= \oint_C \left(F_x \frac{dy}{ds} - F_y \frac{dx}{ds}\right) ds \\ &= \oint_C F_x\, dy - F_y\, dx = \oint_C - F_y\, dx + F_x\, dy \\ &= \iint_D \left[\frac{\partial}{\partial x}\left(\frac{\partial F}{\partial x}\right) - \frac{\partial}{\partial y}\left(-\frac{\partial F}{\partial y}\right)\right] \\ &= \iint_D \left(\frac{\partial^2 F}{\partial x^2} + \frac{\partial^2 F}{\partial y^2}\right).\end{aligned}$$

The quantity $(\partial^2 F/\partial x^2) + (\partial^2 F/\partial y^2)$ is called the *Laplacian* of F, written $\nabla^2 F$, and plays an important role in potential theory, heat conduction, and vibrating-membrane problems, as we shall see later. For now, we have just derived Green's first identity:

$$\oint_C \frac{dF}{d\eta}\, ds = \iint_D \nabla^2 F.$$

A function F is called *harmonic* in D if $\nabla^2 F = 0$ in D. For harmonic F, then,

$$\oint_C \frac{dF}{d\eta}\, ds = 0.$$

Now suppose that G is also continuous with continuous first partials in D. Green's first identity can be generalized as follows. Write

$$\begin{aligned}\oint_C G\frac{dF}{d\eta}\, ds &= \oint_C G(F_x x'(s) - F_y y'(s))\, ds = \oint_C -GF_y\, dx + GF_x\, dy \\ &= \iint_D [(GF_x)_x - (-GF_y)_y] \\ &= \iint_D \left[G\left(\frac{\partial^2 F}{\partial x^2} + \frac{\partial^2 F}{\partial y^2}\right) + \frac{\partial G}{\partial x}\frac{\partial F}{\partial x} + \frac{\partial G}{\partial y}\frac{\partial F}{\partial y}\right] \\ &= \iint_D (G\nabla^2 F + \nabla F \cdot \nabla G).\end{aligned}$$

This gives us

$$\oint_C G\frac{dF}{d\eta}\, ds = \iint_D (G\nabla^2 F + \nabla F \cdot \nabla G),$$

which reduces to Green's first identity upon putting $G(x, y) = 1$ for all (x, y). This last formula is also known as *Green's first identity*.

Note that, by symmetry, we also have

$$\oint_C F\frac{dG}{d\eta}\, ds = \iint_D (F\nabla^2 G + \nabla F \cdot \nabla G).$$

Subtracting the last two gives us

$$\oint_C \left(F\frac{dG}{d\eta} - G\frac{dF}{d\eta}\right) ds = \iint_D (F\nabla^2 G - G\nabla^2 F),$$

which is *Green's second identity*. This gives us the rather surprising fact that, if F and G are both harmonic in D, then

$$\oint_C F\frac{dG}{d\eta}\, ds = \oint_C G\frac{dF}{d\eta}\, ds.$$

Thus far our applications of Green's theorem may have seemed rather pointless. We have derived several identities with no justification other than a statement that they are sometimes useful. We now show several uses for Green's identities in fairly concrete problems.

B. Application to the Uniqueness of Solution to a Boundary-Value Problem

Consider the partial differential equation

$$\frac{\partial u}{\partial t} = k\left(\frac{\partial^2 u}{\partial x^2} + \frac{\partial^2 u}{\partial y^2}\right), \qquad t > 0,\ (x, y) \text{ in some region } D \text{ bounded by a closed path } C.$$

This is a two-dimensional heat equation, with $u(x, y, t)$ the temperature at time t and point (x, y) in a thin homogeneous solid occupying the region D. If u is independent of y, we are back in the one-dimensional case derived in Section 5.2. This two-dimensional equation will be derived in Chapter 8.

We suppose that the initial temperature is given at each point in D:

$$u(x, y, 0) = f(x, y) \qquad \text{for } (x, y) \text{ in } D,$$

and that the temperature is known at all times on the boundary C:

$$u(x, y, t) = g(x, y, t) \qquad \text{for } t > 0,\ (x, y) \text{ on } C.$$

The differential equation, together with the given initial and boundary conditions, constitute a mathematical model for the heat-conduction problem, and we should expect on physical grounds that the model uniquely specifies $u(x, y, t)$. The question we ask now is: Can we show mathematically that the above equations determine u uniquely?

In order to prove that they do, we proceed by supposing that they do not. That is, suppose there are two distinct solutions, φ_1 and φ_2. Then

$$\frac{\partial \varphi_1}{\partial t} = k\,\nabla^2\varphi_1 \text{———} t > 0,\ (x, y) \text{ in } D \text{———} \frac{\partial \varphi_2}{\partial t} = k\,\nabla^2\varphi_2,$$

$$\varphi_1(x, y, 0) = f(x, y) \text{———} (x, y) \text{ in } D \text{———} \varphi_2(x, y, 0) = f(x, y),$$

$$\varphi_1(x, y, t) = g(x, y, t) \text{———} t \geq 0,\ (x, y) \text{ on } C \text{———} \varphi_2(x, y, t) = g(x, y, t).$$

Here we employ a trick. Put $w(x, y, t) = \varphi_1(x, y, t) - \varphi_2(x, y, t)$, and let

$$I(t) = \tfrac{1}{2}\iint_D w^2(x, y, t).$$

Assuming that we can interchange $\iint_D$ and $\partial/\partial t$ (we discuss this in Chapter 7), then

$$I'(t) = \iint_D \tfrac{1}{2} 2w \frac{\partial w}{\partial t} = \iint_D w \frac{\partial w}{\partial t}.$$

But

$$\frac{\partial w}{\partial t} = \frac{\partial \varphi_1}{\partial t} - \frac{\partial \varphi_2}{\partial t} = k\, \nabla^2 \varphi_1 - k\, \nabla^2 \varphi_2 = k\, \nabla^2(\varphi_1 - \varphi_2) = k\, \nabla^2 w.$$

Then

$$I'(t) = k \iint_D w\, \nabla^2 w.$$

Now apply Green's first identity with $F = G = w$. Then

$$k \iint_D w\, \nabla^2 w = k \oint_C w \frac{dw}{d\eta}\, ds - k \iint_D \nabla w \cdot \nabla w.$$

Now, if (x, y) is on C, then $w(x, y, t) = \varphi_1(x, y, t) - \varphi_2(x, y, t) = g(x, y, t) - g(x, y, t) = 0$. Hence $\oint_C w(dw/d\eta)\, ds = 0$. Then $I'(t) = -k \iint_D \nabla w \cdot \nabla w = -k \iint_D \|\nabla w\|^2$. Now, it is easy to check that $\|\nabla w\| \neq 0$ for at least one point of D, hence $\iint_D \|\nabla w\|^2 > 0$ (see Exercises 4 and 5). Then

$$I'(t) < 0 \qquad \text{for } t > 0.$$

Now apply the mean value theorem to I on any interval $[0, t]$, $t > 0$. For some τ, $0 < \tau < t$ and

$$I'(\tau) = \frac{I(t) - I(0)}{t}.$$

But

$$I(0) = \tfrac{1}{2} \iint_D w(x, y, 0)^2 = 0,$$

as $w(x, y, 0) = \varphi_1(x, y, 0) - \varphi_2(x, y, 0) = f(x, y) - f(x, y)$ for (x, y) in D. Then $I'(\tau) = I(t)/t < 0$, implying that $I(t) < 0$ for every $t > 0$. This is impossible, as $\frac{1}{2} \iint_D w^2 > 0$. Thus we must have $\varphi_1(x, y, t) = \varphi_2(x, y, t)$ for every (x, y) in D, and the uniqueness is proved.

C. Application to the Analysis of Harmonic Functions

The two-dimensional potential, or Laplace's, equation is

$$\nabla^2 u = \frac{\partial^2 u}{\partial x^2} + \frac{\partial^2 u}{\partial y^2} = 0.$$

A function satisfying Laplace's equation in a region D is said to be *harmonic* in D, as mentioned before. We shall now show that harmonic functions have a mean value property. That is, if (a, b) is an interior point of D, and C_r is a circle about (a, b) of sufficiently small radius r that C_r and all points within C_r are in D, then $u(a, b)$ is the arithmetic mean of the values $u(x, y)$ for (x, y) on C_r. That is,

$$u(a, b) = \frac{1}{2\pi r} \oint_{C_r} u \, ds,$$

$2\pi r$ being the length of C_r (see Fig. 6-15).

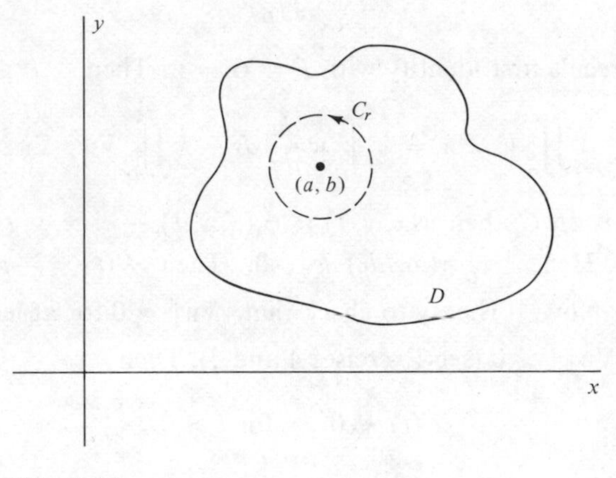

Fig. 6-15

To show that this is true, we first parametrize C_r in a natural way, using the polar angle from (a, b):

$$C_r(\theta) = (a + r\cos(\theta), b + r\sin(\theta)), 0 \leq \theta \leq 2\pi.$$

Next

$$\frac{ds}{d\theta} = \sqrt{x'(\theta)^2 + y'(\theta)^2} = \sqrt{r^2\sin^2(\theta) + r^2\cos^2(\theta)} = r,$$

so, in terms of θ,

$$\frac{1}{2\pi r} \oint_{C_r} u \, ds = \frac{1}{2\pi r} \int_0^{2\pi} u(a + r\cos(\theta), b + r\sin(\theta)) r \, d\theta$$

$$= \frac{1}{2\pi} \int_0^{2\pi} u(a + r\cos(\theta), b + r\sin(\theta)) \, d\theta.$$

What we must show, then, is that

$$u(a, b) = \frac{1}{2\pi} \int_0^{2\pi} u(a + r\cos(\theta), b + r\sin(\theta))\, d\theta.$$

Put $F = u$ in Green's first identity to obtain

$$\oint_{C_r} \frac{du}{d\eta}\, ds = \iint_{D_r} \nabla^2 u,$$

where D_r is the region bounded by C_r. Since $\nabla^2 u = 0$ in D, hence also in D_r, then

$$\oint_{C_r} \frac{du}{d\eta}\, ds = 0.$$

Now, a unit outer normal to C_r is $(\cos(\theta), \sin(\theta))$. Then, by the chain rule at the last part of the equation,

$$\begin{aligned} \frac{du}{d\eta} &= \nabla u \cdot \eta = \frac{\partial u}{\partial x}\cos(\theta) + \frac{\partial u}{\partial y}\sin(\theta) \\ &= \frac{d(u(a + r\cos(\theta), b + r\sin(\theta))}{dr}. \end{aligned}$$

Then

$$\begin{aligned} \oint_{C_r} \frac{du}{d\eta}\, ds &= \oint_{C_r} \frac{d(u(a + r\cos(\theta), b + r\sin(\theta))}{dr}\, ds \\ &= \int_0^{2\pi} \frac{d(u(a + r\cos(\theta), b + r\sin(\theta))}{dr}\, r\, d\theta \\ &= r\int_0^{2\pi} \frac{du(a + r\cos(\theta), b + r\sin(\theta))}{dr}\, d\theta \\ &= 0. \end{aligned}$$

Then

$$\int_0^{2\pi} \frac{du(a + r\cos(\theta), b + r\sin(\theta))}{dr}\, d\theta = 0.$$

Assuming that $\int_0^{2\pi}$ and d/dr can be interchanged, we have

$$\frac{d}{dr}\int_0^{2\pi} u(a + r\cos(\theta), b + r\sin(\theta))\, d\theta = 0.$$

Thinking of the integral expression as a function of r, this tells us that

$$\int_0^{2\pi} u(a + r\cos(\theta), b + r\sin(\theta))\, d\theta = \text{constant}.$$

To evaluate this constant, let $r \to 0$ to obtain:

$$\text{constant} = \int_0^{2\pi} u(a + r\cos(\theta), b + r\sin(\theta))\, d\theta$$

$$\to \int_0^{2\pi} u(a, b)\, d\theta = 2\pi u(a, b).$$

Then

$$2\pi u(a, b) = \int_0^{2\pi} u(a + r\cos(\theta), b + r\sin(\theta))\, d\theta,$$

as we wanted to show. This mean value property of harmonic functions will show up again in Chapter 9, when we discuss Cauchy's integral formula.

As another application of Green's identities to the study of harmonic functions, we shall show that the solution to the equation $\nabla^2 u = 0$ in D is uniquely determined by specifying the values of u on the boundary C of D. To do this, suppose that φ_1 and φ_2 are solutions to the problem

$$\begin{aligned} \nabla^2 u &= 0 && \text{in } D, \\ u(x, y) &= f(x, y) && \text{for } (x, y) \text{ on } C. \end{aligned}$$

Put $w = \varphi_1 - \varphi_2$, and note that

$$\begin{aligned} \nabla^2 w &= 0 && \text{in } D, \\ w &= 0 && \text{on } C. \end{aligned}$$

Let $F = G = w$ in Green's first identity to obtain

$$\oint_C w \frac{dw}{d\eta}\, ds = \iint_D (w\, \nabla^2 w + \nabla w \cdot \nabla w).$$

Now, $\iint_D w\, \nabla^2 w = 0$, as $\nabla^2 w = 0$ in D.

Further, $\oint_C w \dfrac{dw}{d\eta}\, ds = 0$, as $w = 0$ on C. Then

$$\iint_D \nabla w \cdot \nabla w = \iint_D \|\nabla w\|^2 = 0.$$

This is impossible unless $\|\nabla w\|^2 = 0$ in D, hence $w_x = w_y = 0$ in D. Then $w(x, y) =$ constant for (x, y) in D. But $w(x, y) = 0$ on C, so this constant must be zero. Then $\varphi_1(x, y) = \varphi_2(x, y)$ for (x, y) in D, as was to be shown.

The equation $\nabla^2 u = 0$ in D, together with the condition that u is given on the boundary C of D, is called a *Dirichlet problem*. We shall solve Dirichlet problems for various choices of the region D in Chapters 9 and 10.

D. Application to Change of Variables in Double Integrals

One important method of evaluating Riemann integrals $\int_a^b f(x)\,dx$ is to change variables. If $x = \varphi(t)$, for $c \leq t \leq d$, with $\varphi(c) = a$ and $\varphi(d) = b$ and suitable conditions hold for f and φ, then

$$\int_a^b f(x)\,dx = \int_c^d f(\varphi(t))\varphi'(t)\,dt.$$

The effectiveness of the method lies in the ability of the interested party to choose φ cleverly enough that the new integral on the right is easier to evaluate than the original one on the left.

We now want to extend the change of variables method to double integrals (and later to triple integrals). Suppose then that we want to evaluate

$$\iint_D F(x, y)\,dx\,dy.$$

We think of a change of variables here as a pair of functions

$$x = f(u, v), \qquad y = g(u, v),$$

relating points (u, v) in some region D^* of the uv-plane with points (x, y) in D. It is reasonable to demand at the outset that to each (x, y) in D there is exactly one (u, v) in D^* with $x = f(u, v)$, $y = g(u, v)$. That is, we suppose, at least in theory, that the equations $x = f(u, v)$, $y = g(u, v)$ can be inverted and solved for u and v in terms of x and y. This is equivalent to solving the system $A(x, y, u, v) = f(u, v) - x = 0$, $B(x, y, u, v) = g(u, v) - y = 0$, for u and v. By an implicit function theorem, assuming that f and g are continuous with continuous partials in D^*, this can be done if $\partial(A, B)/\partial(u, v) \neq 0$ throughout D^*. A simple calculation shows that

$$\frac{\partial(A, B)}{\partial(u, v)} = \frac{\partial(f, g)}{\partial(u, v)}.$$

We therefore assume that $\partial(f, g)/\partial(u, v) \neq 0$ in D^*.

We now want to see how to go about transforming $\iint_D F$ into an integral of some function of u and v over D^*. Go back to the definition and recall that, in forming $\iint_D F$, we begin by partitioning D into pieces $D_1, \ldots, D_n$ of areas $A(D_1), \ldots, A(D_n)$, respectively, choosing (ξ_i, η_i) in D_i, and forming sums $\sum_{i=1}^{n} F(\xi_i, \eta_i)A(D_i)$. Now, corresponding to the subregions $D_1, \ldots, D_n$ of D, are the subregions $D_1^*, \ldots, D_n^*$ of D^*, with

$$D_i^* = \{(u, v) \mid (f(u, v), g(u, v)) \text{ is in } D\}$$

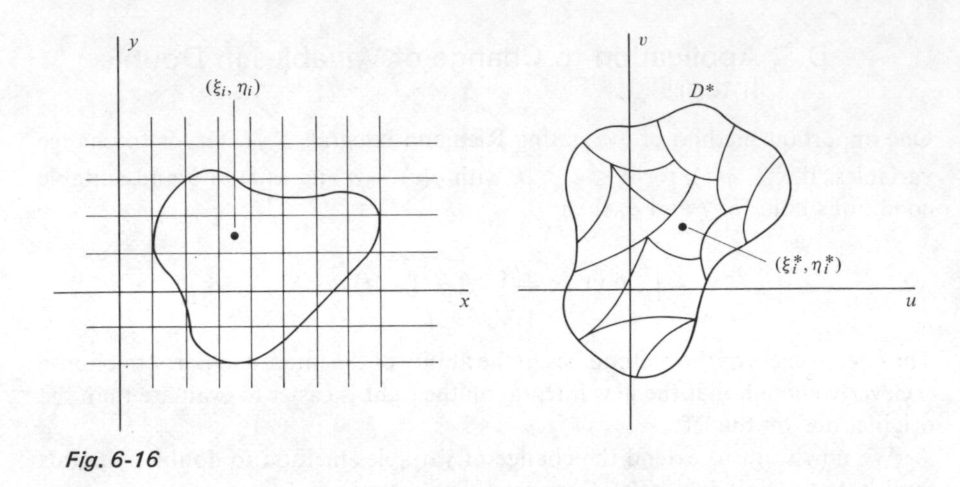

Fig. 6-16

(see Fig. 6-16). In each D_i^* is a point (ξ_i^*, η_i^*) such that $\xi_i = f(\xi_i^*, \eta_i^*)$ and $\eta_i = g(\xi_i^*, \eta_i^*)$. Then, so far

$$\sum_{i=1}^{n} F(\xi_i, \eta_i)A(D_i) = \sum_{i=1}^{n} F(f(\xi_i^*, \eta_i^*), g(\xi_i^*, \eta_i^*))A(D_i).$$

In order to make the sum on the right look like an approximating sum for a double integral in terms of u and v over D^*, we must express $A(D_i)$ in terms of u and v. To do this, we attempt to relate $A(D_i)$, the area of D_i, with $A(D_i^*)$, the area of D_i^*. The problem is that D_i is a rectangle by choice of the partition of D, while D_i^*, which consists of the points (u, v) in D^* corresponding to points (x, y) in D_i under the change of variables, need not be rectangular at all. So we invoke Green's theorem. To be able to do this, we shall suppose that D is the region bounded by a closed path $C: [a, b] \to R^2$, that D^* is bounded by a closed path $C^*: [a^*, b^*] \to R^2$, and that f and g are continuous with continuous first and second partials on D^* and C^*. Since $\partial(f, g)/\partial(u, v)$ is continuous and, by assumption, nonzero, then $\partial(f, g)/\partial(u, v)$ is strictly positive or strictly negative on D^*. As we shall see in the examples later, $\partial(f, g)/\partial(u, v) > 0$ means that orientation is preserved by the change of variables in the sense that, as (u, v) goes around C^* once counterclockwise, the image point $(x, y) = (f(u, v), g(u, v))$ goes once counterclockwise around C. If $\partial(f, g)/\partial(u, v) < 0$, then orientation is reversed, and (x, y) goes around C in a direction opposite that of (u, v) around C^*.

We now get back to the immediate objective, which is to relate $A(D_i)$ with $A(D_i^*)$. Let C_i be the closed path bounding D_i, and C_i^* the corresponding closed path bounding D_i^*. By Green's theorem,

$$A(D_i) = \iint_{D_i} 1 = \iint_{D_i} \frac{\partial x}{\partial x} = \oint_{C_i} x\, dy.$$

Now suppose for the moment that $\partial(f, g)/\partial(u, v) > 0$. Put $x = f(u, v)$ and $dy = g_u\, du + g_v\, dv$ to write

$$\oint_{C_i} x\, dy = \oint_{C_i^*} f(u, v) g_u\, du + f(u, v) g_v\, dv.$$

Now apply Green's theorem to the last line integral:

$$\begin{aligned}\oint_{C_i^*} (fg_u\, du + fg_v\, dv) &= \iint_{D_i^*} \left[\frac{\partial}{\partial u}(fg_v) - \frac{\partial}{\partial v}(fg_u)\right] du\, dv \\ &= \iint_{D_i^*} [f_u g_v - f_v g_u + f(g_{uv} - g_{vu})]\, du\, dv \\ &= \iint_{D_i^*} (f_u g_v - f_v g_u)\, du\, dv \qquad (\text{as } g_{uv} = g_{vu}) \\ &= \iint_{D_i^*} \frac{\partial(f, g)}{\partial(u, v)}\, du\, dv.\end{aligned}$$

Since $\partial(f, g)/\partial(u, v)$ is continuous, we can apply a mean value theorem to write

$$\iint_{D_i^*} \frac{\partial(f, g)}{\partial(u, v)} = \left(\left.\frac{\partial(f, g)}{\partial(u, v)}\right|_{(\alpha_i^*, \beta_i^*)}\right) \iint_{D_i^*} du\, dv,$$

for some (α_i^*, β_i^*) in D_i^*. Since $\iint_{D_i^*} du\, dv = A(D_i^*)$, then

$$A(D_i) = \left(\left.\frac{\partial(f, g)}{\partial(u, v)}\right|_{(\alpha_i^*, \beta_i^*)}\right) A(D_i^*).$$

If $\partial(f, g)/\partial(u, v) < 0$, the argument is similar except that now the change of variables reverses orientation in the line integrals:

$$\oint_{C_i^*} x\, dy = \oint_{C_i^*} fg_u\, du + fg_v\, dv = -\oint_{C_i^*} fg_u\, du + fg_v\, dv.$$

We then obtain

$$A(D_i) = \left(\left.\frac{-\partial(f, g)}{\partial(u, v)}\right|_{(\alpha_i^*, \beta_i^*)}\right) A(D_i^*).$$

Both results are summarized in one by writing absolute values:

$$A(D_i) = \left|\frac{\partial(f, g)}{\partial(u, v)}\right|_{(\alpha_i^*, \beta_i^*)} A(D_i^*).$$

Having succeeded in relating $A(D_i)$ to $A(D_i^*)$, the rest is easy. We now have

$$\sum_{i=1}^{n} F(\xi_i, \eta_i)A(D_i) = \sum_{i=1}^{n} F(f(\xi_i^*, \eta_i^*), g(\xi_i^*, \eta_i^*))\left|\frac{\partial(f, g)}{\partial(u, v)}\right|_{(\alpha_i^*, \beta_i^*)} A(D_i^*).$$

In the limit as $n \to \infty$ and each $A(D_i) \to 0$, so that each $A(D_i^*) \to 0$ as well, the sum of the left approaches $\iint_D F$, while that on the right (with the help of a Duhamel-type theorem) goes to

$$\iint_{D^*} F(f(u, v), g(u, v))\left|\frac{\partial(f, g)}{\partial(u, v)}\right| du\, dv.$$

This is the change of variable formula for double integrals. For convenience, we summarize the conditions we assumed as we went through the argument above: *Suppose that D is bounded by a closed path C in the xy-plane, and D^* by a closed path C^* in the uv-plane. Let $x = f(u, v)$, $y = g(u, v)$ relate each point of D^* with exactly one point of D, and conversely. Suppose that f and g are continuous with continuous first and second partials on D^*. Suppose that $\partial(f, g)/\partial(u, v) \neq 0$ on D^*, and F is continuous on D. Then*

$$\iint_D F(x, y)\, dx\, dy = \iint_{D^*} F(f(u, v), g(u, v))\left|\frac{\partial(f, g)}{\partial(u, v)}\right| du\, dv.$$

At the expense of a much more complicated derivation, the hypotheses just listed can be relaxed a great deal. In many applications, for example, the Jacobian $\partial(f, g)/\partial(u, v)$ may be allowed to vanish at points on C^* without invalidating the change-of-variables formula.

We now give some examples.

EXAMPLE

Evaluate $\iint_D \sqrt{1 + x^2 + y^2}$, where D is the region bounded by $x^2 + y^2 = \delta^2$. Polar coordinates are the natural choice here (see Fig. 6-17). Put

$$x = r\cos(\theta), \qquad 0 \le \theta \le 2\pi,$$
$$y = r\sin(\theta), \qquad 0 \le r \le \delta.$$

Then

$$\sqrt{1 + x^2 + y^2} = \sqrt{1 + r^2}.$$

The Jacobian is

$$\frac{\partial(x, y)}{\partial(r, \theta)} = \begin{vmatrix} \cos(\theta) & -r\sin(\theta) \\ \sin(\theta) & r\cos(\theta) \end{vmatrix} = r.$$

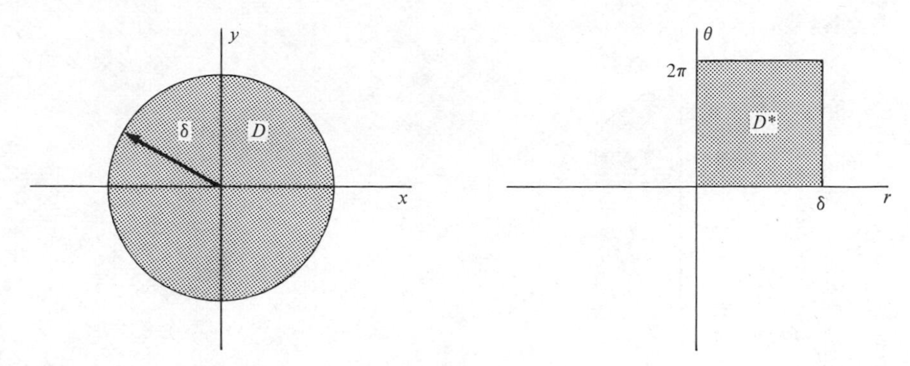

Fig. 6-17

Then

$$\iint_D \sqrt{1 + x^2 + y^2} = \iint_{D^*} \sqrt{1 + r^2}\, r\, dr\, d\theta.$$

Note that D^* is the rectangle $0 \leq \theta \leq 2\pi$, $0 \leq r \leq \delta$. Then

$$\begin{aligned}\iint_{D^*} r\sqrt{1 + r^2}\, dr\, d\theta &= \int_0^{2\pi}\int_0^{\delta} r\sqrt{1 + r^2}\, dr\, d\theta \\ &= 2\pi \int_0^{\delta} r\sqrt{1 + r^2}\, dr = \frac{2\pi}{3}(1 + r^2)^{3/2}\Big|_0^{\delta} \\ &= \frac{2\pi}{3}(1 + \delta^2)^{3/2} - \frac{2\pi}{3}. \qquad \blacksquare\end{aligned}$$

EXAMPLE

Evaluate $\iint_D xy$, where D is the region bounded by the quarter-circle $x^2 + y^2 = 1$, $x \geq 0$, $y \geq 0$ (see Fig. 6-18). Again, try polar coordinates, with $0 \leq \theta \leq \pi/2$, $0 \leq r \leq 1$. We get

$$\begin{aligned}\iint_D xy &= \int_0^{\pi/2}\int_0^1 r\cos(\theta) r\sin(\theta) r\, dr\, d\theta \\ &= \int_0^{\pi/2}\int_0^1 r^3 \cos(\theta)\sin(\theta)\, dr\, d\theta = \tfrac{1}{4}\int_0^{\pi/2} \cos(\theta)\sin(\theta)\, d\theta \\ &= \tfrac{1}{8}\sin^2(\theta)\big|_0^{\pi/2} = \tfrac{1}{8}. \qquad \blacksquare\end{aligned}$$

EXAMPLE

Evaluate $\iint_D (y - x)/(y + x)$, where D is the region bounded by $x + y = 4$, $x + y = 2$, the x-axis and the y-axis. Here we must try something more clever than polar coordinates. The form of the integrand suggests we try something like

$$u = y - x,$$
$$v = y + x.$$

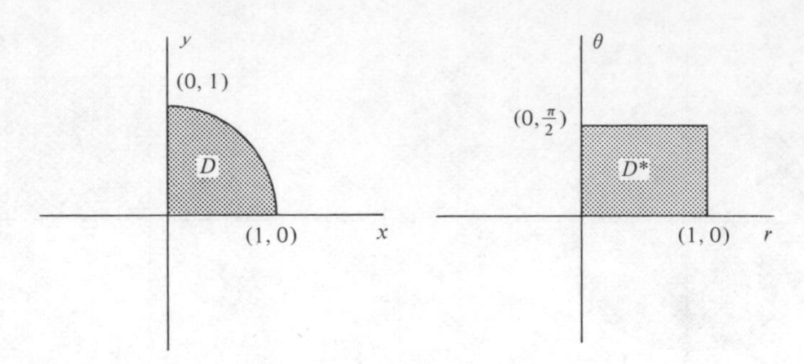

Fig. 6-18

Now we must obtain the corresponding region D^* in the uv-plane. To do this, consider the boundary of D, a piece at a time (Fig. 6-19). On $x + y = 2$, we have $v = 2$, so the change of variables maps $x + y = 2$ to the line $v = 2$. Similarly, the line $x + y = 4$ corresponds to $v = 4$. Next, look at the x-axis. If $y = 0$, then $u = -x$ and $v = x$, so $u = -v$. Finally, on the y-axis, $x = 0$, so $u = y = v$. Then D^* is bounded by the lines $v = 2, v = 4, u = -v$, and $u = v$.

Here, there is a fine point to be recognized. It is reasonable that the boundary of D should correspond to the boundary of D^* under the change of variables. But how do we know that D corresponds to the region *inside* the trapezoid formed by $v = 2$, $v = 4$, $u = -v$, $u = v$, instead of the region *outside* the trapezoid? (Both regions have the same boundary points.) There is an easy, practical way to check. Take an interior point of D, say (1, 2). The image is (1, 3), which is inside the trapezoid. By continuity, all the points inside D must similarly correspond to points inside the trapezoid, and we have D^* as in Fig. 6-19.

Now back to the computation. We can write

$$x = \frac{v - u}{2} \quad \text{and} \quad y = \frac{v + u}{2}.$$

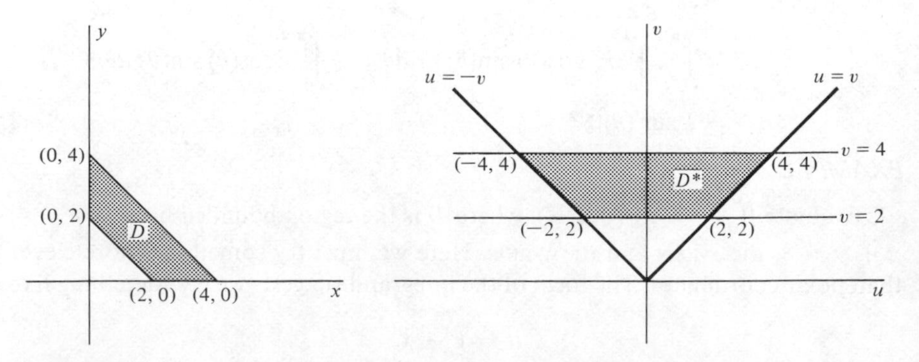

Fig. 6-19

The Jacobian is then

$$\frac{\partial(x, y)}{\partial(u, v)} = \begin{vmatrix} -\frac{1}{2} & \frac{1}{2} \\ \frac{1}{2} & \frac{1}{2} \end{vmatrix} = -\frac{1}{2}.$$

Then

$$\iint_D e^{(y-x)/(y+x)} = \iint_{D*} e^{u/v}|-\tfrac{1}{2}|\, du\, dv = \tfrac{1}{2}\iint_{D*} e^{u/v}\, du\, dv.$$

Since $e^{u/v}$ is difficult to integrate with respect to v, we integrate first with respect to u:

$$\iint_{D*} e^{u/v}\, du\, dv = \int_2^4 \int_{-v}^{v} e^{u/v}\, du\, dv.$$

Now

$$\int_{-v}^{v} e^{u/v}\, du = ve^{u/v}\Big|_{-v}^{v} = ve - \frac{v}{e} = v\left(e - \frac{1}{e}\right).$$

Then

$$\iint_{D*} e^{u/v} = \int_2^4 v\left(e - \frac{1}{e}\right) = \left(e - \frac{1}{e}\right)\frac{v^2}{2} = 6\left(e - \frac{1}{e}\right).$$

Thus

$$\iint_D e^{(y-x)/(y+x)} = 3\left(e - \frac{1}{e}\right). \quad \blacksquare$$

Note that we encountered a negative Jacobian in this example. We said before that this should indicate a reversal of orientation in our change of variables. This is exactly what happens here. The reader can check that, as (x, y) starts at $(2, 0)$ and proceeds to $(4, 0)$, up to $(0, 4)$, down to $(0, 2)$, and back to $(2, 0)$ along the boundary of D, the image point in the uv-plane goes from $(-2, 2)$ to $(-4, 4)$ to $(4, 4)$ to $(2, 2)$ and back to $(-2, 2)$, in the opposite direction, around the boundary of D^*.

Another point to be noticed from the last example is that the change of variables is sometimes more conveniently defined from the statement of the problem in the form $u = p(x, y)$, $v = q(x, y)$, giving u and v in terms of x and y instead of the other way around. In theory, we always suppose that these equations can be inverted to give x and y in terms of u and v. In practice this is not always so easy. The place where this can cause difficulty is in computing the Jacobian $\partial(x, y)/\partial(u, v)$, in which it is nice to have x and y in terms of u and v. Fortunately, this difficulty can be easily circumvented. We shall now show that

$$\frac{\partial(x, y)}{\partial(u, v)} = \frac{1}{\dfrac{\partial(u, v)}{\partial(x, y)}}.$$

Thus, if we have $x = f(u, v)$, $y = g(u, v)$, then go ahead and calculate $\partial(x, y)/\partial(u, v)$ directly. If we have only $u = p(x, y)$ and $v = q(x, y)$, then calculate

$$\frac{\partial(u, v)}{\partial(x, y)} = \begin{vmatrix} p_x & p_y \\ q_x & q_y \end{vmatrix}$$

and its reciprocal is $\partial(x, y)/\partial(u, v)$.

To prove this reciprocity relationship, go back to the basic question of how we would in theory solve for x and y in terms of u and v. We would set up the equations

$$p(x, y) - u = 0,$$

$$q(x, y) - v = 0$$

and apply an implicit function theorem. Once we know that a solution $x = f(u, v)$, $y = g(u, v)$ exists, we have

$$p(f(u, v), g(u, v)) - u = 0,$$

$$q(f(u, v), g(u, v)) - v = 0.$$

Differentiate with respect to u and then v:

$$p_x x_u + p_y y_u - 1 = 0,$$

$$p_x x_v + p_y y_v = 0,$$

$$q_x x_u + q_y y_u = 0,$$

and

$$q_x x_v + q_y y_v - 1 = 0.$$

Solve the first and third for x_u and y_u:

$$x_u = \frac{q_y}{\dfrac{\partial(p, q)}{\partial(x, y)}} \quad \text{and} \quad y_u = \frac{-q_x}{\dfrac{\partial(p, q)}{\partial(x, y)}}.$$

Solve the second and fourth for x_v and y_v:

$$x_v = \frac{-p_v}{\dfrac{\partial(p, q)}{\partial(x, y)}} \quad \text{and} \quad y_v = \frac{p_x}{\dfrac{\partial(p, q)}{\partial(x, y)}}.$$

Writing $J = \partial(p, q)/\partial(x, y)$ for convenience, we have by direct calculation

$$\frac{\partial(x, y)}{\partial(u, v)} = \begin{vmatrix} x_u & x_v \\ y_u & y_v \end{vmatrix} = \begin{vmatrix} \frac{1}{J} q_y & -\frac{1}{J} p_y \\ -\frac{1}{J} q_x & \frac{1}{J} p_x \end{vmatrix}$$

$$= \frac{1}{J^2} \begin{vmatrix} q_y & -p_y \\ -q_x & p_x \end{vmatrix} = \frac{1}{J^2} \begin{vmatrix} p_x & p_y \\ q_x & q_y \end{vmatrix}$$

$$= \frac{1}{J} = \frac{1}{\dfrac{\partial(u, v)}{\partial(x, y)}}.$$

EXAMPLE

In the last example, we had

$$u = y - x, \qquad v = x + y.$$

Then

$$\frac{\partial(u, v)}{\partial(x, y)} = \begin{vmatrix} -1 & 1 \\ 1 & 1 \end{vmatrix} = -2.$$

Solving, we get $x = (v - u)/2$ and $y = (v + u)/2$. Then

$$\frac{\partial(x, y)}{\partial(u, v)} = \begin{vmatrix} -\frac{1}{2} & \frac{1}{2} \\ \frac{1}{2} & \frac{1}{2} \end{vmatrix} = -\frac{1}{2} = \frac{1}{\dfrac{\partial(u, v)}{\partial(x, y)}}.$$ ∎

EXAMPLE

Let $u = x^2 - y^2$ and $v = 2xy$. Then

$$\frac{\partial(u, v)}{\partial(x, y)} = \begin{vmatrix} 2x & -2y \\ 2y & 2x \end{vmatrix} = 4(x^2 + y^2).$$

If $x = x(u, v)$ and $y = y(u, v)$ are the solutions of the above equations for x and y in terms of u and v, then we should have

$$\frac{\partial(x, y)}{\partial(u, v)} = \frac{1}{4[x(u, v)^2 + y(u, v)^2]}.$$

The reader can verify this directly by solving for $x(u, v)$ and $y(u, v)$ and computing $\partial(x, y)/\partial(u, v)$ explicitly. ∎

We conclude this section with an illustration of change-of-variables techniques in multiple integrals applied to evaluation of an improper single integral.

EXAMPLE

$$\text{Evaluate } \int_0^\infty e^{-x^2}\,dx.$$

It is easy to check that this integral converges (compare with $\int_0^\infty e^{-x}\,dx$), but its value is not obvious, as e^{-x^2} has no elementary antiderivative. Here is a trick. Look at

$$\iint_{D_R} e^{-x^2-y^2},$$

where D_R is the quarter-circular region $x^2 + y^2 \le R^2$, $x \ge 0$, $y \ge 0$ (Fig. 6-20). Using polar coordinates, $\iint_{D_R} e^{-x^2-y^2}$ is easy:

$$\iint_{D_R} e^{-x^2-y^2} = \int_0^{\pi/2}\int_0^R e^{-r^2} r\,dr\,d\theta = \frac{\pi}{2}\int_0^R re^{-r^2}\,dr$$

$$= -\frac{\pi}{4}e^{-r^2}\bigg|_0^R = \frac{\pi}{4}(1 - e^{-R^2}).$$

As $R \to \infty$, $\iint_{D_R} e^{-x^2-y^2} \to \iint_D e^{-x^2-y^2}$, where D is the quarter-plane $x \ge 0$, $y \ge 0$. But $(\pi/4)(1 - e^{-R^2}) \to \pi/4$, so we have

$$\iint_D e^{-x^2-y^2}\,dx\,dy = \frac{\pi}{4}.$$

But, in rectangular coordinates,

$$\iint_D e^{-x^2-y^2} = \int_0^\infty e^{-x^2}\,dx \int_0^\infty e^{-y^2}\,dy$$

$$= \left(\int_0^\infty e^{-x^2}\,dx\right)^2.$$

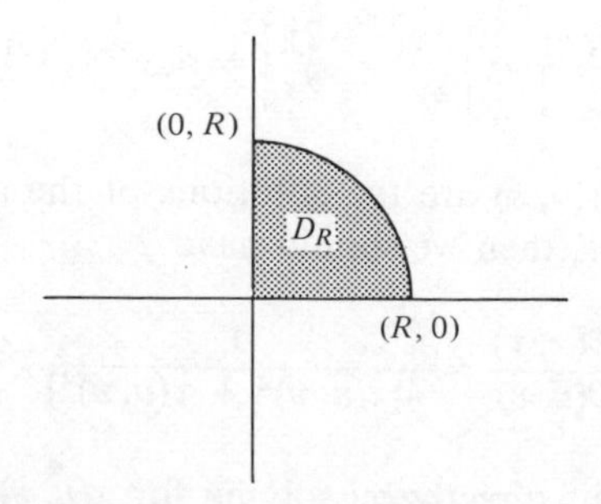

Fig. 6-20

Hence

$$\int_D^\infty e^{-x^2}\,dx = \sqrt{\frac{\pi}{4}} = \frac{\sqrt{\pi}}{2}. \qquad \blacksquare$$

E. Application to Independence of Path

A line integral $\int_C P\,dx + Q\,dy$ generally depends not only on P and Q, but also on the curve C. If it happens that we are working in some region D of the plane, and $\int_C P\,dx + Q\,dy = \int_{C*} P\,dx + Q\,dy$ whenever C and C^* are paths in D having the same initial and terminal points, then we say that $\int_C P\,dx + Q\,dy$ is independent of path in D. We shall spend some time now discussing how such a thing can happen, and some of its ramifications, which are important both mathematically and physically.

First, we make a convention. It will turn out to be convenient to restrict our attention to regions D in the xy-plane having the following two properties: (1) D is connected; (2) for each (a, b) in D, there is an r-neighborhood of (a, b) wholly contained within D. In particular, (2) implies that D contains none of its boundary points. A set D satisfying (1) and (2) will be called a *domain*. Thus, for the rest of this section we shall concentrate on line integrals of the form $\int_C P\,dx + Q\,dy$, with P and Q continuous with continuous first partials in a domain D and C a path in D.

The first theorem we shall prove is that *independence of path of* $\int_C P\,dx + Q\,dy$ *in* D *is equivalent to having* $\oint_K P\,dx + Q\,dy = 0$ *for every closed path* K *in* D.

To prove this, first suppose that $\int_C P\,dx + Q\,dy$ is independent of path in D, and K is a closed path in D. We must show that $\oint_K P\,dx + Q\,dy = 0$. Choose distinct points P_0 and P_1 on K, as in Fig. 6-21, and break K up into paths C_1 from P_0 to P_1 and C_2 along the rest of K back from P_1 to P_0.

By the independence of path, and noting that orientation makes $-C_2$ a path in D from P_0 to P_1, we have

$$\int_{C_1} P\,dx + Q\,dy = \int_{-C_2} P\,dx + Q\,dy = -\int_{C_2} P\,dx + Q\,dy.$$

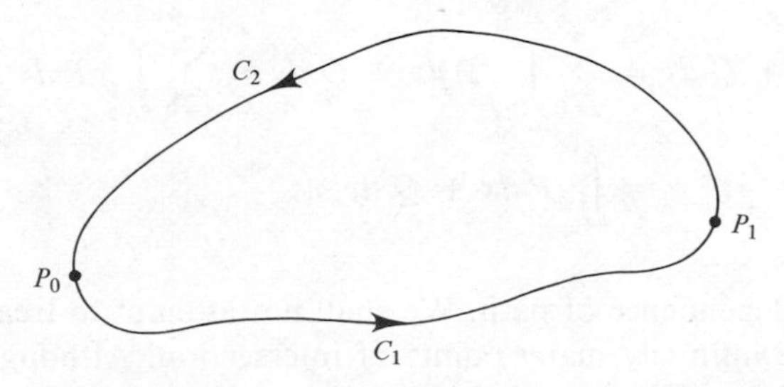

Fig. 6-21

Then

$$\int_{C_1} P\,dx + Q\,dy + \int_{C_2} P\,dx + Q\,dy = \oint_K P\,dx + Q\,dy = 0.$$

Conversely, suppose that $\oint_K P\,dx + Q\,dy = 0$ for every closed path K in D. Let P_0 and P_1 be distinct points of D, and C_1 and C_2 paths in D from P_0 to P_1. We want to show that $\int_{C_1} P\,dx + Q\,dy = \int_{C_2} P\,dx + Q\,dy$.

If C_1 and C_2 have only P_0 and P_1 in common, then C_1 and $-C_2$ form a closed path K in D, so, by assumption, $\oint_K P\,dx + Q\,dy = \int_{C_1} P\,dx + Q\,dy + \int_{-C_2} P\,dx + Q\,dy = 0$ gives us $\int_{C_1} P\,dx + Q\,dy = \int_{C_2} P\,dx + Q\,dy$.

Things are more complicated if C_1 and C_2 intersect in finitely many other points, say $U_1, \ldots, U_n$ ordered from P_0 to P_1, as in Fig. 6-22. For convenience, relabel $P_0 = U_0$, $P_1 = U_{n+1}$.

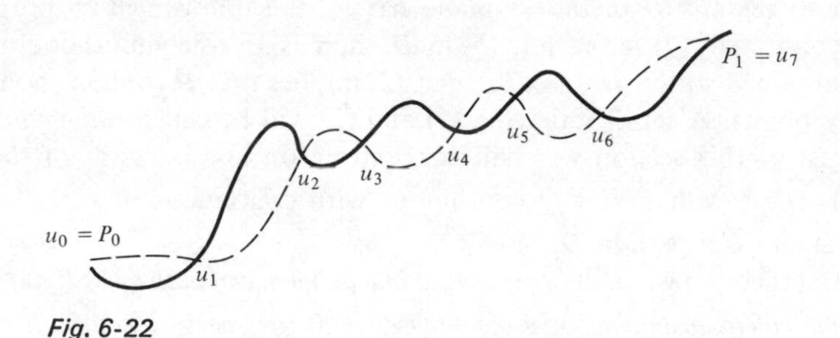

Fig. 6-22

Let C_{1j} be the part of C_1 from U_j to U_{j+1}, and C_{2j} the part of C_2 from U_j to U_{j+1}, $j = 0, \ldots, n$. Then C_{1j} and C_{2j} are paths in D between U_j and U_{j+1} intersecting only at U_j and U_{j+1}, so by the preceding argument,

$$\int_{C_{1j}} P\,dx + Q\,dy = \int_{C_{2j}} P\,dx + Q\,dy.$$

Then

$$\int_{C_1} P\,dx + Q\,dy = \sum_{j=0}^{n} \int_{C_{1j}} P\,dx + Q\,dy = \sum_{j=0}^{n} \int_{C_{2j}} P\,dx + Q\,dy$$
$$= \int_{C_2} P\,dx + Q\,dy$$

and we have independence of path. We shall not attempt to treat the case that C_1 and C_2 have infinitely many points of intersection. Although theoretically interesting, it is also difficult, and is rarely encountered in practice.

Physically, the last few paragraphs may be summarized by saying that the work done by the force (P, Q) in moving a particle from P_0 to P_1, for any P_0 and P_1 in D is independent of the choice of path in D from P_0 to P_1 exactly when the work done around any closed path in D is zero. Such a force is often called *conservative*.

In physics, the student is often told that a conservative force is derivable from a potential function. That is, there is a real-valued $F(x, y)$, called a *potential function* for (P, Q), such that $\nabla F = (P, Q)$. Let us examine this statement mathematically. Noting that $\nabla F = \left(\frac{\partial F}{\partial x}, \frac{\partial F}{\partial y}\right) = (P, Q)$ exactly when $P = \frac{\partial F}{\partial x}$ and $Q = \frac{\partial F}{\partial y}$, we shall now prove the following: *$\int_C P\,dx + Q\,dy$ is independent of path in D if and only if there is a function F such that $P = \frac{\partial F}{\partial x}$ and $Q = \frac{\partial F}{\partial y}$.*

To prove this, suppose first that such F exists. Parametrizing C by t, $a \leq t \leq b$, we have

$$\int_C P\,dx + Q\,dy = \int_C \frac{\partial F}{\partial x}\,dx + \frac{\partial F}{\partial y}\,dy$$

$$= \int_a^b \left(\frac{\partial F}{\partial x}\frac{dx}{dt} + \frac{\partial F}{\partial y}\frac{dy}{dt}\right) dt.$$

Assuming that we can apply the chain rule to F, we recognize that

$$\frac{\partial F}{\partial x}\frac{dx}{dt} + \frac{\partial F}{\partial y}\frac{dy}{dt} = \frac{dF(x(t), y(t))}{dt}.$$

Then

$$\int_C P\,dx + Q\,dy = \int_a^b \frac{dF(x(t), y(t))}{dt}\,dt$$

$$= F(x(b), y(b)) - F(x(a), y(a)) = F(C(b)) - F(C(a)).$$

This is a remarkable result, depending only upon the end points $C(b)$ and $C(a)$ of C and not on C itself. Hence $\int_C P\,dx + Q\,dy$ is independent of path in D. [A note here for the physics students: The last equation says that the work done by a force (P, Q) derivable from a potential function F in moving a particle from P_0 to P_1 is just the difference $F(P_1) - F(P_0)$ in potential energies at the two points.]

We now turn to the converse. Suppose that $\int_C P\,dx + Q\,dy$ is independent of path in D. We must produce a function F such that $P = \partial F/\partial x$ and $Q = \partial F/\partial y$.

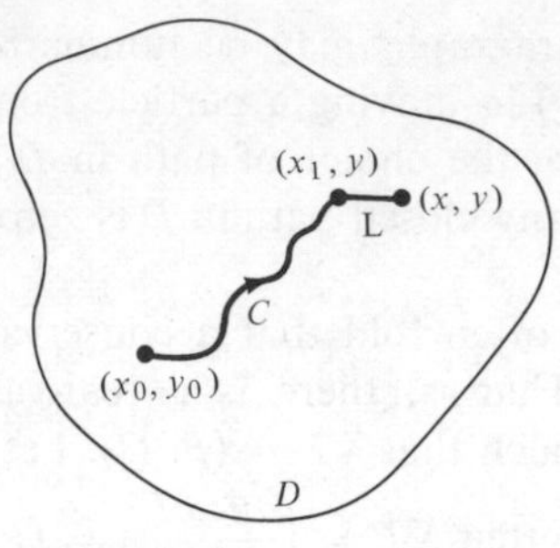

Fig. 6-23

To do this, begin by choosing any point (x_0, y_0) of D. If (x, y) is in D, then the fact that D is a domain implies that there is a path C in D from (x_0, y_0) to (x, y). Put $F(x, y) = \int_C P\,dx + Q\,dy$. This defines a function of (x, y) alone, for, if C^* is any other path in D from (x_0, y_0) to (x, y), then $\int_{C^*} P\,dx + Q\,dy = \int_C P\,dx + Q\,dy$ by the assumption of independence of path in D.

We now claim that $\partial F(x, y)/\partial x = P(x, y)$ for any (x, y) in D. To see why this is the case, pick any (x, y) in D and, using part (2) of the definition of domain, choose some (x_1, y) in D with $x_1 < x$ and each point of the straight line L from (x_1, y) to (x, y) also in D (see Fig. 6-23). Let C be any path in D from (x_0, y_0) to (x_1, y), and form a path K from (x_0, y_0) to (x, y) by stringing C and L together. Then

$$F(x, y) = \int_K P\,dx + Q\,dy = \int_C P\,dx + Q\,dy + \int_L P\,dx + Q\,dy.$$

Now $\int_C P\,dx + Q\,dy$ is independent of x, and, by parametrizing L as $L(\xi) = (\xi, y)$, $\xi: x_1 \to x$, $\int_L P\,dx + Q\,dy = \int_{x_1}^{x} P(\xi, y)\,d\xi$, y being constant on L. Then we have

$$\begin{aligned}\frac{\partial F(x, y)}{\partial x} &= \frac{\partial}{\partial x}\left(\int_C P\,dx + Q\,dy\right) + \frac{\partial}{\partial x}\int_{x_1}^{x} P(\xi, y)\,d\xi \\ &= \frac{\partial}{\partial x}\int_{x_1}^{x} P(\xi, y)\,d\xi = P(x, y),\end{aligned}$$

by the fundamental theorem of calculus.

A similar argument shows that $\partial F/\partial y = Q$. The reader can carry out the details, using an argument like the preceding with a point (x, y_1) in place of (x_1, y) (Fig. 6-24).

Let us pause here to summarize what we have so far. Up to this point we have shown that *the following three conditions are equivalent:*

(1) $\int_C P\,dx + Q\,dy$ *is independent of path in* D.
(2) $\oint_K P\,dx + Q\,dy = 0$ *for every closed path* K *in* D.
(3) $(P, Q) = \nabla F$ *for some* F *defined on* D.

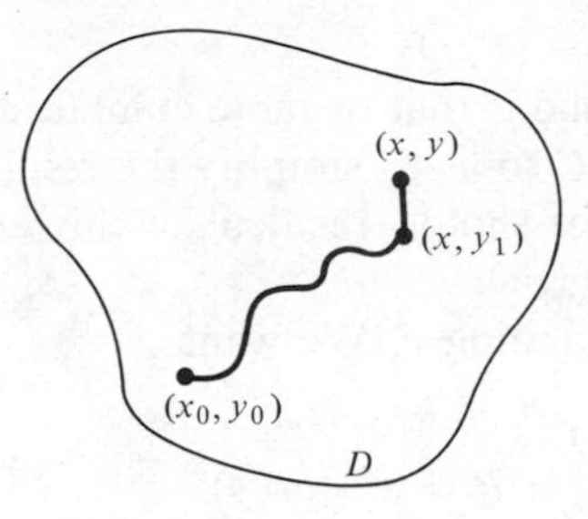

Fig. 6-24

Another way of putting (3) is that $P\,dx + Q\,dy$ is the differential of some function F. In such a case, $P\,dx + Q\,dy$ is sometimes called *exact*.

Ideally, we would like next a simple, practical way of telling whether (1), (2), or (3) holds, given P, Q, and D. Before obtaining such a test, we shall illustrate how one goes about finding the F of (3), assuming existence for the present.

EXAMPLE

Let $P(x, y) = x^2 \cos(y)$ and $Q(x, y) = -(x^3/3)\sin(y) + e^y$. Here D is the whole xy-plane. Assuming that $\int_C P\,dx + Q\,dy$ is independent of path in D, we can find a potential function F for (P, Q) by imitating the proof given before. Begin by choosing a point, say $(0, 0)$, to play the role of (x_0, y_0) in the proof. For any (x, y), put

$$F(x, y) = \int_C P\,dx + Q\,dy$$

for C any path between $(0, 0)$ and (x, y). In order to compute F explicitly, a convenient choice of C is the path consisting of C_1, from $(0, 0)$ to $(x, 0)$ and C_2 from $(x, 0)$ to (x, y). We can parametrize these

$$C_1(t) = (t, 0), \qquad t: 0 \to x$$
$$C_2(t) = (x, t), \qquad t: 0 \to y.$$

To avoid confusion, we shall write the integration variables as u and v. Then

$$F(x, y) = \int_C P(u, v)\,du + Q(u, v)\,dv =$$

$$\int_{C_1} P\,du + Q\,dv + \int_{C_2} P\,du + Q\,dv = \int_0^x t^2 \cos(0)\,dt + \int_0^y \left[\frac{-x^3}{3}\sin(t) + e^t\right] dt$$

$$= \frac{x^3}{3} + \frac{x^3}{3}\cos(y) - \frac{x^3}{3}\cos(0) + e^y - 1$$

$$= \frac{x^3}{3}\cos(y) + e^y - 1.$$

It is easy to check that $\nabla F = (P, Q)$.

A difficulty of this method is that in more complicated examples it may not be obvious how to choose C so as to simplify the resulting integrations. As an example of this, the student should recalculate this example, using as C the straight line from $(0, 0)$ to (x, y).

Here is another way of finding F. We want

$$\frac{\partial F}{\partial x} = P = x^2 \cos(y)$$

and

$$\frac{\partial F}{\partial y} = Q = -\frac{x^3}{3} \sin(y) + e^y.$$

Pick one of these equations, say the first. For $\partial F/\partial x = x^2 \cos(y)$, we need $F(x, y)$ to be of the form $\frac{x^3}{3} \cos(y) + \varphi(y)$. For any function $\varphi(y)$,

$$\frac{\partial}{\partial x}\left(\frac{x^3}{3} \cos(y) + \varphi(y)\right) = x^2 \cos(y).$$

We next use the requirement $\partial F/\partial y = Q$ to try to determine φ:

$$\frac{\partial F}{\partial y} = Q = -\frac{x^3}{3} \sin(y) + e^y = \frac{\partial}{\partial y}\left(-\frac{x^3}{3} \cos(y) + \varphi(y)\right) = -\frac{x^3}{3} \sin(y) + \varphi'(y).$$

Then $\varphi'(y) = e^y$, so we may choose $\varphi(y) = e^y$. This gives us

$$F(x, y) = \frac{x^3}{3} \cos(y) + e^y,$$

as we obtained above by the first method to within a constant.

Obviously F is only determined up to a constant. For example, if $G(x, y) = F(x, y) + k$, for any real number k, then $\nabla G = \nabla F = (P, Q)$. ∎

The second method of the last example actually provides a crude test for independence of path, as we now illustrate.

EXAMPLE

Change the last example a little bit, say with $P(x, y) = x^3 \cos(y)$ and $Q(x, y) = -(x^3/3) \sin(y) + e^y$. Is $\int_C P\,dx + Q\,dy$ independent of path in the xy-plane? If so, then for some F,

$$\frac{\partial F}{\partial x} = x^3 \cos(y) \qquad \text{and} \qquad \frac{\partial F}{\partial y} = -\frac{x^3}{3} \sin(y) + e^y.$$

From the first, $F(x, y) = (x^4/4)\cos(y) + \varphi(y)$ for some function φ. Then

$$\frac{\partial F}{\partial y} = -\frac{x^3}{3}\sin(y) + e^y = -\frac{x^4}{4}\sin(y) + \varphi'(y),$$

so

$$\varphi'(y) = e^y + \left(\frac{x^4}{4} - \frac{x^3}{3}\right)\sin(y).$$

This is impossible if φ is to be a function of y alone. We conclude that no such F exists; hence $\int_C P\,dx + Q\,dy$ is not independent of path in the xy-plane.

Note that, in the previous example, our ability to produce an F such that $\nabla F = (P, Q)$ justified our assumption that $\int_C P\,dx + Q\,dy$ was independent of path for that choice of P and Q. ∎

The last two examples suggest that we can test $\int_C P\,dx + Q\,dy$ for independence of path in D by either producing, or showing that it is impossible to produce, an F whose gradient is (P, Q). There is an easier way, however. Assuming that P and Q have continuous partials, then independence of path implies that $P = \dfrac{\partial F}{\partial x}$ and $Q = \dfrac{\partial F}{\partial y}$ for some F. Then

$$\frac{\partial P}{\partial y} = \frac{\partial^2 F}{\partial y\,\partial x} = \frac{\partial^2 F}{\partial x\,\partial y} = \frac{\partial Q}{\partial x} \qquad \text{at each } (x, y) \text{ in } D.$$

Hence $\partial P/\partial y \neq \partial Q/\partial x$ *for any* (x, y) *in* D *implies that* $\int_C P\,dx + Q\,dy$ *is not independent of path in* D.

This is strictly a negative result so far, and can only be used to show that a given line integral is not independent of path. To illustrate the failure of the converse, consider

$$\int_C \frac{-y\,dx}{x^2 + y^2} + \frac{x\,dy}{x^2 + y^2},$$

with $D = \{(x, y) \mid (x, y) \neq (0, 0)\}$. Choose $C(\theta) = (\cos(\theta), \sin(\theta), 0)$, $0 \leq \theta \leq 2\pi$. Then

$$\int_C \frac{-y\,dx}{x^2 + y^2} + \frac{x\,dy}{x^2 + y^2} = \int_0^{2\pi} d\theta = 2\pi,$$

so the integral around a closed path in D need not be zero. But

$$\frac{\partial}{\partial x}\left(\frac{x}{x^2 + y^2}\right) = \frac{y^2 - x^2}{(x^2 + y^2)^2} = \frac{\partial}{\partial y}\left(\frac{-y}{x^2 + y^2}\right).$$

The difficulty is with the region D in which our functions P and Q are defined. Here P and Q are continuous, with continuous partials, in the plane minus the origin, and the path C we chose in showing nonindependence of path went around the origin. In general, we may encounter difficulties whenever our region D has holes in it. If D has no holes, we say that D is *simply connected.* Another way of saying this is that any closed path in D encloses only points of D.

Note the effect of $\partial P/\partial y = \partial Q/\partial x$ in D when D is simply connected. If K is any closed path in D, then the region B bounded by K is also in D, hence P and Q are continuous with continuous first partials in B and on K, and Green's Theorem applies. Then

$$\oint_K P\,dx + Q\,dy = \iint_B \left(\frac{\partial Q}{\partial x} - \frac{\partial P}{\partial y}\right) = 0,$$

and $\int_C P\,dx + Q\,dy$ is independent of path in D. The definition of simple connectivity, then, is tailormade for the application of Green's theorem over the region in D enclosed by any closed path in D.

This gives us our test. *When D is simply connected, then $\int_C P\,dx + Q\,dy$ is independent of path exactly when $\partial Q/\partial x = \partial P/\partial y$ in D.*

Let

$$P(x, y) = \frac{-y}{x^2 + y^2} \quad \text{and} \quad Q(x, y) = \frac{x}{x^2 + y^2}$$

for (x, y) in the simply connected region $D^* = \{(x, y) \mid y \geq 2\}$ (Fig. 6-25). Then $\partial P/\partial y = \partial Q/\partial x$ in D^*, so we should be able to find a potential function for (P, Q) in D^*. To do this, put

$$\frac{\partial F}{\partial x} = P = \frac{x}{x^2 + y^2}.$$

Then a simple integration gives us $F(x, y) = \arctan(y/x) + \varphi(y)$ for some φ. Finally, we need

$$\frac{\partial F}{\partial y} = Q = \frac{x}{x^2 + y^2} = \frac{\partial}{\partial y}\left[\arctan\left(\frac{y}{x}\right) + \varphi(y)\right]$$

$$= \frac{x}{x^2 + y^2} + \varphi'(y),$$

and we can choose $\varphi(y) = 0$. Thus $F(x, y) = \arctan(y/x)$ gives us our potential function in D^*.

This example demonstrates the importance of the region in path independence. In D^*,

$$\int_C \frac{-y\,dx}{x^2 + y^2} + \frac{x\,dy}{x^2 + y^2}$$

is independent of path; in D, the plane with the origin removed as in Fig. 6-24, this line integral is not independent of path. ∎

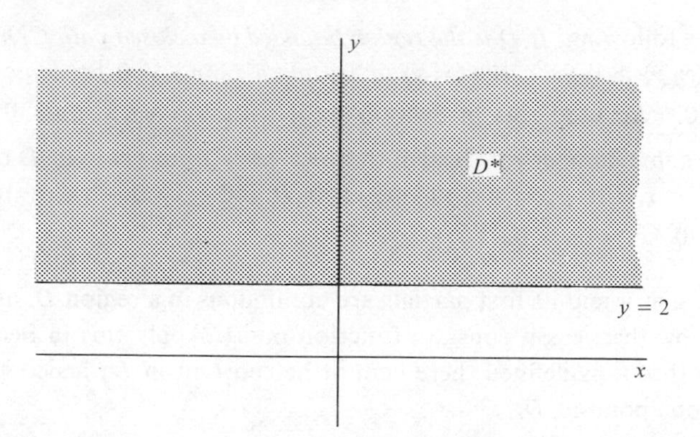

Fig. 6-25

Exercises for Section 6.5

1. For F, G, and C in each of the following, show that

$$\int_C F\frac{dG}{d\eta}\,ds = \iint_D F\,\nabla^2 G + \nabla F\cdot\nabla G$$

by direct computation of both sides.
   (a) $F(x, y) = xy$, $G(x, y) = x^2 + y^2$, $C(t) = (\cos(t), \sin(t))$, $0 \leq t \leq 2\pi$
   (b) $F(x, y) = 1$, $G(x, y) = xy$, C the square with vertices $(0, 0)$, $(1, 0)$, $(0, 1)$, $(1, 1)$
   (c) $F(x, y) = \cos(x)$, $G(x, y) = x + y$, C the triangle with vertices $(0, 0)$, $(1, 0)$, $(1, 1)$
   (d) $F(x, y) = e^{x+y}$, $G(x, y) = x$, C the triangle with vertices $(0, 0)$, $(1, 0)$, $(0, 1)$

2. Let $F(x, y) = x^2 - y^2$ and $G(x, y) = 2xy$, for (x, y) in the region determined by $|x| \leq 1$, $|y| \leq 1$.
   (a) Show that F and G are harmonic in D.
   (b) Show that

$$\oint_C F\frac{dG}{d\eta}\,ds = \oint_C G\frac{dF}{d\eta}\,ds,$$

   with C the boundary of D, by computing both sides.

3. Use the method of Section 6.5B to prove uniqueness of solution to the following problem:

$$\frac{\partial u}{\partial t} = k\,\nabla^2 u + \varphi(x, y)$$

$$[t > 0,\ (x, y) \text{ in some region } D \text{ bounded by a closed path } C];$$

$$u(x, y, 0) = f(x, y) \text{ for } (x, y) \text{ in } D,$$
$$u(x, y, t) = g(x, y, t) \text{ for } t > 0,\ (x, y) \text{ on } C.$$

Here $\varphi(x, y)$ is given for (x, y) on C and in D.

4. Prove the following: *If D is the region bounded by a closed path C, h is continuous on D, $h(x, y) \geq 0$ for every (x, y) in D, and $h(x_0, y_0) \neq 0$ for some (x_0, y_0), then* $\iint_D h > 0$. (We used this in Section 6.5B with $h = w^2$.) *Hint:* By continuity, produce some number $r > 0$ such that $h(x, y) > 0$ for (x, y) in D and $\|(x, y) - (x_0, y_0)\| < r$. Let $D_r = \{(x, y) \mid (x, y)$ in D and $\|(x, y) - (x_0, y_0)\| < r\}$. Then $\iint_{D_r} h > 0$, so $\iint_D h > 0$.

5. Suppose that w and its first partials are continuous in a region D, and $\|\nabla w\| = 0$ in D. Show that w is a constant function on D. Apply this in Section 6.5B by showing that w as defined there cannot be constant in D; hence $\|\nabla w\| \neq 0$ for at least one point in D.

6. Let $u(x, y) = x^2 - y^2$, for all (x, y). Show by direct computation that u has the mean value property.

7. Let $u(x, y) = 2xy$ for all real x and y. Show by direct computation that u has the mean value property.

8. Let $u(x, y) = x + y$ for all real x and y. Show by direct computation that u has the mean value property.

9. Let D be the region in the plane bounded by the closed path C. Suppose that $g(x, y)$ is given on C. Show that the problem $\nabla^2 u = 0$ for (x, y) in D, $\partial u(x, y)/\partial \eta = g(x, y)$ for (x, y) on C, and $\oint_C g\, dc = 0$, can have no more than one solution. (A problem of this type, asking for a harmonic function with prescribed normal derivative on the boundary of the region, is called a *Neumann problem.*)

10. Use polar coordinates to evaluate the following:
   (a) $\iint_D (x^2 + y^2)$, where $D = \{(x, y) \mid x^2 + y^2 \leq 4\}$.
   (b) $\iint_D x^2 y$, $D = \{(x, y) \mid 2 \leq x^2 + y^2 \leq 4$ and $x \geq 0$ and $y \geq 0\}$.
   (c) $\displaystyle\iint_D \frac{x}{y + 10}$, $D = \{(x, y) \mid 2 \leq x^2 + y^2 \leq 4\}$.

11. Evaluate $\iint_D xy$, where D is the region bounded by the ellipse $(x^2/a^2) + (y^2/b^2) = 1$ (a and b are positive constants). *Hint:* First put $u = ax$, $v = by$; then use polar coordinates on the resulting integral.

12. Evaluate $\iint_D \cos[(x - y)/(x + y)]$, D the region bounded by $x + y = 1$, $x + y = 4$, the x-axis, and the y-axis.

13. Evaluate $\iint_D e^{-x^2-y^2}$, where D is bounded by $x^2 + y^2 = 1$, the positive x-axis, and the positive y-axis.

14. Find the mass, center of mass, and moments of inertia about the x- and y-axes of the region bounded by $x^2 - y^2 = 1$, $x^2 - y^2 = 4$, $xy = 1$, $xy = 4$, assuming a density function $\rho(x, y) = xy$. *Hint:* Put $u = x^2 - y^2$, $v = xy$.

15. Evaluate $\iint_D f$, with $D = \{(x, y) \mid x \geq 0, y \geq 0 \text{ and } x + y \leq 4\}$ and

$$f(x, y) = \begin{cases} e^{(y-x)/(y+x)} & \text{if } x \neq -y, \\ e & \text{if } x = -y. \end{cases}$$

16. Evaluate $\int_0^\infty e^{-ax^2}\,dx$, where a is any positive real number.

17. Evaluate $\iint_D \dfrac{x}{y + x^2}$, where D is the region bounded by $y = x^2$, $y = 2 - x^2$.
Hint: Put $x = \sqrt{v - u}$ and $y = u + v$

18. In the following, test to see whether $\int_C P\,dx + Q\,dy$ is independent of path in D first, by attempting to find a function F such that $\nabla F = (P, Q)$ in D, and second, by testing to see if $\partial Q/\partial x = \partial P/\partial y$ in those cases where D is simply connected.
(a) $P = -e^x \cos(y)$, $Q = e^x \sin(y) + y^2 + 2$, D the whole xy-plane
(b) $P = xy^2$, $Q = x^2y$, $D = \{(x, y) \mid |x| \leq 1, |y| \leq 1\}$
(c) $P = \cos(x)\cos(y) + x^2y + e^y$, $Q = xe^y + (x^2/3) - \sin(x)\sin(y)$, $D = \{(x, y) \mid 2 \leq x^2 + y^2 \leq 4\}$
(d) $P = x^2 + y^2$, $Q = 2yx + \sin(x)$, $D = \{(x, y) \mid |x| \geq 2\}$
(e) $P = \dfrac{y}{x^2 + y^2}$, $Q = \dfrac{y}{x^2 + y^2} + \sin(2y)$, $D = \{(x, y) \mid 0 < x^2 + y^2 \leq 1\}$
(f) $P = xe^{xy} + \sin(y)$, $Q = 2ye^{xy} + \cos(x)$, $D =$ the whole xy-plane

19. Compute $\iint_D (x^2 + y^2 - 2xy)\sin(x + y)\,dx\,dy$, with D bounded by $x + y = 2$, $x + y = 4$, and the x and y-axes, in two ways:
(a) Directly.
(b) By changing variables to $u = x + y$, $v = x - y$.

20. Evaluate $\iint_D \sin(x^2 + y^2)\,dx\,dy$ over the region $x^2 + y^2 \leq 4$.

21. Show that the center of mass of a homogeneous plate in the shape of an ellipse is at its center.

22. Show that the center of mass of a homogeneous triangle is at the intersection of the medians.

6.6 Green's Theorem for Non-Simply-Connected Regions

We have seen in Section 6.5E that, for reasonable P and Q, equality of $\partial P/\partial y$ and $\partial Q/\partial x$ in the region D is sufficient for $\int_C P\,dx + Q\,dy$ to be path independent in D, if D is simply connected. We also saw that $\partial P/\partial y = \partial Q/\partial x$ does not guarantee path independence if D is not simply connected. We now ask: What can be said when $\partial P/\partial y = \partial Q/\partial x$ and D is not simply connected?

There is a clever argument which enables us to draw an interesting conclusion in this case. To fix ideas, suppose that D has a single hole in it (as in Fig. 6-26).

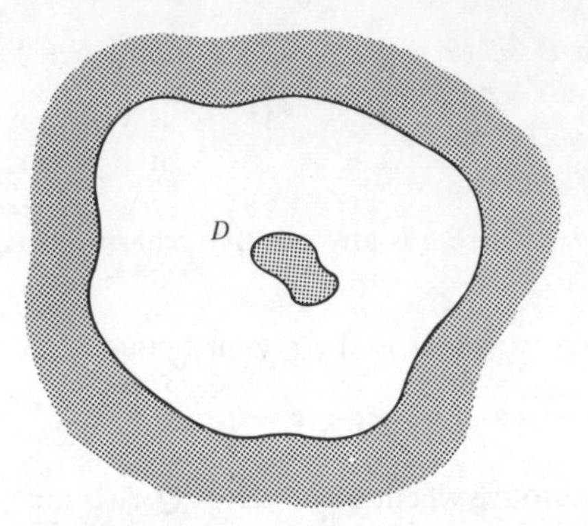

Fig. 6-26

For example, D might consist of the plane with the origin removed. Suppose that P, Q, and their first and second partials are continuous in D, as usual, and suppose that $\partial P/\partial y = \partial Q/\partial x$ in D. Finally, let C be a simple closed path in D. There are two possibilities for C:

(1) C does not enclose the hole (Fig. 6-27a). In this case, $\oint_C P\,dx + Q\,dy = 0$, by the results of Section 6.5E, since we can think of C as being in a simply connected part of D (e.g., inside L in Fig. 6-27).

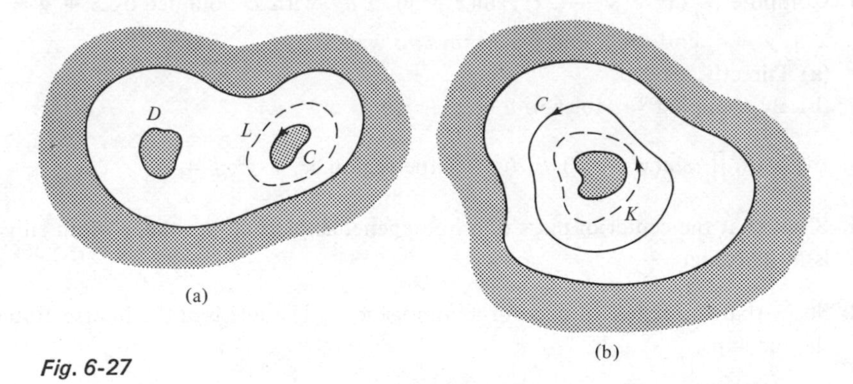

Fig. 6-27

(2) C goes around the hole once (Fig. 6-27b). We cannot argue here that $\oint_C P\,dx + Q\,dy = 0$, as P, Q, or their partials may blow up or be undefined in the hole enclosed by C, so that Green's theorem may not apply in the region interior to C. What we claim is that, if K is any other simple closed path in D lying between C and the hole, and going around the hole once (Fig. 6-27b), then

$$\int_C P\,dx + Q\,dy = \int_K P\,dx + Q\,dy.$$

To prove this rather remarkable result, cut out small pieces of C and K, forming C_1 and K_1, and insert straight-line channels L_1 and L_2 connecting C and K at the edges of the cuts, as shown in Fig. 6-28a. This yields a new closed path $\mathscr{C}$, consisting of C_1 and K_1 together with L_1 and L_2 (Fig. 6-28b). Observe that the subregion of D bounded by $\mathscr{C}$ is simply connected (because of L_1 and L_2, no closed path in the interior of $\mathscr{C}$ can go around the hole in D). Since $\partial P/\partial y = \partial Q/\partial x$ in this region, then

$$\int_{\mathscr{C}} P\,dx + Q\,dy = 0.$$

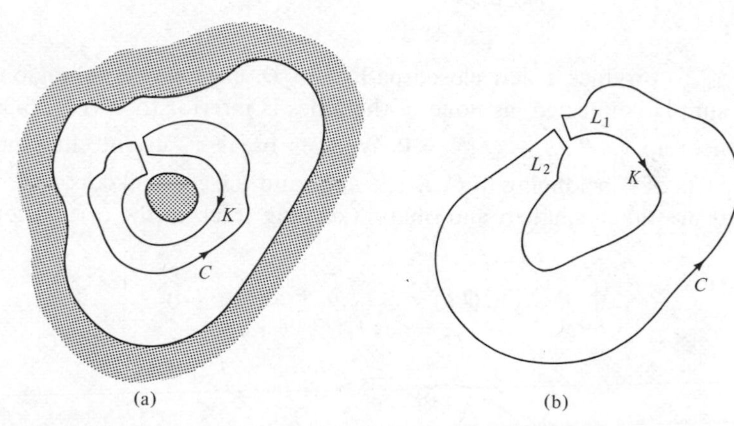

Fig. 6-28

But

$$\oint_{\mathscr{C}} = \int_{\underset{\curvearrowright}{C_1}} + \int_{L_1\downarrow} + \int_{L_2\uparrow} + \int_{\underset{\curvearrowleft}{K_1}}$$

the arrows serving as a reminder of orientation on the various pieces of $\mathscr{C}$. Now take a kind of limit. If we make the pieces cut out of C and K smaller, then L_1 and L_2 merge to a single line segment, and $\int_{L_1\downarrow}$ cancels $\int_{L_2\uparrow}$. Further, $\int_{\underset{\curvearrowright}{C_1}} \to \oint_C$ and $\int_{\underset{\curvearrowleft}{K_1}} \to \oint_K$. Then $\oint_C + \oint_K = 0$. But then

$$\oint_C P\,dx + Q\,dy = -\oint_K P\,dx + Q\,dy = \oint_K P\,dx + Q\,dy,$$

as we wanted to show.

This kind of reasoning can be extended to the case that C encloses n holes in D. Let K_i be a closed path in the interior of C and enclosing the ith hole, and suppose that K_i and K_j do not intersect for $i \neq j$ (Fig. 6-29). Cut a channel from C to K_1, then from K_1 to K_2, and so on, ending with a channel from

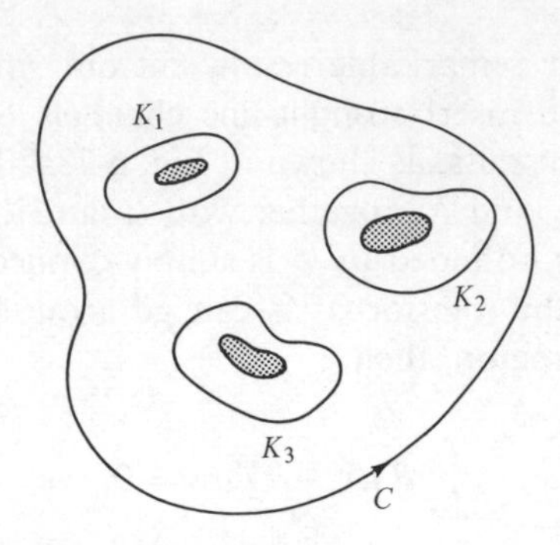

Fig. 6-29

K_{n-1} to K_n to produce a new closed path $\mathscr{C}$ in D, enclosing a subregion of D which is simply connected, as none of the holes is interior to $\mathscr{C}$ (Fig. 6-30). By Green's theorem, $\oint_{\mathscr{C}} P\,dx + Q\,dy = 0$. We then break $\oint_{\mathscr{C}}$ up into line integrals over the parts of $\mathscr{C}$ belonging to $C, K_1, \ldots, K_n$, and the channel cuts, take a limit as the cuts are taken smaller, and obtain (keeping track of the orientations):

$$\oint_C P\,dx + Q\,dy = \sum_{i=1}^{n} \oint_{K_i} P\,dx + Q\,dy.$$

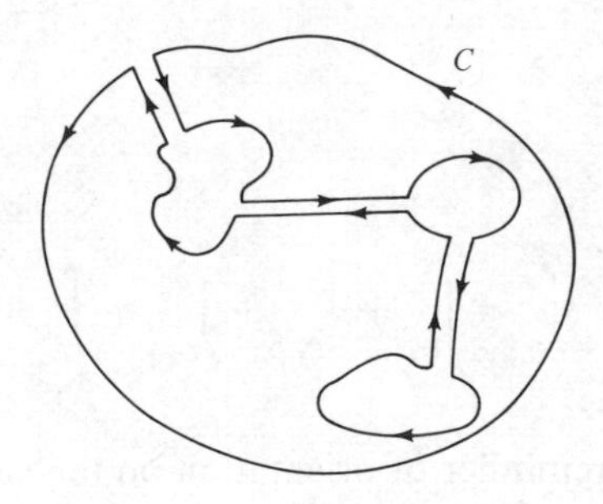

Fig. 6-30

We illustrate these ideas with two examples.

EXAMPLE

Evaluate $\oint_C \dfrac{-y\,dx}{x^2 + y^2} + \dfrac{x\,dy}{xy + y^2}$, where C is *any* simple closed path not passing through the origin. Consider two cases, noting that

$$\frac{\partial}{\partial x}\left(\frac{x}{x^2 + y^2}\right) = \frac{\partial}{\partial y}\left(\frac{-y}{x^2 + y^2}\right)$$

for $(x, y) \neq (0, 0)$:

(1) C does not enclose the origin. Then C lies in a simply connected part of D, so

$$\oint_C \frac{-y\,dx}{x^2 + y^2} + \frac{x\,dy}{x^2 + y^2} = 0.$$

(2) C encloses the origin. Now replace C by a circle K of radius ε sufficiently small that K and C do not intersect. Parametrize:

$$K(\theta) = (\varepsilon \cos(\theta), \varepsilon \sin(\theta)), \qquad 0 \le \theta \le 2\pi.$$

Then

$$\begin{aligned}\oint_C \frac{-y\,dx}{x^2 + y^2} + \frac{x\,dy}{x^2 + y^2} &= \oint_K \frac{-y\,dx}{x^2 + y^2} + \frac{x\,dy}{x^2 + y^2} \\ &= \int_0^{2\pi} \frac{-\varepsilon \sin(\theta)[-\varepsilon \sin(\theta)] + \varepsilon \cos(\theta)[\varepsilon \cos(\theta)]}{\varepsilon^2 \cos^2(\theta) + \varepsilon^2 \sin^2(\theta)}\, d\theta \\ &= \int_0^{2\pi} d\theta = 2\pi.\end{aligned}$$

Above, we have assumed (as indicated by the symbol $\oint$) that C is not only simple, but is taken in a counterclockwise sense. Note that if we drop the requirement that C be simple (but C is still closed), then in case (2) C may be replaced by a circle K going counterclockwise around (0, 0) n times, and the value of the integral is $2n\pi$. If we allow C to have clockwise orientation, then n may be a negative integer. ∎

Exercises for Section 6.6

1. Evaluate

$$\oint_C \frac{-y\,dx}{x^2 + y^2} + \frac{x\,dy}{x^2 + y^2}$$

on the path C shown in Fig. 6-31. *Hint:* Can you replace C by a more convenient curve K so that $\oint_C = \oint_K$?

2. Determine all possible values for

$$\oint_C \frac{y^3\,dx - xy^2\,dy}{(x^2 + y^2)^2},$$

where C is a plane closed curve not passing through the origin. *Hint: C* may or may not enclose (0, 0). Also, C may go around the origin more than once.

3. Determine all possible values for

$$\oint_C \frac{-y\,dx + x\,dy}{x^2 + y^2},$$

where C is a plane closed curve not passing through the origin.

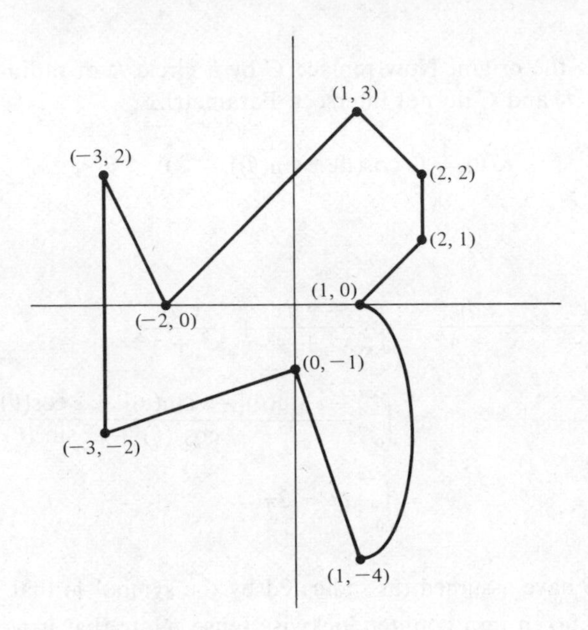

Fig. 6-31

6.7 Surfaces in R^3

We would like to develop a concept of an integral over a surface analogous to that of a line integral over a curve, and from that a three-dimensional analogue of Green's theorem. As with the development of line integrals, this will require a detailed look at the objects over which we want to integrate, in this case, surfaces. In particular, the notion of orientation on a surface must be developed. Unlike curves, where one can readily speak of a sense of direction from one end to the other, it is not immediately obvious how to assign a positive and negative orientation to a surface.

Recalling the definition of a curve as a function mapping an interval to three-vectors, or points in R^3, we are led to define a *surface* as a function mapping a set D of ordered pairs to 3-vectors, or points, in R^3. Often we write $\Sigma: D \to R^3$ if Σ is the function and D the set of ordered pairs mapped by Σ to R^3. For example, we might have

$$\Sigma(u, v) = (\cos(u)\sin(v), \sin(u)\sin(v), \cos(v)),$$

for $0 < u < \pi$, $0 < v < \pi$. Then Σ maps the "rectangle" $0 < u < \pi$, $0 < v < \pi$, to R^3.

The *graph* of a surface $\Sigma: D \to R^3$ is the locus of points $\Sigma(u, v)$ for (u, v) in D. In the last example, the graph of Σ consists of all points $(\cos(u)\sin(v), \sin(u)\sin(v), \cos(v))$ in R^3, with $0 < u < \pi$, $0 < v < \pi$. Putting $x = \cos(u)\sin(v)$, $y = \sin(u)\sin(v)$ and $z = \cos(v)$, we have $-1 < x < 1$, $0 < y < 1$, $-1 < z < 1$, and $x^2 + y^2 + z^2 = 1$. Thus the graph of Σ is a hemisphere, as shown in Fig. 6-32. The points on $x^2 + z^2 = 1, y = 0$ are not in the graph of Σ.

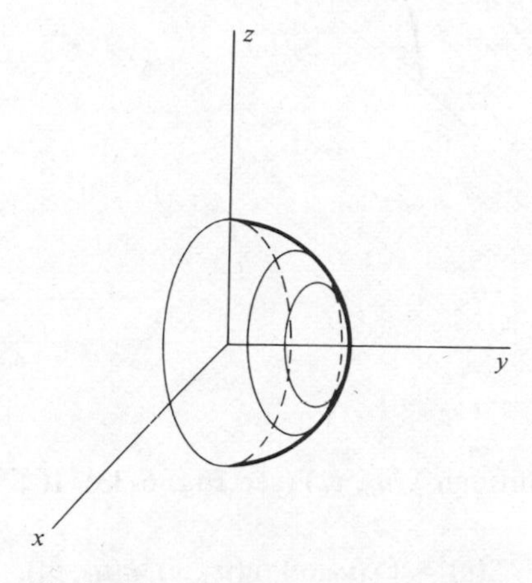

Fig. 6-32

As with curves, there is a distinction between a surface and its graph. However, as a matter of convenience, we often speak of a point on a surface. By this we mean of course a point on the graph of the surface. The reader should also note that in previous chapters we often thought of a surface as a locus $z = \varphi(x, y)$ or $\Phi(x, y, z) = 0$. These are both consistent with present terminology and convention. If $z = \varphi(x, y)$, we think of the surface as a function: $\Sigma(x, y) = (x, y, \varphi(x, y))$ for (x, y) in some specified set. Or, if $\Phi(x, y, z) = 0$, we can think of one of the variables as being implicitly defined in terms of the other two, say $y = \xi(x, z)$, and write the surface as $\Sigma(x, z) = (x, \xi(x, z), z)$ for the appropriate (x, z).

A surface $\Sigma: D \to R^3$ is *simple* if it has no multiple points, that is, if $\Sigma(u, v) \neq \Sigma(\bar{u}, \bar{v})$ whenever $(u, v) \neq (\bar{u}, \bar{v})$. For (u, v) in D, $\Sigma(u, v)$ is a point in R^3, say $\Sigma(u, v) = (f(u, v), g(u, v), h(u, v))$. We call f, g, and h the coordinate functions of Σ and often write $\Sigma = (f, g, h)$.

Suppose now that (u_0, v_0) is in D, and that Σ (i.e., the graph of Σ) has a tangent plane at $\Sigma(u_0, v_0)$. What is the equation of this tangent plane?

One method of approach is this. Think of

$$\gamma_{u_0}(v) = \Sigma(u_0, v)$$

as a curve on Σ through $\Sigma(u_0, v_0)$, obtained by fixing u at u_0 and varying v. Similarly,

$$\gamma_{v_0}(u) = \Sigma(u, v_0)$$

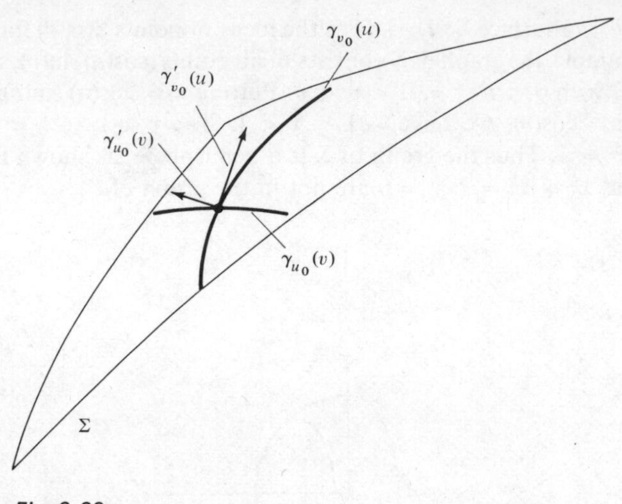

Fig. 6-33

is also a curve on Σ through $\Sigma(u_0, v_0)$ (see Fig. 6-33). If $\Sigma = (f, g, h)$, then

$$\gamma_{u_0}(v) = (f(u_0, v), g(u_0, v), h(u_0, v)),$$

and a tangent to γ_{u_0} at $\gamma_{u_0}(v_0) = \Sigma(u_0, v_0)$ is the vector

$$\gamma'_{u_0}(v_0) = (f_v(u_0, v_0), g_v(u_0, v_0), h_v(u_0, v_0)).$$

Similarly,

$$\gamma'_{v_0}(u_0) = (f_u(u_0, v_0), g_u(u_0, v_0), h_u(u_0, v_0))$$

is tangent to γ_{v_0} at $\Sigma(u_0, v_0)$. Now $\gamma'_{v_0}(u_0) \times \gamma'_{u_0}(v_0)$ is perpendicular to both these tangents, hence normal to the tangent plane at $\Sigma(u_0, v_0)$. This gives us a normal vector:

$$\left(\begin{vmatrix} g_u(u_0, v_0) & h_u(u_0, v_0) \\ g_v(u_0, v_0) & h_v(u_0, v_0) \end{vmatrix}, \begin{vmatrix} h_u(u_0, v_0) & f_u(u_0, v_0) \\ h_v(u_0, v_0) & f_v(u_0, v_0) \end{vmatrix}, \begin{vmatrix} f_u(u_0, v_0) & g_u(u_0, v_0) \\ f_v(u_0, v_0) & g_v(u_0, v_0) \end{vmatrix} \right)$$

to Σ at $\Sigma(u_0, v_0)$. The components of this normal vector are just Jacobians, and we may rewrite the normal vector as:

$$\eta(u_0, v_0) = \left(\left.\frac{\partial(g, h)}{\partial(u, v)}\right|_{(u_0, v_0)}, \left.\frac{\partial(h, f)}{\partial(u, v)}\right|_{(u_0, v_0)}, \left.\frac{\partial(f, g)}{\partial(u, v)}\right|_{(u_0, v_0)} \right).$$

For any (x, y, z) on the tangent plane, the vector $(x - f(u_0, v_0), y - g(u_0, v_0),$

$z - h(u_0, v_0))$ is in the tangent plane, hence perpendicular to the normal to the surface at $\Sigma(u_0, v_0)$, and we have as the equation of the tangent plane:

$$\left.\frac{\partial(g, h)}{\partial(u, v)}\right|_{(u_0,v_0)}(x - f(u_0, v_0)) + \left.\frac{\partial(h, f)}{\partial(u, v)}\right|_{(u_0,v_0)}(y - g(u_0, v_0)) + \left.\frac{\partial(f, g)}{\partial(u, v)}\right|_{(u_0,v_0)}(z - h(u_0, v_0)) = 0.$$

Henceforth, when speaking of the normal to Σ at $\Sigma(u_0, v_0)$, we shall always agree to use the vector

$$\eta(u_0, v_0) = \left(\left.\frac{\partial(g, h)}{\partial(u, v)}\right|_{(u_0,v_0)}, \left.\frac{\partial(h, f)}{\partial(u, v)}\right|_{(u_0,v_0)}, \left.\frac{\partial(f, g)}{\partial(u, v)}\right|_{(u_0,v_0)}\right).$$

The equation we have just derived for the tangent plane is entirely consistent with that obtained in Chapter 5. If our surface is given by $z = \varphi(x, y)$, we write this parametrically, as

$$\Sigma(x, y) = (x, y, \varphi(x, y)).$$

Putting $u = x$, $v = y$, $f(u, v) = x$, $g(u, v) = y$, and $h(u, v) = \varphi(x, y)$, we have

$$\frac{\partial(g, h)}{\partial(u, v)} = \begin{vmatrix} 0 & 1 \\ \dfrac{\partial\varphi}{\partial x} & \dfrac{\partial\varphi}{\partial y} \end{vmatrix} = \frac{-\partial\varphi}{\partial x},$$

$$\frac{\partial(h, f)}{\partial(u, v)} = \begin{vmatrix} \dfrac{\partial\varphi}{\partial x} & \dfrac{\partial\varphi}{\partial y} \\ 1 & 0 \end{vmatrix} = \frac{-\partial\varphi}{\partial y},$$

and

$$\frac{\partial(f, g)}{\partial(u, v)} = \begin{vmatrix} 1 & 0 \\ 0 & 1 \end{vmatrix} = 1.$$

The normal vector at (x_0, y_0) is then

$$\eta(x_0, y_0) = \left(\frac{-\partial\varphi(x_0, y_0)}{\partial x}, \frac{-\partial\varphi(x_0, y_0)}{\partial y}, 1\right),$$

and the tangent plane to $z = \varphi(x, y)$ at $(a, b, \varphi(a, b))$ is

$$\frac{-\partial\varphi(a, b)}{\partial x}(x - a) - \frac{\partial\varphi(a, b)}{\partial y}(y - b) + (z - \varphi(a, b)) = 0,$$

as we had before.

It will also be convenient to derive the normal vector and tangent plane when the surface is given in the form $\Phi(x, y, z) = 0$. We assume that at least one partial is not zero at (a, b, c), say $\varphi_z(a, b, c) = 0$. Then we can (in theory) solve for $z = \varphi(x, y)$ in a neighborhood of (a, b) such that $c = \varphi(a, b)$. In this neighborhood,

$$\frac{\partial \varphi}{\partial x} = \frac{\frac{-\partial \Phi}{\partial x}}{\frac{\partial \Phi}{\partial z}} \quad \text{and} \quad \frac{\partial \varphi}{\partial y} = \frac{\frac{-\partial \Phi}{\partial y}}{\frac{\partial \Phi}{\partial z}}.$$

From the last result, then, the normal vector at (x_0, y_0, z_0) on the surface is

$$\left(\frac{\frac{\partial \Phi}{\partial x}}{\frac{\partial \Phi}{\partial z}}\bigg|_{(a,b,c)}, \frac{\frac{\partial \Phi}{\partial y}}{\frac{\partial \Phi}{\partial z}}\bigg|_{(a,b,c)}, 1 \right).$$

The tangent plane at (a, b, c) is then

$$\frac{\partial \Phi(a, b, c)}{\partial x}(x - a) + \frac{\partial \Phi(a, b, c)}{\partial y}(y - b) + \frac{\partial \Phi(a, b, c)}{\partial z}(z - c) = 0.$$

By analogy with smooth curves, we would like a smooth surface to have a "continuously varying" tangent plane as we move over the surface. This is equivalent to having a continuous normal on the surface. To ensure this, we define a *smooth surface* as one that has a normal at each point, and whose coordinate functions are continuous with continuous first partials. By "has a normal at each point," we mean that the normal vector derived above never degenerates into the zero vector. This is guaranteed by requiring that

$$\left(\frac{\partial(g, h)}{\partial(u, v)}\right)^2 + \left(\frac{\partial(h, f)}{\partial(u, v)}\right)^2 + \left(\frac{\partial(f, g)}{\partial(u, v)}\right)^2 > 0$$

for all (u, v) in D.

As an example, let $\Sigma(u, v) = (\cos(u)\sin(v), \sin(u)\sin(v), \cos(v))$, $0 < u < \pi$, $0 < v < \pi$. Then a simple calculation yields

$$\eta(u, v) = -\sin(v)(\cos(u)\sin(v), \sin(u)\sin(v), \cos(v)).$$

Note that $|\eta(u, v)|^2 = \sin^2(v) > 0$ for $0 < v < \pi$, so Σ has a nondegenerate normal at each point. Since the coordinate functions are continuous with continuous first partials, then Σ is smooth. This is consistent with intuition—as a sphere, at least geometrically, appears to have a continuous normal.

We say that a surface is *piecewise smooth* if it consists of a finite number of smooth surfaces. A cube is piecewise smooth, as each face is smooth, but there is no normal at any point on the edge, and the normal jumps by 90 degrees as we move from any face to an adjacent face.

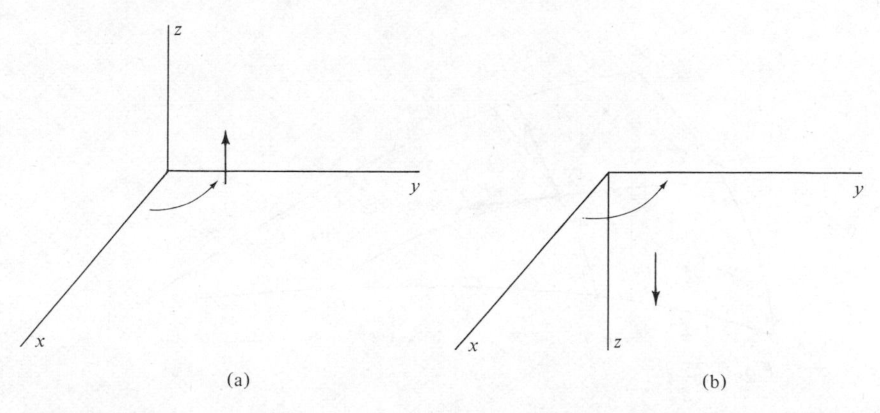

Fig. 6-34

We shall now concern ourselves with orientation. First, consider orientation as it is implicitly determined by a three-dimensional rectangular coordinate system. There are essentially two orientations we can give such a system. In one (Fig. 6-34a), the thumb points up along the z-axis as we place the right hand so that the fingers rotate from the positive x- toward the positive y-axes. In the other (Fig. 6-34b), the positive z-axis is in the opposite direction. We are free to call whichever of these we like the positively oriented system. The other then becomes the negatively oriented system. We usually think of the orientation given in Fig. 6-34a as the positive one, purely as a matter of convention.

Now back to surfaces. Say that we have $\Sigma: D \to R^3$, and we want to assign an orientation to Σ. For any (u_0, v_0) in D, the curves $\gamma_{u_0}(v) = \Sigma(u_0, v)$, $\gamma_{v_0}(u) = \Sigma(u, v_0)$, pass through $\Sigma(u_0, v_0)$ and lie on Σ. We saw above that we can form a normal to Σ at $\Sigma(u_0, v_0)$ by taking the cross products of the tangents to γ_{u_0} and γ_{v_0} at $\Sigma(u_0, v_0)$:

$$\eta(u_0, v_0) = \gamma'_{v_0}(u_0) \times \gamma'_{u_0}(v_0).$$

Now, it is not necessarily the case that $\gamma'_{v_0}(u_0)$ and $\gamma'_{u_0}(v_0)$ are at right angles. Nevertheless, the sense of rotation from $\gamma'_{v_0}(u_0)$ to $\gamma'_{u_0}(v_0)$, together with $\eta(u_0, v_0)$, determine an orientation at $\Sigma(u_0, v_0)$ which we shall designate as the positive orientation (see Fig. 6-35). For the opposite, or negative, orientation, we replace $\eta(u_0, v_0)$ by $-\eta(u_0, v_0)$.

Put another way, we declare the positive side of Σ at $\Sigma(u_0, v_0)$ to be the one away from which $\eta(u_0, v_0)$ points. The other side of Σ at $\Sigma(u_0, v_0)$ is then the negative side.

Note that the choice of a sense of positive rotation from $\gamma'_{v_0}(u_0)$ to $\gamma'_{u_0}(v_0)$ is a natural one in view of the orientation we have adopted in the plane. Since we are writing our ordered pairs as (u, v) as opposed to (v, u), we are thinking of the u-axis as horizontal and the v-axis as vertical. The positive sense of rotation from the u- to the v-axis induces the counterclockwise orientation chosen before as positive for the plane, and it is this sense that we are preserving in

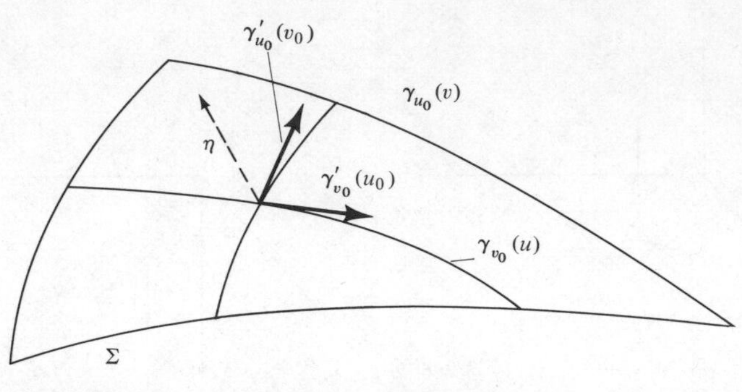

Fig. 6-35

going from $\gamma'_{v_0}(u_0)$, which is tangent to $\gamma_{v_0}(u)$ in which u is the variable, to $\gamma'_{u_0}(v_0)$, tangent to $\gamma_{u_0}(v)$, in which v is the variable.

Thus far we have only oriented Σ at a point $\Sigma(u_0, v_0)$. Conceivably the orientation could change abruptly from point to point on Σ, an obviously unpleasant situation. However, if Σ is smooth, then $\eta(u, v)$ is continuous and must vary continuously as we travel over the graph of Σ. This means that, if we start at a point on the positive size of Σ and move to another point on the same side of Σ, then the new point will also be on the positive side, as determined by the normal at that point. The same remark applies to piecewise smooth surfaces if we consider the surface a smooth piece at a time.

Some examples will help the reader digest these intuitively phrased concepts.

EXAMPLE

Let $\Sigma(u, v) = (\cos(u)\sin(v), \sin(u)\sin(v), \cos(v))$, $0 < u < \pi$, $0 < v < \pi$. The graph of Σ is the hemisphere $x^2 + y^2 + z^2 = 1$, with $-1 < x < 1$, $-1 < z < 1$, $0 < y < 1$. Now

$$\eta(u, v) = -\sin(v)(\cos(u)\sin(v), \sin(u)\sin(v), \cos(v))$$
$$= -\sin(v)\Sigma(u, v).$$

Since $\sin(v) > 0$ for $0 < v < \pi$, then $\eta(u, v)$ is always in the direction opposite $\Sigma(u, v)$, hence points toward the origin and away from the inside of the sphere. The inside of the hemisphere is then the positive side, according to our convention. The outside is the negative side. ▮

EXAMPLE

Let

$$\Sigma(x, y) = (x, y, x^2 + y^2), \qquad x^2 + y^2 \leq 1.$$

The graph of Σ looks like a parabolic bowl (Fig. 6-36). The surface is obviously smooth, and we find

$$\eta(x, y) = (-2x, -2y, 1).$$

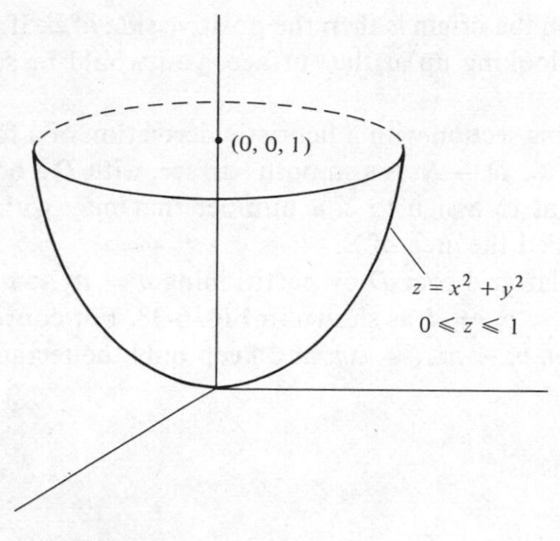

Fig. 6-36

This is an inner normal. For example, $\eta(0, 0) = (0, 0, 1)$, the z-axis being normal to Σ at $\Sigma(0, 0) = (0, 0, 0)$; the inside of the bowl is then the positive side.

EXAMPLE

Let

$$\Sigma(x, y) = (x, y, -x + y + 2), \qquad 0 \leq x \leq 1, 0 \leq y \leq 1.$$

The graph of Σ is shown in Fig. 6-37 and consists of points on the plane $x + y + z = 2$ for $0 \leq x \leq 2$, $0 \leq y \leq 2$. Here

$$\eta(x, y) = (1, 1, 1).$$

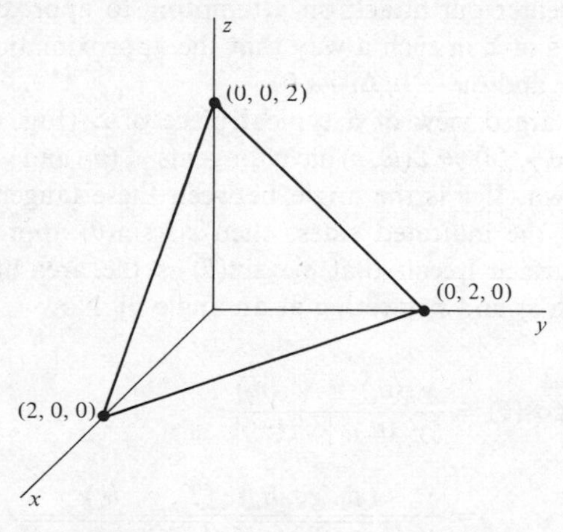

Fig. 6-37

The side away from the origin is then the positive side of Σ. If you were standing at the origin and looking up at the surface, you would be seeing the negative side. ∎

We conclude this section with a heuristic derivation of a formula for surface area. Suppose that $\Sigma: D \to R^3$ is a smooth surface, with D a bounded, connected set in R^2. We want to assign to Σ a number that may with some reasonable justification be called the area of Σ.

Put a rectangular grid over D by partitioning $a = u_0 < u_1 < \cdots < u_n = b$, $c = v_0 < v_1 < \cdots < v_m = d$, as shown in Fig. 6-38. For convenience, standardize $u_i = u_{i-1} = \Delta u$, $v_j - v_{j-1} = \Delta v$, and keep only the rectangles lying entirely within D.

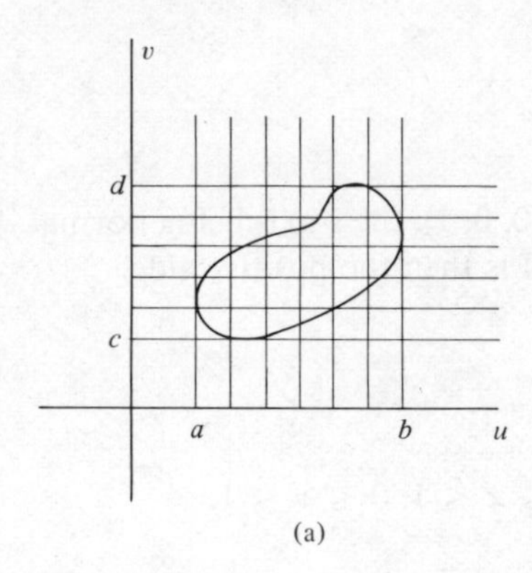

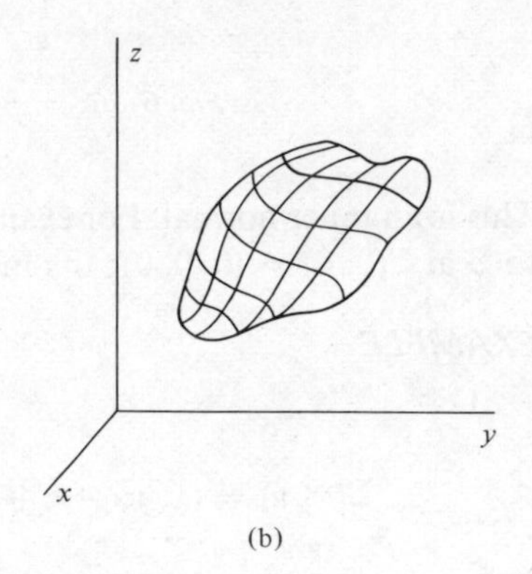

Fig. 6-38

Note that the curves $\Sigma(u, v_j)$, $\Sigma(u_i, v)$, obtained by fixing one or the other of u or v at some partition point, subdivide the surface into curvilinear pieces (Fig. 6-38b). We center our attack on attempting to approximate the area of each of these pieces of Σ in such a way that the approximations become better as $n \to \infty$, $m \to \infty$, and $\Delta u \to 0$, $\Delta v \to 0$.

Look at an enlarged view of a typical piece of Σ (Fig. 6-39). The curves $\gamma_{v_j}(u) = \Sigma(u, v_j)$ and $\gamma_{u_i}(v) = \Sigma(u_i, v)$ have tangents $\gamma'_{v_j}(u_i)$ and $\gamma'_{u_i}(v_j)$, respectively, at $\Sigma(u_i, v_i)$, as shown. If θ is the angle between these tangents, and s_1 and s_2 are the lengths of the indicated sides, then $s_1 s_2 \sin(\theta)$ approximates the area of this piece of surface [recall that $s_1 s_2 \sin(\theta)$ is the area of a parallelogram with sides of length s_1 and s_2 meeting at an angle θ]. Now

$$\cos(\theta) = \frac{\gamma'_{v_j}(u_i) \times \gamma'_{u_i}(v_j)}{\|\gamma'_{v_j}(u_i)\| \, \|\gamma'_{u_i}(v_j)\|}$$

$$= \frac{(f_u, g_u, h_u) \cdot (f_v, g_v, h_v)}{\sqrt{f_u^2 + g_u^2 + h_u^2}\sqrt{f_v^2 + g_v^2 + h_v^2}},$$

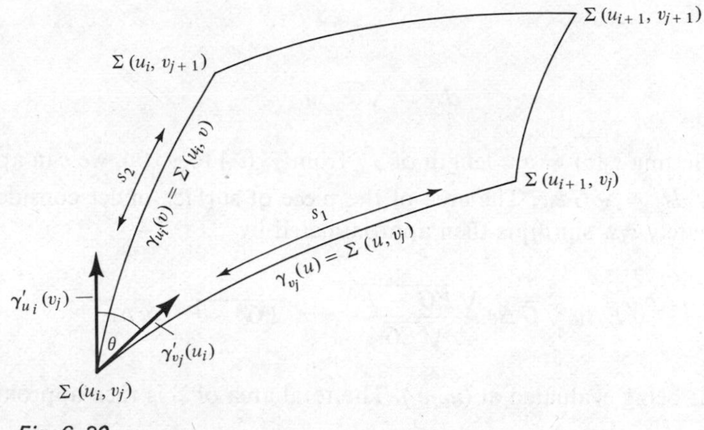

Fig. 6-39

where $\Sigma = (f, g, h)$, and all partials are evaluated at (u_i, v_j). For convenience, denote

$$E = f_u^2 + g_u^2 + h_u^2 \qquad [= \|\gamma'_{v_j}(u)\|^2],$$

$$F = f_u f_v + g_u g_v + h_u h_v \qquad [= \gamma'_{v_j}(u) \cdot \gamma'_{u_i}(v)],$$

and

$$G = f_v^2 + g_v^2 + h_v^2 \qquad [= \|\gamma'_{u_i}(v)\|^2].$$

Then

$$\sin(\theta) = [1 - \cos^2(\theta)]^{1/2} = \frac{\sqrt{EG - F^2}}{\sqrt{EG}},$$

We now seek some useful expressions for s_1 and s_2. Here an approximation by differentials is handy. Let

$$s_1(u) = \text{arc length of } \gamma_{v_j} \text{ from } \gamma_{v_j}(u_i) \text{ to } \gamma_{v_j}(u).$$

We approximate s_1 $[= s_1(u_{i+1})]$ by the differential of $s_1(u)$ at u_i due to the increment Δu. Now

$$\gamma_{v_j}(u) = (f(u, v_j), g(u, v_j), h(u, v_j)),$$

so

$$s_1(u) = \int_{u_i}^{u} \sqrt{f_u^2 + g_u^2 + h_u^2}\, du = \int_{u_i}^{u} \sqrt{E}\, du.$$

Then

$$ds_1 = \sqrt{E}\,\Delta u.$$

Similarly, letting $s_2(v)$ = arc length of γ_{u_i} from $\gamma_{u_i}(v_j)$ to $\gamma_{u_i}(v)$, we can approximate s_2 by $ds_2 = \sqrt{G}\,\Delta v$. The area of the piece of surface under consideration [approximately $s_1 s_2 \sin(\theta)$] is then approximated by

$$\sqrt{E}\,\Delta u\,\sqrt{G}\,\Delta v\,\frac{\sqrt{EG - F^2}}{\sqrt{EG}} = \sqrt{EG - F^2}\,\Delta u\,\Delta v,$$

the partials being evaluated at (u_i, v_j). The total area of Σ is then approximated by

$$\sum_{i,j} \sqrt{EG - F^2}\big|_{(u_i, v_j)} \cdot \Delta u\,\Delta v,$$

extending over all (u_i, v_j) at corners of rectangles contained in D. As $m, n \to \infty$ and $\Delta u, \Delta v \to 0$, these sums approximate more closely the surface area of Σ, leading us to define this surface area as

$$\iint_D \sqrt{EG - F^2}\,du\,dv.$$

The reader can check that $\sqrt{EG - F^2} = \|\eta(u, v)\|$, giving us $\iint_D \|\eta(u, v)\|$ as the area of Σ. This is a somewhat easier expression to remember.

EXAMPLE

Let

$$\Sigma(u, v) = (r\cos(u)\sin(v),\ r\sin(u)\sin(v),\ r\cos(v)), \qquad 0 < u < \pi,\ 0 < v < \pi,$$

with r a positive constant. We compute

$$\begin{aligned} x_u &= -r\sin(u)\sin(v), & x_v &= r\cos(u)\cos(v),\\ y_u &= r\cos(u)\sin(v), & y_v &= r\sin(u)\cos(v),\\ z_u &= 0 & z_v &= -r\sin(v). \end{aligned}$$

Then

$$E = r^2\sin^2(v),$$

$$F = 0,$$

and

$$G = r^2.$$

Then

$$\sqrt{EG - F^2} = \sqrt{r^4 \sin^2(v)} = r^2|\sin(v)|.$$

Since $\sin(v) > 0$ for $0 < v < \pi$, then $r^2|\sin(v)| = r^2 \sin(v)$. Then

$$\text{area of } \Sigma = \iint_D r^2 \sin(v)\, du\, dv = r^2 \int_0^{\pi}\int_0^{\pi} \sin(v)\, du\, dv$$
$$= 2\pi r^2.$$

This is the usual result for the area of a hemisphere of radius r. ∎

Exercises for Section 6.7

1. For each of the following surfaces, determine at each point the tangent plane (if any) and the normal agreed upon in the text. Also determine the positive side of the surface according to the convention of the text, and the surface area.
   (a) $\Sigma(u, v) = (u, 2v, u^2 + 3v^2)$, $u^2 + v^2 \leq 2$
   (b) $\Sigma(u, v) = (v\sin(u), v\cos(u), 2v)$, $0 \leq u \leq \pi/2$, $0 \leq v \leq 1$
   (c) $\Sigma(u, v) = (u, v, u^2 - v^2)$, $0 \leq u \leq 1$, $0 \leq v \leq 1$
   (d) $\Sigma(u, v) = (u, v, \sqrt{u^2 + v^2})$, $u^2 + v^2 \leq 4$
   (e) $\Sigma(u, v) = (u^2\sin(v), u^2\cos(v), u^4/6)$, $0 \leq u \leq 1$, $0 \leq v \leq 2\pi$
   (f) Σ is the surface of a cube with vertices at $(0, 0, 0)$, $(1, 0, 0)$, $(0, 1, 0)$, $(0, 0, 1)$, $(1, 1, 0)$, $(1, 0, 1)$, $(0, 1, 1)$, and $(1, 1, 1)$

2. Let $\Sigma(x, y) = (x, y, x^2 + y^2)$ for $0 \leq x^2 + y^2 \leq 1$.
   (a) Sketch the graph of Σ.
   (b) Draw the curves $\gamma_{1/\sqrt{2}}(x)$ and $\gamma_{1/\sqrt{2}}(y)$ on the graph of Σ.
   (c) Compute $\gamma'_{1/\sqrt{2}}(x)$ at $x = 1/\sqrt{2}$ and $\gamma'_{1/\sqrt{2}}(y)$ at $y = 1/\sqrt{2}$, and take the cross product
   $$\gamma'_{1/\sqrt{2}}(1/\sqrt{2}) \times \gamma'_{1/\sqrt{2}}(1/\sqrt{2}).$$
   (d) Draw the coordinate system obtained by taking one axis along $\gamma'_{1/\sqrt{2}}(x)$, another along $\gamma'_{1/\sqrt{2}}(y)$, the third along their cross product.
   (e) Determine from (d) the positive side of $\sum$.

3. Obtain an integral formula for the area of a surface defined by $z = \varphi(x, y)$ for (x, y) in D. The integrand should be in terms of partials of φ. *Hint:* Parametrize $\Sigma(x, y) = (x, y, \varphi(x, y))$ for (x, y) in D.

4. Obtain an integral formula for the area of a surface given by an equation $\Phi(x, y, z) = 0$, assuming that Φ has nonzero partials. The integrand should be in terms of partials of Φ.

5. In the text we agreed on geometric grounds that $\iint_D \sqrt{EG - F^2}\, du\, dv$ gives a number that we can reasonably call the area of the given surface. This problem

shows that there is more to the problem of defining area than we admitted at the time.

Let S be a right circular cylinder, given by $x^2 + y^2 = 1$, $0 \le z \le 1$. The surface area is 2π (if you slit the cylinder and unfold it into a rectangle 2π by 1 unit in dimension, this follows from the formula for area of a rectangle). Now approximate the surface area of S as follows. First, divide the bottom circle into n equal arcs by placing points $P_1, \ldots, P_n$ as shown by the solid dots in the diagram, and divide the top circle by placing the same number of points, $Q_1, \ldots, Q_n$ (open dots), spaced as shown relative to the points $P_1, \ldots, P_n$. Form $2n$ isosceles triangles with sides P_1P_2, Q_1P_2, Q_1P_1; Q_1Q_2, Q_1P_2, Q_2P_1; and so on. This inscribes a polyhedron inside S. Show that the area of the inscribed polyhedron (obtained by summing the areas of the triangles) is

$$2n \sin\left(\frac{\pi}{n}\right)\sqrt{1 + \left[1 - \cos\left(\frac{\pi}{n}\right)\right]^2}.$$

Now slice the cylinder into m cylinders by planes $z = 0, 1/m, \ldots, 1$ parallel to the xy-plane, and perform the above construction on each slice. Show that the total surface area of the resulting inscribed polyhedron is now given by

$$A(m, n) = 2n \sin\left(\frac{\pi}{n}\right)\sqrt{1 + m^2\left[1 - \cos\left(\frac{\pi}{n}\right)\right]^2}.$$

It is reasonable to think that as $m, n \to \infty$, the inscribed polyhedron fits closer and closer to S, and hence that $A(m, n)$ approaches the surface area of S in the limit. Show that this is incorrect by showing that $A(m, n)$ can be made to approach different values by letting m and n approach infinity at different rates. [For example, compute $\lim_{n\to\infty} A(n, n)$, $\lim_{n\to\infty} A(n^2, n)$, and $\lim_{m\to\infty} A(m, m^2)$.]

Figure 6-40 is an example of this intuitively plausible procedure that fails even for such a simple surface as a cylinder.

6. Use the results of this section to derive a formula for the surface area of a right circular cylinder of height h, radius r.

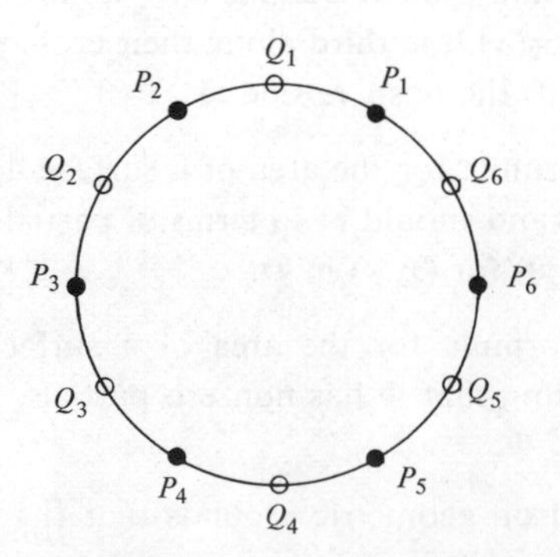

Fig. 6-40

6.8 Surface Integrals of Vector-Valued Functions

With the above background on surfaces, it is possible to develop a notion of a surface integral in a manner that exactly parallels that of a line integral. However, matters can be greatly simplified if we recall that

$$\int_C F = \int_a^b F(C(t)) \cdot C'(t)\, dt$$

whenever F is continuous (i.e., has continuous components) on the graph of C and C is a smooth curve in R^3. By analogy, suppose that F is continuous on the smooth surface $\Sigma: D \to R^3$, with D bounded. Then we put

$$\iint_\Sigma F = \iint_D F(\Sigma(u, v)) \cdot \eta(u, v)\, du\, dv,$$

where

$$\eta(u, v) = \left(\frac{\partial(g, h)}{\partial(u, v)}, \frac{\partial(h, f)}{\partial(u, v)}, \frac{\partial(f, g)}{\partial(u, v)}\right)$$

is the normal to Σ as agreed upon in Section 6.7. We call $\iint_\Sigma F$ the surface integral of F over Σ.

Surface integrals have the usual properties we expect of integrals. If F and G are both continuous on Σ, and α and β are real numbers, then

$$\iint_\Sigma (\alpha F + \beta G) = \alpha \iint_\Sigma F + \beta \iint_\Sigma G.$$

Further, if Σ is partitioned into pieces Σ_1 and Σ_2, then

$$\iint_\Sigma F = \iint_{\Sigma_1} F = \iint_{\Sigma_2} F.$$

This allows us to integrate over piecewise smooth surfaces, by summing the integrals over the individual smooth components.

Finally, as with line integrals, a reversal in orientation results in a negative sign introduced into the calculation. Last section we agreed on how to arrive at a side of Σ that we would call the positive side, and the choice was based on the normal vector η. If we reversed our orientation and used $-\eta$ instead, then we would have

$$\iint_D F(\Sigma(u, v)) \cdot (-\eta(u, v))\, du\, dv,$$

which is $-\iint_\Sigma F$. Unless explicitly stated to the contrary, however, we always understand in the above definition of $\iint_\Sigma F$ that η is the normal agreed on in Section 6.7, and we choose the positive side of Σ as the side away from which η points.

We shall work out some surface integrals by way of illustration.

EXAMPLE

Let $\Sigma(u, v) = (\cos(u)\sin(v), \sin(u)\sin(v), \cos(v))$ for $\pi/4 \le u \le \pi/2, \pi/4 \le v \le \pi/2$. Let $F(x, y, z) = (x, y, z^2)$. Then

$$F(\Sigma(u, v)) = (\cos(u)\sin(v), \sin(u)\sin(v), \cos^2(v))$$

and

$$\eta(u, v) = (-\cos(u)\sin^2(v), -\sin(u)\sin^2(v), -\cos(v)\sin(v)).$$

Then

$$F(\Sigma(u, v)) \cdot \eta(u, v) = -\sin^3(v) - \cos^3(v)\sin(v).$$

Then

$$\iint_\Sigma F = \int_{\pi/4}^{\pi/2}\int_{\pi/4}^{\pi/2} [-\sin^3(v) - \cos^3(v)\sin(v)]\, du\, dv$$

$$= \frac{\pi}{4}\int_{\pi/4}^{\pi/2} [-\sin^3(v) - \cos^3(v)\sin(v)]\, dv$$

$$= \frac{\pi}{12}\cos(v)(\sin^2(v) + 2)\Big|_{\pi/4}^{\pi/2} + \frac{\pi}{16}\cos^4(v)\Big|_{\pi/4}^{\pi/2}$$

$$= -\frac{\pi}{12}\frac{1}{\sqrt{2}}\left(\frac{1}{2} + 2\right) - \frac{\pi}{16}\frac{1}{4} = -\frac{\pi}{8}\left(\frac{5}{3\sqrt{2}} + \frac{1}{8}\right). \quad ▮$$

EXAMPLE

Compute $\iint_\Sigma G$, where $G(x, y, z) = (xy, z, 0)$, and $\Sigma(x, y) = (x, y, x^2 + y^2)$ for $x^2 + y^2 \le 1$. Here $\eta(x, y) = (-2x, -2y, 1)$, and $G(\Sigma(x, y)) = (xy, x^2 + y^2, 0)$. Then

$$\iint_\Sigma G = \iint_D (xy, x^2 + y^2, 0) \cdot (-2x, -2y, 1)\, dx\, dy$$

$$= \iint_D (-2x^2y - 2yx^2 - 2y^3)\, dx\, dy$$

$$= \iint_D (-4x^2y - 2y^3)\, dx\, dy,$$

with D the region in the xy-plane bounded by $x^2 + y^2 = 1$.

A straightforward computation now gives us

$$\iint_{\Sigma} G = \int_{-1}^{1} \int_{-\sqrt{1-x^2}}^{\sqrt{1-x^2}} (-4x^2y - 2y^3)\, dy\, dx = \int_{-1}^{1} 0\, dx = 0. \quad \blacksquare$$

There is an alternative notation which is often seen for surface integrals. Write $F(x, y, z) = (f(x, y, z), g(x, y, z), h(x, y, z))$. Then

$$\begin{aligned}\iint_{\Sigma} F &= \iint_{D} (f, g, h) \cdot \left(\frac{\partial(y, z)}{\partial(u, v)}, \frac{\partial(z, x)}{\partial(u, v)}, \frac{\partial(x, y)}{\partial(u, v)}\right) du\, dv \\ &= \iint_{D} f(x(u, v), y(u, v), z(u, v)) \frac{\partial(y, z)}{\partial(u, v)}\, du\, dv \\ &\quad + \iint_{D} g(x(u, v), y(u, v), z(u, v)) \frac{\partial(z, x)}{\partial(u, v)}\, du\, dv \\ &\quad + \iint_{D} h(x(u, v), y(u, v), z(u, v)) \frac{\partial(x, y)}{\partial(u, v)}\, du\, dv.\end{aligned}$$

These last integrals resemble the integrals appearing in the change-of-variables formula, and suggest the notation

$$\iint_{\Sigma} f\, dy\, dz + g\, dz\, dx + h\, dx\, dy$$

for $\iint_{\Sigma} (f, g, h)$. It is important to understand that this is purely a matter of notation [as with $\int_C f\, dx + g\, dy + h\, dz$ for $\int_C (f, g, h)$], and does not alter at all the basic notion of the surface integral of F over Σ, or the means of evaluating this integral.

Using the additivity of the surface integral, these are often written

$$\iint_{\Sigma} f\, dy\, dz + \iint_{\Sigma} g\, dz\, dx + \iint_{\Sigma} h\, dx\, dy.$$

Or, one might see these integrals appearing individually. For example,

$$\iint_{\Sigma} f\, dy\, dz$$

stands for

$$\iint_{\Sigma} (f, 0, 0);$$

$$\iint_{\Sigma} g\, dz\, dx$$

stands for

$$\iint_\Sigma (0, g, 0);$$

and

$$\iint_\Sigma h\,dx\,dy$$

stands for

$$\iint_\Sigma (0, 0, h).$$

Again, the consistency of this notation is easy to see:

$$\iint_\Sigma f\,dy\,dz + \iint_\Sigma g\,dz\,dx + \iint_\Sigma h\,dx\,dy$$

$$= \iint_\Sigma (f, 0, 0) + \iint_\Sigma (0, g, 0) + \iint_\Sigma (0, 0, h)$$

$$= \iint_\Sigma (f, g, h),$$

as defined above.

EXAMPLE

Let $f(x, y, z) = x^2yz$. Let $\Sigma(u, v) = (u, v, u^2 + v)$, $0 \le u \le 1$, $0 \le v \le 1$. Then

$$\iint_\Sigma f\,dy\,dz = \iint_\Sigma (f, 0, 0) = \iint_D (u^2v(u^2 + v), 0, 0) \cdot (-2u, 1, 1)\,du\,dv$$

$$= \int_0^1 \int_0^1 -2u^3v(u^2 + v)\,du\,dv = -\tfrac{1}{3};$$

$$\iint_\Sigma f\,dz\,dx = \iint_\Sigma (0, f, 0) = \iint_D (0, u^2v(u^2 + v), 0) \cdot (-2u, 1, 1)\,du\,dv$$

$$= \int_0^1 \int_0^1 (u^4v + u^2v^2)\,du\,dv = \tfrac{19}{90};$$

and

$$\iint_\Sigma f\,dx\,dy = \iint_\Sigma (0, 0, f) = \iint_D (0, 0, u^2v(u^2 + v)) \cdot (-2u, 1, 1)\,du\,dv = \tfrac{19}{90}. \quad \blacksquare$$

As with curves and line integrals, it is often the case, particularly in physical applications, that a surface is described in some way [e.g., sphere of radius r about the origin, cube of side length a about (1, 2, 3)], and it is left to the individual to produce his own parametric representation. How will different parametric representations of the same surface affect $\iint_\Sigma F$? As we might expect, the answer is, not at all, with certain assumptions on Σ and the parametric representations.

Suppose that we have

$$\Sigma_1: D_1 \to R^3$$

and

$$\Sigma_2: D_2 \to R^3,$$

both smooth surfaces. For convenience, we shall write ordered pairs in D_1 as (u, v), and D_2 as (s, t). We shall say that Σ_1 and Σ_2 are *equivalent* if there is a change of variables

$$s = \alpha(u, v),\ t = \beta(u, v),$$

which associates with each (u, v) in D_1 exactly one (s, t) in D_2, and conversely, such that α and β are continuous, with continuous partials in D_1, $\dfrac{\partial(\alpha, \beta)}{\partial(u, v)} > 0$ in D_1, and, for each (u, v) in D_1,

$$\Sigma_1(u, v) = \Sigma_2(\alpha(u, v), \beta(u, v)).$$

We leave it for the reader, as an exercise, to show that equivalent surfaces have the same graph, and that

$$\iint_{\Sigma_1} F = \iint_{\Sigma_2} F$$

whenever Σ_1 and Σ_2 are equivalent and F is continuous on Σ_1.

Exercises for Section 6.8

1. Compute $\iint_\Sigma F$ for each of the following:

(a) $F(x, y, z) = (x, y, z)$, $\Sigma\,(u, v) = (\cos(u)\sin(v), \sin(u)\sin(v), \cos(v))$, for $0 < u < \pi/2$, $0 < v < \pi/2$

(b) $F(x, y, z) = (x^2, yz, xyz)$, $\Sigma\,(x, y) = (x, y, x^2 + y^2)$, for $x^2 + y^2 \geq 2$

(c) $F(x, y, z) = (e^x, \sin(y), z)$, $\Sigma\,(u, v) = (u, v, v + u - 3)$, for $0 \leq u \leq 1$, $0 \leq v \leq 1$

(d) $F(x, y, z) = x + y + z$, $\Sigma\,(u, v) = (u, v, u^2 + v)$ for $0 \leq u \leq 1$, $1 \leq v \leq 2$

2. In each of the following, compute $\iint_\Sigma f\,dx\,dy$, $\iint_\Sigma f\,dy\,dz$, and $\iint_\Sigma f\,dz\,dx$.

(a) $f(x, y, z) = 2x + yz^2$; Σ as in 1(a)

(b) $f(x, y, z) = \sin(x)yz$; Σ as in 1(b)

(c) $f(x, y, z) = ax + by + cz$; Σ as in 1(c), a, b, and c constant

(d) $f(x, y, z) = xy$; Σ as in 1(d)

3. Let Σ_1 and Σ_2 be equivalent smooth surfaces.
(a) Show that Σ_1 and Σ_2 have the same graph.
(b) Let F be a 3-vector-valued function whose components are continuous on Σ_1 [hence also on Σ_2, by (a)]. Show that

$$\iint_{\Sigma_1} F = \iint_{\Sigma_2} F.$$

4. Let $\Sigma_1(x, y) = (x, \sqrt{1 - x^2 - z^2}, z)$, $0 < x^2 + z^2 < 1$, and

$$\Sigma_2(u, v) = (\cos(u)\sin(v), \sin(u)\sin(v), \cos(v)),\ 0 < u < \pi,\ 0 < v < \pi.$$

(a) Show that Σ_1 and Σ_2 are equivalent.
(b) Let $F(x, y, z) = (2x, y^2, 3xyz)$. Show by direct computation that

$$\iint_{\Sigma_1} F = \iint_{\Sigma_2} F.$$

(c) Show by direct computation that Σ_1 and Σ_2 have the same area.

5. Show that equivalent smooth surfaces have the same area.

6.9 Application of Surface Integrals to Fluid Flow, Divergence, and Gradient

In this section we shall give several elementary applications of surface integrals. More substantial applications will be treated later, after we have covered the theorems of Gauss and Stokes.

A. Fluid Flow

The term "fluid" is understood in a rather technical sense by an expert in this area, but it will suffice for our purposes to think of a fluid as water or oil or the like, flowing in a three-dimensional region, inside a pipeline, for example.

Suppose that we have a fixed coordinate system as a frame of reference, and $v(x, y, z, t)$ at time t gives the velocity of the fluid at (x, y, z). Let $\rho(x, y, z, t) =$ density of fluid at (x, y, z) at time t. The total mass at time t within a given volume V of space is then $\iiint_V \rho(x, y, z, t)\, dx\, dy\, dz$.

Suppose now that V is bounded by a smooth surface $\Sigma: D \to R^3$. Consider a small piece of Σ, say $\Sigma_0: D_0 \to R^3$. The area of Σ_0 is $\iint_{D_0} \|\eta\|\, du\, dv$, where η is the normal to Σ_0. By a mean value argument, there is some point (u_0, v_0) in D_0 such that

$$\iint_{D_0} \|\eta(u, v)\|\, du\, dv = \|\eta(u_0, v_0)\| \cdot (\text{area of } D_0).$$

Writing $\Sigma(u_0, v_0) = (x_0, y_0, z_0)$, we may think of

$$\rho(x_0, y_0, z_0, t) \cdot (v(x_0, y_0, z_0, t) \cdot \eta(u_0, v_0)) \cdot (\text{area of } D_0)$$

as approximating the mass of fluid per unit time flowing across Σ_0. By partitioning D into small pieces, whose areas go to zero, we have in the limit $\iint_D \rho(v \cdot \eta)$ as the total mass of fluid per unit time flowing across the entire boundary Σ of V. This is exactly the surface integral $\iint_\Sigma \rho v$.

Note that, if η is an outer normal (i.e., pointing away from V), then the outside of V is the positive side, and $\iint_\Sigma \rho v$ is positive for fluid flowing out of V, negative for fluid flowing into V. With this orientation,

$$\iint_\Sigma \rho v = \text{total mass of fluid per unit time flowing out of } V \text{ across } \Sigma.$$

Omission of ρ gives $\iint_\Sigma v$, or $\iint_D v \cdot \eta$. This is the volume of fluid flowing out of V across Σ per unit time. In this sense, $\iint_\Sigma v$ represents the total strength of sources of fluid interior to Σ.

B. Application to the Divergence of a Vector

Given a vector $F(x, y, z) = (f(x, y, z), g(x, y, z), h(x, y, z))$, the divergence is defined by

$$\operatorname{div} F = \frac{\partial f}{\partial x} + \frac{\partial g}{\partial y} + \frac{\partial h}{\partial z}.$$

We shall now discuss a physical problem in which the divergence arises naturally in the guise of a surface integral.

Consider the setting of Section 6.9A and suppose that we want to obtain some measure of the magnitude of a source of fluid at some point (x_0, y_0, z_0) and time t. If we imagine a volume V of fluid, with boundary surface Σ, then $\iint_\Sigma v$ measures the total strength of all sources interior to V. Imagine that (x_0, y_0, z_0) is interior to V. If we "normalize" by dividing by the volume of V (which we denote $|V|$), and take the limit as V shrinks to the point (x_0, y_0, z_0), we obtain in $\lim_{|V| \to 0} (1/|V|) \iint_\Sigma v$ a measure of the strength of the source at (x_0, y_0, z_0) itself. Since we are envisioning fluid flowing out from this source, this strength or magnitude of the source at (x_0, y_0, z_0) is often called the *divergence* of v at (x_0, y_0, z_0). We shall now show (or at least give a plausibility argument) that this physically based concept of divergence agrees with the mathematical definition given above. That is, we claim that, if $v(x, y, z, t) = (A(x, y, z, t), B(x, y, z, t), C(x, y, z, t))$, then

$$\begin{aligned}\lim_{|V!| \to 0} \frac{1}{|V|} \iint_\Sigma v &= \left.\frac{\partial A}{\partial x}\right|_{(x_0, y_0, z_0, t)} + \left.\frac{\partial B}{\partial y}\right|_{(x_0, y_0, z_0, t)} + \left.\frac{\partial C}{\partial z}\right|_{(x_0, y_0, z_0, t)} \\ &= \operatorname{div} F|_{(x_0, y_0, z_0, t)}.\end{aligned}$$

It is understood by the symbol $\lim_{|V|\to 0}$ that V is shrunk to the point (x_0, y_0, z_0) in such a way as to at each stage always contain (x_0, y_0, z_0) in its interior, and that Σ always denotes the boundary of V, so that Σ as well as V changes in the limit process.

To calculate a limit of this type, we shall choose for V a cube of side length a centered at (x_0, y_0, z_0). Then $|V| = a^3$, and Σ is the piecewise smooth surface consisting of the sides of the cube, which lie in the planes $x = x_0 \pm a/2$, $y = y_0 \pm a/2$, $z = z_0 \pm a/2$. Label the faces of the cube as follows according to the plane each is in:

$$\Sigma_1 \colon x = x_0 + \frac{a}{2},$$

$$\Sigma_2 \colon x = x_0 - \frac{a}{2},$$

$$\Sigma_3 \colon y = y_0 + \frac{a}{2},$$

$$\Sigma_4 \colon y = y_0 - \frac{a}{2},$$

$$\Sigma_5 \colon z = z_0 + \frac{a}{2},$$

$$\Sigma_6 \colon z = z_0 - \frac{a}{2}.$$

The outer normal η_i to Σ_i for $i = 1, \ldots, 6$ is given by

$$\begin{aligned} \eta_1 &= (1, 0, 0), & \eta_2 &= (-1, 0, 0), \\ \eta_3 &= (0, 1, 0), & \eta_4 &= (0, -1, 0), \\ \eta_5 &= (0, 0, 1), & \eta_6 &= (0, 0, -1). \end{aligned}$$

We now want to consider $\iint_{\Sigma_i} v$, using a mean value type of argument for each. Complete details will be given only for $i = 1$ and $i = 2$, the other $\iint_{\Sigma_i} v$ being treated similarly.

Parametrize Σ_1 by

$$\Sigma_1(y, z) = \left(x_0 + \frac{a}{2}, y, z\right)$$

for

$$y_0 - \frac{a}{2} \leq y \leq y_0 + \frac{a}{2}, \qquad z_0 - \frac{a}{2} \leq z \leq z_0 + \frac{a}{2}.$$

Then

$$\iint_{\Sigma_1} v = \int_{z_0 - a/2}^{z_0 + a/2} \int_{y_0 - a/2}^{y_0 + a/2} A\left(x_0 + \frac{a}{2}, y, z, t\right) dy\, dz$$
$$= A\left(x_0 + \frac{a}{2}, \bar{y}, \bar{z}, t\right) a^2,$$

for some y and z with $y_0 - a/2 \le \bar{y} \le y_0 + a/2$, $z_0 - a/2 \le \bar{z} \le z_0 + a/2$. The factor a^2 is simply the area of the region over which we are integrating.

Similarly, we may parametrize

$$\Sigma_2(y, z) = \left(x_0 - \frac{a}{2}, y, z\right)$$

for

$$y_0 - \frac{a}{2} \le y \le y_0 + \frac{a}{2}, \qquad z_0 - \frac{a}{2} \le z \le z_0 + \frac{a}{2}.$$

Then

$$\iint_{\Sigma_2} v = \int_{z_0 - a/2}^{z_0 + a/2} \int_{y_0 - a/2}^{y_0 + a/2} (A, B, C) \cdot (-1, 0, 0)\, dy\, dz$$
$$= \int_{z_0 - a/2}^{z_0 + a/2} \int_{y_0 - a/2}^{y_0 + a/2} -A\left(x_0 - \frac{a}{2}, y, z, t\right) dy\, dz = -A\left(x_0 - \frac{a}{2}, \bar{\bar{y}}, \bar{\bar{z}}, t\right) a^2$$

for some $\bar{\bar{y}}$ and $\bar{\bar{z}}$ with $y_0 - a/2 \le \bar{\bar{y}} \le y_0 + a/2$, $z_0 - a/2 \le \bar{\bar{z}} \le z_0 + a/2$. Then

$$\frac{1}{|V|}\left(\iint_{\Sigma_1} v + \iint_{\Sigma_2} v\right) = \frac{1}{a^3}\left(\iint_{\Sigma_1} v + \iint_{\Sigma_2} v\right)$$
$$= \frac{A(x_0 + a/2, \bar{y}, \bar{z}, t) - A(x_0 - a/2, \bar{\bar{y}}, \bar{\bar{z}}, t)}{a}.$$

In similar fashion,

$$\frac{1}{|V|}\left(\iint_{\Sigma_3} v + \iint_{\Sigma_4} v\right) = \frac{B(x^*, y_0 + a/2, z^*, t) - B(x^{**}, y_0 - a/2, z^{**}, t)}{a}$$

for some x^*, x^{**} in $[x_0 - a/2, x_0 + a/2]$, and some z^*, z^{**} in $[z_0 - a/2, z_0 + a/2]$; and

$$\frac{1}{|V|}\left(\iint_{\Sigma_5} v + \iint_{\Sigma_6} v\right) = \frac{C(\hat{x}, \hat{y}, z_0 + a/2, t) - C(\hat{\hat{x}}, \hat{\hat{y}}, z_0 - a/2, t)}{a}$$

for some $\hat{x}$, $\hat{\hat{x}}$ in $[x_0 - a/2, x_0 + a/2]$ and $\hat{y}$, $\hat{\hat{y}}$ in $[y_0 - a/2, y_0 - a/2]$.

Now take a limit as $a \to 0$. Then x^*, x^{**}, $\hat{x}$, and $\hat{\hat{x}} \to x_0$, $\bar{y}$, $\bar{\bar{y}}$, $\hat{y}$, and $\hat{\hat{y}} \to y_0$, and $\bar{z}$, $\bar{\bar{z}}$, z^*, and $z^{**} \to z_0$. Hence, as in the derivation of the heat equation in Section 5.2, we have

$$\frac{A(x_0 + a/2, \bar{y}, \bar{z}, t) - A(x_0 - a/2, \bar{\bar{y}}, \bar{\bar{z}}, t)}{a} \to \frac{\partial A(x_0, y_0, z_0, t)}{\partial x},$$

$$\frac{B(x^*, y_0 + a/2, z^*, t) - B(x^{**}, y_0 - a/2, z^{**}, t)}{a} \to \frac{\partial B(x_0, y_0, z_0, t)}{\partial y},$$

and

$$\frac{C(\hat{x}, \hat{y}, z_0 + a/2, t) - C(\hat{\hat{x}}, \hat{\hat{y}}, z_0 - a/2, t)}{a} \to \frac{\partial C(x_0, y_0, z_0, t)}{\partial z}.$$

This gives us

$$\operatorname{div} v|_{(x_0, y_0, z_0, t)} = \lim_{|V| \to 0} \frac{1}{|V|} \iint_{\Sigma} v,$$

as we wanted.

C. Application to the Gradient of a Scalar

We have already seen the gradient of a function of two variables:

$$\nabla\varphi(x, y) = \left(\frac{\partial\varphi}{\partial x}, \frac{\partial\varphi}{\partial y}\right).$$

This extends easily to three dimensions:

$$\nabla\varphi(x, y, z) = \left(\frac{\partial\varphi}{\partial x}, \frac{\partial\varphi}{\partial y}, \frac{\partial\varphi}{\partial z}\right).$$

According to Section 6.7, this vector is normal to any surface $\varphi(x, y, z) = C$, where C is any real number. We shall now give a heuristic argument to support the claim that

$$\nabla\varphi|_{(x_0, y_0, z_0)} = \lim_{|V| \to 0} \left(\frac{1}{|V|} \iint_{\Sigma} \varphi \, dy \, dz, \frac{1}{|V|} \iint_{\Sigma} \varphi \, dz \, dx, \frac{1}{|V|} \iint_{\Sigma} \varphi \, dx \, dy\right),$$

where V is a region in R^3 containing (x_0, y_0, z_0), bounded by the (piecewise) smooth surface Σ, and $|V|$ denotes the volume of V. As in Section 6.9B, $\lim\limits_{|V| \to 0}$ denotes a limit as V is shrunk to the point (x_0, y_0, z_0) in such a way that (x_0, y_0, z_0) always remains interior to V.

For ease in calculation, we choose for V a rectangular box of volume $\Delta x \cdot \Delta y \cdot \Delta z$, centered at (x_0, y_0, z_0), bounded by the planes $x = x_0 \pm \Delta x/2$, $y = y_0 \pm \Delta y/2$, $z = z_0 \pm \Delta z/2$.

For $\iint_\Sigma \varphi \, dy \, dz$, we need only consider the parts of Σ in the planes $x = x_0 \pm \Delta x/2$, as y or z is constant on the other four faces. We have

$$\iint_\Sigma \varphi \, dy \, dz = \int_{z_0-\Delta z/2}^{z_0+\Delta z/2} \int_{y_0-\Delta y/2}^{y_0+\Delta y/2} \varphi(x_0 + \Delta x/2, y, z) \, dy \, dz$$
$$- \int_{z_0-\Delta z/2}^{z_0+\Delta z/2} \int_{y_0-\Delta y/2}^{y_0+\Delta y/2} \varphi(x_0 - \Delta x/2, y, z) \, dy \, dz.$$

The minus sign in front of the second integral arises from our orientation based on the outer normal, and is similar in this respect to the calculation in Section 6.9B.

By a mean value argument,

$$\int_{z_0-\Delta z/2}^{z_0+\Delta z/2} \int_{y_0-\Delta y/2}^{y_0+\Delta y/2} \varphi(x_0 + \Delta x/2, y, z) \, dy \, dz = \varphi(x_0 + \Delta x/2, \bar{y}, \bar{z}) \, \Delta y \, \Delta z$$

and
$$\int_{z_0-\Delta z/2}^{z_0+\Delta z/2} \int_{y_0-\Delta y/2}^{y_0+\Delta y/2} \varphi(x_0 - \Delta x/2, y, z) \, dy \, dz = -\varphi(x_0 - \Delta x/2, \bar{\bar{y}}, \bar{\bar{z}}) \, \Delta y \, \Delta z,$$

for some $\bar{y}$ and $\bar{\bar{y}}$ in $[y_0 - \Delta y/2, y_0 + \Delta y/2]$, and some $\bar{z}$ and $\bar{\bar{z}}$ in $[z_0 - \Delta z/2, z_0 + \Delta z/2]$. Then

$$\frac{1}{|V|} \iint_\Sigma \varphi \, dy \, dz = \frac{\varphi(x_0 + \Delta x/2, \bar{y}, \bar{z}) - \varphi(x_0 - \Delta x/2, \bar{\bar{y}}, \bar{\bar{z}})}{\Delta x} \to \frac{\partial \varphi(x_0, y_0, z_0)}{\partial x}$$

as $\Delta x \to 0$. Similarly,

$$\frac{1}{|V|} \iint_\Sigma \varphi \, dz \, dx \to \frac{\partial \varphi(x_0, y_0, z_0)}{\partial y} \qquad \text{as } \Delta y \to 0,$$

and
$$\frac{1}{|V|} \iint_\Sigma \varphi \, dx \, dy \to \frac{\partial \varphi(x_0, y_0, z_0)}{\partial z} \qquad \text{as } \Delta z \to 0.$$

This gives us the result we wanted.

In concluding this section, we add a word about the sense in which Sections 6.9B and C might be called applications, since we did nothing but express familiar quantities in terms of surface integrals in a somewhat unusual way.

One point is demonstrated in Section 6.9B, where the physical significance of div v is much clearer from the surface integral formulation than from the derivative formula, although the latter is certainly easier to compute with. Another point is that the usual derivative formulas for div F and $\nabla\varphi$ are in rectangular coordinates, while the surface integral formulations may be used in any kind of coordinate system convenient to the problem at hand. This is important in a number of nontrivial physical applications, besides being of theoretical interest.

Exercises for Section 6.9

1. Redo Section 6.9B, using a sphere about (x_0, y_0, z_0) instead of a cube.

2. Redo Section 6.9C, using a sphere about (x_0, y_0, z_0) instead of a rectangular box.

3. Let φ be a differentiable function of three variables. Change variables to cylindrical coordinates:

$$x = r\cos(\theta), \qquad y = r\sin(\theta), \qquad z = z.$$

Put $\Phi(r, \theta, z) = \varphi(r\cos(\theta), r\sin(\theta), z)$.

(a) Derive a formula in terms of partial derivatives of Φ for $\nabla\Phi$ in cylindrical coordinates, using chain-rule differentiation. *Hint:*

$$\nabla\varphi = \left(\frac{\partial\varphi}{\partial x}, \frac{\partial\varphi}{\partial y}, \frac{\partial\varphi}{\partial z}\right).$$

Express $\partial\varphi/\partial x$, $\partial\varphi/\partial y$, and $\partial\varphi/\partial z$ in terms of partials of Φ with respect to r, θ, and z.

(b) Rederive the formula of (a) for $\nabla\Phi$ in cylindrical coordinates by using the surface integral formulation of *C*. *Hint:* Use a cylinder for V, centered at the point at which you want to compute $\nabla\Phi$.

4. Let $F(x, y, z) = (A(x, y, z), B(x, y, z), C(x, y, z))$, with A, B, and C differentiable. Switch to cylindrical coordinates, as defined in Exercise 3, and put

$$\begin{aligned} G(r, \theta, z) &= (A(r\cos(\theta), r\sin(\theta), z), B(r\cos(\theta), r\sin(\theta), z), C(r\cos(\theta), r\sin(\theta), z)) \\ &= (a(r, \theta, z), b(r, \theta, z), e(r, \theta, z)). \end{aligned}$$

(a) Use the chain rule to express div G as a sum of terms involving partials of a, b, and e with respect to r, θ, and z.

(b) Rederive the solution to (a) by using the surface integral formulation of *B*.

5. Let ξ be a differentiable function of three variables. Switch to spherical coordinates:

$$x = \rho\cos(\theta)\sin(\varphi), \qquad y = \rho\sin(\theta)\sin(\varphi), \qquad z = \rho\cos(\varphi).$$

Put

$$\Phi(\rho, \theta, \varphi) = \xi(\rho\cos(\theta)\sin(\varphi), \rho\sin(\theta)\sin(\varphi), \rho\cos(\varphi)).$$

(a) Compute $\nabla\Phi$ in terms of partials of Φ with respect to ρ, θ, and φ, using the chain rule.

(b) Check the answer to (a) by using the integral formulation given in C. *Hint:* Use a sphere for V.

6. Let F be as in Exercise 4. Using spherical coordinates as defined in Exercise 5, put

$$\begin{aligned} H(\rho, \theta, \varphi) &= (A(\rho \cos(\theta) \sin(\varphi), \rho \sin(\theta) \sin(\varphi), \rho \cos(\varphi)), \\ &\quad B(\rho \cos(\theta) \sin(\varphi), \rho \sin(\theta) \sin(\varphi), \rho \cos(\varphi)), \\ &\quad C(\rho \cos(\theta) \sin(\varphi), \rho \sin(\theta) \sin(\varphi), \rho \cos(\varphi))) \\ &= (a(\rho, \theta, \varphi), b(\rho, \theta, \varphi), e(\rho, \theta, \varphi)). \end{aligned}$$

   (a) Use the chain rule to express div H in terms of partials of a, b, and e with respect to ρ, θ, and φ.
   (b) Redo (a), using the surface integral formulation of B.

6.10 Surface Integrals of Scalar-Valued Functions and Applications

It is convenient to have a concept of $\iint_\Sigma f$ when f is a real-valued function of three variables, defined at least for (x, y, z) on Σ. Without further ado, we define

$$\iint_\Sigma f = \iint_D f(\Sigma(u, v)) \|\eta(u, v)\| \, du \, dv$$

if $\Sigma\colon D \to R^3$ is a smooth surface and f is continuous on Σ. Note that, with this definition, the area of Σ is just $\iint_\Sigma 1$.

In physical applications, it is often handy to observe that

$$\iint_D f(\Sigma(u, v)) \|\eta(u, v)\| \, du \, dv$$

is a limit, as $n \to \infty$ and $\Delta u, \Delta v \to 0$, of sums of the form

$$\sum_{i=1}^{n} f(\Sigma(\xi_i, \varphi_i)) \cdot \|\eta(\xi_i, \varphi_i)\| \, \Delta u \, \Delta v,$$

which often arise in approximating such quantities as mass and center of mass, as we shall see shortly.

It will be convenient to see what the defining equation for $\iint_\Sigma f$ looks like when Σ has the special forms $z = \varphi(x, y)$ or $\Phi(x, y, z) = 0$.

Suppose first that Σ is given by $z = \varphi(x, y)$. That is,

$$\Sigma(x, y) = (x, y, \varphi(x, y))$$

for (x, y) in some region D of the xy-plane. Then

$$\|\eta(x, y)\| = \left[\left(\frac{\partial \varphi}{\partial x}\right)^2 + \left(\frac{\partial \varphi}{\partial y}\right)^2 + 1\right]^{1/2}$$

and we have

$$\iint_\Sigma f = \iint_D f(x, y, \varphi(x, y))\sqrt{1 + \left(\frac{\partial\varphi}{\partial x}\right)^2 + \left(\frac{\partial\varphi}{\partial y}\right)^2}.$$

If the surface is described by an equation $\Phi(x, y, z) = 0$, then the assumption that the surface is smooth means that the normal, which we saw in Section 6.7 is $(\partial\Phi/\partial x, \partial\Phi/\partial y, \partial\Phi/\partial z)$, is never the zero vector. Let us suppose that $\partial\Phi/\partial z \neq 0$. Assume continuity of Φ and its partials; we can then solve for $z = \varphi(x, y)$, at least in some neighborhood of any given point. Then

$$\frac{\partial\varphi}{\partial x} = \frac{-\dfrac{\partial\Phi}{\partial x}}{\dfrac{\partial\Phi}{\partial z}}$$

and

$$\frac{\partial\varphi}{\partial y} = \frac{-\dfrac{\partial\Phi}{\partial y}}{\dfrac{\partial\Phi}{\partial z}}.$$

Then, by the last result,

$$\iint_\Sigma f = \iint_D \frac{\Phi(x, y, \varphi(x, y))\sqrt{\left(\dfrac{\partial\Phi}{\partial x}\right)^2 + \left(\dfrac{\partial\Phi}{\partial y}\right)^2 + \left(\dfrac{\partial\Phi}{\partial z}\right)^2}}{\left|\dfrac{\partial\Phi}{\partial z}\right|},$$

it being understood that $\partial\Phi/\partial z$ is evaluated at $(x, y, \varphi(x, y))$.

There is a fine point that we have glossed over here. We have used the implicit function theorem, which only guarantees local solutions of $\Phi(x, y, z) = 0$ under the conditions assumed above. It may happen, for example, that $\Phi(x_0, y_0, z_0) = 0$, $\partial\Phi(x_0, y_0, z_0)/\partial z \neq 0$, and we can solve for $z = \varphi(x, y)$ near (x_0, y_0). But possibly $\Phi(x_1, y_1, z_1) = 0$, and, say $\partial\Phi(x_1, y_1, z_1)/\partial y \neq 0$, so near (x_1, y_1) we solve for y in terms of x and z. The point is that when Σ is defined implicitly by $\Phi(x, y, z) = 0$, we may have to break Σ up into individual pieces, on each of which one (but possibly not two or three) of the partials $\partial\Phi/\partial x$, $\partial\Phi/\partial y$, and $\partial\Phi/\partial z$ is nonzero. We must then write $\iint_\Sigma f$ as a sum of integrals over each piece, making sure that the partial of Φ appearing in the denominator is nonzero on that piece.

We now give a computational example and then some applications.

EXAMPLE

Evaluate $\iint_\Sigma 3x^2$, where Σ is given by $z = \sqrt{1 - x^2 - y^2}$, with $x^2 + y^2 \leq 1$. Here Σ is parametrized by x and y:

$$\Sigma(x, y) = (x, y, \sqrt{1 - x^2 - y^2}), \qquad x^2 + y^2 \leq 1.$$

We have

$$\iint_{\Sigma} 3x^2 = \iint_{D} 3x^2 \sqrt{1 + \frac{x^2}{1 - x^2 - y^2} + \frac{y^2}{1 - x^2 - y^2}}\, dx\, dy$$

$$= \iint_{D} \frac{3x^2}{(1 - x^2 - y^2)^{1/2}}\, dx\, dy,$$

where $D = \{(x, y) \mid x^2 + y^2 \leq 1\}$. This is an improper double integral, which we can treat as follows. Consider

$$\iint_{D_\varepsilon} \frac{3x^2}{(1 - x^2 - y^2)^{1/2}}\, dx\, dy, \qquad \text{where } D_\varepsilon = \{(x, y) \mid x^2 + y^2 \leq \varepsilon\}$$

(see Fig. 6-41), with ε chosen so that $0 < \varepsilon < 1$. We evaluate this integral, then let $\varepsilon \to 1$. Switching to polar coordinates, we have

$$\iint_{D_\varepsilon} \frac{3x^2}{(1 - x^2 - y^2)^{1/2}}\, dx\, dy = \int_0^{2\pi}\int_0^{\varepsilon} \frac{3r^2\cos^2(\theta) r\, dr\, d\theta}{\sqrt{1 - r^2}}$$

$$= \int_0^{2\pi}\int_0^{\varepsilon} \frac{3r^3\cos^2(\theta)\, dr\, d\theta}{\sqrt{1 - r^2}}$$

$$= 3\pi \int_0^{\varepsilon} \frac{r^3}{\sqrt{1 - r^2}}\, dr$$

$$= 3\pi\left[\tfrac{1}{3}(1 - r^2)^{3/2} - \sqrt{1 - r^2}\right]_0^{\varepsilon}$$

$$= 3\pi\left[\tfrac{1}{3}(1 - \varepsilon^2)^{3/2} - \sqrt{1 - \varepsilon^2}\right] - 3\pi\left(\tfrac{1}{3} - 1\right).$$

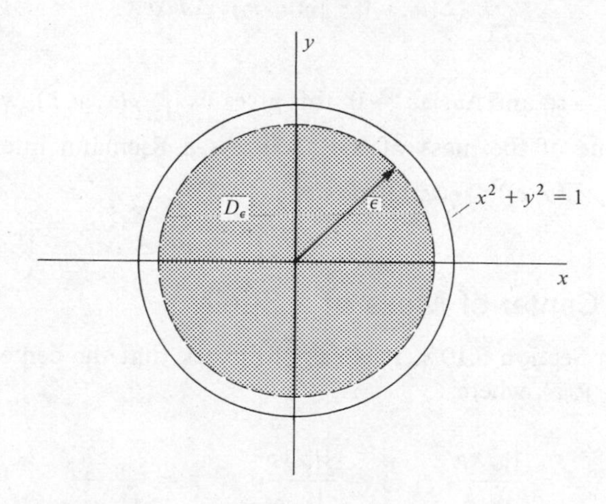

Fig. 6-41

As $\varepsilon \to 1$, then

$$\iint_{D_\varepsilon} \frac{3x^2\,dx\,dy}{(1 - x^2 - y^2)^{1/2}} \to 2\pi.$$

Thus

$$\iint_D \frac{3x^2\,dx\,dy}{\sqrt{1 - x^2 - y^2}}$$

converges to 2π, and $\iint_\Sigma 3x^2 = 2\pi$. ▮

We conclude this section with some physical applications.

A. Mass of a Shell

Suppose that we have a shell which can be thought of as the graph of a smooth surface $\Sigma: D \to R^3$. Let $\rho(x, y, z)$ be the mass per unit area at (x, y, z) on the shell. We want the mass of the shell. As usual, we approach the problem by first forming an approximating sum. Subdivide D by small rectangular pieces $D_1, \ldots, D_n$ of dimensions Δu by Δv, keeping only the rectangles entirely within D. Let Σ_i be the part of the surface corresponding to D_i under the function Σ. [That is, $\Sigma_i(u, v) = \Sigma(u, v)$ for (u, v) in D_i.] Consider $\rho(\xi_i, \varphi_i, \gamma_i)$ (area of Σ_i) as an approximation to the mass of Σ_i, where $(\xi_i, \varphi_i, \gamma_i)$ is any point on Σ_i. Now the area of Σ_i is $\iint_D \|\eta\|\,du\,dv$, which is approximated by $\|\eta(u_i, v_i)\|\,\Delta u\,\Delta v$ for some (u_i, v_i) in D_i. If we choose $(\xi_i, \varphi_i, \gamma_i)$ above as the point $\Sigma_i(u_i, v_i)$, then the mass of Σ_i is approximated by $\rho(\Sigma(u_i, v_i))\|\eta(u_i, v_i)\|\,\Delta v\,\Delta v$. The mass of the shell is then approximated by

$$\sum_{i=1}^{n} \rho(\Sigma(u_i, v_i)) \cdot \|\eta(u_i, v_i)\|\,\Delta u\,\Delta v.$$

In the limit, as $n \to \infty$ and $\Delta u, \Delta v \to 0$, this gives us $\iint_\Sigma \rho(x, y, z)$, which we take as the exact value of the mass of the shell. As a Riemann integral, $\iint_\Sigma \rho$ is evaluated as $\iint_D \rho(\Sigma(u, v)) \cdot \|\eta(u, v)\|\,du\,dv$.

B. Center of Mass of a Shell

Continuing from Section 6.10A, it is easy to check that the center of mass of the shell is at $(\bar{x}, \bar{y}, \bar{z})$, where

$$\bar{x} = \frac{\iint_\Sigma x\rho}{\iint_\Sigma \rho}, \qquad \bar{y} = \frac{\iint_\Sigma y\rho}{\iint_\Sigma \rho}, \qquad \bar{z} = \frac{\iint_\Sigma z\rho}{\iint_\Sigma \rho}.$$

C. Moments of Inertia

Continuing again from Section 6.10A, the moments of inertia of the shell about the x, y, and z axes, respectively, are given by

$$I_x = \iint_\Sigma \rho(z^2 + y^2),$$

$$I_y = \iint_\Sigma \rho(z^2 + x^2),$$

$$I_z = \iint_\Sigma \rho(x^2 + y^2).$$

Exercises for Section 6.10

1. Evaluate the following:

(a) $\iint_\Sigma y^2x^2z^2$, $\Sigma(u, v) = (u, v, u^2 + v^2)$, $u^2 + v^2 \leq 1$

(b) $\iint_\Sigma \cos(x)z^3$, $\Sigma(x, y) = (x, y, 2 - x - y)$, $0 \leq x \leq 2$, $0 \leq y \leq 2$

(c) $\iint_\Sigma (x + y^2 + z)$, $\Sigma(u, v) = (2\cos(u)\sin(v), 2\sin(u)\sin(v), 2\cos(v))$, $0 < u < \pi$, $0 < v < \pi$

2. For each of the following, find the mass, center of mass, and moments of inertia about the x-, y-, and z-axes:

(a) Sphere of radius 3 about (8, 1, 7), with $\rho(x, y, z) = 2$.

(b) Parabolic bowl $z = x^2 + y^2$, $x^2 + y^2 \leq 3$, with $\rho(x, y, z) = z$.

(c) Cube with vertices at (0, 0, 0), (1, 0, 0), (0, 1, 0), (0, 0, 1), (1, 1, 0), (1, 0, 1), (0, 1, 1), (1, 1, 1), with $\rho(x, y, z) = 1$ on the faces parallel to the xy-plane, $\rho(x, y, z) = 2$ on the faces parallel to the yz-plane, and $\rho(x, y, z) = 3$ on the faces parallel to the xz-plane.

(d) Cylinder $x^2 + y^2 = a^2$, $0 \leq z \leq 1$, with $\rho(x, y, z) = 1$ (a is a positive constant).

(e) Hemisphere $x^2 + y^2 + z^2 = 2$, $z \geq 0$, with $\rho(x, y, z) = 1$.

3. Fill in the details for a derivation of the formulas for center of mass in Section 6.10B and moments of inertia in Section 6.10C.

4. Derive a formula for the moment of inertia about a line

$$\frac{x - x_0}{a} = \frac{y - y_0}{b} = \frac{z - z_0}{c}$$

of a shell with density function ρ. Here a, b, and c are nonzero constants.

6.11 Vector Forms of Green's Theorem

This section paves the way for a generalization of Green's theorem to three dimensions. From the formula $\oint_C P\,dx + Q\,dy = \iint_D (Q_x - P_y)$ generalizations do not readily come to mind. This situation will be remedied when we rewrite Green's theorem in vector form.

There are two fairly simple ways to do this. First, recall that, if we write $C(t) = (\alpha(t), \beta(t))$ for $a \leq t \leq b$, then under suitable conditions on α, β, P, and Q we have

$$\begin{aligned}\oint_C P\,dx + Q\,dy &= \oint_C (P(\alpha(t), \beta(t))\alpha'(t) + Q(\alpha(t), \beta(t))\beta'(t)))\,dt \\ &= \oint_C (P, Q) \cdot (\alpha'(t), \beta'(t))\,dt \\ &= \oint_C (P, Q) \cdot C'(t)\,dt = \oint_C (P, Q).\end{aligned}$$

Now, what does $Q_x - P_y$ have to do with (P, Q)? Observe that $Q_x - P_y$ is the third component of $\operatorname{curl}(P, Q, 0)$. That is, $\operatorname{curl}(P, Q, 0) \cdot (0, 0, 1) = Q_x - P_y$. Now think of C as a curve in three space by writing $C(t) = (\alpha(t), \beta(t), 0)$ for $a \leq t \leq b$. We can also envision D, the region enclosed by C, as the graph of a surface Σ, given by $\Sigma(x, y) = (x, y, 0)$ for (x, y) in D. Noting that $\eta(x, y) = (0, 0, 1)$ is the normal to Σ, then

$$Q_x - P_y = \operatorname{curl}(P, Q, 0) \cdot \eta(\Sigma).$$

Then Green's theorem becomes

$$\begin{aligned}\iint_D (Q_x - P_y) &= \iint_D \operatorname{curl}(P, Q, 0) \cdot \eta(\Sigma) \\ &= \iint_\Sigma \operatorname{curl}(P, Q, 0) = \oint_C P\,dx + Q\,dy = \oint_C (P, Q, 0).\end{aligned}$$

We can attempt to generalize this to three dimensions by

(1) Dropping the restrictions that C and D lie in a plane.
(2) Allowing 3-vectors whose third component is nonzero.

Thus put $u = (P, Q, R)$, and suppose that Σ is a surface in 3-space bounded (in some sense yet to be discussed) by a closed curve C. We conjecture, under reasonable conditions on Σ, C, P, Q and R, that

$$\iint_\Sigma \operatorname{curl} u = \oint_C u.$$

This will turn out to be Stokes's theorem.

Now leave this until later and start out again from Green's Theorem on a new track. Parametrize C by arc length s. A unit tangent to C at $C(s)$ is $C'(s) = (\alpha'(s), \beta'(s))$. The vector $(\beta'(s), -\alpha'(s))$ is then a unit vector normal to C at

$C(s)$. That is, $(\beta'(s), -\alpha'(s))$ is orthogonal to the tangent $C'(s)$, and points away from the region D enclosed by C (see Fig. 6-42).

Now observe that

$$(Q, -P) \cdot (\beta'(s), -\alpha'(s)) = P\alpha'(s) + Q\beta'(s),$$

so

$$\oint_C (Q, -P) \cdot (\beta'(s), -\alpha'(s))\, ds = \oint_C P\, dx + Q\, dy.$$

But also $\operatorname{div}(Q, -P) = Q_x - P_y$, so Green's theorem can be written

$$\iint_D \operatorname{div}(Q, -P) = \oint_C (Q, -P) \cdot (\beta'(s), -\alpha'(s))\, ds = \oint_C P\, dx + Q\, dy.$$

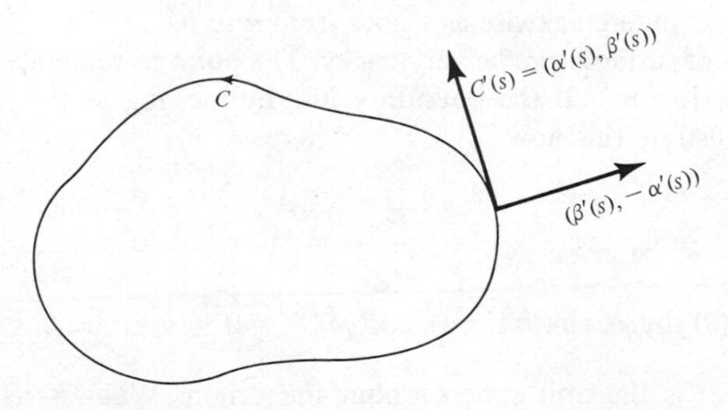

Fig. 6-42

We attempt to generalize this by

(1) Replacing the arbitrary 2-vector $(Q, -P)$ by a 3-vector, say (u, v, w).
(2) Replacing D by a volume V in R^3, bounded by a surface Σ, and the double integral over D by a triple integral over V.
(3) Replacing the normal to the curve C bounding D by the appropriate normal η to Σ, and the line integral over C by a surface integral over Σ.

We then conjecture that

$$\iiint_V \operatorname{div}(u, v, w) = \iint_\Sigma (u, v, w).$$

This will turn out to be Gauss's divergence theorem.

6.12 Stokes's Theorem

We want now to justify one of our conjectured generalizations of Green's theorem. Stokes's theorem says that

$$\iint_{\Sigma} \operatorname{curl} u = \oint_{C} u,$$

with C the boundary of the surface Σ, and under suitable conditions on u.

First we must clarify what we mean by the boundary of a surface. Let $\Sigma: D \to R^3$ be a surface, and suppose that D is a compact set of points in R^2 bounded by a closed path $C: [a, b] \to R^2$. We then define the *boundary* of Σ to be the curve $\partial\Sigma$ given by

$$\partial\Sigma(t) = \Sigma(C(t)), \qquad a \leq t \leq b.$$

We consider $\partial\Sigma$ to have positive orientation if C has been oriented so that $C(t)$ goes around C counterclockwise as t goes from a to b.

Boundaries of surfaces can be very tricky. The point to remember is that it is not always possible to tell the boundary just by looking at the graph of the surface. We illustrate this now.

EXAMPLE

Let $\Sigma: D \to R^3$ be given by

$$\Sigma(\theta, \varphi) = (\cos(\theta)\sin(\varphi), \sin(\theta)\sin(\varphi), \cos(\varphi)), \qquad 0 \leq \theta \leq 2\pi, 0 \leq \varphi \leq \pi.$$

The graph of Σ is the unit sphere about the origin. What is its boundary? Intuitively, it is not clear what curve on Σ should be called the boundary of Σ. So we proceed to compute somewhat mechanically directly from the definition.

Here D is a rectangle in the $\theta\varphi$-plane, and the boundary C of D is made up of four pieces, which we parametrize (keeping orientation in mind) as follows (see Fig. 6-43):

$$C_1(\theta) = (\theta, 0), \theta: 0 \to 2\pi,$$
$$C_2(\varphi) = (2\pi, \varphi), \varphi: 0 \to \pi,$$
$$C_3(\theta) = (\theta, \pi), \theta: 2\pi \to 0,$$
$$C_4(\varphi) = (0, \varphi), \varphi: \pi \to 0.$$

We must compose C with Σ in four pieces:

$$\partial\Sigma_1(\theta) = \Sigma(C_1(\theta)) = \Sigma(\theta, 0) = (0, 0, 1), \theta: 0 \to 2\pi,$$
$$\partial\Sigma_2(\varphi) = \Sigma(C_2(\varphi)) = \Sigma(2\pi, \varphi) = (\sin(\varphi), 0, \cos(\varphi)), \varphi: 0 \to \pi,$$
$$\partial\Sigma_3(\theta) = \Sigma(C_3(\theta)) = \Sigma(\theta, \pi) = (0, 0, -1), \theta: 2\pi \to 0,$$

and $$\partial\Sigma_4(\varphi) = \Sigma(C_4(\varphi)) = \Sigma(0, \varphi) = (\sin(\varphi), 0, \cos(\varphi)), \varphi: \pi \to 0.$$

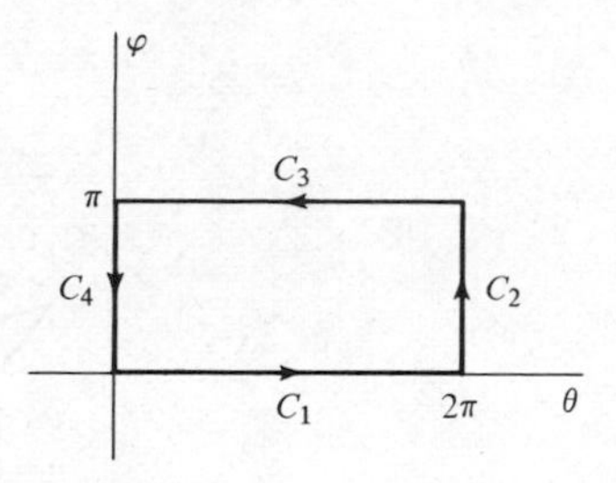

Fig. 6-43

Then $\partial\Sigma$ is the result of stringing the four curves $\partial\Sigma_1, \ldots, \partial\Sigma_4$ together on Σ. Graphically, $\partial\Sigma$ starts out at (0, 0, 1), sits there awhile (as θ varies from 0 to 2π), then slides down the curve $x^2 + z^2 = 1$ in the $x \geq 0$ region until it reaches $(0, 0, -1)$, sits there awhile (as $\theta: 2\pi \to 0$), then moves up $x^2 + z^2 = 1$ in the $x \leq 0$ region until it ends at (0, 0, 1). The orientation on $\partial\Sigma$ induced by the positive orientations on C is indicated by the arrows in Fig. 6-44.

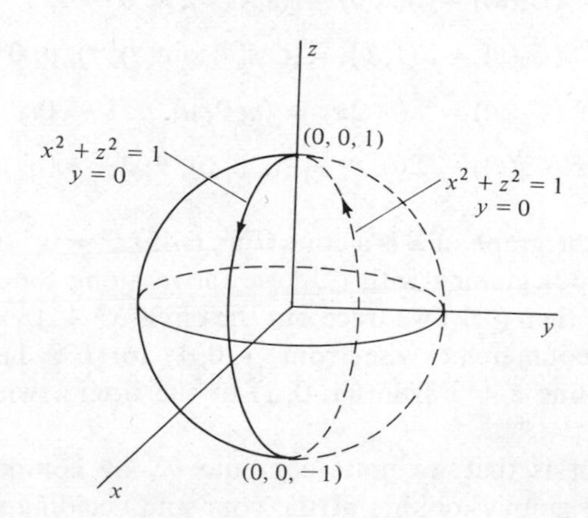

Fig. 6-44

EXAMPLE

Let

$$\Sigma(u, \theta) = (u\cos(\theta), u\sin(\theta), u), \qquad 0 \leq u \leq 1, 0 \leq \theta \leq 2\pi.$$

Here $\Sigma: D \to R^3$, where D is a rectangle in the $u\theta$-plane (Fig. 6-45a) bounded by a curve C which consists of four straight-line segments. Parametrize the pieces of C:

$$C_1(u) = (u, 0), u: 0 \to 1,$$
$$C_2(\theta) = (1, \theta), \theta: 0 \to 2\pi,$$
$$C_3(u) = (u, 2\pi), u: 1 \to 0,$$
$$C_4(\theta) = (0, \theta), \theta: 2\pi \to 0.$$

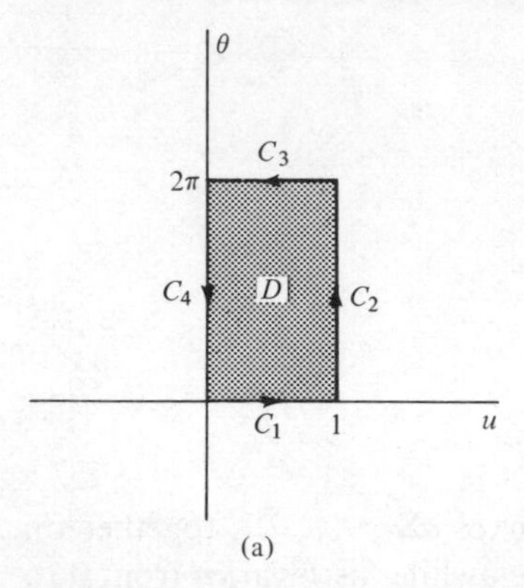

Fig. 6-45

We can then compute $\partial\Sigma$ in four pieces:

$$\partial\Sigma_1(u) = \Sigma(C_1(u)) = \Sigma(u, 0) = (u, 0, u),\ u\colon 0 \to 1,$$
$$\partial\Sigma_2(\theta) = \Sigma(C_2(\theta) = \Sigma(1, \theta) = (\cos(\theta), \sin(\theta), 1),\ \theta\colon 0 \to 2\pi,$$
$$\partial\Sigma_3(u) = \Sigma(C_3(u)) = \Sigma(u, 2\pi) = (u, 0, u),\ u\colon 1 \to 0,$$
$$\partial\Sigma_4(\theta) = \Sigma(C_4(\theta)) = \Sigma(0, \theta) = (0, 0, 0),\ \theta\colon 2\pi \to 0.$$

Geometrically, the graph of Σ is a cone (Fig. 6-45) $z^2 = x^2 + y^2, 0 \le z \le 1$. Trace out $\partial\Sigma$ on Σ. Beginning with $\partial\Sigma_1$, we move along the line $z = x$ from (0, 0, 0) to (1, 0, 1); then ($\partial\Sigma_2$) we trace out the circle $x^2 + y^2 = 1$ in the $z = 1$ plane, proceeding counterclockwise from (1, 0, 1) to (1, 0, 1); then ($\partial\Sigma_3$) we proceed down the line $z = x$ from (1, 0, 1) to the origin, where we sit ($\partial\Sigma_4$) for awhile.

Again, the point is that we must compute $\partial\Sigma$ by composing C with Σ. There is no way of simply looking at the cone and deciding what the boundary is. ∎

We can now state the main result of this section.

Stokes's Theorem

Suppose that $\Sigma = (f, g, h)\colon D \to R^3$ is a piecewise smooth surface, with D a compact region in R^2 bounded by a closed path. Suppose that f, g, and h are continuous with continuous first and second partials in D. Let $F(x, y, z) = (A(x, y, z), B(x, y, z), C(x, y, z))$, and suppose that A, B and C are continuous with continuous first and second partials at least on the graph of Σ. Then

$$\iint_\Sigma \operatorname{curl} F = \oint_{\partial\Sigma} F.$$

To verify the formula, we begin by calculating $\oint_{\partial\Sigma} F$. Write $\mathscr{C}(t) = (\alpha(t), \beta(t))$,

$a \le t \le b$, for the boundary of D, positively oriented as $t: a \to b$. Then $\partial\Sigma(t) = \Sigma(\mathscr{C}(t))$ for $a \le t \le b$, and

$$\oint_{\partial\Sigma} F = \int_a^b F(\partial\Sigma(t)) \cdot (\partial\Sigma)'(t)\, dt.$$

In order to carry out this integration, we must first compute $F(\partial\Sigma(t))$ and $(\partial\Sigma)'(t)$. First,

$$(\partial\Sigma)(t) = \Sigma(\mathscr{C}(t)) = (f(\alpha(t), \beta(t)), g(\alpha(t), \beta(t)), h(\alpha(t), \beta(t))), \qquad a \le t \le b,$$

so by the chain rule:

$$(\partial\Sigma)'(t) = (f_u\alpha'(t) + f_v\beta'(t), g_u\alpha'(t) + g_v\beta'(t), h_u\alpha'(t) + h_v\beta'(t)).$$

Next

$$\begin{aligned} F(\partial\Sigma(t)) &= (A(\partial\Sigma(t)), B(\partial\Sigma(t)), C(\partial\Sigma(t))) \\ &= (A(f(\alpha(t), \beta(t)), g(\alpha(t), \beta(t)), h(\alpha(t), \beta(t))), \\ &\quad\; B(f(\alpha(t), \beta(t)), g(\alpha(t), \beta(t)), h(\alpha(t), \beta(t))), \\ &\quad\; C(f(\alpha(t), \beta(t)), g(\alpha(t), \beta(t)), h(\alpha(t), \beta(t)))). \end{aligned}$$

Then

$$\begin{aligned} &\int_a^b F(\partial\Sigma(t)) \cdot (\partial\Sigma)'(t)\, dt \\ &\quad = \int_a^b [A(f(\alpha(t), \beta(t)), g(\alpha(t), \beta(t)), h(\alpha(t), \beta(t)))(f_u\alpha'(t) + f_v\beta'(t)) \\ &\qquad + B(f(\alpha(t), \beta(t)), g(\alpha(t), \beta(t)), h(\alpha(t), \beta(t)))(g_u\alpha'(t) + g_v\beta'(t)) \\ &\qquad + C(f(\alpha(t), \beta(t)), g(\alpha(t), \beta(t)), h(\alpha(t), \beta(t)))(h_u\alpha'(t) + h_v\beta'(t))]\, dt. \end{aligned}$$

As a bookkeeping convenience, consider the last as a sum of three integrals and take each one separately. We also abbreviate $u = \alpha(t)$, $v = \beta(t)$, $x = f(u, v)$, $y = g(u, v)$, and $z = h(u, v)$. Applying Green's theorem in the uv-plane, we have

$$\begin{aligned} \int_a^b A(x, y, z)(x_u\alpha'(t) + x_v\beta'(t))\, dt &= \oint_{\mathscr{C}} Ax_u\, du + Ax_v\, dv \\ &= \iint_D \left[\frac{\partial(Ax_v)}{\partial u} - \frac{\partial(Ax_u)}{\partial v}\right] du\, dv. \end{aligned}$$

Now

$$\begin{aligned} \frac{\partial}{\partial u}(Ax_v) - \frac{\partial}{\partial v}(Ax_u) &= A(x_{uv} - x_{vu}) + A_u x_v - A_v x_u \\ &= A_u x_v - A_v x_u, \qquad \text{as } x_{uv} = x_{vu}. \end{aligned}$$

Then we have

$$\iint_D \left[\frac{\partial(Ax_v)}{\partial u} - \frac{\partial(Ax_u)}{\partial v}\right] du\, dv = \iint_D (A_u x_v - A_v x_u)\, du\, dv.$$

Remembering that x, y, and z are functions of u and v, we have, by the chain rule,

$$\begin{aligned}\iint_D &(A_u x_v - A_v x_u)\, du\, dv \\ &= \iint_D [(A_x x_u + A_y y_u + A_z z_u)x_v - (A_x x_v + A_y y_v + A_z z_v)x_u]\, du\, dv \\ &= \iint_D [A_z(x_v z_u - x_u z_v) - A_y(x_u y_v - x_v y_u)]\, du\, dv \\ &= \iint_D \left[A_z \frac{\partial(z, x)}{\partial(u, v)} - A_y \frac{\partial(x, y)}{\partial(u, v)}\right] du\, dv.\end{aligned}$$

In summary,

$$\int_a^b A(x_u\alpha'(t) + x_v\beta'(t))\, dt = \iint_D \left[A_z \frac{\partial(z, x)}{\partial(u, v)} - A_y \frac{\partial(x, y)}{\partial(u, v)}\right] du\, dv.$$

In similar fashion, we can obtain

$$\int_a^b B(x, y, z)(y_u\alpha'(t) + y_v\beta'(t))\, dt = \iint_D \left[B_x \frac{\partial(x, y)}{\partial(u, v)} - B_z \frac{\partial(y, z)}{\partial(u, v)}\right] du\, dv,$$

and

$$\int_a^b C(x, y, z)(z_u\alpha'(t) + z_v\beta'(t))\, dt = \iint_D \left[C_y \frac{\partial(x, y)}{\partial(u, v)} - C_x \frac{\partial(z, x)}{\partial(u, v)}\right] du\, dv.$$

Adding up the left sides of the last three equations, and equating to the sum of the right sides, we have

$$\oint_{\partial\Sigma} F = \iint_D \left[(C_y - B_z)\frac{\partial(y, z)}{\partial(u, v)} + (A_z - C_x)\frac{\partial(z, x)}{\partial(u, v)} + (B_x - A_y)\frac{\partial(x, y)}{\partial(u, v)}\right] du\, dv.$$

But

$$\eta(u, v) = \left(\frac{\partial(y, z)}{\partial(u, v)}, \frac{\partial(z, x)}{\partial(u, v)}, \frac{\partial(x, y)}{\partial(u, v)}\right)$$

and

$$\text{curl}(A, B, C) = (C_y - B_z, A_z - C_x, B_x - A_y).$$

Then the integral on the right is $\iint_\Sigma (\text{curl } F) \cdot \eta \, du \, dv$, which is $\iint_\Sigma \text{curl } F$. Thus $\int_{\partial\Sigma} F = \iint_\Sigma \text{curl } F$, as we wanted to show.

As the reader might have noticed, the above proof, although long and tedious, was purely a matter of calculation. The pedagogical value of going through such an argument is debatable, but it does make use of a number of the definitions and theorems developed to this point and should give the tenacious student a good test of his understanding of line and surface integrals, curl, normals, chain-rule differentiation, boundary of a surface, and Green's theorem in the plane.

The student can review the notation of Section 6.8 to verify that an entirely equivalent form of Stokes's formula is

$$\iint_\Sigma (C_y - B_z) \, dy \, dz + (A_z - C_x) \, dz \, dx + (B_x - A_y) \, dx \, dy = \oint_{\partial\Sigma} A \, dx + B \, dy + C \, dz.$$

For illustration, we shall compute some numerical examples.

EXAMPLE

Let Σ be given by $\Sigma(\theta, \varphi) = (\cos(\theta) \sin(\varphi), \sin(\theta) \sin(\varphi), \cos(\varphi))$, $0 \leq \theta \leq 2\pi$, $0 \leq \varphi \leq \pi$. We have already computed $\partial\Sigma$, in the first example of this section. Let $F(x, y, z) = (xyz, x^2, y^2)$. We want to show that

$$\iint_\Sigma \text{curl } F = \oint_{\partial\Sigma} F$$

by calculating each side separately. First,

$$\eta(\theta, \varphi) = -\sin(\varphi) \, \Sigma \, (\theta, \varphi).$$

Next,

$$\text{curl } F = (2y, xy, 2x - xz).$$

Then

$$\begin{aligned} \text{curl } F \cdot \eta &= -2y \cos(\theta) \sin^2(\varphi) - xy \sin(\theta) \sin^2(\varphi) - (2x - xz) \sin(\varphi) \cos(\varphi) \\ &= -2 \sin(\theta) \cos(\theta) \sin^3(\varphi) - \cos(\theta) \sin^2(\theta) \sin^4(\varphi) \\ &\quad - 2 \cos(\theta) \sin^2(\varphi) \cos(\varphi) + \cos(\theta) \sin^2(\varphi) \cos^2(\varphi). \end{aligned}$$

Then

$$\iint_{\Sigma} \text{curl } F = \iint_{D} (\text{curl } F)(\Sigma(\theta, \varphi)) \cdot \eta(\theta, \varphi)\, d\theta\, d\varphi$$

$$= \int_0^{\pi} \int_0^{2\pi} -2 \sin(\theta) \cos(\theta) \sin^3(\varphi)\, d\theta\, d\varphi$$

$$- \int_0^{\pi} \int_0^{2\pi} \cos(\theta) \sin^2(\theta) \sin^4(\varphi)\, d\theta\, d\varphi$$

$$- \int_0^{\pi} \int_0^{2\pi} 2 \cos(\theta) \sin^2(\varphi) \cos(\varphi)\, d\theta\, d\varphi$$

$$+ \int_0^{\pi} \int_0^{2\pi} \cos(\theta) \sin^2(\varphi) \cos^2(\varphi)\, d\theta\, d\varphi = 0.$$

Next, we compute $\int_{\partial\Sigma} F$. Split $\partial\Sigma$ into four parts, $\partial\Sigma_1, \ldots, \partial\Sigma_4$, as in the previous example. Clearly $\int_{\partial\Sigma_1} F = \int_{\partial\Sigma_3} F = 0$, as $\partial\Sigma_1(\theta) = (0, 0, 1)$ for $\theta: 0 \to 2\pi$, and $\partial\Sigma_3(\theta) = (0, 0, -1)$ for $\theta: 2\pi \to 0$. Next, we can see by inspection that

$$\int_{\partial\Sigma_2} F = -\int_{\partial\Sigma_4} F.$$

Thus

$$\int_{\partial\Sigma_2} F + \int_{\partial\Sigma_4} F = 0, \quad \text{so} \int_{\partial\Sigma} F = 0. \qquad \blacksquare$$

EXAMPLE

Let $\Sigma(u, \theta) = (u \cos(\theta), u \sin(\theta), u)$, $0 \le u \le 1$, $0 \le \theta \le 2\pi$, as in the second example of this section. Let $F(x, y, z) = (xyz, x^2, y^2)$, as above. The normal to Σ is $\eta(u, \theta) = (-u \cos(\theta), -u \sin(\theta), u)$. Then, on Σ,

$$(\text{curl } F) \cdot \eta = -2u^2 \sin(\theta) \cos(\theta) - u^3 \cos(\theta) \sin^2(\theta) + 2u^3 \cos(\theta) - u^3 \cos(\theta).$$

Then

$$\iint_{\Sigma} \text{curl } F = \int_0^{2\pi} \int_0^1 -2u^2 \sin(\theta) \cos(\theta)\, du\, d\theta$$

$$- \int_0^{2\pi} \int_0^1 u^3 \cos(\theta) \sin^2(\theta)\, du\, d\theta$$

$$+ \int_0^{2\pi} \int_0^1 2u^3 \cos(\theta)\, du\, d\theta - \int_0^{2\pi} \int_0^1 u^3 \cos(\theta)\, du\, d\theta = 0.$$

Next we compute $\int_{\partial\Sigma} F$. Split $\partial\Sigma$ into $\partial\Sigma_1, \ldots, \partial\Sigma_4$, as we found previously. Clearly $\int_{\partial\Sigma_4} F = 0$. Next

$$\int_{\partial\Sigma_1} F = \int_0^1 F(\partial\Sigma_1(u)) \cdot (\partial\Sigma_1')(u)\, du = \int_0^1 F(u, 0, u) \cdot (1, 0, 1)\, du$$

$$= \int_0^1 (0, u^2, 0) \cdot (1, 0, 1)\, du = 0 = -\int_{\partial\Sigma_3} F.$$

Finally,

$$\int_{\partial\Sigma_2} F = \int_0^{2\pi} F(\cos(\theta), \sin(\theta), 1) \cdot (-\sin(\theta), \cos(\theta), 0)\, d\theta$$

$$= \int_0^{2\pi} (-\cos(\theta)\sin^2(\theta) + \cos^3(\theta))\, d\theta = 0. \qquad \blacksquare$$

EXAMPLE

Lest it appear that these integrals are always zero, we compute an example that gives a nonzero result. Let $\Sigma(x, y) = (x, y, x^2 + y^2)$, $x^2 + y^2 \le 1$. Let $F(x, y, z) = (z, x, y)$. Then

$$\operatorname{curl} F = (1, 1, 1) \qquad \text{and} \qquad \eta(x, y) = (-2x, -2y, 1).$$

Then $(\operatorname{curl} F) \cdot \eta = -2x - 2y + 1$, so

$$\iint_\Sigma \operatorname{curl} F = \iint_D (-2x - 2y + 1)\, dx\, dy$$

where $D = \{(x, y) \mid x^2 + y^2 \le 1\}$. To evaluate this integral, go to polar coordinates to obtain

$$\int_0^{2\pi}\int_0^1 [-2r\cos(\theta) - 2r\sin(\theta) + 1]r\, dr\, d\theta = \pi.$$

Thus $\iint_\Sigma \operatorname{curl} F = \pi$.

To evaluate $\oint_{\partial\Sigma} F$, we first have to find $\partial\Sigma$. Write the boundary C of D in two pieces:

$$C_1(x) = (x, \sqrt{1 - x^2}), \qquad x: 1 \to -1,$$

$$C_2(x) = (x, -\sqrt{1 - x^2}), \qquad x: -1 \to 1.$$

Then $\partial\Sigma$ is in two pieces:

$$\partial\Sigma_1(x) = \Sigma(C_1(x)) = (x, \sqrt{1 - x^2}, 1), \qquad x: 1 \to -1;$$

$$\partial\Sigma_2(x) = \Sigma(C_2(x)) = (x, -\sqrt{1 - x^2}, 1), \qquad x: -1 \to 1.$$

Now

$$\int_{\partial\Sigma_1} F = \int_1^{-1} F(\partial\Sigma_1(x)) \cdot (\partial\Sigma_1')(x)\,dx$$
$$= \int_1^{-1} (1, x, \sqrt{1-x^2}) \cdot \left(1, \frac{-x}{\sqrt{1-x^2}}, 0\right) dx$$
$$= \int_{-1}^{1} \left(-1 + \frac{x^2}{\sqrt{1-x^2}}\right) dx.$$

And

$$\int_{\partial\Sigma_2} F = \int_{-1}^{1} F(\partial\Sigma_2(x)) \cdot (\partial\Sigma_2'(x))\,dx$$
$$= \int_{-1}^{1} (1, x, \sqrt{1-x^2}) \cdot \left(1, \frac{x}{\sqrt{1-x^2}}, 0\right) dx = \int_{-1}^{1} \left(1 + \frac{x^2}{\sqrt{1-x^2}}\right) dx.$$

We would therefore expect that $\oint_{\partial\Sigma} F = 2\int_{-1}^{1} (x^2/\sqrt{1-x^2})\,dx$ if this integral converges (it is improper at 1 and -1).

Put $x = \cos(\theta)$. Then $\theta = \pi$ for $x = -1$ and $\theta = 0$ for $x = 1$, so

$$\int_{-1}^{1} \frac{x^2\,dx}{\sqrt{1-x^2}} = \int_{\pi}^{0} \frac{\cos^2(\theta)(-\sin(\theta))\,d\theta}{\sqrt{1-\cos^2(\theta)}}$$
$$= \int_0^{\pi} \cos^2(\theta)\,d\theta = \frac{\pi}{2},$$

giving us $\oint_{\partial\Sigma} F = \pi$, as we should have.

Strictly speaking, changing variables this way in an improper integral can be justified only after convergence has been established. We leave it to the reader to show that $\int_{-1}^{1} (x^2/\sqrt{1-x^2})\,dx$ converges. [*Hint:* For $-1 < \varepsilon_1 < \varepsilon_2 < 1$, calculate $\int_{\varepsilon_1}^{\varepsilon_2} (x^2/\sqrt{1-x^2})\,dx$ and then let $\varepsilon_1 \to -1$ from the right and $\varepsilon_2 \to 1$ from the left. It will help to note that

$$\int \frac{x^2\,dx}{\sqrt{1-x^2}} = -\frac{x}{2}\sqrt{1-x^2} + \tfrac{1}{2}\arcsin(x).\Big]$$

∎

Exercises for Section 6.12

1. For each F, and for each Σ, show that $\iint_{\Sigma} \operatorname{curl} F = \oint_{\partial\Sigma} F$, by computing both sides separately.

$F(x, y, z) =$	$\Sigma(u, v) =$	
(x, x, xyz)	$(u, v, u^2 + v^2),$	$0 \leq u \leq 1, 0 \leq v \leq 1$
(y^2, xy, z)	$(u, v, \sin(u + v)),$	$0 \leq u \leq \pi/2, 0 \leq v \leq \pi/2$
$(x + y + z, xz, y^2)$	$(u, v, 2 - u^2 + v),$	$0 \leq u \leq 1, 0 \leq v \leq 1$
(z, x, y)	$(u, v, 2 - u - v),$	$0 \leq u \leq 2, 0 \leq v \leq 2$
(x, y, z)	$(u, v, uv),$	$0 \leq u \leq 1, 0 \leq v \leq 1$

(There are twenty-five problems in all here.)

2. Verify Stokes's theorem when $F(x, y, z) = (xy, xz, yz)$ and

$$\Sigma(x, y) = (x, y, x^2 + y^2), \qquad x^2 + y^2 \leq 1.$$

6.13 Two Applications of Stokes's Theorem

A. Curl of a Vector

In Section 6.2 we gave a lengthy and very loose argument to the effect that

$$(\text{curl } F|_{(x_0, y_0, z_0)}) \cdot \eta = \lim_{A \to 0} \frac{1}{A} \oint_C F,$$

where η is the normal to the plane curve C bounding a region D of area A, containing (x_0, y_0, z_0), and $\lim_{A \to 0}$ means a limit as D shrinks to the point (x_0, y_0, z_0). A much neater argument can be based on Stokes's theorem. Write

$$\iint_\Sigma \text{curl } F = \iint_D (\text{curl } F) \cdot \eta \, du \, dv = \oint_{\partial\Sigma} F.$$

Choose Σ as the plane region D. Then $\partial\Sigma$ is the curve C bounding D. By a mean value theorem,

$$\iint_D (\text{curl } F) \cdot \eta \, du \, dv = (\text{curl } F|_{(\bar{x}, \bar{y}, \bar{z})}) \cdot \eta(\bar{x}, \bar{y}, \bar{z})(A(D))$$

for some $(\bar{x}, \bar{y}, \bar{z})$ in D. Then

$$((\text{curl } F) \cdot \eta)(\bar{x}, \bar{y}, \bar{z}) = \frac{1}{A} \oint_C F.$$

Since Σ is in a plane, η is a constant vector on Σ, and

$$\eta(\bar{x}, \bar{y}, \bar{z}) = \eta(x_0, y_0, z_0) = \eta.$$

Further, as $A \to 0$, $(\bar{x}, \bar{y}, \bar{z}) \to (x_0, y_0, z_0)$. Then

$$(\text{curl } F|_{(x_0, y_0, z_0)}) \cdot \eta = \lim_{A \to 0} \frac{1}{A} \oint_C F.$$

B. Potential Theory and Independence of Path in R^3

In physics it is often useful to know whether or not a given force is the gradient of some potential function. We have already seen conditions for this in the plane (Section 6.5E). In particular, if D is a connected region in the xy-plane, then $(A, B) = \nabla\varphi$ is equivalent to having $\oint_C A\,dx + B\,dy = 0$ for every closed path in D, or, equivalently, $\int_C A\,dx + B\,dy$ independent of path in D. When D is simply connected, these are in turn equivalent to assuming that A and B satisfy the condition $B_x = A_y$.

The first part of the above result extends immediately to three dimensions. If D is connected in R^3, then the conditions—$F = \nabla\varphi$, $\int_C F$ is independent of path in D, and $\oint_C F = 0$ for every closed path C in D—are equivalent. The proofs mirror the two-dimensional arguments.

It is somewhat more difficult to generalize the test $\dfrac{\partial B}{\partial x} = \dfrac{\partial A}{\partial y}$ for independence of path of $\int_C A\,dx + B\,dy$. The first step is to generalize the notion of simple connectedness to three dimensions in such a way as to be able to exploit Stokes's Theorem. A region D in R^3 will be called *simply connected* if every closed (not necessarily plane) curve in D is the boundary of a surface whose graph is in D. For example, the interior of a sphere is simply connected, while the interior of a doughnut is not (take a curve going around the hole).

Now suppose that D is a simply connected region in R^3. Let $F(x, y, z)$ be a 3-vector for each (x, y, z) in D, and suppose that F has coordinate functions satisfying the hypotheses of Stokes's Theorem in D. We shall examine the condition curl $F = (0, 0, 0)$ in D.

First, suppose that curl $F = (0, 0, 0)$ in D. If C is any closed path in D, invent a surface Σ in D having C as boundary. Then we have, by Stokes's Theorem,

$$\int_C F = \iint_\Sigma \text{curl } F = 0,$$

and we conclude that $\int_C F$ is independent of path in D. Hence also F is the gradient of a scalar function φ defined in D. [In fact, we can choose any (x_0, y_0, z_0) in D, define $\varphi(x, y, z) = \int_C F$ for any path C in D from (x_0, y_0, z_0) to (x, y, z), and show that $\nabla\varphi = F$ in much the same way that we argued in two dimensions in Section 6.5E.]

Conversely, if $F = \nabla\varphi$ in D, then a straightforward calculation gives curl $F =$ curl$(\nabla\varphi) = (0, 0, 0)$.

To summarize the last two paragraphs: *In a simply connected region D, independence of path of $\int_C F$ is equivalent to the condition that* curl $F = (0, 0, 0)$, *provided that F satisfies the conditions of Stokes's Theorem in D.*

Exercises for Section 6.13

1. Show that the condition curl $F = (0, 0, 0)$ given here for path independence reduces to the condition of Section 6.5E in R^2. *Hint:* Put $F = (P, Q, 0)$ and compute curl F.

2. Suppose that $F = \nabla\varphi$ in D, a region in R^3. If C is a path in D from P_0 to P_1, show that

$$\int_C F = \varphi(P_1) - \varphi(P_0).$$

3. Carry out the details of showing that $F = \nabla\varphi$ if φ is defined as $\varphi(x, y, z) = \int_K F$ as discussed in *B*.

4. Test the following for independence of path in the given region. If $\int_C F$ is independent of path in D, find a potential function for F.
   (a) $F(x, y, z) = (1, 2y, xy)$ in $D = \{(x, y, z) \mid x^2 + y^2 + z^2 \leq 1\}$
   (b) $F(x, y, z) = \left(yz + \dfrac{1}{x + y + z}, e^x z + \dfrac{1}{x + y + z}, e^x y + \dfrac{1}{x + y + z}\right)$, $D = \{(x, y, z) \mid x > 0, y > 0 \text{ and } z > 0\}$
   (c) $F(x, y, z) = (1, z\cos(yz), e^z + y\cos(yz))$, $D = \{(x, y, z) \mid 2 \leq x^2 + y^2 + z^2 \leq 4\}$
   (d) $F(x, y, z) = (y, x, z^2)$, $D = R^3$
   (e) $F(x, y, z) = (2xy^2z^2, 2x^2yz^2, 2x^2y^2z)$, $D = R^3$

6.14 Divergence Theorem

In this section we complete the program started in Section 6.11 and generalize the second vector form of Green's theorem. As a preliminary, we need the notion of a *closed surface*, by which we mean a surface in R^3 whose graph bounds a compact, simply connected region of finite, nonzero volume. In this case we denote by $V(\Sigma)$ the region enclosed by Σ, together with the graph of Σ itself. We assume that Σ is always oriented so that the outer normal determines the positive side of Σ.

We can now state the main result of this section.

Divergence (Gauss's) Theorem

Let Σ be a closed, piecewise smooth surface in R^3. Let $F(x, y, z)$ be a vector defined at each (x, y, z) in $V(\Sigma)$, with continuous component functions having continuous partials in $V(\Sigma)$. Then

$$\iiint_{V(\Sigma)} \operatorname{div} F = \iint_{\Sigma} F.$$

Before going into a mathematical argument, look at the equation from a physical point of view. Think of $F(x, y, z)$, or $F(x, y, z, t)$, as a velocity of a fluid. We have already seen an interpretation of div F as the strength of a source of fluid at each point (x, y, z). Then $\iiint_{V(\Sigma)} \operatorname{div} F$ represents at any time t the rate of fluid flow out of $V(\Sigma)$. This must equal the rate of flow across the bounding surface Σ, this rate being exactly $\iint_\Sigma F$. In this context $\iiint_{V(\Sigma)} \operatorname{div} F = \iint_\Sigma F$ is a thoroughly plausible equation.

A mathematical justification can be based on an argument like that used for Green's Theorem, assuming that $V(\Sigma)$ and Σ can be represented in various particularly convenient ways. Such an argument is a terrible thing to behold, so we shall be content to demonstrate the theorem when the region is a rectangular box. This does not constitute much of a proof, but may give the reader an idea of what is involved.

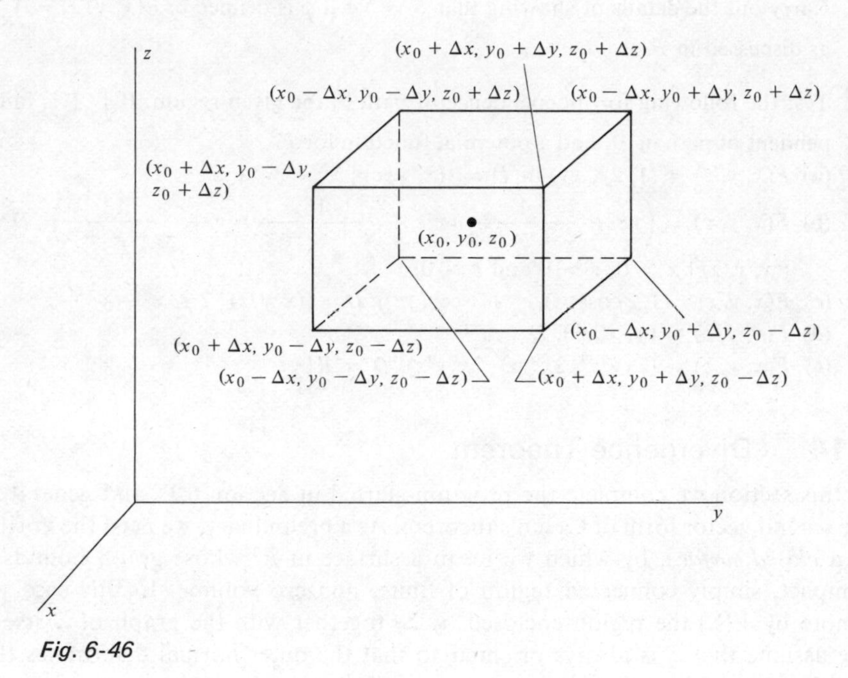

Fig. 6-46

Write $F(x, y, z) = (A(x, y, z), B(x, y, z), C(x, y, z))$, and let $V(\Sigma)$ be the box-shaped region centered at (x_0, y_0, z_0) as shown in Fig. 6-46. Write

$$\iiint_{V(\Sigma)} \operatorname{div} F = \iiint_{V(\Sigma)} A_x + \iiint_{V(\Sigma)} B_y + \iiint_{V(\Sigma)} C_z,$$

and compute each integral separately:

$$\iiint_{V(\Sigma)} A_x = \int_{z_0-\Delta z}^{z_0+\Delta z} \int_{y_0-\Delta y}^{y_0+\Delta y} \int_{x_0-\Delta x}^{x_0+\Delta x} A_x \, dx\, dy\, dz$$

$$= \int_{z_0-\Delta z}^{z_0+\Delta z} \int_{y_0-\Delta y}^{y_0+\Delta y} [A(x_0 + \Delta x, y, z) - A(x_0 - \Delta x, y, z)]\, dy\, dz,$$

$$\iiint_{V(\Sigma)} B_y = \int_{z_0-\Delta z}^{z_0+\Delta z} \int_{x_0-\Delta x}^{x_0+\Delta x} \int_{y_0-\Delta y}^{y_0+\Delta y} B_y \, dy \, dx \, dz$$
$$= \int_{z_0-\Delta z}^{z_0+\Delta z} \int_{x_0-\Delta x}^{x_0+\Delta x} [B(x, y_0 + \Delta y, z) - B(x, y_0 - \Delta y, z)] \, dx \, dz,$$

and

$$\iiint_{V(\Sigma)} C_z = \int_{y_0-\Delta y}^{y_0+\Delta y} \int_{x_0-\Delta x}^{x_0+\Delta x} \int_{z_0-\Delta z}^{z_0+\Delta z} C_z \, dz \, dx \, dy$$
$$= \int_{y_0-\Delta y}^{y_0+\Delta y} \int_{x_0-\Delta x}^{x_0+\Delta x} [C(x, y, z_0 + \Delta z) - C(x, y, z_0 - \Delta z)] \, dx \, dy.$$

Now for the surface integral side of the equation. Break Σ into the six faces $\Sigma_1, \ldots, \Sigma_6$ of the box. Parametrize:

$$\begin{aligned} \Sigma_1(x, z) &= (x, y_0 - \Delta y, z), \\ \Sigma_2(x, z) &= (x, y_0 + \Delta y, z), \end{aligned} \qquad \begin{cases} x_0 - \Delta x \le x \le x_0 + \Delta x, \\ z_0 - \Delta z \le z \le z_0 + \Delta z. \end{cases}$$

$$\begin{aligned} \Sigma_3(x, y) &= (x, y, z_0 - \Delta z), \\ \Sigma_4(x, y) &= (x, y, z_0 + \Delta z), \end{aligned} \qquad \begin{cases} x_0 - \Delta x \le x \le x_0 + \Delta x, \\ y_0 - \Delta y \le y \le y_0 + \Delta y. \end{cases}$$

$$\begin{aligned} \Sigma_5(y, z) &= (x_0 - \Delta x, y, z), \\ \Sigma_6(y, z) &= (x_0 + \Delta x, y, z), \end{aligned} \qquad \begin{cases} y_0 - \Delta y \le y \le y_0 + \Delta y, \\ z_0 - \Delta z \le z \le z_0 + \Delta z. \end{cases}$$

The normal η_i to Σ_i, $i = 1, \ldots, 6$, is given by

$$\begin{aligned} \eta_1 &= (0, -1, 0) = \eta_2, \\ \eta_3 &= (0, 0, 1) = \eta_4, \\ \eta_5 &= (1, 0, 0) = \eta_6. \end{aligned}$$

But these are not in agreement with our previous declaration to use the outer normals, since η_2, η_3, and η_5 are inner normals to Σ_2, Σ_3, and Σ_5, respectively. So, we readjust our parametrizations and rewrite Σ_i as S_i, $i = 1, \ldots, 6$, where

$$S_1 = \Sigma_1, \qquad S_4 = \Sigma_4, \qquad S_6 = \Sigma_6,$$

and

$$S_2(z, x) = (x, y_0 + \Delta y, z), \qquad \begin{cases} x_0 - \Delta x \le x \le x_0 + \Delta x, \\ z_0 - \Delta z \le z \le z_0 + \Delta z. \end{cases}$$

$$S_3(y, x) = (x, y, z_0 - \Delta z), \qquad \begin{cases} x_0 - \Delta x \le x \le x_0 + \Delta x, \\ y_0 - \Delta y \le y \le y_0 + \Delta y. \end{cases}$$

$$S_5(z, y) = (x_0 - \Delta x, y, z), \qquad \begin{cases} y_0 - \Delta y \le y \le y_0 + \Delta y, \\ z_0 - \Delta z \le z \le z_0 - \Delta z. \end{cases}$$

This change in order of the parameters reverses the direction of the normals, so that $S_1, \ldots, S_6$ all have outer normals now. We now compute $\sum_{i=1}^{6} \iint_{S_i} F$.

We have

$$\iint_{S_1} F = \int_{z_0-\Delta z}^{z_0+\Delta z} \int_{x_0-\Delta x}^{x_0+\Delta x} (A(S_1(x, z)), B(S_1(x, z)), C(S_1(x, z))) \cdot (0, -1, 0)\, dx\, dz$$

$$= \int_{z_0-\Delta z}^{z_0+\Delta x} \int_{x_0-\Delta x}^{x_0+\Delta x} -B(x, y_0 - \Delta y, z)\, dx\, dz$$

and

$$\iint_{S_2} F = \int_{z_0-\Delta z}^{z_0+\Delta z} \int_{x_0-\Delta x}^{x_0+\Delta x} (A(S_2(z, x)), B(S_2(z, x)), C(S_2(z, x))) \cdot (0, 1, 0)\, dx\, dz$$

$$= \int_{z_0-\Delta z}^{z_0+\Delta z} \int_{x_0-\Delta x}^{x_0+\Delta x} B(x, y_0 + \Delta y, z)\, dx\, dz.$$

Then

$$\iint_{S_1} F + \iint_{S_2} F = \int_{z_0-\Delta z}^{z_0+\Delta z} \int_{x_0-\Delta x}^{x_0+\Delta x} [B(x, y_0 + \Delta y, z) - B(x, y_0 - \Delta y, z)]\, dx\, dz.$$

Similarly,

$$\iint_{S_3} F + \iint_{S_4} F = \int_{y_0-\Delta y}^{y_0+\Delta y} \int_{x_0-\Delta x}^{x_0+\Delta x} [C(x, y, z_0 + \Delta z) - C(x, y, z_0 - \Delta z)]\, dx\, dy,$$

and

$$\iint_{S_5} F + \iint_{S_6} F = \int_{z_0-\Delta z}^{z_0+\Delta z} \int_{y_0-\Delta y}^{y_0+\Delta y} [A(x_0 + \Delta x, y, z) - A(x_0 - \Delta x, y, z)]\, dy\, dz.$$

This completes the proof for the kind of region under consideration.

Sometimes the divergence theorem is seen in a slightly different form. Using the alternative notation defined in Section 6.8, we often see it as

$$\iiint_{V(\Sigma)} \left(\frac{\partial A}{\partial x} + \frac{\partial B}{\partial y} + \frac{\partial C}{\partial z}\right) dx\, dy\, dz = \iint_{\Sigma} A\, dy\, dz + B\, dz\, dx + C\, dx\, dy.$$

This is purely a matter of notation.

We conclude this section with some computational examples, and devote the next two sections to applications.

EXAMPLE

Let $\Sigma(\theta, \varphi) = (\cos(\theta)\sin(\varphi), \sin(\theta)\sin(\varphi), \cos(\varphi))$, $0 \le \theta \le 2\pi$, $0 \le \varphi \le \pi$. As before, $\eta(\theta, \varphi) = -\sin(\varphi)\,\Sigma(\theta, \varphi)$. Let $F(x, y, z) = (x^2, yz, 2xyz)$. Here $V(\Sigma)$

is the region bounded by the sphere $x^2 + y^2 + z^2 = 1$ and may be written as the set of points (x, y, z) with

$$-\sqrt{1 - x^2 - y^2} \leq z \leq \sqrt{1 - x^2 - y^2},$$

$$-\sqrt{1 - x^2} \leq y \leq \sqrt{1 - x^2},$$

$$-1 \leq x \leq 1.$$

Then

$$\iiint_{V(\Sigma)} \operatorname{div} F = \int_{-1}^{1} \int_{-\sqrt{1-x^2}}^{\sqrt{1-x^2}} \int_{-\sqrt{1-x^2-y^2}}^{\sqrt{1-x^2-y^2}} (2x + z + 2xy)\, dz\, dy\, dx.$$

First,

$$\int_{-\sqrt{1-x^2-y^2}}^{\sqrt{1-x^2-y^2}} (2x + z + 2xy)\, dz$$

$$= \left(2xz + \frac{z^2}{2} + 2xyz\right)\Bigg|_{-\sqrt{1-x^2-y^2}}^{\sqrt{1-x^2-y^2}}$$

$$= 2x(1 - x^2 - y^2)^{1/2} + \frac{1 - x^2 - y^2}{2} + 2xy(1 - x^2 - y^2)^{1/2}$$

$$- 2x(-\sqrt{1 - x^2 - y^2}) - \frac{1 - x^2 - y^2}{2} + 2xy\sqrt{1 - x^2 - y^2}$$

$$= 4x\sqrt{1 - x^2 - y^2} + 4xy\sqrt{1 - x^2 - y^2}.$$

Thus far we have

$$\iiint_{V(\Sigma)} \operatorname{div} F = \int_{-1}^{1} \int_{-\sqrt{1-x^2}}^{\sqrt{1-x^2}} (4x\sqrt{1 - x^2 - y^2} + 4xy\sqrt{1 - x^2 - y^2})\, dx\, dy.$$

This integral is most easily handled by switching to polar coordinates. We then have

$$\int_0^{2\pi} \int_0^1 [4r\cos(\theta)\sqrt{1 - r^2} + 4r^2\cos(\theta)\sin(\theta)\sqrt{1 - r^2}]r\, dr\, d\theta$$

$$= \int_0^{2\pi} \int_0^1 [4r^2\sqrt{1 - r^2}\cos(\theta)\, dr\, d\theta + \int_0^{2\pi} \int_0^1 4r^3\sqrt{1 - r^2}\cos(\theta)\sin(\theta)\, dr\, d\theta.$$

Since $\int_0^{2\pi} \cos(\theta)\, d\theta = \int_0^{2\pi} \cos(\theta)\sin(\theta)\, d\theta = 0$, then $\iiint_{V(\Sigma)} \operatorname{div} F = 0$. Next we compute

$$\begin{aligned}\iint_\Sigma F &= \int_0^\pi \int_0^{2\pi} [(\cos^2(\theta)\sin^2(\varphi), \sin(\theta)\sin(\varphi)\cos(\varphi), 2\cos(\theta)\sin(\theta)\sin^2(\varphi)\cos(\varphi))\\ &\qquad \cdot (-\cos(\theta)\sin^2(\varphi), -\sin(\theta)\sin^2(\varphi), -\sin(\varphi)\cos(\varphi))]\, d\theta\, d\varphi\\ &= \int_0^\pi \int_0^{2\pi} [-\cos^3(\theta)\sin^4(\varphi) - \sin^2(\theta)\sin^3(\varphi)\cos(\varphi)\\ &\qquad - 2\cos(\theta)\sin(\theta)\sin^3(\varphi)\cos^2(\varphi)]\, d\theta\, d\varphi\\ &= 0.\end{aligned}$$

Strictly speaking, we should multiply the last answer by -1, since $\eta(\theta, \varphi)$ as computed above is an inner normal, making $-\eta(\theta, \varphi)$ the outer normal. In this case it makes no difference, as $-0 = 0$. ∎

EXAMPLE

Let Σ consist of the two pieces Σ_1 and Σ_2:

$$\begin{aligned}\Sigma_1(x, y) &= (x, y, x^2 + y^2),\\ \Sigma_2(x, y) &= (x, y, 1),\end{aligned} \qquad x^2 + y^2 \le 1.$$

Then $V(\Sigma)$ is the region shown in Fig. 6-47.

Let $F(x, y, z) = (x, y, z)$. Then

$$\iiint_{V(\Sigma)} \operatorname{div} F = \iiint_{V(\Sigma)} 3 = 3\int_{-1}^1 \int_{-\sqrt{1-x^2}}^{\sqrt{1-x^2}} \int_{x^2+y^2}^1 dz\, dy\, dx.$$

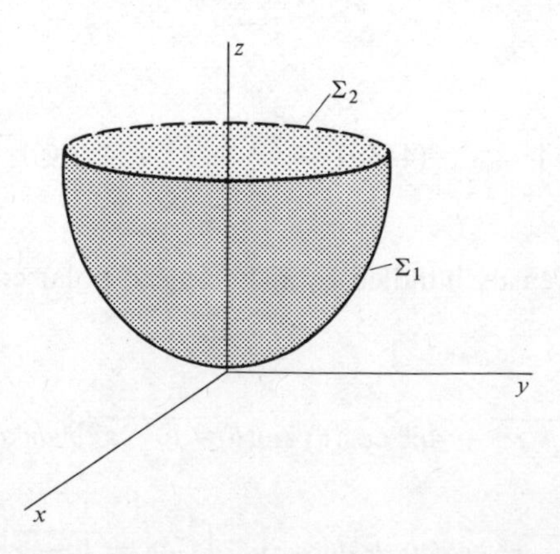

Fig. 6-47

Using polar coordinates, we have

$$3\int_0^{2\pi}\int_0^1 (1 - r^2)r\,dr\,d\theta = 6\pi\int_0^1 (r - r^3)\,dr = \frac{3\pi}{2}.$$

Next we compute $\iint_{\Sigma_1} F + \iint_{\Sigma_2} F$.

On Σ_2, the outer normal is $(0, 0, 1)$, and we have

$$\iint_{\Sigma_2} F = \iint_D (x, y, 1)\cdot(0, 0, 1)\,dx\,dy = \iint_D dx\,dy = A(D) = \pi,$$

since D here is the region bounded by the circle $x^2 + y^2 = 1$.

Next, on Σ_1, the normal is $\eta = (-2x, -2y, 1)$. Again, this is an inner normal. But we will use it anyway, then multiply the final result by -1. (Alternatively, we could use $-\eta$ in place of η.) Anyway, using η, we have

$$\begin{aligned}\iint_{\Sigma_1} F &= \iint_D (x, y, x^2 + y^2)\cdot(-2x, -2y, 1)\,dx\,dy\\ &= \iint_D (-2x^2 - 2y^2 + x^2 + y^2)\,dx\,dy\\ &= \int_0^1\int_0^{2\pi} -r^3\,d\theta\,dr = -2\pi\int_0^1 r^3\,dr = -\frac{\pi}{2}.\end{aligned}$$

Then

$$\iint_\Sigma F = \pi - \left(\frac{-\pi}{2}\right) = \frac{3\pi}{2}. \qquad \blacksquare$$

Exercises for Section 6.14

1. For each F, and each Σ below, show by direct computation that

$$\iiint_{V(\Sigma)} \operatorname{div} F = \iint_\Sigma F.$$

$F(x, y, z)$	Σ
(x, y, z^2)	sphere of radius 2 about $(0, 0, 0)$
(x^2, ye^x, z)	cube with vertices at $(0, 0, 0)$, $(1, 0, 0)$, $(0, 1, 0)$, $(0, 0, 1)$, $(1, 1, 0)$, $(1, 0, 1)$, $(0, 1, 1)$, $(1, 1, 1)$
(x, y, z)	surface consisting of $z^2 = x^2 + y^2$, and $z = 1$ for $x^2 + y^2 \le 1$
(x^3, y^3, z)	the surface of a tetrahedron formed from the planes $x = 0$, $y = 0$, $z = 0$, and $x + y + z = 1$

(There are 16 problems here in all.)

2. Let Σ be a smooth surface bounding a volume $V(\Sigma)$ in R^3. Let f, g, and h be continuous with continuous first, second, and third partials on $V(\Sigma)$. Show that

$$\iint_\Sigma \operatorname{curl}(f, g, h) = 0.$$

3. Prove the divergence theorem for the special case that $V(\Sigma)$ is the region enclosed by $x^2 + y^2 + z^2 = 1$.

4. Verify the divergence theorem for $F(x, y, z) = (xy, yz, xz)$, when $V(\Sigma)$ is bounded by $x = 0$, $y = 0$, $z = 0$, and $x + y + z = 1$.

6.15 Application of the Divergence Theorem to Change of Variables in Triple Integrals

Suppose that we want to evaluate $\iiint_V F(x, y, z)\,dx\,dy\,dz$, and things appear to simplify if we change variables to $x = f(u, v, w)$, $y = g(u, v, w)$, $z = h(u, v, w)$ for (u, v, w) in some corresponding region V^* of uvw-space. Experience with the two-variable case suggests that we might expect

$$\iiint_V F(x, y, z)\,dx\,dy\,dz$$

$$= \iiint_{V^*} F(f(u, v, w), g(u, v, w), h(u, v, w)) \left|\frac{\partial(f, g, h)}{\partial(u, v, w)}\right| du\,dv\,dw.$$

This is exactly the case, with reasonable restrictions on V, F, f, g, and h. The proof of this is quite long, but we shall sketch the idea of the argument so

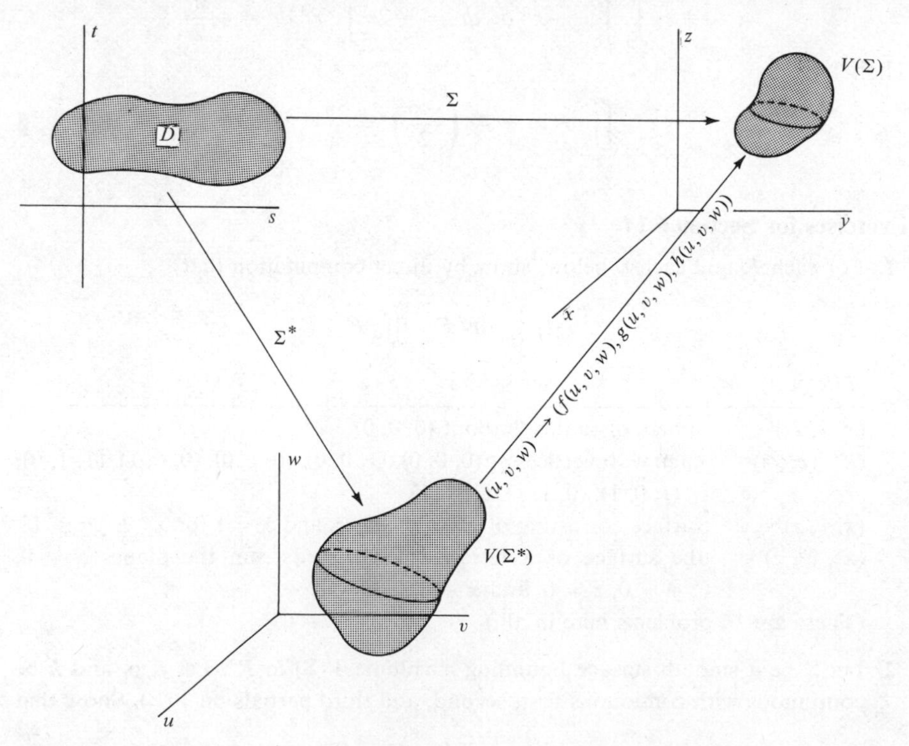

Fig. 6-48

that the role of the divergence theorem will be clear (see Fig. 6-48). Some details of calculation will be left to the reader.

To make the setting precise, *suppose that* Σ *is a closed smooth surface in* R^3 *bounding a volume* $V(\Sigma)$. *Put* $\Sigma(s, t) = (a(s, t), b(s, t), c(s, t))$ *for* (s, t) *in* D, *a compact connected set in the st-plane. Suppose that* $x = f(u, v, w)$, $y = g(u, v, w)$, $z = h(u, v, w)$ *define a mapping from uvw-space to xyz-space, so that to each point in a volume* $V(\Sigma^*)$ *of uvw-space, there corresponds exactly one point in* $V(\Sigma)$, *and conversely. Suppose that* $V(\Sigma^*)$ *is bounded by the closed smooth surface* Σ^*. *For convenience, parametrize* $\Sigma^*(s, t) = (\alpha(s, t), \beta(s, t), \gamma(s, t))$ *for* (s, t) *in* D. Note that

$$a(s, t) = f(\alpha(s, t), \beta(s, t), \gamma(s, t)),$$
$$b(s, t) = g(\alpha(s, t), \beta(s, t), \gamma(s, t)),$$

and

$$c(s, t) = h(\alpha(s, t), \beta(s, t), \gamma(s, t)).$$

Finally, *suppose that* $\partial(f, g, h)/\partial(u, v, w) > 0$ *in* $V(\Sigma^*)$, *and that* f, g, *and* h *have continuous first and second partials in* $V(\Sigma^*)$. We first concentrate our efforts on showing that

$$\text{volume of } V(\Sigma) = \iiint_{V(\Sigma^*)} \frac{\partial(f, g, h)}{\partial(u, v, w)}\, du\, dv\, dw.$$

To do this, let us denote the normals to Σ and Σ^* as $\eta(\Sigma)$ and $\eta(\Sigma^*)$, respectively. Then

$$\eta(\Sigma) = \left(\frac{\partial(y, z)}{\partial(s, t)}, \frac{\partial(z, x)}{\partial(s, t)}, \frac{\partial(x, y)}{\partial(s, t)}\right)$$

and

$$\eta(\Sigma^*) = \left(\frac{\partial(v, w)}{\partial(s, t)}, \frac{\partial(w, u)}{\partial(s, t)}, \frac{\partial(u, v)}{\partial(s, t)}\right).$$

Now, by the divergence theorem,

$$\begin{aligned} \text{volume of } V(\Sigma) &= \iiint_{V(\Sigma)} 1\, dx\, dy\, dz \\ &= \iiint_{V(\Sigma)} \operatorname{div}(0, 0, z)\, dx\, dy\, dz = \iint_{\Sigma} (0, 0, z) \\ &= \iint_{D} (0, 0, z) \cdot \eta(\Sigma)\, ds\, dt = \iint_{D} z \frac{\partial(x, y)}{\partial(s, t)}\, ds\, dt. \end{aligned}$$

A long but straightforward calculation will show that

$$\frac{\partial(x, y)}{\partial(s, t)} = \frac{\partial(x, y)}{\partial(v, w)}\frac{\partial(v, w)}{\partial(s, t)} + \frac{\partial(x, y)}{\partial(w, u)}\frac{\partial(w, u)}{\partial(s, t)} + \frac{\partial(x, y)}{\partial(u, v)}\frac{\partial(u, v)}{\partial(s, t)}.$$

Thus

$$\begin{aligned}
&\iint_D z\,\frac{\partial(x, y)}{\partial(s, t)}\,ds\,dt \\
&\quad= \iint_D h(u, v, w)\left[\frac{\partial(x, y)}{\partial(v, w)}\frac{\partial(v, w)}{\partial(s, t)} + \frac{\partial(x, y)}{\partial(w, u)}\frac{\partial(w, u)}{\partial(s, t)} + \frac{\partial(x, y)}{\partial(u, v)}\frac{\partial(u, v)}{\partial(s, t)}\right] ds\,dt \\
&\quad= \iint_D \left(h\,\frac{\partial(x, y)}{\partial(v, w)}, h\,\frac{\partial(x, y)}{\partial(w, u)}, h\,\frac{\partial(x, y)}{\partial(u, v)}\right)\cdot\left(\frac{\partial(v, w)}{\partial(s, t)}, \frac{\partial(w, u)}{\partial(s, t)}, \frac{\partial(u, v)}{\partial(s, t)}\right) ds\,dt \\
&\quad= \iint_D \left(h\,\frac{\partial(x, y)}{\partial(v, w)}, h\,\frac{\partial(x, y)}{\partial(w, u)}, h\,\frac{\partial(x, y)}{\partial(u, v)}\right)\cdot\eta(\Sigma^*) \\
&\quad= \iint_{\Sigma^*} \left(h\,\frac{\partial(x, y)}{\partial(v, w)}, h\,\frac{\partial(x, y)}{\partial(w, u)}, h\,\frac{\partial(x, y)}{\partial(u, v)}\right).
\end{aligned}$$

In these integrals it is understood that u, v, and w are functions of s and t through Σ^*, since, on Σ^*, $u = \alpha(s, t)$, $v = \beta(s, t)$, and $w = \gamma(s, t)$.

Now use the divergence theorem on the last integral to obtain, finally,

volume of $V(\Sigma)$

$$\begin{aligned}
&= \iiint_{V(\Sigma^*)} \operatorname{div}\left(h\,\frac{\partial(x, y)}{\partial(v, w)}, h\,\frac{\partial(x, y)}{\partial(w, u)}, h\,\frac{\partial(x, y)}{\partial(u, v)}\right) du\,dv\,dw \\
&= \iiint_{V(\Sigma^*)} \left[\frac{\partial}{\partial u}\left(h\,\frac{\partial(x, y)}{\partial(v, w)}\right) + \frac{\partial}{\partial v}\left(h\,\frac{\partial(x, y)}{\partial(w, u)}\right) + \frac{\partial}{\partial w}\left(h\,\frac{\partial(x, y)}{\partial(u, v)}\right)\right] du\,dv\,dw.
\end{aligned}$$

Another long calculation at this point shows that the integrand is exactly the Jacobian $\partial(f, g, h)/\partial(u, v, w)$. If this Jacobian is everywhere negative instead of positive, then the transformation has the effect of reversing orientation, changing an outer normal on Σ to an inner normal on Σ^*. In this case we must multiply the last integral obtained from the divergence theorem by -1. In summary, then, whether $\partial(f, g, h)/\partial(u, v, w)$ is strictly positive or strictly negative, we have

$$\text{volume of } V(\Sigma) = \iiint_{V(\Sigma^*)} \left|\frac{\partial(f, g, h)}{\partial(u, v, w)}\right| du\,dv\,dw.$$

It is now a simple matter to decide on a formula for transforming

$$\iiint_{V(\Sigma)} F(x, y, z)\,dx\,dy\,dz.$$

A typical Riemann sum for this integral is $\sum_{i=1}^{N} F(\xi_i, \zeta_i, \varphi_i)\,\Delta V_i$, where ΔV_i is the volume of a subdivided piece $V_i(\Sigma)$ of $V(\Sigma)$ and $(\xi_i, \zeta_i, \varphi_i)$ is in $V_i(\Sigma)$. Now, corresponding to $V_i(\Sigma)$ under the change of variables is a subdivided piece $V_i(\Sigma^*)$ of $V(\Sigma^*)$. The volume ΔV_i^* of $V_i(\Sigma^*)$ is

$$\iiint_{V_i(\Sigma^*)} \left| \frac{\partial(f, g, h)}{\partial(u, v, w)} \right| du\, dv\, dw,$$

which by a mean value theorem can be written

$$\left| \frac{\partial(f, g, h)}{\partial(u, v, w)} \right|_{(\bar{u}_i, \bar{v}_i, \bar{w}_i)} \Delta V_i^* \qquad \text{for some } (\bar{u}_i, \bar{v}_i, \bar{w}_i) \text{ in } V_i(\Sigma^*).$$

If we put $\xi_i = f(\bar{u}_i, \bar{v}_i, \bar{w}_i)$, $\zeta_i = g(\bar{u}_i, \bar{v}_i, \bar{w}_i)$, and $\varphi_i = h(\bar{u}_i, \bar{v}_i, \bar{w}_i)$, then our Riemann sum becomes

$$\sum_{i=1}^{N} F(f(\bar{u}_i, \bar{v}_i, \bar{w}_i), g(\bar{u}_i, \bar{v}_i, \bar{w}_i), h(\bar{u}_i, \bar{v}_i, \bar{w}_i)) \left| \frac{\partial(f, g, h)}{\partial(u, v, w)} \right|_{(\bar{u}_i, \bar{v}_i, \bar{w}_i)} \Delta V_i^*,$$

and this converges to

$$\iiint_{V(\Sigma^*)} F(f(u, v, w), g(u, v, w), h(u, v, w)) \left| \frac{\partial(f, g, h)}{\partial(u, v, w)} \right| du\, dv\, dw$$

as $N \to \infty$ and each $\Delta V_i^* \to 0$.

As is usually the case, the conditions of the theorem can generally be weakened considerably if one wants to spend a great deal more time on the proof. In practice, for example, vanishing of the Jacobian at finitely many points is tolerable.

EXAMPLE

Evaluate

$$\iiint_{V(\Sigma)} \sqrt{x^2 + y^2 + z^2}\, dx\, dy\, dz,$$

where $V(\Sigma)$ is the region bounded by the unit sphere about the origin.

Here we switch to spherical coordinates. Put

$$x = \rho \cos(\theta) \sin(\varphi), \qquad y = \rho \sin(\theta) \sin(\varphi), \qquad z = \rho \cos(\varphi),$$
$$0 \le \rho \le 1,\ 0 \le \theta \le 2\pi,\ 0 \le \varphi \le \pi.$$

Now

$$\frac{\partial(x, y, z)}{\partial(\rho, \theta, \varphi)} = \begin{vmatrix} \cos(\theta)\sin(\varphi) & -\rho\sin(\theta)\sin(\varphi) & \rho\cos(\theta)\cos(\varphi) \\ \sin(\theta)\sin(\varphi) & \rho\cos(\theta)\sin(\varphi) & \rho\sin(\theta)\cos(\varphi) \\ \cos(\varphi) & 0 & -\rho\sin(\varphi) \end{vmatrix}$$

$$= -\rho^2\sin(\varphi).$$

Further, $x^2 + y^2 + z^2 = \rho^2$, so our integral becomes

$$\iiint_{V(\Sigma)} \sqrt{x^2 + y^2 + z^2}\, dx\, dy\, dz = \int_0^{\pi}\int_0^{2\pi}\int_0^1 \sqrt{\rho^2}|-\rho^2\sin(\varphi)|\, d\rho\, d\theta\, d\varphi$$

$$= \int_0^{\pi}\int_0^{2\pi}\int_0^1 \rho^3\sin(\varphi)\, d\rho\, d\theta\, d\varphi$$

$$= \frac{1}{4}(2\pi)\int_0^{\pi}\sin(\varphi)\, d\varphi$$

$$= \frac{\pi}{2}(-\cos(\varphi))\Big|_0^{\pi} = \pi. \qquad \blacksquare$$

Exercises for Section 6.15

1. Use suitable changes of variables in the following:

(a) $\iiint_V \frac{1}{\sqrt{x^2 + y^2 + z^2}}$, V the region $x^2 + y^2 + z^2 \leq 1$ *Hint:* This is an improper integral. Let V_ε be a sphere of radius $\varepsilon < 1$ about the origin, and use spherical coordinates to evaluate $\iiint_{W_\varepsilon} \frac{1}{\sqrt{x^2 + y^2 + z^2}}$, where W_ε is the region inside V and exterior to V_ε. Then take a limit as $\varepsilon \to 0$.

(b) Find the volume of a sphere of radius R about (x_0, y_0, z_0). *Hint:* Put $x - x_0 = \rho\cos(\theta)\sin(\varphi)$, $y - y_0 = \rho\sin(\theta)\sin(\varphi)$, $z - z_0 = \rho\cos(\varphi)$, $0 \leq \theta \leq 2\pi$, $0 \leq \varphi \leq \pi$, $0 \leq \rho \leq R$.

(c) Find the volume of a right circular cylinder of height h and circular radius R. *Hint:* Use cylindrical coordinates.

(d) Find the mass, center of mass, and moments of inertia about the axes of

$$V = \{(x, y, z)\} \mid z^2 \geq x^2 + y^2, 0 \leq z \leq 1\} \quad \text{if } \rho(x, y, z) = x + y.$$

(e) Find the mass, center of mass, and moments of inertia about the x, y, and z axes of

$$V = \{(x, y, z) \mid (x - x_0)^2 + (y - y_0)^2 + (z - z_0)^2 \leq R^2\}$$

if $\rho(x, y, z) = 3z$.

(f) Find the mass, center of mass, and moments of inertia about the axes of $V = \{(x, y, z) \mid z \geq x^2 + y^2, 0 \leq z \leq 1\}$, if $\rho(x, y, z) = 1$.

2. Show that, if we switch to spherical coordinates, then

$$\iiint_V f(x, y, z)\, dx\, dy\, dz$$
$$= \iiint_{V^*} f(\rho \cos(\theta) \sin(\varphi), \rho \sin(\theta) \sin(\varphi), \rho \cos(\varphi))\rho^2|\sin(\varphi)|\, d\rho\, d\theta\, d\varphi,$$

where V^* denotes the set of points in V described in terms of (ρ, θ, φ) coordinates.

3. Show that, if we switch to cylindrical coordinates, then

$$\iiint_V f(x, y, z)\, dx\, dy\, dz = \iiint_{V^*} f(r \cos(\theta), r \sin(\theta), z) r\, dr\, d\theta\, dz,$$

where V^* denotes the region V described in (r, θ, z) coordinates.

4. Evaluate $\iiint_V z$, where V is the region inside $x^2 + y^2 + z^2 = 1$, $z \geq 0$ and outside the cone $z^2 = x^2 + y^2$.

5. Compute $\iiint_V \cos[(\pi/3)(x + y + z)]\, dx\, dy\, dz$ if V is the region given by $x^2 + y^2 + z^2 \leq 1$. *Hint:* Rotate xyz-space to $\bar{x}\bar{y}\bar{z}$-space so that the plane $x + y + z = 0$ becomes $\bar{z} = 0$, and then use cylindrical coordinates. The answer is

$$\frac{4(3)^{3/2}}{\pi^2}\left[\sin\left(\frac{\pi}{\sqrt{3}}\right) - \frac{\pi}{\sqrt{3}}\cos\left(\frac{\pi}{\sqrt{3}}\right)\right].$$

6. The volume of a right circular cylinder of height h and radius r is $\pi r^2 h$. Derive this result by integrating $\iiint_V dx\, dy\, dz$ over the region V bounded by such a cylinder and using a change of variables to cylindrical coordinates.

7. The volume of a sphere of radius r is $\frac{4}{3}\pi r^3$. Derive this formula by integrating $\iiint_V dx\, dy\, dz$ over the region V bounded by a sphere of radius r centered about the origin, and making a change of variables to spherical coordinates.

8. Find the mass, center of mass, and moments of inertia about the x-, y-, and z-axes, of a right circular cylinder $x^2 + y^2 \leq r^2$, $0 \leq z \leq h$, if the density function is given by $\rho(x, y, z) = z(x^2 + y^2)$.

9. Evaluate $\iiint_V \cos(x^2 + y^2 + z)\, dx\, dy\, dz$, with V the region given by $2 \leq x^2 + y^2 \leq 4$, $0 \leq z \leq 4$.

6.16 Additional Applications of the Divergence Theorem

A. Divergence of a Vector

Before (in Section 6.9B) we argued that

$$(\operatorname{div} F)|_{(x_0, y_0, z_0)} = \lim_{V \to 0} \frac{1}{|V|} \iint_\Sigma F,$$

where Σ is a closed surface bounding the region V of volume $|V|$, and (x_0, y_0, z_0) is in V. We now give a shorter justification of this based on the divergence theorem. Write

$$\iiint_{V(\Sigma)} \operatorname{div} F = \iint_{\Sigma} F.$$

Use a mean value theorem on the triple integral to write

$$\iiint_{V(\Sigma)} \operatorname{div} F = (\operatorname{div} F)\Big|_{(\bar{x},\bar{y},\bar{z})} [\text{volume of } V(\Sigma)]$$

for some $(\bar{x}, \bar{y}, \bar{z})$ in $V(\Sigma)$. Then

$$(\operatorname{div} F)|_{(x_0,y_0,z_0)} = \frac{1}{\text{volume of } V(\Sigma)} \iint_{\Sigma} F.$$

Now let $V(\Sigma)$ collapse to (x_0, y_0, z_0). Then also $(\bar{x}, \bar{y}, \bar{z}) \to (x_0, y_0, z_0)$ and we obtain

$$(\operatorname{div} F)|_{(x_0,y_0,z_0)} = \lim_{V\to 0} \frac{1}{|V|} \iint_{\Sigma} F.$$

B. Applications in Fluid Mechanics

Suppose that we have a fluid which has velocity $v(x, y, z, t)$ and density $\rho(x, y, z, t)$ at points (x, y, z) (measured with respect to some fixed frame) and time t. We have already noted that $\iint_{\Sigma} \rho v =$ total mass of fluid flowing out of a volume $V(\Sigma)$ bounded by Σ. Now note that $\iiint_{V(\Sigma)} \rho\, dx\, dy\, dz$ gives us the mass of fluid in $V(\Sigma)$. But we must have [in the absence of a source or sink in $V(\Sigma)$], mass of fluid flowing out of $V(\Sigma)$ at time $t =$ decrease of mass of fluid in $V(\Sigma)$. Thus

$$\frac{-\partial}{\partial t} \iiint_{V(\Sigma)} \rho = \iint_{\Sigma} \rho v.$$

Assume that $\dfrac{\partial}{\partial t}$ and $\iiint_{(V\Sigma)}$ interchange; we then have

$$\iiint_{V(\Sigma)} \frac{\partial}{\partial t} \rho + \iint_{\Sigma} \rho v = 0.$$

By the divergence theorem,

$$\iint_{\Sigma} \rho v = \iiint_{V(\Sigma)} \operatorname{div}(\rho v).$$

Hence

$$\iiint_{V(\Sigma)} \left[\frac{\partial \rho}{\partial t} + \operatorname{div}(\rho v)\right] = 0.$$

This equation must hold for *any* volume $V(\Sigma)$ within the fluid. The only way this can happen is if the integrand is zero. This gives us

$$\frac{\partial \rho}{\partial t} + \operatorname{div}(\rho v) = 0,$$

which is called the *equation of continuity*.

C. Green's Identities

We have seen Green's identities in the plane derived from Green's Theorem. There are three-dimensional analogues which can be readily derived from Gauss's theorem.

Let f and g have continuous first and second partials throughout a region D of 3-space. Suppose that the closed surface Σ and its interior $V(\Sigma)$ are within D. Denote

$$\nabla^2 f = \frac{\partial^2 f}{\partial x^2} + \frac{\partial^2 f}{\partial y^2} + \frac{\partial^2 f}{\partial z^2}.$$

(This is the three-dimensional Laplacian, also often denoted Δf.)

Green's first identity in R^3 then says that

$$\iint_\Sigma f\,\nabla g = \iiint_{V(\Sigma)} (f\,\nabla^2 g + \nabla f \cdot \nabla g).$$

To prove this, note by a simple calculation that

$$\operatorname{div}(f\,\nabla g) = \nabla f \cdot \nabla g + f\,\nabla^2 g.$$

By Gauss's Theorem, $\iint_\Sigma f\,\nabla g = \iiint_{V(\Sigma)} \operatorname{div}(f\,\nabla g)$, hence

$$\iint_\Sigma f\,\nabla g = \iiint_{V(\Sigma)} (f\,\nabla^2 g + \nabla f \cdot \nabla g).$$

Green's second identity is immediate from the first, by interchanging f and g and subtracting:

$$\iint_\Sigma (f\,\nabla g - g\,\nabla f) = \iiint_{V(\Sigma)} (f\,\nabla^2 g - g\,\nabla^2 f).$$

We call f harmonic in $V(\Sigma)$ if $\nabla^2 f = 0$ in $V(\Sigma)$. For such f, we have immediately

$$\iint_\Sigma f \nabla f = \iiint_{V(\Sigma)} \nabla f \cdot \nabla f = \iiint_{V(\Sigma)} \|\nabla f\|^2$$

and, for harmonic f and g,

$$\iint_\Sigma (f \nabla g - g \nabla f) = 0.$$

D. Is the Dirichlet Problem Well Posed?

A Dirichlet problem for a region $V(\Sigma)$ bounded by Σ consists of the differential equation $\nabla^2 f = 0$, together with data giving values of f on Σ. The question is whether or not this is sufficient to imply uniqueness of solution (assuming existence).

To see that it is, suppose that $\nabla^2 f = \nabla^2 g = 0$ in $V(\Sigma)$ and $f(x, y, z) = g(x, y, z)$ on Σ. Let $h = f - g$; then $\nabla^2 h = 0$ in $V(\Sigma)$ and $h(x, y, z) = 0$ on Σ. From Section 6.16C we have

$$\iint_\Sigma h \nabla h = \iiint_{V(\Sigma)} \|\nabla h\|^2.$$

But $\iint_\Sigma h \nabla h = 0$, as $h = 0$ on Σ, so $\iiint_{V(\Sigma)} \|\nabla h\|^2 = 0$. Then $\|\nabla h\| = 0$, in $V(\Sigma)$, hence $h_x = h_y = h_z = 0$ in $V(\Sigma)$. Then h is constant in $V(\Sigma)$. But $h = 0$ on Σ, so $h = 0$ throughout $V(\Sigma)$. Then $f = g$ in $V(\Sigma)$.

E. Uniqueness of Solution to a Boundary-Value Problem

Consider the problem

$$\frac{\partial u}{\partial t} = k \nabla^2 u + \varphi(x, y, z, t) \qquad \text{for } t > 0, (x, y, z) \text{ in } V(\Sigma),$$

$$u(x, y, z, 0) = f(x, y, z) \qquad \text{for } (x, y, z) \text{ in } V(\Sigma),$$

$$u(x, y, z, t) = g(x, y, z) \qquad \text{for } (x, y, z) \text{ on } \Sigma, \qquad t > 0.$$

This is a heat problem, with f giving initial temperature inside the solid $V(\Sigma)$ and g the temperature at time t on the surface Σ. Physically, this should be enough information to specify the temperature $u(x, y, z, t)$ at any point on the surface and in the interior, and at any time. It is therefore important that the differential equation and initial and boundary values, which purport to describe the heat flow, specify u uniquely. To show that this is the case, suppose that

u_1 and u_2 are solutions, and put $w = u_1 - u_2$. Then $\dfrac{\partial w}{\partial t} = k\,\nabla^2 w$ for (x, y, z) in $V(\Sigma)$ and $t > 0$,

$$w(x, y, z, 0) = 0 \text{ in } V(\Sigma)$$

and

$$w(x, y, z, t) = 0 \text{ on } \Sigma.$$

Now put

$$I(t) = \tfrac{1}{2} \iiint_{V(\Sigma)} w^2.$$

Then

$$I'(t) = \iiint_{V(\Sigma)} w \frac{\partial w}{\partial t} = k \iiint_{V(\Sigma)} w\,\nabla^2 w$$

$$= k \iint_{\Sigma} w\,\nabla w - k \iiint_{V(\Sigma)} \|\nabla w\|^2.$$

Since $w = 0$ on Σ, then $\iint_\Sigma w\,\nabla w = 0$. Hence $I'(t) = -k \iiint_{V(\Sigma)} \|\nabla w\|^2 \leq 0$, for $t > 0$.

Now, for any $t > 0$, there is, by the mean value theorem, some $\bar{t}$ such that $0 < \bar{t} < t$ and

$$I(t) - I(0) = tI'(\bar{t}) \leq 0.$$

But $I(0) = \frac{1}{2}\iiint_{V(\Sigma)} w(x, y, z, 0)^2 = 0$. Then $I(t) \leq 0$ for $t > 0$. But clearly $I(t) \geq 0$, hence $I(t) = 0$. Then $w^2 = 0$, hence $w = 0$ throughout $V(\Sigma)$, and $u_1 = u_2$, giving uniqueness of solution.

F. Gauss's Law

Suppose that we have a point electric charge q at the origin of our coordinate system. The electric field produced by q is given by the vector

$$E = \frac{q}{4\pi\varepsilon_0} \frac{(x, y, z)}{(x^2 + y^2 + z^2)^{3/2}},$$

where ε_0 is a constant depending on the medium.

If Σ is a smooth closed surface, then Gauss's law says that

$$E = \begin{cases} \dfrac{q}{\varepsilon_0} & \text{if } (0, 0, 0) \text{ is within the interior of } \Sigma, \\ 0 & \text{if } (0, 0, 0) \text{ is exterior to } \Sigma. \end{cases}$$

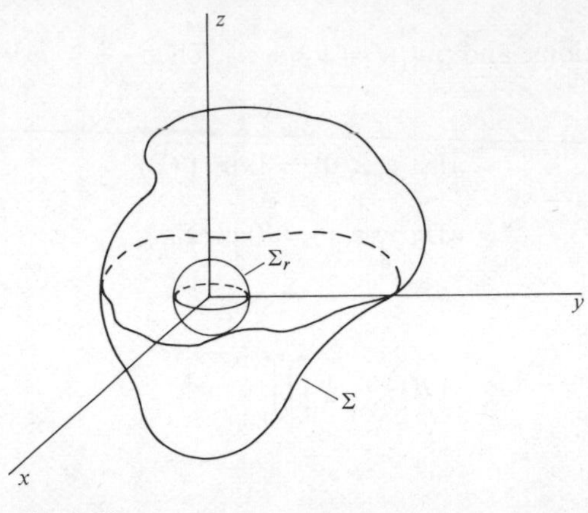

Fig. 6-49

To prove this, we use Gauss's theorem. Suppose first that Σ does not enclose the origin. Then a simple calculation shows that div $E = 0$ in $V(\Sigma)$, so by Gauss's theorem,

$$\iint_{\Sigma} E = \iiint_{V(\Sigma)} \operatorname{div} E = 0.$$

If Σ encloses the origin, however, then div E is not 0 in $V(\Sigma)$, as E is not defined at the origin.

Enclose (0, 0, 0) by a sphere Σ_r of sufficiently small radius r that Σ_r lies entirely inside Σ (Fig. 6-49). Note that div $E = 0$ in the region $V(\Sigma - \Sigma_r)$ between Σ and Σ_r. By cutting a channel between Σ and Σ_r and arguing as in Section 6.6 (using Gauss's instead of Green's theorem), we have

$$\iint_{\Sigma} E = \iint_{\Sigma_r} E.$$

To evaluate $\iint_{\Sigma_r} E$, parametrize

$$\Sigma_r(\theta, \varphi) = (r\cos(\theta)\sin(\varphi), r\sin(\theta)\sin(\varphi), r\cos(\varphi)), \qquad 0 \le \theta \le 2\pi, 0 \le \varphi \le \pi.$$

Then

$$\begin{aligned} E(\Sigma(\theta, \varphi)) &= \frac{q}{4\pi\varepsilon_0} \frac{r(\cos(\theta)\sin(\varphi), \sin(\theta)\sin(\varphi), \cos(\varphi))}{(r^2)^{3/2}} \\ &= \frac{q}{4\pi\varepsilon_0 r^2}(\cos(\theta)\sin(\varphi), \sin(\theta)\sin(\varphi), \cos(\varphi)). \end{aligned}$$

Next

$$\eta(\theta, \varphi) = (-r^2\cos(\theta)\sin^2(\varphi), -r^2\sin(\theta)\sin^2(\varphi), -r^2\sin(\varphi)\cos(\varphi)).$$

This is an inner normal to Σ_r, and Gauss's theorem calls for an outer normal. Hence we shall use $-\eta$ in place of η to obtain

$$\iint_{\Sigma} E = \iint_{\Sigma_r} E$$

$$= \frac{q}{4\pi\varepsilon_0}\int_0^{\pi}\int_0^{2\pi} [\cos^2(\theta)\sin^3(\varphi) + \sin^2(\theta)\sin^3(\varphi) + \sin(\varphi)\cos^2(\varphi)]\, d\theta\, d\varphi$$

$$= \frac{q}{4\pi\varepsilon_0}\int_0^{\pi}\int_0^{2\pi} [\sin^3(\varphi) + \sin(\varphi)\cos^2(\varphi)]\, d\theta\, d\varphi$$

$$= \frac{2\pi q}{4\pi\varepsilon_0}\int_0^{\pi} [\sin^3(\varphi) + \sin(\varphi)\cos^2(\varphi)]\, d\varphi$$

$$= \frac{q}{\varepsilon_0},$$

as we wanted to show.

G. Three-Dimensional Heat Equation

In this section we shall derive the heat equation in three dimensions. Suppose that we have a medium with density $\rho(x, y, z)$, specific heat $C(x, y, z)$, and coefficient of thermal conductivity $k(x, y, z)$. Let $I(x, y, z, t)$ denote the intensity of the heat source at (x, y, z) at time t, and $u(x, y, z, t)$ the temperature at (x, y, z) at time t.

Now imagine a closed surface Σ within the medium, enclosing a region $V(\Sigma)$. Fourier's law says that the amount of heat entering $V(\Sigma)$ across Σ in a time interval Δt is equal to $(\iint_{\Sigma} k\, \nabla u)\, \Delta t$. Next, observe that $(\iiint_{V(\Sigma)} I(x, y, z, t))\, \Delta t$ is the amount of heat originating from sources in $V(\Sigma)$ in the interval Δt. Further, the increase in temperature in $V(\Sigma)$ in the interval Δt is $u(x, y, z, t + \Delta t) - u(x, y, z, t)$, which is approximately $\frac{\partial u}{\partial t}\Delta t$. This necessitates a heat expenditure of

$$\left(\iiint_{V(\Sigma)} C\rho\, \frac{\partial u}{\partial t}\right) \Delta t.$$

Now

amount of heat expended

= amount entering $V(\Sigma)$ across Σ + amount produced by sources within $V(\Sigma)$.

Then

$$\left(\iiint_{V(\Sigma)} C\rho \frac{\partial u}{\partial t}\right) \Delta t = \left(\iint_{\Sigma} k \nabla u\right) \Delta t + \left(\iiint_{V(\Sigma)} I\right) \Delta t.$$

By Gauss's theorem,

$$\iint_{\Sigma} k \nabla u = \iiint_{V(\Sigma)} \operatorname{div}(k \nabla u).$$

Then

$$\iiint_{V(\Sigma)} \left[\operatorname{div}(k \nabla u) + I - C\rho \frac{\partial u}{\partial t}\right] = 0.$$

Since this must hold for every solid region $V(\Sigma)$ within the medium, then we must have

$$\operatorname{div}(k \nabla u) + I - C\rho \frac{\partial u}{\partial t} = 0.$$

This is the three-dimensional heat equation. Noting that

$$\begin{aligned}\operatorname{div}(k \nabla u) &= \frac{\partial}{\partial x}\left(k \frac{\partial u}{\partial x}\right) + \frac{\partial}{\partial y}\left(k \frac{\partial u}{\partial y}\right) + \frac{\partial}{\partial z}\left(k \frac{\partial u}{\partial z}\right) \\ &= k \nabla^2 u + \nabla u \cdot \nabla k,\end{aligned}$$

we can write the equation as

$$k \nabla^2 u + \nabla u \cdot \nabla k + I = C\rho \frac{\partial u}{\partial t}.$$

We briefly mention some special cases of interest:

(1) C, ρ, and k are constant and $I = 0$. Then

$$\frac{k}{C\rho} \nabla^2 u = \frac{\partial u}{\partial t}.$$

If u is independent of y and z, this becomes

$$\frac{\partial u}{\partial t} = \frac{k}{C\rho} \frac{\partial^2 u}{\partial x^2},$$

which we derived in Section 5.2.

(2) $\partial u/\partial t = 0$. This is called the *steady-state case*. Then

$$k \nabla^2 u + \nabla k \cdot \nabla u = -I.$$

If also k is a constant, this is

$$\nabla^2 u = -\frac{1}{k} I,$$

which is called *Poisson's equation*. If, in addition, $I = 0$, we have Laplace's equation,

$$\nabla^2 u = 0.$$

7

Functions Defined by Integrals

7.1 Introduction

We have already encountered several instances in which the function under consideration was defined by an integral. For example, in deriving the continuity equation of fluid mechanics, we encountered the problem of calculating $\partial/\partial t \iiint_V \rho(x, y, z, t)\, dx\, dy\, dz$. Here x, y, and z are integration variables, and $\iiint_V \rho\, dx\, dy\, dz$ depends only on t. It seems worthwhile, then, to devote a short chapter to such matters. A comprehensive treatment would necessarily be very long and cumbersome, but in a few pages we can give the reader an idea of what to expect and how to cope with such situations as he meets them.

7.2 Leibniz's Rule for Differentiation

We begin with a simple case. Suppose that

$$F(x) = \int_a^b f(x, t)\, dt,$$

where f is continuous on a rectangle $D: c \le x \le d, a \le t \le b$. We shall show that

$$F'(x) = \int_a^b \frac{\partial f}{\partial x}(x, t)\, dt \qquad \text{for } c \le x \le d$$

if $\partial f/\partial x$ is also continuous on D.

To begin, look at a difference quotient

$$\frac{F(x + \Delta x) - F(x)}{\Delta x} = \int_a^b \frac{\{f(x + \Delta x, t) - f(x, t)\}}{\Delta x} \, dt.$$

By a mean value theorem, we can write

$$f(x + \Delta x, t) - f(x, t) = \frac{\partial f(x + \theta \, \Delta x, t)}{\partial x} \Delta x \text{ for some } \theta \text{ between 0 and 1.}$$

Then

$$\left| \frac{F(x + \Delta x) - F(x)}{\Delta x} - \int_a^b \frac{\partial f(x, t)}{\partial x} \, dt \right|$$

$$= \left| \int_a^b \frac{f(x + \Delta x, t) - f(x, t)}{\Delta x} \, dt - \int_a^b \frac{\partial f(x, t)}{\partial x} \, dt \right|$$

$$= \left| \int_a^b \left[\frac{\partial f(x + \theta \, \Delta x, t)}{\partial x} - \frac{\partial f(x, t)}{\partial x} \right] dt \right| \leq \int_a^b \left| \frac{\partial f(x + \theta \, \Delta x, t)}{\partial x} - \frac{\partial f(x, t)}{\partial x} \right| dt.$$

As $\Delta x \to 0$,

$$\frac{\partial f(x + \theta \, \Delta x, t)}{\partial x} - \frac{\partial f(x, t)}{\partial x} \to 0,$$

by continuity of $\partial f / \partial x$ on D. Then

$$\lim_{\Delta x \to 0} \left| \frac{F(x + \Delta x) - F(x)}{x} - \int_a^b \frac{\partial f(x, t)}{\partial x} \, dt \right| = 0,$$

implying that

$$F'(x) = \int_a^b \frac{\partial f(x, t)}{\partial x} \, dt.$$

Leibniz's rule, stated above, does not cover all possible situations in which we want to differentiate a function defined by an integral. For example, suppose that

$$h(x) = \int_{a(x)}^{b(x)} f(t) \, dt,$$

where a and b are differentiable on $[\alpha, \beta]$ and f is continuous on $[a(x), b(x)]$ for $\alpha \leq x \leq \beta$.

Put

$$H(u, v) = \int_u^v f(t) \, dt.$$

Then $H(a(x), b(x)) = h(x)$. By the chain rule,

$$h'(x) = H_u u_x + H_v v_x = H_u a'(x) + H_v b'(x).$$

There remains to calculate H_u and H_v.

Choose a convenient number c and put

$$H(u, v) = \int_u^c f(t)\,dt + \int_c^v f(t)\,dt.$$

By the fundamental theorem of calculus,

$$\frac{d}{dv}\int_c^v f(t)\,dt = f(v)$$

and

$$\frac{d}{du}\int_u^c f(t)\,dt = \frac{d}{du}\left[-\int_c^u f(t)\,dt\right] = -f(u).$$

Then $H_v = f(v)$ and $H_u = -f(u)$. So

$$h'(x) = -a'(x)f(u) + b'(x)f(v) = f(b(x))b'(x) - f(a(x))a'(x).$$

Obviously, similar theorems can be developed for a number of situations like these. For example, going back again to the continuity equation of fluids, we can be sure that

$$\frac{\partial}{\partial t}\iiint_{V(\Sigma)} \rho(x, y, z, t)\,dx\,dy\,dz = \iiint_{V(\Sigma)} \frac{\partial \rho}{\partial t}$$

if ρ and $\partial\rho/\partial t$ are continuous at all (x, y, z, t) with (x, y, z) in $V(\Sigma)$ and $t > 0$. In most fluid problems, this condition will be satisfied, as density is generally continuous and varies continuously with time. Even in cases where ρ or $\partial\rho/\partial t$ is not continuous, so that the theorem fails, one may still be able to interchange $\partial/\partial t$ and $\iiint_{V(\Sigma)}$, as the theorem gives sufficient, not necessary, conditions.

Exercises for Section 7.2

1. Show that, if

$$F(x) = \int_{\alpha(x)}^{\beta(x)} f(x, t)\,dt,$$

then under suitable conditions on f, α, and β:

$$F'(x) = \int_{\alpha(x)}^{\beta(x)} \frac{\partial f}{\partial x}\,dt + \beta'(x)f(x, \beta(x)) - \alpha'(x)f(x, \alpha(x)).$$

What did you have to assume about f, α, and β to derive the conclusion?

2. Obtain derivatives of the following:

(a) $F(x) = \int_0^1 e^{-xt^2}\,dt,\ x > 0$

(b) $F(x) = \int_{x^2}^{x^4} \ln(xt)\,dt,\ x > 0$

(c) $F(x) = \cosh(x)\left[\int_x^{x+1} e^{xt}\cos(x^2 + t^2)\,dt\right],\ x$ real

7.3 Pointwise and Uniform Convergence of Improper Integrals

A number of important functions are of the form $F(x) = \int_a^\infty f(x, t)\,dt$. Examples, which we shall see shortly, are the gamma and beta functions. If we think of $\int_a^\infty$ as analogous to $\sum_{n=1}^{\infty}$, then the means of treating such functions becomes fairly obvious. The key is to develop concepts analogous to pointwise and uniform convergence of series.

To do this, let us make the setting more precise. Suppose that f is continuous for (x, t) in the "infinite rectangle" $c \leq x \leq d, t \geq a$. If, for $c \leq x \leq d$, $\int_a^\infty f(x, t)\,dt$ converges to some number $F(x)$, then we say that $\int_a^\infty f(x, t)\,dt$ *converges pointwise* to $F(x)$ on $[c, d]$. This means that $\lim_{r\to\infty} \int_a^r f(x, t)\,dt = F(x)$ for each x in $[c, d]$. Or, put another way, given $\varepsilon > 0$ and x in $[c, d]$, there is some δ such that $\left|\int_a^r f(x, t)\,dt - F(x)\right| < \varepsilon$ whenever $r \geq \delta$. Here, of course, δ probably depends upon both x and ε. If the same δ can be used for every x in $[c, d]$, we say that the convergence is uniform. Thus: $\int_a^\infty f(x, t)\,dt$ *converges to* $F(x)$ *uniformly on* $[c, d]$ *if, given* $\varepsilon > 0$, *there is some* δ *such that* $\left|\int_a^r f(x, t)\,dt - F(x)\right| < \varepsilon$ *whenever* $r \geq \delta$ *and* $c \leq x \leq d$.

The following analogue of the Weierstrass M-test is often useful in establishing uniform convergence of an integral.

Theorem

Let g *be continuous on* (a, ∞), *and suppose that* $|f(x, t)| \leq g(t)$ *for* $t \geq a$, $c \leq x \leq d$. *If* $\int_a^\infty g(t)\,dt$ *converges, then* $\int_a^\infty f(x, t)\,dt$ *converges uniformly on* $[c, d]$.

The proof is similar to that of the M-test, and is left to the reader.

We now state for integrals the analogues of the theorems on continuity, integrals, and derivatives of uniformly convergent series.

Theorem

Suppose that f *is continuous for* $c \leq x \leq d$, $t \geq a$, *and that* $\int_a^\infty f(x, t)\,dt$ *converges uniformly to* $F(x)$ *on* $[c, d]$. *Then*

(1) *F is continuous on* $[c, d]$.

(2) $\displaystyle\int_c^d F(x)\,dx = \int_a^\infty \left[\int_c^d f(x, t)\,dx\right] dt.$

(3) *If $\partial f/\partial x$ is continuous for $t \geq a$ and $c \leq x \leq d$, and*

$$\int_a^\infty \frac{\partial f(x, t)}{\partial x}\, dt$$

converges uniformly on $[c, d]$, then

$$F'(x) = \int_a^\infty \frac{\partial f(x, t)}{\partial x}\, dt \text{ on } [c, d].$$

Again, the proofs are a reworking of the corresponding proofs for series and are left to the reader. We also leave it to the reader to formulate the corresponding notion of uniform convergence for improper integrals of type II, and to prove a theorem like the last for such integrals.

We shall give some examples to show how the operations with $\int_a^\infty f(x, t)\, dt$ permitted by the last theorem can be useful.

EXAMPLE

Evaluate $\int_0^\infty e^{-t^2} \cos(xt)\, dt$. A standard trick with such problems is to put $F(x) = \int_0^\infty e^{-t^2} \cos(xt)\, dt$ and use calculus operations to find out what we can about F.

First, note that, for any x, $|e^{-t^2} \cos(xt)| \leq e^{-t^2}$, and $\int_0^\infty e^{-t^2}\, dt$ converges (to $\sqrt{\pi}/2$; see Section 6.5D). Thus $\int_0^\infty e^{-t^2} \cos(xt)\, dt$ converges uniformly on every interval $[c, d]$. Now $\dfrac{\partial(e^{-t^2} \cos(xt))}{\partial x} = -te^{-t^2} \sin(xt)$. Since $|-te^{-t^2} \sin(x, t)| \leq te^{-t^2}$ and $\int_0^\infty te^{-t^2}$ converges, then also

$$\int_0^\infty \frac{\partial(e^{-t^2} \cos(xt))}{\partial x}\, dt$$

converges uniformly on any interval $[c, d]$. We now have $F'(x) = \int_0^\infty -te^{-t^2} \sin(xt)\, dt$. Integrate by parts: for any $r > 0$,

$$\int_0^r -te^{-t^2} \sin(xt)\, dt = \frac{\sin(xt)e^{-t^2}}{2}\bigg|_0^r - \int_0^r \frac{e^{-t^2} x \cos(xt)}{2}\, dt.$$

Letting $r \to \infty$, we have

$$F'(x) = \int_0^\infty -te^{-t^2} \sin(xt)\, dt = -\tfrac{1}{2}x \int_0^\infty e^{-t^2} \cos(xt)\, dt = -\tfrac{1}{2}xF(x).$$

If $F(x) \neq 0$, then $F'(x)/F(x) = -x/2$, so that $\ln(F(x)) = -x^2/4 + C$, where C

is a constant of integration. Then $F(x) = Ke^{-x^2/4}$, K a constant to be determined. But $F(0) = K = \int_0^\infty e^{-t^2}\,dt = \sqrt{\pi}/2$. Then

$$F(x) = \int_0^\infty e^{-t^2}\cos(xt)\,dt = \frac{\sqrt{\pi}}{2}e^{-x^2/4}. \qquad \blacksquare$$

EXAMPLE

Show that

$$\int_0^\infty e^{-px}\frac{\cos(\alpha x) - \cos(\beta x)}{x}\,dx = \frac{1}{2}\ln\left(\frac{p^2 + \beta^2}{p^2 + \alpha^2}\right)$$

for $p > 0$ (α and β are positive constants).

Note first that the integrand is bounded at 0, since

$$\lim_{x\to 0}\frac{\cos(\alpha x) - \cos(\beta x)}{x} = 0.$$

One way to see this is to use L'Hospital's Rule; another is to expand the cosine terms in power series and look directly at the quotient involved in the limit.

Now observe that

$$\frac{\cos(\alpha x) - \cos(\beta x)}{x} = \int_\alpha^\beta \sin(xt)\,dt,$$

hence

$$\int_\alpha^\beta e^{-px}\sin(xt)\,dt = \frac{e^{-px}[\cos(\alpha x) - \cos(\beta x)]}{x}.$$

Further $|e^{-px}\sin(xt)| \le e^{-px}$ and $\int_0^\infty e^{-px}\,dx$ converges for $p > 0$. Thus $\int_0^\infty e^{-px}\sin(xt)\,dx$ converges uniformly on any interval, and we can interchange $\int_0^\infty dx$ and $\int_\alpha^\beta dt$, enabling us to write

$$\int_0^\infty e^{-px}\frac{\cos(\alpha x) - \cos(\beta x)}{x}\,dx = \int_0^\infty \left[\int_\alpha^\beta e^{-px}\sin(xt)\,dt\right]dx$$

$$= \int_\alpha^\beta \left[\int_0^\infty e^{-px}\sin(xt)\,dx\right]dt.$$

The advantage in doing this is that we can evaluate $\int_0^\infty e^{-px}\sin(xt)\,dx$ directly. Since

$$\int e^{-px}\sin(xt)\,dx = \frac{e^{-px}[t\sin(xt) - t\cos(xt)]}{t^2 + p^2},$$

then

$$\int_0^\infty e^{-px}\sin(xt)\,dx = \lim_{r\to\infty}\int_0^r e^{-px}\sin(xt)\,dx$$

$$= \lim_{r\to\infty}\left[\frac{e^{-pr}[t\sin(rt) - t\cos(rt)]}{t^2+p^2} + \frac{t}{t^2+p^2}\right] = \frac{t}{t^2+p^2}$$

since $e^{-pr} \to 0$ as $r \to \infty$ for $p > 0$. Then

$$\int_0^\infty \frac{e^{-px}[\cos(\alpha x) - \cos(\beta x)]}{x}\,dx = \int_\alpha^\beta \frac{t}{t^2+p^2}$$

$$= \tfrac{1}{2}\ln(p^2+t^2)|_\alpha^\beta = \tfrac{1}{2}\ln(p^2+\beta^2) - \tfrac{1}{2}\ln(p^2+\alpha^2)$$

$$= \tfrac{1}{2}\ln\left(\frac{p^2+\beta^2}{p^2+\alpha^2}\right). \qquad \blacksquare$$

The reader should note the similarity between these tricks to evaluate improper integrals and the integration–differentiation methods used in Chapter 1 to sum power series. We conclude this section with one more example, which will be useful in treating the convergence of Fourier series in Chapter 10.

EXAMPLE

Evaluate $\int_0^\infty [\sin(x)/x]\,dx$. Note that $\sin(x)/x$ is well defined for $x > 0$, and $\lim_{x\to 0}[\sin(x)/x] = 1$, so we assign to $\sin(x)/x$ the value 1 at 0. The integral is improper only of Type I. The trick now is to start with the apparently unrelated integral

$$f(t) = \int_0^\infty \frac{e^{-xt}\sin(x)}{x}\,dx, \qquad t > 0.$$

The connection with $\int_0^\infty [\sin(x)/x]\,dx$ is seen by letting $t \to 0$ from the right under the integral sign. To see that we can do this, note that $|\sin(x)/x| \le 1$ for $0 \le x < \infty$, so

$$\left|\frac{e^{-xt}\sin(x)}{x}\right| \le e^{-xt}$$

and $\int_0^\infty e^{-xt}\,dx$ converges on any interval $0 < t_0 \le t < \infty$. Thus

$$\int_0^\infty \frac{e^{-xt}\sin(x)}{x}\,dx$$

converges uniformly on $t > 0$, and

$$\lim_{t \to 0+} f(t) = \int_0^\infty \left[\lim_{t \to 0+} \left(\frac{e^{-xt} \sin(x)}{x} \right) \right] dx = \int_0^\infty \frac{\sin(x)}{x}\, dx.$$

Thus far, of course, these observations are of use to us only if we can do something helpful with

$$\int_0^\infty \frac{e^{-xt} \sin(x)}{x}\, dx.$$

If we differentiate under the integral sign with respect to t, we have

$$\int_0^\infty e^{-xt} \sin(x)\, dx.$$

In any interval $0 < t_0 \le t$, $|e^{-xt} \sin(x)| \le e^{-xt_0}$, and $\int_0^\infty e^{-xt_0}\, dx$ converges. Thus $\int_0^\infty e^{-xt} \sin(x)\, dx$ converges uniformly in $[t_0, \infty)$ for any $t_0 > 0$, so

$$f'(t) = \int_0^\infty e^{-xt} \sin(x)\, dx \qquad \text{for } t > 0.$$

But this is something we can integrate explicitly:

$$\begin{aligned}
\int_0^\infty e^{-xt} \sin(x)\, dx &= \lim_{b \to \infty} \int_0^b e^{-xt} \sin(x)\, dx \\
&= \lim_{b \to \infty} \left[\frac{-e^{-xt}(-t \sin(x) - \cos(x))}{1 + t^2} \right]_0^b \\
&= \lim_{b \to \infty} - \left[\frac{e^{-bt}[-t \sin(b) - \cos(b)]}{1 + t^2} + \frac{1}{1 + t^2} \right] = \frac{-1}{1 + t^2}.
\end{aligned}$$

Thus

$$f'(t) = \frac{-1}{1 + t^2} \qquad \text{for } t > 0.$$

Then, for some constant of integration K,

$$f(t) = -\arctan(t) + K.$$

To find K, let $t \to \infty$. Note that

$$0 \le |f(t)| \le \int_0^\infty \left| \frac{e^{-xt} \sin(x)}{x} \right| dx \le \int_0^\infty e^{-xt}\, dx = \frac{1}{t}$$

for any $t > 0$.

As $t \to \infty$, $1/t \to 0$, so $f(t) \to 0$ also. Thus

$$\lim_{t\to\infty} f(t) = 0 = \lim_{t\to\infty} (-\arctan(t) + K)$$

$$= -\frac{\pi}{2} + K,$$

giving us $K = \pi/2$. Then

$$f(t) = \int_0^\infty \frac{e^{-xt}\sin(x)}{x}\,dx = -\arctan(t) + \frac{\pi}{2}.$$

Taking a limit as $t \to 0+$, we have finally

$$\int_0^\infty \frac{\sin(x)}{x}\,dx = \lim_{t\to 0+}\left[-\arctan(t) + \frac{\pi}{2}\right] = \frac{\pi}{2}. \quad \blacksquare$$

Exercises for Section 7.3

1. Show that the following converge uniformly on the given interval:

(a) $\int_0^\infty \frac{\cos(xt)}{1+t^2}\,dt$, $-\infty < x < \infty$

(b) $\int_0^\infty e^{-xt}\,dt$, $0 < \varepsilon \le x < \infty$

(c) $\int_0^\infty e^{-xt}\cos(xt)\,dt$, $0 < \varepsilon \le x < \infty$

(d) $\int_1^\infty \frac{t\tanh(x)}{x^4+t^3}\,dt$, $-\infty < x < \infty$

2. Evaluate

$$\int_0^\infty \frac{e^{-at} - e^{-bt}}{t}\,dt, \qquad \text{where } 0 < a < b.$$

Hint: Let $f(x) = \int_0^\infty e^{-xt}\,dt$, for $x > 0$. Show that $f(x) = 1/x$, and consider $\int_a^b f(x)\,dx$.

3. Show that $\int_0^\infty e^{-t^2 - t^2/x^2}\,dt = (\sqrt{\pi}/2)e^{-2|x|}$ for $x \neq 0$. *Hint:* First suppose that $x > 0$ and derive the result, then worry about the case that $x < 0$.

4. Evaluate

$$\int_0^\infty \frac{\arctan(bt) - \arctan(at)}{t}\,dt \qquad (a, b > 0).$$

Hint: Show that

$$\int_a^b \int_0^\infty \frac{1}{1 + x^2t^2}\, dt\, dx = \int_0^\infty \int_a^b \frac{1}{1 + x^2t^2}\, dx\, dt.$$

5. Show that $\int_0^\infty te^{-t} \sin(xt)\, dt = 2x/(1 + x^2)^2$. *Hint:* Show first that

$$\frac{1}{1 + x^2} = \int_0^\infty e^{-t} \cos(xt)\, dt.$$

6. Develop a notion of uniform convergence for integrals of the form $\int_c^d f(x, y)\, dy$, $a \leq x \leq b$, where $f(x, y)$ is defined on the "rectangle" $a \leq x \leq b, c < y \leq d$. Prove theorems on interchange of $\int_a^b$ and d/dx with $\int_c^d$, and on continuity of the function $g(x) = \int_c^d f(x, y)\, dy$ on $[a, b]$.

7. Evaluate $\int_0^1 x^y \ln(x)\, dx$, $y > -1$: *Hint:* Start with $\int_0^1 x^y\, dx = 1/(y + 1)$.

7.4 Gamma and Beta Functions

Two important functions defined by improper integrals are the *gamma* and *beta functions*, given by

$$\Gamma(x) = \int_0^\infty e^{-t}t^{x-1}\, dt \qquad \text{for } x > 0$$

and

$$\mathrm{B}(x, y) = \int_0^1 t^{x-1}(1 - t)^{y-1}\, dt \qquad \text{for } x > 0 \text{ and } y > 0.$$

We treat $\Gamma(x)$ first. The definition involves a mixed type I–type II improper integral, as by definition $e^{-t}t^{x-1} = e^{-t}e^{(x-1)\ln(t)}$, and $\ln(t)$ is no good at the lower limit of integration, 0. We shall prove convergence of $\int_1^\infty e^{-t}t^{x-1}\, dt$ and of $\int_0^1 e^{-t}t^{x-1}\, dt$ for $x > 0$. We can then paste these together to form $\Gamma(x)$.

Suppose that $0 < a \leq x \leq b$. Then $|e^{-t}t^{x-1}| \leq e^{-t}t^{b-1}$ for $t \geq 1$, so $\int_1^\infty e^{-t}t^{x-1}\, dt$ converges uniformly on $[a, b]$. If $0 < t \leq 1$, then $|e^{-t}t^{x-1}| \leq t^{a-1}$, and $\int_0^1 t^{a-1}\, dt$ converges, so $\int_0^1 e^{-t}t^{x-1}\, dt$ converges uniformly on $[a, b]$. We then say that $\int_0^\infty e^{-t}t^{x-1}\, dt$ converges uniformly on $[a, b]$.

In such a case of a uniformly convergent improper integral of mixed type, one can prove theorems on continuity, integration, and differentiation just like those of the last section. Thus, for example, Γ is continuous on any closed interval $[a, b]$ with $a > 0$.

The gamma function is often called the *factorial function*, owing to the fact that $\Gamma(n + 1) = n!$ for any positive integer n. This will follow from the more general result:

$$\Gamma(x + 1) = x\Gamma(x) \qquad \text{for } x > 0.$$

A proof of this consists of a simple integration by parts:

$$\Gamma(x + 1) = \int_0^\infty t^x e^{-t}\, dt = -t^x e^{-t}\Big|_0^\infty - \int_0^\infty -e^{-t} t^{x-1} x\, dt$$

$$= x\int_0^\infty e^{-t} t^{x-1}\, dt = x\Gamma(x).$$

In particular, $\Gamma(n + 1) = n\Gamma(n)$ for n any positive integer, from which $\Gamma(n + 1) = n!$ follows by an easy induction argument.

Gamma functions are useful in a variety of different connections, of which we shall mention just two. By various changes of variables, it is often the case that an integral encountered in a problem may turn out to be a gamma function. In this event the integral in question can be approximated very closely numerically from the rather extensive tabulations that have been made of $\Gamma(x)$. As examples of two different forms of the gamma function, the reader can show that

$$\Gamma(x) = \int_0^1 [\ln(1/u)]^{x-1}\, du \qquad [\text{put } t = -\ln(u)]$$

and

$$\Gamma(x) = \int_{-\infty}^\infty e^{xu} e^{-e^u}\, du \qquad (\text{put } t = e^u).$$

Gamma functions also arise frequently in the solution of important differential equations. Bessel's equation with index ν is

$$x^2 y'' + xy' + (x^2 - \nu^2)y = 0,$$

where ν is assumed to be real. A solution is given by the ν-order Bessel function

$$J_\nu(x) = \sum_{k=0}^\infty \frac{(-1)^k}{k!\,\Gamma(k + \nu + 1)} \left(\frac{x}{2}\right)^{2k+\nu},$$

provided that ν is not a negative integer. Such functions have been studied and tabulated extensively, as Bessel's equation arises in the study of such diverse phenomena as vibration in membranes, heat conduction, alternating current in a wire, critical length of a rod in elasticity theory, and planetary motion.

The alert reader will note that we have not explicitly determined the value of $\Gamma(x)$ for any noninteger x yet. Depending upon x, this is a very difficult problem. However, there is a clever trick by which we can calculate $\Gamma(\frac{1}{2})$. By definition,

$$\Gamma(\tfrac{1}{2}) = \int_0^\infty t^{-1/2}e^{-t}\,dt.$$

Change variables by putting $t = u^2$. Then

$$\Gamma(\tfrac{1}{2}) = \int_0^\infty u^{-1}e^{-u^2}2u\,du = 2\int_0^\infty e^{-u^2}\,du.$$

Now, we have already computed $\int_0^\infty e^{-u^2}\,du$ in Section 6.5D, and found the value $\sqrt{\pi}/2$. Hence $\Gamma(\frac{1}{2}) = \sqrt{\pi}$. From $\Gamma(x+1) = x\Gamma(x)$ it follows that

$$\Gamma(n + \tfrac{1}{2}) = \frac{(2n)!\,\sqrt{\pi}}{4^n n!} \qquad \text{for } n = 1, 2, \ldots$$

We now take a brief look at the beta function, given by

$$B(x, y) = \int_0^1 t^{x-1}(1-t)^{y-1}\,dt \qquad x > 0,\ y > 0.$$

It is not immediately obvious that this integral converges. However, we shall show that

$$B(x, y) = \frac{\Gamma(x)\Gamma(y)}{\Gamma(x+y)}.$$

Write

$$\Gamma(x) = \int_0^\infty t^{x-1}e^{-t}\,dt = 2\int_0^\infty u^{2x-1}e^{-u^2}\,du$$

by putting $t = u^2$. Similarly, $\Gamma(y) = 2\int_0^\infty u^{2y-1}e^{-u^2}\,du$. Then

$$\begin{aligned}\Gamma(x)\Gamma(y) &= 4\int_0^\infty u^{2x-1}e^{-u^2}\,du\int_0^\infty u^{2y-1}e^{-u^2}\,du \\ &= 4\int_0^\infty u^{2x-1}e^{-u^2}\,du\int_0^\infty v^{2y-1}e^{-v^2}\,dv.\end{aligned}$$

Now use the same trick as that used previously in evaluating $\int_0^\infty e^{-x^2}\,dx$. Think of

$$\begin{aligned}&\int_0^\infty u^{2x-1}e^{-u^2}\,du\int_0^\infty v^{2y-1}e^{-v^2}\,dv \\ &\qquad = \int_0^\infty\int_0^\infty u^{2x-1}v^{2y-1}e^{-u^2-v^2}\,du\,dv = \lim_{R\to\infty}\iint_{D_R} u^{2x-1}v^{2y-1}e^{-u^2-v^2}\,du\,dv,\end{aligned}$$

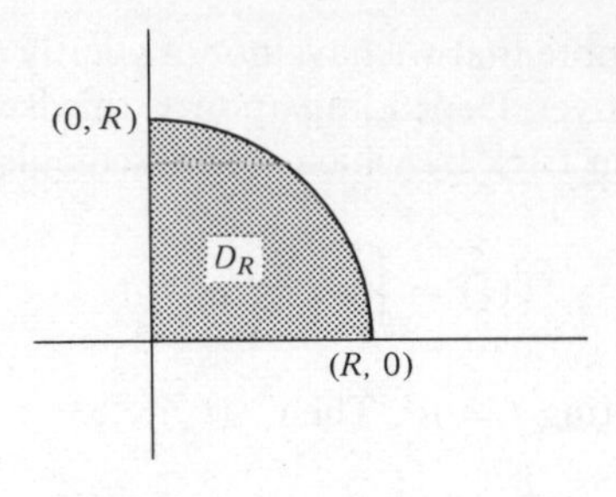

Fig. 7-1

and use polar coordinates, with D_R the region $x^2 + y^2 \leq R^2$, $x \geq 0$ and $y \geq 0$, as shown in Fig. 7-1. We obtain

$$\iint_{D_R} u^{2x-1}v^{2y-1}e^{-u^2-v^2}\,du\,dv$$

$$= \int_0^{\pi/2}\int_0^R r^{2x-1}[\cos(\theta)]^{2x-1}r^{2y-1}[\sin(\theta)]^{2y-1}e^{-r^2}r\,dr\,d\theta$$

$$= \int_0^R r^{2x+2y-1}e^{-r^2}\,dr \int_0^{\pi/2} \cos^{2x-1}(\theta)\sin^{2y-1}(\theta)\,d\theta.$$

We work these out individually. In the integral involving r, put $z = r^2$. Then

$$\int_0^R r^{2x+2y-1}e^{-r^2}\,dr = \int_0^{R^2} \frac{z^{x+y-1/2}e^{-z}}{2z^{1/2}}\,dz$$

$$= \frac{1}{2}\int_0^{R^2} z^{x+y-1}e^{-z}\,dz.$$

In the other integral, put $w = \sin^2(\theta)$. Then

$$\int_0^{\pi/2} \cos^{2x-1}(\theta)\sin^{2y-1}(\theta)\,d\theta = \int_0^{\pi/2} [1 - \sin^2(\theta)]^{x-1/2}\sin^{2y-1}(\theta)\,d\theta$$

$$= \int_0^1 (1 - w)^{x-1/2}w^{y-1/2}(\tfrac{1}{2})w^{-1/2}(1 - w)^{-1/2}\,dw$$

$$= \tfrac{1}{2}\int_0^1 w^{y-1}(1 - w)^{x-1}\,dw = \tfrac{1}{2}\,\mathrm{B}(x, y).$$

Then

$$\iint_{D_R} u^{2x-1}v^{2y-1}e^{-u^2-v^2}\,du\,dv = \tfrac{1}{4}\int_0^{R^2} z^{x+y-1}e^{-z}\,dz\,\mathrm{B}(x, y).$$

Finally, take a limit as $R \to \infty$. The left side goes to $\Gamma(x)\Gamma(y)$, and we have

$$\Gamma(x)\Gamma(y) = \int_0^\infty z^{x+y-1}e^{-z}\,dz\,\mathrm{B}(x, y)$$
$$= \Gamma(x + y)\mathrm{B}(x, y),$$

which is what we wanted to show.

The beta-function is often handy in problems involving nasty-looking improper integrals. For example, put $u = \sin^2(\theta)$ in $\int_0^{\pi/2} \sqrt{\sin(\theta)}\,d\theta$ to obtain

$$\int_0^{\pi/2} \sqrt{\sin(\theta)}\,d\theta = \int_0^1 \frac{u^{1/4}}{2u^{1/2}(1 - u)^{1/2}}\,du$$
$$= \tfrac{1}{2}\int_0^1 u^{-1/4}(1 - u)^{1/2}\,du$$
$$= \tfrac{1}{2}\mathrm{B}(\tfrac{3}{4}, \tfrac{1}{2}) = \frac{\Gamma(\frac{3}{4})\Gamma(\frac{1}{2})}{2\Gamma(\frac{5}{4})} = \frac{2\Gamma(\frac{3}{4})\Gamma(\frac{1}{2})}{\Gamma(\frac{1}{4})}.$$

Exercises for Section 7.4

1. Prove the following:

(a) $\Gamma(x)\Gamma(1 - x) = \pi \csc(\pi x)$ for $0 < x < 1$

(b) $\Gamma(n/2)\Gamma(2n) = 2^{2n-1}\Gamma(n)\Gamma(n + \frac{1}{2})$ for $n = 1, 2, \ldots$

(c) $\Gamma(x) = \alpha^x \int_0^\infty e^{-\alpha t}t^{x-1}\,dt$ for $\alpha > 0$ and $x > 0$

(d) $\Gamma(x) = 2\int_0^\infty e^{-t^2}t^{2x-1}\,dt$ for $x > 0$

(e) $\Gamma(x) = \dfrac{1}{s^x}\int_0^1 t^{s-1}[\ln(1/t)]^{x-1}\,dt$ for $s > 0$, $x > 0$

(f) $B(x, y) = \displaystyle\int_0^\infty \frac{t^{x-1}}{(1 + t)^{x+y}}\,dt$ *Hint:* Put $t = u/(1 + u)$.

(g) $B(x, y) = \displaystyle\int_0^{\pi/2} 2\sin^{2x-1}(\theta)\cos^{2y-1}(\theta)\,d\theta$

(h) $\displaystyle\int_0^{\pi/2} \sin^{2n}(x)\,dx = \frac{\sqrt{\pi}\,\Gamma(n + \frac{1}{2})}{2(n!)}$, $n = 1, 2, \ldots$

(i) $\displaystyle\int_0^{\pi/2} \sin^{2n+1}(x)\,dx = \frac{\sqrt{\pi}\,n!}{2\Gamma(n + \frac{3}{2})}$, $n = 1, 2, \ldots$

2. Evaluate the following in terms of gamma functions:

(a) $\displaystyle\int_0^\infty [t^{x-1}/(1 + t)]\,dt$, $0 < x < 1$

(b) $\displaystyle\int_0^\infty \sqrt{x}e^{-x^3}\,dx$

(c) $\int_0^\infty x^\alpha e^{-\beta x}\, dx$, α, β, γ positive constants

(d) $\int_0^\infty te^{-t}\, dt$

(e) $\int_0^{\pi/4} \sqrt{\tan(x)}\, dx$

(f) $\int_0^1 t^{x-1}[\ln(t)](1-t)^{y-1}\, dt$

3. Show that

$$J_\nu(x) = \sum_{k=0}^{\infty} \frac{(-1)^k (x/2)^{2k+\nu}}{k!\,\Gamma(k+\nu+1)}$$

is a solution to $x^2y'' + xy' + (x^2 + \nu^2)y = 0$.

4. Let a, b, c, α, β, γ, P, Q, and R be positive constants, and V the region in R^3 given by $x \geq 0$, $y \geq 0$, $z \geq 0$, and inside the ellipsoid $(x/a)^\alpha + (y/b)^\beta + (z/c)^\gamma = 1$. Prove that

$$\iiint_V x^{P-1}y^{Q-1}z^{R-1} = \frac{a^P b^Q c^R}{\alpha\beta\gamma} \frac{\Gamma(P/\alpha)\Gamma(Q/\beta)\Gamma(R/\gamma)}{\Gamma(1 + P/\alpha + Q/\beta + R/\gamma)}.$$

5. Derive Wallis's product formula for π:

$$\frac{\pi}{2} = \frac{2}{1}\frac{2}{3}\frac{4}{3}\frac{4}{5}\frac{6}{5}\frac{6}{7}\cdots\frac{2n}{2n-1}\frac{2n}{2n+1}\cdots.$$

(Here, of course, we must make clear what we mean by an infinite product. If P_n is the product of the first n terms of the product indicated above, we mean that $\lim_{n\to\infty} P_n = \pi/2$.) *Hint:* Carry out the following steps:

Step 1

$$\left(\frac{1}{P_{2n}}\right)\left(\frac{\pi}{2}\right) = \int_0^{\pi/2} \sin^{2n}(x)\, dx \Big/ \int_0^{\pi/2} \sin^{2n+1}(x)\, dx \qquad \text{for } n = 1, 2, 3, \ldots.$$

To prove this, use Exercise 1(h) and (i), to directly compute the quotient on the right.

Step 2

$$\lim_{n\to\infty} \frac{\int_0^{\pi/2} \sin^{2n}(x)\, dx}{\int_0^{\pi/2} \sin^{2n+1}(x)\, dx} = 1.$$

To prove this, begin with $0 < \sin^{2n+1}(x) < \sin^{2n}(x) < \sin^{2n-1}(x)$ for $0 < x < \pi/2$. Conclude from this that

$$0 < \frac{\int_0^{\pi/2} \sin^{2n}(x)\, dx}{\int_0^{\pi/2} \sin^{2n+1}(x)\, dx} < \frac{\int_0^{\pi/2} \sin^{2n-1}(x)\, dx}{\int_0^{\pi/2} \sin^{2n+1}(x)\, dx}.$$

Use Exercise 1(h) and (i) to evaluate the quotient on the right and the fact that $\Gamma(1 + x) = x\Gamma(x)$; obtain

$$1 < \frac{\int_0^{\pi/2} \sin^{2n}(x)\, dx}{\int_0^{\pi/2} \sin^{2n+1}(x)\, dx} \leq \frac{2n+1}{2n} \to 1 \qquad \text{as } n \to \infty.$$

Step 3. $P_{2n} \to \pi/2$ as $n \to \infty$. This should be immediate from previous steps.

Step 4

$$P_{2n+1} = \frac{(2n + 2)P_{2n}}{2n + 1}.$$

Step 5. $P_{2n+1} \to \pi/2$ as $n \to \infty$.

Step 6. $\lim_{n\to\infty} P_n = \pi/2$.

6. Derive Stirling's formula:

$$\lim_{n\to\infty} \frac{(n/e)^n \sqrt{2\pi n}}{n!} = 1.$$

(The value of such a limit is that, for large n, the numerator of the quotient in the limit is easier to compute than the denominator, and the fact that the limit is 1 means that the numerator becomes a better approximation to $n!$ as n is taken larger.) *Hint:* Put $S_n = n!/(n/e)^n \sqrt{n}$.

Step 1. Show that it suffices to prove that $\lim_{n\to\infty} S_n = \sqrt{2\pi}$.

Step 2. $\{S_n\}$ converges. To prove this, note first that $\{S_n\}$ is bounded below. It now suffices to show that $\{S_n\}$ is monotone decreasing. To do this, show first that

$$\frac{S_n}{S_{n+1}} = \frac{(1 + 1/n)^{n+1/2}}{e}.$$

Hence it suffices to show that $\ln(1 + 1/n) > 2/(2n + 1)$. Here write $\ln(1 + 1/n) = \int_1^{1+1/n} \frac{1}{t}\,dt$, which is the area under the graph of $y = 1/t$ from $t = 1$ to $t = 1 + 1/n$. Compare this area with the area of the trapezoid shown in Fig. 7-2.

Step 3. Each $S_n \geq 1$. To prove this, compare the area under the curve $y = \ln(x)$, $1 \leq x \leq n$, with the sum of the areas of the rectangles and trapezoids shown in Fig. 7-3.

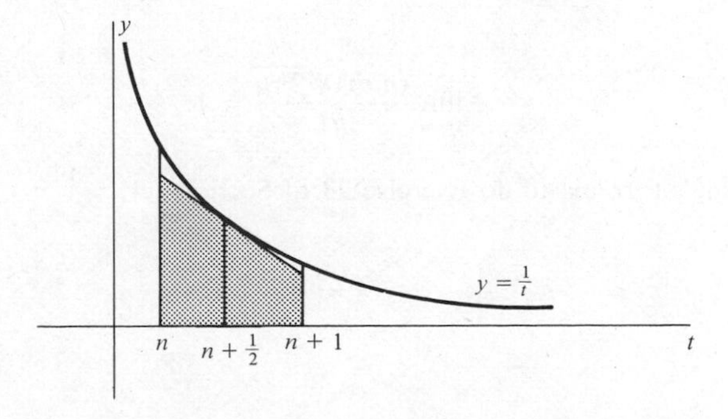

Fig. 7-2

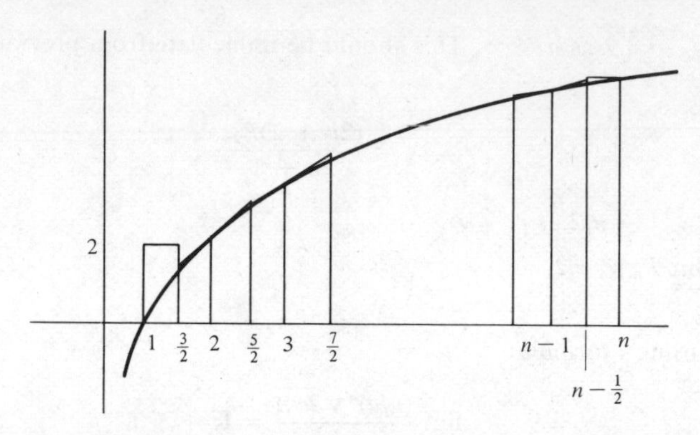

Fig. 7-3

Step 4. $\lim\limits_{n\to\infty} S_n \neq 0.$

Step 5.

$$\lim_{n\to\infty} \frac{(n!)^2 2^{2n}}{(2n)!\,\sqrt{n}} = \sqrt{\pi}.$$

Multiply numerator and denominator of the product for P_{2n} in Exercise 5 by $2\cdot 2\cdot 4\cdot 4\cdot 6\cdot 6\cdots(2n)(2n)$ to obtain

$$P_{2n} = \frac{(n!)^4 2^{4n}}{(2n)!\,(2n+1)!}.$$

Show from this, and from Exercise 5, that

$$\frac{(n!)^2 2^{2n}}{(2n)!\,\sqrt{2n+1}} \to \sqrt{\pi}.$$

From this, show the limit stated as Step 5.

Step *6.* $\lim\limits_{n\to\infty} S_n = \sqrt{2\pi}$. Use Steps 5 and 4, and the fact that

$$\frac{(n!)^2 2^{2n}}{(2n)!\,\sqrt{n}} = \frac{S_n^2}{S_{2n}\sqrt{2}}.$$

Step 7.

$$\lim_{n\to\infty} \frac{(n/e)^n\sqrt{2\pi n}}{n!} = 1.$$

7. Use Stirling's formula to do Exercise 23 of Section 1.1.

8

Calculus of Variations

8.1 Some Historical Remarks

The calculus of variations, which deals with certain kinds of extremal problems, has had a particularly fascinating history, especially if you believe everything you read. According to legend, the Phoenician princess Dido was awarded by an African chief as much land as she could enclose in an oxhide. In order to make the most of her opportunity, she cut the oxhide into thin strips and enclosed a circular area in which she founded Carthage. In modern terminology, the problem before Dido was to determine a closed curve in the plane enclosing a maximum area, subject to the constraint that the length was fixed.

From very early times, it was apparent to philosophers and scientists that nature often tends to act in such a way as to produce a maximum or minimum. Heron of Alexandria, who was aware that the angle of incidence equals the angle of reflection for a ray of light, observed that this law of nature actually guarantees that the beam will takc the shortest path from source to reflection point by way of the mirror. This is similar to the optical principle of Fermat, which says that whenever light travels between two points in different media, the path traveled is the one ensuring the least time of passage.

A problem that gave major impetus to the development of a calculus of variations was the brachistochrone problem of James and John Bernoulli in the seventeenth and eighteenth centuries. What path should we draw between the points P_0 and P_1 so that a ball rolling from P_0 without friction on the path will reach P_1 in minimum time? One might be tempted to think of a straight line first, but Galileo had already observed that some other paths give shorter times.

Using an argument from optics, the Bernoullis obtained the solution, the resulting curve being called a *brachistochrone* (brachos = shortest, chronos = time).

Today max–min problems and principles abound in both physics and mathematics. The third law of thermodynamics states that the universe tends toward a state of maximum entropy. Hamilton's principle in mechanics says that a mechanical system will move in such a way as to minimize a certain integral. As a final introductory illustration, a tantalizing problem for many years was Plateau's soap-bubble problem: Given a closed curve C in 3-space, find a surface of minimum area spanning C. Long before a mathematical solution was achieved, solutions could be obtained experimentally by passing through a soap mixture a thin wire bent into the shape of C. The resulting soap bubble assumes the shape of the surface of minimum area.

8.2 Extremizing $\int_a^b f(x, y, y')\,dx$

The first problem we shall consider in the calculus of variations is that of choosing $y(x)$ so as the minimize or maximize

$$\int_a^b f(x, y, y')\,dx,$$

given f, a, and b. First, why look at problems of this form? As an illustration, we shall phrase the brachistochrone problem in these terms. As a matter of convenience put P_0 at the origin, with the positive y-axis directed downward. Let $y = \varphi(x)$ denote the path of least time which we seek.

Now, at any time t, the particle is moving with a velocity $v(t) = ds/dt$, where s = distance traveled along $y = \varphi(x)$ from P_0. According to the conservation-of-energy principle, change in kinetic energy = change in potential energy. This, if the particle is at (x, y) at time t, then

$$mgy = \tfrac{1}{2}mv(t)^2 - \tfrac{1}{2}mv(0)^2.$$

Here $v(0)$ is the initial velocity, which for convenience we take to be 0. Then

$$v(t) = \frac{ds}{dt} = \sqrt{2gy}.$$

Then

$$dt = \frac{ds}{\sqrt{2gy}} = \frac{ds}{[2g\varphi(x)]^{1/2}}.$$

Now ds = differential of arc length $= \sqrt{1 + \varphi'(x)^2}\,dx$. So

$$dt = \frac{\sqrt{1 + \varphi'(x)^2}}{\sqrt{2g\varphi}}\,dx.$$

Total time in traveling from P_0 to P_1 is then

$$\int_0^{x_1} dt = \int_0^{x_1} \frac{\sqrt{1 + \varphi'(x)^2}}{\sqrt{2g\varphi(x)}}\,dx.$$

The problem, then, is to determine the path $y = \varphi(x)$ so as to minimize

$$\int_{x_0}^{x_1} f(x, y, y')\,dx \qquad \text{where } f(x, y, y') = \sqrt{\frac{1 + (y')^2}{2gy}}.$$

Now let us consider the problem in general with some attention to detail. In many max–min problems, it is clear from the setting of the problem that a max or min exists, and the main source of difficulty lies in finding it. Suppose then that y_m is the solution to our problem. That is, y_m extremizes $\int_a^b f(x, y, y')\,dx$, and the curve $y = y_m(x)$ passes through the prescribed end points (a, α), (b, β), so that $y_m(a) = \alpha$ and $y_m(b) = \beta$. We proceed to derive a differential equation for y_m as follows. Consider "test" or "comparison" functions

$$y(x) = y_m(x) + \varepsilon\eta(x),$$

where ε is real. By choosing η so that $\eta(a) = \eta(b) = 0$, we ensure that each such function passes through (a, α) and (b, β). We shall also require that η be differentiable. Define

$$I(\varepsilon) = \int_a^b f(x, y, y')\,dx.$$

We think of y as a variation from y_m by $\varepsilon\eta(x)$. Note that, for each admissible choice of η, the value of ε which extremizes $I(\varepsilon)$ is $\varepsilon = 0$. Thinking back to max–min tests for functions of one variable, we are led to the conclusion that we should look at the equation $I'(\varepsilon) = 0$ at the value $\varepsilon = 0$, which we know beforehand gives the extremal value by choice of y_m. Now

$$I'(\varepsilon) = \frac{d}{d\varepsilon}\int_a^b f(x, y, y')\,dx = \int_a^b \frac{\partial f(x, y_m + \varepsilon\eta, y_m' + \varepsilon\eta')}{\partial\varepsilon}\,dx$$

$$= \int_a^b (f_y y_\varepsilon + f_{y'} y_\varepsilon')\,dx = \int_a^b (f_y\eta + f_{y'}\eta')\,dx.$$

Since $y = y_m$ when $\varepsilon = 0$, then $I'(0) = 0$ is the equation

$$I'(0) = \int_a^b (f_{y_m}\eta + f_{y_m'}\eta')\,dx = 0.$$

Integrate $\int_a^b f_{y'_m}\eta') \, dx$ by parts:

$$\int_a^b f_{y'_m}\eta' \, dx = f_{y'_m}\eta \Big|_a^b - \int_a^b \frac{d(f_{y'_m})}{dx} \eta \, dx = -\int_a^b \eta \frac{d}{dx}(f_{y'_m}) \, dx,$$

as by choice $\eta(a) = \eta(b) = 0$. This gives us

$$\int_a^b \left(\frac{\partial f}{\partial y_m} - \frac{d}{dx} \frac{\partial f}{\partial y'_m} \right) \eta \, dx = 0,$$

for *every* choice of η such that η' exists and $\eta(a) = \eta(b) = 0$.

We claim that this forces

$$\frac{\partial f}{\partial y_m} - \frac{d}{dx} \frac{\partial f}{\partial y'_m} = 0.$$

This will follow from the following lemma:

Lemma

If g is continuous on $[a, b]$ and $\int_a^b g(x)\eta(x) \, dx = 0$ whenever η is a differentiable function on $[a, b]$ with $\eta(a) = \eta(b) = 0$, then $g(x) = 0$ for every x in $[a, b]$.

To prove the lemma, suppose that $g(x) \neq 0$ at some x_0 in $[a, b]$. Say that $g(x_0) > 0$. By continuity of g, there is some subinterval $[u, v]$ of $[a, b]$ containing x_0 on which $g(x)$ is positive. Define

$$\eta(x) = \begin{cases} 0, & a \leq x \leq u, \\ (x - u)^2(x - v)^2, & u \leq x \leq v, \\ 0, & v \leq x \leq b. \end{cases}$$

Then η is differentiable on $[a, b]$ and $\eta(a) = \eta(b) = 0$. But $\int_a^b g(x)\eta(x) \, dx = \int_a^b g(x)\eta(x) \, dx > 0$, since $g(x) > 0$ on $[u, v]$ and $\eta(x) > 0$ on (u, v). This contradiction implies that $g(x) = 0$ for $a \leq x \leq b$. ∎

Getting back to our problem, we now have

$$\frac{\partial f}{\partial y_m} - \frac{d}{dx} \left(\frac{\partial f}{\partial y'_m} \right) = 0.$$

This gives a necessary condition for y_m to be our extremizing function. The condition is not sufficient, but in most problems arising out of a physical situation the solution to this equation is the solution to the extremal problem.

Our procedure, then, for finding y to extremize $\int_a^b f(x, y, y')\,dx$ is to solve the Euler–Lagrange differential equation

$$\frac{\partial f}{\partial y} - \frac{d}{dx}\left(\frac{\partial f}{\partial y'}\right) = 0,$$

subject to the specified end-point conditions $y(a) = \alpha$, $y(b) = \beta$.

EXAMPLE

We shall solve the brachistochrone problem explicitly. Here

$$f(x, y, y') = \frac{\sqrt{1 + (y')^2}}{\sqrt{2gy}}.$$

We can certainly calculate

$$\frac{\partial f}{\partial y} - \frac{d}{dx}\frac{\partial f}{\partial y'} = 0$$

to obtain the Euler–Lagrange equation directly. However, observe that f in this case depends upon x only through y and y', and x does not explicitly appear. In this event, we invoke the identity:

$$\frac{d}{dx}\left(y'\frac{\partial f}{\partial y'} - f\right) = -y'\left(\frac{\partial f}{\partial y} - \frac{d}{dx}\frac{\partial f}{\partial y'}\right) - \frac{\partial f}{\partial x},$$

and the fact that $\partial f/\partial x = 0$, to rewrite the Euler–Lagrange equation as

$$\frac{d}{dx}\left(y'\frac{\partial f}{\partial y'} - f\right) = 0.$$

This is equivalent to

$$y'\frac{\partial f}{\partial y'} - f = C$$

for some constant C.

Computing the last equation for

$$f(x, y, y') = \sqrt{\frac{1 + (y')^2}{2gy}}$$

gives us

$$\frac{(y')^2}{\sqrt{1 + (y')^2}\,\sqrt{2gy}} - \frac{\sqrt{1 + (y')^2}}{\sqrt{2gy}} = C.$$

After some simplification, this becomes

$$(y')^2 = \frac{1}{2C^2gy} - 1,$$

or

$$\frac{dy}{dx} = \sqrt{\frac{1 - 2C^2gy}{2C^2gy}}.$$

For ease of integration, write this as

$$dx = \frac{\sqrt{2C^2gy}\,dy}{\sqrt{1 - 2C^2gy}} = \frac{\sqrt{y}}{\sqrt{\alpha^2 - y}}\,dy,$$

where $\alpha^2 = 1/2C^2g$. Then

$$x = \int \frac{\sqrt{y}}{\sqrt{\alpha^2 - y}}\,dy.$$

Put $y = \alpha^2 \sin^2(\theta/2)$ to write

$$x = \int \frac{\alpha \sin(\theta/2)\, \alpha^2 \sin(\theta/2) \cos(\theta/2)}{\sqrt{\alpha^2(1 - \sin^2(\theta/2))}}\,d\theta$$

$$= \int \alpha^2 \sin^2\left(\frac{\theta}{2}\right) d\theta = \frac{\alpha^2}{2}\,[\theta - \sin(\theta)] + K,$$

where K is the constant of integration.

This gives us two equations:

$$x(\theta) = \frac{\alpha^2}{2}\,[\theta - \sin(\theta)] + K$$

and

$$y(\theta) = \frac{\alpha^2}{2} \sin^2\left(\frac{\theta}{2}\right) = \frac{\alpha^2}{2}\,[1 - \cos(\theta)].$$

These may be thought of as parametric equations for the path in terms of parameter θ. These equations also represent an arc of the cycloid, which is the locus swept out by a point on the circumference of a circle of radius $\alpha/\sqrt{2}$, rolling without slipping along the x-axis (Fig. 8-1). By adjusting α and K, the cycloid can be set to pass through $(0, 0)$ and (x_1, y_1), solving the brachistochrone problem.

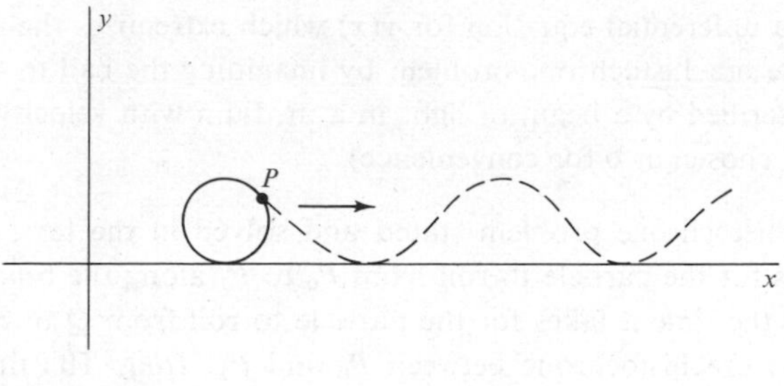

Fig. 8-1

Exercises for Section 8.2

1. Let (x_0, y_0) and (x_1, y_1) be given, distinct points, $x_0 < x_1$ and $y_0 > 0$, $y_1 > 0$. Determine a curve $y = f(x)$, $x_0 \leq x \leq x_1$, with $f(x_0) = y_0$, $f(x_1) = y_1$, and whose rotation about the x-axis produces a surface of minimum area. *Hint:* The area of such a surface is

$$2\pi \int_{x_0}^{x_1} f(x)\sqrt{1 + (f'(x))^2}\,dx.$$

2. Derive the Euler–Lagrange equation for extremizing

$$\int_a^b f(x, y, y', y'')\,dx,$$

where $y(a) = \alpha$ and $y(b) = \beta$ are given.

3. Consider the problem of extremizing $\int_a^b f(x, y, y')\,dx$ subject only to $y(a) = \alpha$ [so $y(b)$ is not prescribed]. Derive the Euler–Lagrange equations:

$$\frac{\partial f}{\partial y} - \frac{d}{dx}\left(\frac{\partial f}{\partial y'}\right) = 0,$$

$$\left.\frac{\partial f}{\partial y'}\right|_{x=b} = 0.$$

Hint: Suppose that the extremizing function is y_m, and proceed as in the section, with $y = y_m + \varepsilon\eta$ and $\eta(a) = 0$ only.

4. Using the result of Exercise 3, find the path of quickest descent (brachistochrone) from (0, 0) to a point on the vertical line $x = b > 0$.

5. Suppose we have a beam of light traveling in an inhomogeneous medium, in which the velocity is a function of the vertical coordinate, say $v = v(y)$. Fermat's principle says that the beam between given points (x_1, y_1) and (x_2, y_2) will describe a path that extremizes

$$\int_{x_1}^{x_2} \frac{\sqrt{1 + (y')^2}}{v(y)}\,dx.$$

(a) Obtain a differential equation for $y(x)$ which extremizes the above integral.
(b) Solve the brachistochrone problem by imagining the ball to be rolling on a path described by a beam of light in a medium with velocity $v(y) = \sqrt{2gy}$ (with x_1 chosen as 0 for convenience).

6. In the brachistochrone problem stated and solved in the text, show that the time it takes for the particle to roll from P_0 to P_1 along the brachistochrone is the same as the time it takes for the particle to roll from Q to P_1, with Q any point on the brachistochrone between P_0 and P_1. *Hint:* The time from P_0 to P_1 can be computed directly as $\int_0^{x_1} dt$, using the solution obtained in the text. In computing the time from Q, be sure to take into account the fact that, though the particle is still presumed to start from rest (i.e., zero initial velocity), the potential energy equation is different than that used when the particle started from P_0, because Q has a different vertical coordinate than P_0.

8.3 Extremizing $\int_{t_1}^{t_2} f(x_1(t), \ldots, x_n(t), \dot{x}_1(t), \ldots, \dot{x}_n(t), t)\,dt$

We now generalize the result of the last section. Suppose that f is a function of $2n + 1$ variables, with continuous first and second partials. Imagine that we want to choose $x_1(t), \ldots, x_n(t)$, each twice-differentiable and having given values at t_1 and t_2, say

$$x_i(t_1) = \alpha_i,$$

$$x_i(t_2) = \beta_i$$

in such a way as to render $\int_{t_1}^{t_2} f(x_1, \ldots, x_n, \dot{x}_1, \ldots, \dot{x}_n, t)\,dt$ an extremum. Here $\dot{x}_i$ denotes dx_i/dt (in most applications, t is time, so $\dot{x}_i$ denotes a velocity).

Proceed as in Section 8.2, denoting the functions we seek as $x_1, \ldots, x_n$, and varying these by putting

$$X_i(t) = x_i(t) + \varepsilon\eta_i(t),$$

where each η_i has two derivatives and $\eta_i(t_1) = \eta_i(t_2) = 0$. Put

$$I(\varepsilon) = \int_{t_1}^{t_2} f(X_1, \ldots, X_n, \dot{X}_1, \ldots, \dot{X}_n, t)\,dt.$$

Then I is a function of ε for any given $\eta_1, \ldots, \eta_n$. Put $I'(\varepsilon) = 0$ at $\varepsilon = 0$:

$$I'(0) = 0 = \int_{t_1}^{t_2} \left[\frac{\partial f}{\partial x_1}\eta_1 + \cdots + \frac{\partial f}{\partial x_n}\eta_n + \frac{\partial f}{\partial \dot{x}_1}\dot{\eta}_1 + \cdots + \frac{\partial f}{\partial \dot{x}_n}\dot{\eta}_n\right] dt.$$

Integrate the terms

$$\int_{t_1}^{t_2} \frac{\partial f}{\partial \dot{x}_i} \dot{\eta}_i \, dt$$

by parts to obtain

$$\int_{t_1}^{t_2} \frac{\partial f}{\partial \dot{x}_i} \dot{\eta}_i \, dt = \frac{\partial f}{\partial \dot{x}_i} \eta_i \Big|_{t_1}^{t_2} - \int_{t_1}^{t_2} \eta_i \frac{d}{dt} \frac{\partial f}{\partial \dot{x}_i} \, dt$$

$$= -\int_{t_1}^{t_2} \eta_i \frac{d}{dt} \frac{\partial f}{\partial \dot{x}_i} \, dt, \qquad \text{since } \eta_i(t_1) = \eta_i(t_2) = 0.$$

Then

$$\sum_{i=1}^{n} \int_{t_1}^{t_2} \left(\frac{\partial f}{\partial x_i} - \frac{d}{dt} \frac{\partial f}{\partial \dot{x}_i} \right) \eta_i = 0.$$

For any given j, $1 \le j \le n$, we may choose the η_i's to be identically zero for $i \ne j$, giving us

$$\int_{t_1}^{t_2} \left(\frac{\partial f}{\partial x_j} - \frac{d}{dt} \frac{\partial f}{\partial \dot{x}_j} \right) \eta_j \, dt = 0$$

for $j = 1, 2, \ldots, n$. Applying the lemma of the last section now gives us the system of n partial differential equations:

$$\frac{\partial f}{\partial x_j} - \frac{d}{dt} \frac{\partial f}{\partial \dot{x}_j} = 0, \qquad j = 1, \ldots, n.$$

These are the Euler–Lagrange equations for the problem of extremizing $\int_{t_1}^{t_2} f(x_1, \ldots, x_n, \dot{x}_1, \ldots, \dot{x}_n, t)\, dt$, and we solve them subject to the given boundary conditions:

$$x_j(t_1) = \alpha_j,$$
$$x_j(t_2) = \beta_j, \qquad j = 1, \ldots, n.$$

Exercises for Section 8.3

1. A geodesic on a surface Σ is a curve having the property that any other curve on the surface and having the same end points is at least as long (i.e., a geodesic on Σ is a shortest length path on Σ between two points on Σ).

(a) Show that geodesics in the plane are straight lines.

(b) Let Σ be a right circular cone:

$$\Sigma(x, y) = (x, y, \sqrt{x^2 + y^2}), \qquad x^2 + y^2 \le R^2, x \ge 0, y \ge 0.$$

Find a differential equation for geodesics on Σ. *Hint:* Let P_0, P_1 be points on Σ. If $C: [a, b] \to R^3$ is a curve on Σ between P_0 and P_1, then we can write $C(t)$ in spherical coordinates as

$$C(t) = (\rho \cos(\theta) \sin(\varphi), \rho \sin(\theta) \sin(\varphi), \rho \cos(\varphi)),$$

with ρ and θ functions of t and $\varphi = \alpha =$ constant. The length of C is then $\int_a^b \sqrt{(x'(t))^2 + (y'(t))^2 + (z'(t))^2}\, dt$. Show that

$$x'(t)^2 + y'(t)^2 + z'(t)^2 = \left(\frac{d\rho}{dt}\right)^2 + \rho^2 \sin^2(\alpha)\left(\frac{d\theta}{dt}\right)^2.$$

The problem is then to minimize $\int_a^b [(\rho')^2 + \rho^2 \sin^2(\alpha)(\theta')^2]^{1/2}\, dt$, with integrand of the form $f(\rho, \theta, \dot{\rho}, \dot{\theta}, t)$.

(c) Derive a differential equation for geodesics on a sphere. *Hint:* Let the sphere be $x^2 + y^2 + z^2 = 1$, switch to spherical coordinates and proceed as in (b), noting that here $\rho = 1$.

8.4 Application to Hamilton's Principle and Lagrange's Equations of Motion

Suppose that we have a mechanical system in which the forces are conservative, that is, derivable from a potential function. Suppose the kinetic energy T and the potential energy V are expressed in terms of coordinates $q_1, \ldots, q_n$, which we suppose are conveniently chosen to suit the setting of the problem, and their time derivatives $\dot{q}_1, \ldots, \dot{q}_n$. We call $L = T - V$ the *Lagrangian of the system. Hamilton's Principle* then says: *The motion of the system is such that, for arbitrary times t_1 and t_2,*

$$\int_{t_1}^{t_2} L(q_1, \ldots, q_n, \dot{q}_1, \ldots, \dot{q}_n, t)\, dt$$

is an extremum when considered over all twice-differentiable functions $q_1(t), \ldots, q_n(t)$ having specified values at t_1 and t_2.

Lagrange's equations of motion are nothing more than the Euler-Lagrange equations of Section 8.3 applied to the integral of Hamilton's Principle:

$$\frac{\partial L}{\partial q_i} - \frac{d}{dt}\left(\frac{\partial L}{\partial \dot{q}_i}\right) = 0.$$

As an example, consider the pulley system shown in Fig. 8-2, with masses M_1 and M_2 at the ends of the rope. Assuming no friction, the potential energy is $V = -M_1 g x - M_2 g(R - x)$, and the kinetic energy is

$$T = \tfrac{1}{2}(M_1 + M_2)\dot{x}^2,$$

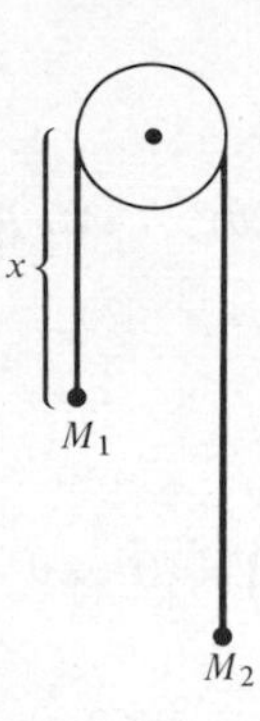

Fig. 8-2

where we have used x as indicated in Fig. 8-2 for our coordinate and R is the length of the rope connecting the two masses. Then

$$L = T - V = \tfrac{1}{2}(M_1 + M_2)\dot{x}^2 + M_1 g x + M_2 g(R - x).$$

The equation of motion is

$$\frac{\partial L}{\partial x} - \frac{d}{dt}\left(\frac{\partial L}{\partial \dot{x}}\right) = 0 = (M_1 - M_2)g - (M_1 + M_2)\ddot{x}.$$

This becomes

$$\ddot{x} = \left(\frac{M_1 - M_2}{M_1 + M_2}\right)g.$$

As a second example, consider a simple pendulum. Use θ as the angle of deviation from the vertical. Since $x = R\sin(\theta)$ (Fig. 8-3) and $y = -R\cos(\theta)$, then $\dot{x}^2 + \dot{y}^2 = R^2\sin^2(\theta)\dot{\theta} + R^2\cos^2(\theta)\dot{\theta} = R^2\dot{\theta}^2$. Thus the kinetic energy is given by $T = \frac{1}{2}M(R\dot{\theta})^2$.

To find the potential energy, note that the work done by gravity in lifting M from the equilibrium position ($\theta = 0$) to some other position is $-MgR - [-MgR\cos(\theta)] = -MgR[1 - \cos(\theta)]$. Thus the potential energy is of the form

$$V = MgR[1 - \cos(\theta)] + \text{constant}.$$

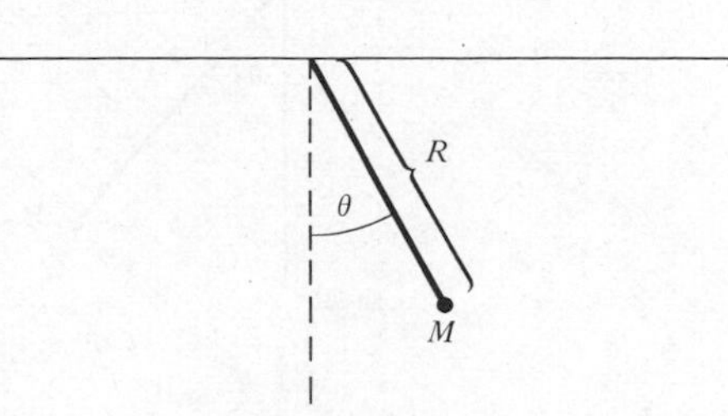

Fig. 8-3

Then

$$L = T - V = \tfrac{1}{2}M(R\dot{\theta})^2 - MgR[1 - \cos(\theta)] - K,$$

where K is a constant.

The equation of motion is then

$$\frac{\partial L}{\partial \theta} - \frac{d}{dt}\left(\frac{\partial L}{\partial \dot{\theta}}\right) = 0 = \ddot{\theta} + \frac{g}{R}\sin(\theta).$$

Things are a little more complicated with the double pendulum shown below in Fig. 8-4. Here it is natural to use θ_1 and θ_2 as the coordinates to describe the motion. Now M_1 is at (x_1, y_1), with $x_1 = L_1 \sin(\theta_1)$, $y_1 = -L_1 \cos(\theta_1)$, and M_2 is at (x_2, y_2) with

$$x_2 = L_1 \sin(\theta_1) + L_2 \sin(\theta_2)$$
$$y_2 = -L_1 \cos(\theta_1) - L_2 \cos(\theta_2).$$

The kinetic energy is

$$T = \tfrac{1}{2}M_1(\dot{x}_1^2 + \dot{y}_1^2) + \tfrac{1}{2}M_2(\dot{x}_2^2 + \dot{y}_2^2).$$

This works out to

$$T = \tfrac{1}{2}(M_1 + M_2)(L_1\dot{\theta}_1)^2 + M_2L_1L_2\dot{\theta}_1\dot{\theta}_2\cos(\theta_1 - \theta_2) + \tfrac{1}{2}M_2(L_2\dot{\theta}_2)^2.$$

The potential energy is of the form

$$V = M_1gy_1 + M_2gy_2 + K, \qquad \text{for some } K,$$

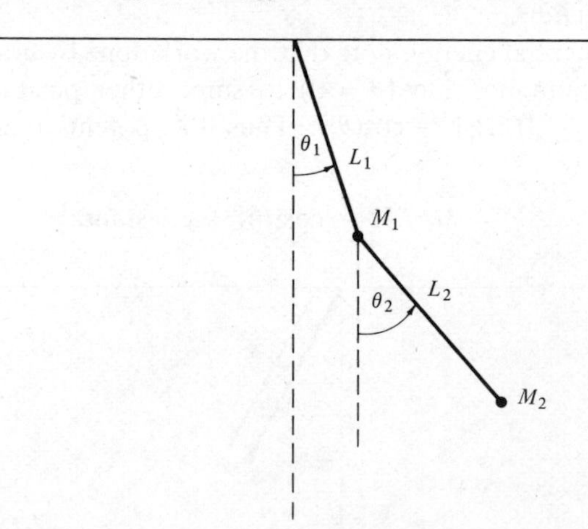

Fig. 8-4

so

$$V = -(M_1 + M_2)gL_1 \cos(\theta_1) - M_2 g L_2 \cos(\theta_2) + K.$$

Then

$$0 = \frac{\partial L}{\partial \theta_i} - \frac{d}{dt}\left(\frac{\partial L}{\partial \dot{\theta}_i}\right), \qquad i = 1, 2,$$

gives us two equations:

$$(M_1 + M_2)L_1\ddot{\theta}_1 + M_2 L_2[\ddot{\theta}_2 \cos(\theta_1 - \theta_2) + \dot{\theta}_2^2 \sin(\theta_1 - \theta_2)] + (M_1 + M_2)g \sin(\theta_1) = 0$$

and

$$L_1\ddot{\theta}_1 \cos(\theta_1 - \theta_2) + L_2\ddot{\theta}_2 - L_1\dot{\theta}_1^2 \sin(\theta_1 - \theta_2) + g \sin(\theta_1 - \theta_2) = 0.$$

These are too difficult for us to attempt to solve here.

Exercises for Section 8.4

1. The equation of motion for a particle of mass m falling under only the influence of gravity is $\ddot{x}$ = constant, with x the distance measured downward. Obtain this equation by writing the Lagrangian for this problem and using Lagrange's equations of motion.

2. Suppose that a particle of mass m is falling under the influence of gravity and a resistive force proportional to the distance $x(t)$ traveled. Write the Lagrangian and obtain the Lagrange equation of motion.

3. Masses $m_1, \ldots, m_n$ are connected in a spring system as shown in Fig. 8-5. Let x_i = displacement of m_i from equilibrium, and assume as usual that the force of a spring is proportional to the displacement. Derive the equations of motion.

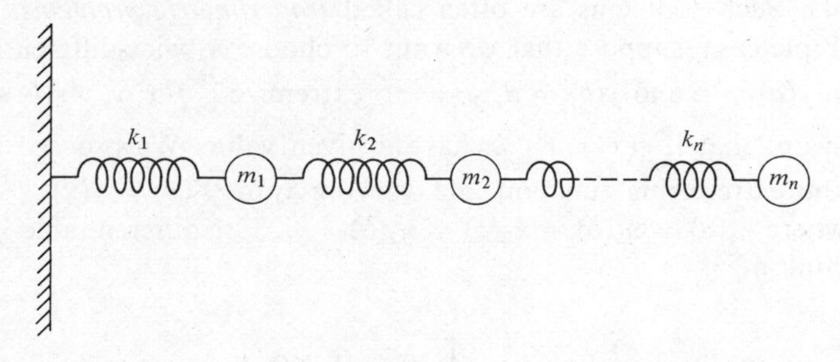

Fig. 8-5

Hint: If k_i is the force constant of proportionality for the spring, show that the Lagrangian is

$$L = \frac{1}{2}\sum_{i=1}^{n}(m_i \dot{x}_i^2) - \frac{1}{2}k_1 x_1^2 - \frac{1}{2}k_2(x_2 - x_1)^2 - \cdots - \frac{1}{2}k_n(x_n - x_{n-1})^2.$$

4. Obtain the equations of motion for the triple pendulum shown in Fig. 8-6.

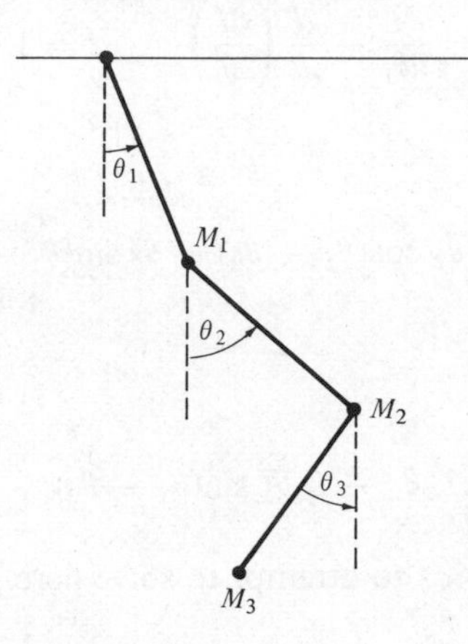

Fig. 8-6

5. This exercise is for those familiar with some theory of electricity and magnetism. Starting with the Lagrangian

$$L = \frac{1}{2}\left(\varepsilon_0 E^2 - \frac{B^2}{\mu_0}\right) - \rho\varphi + \rho(V \cdot A)$$

for an electromagnetic field with charge density ρ, use Lagrange's equations to derive Maxwell's equations.

8.5 Extremal Problems with Constraints

We want now to consider problems like those of Sections 8.2 and 8.3, but with constraints. Such problems are often called *isoperimetric problems.* To begin with a simple case, suppose that we want to choose a twice-differentiable $y = y(x)$, with $y(a) = \alpha$ and $y(b) = \beta$, so as to extremize $\int_a^b f(x, y, y')\,dx$ subject to the constraint that $\int_a^b g(x, y, y')\,dx$ have a given value. We proceed by letting $y(x)$ be the extremizing function and varying y to $Y(x) = y(x) + \varepsilon_1\eta_1(x) + \varepsilon_2\eta_2(x)$, where $\eta_1(a) = \eta_1(b) = \eta_2(a) = \eta_2(b) = 0$. For differentiable η_1 and η_2, we can think of

$$I(\varepsilon_1, \varepsilon_2) = \int_a^b f(x, Y, Y')\,dx$$

as a function of ε_1 and ε_2, to be extremized subject to the condition

$$J(\varepsilon_1, \varepsilon_2) = \int_a^b g(x, Y, Y')\, dx = K, \qquad K \text{ a given constant.}$$

We already know how to handle such a problem. Consider $I(\varepsilon_1, \varepsilon_2) + \lambda J(\varepsilon_1, \varepsilon_2)$, where λ is a Lagrange multiplier to be determined, and solve for the critical points. Thus consider

$$\frac{\partial(I + \lambda J)}{\partial \varepsilon_1} = 0$$

and

$$\frac{\partial(I + \lambda J)}{\partial \varepsilon_2} = 0.$$

Since we already know that the extremum occurs when $\varepsilon_1 = \varepsilon_2 = 0$, we have

$$\left.\frac{\partial(I + \lambda J)}{\partial \varepsilon_1}\right|_{\varepsilon_1 = 0} = 0 = \int_a^b \left[\frac{\partial(f + \lambda g)}{\partial y}\eta_1 + \frac{\partial(f + \lambda g)}{\partial y'}\eta_1'\right] dx$$

and

$$\left.\frac{\partial(I + \lambda J)}{\partial \varepsilon_2}\right|_{\varepsilon_2 = 0} = 0 = \int_a^b \left[\frac{\partial(f + \lambda g)}{\partial y}\eta_2 + \frac{\partial(f + \lambda g)}{\partial y'}\eta_2'\right] dx.$$

After the usual integration by parts, we obtain

$$\int_a^b \left[\frac{\partial(f + \lambda g)}{\partial y} - \frac{d}{dx}\frac{\partial(f + \lambda g)}{\partial y'}\right] \eta_1\, dx = 0$$

and

$$\int_a^b \left[\frac{\partial(f + \lambda g)}{\partial y} - \frac{d}{dx}\frac{\partial(f + \lambda g)}{\partial y'}\right] \eta_2\, dx = 0.$$

Because of the freedom of choice in selecting η_1 and η_2, we obtain the single equation

$$\frac{\partial(f + \lambda g)}{\partial y} - \frac{d}{dx}\frac{\partial(f + \lambda g)}{\partial y'} = 0.$$

In effect, then, we incorporate the constraint condition into the Euler–Lagrange equation through the Lagrange multiplier λ. We have enough information to solve the problem, as $y(x_1) = \alpha$ and $y(x_2) = \beta$ give two pieces of information, and $\int_a^b g(x, y, y')\, dx = K$ a third.

There are two directions in which this result may be extended. First, we may have more than one constraint. That is, we may want to extremize $\int_a^b f(x, y, y')\,dx$ with $y(a) = \alpha$, $y(b) = \beta$, subject to $\int_a^b g_1(x, y, y')\,dx = K_1, \ldots, \int_a^b g_n(x, y, y')\,dx = K_n$. We then simply introduce enough Lagrange multipliers to cover all the constraints and consider the Euler–Lagrange equation:

$$\frac{\partial(f + \lambda_1 g_1 + \cdots + \lambda_n g_n)}{\partial y} - \frac{d}{dx}\frac{\partial(f + \lambda_1 g_1 + \cdots + \lambda_n g_n)}{\partial y'} = 0.$$

We may, on the other hand, want to increase the number of variables, and extremize

$$\int_a^b f(x_1, \ldots, x_n, \dot{x}_1, \ldots, \dot{x}_n, t)\,dt,$$

$$x_i(a) = \alpha_i, \qquad x_i(b) = \beta_i, \qquad i = 1, \ldots, n,$$

subject to

$$\int_a^b g_i(x_1, \ldots, x_n, \dot{x}_1, \ldots, \dot{x}_n, t)\,dt = K_i, \qquad i = 1, \ldots, N.$$

We then obtain a system of n equations:

$$\frac{\partial(f + \lambda_1 g_1 + \cdots + \lambda_N g_N)}{\partial x_i} - \frac{d}{dt}\frac{\partial(f + \lambda_1 g_1 + \cdots + \lambda_N g_N)}{\partial \dot{x}_i} = 0, \qquad i = 1, \ldots, n.$$

EXAMPLE

Suppose that we want to find a plane closed curve $x = x(t)$, $y = y(t)$ of given length L enclosing the maximum area. Thus we want to maximize

$$\int_a^b \frac{1}{2}(x\dot{y} - y\dot{x})\,dt$$

subject to $\int_a^b \sqrt{(\dot{x})^2 + (\dot{y})^2}\,dt = L$. Consider $f + \lambda g = \frac{1}{2}(x\dot{y} - y\dot{x}) + \lambda\sqrt{\dot{x}^2 + \dot{y}^2}$. We must calculate

$$\frac{\partial}{\partial x}(f + \lambda g) - \frac{d}{dt}\frac{\partial(f + \lambda g)}{\partial \dot{x}} = \frac{1}{2}\dot{y} - \frac{d}{dt}\left(-\frac{1}{2}y + \frac{\lambda\dot{x}}{\sqrt{\dot{x}^2 + \dot{y}^2}}\right) = 0,$$

and

$$\frac{\partial}{\partial y}(f + \lambda g) - \frac{d}{dt}\frac{\partial(f + \lambda g)}{\partial \dot{y}} = -\frac{1}{2}\dot{x} - \frac{d}{dt}\left(\frac{1}{2}x + \frac{\lambda\dot{y}}{\sqrt{\dot{x}^2 + \dot{y}^2}}\right) = 0.$$

Integrate directly with respect to t:

$$\frac{1}{2}y - \left(-\frac{1}{2}y + \frac{\lambda\dot{x}}{\sqrt{\dot{x}^2 + \dot{y}^2}}\right) = C$$

and

$$-\frac{1}{2}x - \left(\frac{1}{2}x + \frac{\lambda\dot{y}}{\sqrt{\dot{x}^2 + \dot{y}^2}}\right) = K,$$

where C and K are constants of integration. Then

$$y - \frac{\lambda\dot{x}}{\sqrt{\dot{x}^2 + \dot{y}^2}} = C$$

and

$$x + \frac{\lambda\dot{y}}{\sqrt{\dot{x}^2 + \dot{y}^2}} = -K.$$

Note that

$$(y - C)^2 + (x + K)^2 = \frac{\lambda^2\dot{x}^2}{\dot{x}^2 + \dot{y}^2} + \frac{\lambda^2\dot{y}^2}{\dot{x}^2 + \dot{y}^2} = \lambda^2.$$

Thus we have a circle. Knowing $x(a)$, $x(b)$, $y(a)$, $y(b)$, and L, we can determine C, K, and λ. [Note that the fact that our curve is closed means that $x(a) = x(b)$ and $y(a) = y(b)$, so we have exactly three pieces of data to find the three constants C, K, and λ.] ∎

Exercises for Section 8.5

1. Suppose that a particle moves from P_0 to P_1 on the surface $g(x, y, z) = 0$, in such a way that the integral of the kinetic energy over the total time τ of travel is a minimum. Show that the coordinates $(x(t), y(t), z(t))$ of the particle satisfy

$$\frac{\ddot{x}}{\dfrac{\partial g}{\partial x}} = \frac{\ddot{y}}{\dfrac{\partial g}{\partial y}} = \frac{\ddot{z}}{\dfrac{\partial g}{\partial z}}.$$

Hint: Minimize $\int_0^\tau (\dot{x}^2 + \dot{y}^2 + \dot{z}^2)\, dt$.

2. Derive the Euler–Lagrange equations for extremizing

$$\int_a^b f(x, y, y', y'')\, dx$$

subject to $y(a) = \alpha$, $y(b) = \beta$, and the constraint

$$\int_a^b g(x, y, y', y'')\, dx = \gamma.$$

8.6 Extremizing Multiple Integrals and Applications

Suppose that the integral we want to extremize is a multiple integral

$$\iint_D f(x, y, z, z_x, z_y)\, dx\, dy,$$

where D is a region of the xy-plane bounded by a smooth simple closed curve C and f and its first and second partials are continuous on D. We seek our extremizing function from among continuous functions $z = z(x, y)$ having continuous first and second partials in D and prescribed values along C.

Let $z(x, y)$ be the desired function, and $Z(x, y) = z(x, y) + \varepsilon\eta(x, y)$, where η has continuous first and second partials in D and vanishes on C. Consider, as usual,

$$I(\varepsilon) = \iint_D f(x, y, Z, Z_x, Z_y)\, dx\, dy$$

as a function of ε for any given admissible η, and look at the equation $I'(0) = 0$, since we know that the extremum occurs when $\varepsilon = 0$. Now

$$I'(0) = 0 = \iint_D \left(\frac{\partial f}{\partial z}\eta + \frac{\partial f}{\partial z_x}\eta_x + \frac{\partial f}{\partial z_y}\eta_y\right) dx\, dy.$$

Here we have a new twist. There is no integration by parts to reduce the terms involving η_x and η_y to integrals involving just η. However, Green's theorem comes to the rescue. Put

$$A = -\frac{\partial f}{\partial z_y}\eta \qquad \text{and} \qquad B = \frac{\partial f}{\partial z_x}\eta$$

into the formula $\iint_D (B_x - A_y) = \oint_C A\, dx + B\, dy$ to obtain

$$\begin{aligned}
\iint_D &\left[\frac{\partial}{\partial x}\left(\eta\frac{\partial f}{\partial z_x}\right) + \frac{\partial}{\partial y}\left(\eta\frac{\partial f}{\partial z_y}\right)\right] dx\, dy \\
&= \iint_D \left\{\eta\left[\frac{\partial}{\partial x}\left(\frac{\partial f}{\partial z_x}\right) + \frac{\partial}{\partial y}\left(\frac{\partial f}{\partial z_y}\right)\right] + \frac{\partial f}{\partial z_x}\eta_y + \frac{\partial f}{\partial z_y}\eta_y\right\} dx\, dy \\
&= \oint_C -\frac{\partial f}{\partial z_y}\eta\, dx + \frac{\partial f}{\partial z_x}\eta\, dy.
\end{aligned}$$

Now observe that the line integral is zero, as $\eta(x, y) = 0$ for (x, y) on C. Then

$$\iint_D \left(\frac{\partial f}{\partial z_x}\eta_x + \frac{\partial f}{\partial z_y}\eta_y\right) = -\iint_D \eta\left[\frac{\partial}{\partial x}\frac{\partial f}{\partial z_x} + \frac{\partial}{\partial y}\frac{\partial f}{\partial z_y}\right].$$

Then

$$I'(0) = 0 = \iint_D \left(\frac{\partial f}{\partial z} - \frac{\partial}{\partial x}\frac{\partial f}{\partial z_x} - \frac{\partial}{\partial y}\frac{\partial f}{\partial z_y}\right) \eta \, dx \, dy,$$

for every admissible η. We then conclude that

$$\frac{\partial f}{\partial z} - \frac{\partial}{\partial x}\frac{\partial f}{\partial z_x} - \frac{\partial}{\partial y}\frac{\partial f}{\partial z_y} = 0.$$

This is the *Euler–Lagrange equation* for the current situation.

If we have a constraint, say

$$\iint_D g(x, y, z, z_x, z_y) \, dx \, dy = A,$$

then we apply the Euler–Lagrange condition to $f + \lambda g$, with λ a Lagrange multiplier. Similarly, we can handle more constraints by throwing in more Lagrange multipliers.

We devote the remainder of this section to some applications.

A. Vibrating String

We derived the one-dimensional wave equation $y_{tt} = (\rho/\alpha) y_{xx}$ in Section 5.2. We now give an alternative derivation. Think of the tension T as constant, as before, and let the mass per unit length be $\rho(x)$ at x. The potential energy may be thought of as derivable from the amount of work done in stretching the string from its initial length R to its length at time t. If $y = y(x, t)$ gives the shape of the string at time t, the potential energy is then

$$V = T\left(\int_0^R \sqrt{1 + (y_x)^2} \, dx - R\right).$$

Now

$$\sqrt{1 + u} = 1 + \tfrac{1}{2}u - \tfrac{1}{4}u^2 + \cdots,$$

so

$$\sqrt{1 + (y_x)^2} = 1 + \tfrac{1}{2}y_x - \tfrac{1}{4}(y_x)^2 + \cdots.$$

If $|y_x|$ is much less than 1, then we can neglect terms $(y_x)^n$ for $n \geq 4$, and reasonably approximate $\sqrt{1 + (y_x)^2}$ by $1 + \frac{1}{2}(y_x)^2$. We then write

$$V = T\left\{\int_0^R \left[1 + \tfrac{1}{2}(y_x)^2\right] dx - R\right\} = \tfrac{1}{2}T\int_0^R \left(\frac{\partial y}{\partial x}\right)^2 dx.$$

Next, the kinetic energy of the string is

$$T^* = \tfrac{1}{2}\int_0^R \rho(x) y_t^2 \, dx.$$

The Lagrangian is then

$$L = T^* - V = \tfrac{1}{2}\int_0^R \left[T\left(\frac{\partial y}{\partial x}\right)^2 - \rho\left(\frac{\partial y}{\partial t}\right)^2 \right] dx.$$

By Hamilton's Principle, the motion should be such as to extremize this integral on any interval of time $[t_1, t_2]$. Thus we want to extremize

$$\int_{t_1}^{t_2}\int_0^R \tfrac{1}{2}\left[T\left(\frac{\partial y}{\partial x}\right)^2 - \rho\left(\frac{\partial y}{\partial t}\right)^2 \right] dx\, dt.$$

Think of this as a double integral of $f(x, t, y, y_x, y_t)$ over the rectangular region $t_1 \leq t \leq t_2$, $0 \leq x \leq L$, in xt-space. The extremizing function is then found as a solution to the Euler–Lagrange equation:

$$\begin{aligned} \frac{\partial f}{\partial y} - \frac{\partial}{\partial x}\left(\frac{\partial f}{\partial y_x}\right) - \frac{\partial}{\partial t}\left(\frac{\partial f}{\partial y_t}\right) &= 0 \\ &= -\frac{\partial}{\partial x}\left(T\frac{\partial y}{\partial x}\right) - \frac{\partial}{\partial t}\left(-\rho\frac{\partial y}{\partial t}\right) \\ &= T\frac{\partial^2 y}{\partial x^2} - \frac{\partial}{\partial t}\left(\rho\frac{\partial y}{\partial t}\right). \end{aligned}$$

The wave equation is then

$$\frac{\partial}{\partial t}\left(\rho\frac{\partial y}{\partial t}\right) = \rho(x)\frac{\partial^2 y}{\partial t^2} = T\frac{\partial^2 y}{\partial x^2},$$

which can be written

$$\frac{\rho}{T}\frac{\partial^2 y}{\partial t^2} = \frac{\partial^2 y}{\partial x^2}.$$

This agrees with the result derived in Section 5.2.

Once the method used here to treat $\iint_D f(x, y, z, z_x, z_y)\, dx\, dy$ has been digested, it is not difficult to go on to triple integrals. For

$$\iiint_{V(\Sigma)} f(x, y, z, u, u_x, u_y, u_z)\, dx\, dy\, dz$$

with u given on the boundary Σ of $V(\Sigma)$, we obtain

$$\frac{\partial f}{\partial u} - \frac{\partial}{\partial x}\left(\frac{\partial f}{\partial u_x}\right) - \frac{\partial}{\partial y}\left(\frac{\partial f}{\partial u_y}\right) - \frac{\partial}{\partial z}\left(\frac{\partial f}{\partial u_z}\right) = 0.$$

The derivation is left to the reader, with the warning that, instead of Green's theorem, one must use the divergence theorem in the appropriate place.

B. Two-Dimensional Wave Equation

Imagine that we have a perfectly elastic membrane stretched over a framework that forms a smooth, simple closed curve C in the xy-plane. The membrane is displaced and allowed to vibrate freely. At time t, let $z(x, y, t)$ denote the z-coordinate of the particle initially at position $(x, y, 0)$. We want a differential equation for z.

Let $\rho(x, y)$ = mass/unit area. The kinetic energy at time t is

$$T^* = \tfrac{1}{2}\iint_D \rho\left(\frac{\partial z}{\partial t}\right)^2 dx\, dy.$$

The potential energy at time t is due to the work done in stretching the membrane from its initial area A to its surface area at time t. Thus

$$V = T\left(\iint_D \sqrt{1 + (z_x)^2 + (z_y)^2}\, dx\, dy - A\right),$$

where T = tension (=force/unit area), assumed to be constant. Here D is the region bounded by C, and we recall the previous result that

$$\iint_D \sqrt{1 + (z_x)^2 + (z_y)^2}\, dx\, dy = \text{area of surface } z = z(x, y, t) \qquad \text{at time } t.$$

Approximate

$$\sqrt{1 + (z_x)^2 + (z_y)^2} \approx 1 + \tfrac{1}{2}[(z_x)^2 + (z_y)^2]$$

by expanding the radical in a Taylor series and neglecting higher powers of z_x and z_y. We are assuming then that $|z_x|$ and $|z_y|$ are much smaller than 1 (i.e., the amplitudes of the vibrations are small). Then, putting $A = \iint_D dx\, dy$, we have

$$\begin{aligned} V &= \tfrac{1}{2}T\iint_D \left[\left(\frac{\partial z}{\partial x}\right)^2 + \left(\frac{\partial z}{\partial y}\right)^2\right] dx\, dy \\ &= \tfrac{1}{2}T\iint_D \nabla^2 z. \end{aligned}$$

The Lagrangian of the system is then

$$L = T^* - V = \frac{1}{2}\iint_D \left[\rho\left(\frac{\partial z}{\partial t}\right)^2 - T\,\nabla^2 z\right] dx\, dy.$$

According to Hamilton's Principle, we must extremize

$$\int_{t_1}^{t_2}\iint_D \frac{1}{2}\left[\rho\left(\frac{\partial z}{\partial t}\right)^2 - T\,\nabla^2 z\right] dx\, dy\, dt.$$

Think of this as a triple integral over the region in xyt-space determined by (x, y) in D and $t_1 \leq t \leq t_2$. The Euler–Lagrange equation is then

$$\frac{\partial L}{\partial z} - \frac{\partial}{\partial x}\left(\frac{\partial L}{\partial z_x}\right) - \frac{\partial}{\partial y}\left(\frac{\partial L}{\partial z_y}\right) - \frac{\partial}{\partial t}\left(\frac{\partial L}{\partial z_t}\right) = 0.$$

After some computations, this becomes

$$\nabla^2 z = \frac{\rho}{T}\frac{\partial^2 z}{\partial t^2},$$

which is the two-dimensional analogue of the vibrating string equation.

C. Schrödinger's Equation for the Hydrogen Atom

We conclude this section with an application to quantum mechanics. Consider a single particle of mass M moving under the influence of a force field determined by a potential function $V(x, y, t)$. Let E be the total energy of the particle, and, as usual, V the potential energy. Let $\psi(x, y, z, t)$ denote the particle's wave function. That is, over any volume $V(\Sigma)$ of 3-space, $\iiint_{V(\Sigma)} |\psi(x, y, z, t)|^2\, dx\, dy\, dz$ gives at time t the probability that the particle is in $V(\Sigma)$. Schrödinger determined that ψ must be a function with continuous first and second partials which extremizes

$$\iiint_{R^3}\left[\frac{K^2}{2M}\left(\frac{\partial^2\psi}{\partial x^2} + \frac{\partial^2\psi}{\partial y^2} + \frac{\partial^2\psi}{\partial z^2}\right) + (V - E)\psi\right] dx\, dy\, dz,$$

the integral being an improper one over all of 3-space. Here K is a constant that must be determined experimentally.

The Euler–Lagrange equation for ψ is

$$\frac{K^2}{2M}\nabla^2\psi + (E - V)\psi = 0,$$

and this is Schrödinger's wave equation for the hydrogen atom.

Exercises for Section 8.6

1. Derive the Euler–Lagrange equation for extremizing

$$\iint_D F(x, y, u, u_x, u_y, u_{xx}, u_{yy}, u_{xy})$$

with u given on the boundary of D.

2. Derive the Euler–Lagrange conditions for extremizing the integral of Exercise 1 subject to the constraint $\iint_D G(x, y, u, u_x, u_y, u_{xx}, u_{yy}, u_{xy}) = K$.

9

Analysis of Complex-Valued Functions

9.1 Complex Numbers

Before beginning the study of complex-valued functions, we naturally have to know some things about complex numbers.

A *complex number* may be thought of as a symbol $a + bi$ (or $a + ib$) where a and b are real and, in carrying out arithmetic computations, $i^2 = -1$. Addition and multiplication are given by

$$(a + bi) + (c + di) = (a + c) + i(b + d)$$

and

$$\begin{aligned}(a + bi) \cdot (c + di) &= ac + adi + cbi + bdi^2 \\ &= (ac - bd) + (ad + cb)i.\end{aligned}$$

We call a the *real part* of $a + bi$, and b the *imaginary part*. These are denoted $\mathrm{Re}(a + bi)$ and $\mathrm{Im}(a + bi)$, and are of course themselves real.

It is easy to check that the usual rules of arithmetic, such as commutativity and associativity, hold for addition and multiplication of complex numbers. In particular, if $a \neq 0$ or $b \neq 0$, then $a + bi$ has a multiplicative inverse. An easy way to distinguish its real and imaginary parts is to write

$$\frac{1}{a + bi} = \frac{1}{a + bi}\frac{a - bi}{a + bi} = \frac{a}{a^2 + b^2} - \frac{bi}{a^2 + b^2}.$$

Thus $\mathrm{Re}(1/(a + bi)) = a/(a^2 + b^2)$ and $\mathrm{Im}(1/(a + bi)) = -b/(a^2 + b^2)$.

The number $a - bi$ is called the *complex conjugate* of $a + bi$. In general, the conjugate of z is denoted $\bar{z}$. For any complex numbers z_1 and z_2, it is easy to check that

$$\overline{z_1 + z_2} = \overline{z_1} + \overline{z_2},$$

$$\overline{z_1 z_2} = \overline{z_1} \cdot \overline{z_2},$$

and

$$\overline{\left(\frac{z_1}{z_2}\right)} = \frac{\bar{z}_1}{\bar{z}_2}, \qquad \text{whenever } z_2 \neq 0.$$

The magnitude (or modulus) of $a + bi$ is $\sqrt{a^2 + b^2}$. In general, the magnitude of z is denoted $|z|$. As we might expect, $|z_1 z_2| = |z_1||z_2|$, and for addition we have the Schwarz inequality, $|z_1 + z_2| \leq |z_1| + |z_2|$. Schwarz's inequality yields a second inequality which is often useful:

$$\big||z_1| - |z_2|\big| \leq |z_1 - z_2|.$$

To prove this, note that $|z_1| = |z_2 + (z_1 - z_2)| \leq |z_2| + |z_1 - z_2|$, so $|z_1| - |z_2| \leq |z_1 - z_2|$. Similarly, $|z_2| - |z_1| \leq |z_1 - z_2|$, so $-(|z_1| - |z_2|) \leq |z_1 - z_2|$. Then $\big||z_1| - |z_2|\big| \leq |z_1 - z_2|$, as we wanted to show.

Conjugacy and magnitude are related by

$$|z|^2 = z\bar{z} \qquad \text{and} \qquad |\bar{z}| = |z|.$$

A different point of view is achieved by thinking of $a + bi$ as an ordered pair or 2-vector (a, b). This involves a slight loss in computational facility, as, for example, $(a, b) \cdot (c, d) = (ac - bd, ad + bc)$ is not as obvious as $(a + bi) \cdot (c + di) = (ac - bd) + (ad + bc)i$ obtained by direct multiplication. The gain is in geometric insight, as 2-vectors can be thought of as points in the plane. In this context the plane is often called the *complex plane*, with $\text{Re}(z)$ plotted along the horizontal (real) axis and $\text{Im}(z)$ along the vertical (imaginary) axis. Now $|z|$ is interpreted as the distance from the origin to z, and $|z_1 - z_2|$ is the distance between z_1 and z_2. The conjugate $\bar{z}$ of z can be thought of as the reflection of z about the real axis (Fig. 9-1).

It is sometimes convenient to go to polar coordinates. Putting $a = r\cos(\theta)$ and $b = r\sin(\theta)$, then $r = |a + bi|$ and θ is the angle of rotation from the positive real axis to the line from the origin to $a + bi$. Obviously, θ is determined only to within integer multiples of 2π. The numbers $\theta + 2n\pi$, $n = 0, \pm 1, \pm 2, \ldots$, are called *arguments* of $a + bi$. Often we shall specify a certain argument for use in calculations (e.g., by restricting $0 \leq \theta < 2\pi$ or $-\pi < \theta \leq \pi$).

Sequences of complex numbers can be handled as sequences of 2-vectors. Then $\{a_n + ib_n\}$ converges to $\alpha + i\beta$ when $a_n \to \alpha$ and $b_n \to \beta$, and we can use our theorems on real sequences separately on $\{\text{Re}(a_n + b_n i)\}_{n=1}^{\infty}$ and

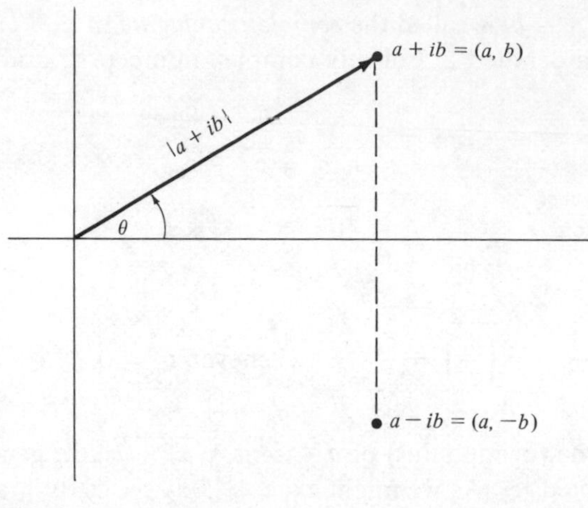

Fig. 9-1

$\{\text{Im}(a_n + b_n i)\}_{n=1}^{\infty}$. There is also a Cauchy theorem for complex sequences: *$\{z_n\}$ converges if and only if, given $\varepsilon > 0$, there is some N such that $|z_n - z_m| < \varepsilon$ whenever $n, m \geq N$.*

Since an infinite series may be treated by looking at the sequence of partial sums, we also have available some machinery for testing series $\sum_{n=1}^{\infty} z_n$ when the z_n's are complex. In particular, if $\sum_{n=1}^{\infty} a_n = \alpha$ and $\sum_{n=1}^{\infty} b_n = \beta$, then $\sum_{n=1}^{\infty} a_n + ib_n = \alpha + i\beta$. Finally, convergence of $\sum_{n=1}^{\infty} |z_n|$ implies that of $\sum_{n=1}^{\infty} z_n$.

We remark in closing this section that it is often convenient to adopt the point of view that every real number is a complex number with zero imaginary part. That is, we identify the real number α with the complex number $\alpha + 0i$. In the geometric view of complex numbers as 2-vectors, this is equivalent to thinking of the real number α as the 2-vector $(\alpha, 0)$, which is on the horizontal (real) axis.

Exercises for Section 9.1

1. Let $p(z)$ be a polynomial of positive degree

$$p(z) = a_0 + a_1 z + \cdots + a_n z_n, \qquad a_n \neq 0.$$

Suppose that ξ is a root $[p(\xi) = 0]$ and that all the a_i's are real. Show that $\bar{\xi}$ is also a root. What happens if one or more a_i is complex?

2. Show that

$$|z_1 + z_2|^2 = |z_1|^2 + |z_2|^2 + 2\,\text{Re}(z_1\bar{z}_2)$$

and

$$|z_1 - z_2|^2 = |z_1|^2 + |z_2|^2 - 2\,\text{Re}(z_1\bar{z}_2).$$

3. Prove Lagrange's identity:

$$\left|\sum_{j=1}^{n} z_j Z_j\right| = \left(\sum_{j=1}^{n} |z_j|^2\right)\left(\sum_{j=1}^{n} |Z_j|^2\right) - \sum_{1 \le j < k \le n} |z_j Z_k - z_k Z_j|^2.$$

Use this to prove Cauchy's inequality:

$$\left|\sum_{j=1}^{n} z_j Z_j\right|^2 \le \left(\sum_{j=1}^{n} |z_j|^2\right)\left(\sum_{j=1}^{n} |Z_j|^2\right).$$

4. Suppose that $\operatorname{Im} z \neq 0$ and $z \neq \pm i$. Show that $z/(1 + z^2)$ is real only if $|z|^2 = 1$.

5. Suppose that $|z| \le 1$ and $|Z| \le 1$. Show that

$$|z + Z| \le |1 + \bar{z}Z|.$$

6. Let $|z_i| < 1$ for $i = 1, \ldots, n$, and $\alpha_i \ge 0$ such that $\sum_{j=1}^{n} \alpha_j = 1$. Prove that $\left|\sum_{j=1}^{n} \alpha_j z_j\right| < 1$.

7. Prove the Cauchy theorem for complex sequences: $\{z_n\}_{n=1}^{\infty}$ converges if and only if, given $\varepsilon > 0$, there is some N such that $|z_n - z_m| < \varepsilon$ whenever $n, m \ge N$.

8. Sketch in the complex plane the locus of points z satisfying each of the following:
(a) $|z + i| = |z - i|$.
(b) $\bar{z} = \dfrac{1}{z + 2i}$
(c) $2 \le \operatorname{Re}(z + i) < 4$
(d) $\operatorname{Im}(z) > 5$
(e) $\operatorname{Im}(z + 2i) = \operatorname{Re}(z + 8)$.
(f) $\operatorname{Re}(z^3) = \operatorname{Im}(z^3)$.
(g) $\operatorname{Re}(z^2 + 2) = 14$.
(h) $|z|^2 = \operatorname{Im}(z)$.
(i) $\left|\dfrac{z - a}{1 - \bar{a}z}\right| \le 1$, where a is any complex number with $|a| < 1$.

9.2 Complex Functions, Limits, and Continuity

We shall want to consider functions that assign complex numbers to complex numbers. Such functions are often called *complex functions*. Examples of such functions are

$$f(z) = z^2 \qquad \text{for each complex } z,$$

and

$$g(z) = \frac{8z^2 + 2}{1 - z} \qquad \text{for } z \neq 1.$$

Often complex functions can be thought of in terms of real-valued functions of two variables as follows. If f is defined on a set D of complex numbers $x + iy$, write $u(x, y) = \operatorname{Re} f(x + iy)$ and $v(x, y) = \operatorname{Im} f(x + iy)$. Then

$$f(x + iy) = u(x, y) + iv(x, y),$$

with u and v real-valued functions defined on the set D in the plane. For example, with $f(z) = z^2$, we have $f(z) = f(x + iy) = (x + iy)^2 = x^2 - y^2 + i(2xy)$, so $u(x, y) = x^2 - y^2$ and $v(x, y) = 2xy$. Although complex functions will be seen to have properties surprisingly different from those of real-valued functions of real variables, nevertheless the student should note how often it is useful in analyzing complex-valued f to write f as $u + iv$, and apply standard theorems on real-valued functions of two real variables to u and v.

The notion of limit is defined exactly as we might reasonable expect. We say that f has the *limit* L at z_0, and write $\lim_{z \to z_0} f(z) = L$, if, given $\varepsilon > 0$, there is some $\delta > 0$ such that $|f(z) - L| < \varepsilon$ whenever $0 < |z - z_0| < \delta$, and $f(z)$ is defined. Generally, in order to make the statement nonvacuous, it is assumed in speaking of a limit of f at z_0 that $f(z)$ is defined at points arbitrarily close to z_0, that is, at least on some sequence of points converging to z_0.

It is easy to copy arguments from elementary calculus to show that

(1) $\lim_{z \to z_0} (f(z) + g(z)) = L + K$,

(2) $\lim_{z \to z_0} (f(z) \cdot g(z)) = L \cdot K$, and

(3) $\lim_{z \to z_0} (f(z)/g(z)) = L/K$

whenever $\lim_{z \to z_0} f(z) = L$, $\lim_{z \to z_0} g(z) = K$, and, in (3), $K \neq 0$.

Next, we say that f is *continuous* at z_0 if $\lim_{z \to z_0} f(z) = f(z_0)$, and that f is continuous on a set D if f is continuous at each point in D. Sums, products, and quotients (where defined) of continuous functions are continuous.

The conjugate function $\bar{f}$ of f is defined by $\bar{f}(z) = \overline{f(z)}$. Since

$$|\bar{f}(z) - \bar{f}(z_0)| = |\overline{f(z)} - \overline{f(z_0)}| = |\overline{f(z) - f(z_0)}| = |f(z) - f(z_0)|,$$

then f is continuous at z_0 exactly when $\bar{f}$ is.

We also define the functions $\operatorname{Re} f$, $\operatorname{Im} f$, and $|f|$ by

$$(\operatorname{Re} f)(z) = \operatorname{Re} f(z),$$

$$(\operatorname{Im} f)(z) = \operatorname{Im} f(z),$$

$$|f|(z) = |f(z)|.$$

Since $\operatorname{Re} f = \frac{1}{2}(f + \bar{f})$, $\operatorname{Im} f = \frac{1}{2}(f - \bar{f})$, and $|f|^2 = (f \cdot \bar{f})$, these three functions are continuous whenever f is.

In concluding this section, we note that *a complex-valued function continuous on a compact set is bounded.* For, when we write $f(x + iy) = u(x, y) + iv(x, y)$ for $x + iy$ in compact D, then u and v are continuous on D by the preceding paragraph, hence bounded on D. If $|u(x, y)| \leq M$ and $|v(x, y)| \leq N$ on D, then

$$|f(x + iy)| = \sqrt{u(x, y)^2 + v(x, y)^2} \leq \sqrt{M^2 + N^2} \qquad \text{for } (x, y) \text{ in } D.$$

Exercises for Section 9.2

1. For each of the following, find u and v such that $f(x + iy) = u(x, y) + iv(x, y)$ for $z = x + iy$. Also, determine where f is continuous.

(a) $f(z) = z^3$

(b) $f(z) = \dfrac{z^3}{|z|}$, $z \neq 0$

(c) $f(z) = \dfrac{1}{1 - z}$, $z \neq 1$

(d) $f(z) = z^{20}$. *Hint:* Use the binomial formula $(a + b)^n = \sum_{j=0}^{n} \binom{n}{j} a^j b^{n-j}$.

(e) $f(z) = \dfrac{z^2 - 2z + 1}{\bar{z} + 3 + i}$, $\bar{z} + 3 + i \neq 0$

2. For each of the functions of Exercise 1, determine $\bar{f}$ and $|f|$, and determine in each case $\alpha(x, y)$ and $\beta(x, y)$ such that $\bar{f}(x + iy) = \alpha(x, y) + i\beta(x, y)$.

3. Write $z = x + iy$ and $f(z) = u(x, y) + iv(x, y)$. Suppose that $\lim_{(x,y)\to(a,b)} u(x, y) = L$ and $\lim_{(x,y)\to(a,b)} v(x, y) = M$. Does it follow that $\lim_{z\to a+ib} f(z) = L + iM$? If so, give a proof. If not, give a counterexample.

9.3 Analyticity

We would now like to define a concept of derivative for functions of a complex variable. It is natural to consider

$$\lim_{z\to z_0} \frac{f(z) - f(z_0)}{z - z_0}.$$

If this limit exists finite, we call it the *derivative* of f at z_0, denoted $f'(z_0)$ or

$$\left.\frac{df}{dz}\right|_{z_0} \qquad \text{or} \qquad \frac{df}{dz}(z_0).$$

Another way of writing the same limit is

$$\lim_{h\to 0} \frac{f(z_0 + h) - f(z_0)}{h},$$

with the understanding here that h must be allowed to go to zero through complex values.

It is easy to show that the usual rules hold for differentiating sums, products, and quotients:

$$(f+g)'(z_0) = f'(z_0) + g'(z_0),$$
$$(fg)'(z_0) = f(z_0)g'(z_0) + f'(z_0)g(z_0),$$

and

$$(f/g)'(z_0) = \frac{g(z_0)f'(z_0) - f(z_0)g'(z_0)}{g(z_0)^2}$$

whenever $f'(z_0)$ and $g'(z_0)$ exist and, in the quotient formula $g(z_0) \neq 0$. Similarly, under suitable conditions we have chain-rule differentiation:

$$f(g(z))' = f'(g(z)) \cdot g'(z).$$

Differentiability at a point obviously implies continuity there, as in the real case.

Derivatives of familiar functions are computed in the usual way. For example, $d(z^n)/dz = nz^{n-1}$ when n is a nonzero integer. We shall also scc that such formulas as $d(e^z)/dz = e^z$ and $d(\cos(z))/dz = -\sin(z)$ hold, after we have defined the complex exponential and trigonometric functions.

We shall now take a closer look at the concept of complex differentiation and place it in the setting that will be most useful for later developments.

To begin, let us write $z = x + iy$ and $f(z) = f(x + iy) = u(x, y) + iv(x, y)$. We shall assume in the following discussion that $f(z)$ is defined in some neighborhood of $z_0 = x_0 + iy_0$, so $f(z_0 + h)$ is defined for $|h|$ sufficiently small, and $f'(z_0)$ exists. Under these conditions we can compute a number of useful expressions for $f'(z)$ in terms of partials of u and v at (x_0, y_0).

First, if h is chosen real (so $z_0 + h$ is on a horizontal line segment through z_0), then

$$\frac{f(z_0+h) - f(z_0)}{h} = \frac{u(x_0+h, y_0) + iv(x_0+h, y_0) - u(x_0, y_0) - iv(x_0, y_0)}{h}$$

$$= \frac{u(x_0+h, y_0) - u(x_0, y_0)}{h} + i\left(\frac{v(x_0+h, y_0) - v(x_0, y_0)}{h}\right).$$

Letting $h \to 0$, we have

$$f'(z_0) = \frac{\partial u(x_0, y_0)}{\partial x} + i\,\frac{\partial v(x_0, y_0)}{\partial x}.$$

But, we may also choose h to be purely imaginary, say $h = ik$, where k is real (now $z_0 + h$ is on a vertical line segment through z_0). Using the fact that $1/i = -i$, we have

$$\frac{f(z_0 + h) - f(z_0)}{h} = \frac{u(x_0, y_0 + k) + iv(x_0, y_0 + k) - u(x_0, y_0) - iv(x_0, y_0)}{ik}$$

$$= \frac{1}{i}\left[\frac{u(x_0, y_0 + k) - u(x_0, y_0)}{k} + i\left(\frac{v(x_0, y_0 + k) - v(x_0, y_0)}{k}\right)\right].$$

As $k \to 0$, we then get

$$f'(z_0) = \frac{\partial v(x_0, y_0)}{\partial y} - i\frac{\partial u(x_0, y_0)}{\partial y}.$$

Equating these expressions for $f'(z_0)$, we have

$$\frac{\partial u(x_0, y_0)}{\partial x} + i\frac{\partial v(x_0, y_0)}{\partial x} = \frac{\partial v(x_0, y_0)}{\partial y} - i\frac{\partial u(x_0, y_0)}{\partial y}.$$

Since the real (imaginary) part of the left side must equal the real (imaginary) part of the right, we conclude that

$$\frac{\partial u(x_0, y_0)}{\partial x} = \frac{\partial v(x_0, y_0)}{\partial y}$$

and

$$\frac{\partial v(x_0, y_0)}{\partial x} = -\frac{\partial u(x_0, y_0)}{\partial y}.$$

These are the Cauchy–Riemann equations for f, and are satisfied at any point z_0 at which $f'(z_0)$ exists and about which $f(z)$ is defined.

We now consider the converse. Suppose that $f(z)$ is defined in some disc about z_0, and the Cauchy–Riemann equations are satisfied by $\operatorname{Re} f$ and $\operatorname{Im} f$ at (x_0, y_0). Does $f'(z_0)$ necessarily exist? The answer is no. As an example, put

$$f(z) = \begin{cases} \dfrac{z^5}{|z|^4} & \text{for } z \neq 0, \\ 0 & \text{for } z = 0. \end{cases}$$

A straightforward (but long) calculation yields $u_x(0, 0) = v_y(0, 0) = 1$, and $u_y(0, 0) = -v_x(0, 0) = 0$. But f cannot have a derivative at 0. For, $f(h)/h = 1$ when h is real, so $f'(0)$ must be 1 if $f'(0)$ exists. On the other hand, along $x = y$ we would get $f'(0) = \lim_{x \to 0} (x(1 + i))^5/|x|^4|1 + i|^4 x(1 + i) = (1 + i)^4/4 \neq 1$.

So we need something stronger than just the Cauchy–Riemann equations to ensure differentiability. Suppose that (1) the Cauchy–Riemann equations hold throughout a disc $|z - \zeta| < \delta$ centered at ζ, and (2) the partials u_x, u_y, v_x, and v_y are continuous in this disc. Choose any $z_0 = x_0 + iy_0$ in this disc (so $|z_0 - \zeta| < \delta$). We shall show that $f'(z_0)$ exists.

Write $h = \Delta x + i\,\Delta y$ and note that

$$\begin{aligned} &f(z_0 + h) - f(z_0) \\ &\quad = u(x_0 + \Delta x, y_0 + \Delta y) - u(x_0, y_0) + i(v(x_0 + \Delta x, y_0 + \Delta y) - v(x_0, y_0)). \end{aligned}$$

Now, continuity of the partials in the disc ensures differentiability of u and v at (x_0, y_0). Then we may write

$$\begin{aligned} &u(x_0 + \Delta x, y_0 + \Delta y) - u(x_0, y_0) \\ &\qquad = u_x(x_0, y_0)\,\Delta x + u_y(x_0, y_0)\,\Delta y + \varepsilon_1\sqrt{(\Delta x)^2 + (\Delta y)^2} \end{aligned}$$

and

$$\begin{aligned} &v(x_0 + \Delta x, y_0 + \Delta y) - v(x_0, y_0) \\ &\qquad = v_x(x_0, y_0)\,\Delta x + v_y(x_0, y_0)\,\Delta y + \varepsilon_2\sqrt{(\Delta x)^2 + (\Delta y)^2}, \end{aligned}$$

where $\varepsilon_1, \varepsilon_2 \to 0$ as $(\Delta x, \Delta y) \to (0, 0)$.

Since $\sqrt{(\Delta x)^2 + (\Delta y)^2} = |h|$, we now have

$$\begin{aligned} &\frac{f(z_0 + h) - f(z_0)}{h} \\ &= u_x(x_0, y_0)\frac{\Delta x}{h} + u_y(x_0, y_0)\frac{\Delta y}{h} + \varepsilon_1\frac{|h|}{h} \\ &\quad + i\left[v_x(x_0, y_0)\frac{\Delta x}{h} + v_y(x_0, y_0)\frac{\Delta y}{h} + \varepsilon_2\frac{|h|}{h}\right] \\ &= (u_x(x_0, y_0) + iv_x(x_0, y_0))\frac{\Delta x}{h} + (u_y(x_0, y_0) + iv_y(x_0, y_0))\frac{\Delta y}{h} + (\varepsilon_1 + i\varepsilon_2)\frac{|h|}{h} \\ &= (u_x(x_0, y_0) + iv_x(x_0, y_0))\frac{\Delta x}{h} + (-v_x(x_0, y_0) + iu_x(x_0, y_0))\frac{\Delta y}{h} + (\varepsilon_1 + i\varepsilon_2)\frac{|h|}{h} \\ &= (u_x(x_0, y_0) + iv_x(x_0, y_0))\frac{\Delta x + i\,\Delta y}{h} + (\varepsilon_1 + i\varepsilon_2)\frac{|h|}{h} \\ &= u_x(x_0, y_0) + iv_x(x_0, y_0) + (\varepsilon_1 + i\varepsilon_2)\frac{|h|}{h}. \end{aligned}$$

Finally, note that $||h|/h| = |h|/|h| = 1$, so that as $(\Delta x, \Delta y) \to (0, 0)$, $(\varepsilon_1 + i\varepsilon_2)(|h|/h) \to 0$. Thus, as $h \to 0$,

$$\frac{f(z_0 + h) - f(z_0)}{h}$$

has the limit $u_x(x_0, y_0) + iv_x(x_0, y_0)$, implying that $f'(z_0)$ exists. Since z_0 was any point of the disc $|z - \zeta| < \delta$, we have shown under conditions (1) and (2) that f is differentiable at each point of the disc.

It is standard terminology to say that f is analytic at z_0 if $f'(z_0)$ exists in some disc $|z - z_0| < \delta$ about z_0. Rephrasing the preceding argument gives us the following:

Theorem

f is analytic at z_0 if $\text{Re} f$ *and* $\text{Im} f$ *have continuous first partials that satisfy the Cauchy–Riemann equations throughout some disc about z_0.*

Exercises for Section 9.3

1. For each of the following, determine whether or not f is analytic at the indicated point (1) by checking for existence of the derivative directly from the definitions and (2) by computing $\text{Re}(f)$ and $\text{Im}(f)$, and checking the Cauchy–Riemann equations about the point.
(a) $f(z) = z^2$, z any complex number
(b) $f(z) = \bar{z}$, z arbitrary
(c) $f(z) = |z|$, z arbitrary
(d) $f(z) = \dfrac{1}{1 - z}$, $z = 1 + i$
(e) $f(z) = z^2 + i$, $z = 3i$

2. Let $f = u + iv$ be analytic in $|z - z_0| < \delta$ and suppose that u and v have continuous second partials. Show that u and v are harmonic [i.e., $\nabla^2 u = \nabla^2 v = 0$) for (x, y) within distance δ of (x_0, y_0)].

3. Let

$$f(z) = \begin{cases} \dfrac{z^5}{|z|^4}, & z \neq 0, \\ 0, & z = 0. \end{cases}$$

Compute $\text{Re}(f)$ and $\text{Im}(f)$ and show that the Cauchy–Riemann equations are satisfied at $(0, 0)$.

4. Let u be harmonic and have continuous first and second partials in a region D interior to a simple closed curve C in the plane. Let

$$f(z) = u_y(x, y) + iu_x(x, y)$$

for $z = x + iy$ in D. Show that f is analytic in D.

9.4 Important Analytic Functions

Having defined the concept of an analytic function, it is natural to look around for some examples (other than, say, polynomials). At the same time, this affords a good opportunity to develop complex analogues of some of the more important functions from real analysis.

We shall begin with the exponential function. There are several very elegant approaches to choose from, but we shall be as direct as possible and define

$$E(x + iy) = e^x[\cos(y) + i\sin(y)].$$

Thus

$$\operatorname{Re} E(x + iy) = e^x \cos(y)$$

and

$$\operatorname{Im} E(x + iy) = e^x \sin(y).$$

Since

$$\frac{\partial\,(e^x \cos(y))}{\partial x} = e^x \cos(y) = \frac{\partial\,(e^x \sin(y))}{\partial y},$$

and

$$\frac{\partial\,(e^x \sin(y))}{\partial x} = e^x \sin(y) = -\frac{\partial\,(e^x \cos(y))}{\partial y},$$

then the Cauchy–Riemann equations hold throughout the complex plane. Further, these partials are everywhere continuous, so E is analytic everywhere. Such a function is said to be *entire*.

Note that

$$\begin{aligned} E'(x + iy) &= \frac{\partial(e^x \cos(y))}{\partial x} + i\frac{\partial(e^x \sin(y))}{\partial x} \\ &= e^x \cos(y) + ie^x \sin(y) = E(x + iy). \end{aligned}$$

Further, $E(x) = e^x$ when $y = 0$ (so E agrees with the usual exponential function on the reals). Finally,

$$\begin{aligned} &E(x_1 + iy_1)E(x_2 + iy_2) \\ &\quad = e^{x_1}[\cos(y_1) + i\sin(y_1)]e^{x_2}[\cos(y_2) + i\sin(y_2)] \\ &\quad = e^{x_1+x_2}\{[\cos(y_1)\cos(y_2) - \sin(y_1)\sin(y_2)] \\ &\qquad\qquad + i[\cos(y_1)\sin(y_2) + \sin(y_1)\cos(y_2)]\} \\ &\quad = e^{x_1+x_2}[\cos(y_1 + y_2) + i\sin(y_1 + y_2)] \\ &\quad = E((x_1 + x_2) + i(y_1 + y_2)) = E((x_1 + iy_1) + (x_2 + iy_2)). \end{aligned}$$

Put more briefly,

$$E(z_1 + z_2) = E(z_1)E(z_2).$$

All this suggests that we call E the complex exponential function, and henceforth write $E(z)$ as e^z.

In magnitude,

$$|e^z| = |e^x \cos(y) + ie^x \sin(y)| = \sqrt{e^{2x}\cos^2(y) + e^{2x}\sin^2(y)} = e^x,$$

which is never zero. Thus e^z is also never zero. Further, $|e^{iy}| = 1$ for all real y, as $|e^{iy}| = \sqrt{\cos^2(y) + \sin^2(y)} = 1$ if y is real.

In order to verify that $e^{-z} = 1/e^z$, write

$$e^{-z} = e^{-x-iy} = e^{-x}e^{-iy} = \frac{e^{-x}e^{-iy}e^{iy}}{e^{iy}}$$

$$= \frac{1}{e^x e^{iy}}(e^{-iy}e^{iy}) = \frac{1}{e^x e^{iy}}|e^{iy}|^2 = \frac{1}{e^z}.$$

A perhaps surprising difference between the real and complex exponential functions is the periodicity of the latter, with period $2\pi i$. To see this, write

$$e^{z+2\pi i} = e^{x+iy+2\pi i} = e^{x+i(y+2\pi)} = e^x[\cos(y + 2\pi) + i\sin(y + 2\pi)]$$

$$= e^x[\cos(y) + i\sin(y)] = e^z.$$

In general, not only does $e^{2n\pi i} = 1$ for any integer n, but $e^z = 1$ only if $z = 2n\pi i$. For suppose that $z = x + iy$ and $e^z = 1$. Then

$$e^x \cos(y) + ie^x \sin(y) = 1.$$

Then

$$e^x \cos(y) = 1 \quad \text{and} \quad e^x \sin(y) = 0.$$

Since $e^x \neq 0$, $\sin(y) = 0$, so $y = k\pi$ for some integer k. But then $e^x \cos(y) = e^x \cos(k\pi) = 1$. Then $x = 0$ and k is even, say $k = 2n$, so $z = x + iy = 2n\pi i$.

Note that this is all consistent with the representation of z in polar form, $re^{i\theta}$, where $r = |z|$ and θ is any argument of z. If we increase the argument by $2n\pi$, we should not change the complex number represented. We can also see the addition rule for arguments. If $z_1 = r_1e^{i\theta_1}$ and $z_2 = r_2e^{i\theta_2}$, then $z_1z_2 = r_1r_2e^{i(\theta_1+\theta_2)}$, which means that $|z_1z_2| = r_1r_2 = |z_1||z_2|$ and $\arg(z_1z_2) - [\arg(z_1) + \arg(z_2)] = 2n\pi$. Thus the *argument of a product is the sum of the individual arguments, plus an arbitrary* $2n\pi$, *n integer*, taking into account the periodicity of e^z.

We now turn to complex analogues of sine and cosine. Observe that

$$e^{iy} = \cos(y) + i\sin(y)$$

and

$$e^{-iy} = \cos(y) - i\sin(y).$$

Then

$$\frac{e^{iy} - e^{-iy}}{2i} = \sin(y)$$

and

$$\frac{e^{iy} + e^{-iy}}{2} = \cos(y) \qquad \text{for any real } y.$$

Using these as a guide, we define

$$\sin(z) = \frac{1}{2i}(e^{iz} - e^{-iz})$$

and

$$\cos(z) = \frac{1}{2}(e^{iz} + e^{-iz}).$$

Writing $z = x + iy$, we have

$$\begin{aligned} \sin(z) &= \frac{1}{2i}(e^{ix-y} - e^{-ix+y}) \\ &= \frac{1}{2i}[e^{-y}\cos(x) + ie^{iy}\sin(x) - e^{y}\cos(x) + ie^{y}\sin(x)] \\ &= \frac{1}{2}(e^{y} + e^{-y})\sin(x) + \frac{1}{2}(e^{y} - e^{-y})\cos(x) \\ &= \cosh(y)\sin(x) + i\sinh(y)\cos(x). \end{aligned}$$

These give Re $\sin(z)$ and Im $\sin(z)$, and it is easy to check from their partials that $\sin(z)$ is analytic in the whole complex plane.

A similar calculation gives

$$\cos(z) = \cosh(y)\cos(x) - i\sinh(y)\sin(x).$$

From this it follows that $\cos(z)$ is also an entire function. The derivative formulas

$$\frac{d\sin(z)}{dz} = \cos(z)$$

and

$$\frac{d\cos(z)}{dz} = -\sin(z)$$

are just like the formulas for the real sine and cosine functions. Further, $\sin^2(z) + \cos^2(z) = 1$ for all z, with the other identities [such as $\sin(2z) = 2\sin(z)\cos(z)$]

also following rather easily. Here the similarities between real and complex sine and cosine are interrupted, however, and we see, as with the periodicity of the exponential function, that something unusual happens when we go from the real to the complex. While $|\sin(x)| \leq 1$ and $|\cos(x)| \leq 1$ for real x, $\sin(z)$ and $\cos(z)$ are unbounded for z complex. For example, putting $z = iy$, we have $\cos(z) = \cos(iy) = \cosh(y) \to \infty$ as $y \to \infty$. As we shall see later, it is impossible for a nonconstant function to be both bounded and analytic in the whole plane.

It is interesting to note that $\sin(x)$ and $\cos(x)$ keep the same zeros when real x is replaced by complex z. For suppose that $\sin(z) = 0$. Then $\cosh(y)\sin(x) = 0$ and $\sinh(y)\cos(x) = 0$. Now, $\cosh(y)$ is never zero, so $\sin(x) = 0$, and $x = n\pi$ for integer n. But then $\cos(x) \neq 0$, so $\sinh(y) = 0$, and we have $y = 0$. Thus $z = n\pi$.

Similarly, if $\cos(z) = 0$, then we arrive at $z = (2n + 1)\pi/2$ for integer n.

Having defined sine and cosine, it is a simple matter to put

$$\tan(z) = \frac{\sin(z)}{\cos(z)}, \qquad \cot(z) = \frac{\cos(z)}{\sin(z)}, \qquad \sec(z) = \frac{1}{\cos(z)}, \qquad \csc(z) = \frac{1}{\sin(z)},$$

whenever the denominators are not zero.

The derivative formulas for these are the same as for the corresponding real functions. The *hyperbolic functions* are defined for complex z much as they are for real x:

$$\sinh(z) = \frac{e^z - e^{-z}}{2}$$

and

$$\cosh(z) = \frac{e^z + e^{-z}}{2}.$$

From the definitions it follows immediately that

$$\sinh(z) = -i\sin(iz)$$

and

$$\cosh(z) = \cos(iz).$$

This means that sinh and cosh are entire functions, and

$$\sinh'(z) = \cosh(z), \qquad \cosh'(z) = \sinh(z).$$

Thus far we have extended the exponential, trigonometric, and hyperbolic functions from the real line to the complex plane, each extension resulting in an entire function. We would like now to extend the logarithm and power functions. However, these present certain difficulties which must be appreciated if the functions are to be used properly in later work.

We begin with the logarithm. We would like to define $\log(z)$ so that $\log(z_1 \cdot z_2) = \log(z_1) + \log(z_2)$ and $\log(z^a) = a\log(z)$. One method of doing this is suggested by the definition of the complex exponential function. If we write $z = x + iy = r\cos(\theta) + ir\sin(\theta) = r[\cos(\theta) + i\sin(\theta)] = re^{i\theta}$, then we would expcct to have $\log(z) = \log(re^{i\theta}) = \log(r) + \log(e^{i\theta}) = \log(r) + i\theta\log(e) = \log(r) + i\theta$. Since $r = |z|$ and is real, and θ is an argument of z, we are led to define $\log(z) = \ln(|z|) + i\arg(z)$, where ln is the usual real logarithm. The trouble with this is that z determines $\arg(z)$ only to within integer multiples of 2π. We therefore agree that any number $\ln(|z|) + i\arg(z)$, for $z \neq 0$, will be called a *logarithm* of z, and that the symbol $\log(z)$ will be taken to mean any one of these numbers. Thus $\log(z)$ is not a function in the usual sense.

Often it is convenient to define a principal value of the argument, called Arg, by restricting, say $-\pi < \text{Arg}(z) \leq \pi$. Then

$$\log(z) = \ln(|z|) + i\,\text{Arg}(z) + 2n\pi i, \qquad n = 0, \pm 1, \pm 2, \ldots$$

The fact that the symbol $\log(z)$ has infinitely many different meanings calls for extreme care in attempting to establish the sense in which such formulas as $\ln(xy) = \ln(x) + \ln(y)$ are valid in the complex case. We shall examine this now.

First, note that, since $e^{2n\pi i} = 1$ for integer n, we have

$$\begin{aligned} e^{\log(z)} &= e^{\ln(|z|) + i\text{Arg}(z) + 2n\pi i} \\ &= e^{\ln(|z|)}e^{i\text{Arg}(z)} \\ &= |z|\{\cos[\text{Arg}(z)] + i\sin[\text{Arg}(z)]\} = z, \end{aligned}$$

as in the real case. In the real case, however, we expect not only $e^{\ln(x)} = x$, but $x = \ln(e^x)$. The latter does not go over to the complex case as does the former. In fact, write $z = x + iy$, and note that

$$e^z = e^x[\cos(y) + i\sin(y)].$$

Then $|e^z| = e^x$ and $\arg(e^z) = y + 2n\pi$, so

$$\begin{aligned} \log(e^z) &= \ln(|e^z|) + i\arg(e^z) = \ln(e^x) + i(y + 2n\pi) \\ &= x + iy + 2n\pi i = z + 2n\pi i. \end{aligned}$$

Instead of $\log(e^z) = z$, then we have only $\log(e^z) = z + 2n\pi i$, where n is any integer.

We can define a logarithm function [as opposed to the many-valued $\log(z)$ above] by suitably restricting the argument. If, say, we denote by $\text{Arg}(z)$ the value of the argument in the range $(-\pi, \pi]$, then $\text{Log}(z) = \ln(|z|) + i\,\text{Arg}(z)$

for $z \neq 0$ is a legitimate function, giving only one value of Log(z) for each $z \neq 0$. We now have, for $z = x + iy \neq 0$,

$$\begin{aligned}\text{Log}(e^z) &= \ln(e^z) + i\,\text{Arg}(e^z)\\ &= \ln(e^x) + iy = x + iy = z.\end{aligned}$$

As an illustration of all this, put $z = 1 + i$. By direct computation, $\log(z) = \ln(\sqrt{2}) + i\arg(1 + i) = \ln(\sqrt{2}) + i(\pi/4) + 2n\pi i$. Then

$$\begin{aligned}e^{\log(z)} &= e^{\ln(\sqrt{2})}e^{i(\pi/4)}e^{2n\pi i}\\ &= e^{\ln(\sqrt{2})}e^{i\pi/4} = \sqrt{2}[\cos(\pi/4) + i\sin(\pi/4)]\\ &= \sqrt{2}\left(\frac{\sqrt{2}}{2} + i\frac{\sqrt{2}}{2}\right) = 1 + i = z,\end{aligned}$$

as we showed should be the case. But

$$\log(e^z) = \log(e^{1+i}) = \ln(|e^{1+i}|) + i\arg(e^{1+i}).$$

Now $e^{1+i} = ee^i$, so $|e^{1+i}| = e$. Then $\ln(|e^{1+i}|) = \ln(e) = 1$. Next, $\arg(e^{1+i}) = 1 + 2n\pi$, with n any integer. Then

$$\log(e^{1+i}) = 1 + i(1 + 2n\pi) = 1 + i + 2n\pi i.$$

We now ask whether it is true that $\log(z_1z_2) = \log(z_1) + \log(z_2)$, as in the real case. From the above we have

$$e^{\log(z_1z_2)} = z_1z_2 = e^{\log(z_1)}e^{\log(z_2)}.$$

However, we cannot conclude from this that $\log(z_1z_2) = \log(z_1) + \log(z_2)$, since the complex exponential is periodic of period $2\pi i$. We can conclude that

$$\frac{e^{\log(z_1z_2)}}{e^{\log(z_1)+\log(z_2)}} = 1 = e^{\log(z_1z_2)-\log(z_1)-\log(z_2)},$$

hence that $\log(z_1z_2) - \log(z_1) - \log(z_2) = 2n\pi i$ for integer n. The extra $2n\pi i$ term should not be too surprising in view of the fact that $\log(z)$ in general denotes infinitely many complex numbers differing by $2n\pi i$.

As an example of this, consider $\log(i(1 + i)) = \log(-1 + i) = \ln(\sqrt{2}) + i\frac{3}{4} + 2n\pi i$, where n is any integer.

Next, compute

$$\log(i) + \log(1 + i) = \ln(1) + \frac{i\pi}{2} + 2k\pi i + \ln(\sqrt{2}) + \frac{i\pi}{4} + 2r\pi i,$$

where k and r are any integers. Now

$$\begin{aligned}
\ln(1) + \frac{i\pi}{2} + 2k\pi i + \ln(\sqrt{2}) + i\frac{\pi}{4} + 2r\pi i \\
&= \ln(\sqrt{2}) + \frac{3\pi i}{4} + 2(k + r)\pi i \\
&= \ln(\sqrt{2}) + \frac{3\pi i}{4} + 2m\pi i \qquad \text{(here we put } m = k + r\text{)}.
\end{aligned}$$

Since k and r are arbitrary integers, so is m. Thus

$$\log(i(1 + i)) = \log(i) + \log(1 + i) + 2m\pi i,$$

where m is any integer.

Since $\log(z)$ is not a function, but a symbol for a lot of numbers, it makes no sense to try to take a derivative. However, if we produce a logarithm function by suitably restricting $\arg(z)$, say, as above, to the interval $(-\pi, \pi]$, then $\text{Log}(z) = \ln(|z|) + i\,\text{Arg}(z)$, $z \neq 0$, defines a function, which we shall call the *principal logarithm function*. To differentiate this, note that for $z \neq 0$,

$$e^{\text{Log}(z)} = z.$$

By the chain rule,

$$\frac{d}{dz}(e^{\text{Log}(z)}) = e^{\text{Log}(z)}\frac{d}{dz}(\text{Log}(z)) = \frac{d}{dz}(z) = 1.$$

Then

$$\frac{d}{dz}(\text{Log}(z)) = \frac{1}{e^{\text{Log}(z)}} = \frac{1}{z}.$$

For $z \neq 0$, the α power of z is given by

$$z^\alpha = e^{\alpha \log(z)}.$$

Since $\text{Log}(z) = \ln(|z|) + i\,\text{Arg}(z) + 2n\pi i$, then

$$z^\alpha = e^{\alpha \ln(|z|)}e^{i\alpha \text{Arg}(z)}e^{2n\pi i\alpha},$$

so that for $\alpha \neq 0$, the symbol z^α generally denotes infinitely many different complex numbers.

As an example

$$\begin{aligned}
i^i = e^{i\log(1)} &= e^{i[\ln(|i|) + i\text{Arg}(i) + 2n\pi i]} \\
&= e^{i(i\pi/2 + 2n\pi i)} = e^{-\pi/2}e^{-2n\pi}, \qquad n = 0, \pm 1, \pm 2, \ldots.
\end{aligned}$$

Surprisingly, all the values of i^i are real.

If we use Log instead of log, we can obtain a principal value of z^α, given by

$$[z^\alpha]_P = e^{\alpha \mathrm{Log}(z)} = e^{\alpha(\ln(|z|) + i\mathrm{Arg}(z))}.$$

As with logarithms, we must investigate the effect of having multiple values for powers on the usual arithmetic calculations. For example, does

$$z^{\alpha_1} z^{\alpha_2} = z^{\alpha_1 + \alpha_2}?$$

To check this, write $z^{\alpha_1+\alpha_2} = e^{(\alpha_1+\alpha_2)\log(z)} = e^{\alpha_1 \log(z)} e^{\alpha_2 \log(z)} = z^{\alpha_1} z^{\alpha_2}$, so this rule still holds.

Next, does $\log(z^\alpha) = \alpha \log(z)$? From previous considerations, the answer is: not exactly. Recall that $\log(e^z) = z + 2n\pi i$. Thus

$$\log(z^\alpha) = \log(e^{\alpha \log(z)}) = \alpha \log(z) + 2n\pi i.$$

Note, however, that $\mathrm{Log}\,(z^\alpha) = \alpha\,\mathrm{Log}(z)$, since we obtained before that $\mathrm{Log}(e^z) = z$; hence $\mathrm{Log}(z^\alpha) = \mathrm{Log}(e^{\alpha \mathrm{Log}(z)}) = \alpha\,\mathrm{Log}(z)$ if we use principal values.

Using this we can determine whether or not $(z^\alpha)^\beta = z^{\alpha\beta}$. Write

$$(z^\alpha)^\beta = e^{\beta \log(z^\alpha)} = e^{\alpha\beta[\log(z) + 2n\pi i]} = e^{\alpha\beta(\log(z)} e^{2n\pi i\beta} = z^{\alpha\beta} e^{2n\pi i \beta}.$$

Again, if we restrict ourselves to a principal value things work out better:

$$[(z^\alpha)^\beta]_P = e^{\beta \mathrm{Log}(z^\alpha)} = e^{\alpha\beta \mathrm{Log}(z)} = [z^{\alpha\beta}]_P.$$

As with logarithms, it makes no sense to attempt to differentiate z^α until we settle upon a specific unambiguous meaning for the symbol. Once this is done, the usual derivative formula results. Thus we have

$$\begin{aligned} \frac{d}{dz}[z^\alpha]_P &= \frac{d}{dz} e^{\alpha \mathrm{Log}(z)} \\ &= e^{\alpha \mathrm{Log}(z)} \frac{d}{dz}(\alpha\,\mathrm{Log}(z)) = e^{\alpha \mathrm{Log}(z)} \left(\frac{\alpha}{z}\right) = \alpha e^{\alpha \mathrm{Log}(z)} z^{-1} \\ &= \alpha e^{\alpha \mathrm{Log}(z)} e^{-\mathrm{Log}(z)} = \alpha e^{(\alpha - 1)\mathrm{Log}(z)} = \alpha [z^{\alpha - 1}]_P. \end{aligned}$$

Here we have used $z^{-1} = 1/z$. In general,

$$[z^{-\alpha}]_P = e^{-\alpha \mathrm{Log}(z)} = \frac{1}{e^{\alpha \mathrm{Log}(z)}} = \frac{1}{[z^\alpha]_P}.$$

It is interesting to note that when α is rational, z^α takes on only a finite number of different values. For example, consider the values of $(1 + i)^{3/4}$. By definition,

$$(1 + i)^{3/4} = e^{(3/4)\log(1+i)} = e^{(3/4)(\ln(\sqrt{2}) + i\pi/4 + 2n\pi i)} = e^{(3/4)\ln(\sqrt{2})} e^{i\pi/4} e^{3n\pi i/2}.$$

When $n = 0, 1, 2$, or 3, we obtain distinct values. They are

$$n = 0: \quad e^{(3/4)\ln(\sqrt{2})}e^{i\pi/4},$$

$$n = 1: \quad e^{(3/4)\ln(\sqrt{2})}e^{i\pi/4}e^{3\pi i/2} = -ie^{(3/4)\ln(\sqrt{2})}e^{i\pi/4},$$

$$n = 2: \quad e^{(3/4)\ln(\sqrt{2})}e^{i\pi/4}e^{3\pi i} = -e^{(3/4)\ln(\sqrt{2})}e^{i\pi/4},$$

$$n = 3: \quad e^{(3/4)\ln(\sqrt{2})}e^{i\pi/4}e^{9\pi i/2} = ie^{(3/4)\ln(\sqrt{2})}e^{i\pi/4}.$$

However, when $n = 4$, $e^{3n\pi i/2} = e^{6\pi i} = 1$, so we repeat the value obtained when $n = 0$; when $n = 5$,

$$e^{3n\pi i/2} = e^{15\pi i/2} = e^{7\pi i}e^{\pi i/2} = -e^{\pi i/2} = -i,$$

and we repeat the value obtained when $n = 1$, and so on. The pattern of repetitions for this example is as follows:

$$n = 0: \quad 4, 8, 12, \ldots,$$

$$n = 1: \quad 5, 9, 13, \ldots, -3, -7, \ldots,$$

$$n = 2: \quad 6, 10, \ldots, -2, -6, -10, \ldots,$$

$$n = 3: \quad 7, 11, 15, \ldots, -1, -5, -9, \ldots.$$

As another example, we shall compute the sixth roots of 8. That is, we want all numbers attached to the symbol $8^{1/6}$. We have

$$8^{1/6} = e^{(1/6)[\ln(8) + i\mathrm{Arg}(8) + 2n\pi i]}.$$

Now, $\mathrm{Arg}(8) = 0$, so

$$8^{1/6} = e^{(1/6)\ln(8)}e^{n\pi i/3}, \qquad n = 0, 1, 2, 3, \ldots.$$

We therefore must look at $e^{n\pi i/3}$, or, equivalently, $\cos(n\pi/3) + i\sin(n\pi/3)$ for integer values of n. We make a table:

n	$\cos(n\pi/3) + i\sin(n\pi/3)$
0	1
1	$\frac{1}{2} + \frac{i\sqrt{3}}{2}$
2	$-\frac{1}{2} + \frac{i\sqrt{3}}{2}$
3	-1
4	$-\frac{1}{2} - \frac{i\sqrt{3}}{2}$
5	$\frac{1}{2} - \frac{i\sqrt{3}}{2}$

For $n = 6$, $e^{n\pi i/3} = e^{2\pi i} = 1$, the same value as for $n = 0$. For $n = 7$, we repeat the value for $n = 1$, and so on. We therefore have, in all, exactly six sixth roots of eight. They are the numbers obtained by multiplying the numbers in the right-hand column by $e^{(1/6)\ln(8)}$.

Exercises for Section 9.4

1. Find the real and imaginary parts of
(a) $\sin(1 + i)$
(b) $\cos(i)$
(c) $\cosh(-3 - 2i)$
(d) e^{8+4i}

2 Compute all possible values of
(a) $\log(i^{i+1})$
(b) $(2 + i)^{3+i}$
(c) $\left(\frac{1}{1-i}\right)^{1/4}$
(d) $\sin(i^{2i})$
(e) $\cosh((1 + i)^{1/2})$

3. Find the real and imaginary parts of $\tan(z)$, $\cot(z)$, $\sec(z)$, $\csc(z)$, $\sinh(z)$, and $\cosh(z)$.

4. If $z = \sinh(w)$, show that

$$w = \log(z + (z^2 + 1)^{1/2}).$$

5. If $w = \sin(z)$, show that $z = -i\log(iz + (1 - z^2)^{1/2})$.

6. If $w = \cos(z)$, show that $z = -i\log(z + (z^2 - 1)^{1/2})$.

7. Use the definitions of $\sin(z)$ and $\cos(z)$ to show that, for $n = 1, 2, 3, \ldots$,

(a) $$\frac{\sin((n + \frac{1}{2})\theta)}{2\sin(\theta/2)} = \frac{1}{2} + \sum_{j=1}^{n} \cos(j\theta)$$

(b) $$\frac{\sin(2n\theta)}{2\sin(\theta)} = \sum_{j=1}^{n} \cos((2j - 1)\theta)$$

(c) $$\frac{\sin(n\theta/2)\sin((n + 1)\theta/2)}{\sin(\theta/2)} = \sum_{j=1}^{n} \sin(j\theta)$$

8. Write each of the following functions in the form $u(x, y) + iv(x, y)$, with $z = x + iy$.
(a) $\tan(z)$
(b) $\tanh(z)$
(c) e^{z^2}
(d) $\sin(\cos(z))$
(e) $\cosh(\sin(z))$

9.5 Integration of Complex Functions

We want now to develop a notion of integration of complex functions over paths in the plane. Here we can set things up in such a way as to be able to fall back heavily upon results already established for line integrals.

If γ is a piecewise smooth curve (path) in the plane, we shall often write $\gamma(t) = (x(t), y(t))$ as $\gamma(t) = x(t) + iy(t)$. Similarly, $\gamma'(t) = x'(t) + iy'(t)$ whenever γ has a derivative.

If γ: $[a, b] \to R^2$ is a smooth curve, and f is continuous on γ, then we put

$$\int_\gamma f(z)\,dz = \int_a^b f(\gamma(t)) \cdot \gamma'(t)\,dt.$$

If γ is piecewise smooth, say γ is a union of smooth $\gamma_1, \ldots, \gamma_n$, then we put $\int_\gamma f(z)\,dz = \sum_{j=1}^{n} \int_{\gamma_j} f(z)\,dz$. We also put $\int_\gamma f(z)\,dz = -\int_{-\gamma} f(z)\,dz$, where $-\gamma$ denotes the path γ with reversed orientation. If γ is a closed curve, we often write $\int_\gamma$ as $\oint_\gamma$, the arrow indicating orientation on γ. At least formally, then, things so far are very much as they were for line integrals of 2-vector-valued functions over curves in the plane in Chapter 6.

In order to clarify the connection between $\int_\gamma f(z)\,dt$ and previous work on line integrals, write $\gamma(t) = \alpha(t) + i\beta(t)$, $z = x + iy$, and $f(z) = u(x, y) + iv(x, y)$. Then

$$\int_\gamma f(z)\,dz = \int_a^b f(\gamma(t)) \cdot \gamma'(t)\,dt = \int_a^b [u(\gamma(t)) + iv\,(\gamma(t))][\alpha'(t) + i\beta'(t)]\,dt$$

$$= \int_a^b \{[u(\gamma(t))\alpha'(t) - v(\gamma(t))\beta'(t)] + i\,[u(\gamma(t))\beta'(t) + v(\gamma(t))\alpha'(t)]\}\,dt$$

$$= \int_\gamma u\,dx - v\,dy + i\int_\gamma v\,dx + u\,dy.$$

Both integrals at the end are of course just ordinary line integrals over a curve in the plane. From this it is obvious that

$$\int_\gamma (\xi f(z) + \zeta\,g(z))\,dz = \xi\int_\gamma f(z)\,dz + \zeta\int_\gamma g(z)\,dz$$

whenever f and g are continuous on γ and ξ and ζ are complex numbers.

We can also use the above calculation to establish a *complex analogue of the fundamental theorem of calculus. Suppose that γ lies in the interior D of some simple closed curve Γ, and that $F'(z) = f(z)$ and F' is continuous there. We claim that*

$$\int_\gamma f(z)\,dz = F(\gamma(b)) - F(\gamma(a)).$$

To place this in a more familiar context, put $F(z) = U(x, y) + iV(x, y)$. Now $F'(z) = U_x + iV_x = f(z) = u + iv$, so we must have $u = U_x$ and $v = V_x$. Now recall that, wherever F has a derivative, U and V must satisfy the Cauchy–Riemann equations. Hence $V_x = -U_y$. Then $u = U_x$ and $-v = U_y$. We now have

$$u\,dx - v\,dy = U_x\,dx + U_y\,dy = dU,$$

hence

$$\int_\gamma u\,dx - v\,dy = \int_\gamma dU = U(\gamma(b)) - U(\gamma(a)).$$

In previous terminology, $u\,dx - v\,dy$ is an exact differential in D, and $\int_\gamma u\,dx - v\,dy$ is independent of path there.

Next, note that we also have again by the Cauchy–Riemann equations, $U_x = V_y$. Hence $v\,dx + u\,dy = V_x\,dx + V_y\,dy = dV$, so

$$\int_\gamma v\,dx + u\,dy = \int_\gamma dV = V(\gamma(b)) - V(\gamma(a)).$$

That is, $\int_\gamma v\,dx + u\,dy$ is also independent of path in D.

Putting these two results together yields

$$\begin{aligned}\int_\gamma f(z)\,dz &= \int_\gamma u\,dx - v\,dy + i\int_\gamma v\,dx + u\,dy\\ &= U(\gamma(b)) - U(\gamma(a)) + i(V(\gamma(b)) - V(\gamma(a)))\\ &= (U(\gamma(b)) + iV(\gamma(b))) - (U(\gamma(a)) + iV(\gamma(a)))\\ &= F(\gamma(b)) - F(\gamma(a)).\end{aligned}$$

EXAMPLE

Let $f(z) = z^2$ and $\gamma(t) = 2e^{it}$, $t\colon 0 \to \pi$. Thus γ is a semicircle from $(2, 0)$ to $(-2, 0)$, since $\gamma(t) = \cos(t) + i\sin(t) = (\cos(t), \sin(t))$, $t\colon 0 \to \pi$. Using the direct approach, we have

$$\begin{aligned}\int_\gamma f(z)\,dz &= \int_0^\pi f(2e^{it})2ie^{it}\,dt = \int_0^\pi 8ie^{3it}\,dt\\ &= 8i\int_0^\pi (\cos(3t) + i\sin(3t))\,dt = 8i\int_0^\pi \cos(3t)\,dt - 8\int_0^\pi \sin(3t)\,dt\\ &= \frac{-8}{3}(1 + 1) = \frac{-16}{3}.\end{aligned}$$

Alternatively, note that $d(z^3/3)/dz = z^2$, so

$$\int_\gamma z^2\, dz = \frac{z^3}{3}\bigg|_{\gamma(0)}^{\gamma(\pi)} = \frac{z^3}{3}\bigg|_{2}^{-2} = \frac{-8}{3} - \frac{8}{3} = \frac{-16}{3}. \quad \blacksquare$$

EXAMPLE

Let $f(z) = 1/z$, $\gamma(t) = 3e^{it}$, $t: 0 \to 2\pi$. Then

$$\int_\gamma f(z)\, dz = \int_\gamma \frac{1}{z}\, dz = \int_0^{2\pi} \frac{1}{3e^{it}}(3ie^{it})\, dt = i\int_0^{2\pi} dt = 2\pi i. \quad \blacksquare$$

Be careful with integrals like this. It is tempting to say that $\dfrac{d}{dz}(\mathrm{Log}(z)) = \dfrac{1}{z}$, hence that

$$\int_\gamma \frac{1}{z}\, dz = \mathrm{Log}(z)\bigg|_{\gamma(0)}^{\gamma(2\pi)} = \mathrm{Log}(\gamma(2\pi)) - \mathrm{Log}(\gamma(0)) = \mathrm{Log}(3) - \mathrm{Log}(3) = 0.$$

The error here is that Log is not defined in the whole interior of γ (0 is interior to γ), as required in the argument given above.

The last example shows why the complex analogue of the fundamental theorem of calculus is somewhat more elaborate than one might expect, requiring as we did that $F'(z) = f(z)$ in the entire region D interior to some simple closed curve, and not just along γ itself. We shall save ourselves some writing in what follows by agreeing to call any set of points in the plane bounded by a simple closed curve (excluding points on the curve itself) a domain. If D is a domain, the simple closed curve bounding D will be noted ∂D. Note that the set consisting of all points in D together with those in ∂D is closed, bounded (hence compact), connected, and simply connected.

Simple connectedness is important in complex integration in much the same way that it was important in considering exact differentials in Chapter 6. Suppose that $f = u + iv$, where u and v have continuous first partials in a domain D. Choose any z_0 in D, and define $F(z) = \int_{\gamma_z} f(\zeta)\, d\zeta$, where z is any point of D and γ_z is any smooth path in D from z_0 to z (Fig. 9-2). Suppose

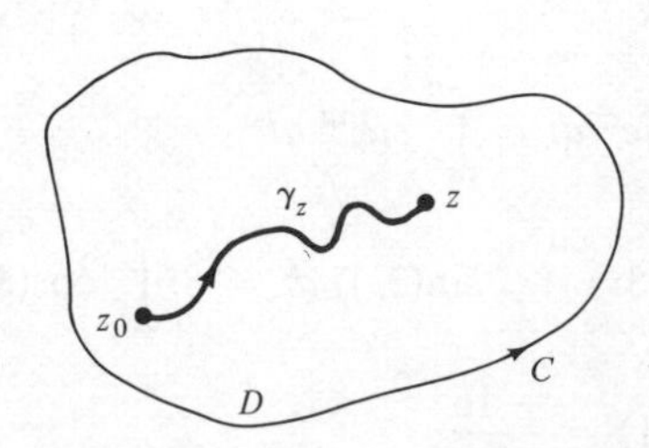

Fig. 9-2

also that the Cauchy–Riemann equations $u_x = v_y, v_x = -u_y$ hold in D. Then $\int_\gamma u\,dx - v\,dy$ and $\int_\gamma v\,dx + u\,dy$ are both independent of path in D. Hence F is a legitimate function of z defined on D. Further, by the simple connectivity of D, there are functions U and V with

$$dU = u\,dx - v\,dy \quad \text{and} \quad dV = v\,dx + u\,dy.$$

Then

$$\int_{\gamma_z} u\,dx - v\,dy = U(z) - U(z_0)$$

and

$$\int_{\gamma_z} v\,dx + u\,dy = V(z) - V(z_0).$$

Then

$$F(z) = U(z) + iV(z) - [U(z_0) + iV(z_0)].$$

For convenience, put $k = U(z_0) + iV(z_0)$. Then

$$F(z) = U(z) + iV(z) - k.$$

Now observe that

$$U_x = u = V_y \quad \text{and} \quad V_x = v = -U_y.$$

Also, U and V are continuous (as both are differentiable by continuity of their partials). Then F is analytic on D. Finally, for z in D,

$$F'(z) = U_x + iV_x = u + iv = f(z).$$

We have therefore proved the following:

Theorem

Suppose that u and v have continuous first partials satisfying the Cauchy–Riemann equations in a domain D. Then there is some F analytic in D such that $F'(z) = u(x, y) + iv(x, y)$ for each $z = x + iy$ in D.

As might be expected, F is not uniquely determined, but comes along with a constant (k in the above discussion), which may be chosen to be 0 by suitably adjusting the arbitrary constants that may be attached to U and V. Combining the last theorem with the results of Section 9.2 gives us the following:

Theorem

Suppose that f is analytic and f' continuous in a domain D. Then for some F, $F'(z) = f(z)$ for each z in D.

This result will be of fundamental importance in the next section.

EXAMPLE

As an example, let us consider a problem similar to one done before, but with an important difference. Consider $\int_\gamma (1/z)\,dz$, where $\gamma: [a, b] \to R^2$ is a closed path not having 0 in its interior or on its graph. We can use $d(\mathrm{Log}(z))/dz = 1/z$ in a domain containing γ, so $\int_\gamma (1/z)\,dz = \mathrm{Log}(\gamma(b)) - \mathrm{Log}(\gamma(a))$. Even if we did not know that $d(\mathrm{Log}(z))/dz = 1/z$, we would still have $\int_\gamma (1/z)\,dz = 0$, since by the previous theorem there would be some analytic F with $F'(z) = 1/z$ in a domain containing γ, and then we would have

$$\int_\gamma \frac{1}{z}\,dz = F(\gamma(b)) - F(\gamma(a)) = 0, \qquad \text{as } \gamma(b) = \gamma(a).$$ ∎

We conclude this section with a simple but often useful estimate. Suppose that f is continuous on a smooth curve $\gamma: [a, b] \to R^2$. By writing $f = u + iv$, it is immediate that $|f|$ is bounded on γ, say $|f(\gamma(t))| \le M$ for $a \le t \le b$. Then

$$\begin{aligned}\left|\int_\gamma f(z)\,dz\right| &= \left|\int_a^b f(\gamma(t))\cdot\gamma'(t)\,dt\right| \\ &\le \int_a^b |f(\gamma(t))|\,|\gamma'(t)|\,dt \\ &\le M\int_a^b |\gamma'(t)|\,dt = M\,(\text{length of } \gamma).\end{aligned}$$

Exercises for Section 9.5

1. Evaluate the following, first by putting $f = u + iv$ and $\int_\gamma f(z)\,dz = \int_\gamma u\,dx - v\,dy + i\int_{\gamma_0} v\,dx + u\,dy$, and, second, by finding $F(z)$ such that $F'(z) = f(z)$.
   (a) $\int_\gamma (z^2 + 2z + 1)$, $\gamma(t) = (1 + i)t$, $t: 0 \to 1$
   (b) $\int_\gamma \sin(z)\,dz$, $\gamma(t) = it$, $t: 0 \to \pi/2$
   (c) $\int_\gamma e^{iz}\,dz$, $\gamma(t) = (1 + i)t$, $t: 1 \to 2$
   (d) $\displaystyle\int_\gamma \cosh((1 + i)z)\,dz$, $\gamma(t) = \begin{cases} it, & t: 0 \to 1 \\ t - 1 + i, & t: 1 \to 2 \end{cases}$
   (e) $\displaystyle\int_\gamma \frac{1}{z}\,dz$, $\gamma(t) = \cos(t) + 2i\sin(t)$, $t: 0 \to 2\pi$
   (f) $\displaystyle\int_\gamma \frac{1}{z - a}\,dz$, where a is any complex number and γ is a circle of radius R centered at a.
   (g) $\int_\gamma \mathrm{Re}(z^2)\,dz$, $\gamma(t) = 2e^{it}$, $t: 0 \to \pi/2$
   (h) $\int_\gamma z^n$, n any positive integer, γ any closed path in the plane.

2. Let f be analytic in a domain D and suppose that f' is continuous on D. Show that $\oint_\gamma f(z)\overline{f'(z)}\,dz$ is pure imaginary for any closed path in D [i.e., show that $\operatorname{Re}\oint_\gamma f(z)\overline{f'(z)}\,dz = 0$].

9.6 Cauchy Integral Theorem and Consequences

We are now in a position to state a version of the Cauchy integral theorem, which forms the basis for most of complex function theory.

Cauchy's Theorem

Suppose that f is analytic in a domain D, and that γ is a closed path in D. Then $\int_\gamma f(z)\,dz = 0$.

The proof of this theorem as stated is extremely delicate and somewhat beyond the purpose of this exposition. We shall give a proof under the added hypothesis that $f'(z)$ is continuous in D. It is important for the reader to realize that this is a serious defect in our treatment, as we are adding to the theorem a strong hypothesis which turns out to be unnecessary. In fact, it is true that a complex function analytic at z_0 has all higher derivatives at z_0, something that fails for real-valued functions differentiable at a point. However, this defect is the price paid for efficiency and, to some extent, simplicity.

Suppose, then, that $f'(z)$ is continuous on D. If we write $f = u + iv$, then u and v have continuous partials on D. We can now use a theorem from Section 9.5 to assert the existence of F such that $F'(z) = f(z)$ on D. Since $\gamma\colon [a, b] \to R^2$ is a closed path, we then have $\gamma(b) = \gamma(a)$, hence

$$\int_\gamma f(z)\,dz = F(\gamma(b)) - F(\gamma(a)) = 0.$$

Another way of looking at essentially the same argument is to write

$$\int_\gamma f(z)\,dz = \int_\gamma u\,dx - v\,dy + i\int_\gamma v\,dx + u\,dy.$$

Again using the assumption that u and v have continuous first partials in D, we can apply Green's theorem to each of these line integrals to obtain

$$\oint_\gamma f(z)\,dz = \iint_D (-v_x + u_y) + i\iint_D (u_x - v_y),$$

where D is the domain bounded by γ. By the Cauchy–Riemann equations for u and v, both double integrals on the right vanish.

A version of Cauchy's theorem can also be proved for non-simply-connected regions. Suppose that f is analytic in a domain D, except at a finite number of

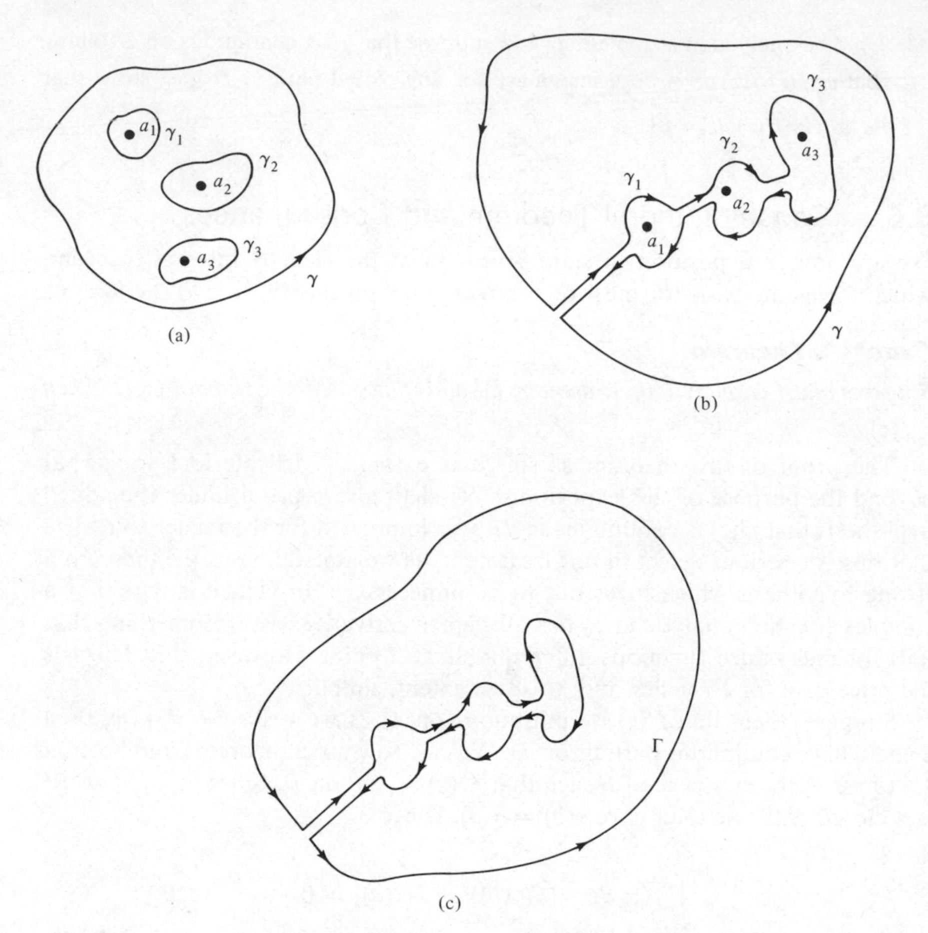

Fig. 9-3

points, and that γ is a closed path in D enclosing points $a_1, \ldots, a_n$ at which f is not analytic, but not passing through any points at which f is not analytic. Let $\gamma_1, \ldots, \gamma_n$ be closed paths, with a_i interior only to γ_i and all γ_j's lying in the region bounded by γ. We suppose that no two of $\gamma, \gamma_1, \gamma_2, \ldots, \gamma_n$ intersect (Fig. 9-3a). *The claim is that*

$$\oint_\gamma f(z)\,dz = \sum_{j=1}^{n} \oint_{\gamma_j} f(z)\,dz.$$

The argument is like that used in Section 6.6. Cut channels from γ to γ_1 to γ_2 to $\cdots$ to γ_n (Fig. 9-3b), and let Γ be the closed path formed from the channel cuts and $\gamma_1, \gamma_2, \ldots, \gamma_n, \gamma$ as in Fig. 9-3c. Since f is analytic in the domain bounded by Γ, then by Cauchy's theorem, $\oint_\gamma f(z)\,dz = 0$. Keep in mind the positive orientation with respect to the interior of Γ (counterclockwise on γ, clockwise on the γ_j's), and let the channel cuts narrow to single lines. The integrals over these line segments, which are taken both ways, then cancel, and we are left with

$$0 = \oint_\gamma f(z)\,dz - \sum_{j=1}^{n} \oint_{\gamma_j} f(z)\,dz,$$

or, equivalently,

$$\oint_\gamma f(z)\,dz = \sum_{j=1}^{n} \oint_{\gamma_j} f(z)\,dz,$$

as we wanted.

EXAMPLE

Evaluate

$$\oint_\gamma \frac{dz}{z^2 + 2z + 1},$$

where $\gamma(t) = -1 + e^{it}$, $t: 0 \to 2\pi$. Immediately, the integral is zero by Cauchy's theorem, as $1/(z^2 + 2z + 1)$ is analytic in a domain containing γ [which is a circle of radius 1 about $(-1, 0)$]. ▮

EXAMPLE

Evaluate

$$\oint_\gamma \frac{dz}{z^2 + 1},$$

where γ is any closed path having i and $-i$ in its interior. Write

$$\frac{1}{z^2 + 1} = \frac{1}{(z + i)(z - i)} = \frac{-1/2i}{z + i} + \frac{1/2i}{z - i}.$$

Then

$$\oint_\gamma \frac{dz}{z^2 + 1} = \frac{-1}{2i} \oint_\gamma \frac{1}{z + i}\,dz + \frac{1}{2i} \oint_\gamma \frac{dz}{z - i}.$$

Put small circles about i and $-i$, say of radius ε, where ε is sufficiently small that neither of these circles intersects γ or each other. This is shown in Fig. 9-4. Parametrize, say,

$$C_\varepsilon(t) = -i + \varepsilon e^{it}, \qquad t: 0 \to 2\pi,$$
$$K_\varepsilon(t) = i + \varepsilon e^{it}, \qquad t: 0 \to 2\pi.$$

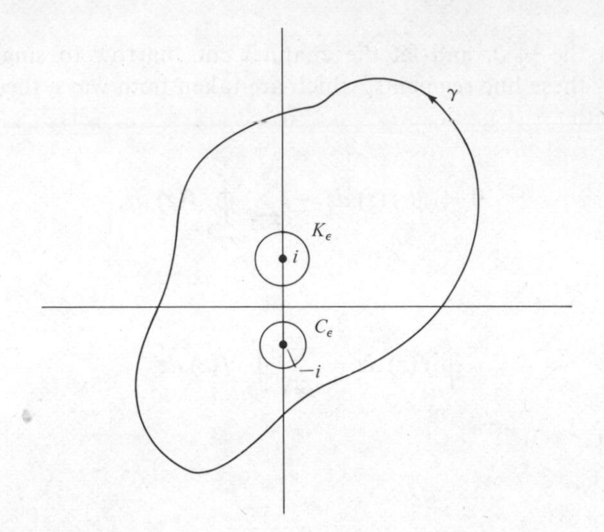

Fig. 9-4

Then

$$\oint_\gamma \frac{dz}{z^2+1} = \frac{-1}{2i}\left(\oint_{C_\varepsilon} \frac{dz}{z+i} + \oint_{K_\varepsilon} \frac{dz}{z+i}\right) + \frac{1}{2i}\left(\oint_{C_\varepsilon} \frac{dz}{z-i} + \oint_{K_\varepsilon} \frac{dz}{z-i}\right).$$

Now

$$\oint_{C_\varepsilon} \frac{dz}{z-i} = \oint_{K_\varepsilon} \frac{dz}{z+i} = 0,$$

by Cauchy's theorem, as $1/(z-i)$ is analytic on and within C_ε, and $1/(z+i)$ is analytic on and within K_ε. Next

$$\oint_{C_\varepsilon} \frac{dz}{z+i} = \int_0^{2\pi} \frac{i\varepsilon e^{it}}{\varepsilon e^{it}}\, dt = 2\pi i$$

and

$$\oint_{K_\varepsilon} \frac{dz}{z-i} = \int_0^{2\pi} \frac{i\varepsilon e^{it}}{\varepsilon e^{it}}\, dt = 2\pi i.$$

Thus

$$\oint_\gamma \frac{dz}{1+z^2} = \frac{-1}{2i}(2\pi i) + \frac{1}{2i}(2\pi i) = 0. \quad \blacksquare$$

Later we shall see methods for working out integrals such as those of the last example without the necessity of parametrizing the curves. For now, we shall develop several important consequences of Cauchy's Theorem.

A. Cauchy Integral Formula

Suppose that f is analytic in a domain D. Let z_0 be a point in D, and γ a circle about z_0 and lying entirely within D. Cauchy's integral formula says that

$$f(z) = \frac{1}{2\pi i} \oint_\gamma \frac{f(\xi)\, d\xi}{\xi - z}$$

for any z inside γ. This says, among other things, that the value of $f(z)$ for any z inside γ is completely determined by the values of $f(\xi)$ for ξ on γ. This is similar in basic idea to the mean value property for harmonic functions (Section 6.5C).

To begin the proof, choose any $\varepsilon > 0$. We shall show that

$$\left| \oint_\gamma \frac{f(\xi)\, d\xi}{\xi - z} - 2\pi i f(z) \right| < \varepsilon.$$

The trick is to enclose z in a circle C of sufficiently small radius ρ as to lie entirely within γ (Fig. 9-5). By Cauchy's theorem for non-simply-connected regions [noting that $f(\xi)/(\xi - z)$ may be undefined at $\xi = z$ inside γ], we have

$$\oint_\gamma \frac{f(\xi)\, d\xi}{\xi - z} = \oint_C \frac{f(\xi)\, d\xi}{\xi - z}.$$

Now write

$$\oint_C \frac{f(\xi)\, d\xi}{\xi - z} = f(z) \oint_C \frac{d\xi}{\xi - z} + \oint_C \frac{f(\xi) - f(z)}{\xi - z}\, d\xi.$$

Parametrize C by $C(t) = z + \rho e^{it}$, $t: 0 \to 2\pi$. Then

$$\oint_C \frac{d\xi}{\xi - z} = \int_0^{2\pi} \frac{i\rho e^{it}\, dt}{\rho e^{it}} = 2\pi i.$$

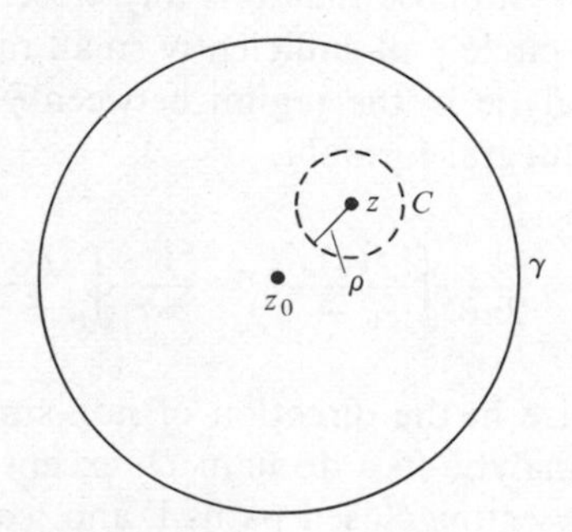

Fig. 9-5

Now, f is continuous at z, so there exists $\delta > 0$ such that $|f(\xi) - f(z)| < \varepsilon/2\pi$ if $|\xi - z| < \delta$. By readjusting δ, if necessary, so that $\rho < \delta$, we can be sure that $|\xi - z| < \delta$ for ξ on C, hence

$$\left|\frac{f(\xi) - f(z)}{\xi - z}\right| = \frac{1}{\rho}|f(\xi) - f(z)| < \frac{\varepsilon}{2\rho\pi},$$

for ξ on C. Then

$$\left|\oint_C \frac{f(\xi) - f(z)}{\xi - z}\,d\xi\right| \leq \frac{\varepsilon}{2\pi\rho}\,(\text{length of } C) = \varepsilon.$$

Hence

$$\begin{aligned}\left|\oint_\gamma \frac{f(\xi)\,d\xi}{\xi - z} - 2\pi i f(z)\right| &= \left|\oint_C \frac{f(\xi)\,d\xi}{\xi - z} - 2\pi i f(z)\right| \\ &= \left|f(z)\oint_C \frac{d\xi}{\xi - z} + \oint_C \frac{f(\xi) - f(z)}{\xi - z}\,d\xi - 2\pi i f(z)\right| \\ &= \left|\oint_C \frac{f(\xi) - f(z)}{\xi - z}\,d\xi\right| \leq \varepsilon.\end{aligned}$$

Since ε was an arbitrary positive number, we conclude that

$$\oint_\gamma \frac{f(\xi)\,d\xi}{\xi - z} = 2\pi i f(z).$$

In particular,

$$f(z_0) = \frac{1}{2\pi i}\oint_\gamma \frac{f(\xi)\,d\xi}{\xi - z_0}.$$

We can generalize Cauchy's integral formula in two directions. First, γ need not be a circle. For, suppose that Γ is any closed path about z and lying inside D. Enclose z in a circle γ of sufficiently small radius that γ lies inside Γ. Then $f(\xi)/(\xi - z)$ is analytic in the region between γ and Γ, so by Cauchy's theorem and Cauchy's integral formula,

$$f(z) = \frac{1}{2\pi i}\oint_\gamma \frac{f(\xi)\,d\xi}{\xi - z} = \frac{1}{2\pi i}\oint_\Gamma \frac{f(\xi)\,d\xi}{\xi - z}.$$

We can also generalize in the direction of non-simply-connected domains. Suppose, say, that f is analytic in a domain D, except at ζ. Let z be in D, but $z \neq \zeta$. Consider nonintersecting closed paths Γ and γ about ζ, as in Fig. 9-6(a), with z between γ and Γ and γ inside Γ. Employing the old channel argument

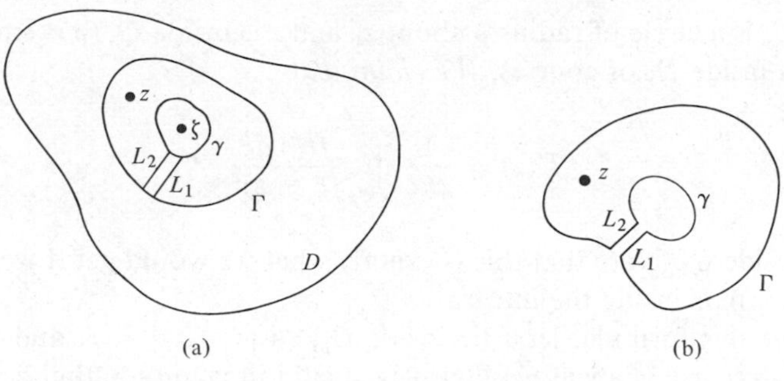

Fig. 9-6

again, let C consist of the pieces of Γ and γ minus the nicks between the channels, together with the channel lines L_1 and L_2 (Fig. 9-6b); then C is a closed path in D, within which f is analytic. Since z is inside C, we have

$$f(z) = \frac{1}{2\pi i} \oint_C \frac{f(\xi)\, d\xi}{\xi - z}.$$

Letting L_1 and L_2 merge, we obtain in the usual way (again, keeping orientation straight)

$$f(z) = \frac{1}{2\pi i} \oint_\Gamma \frac{f(\xi)\, d\xi}{\xi - z} - \frac{1}{2\pi i} \oint_\gamma \frac{f(\xi)\, d\xi}{\xi - z}.$$

Note that, if f were analytic at ζ also, then we would have $\oint_\gamma f(\xi)\, d\xi/(\xi - z) = 0$, and the above would reduce again to Cauchy's integral formula.

EXAMPLE

Sometimes Cauchy's integral formula can be used to evaluate integrals. As an example, consider $\oint_\gamma \cos(2z)\, dz/z$, with γ the circle $|z| = \frac{1}{2}$. Writing this as

$$\oint_\gamma \frac{\cos(2z)\, dz}{z - 0} = \oint_\gamma \frac{f(z)\, dz}{z - 0},$$

we recognize from the integral formula that the value of the integral is $2\pi i f(0)$. With $f(z) = \cos(2z)$ here, the integral is $2\pi i$. ∎

B. Formulas for Higher Derivatives

Using Cauchy's integral formula, we can derive integral representations for all derivatives of f at a point. For, suppose the conditions of Cauchy's integral formula hold, and let z_0 be in D. Then

$$f(z) = \frac{1}{2\pi i} \oint_{C_\rho} \frac{f(\xi)\, d\xi}{\xi - z}$$

whenever C_ρ is a circle of radius ρ about z_0 and z is inside C_ρ (ρ is small enough that C_ρ lies inside D, of course). *We claim that*

$$f'(z) = \frac{1}{2\pi i} \oint_{C_\rho} \frac{f(\xi)\,d\xi}{(\xi - z)^2}$$

for any z inside C_ρ. Note that this is exactly what we would get if we calculated $d(f(\xi)/(\xi - z))/dz$ inside the integral.

To prove this formula, let z lie inside C_ρ, so $|z - z_0| < \rho$, and suppose in the following that h is always sufficiently small in magnitude that $z + h$ is also in C_ρ (so $|z + h - z_0| < \rho$). Then, by Cauchy's integral formula,

$$\begin{aligned}
\frac{f(z+h) - f(z)}{h} &= \frac{1}{2\pi ih} \oint_{C_\rho} \frac{f(\xi)\,d\xi}{\xi - (z+h)} - \frac{1}{2\pi ih} \oint_{C_\rho} \frac{f(\xi)\,d\xi}{\xi - z} \\
&= \frac{1}{2\pi ih} \oint_{C_\rho} f(\xi)\left[\frac{1}{\xi - (z+h)} - \frac{1}{\xi - z}\right] d\xi \\
&= \frac{1}{2\pi ih} \oint_{C_\rho} \frac{hf(\xi)\,d\xi}{(\xi - z + h)(\xi - z)} \\
&= \frac{1}{2\pi i}\left[\oint_{C_\rho} \frac{f(\xi)\,d\xi}{(\xi - z)^2} + \oint_{C_\rho} \frac{hf(\xi)\,d\xi}{(\xi - z)^2(\xi - z - h)}\right].
\end{aligned}$$

The last line is just some algebraic sleight-of-hand, which the reader can show equals the previous line by direct algebraic manipulation.

We now claim that

$$\oint_{C_\rho} \frac{hf(\xi)\,d\xi}{(\xi - z)^2(\xi - z - h)} \to 0 \qquad \text{as } h \to 0.$$

To show this first let 2δ denote the smallest distance between z and a point on C_ρ (Fig. 9-7). Then $|\xi - z| \geq 2\delta$ for ξ on C_ρ, so $1/|\xi - z| \leq 1/2\delta$. Next, restrict $|h| < \delta$ to obtain $|\xi - z - h| \geq |\xi - z| - |h| \geq 2\delta - \delta = \delta$, for ξ on C_ρ.

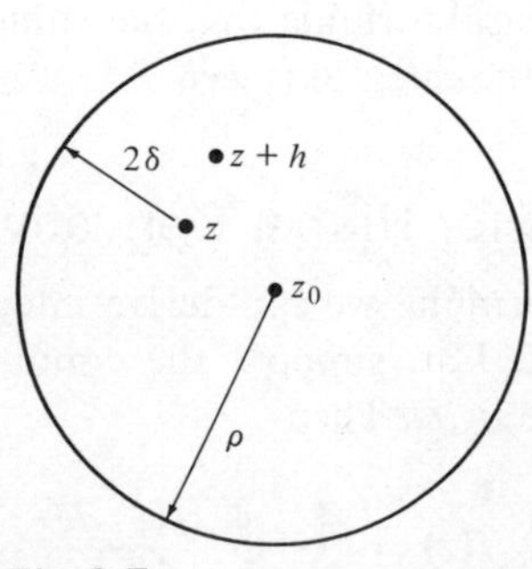

Fig. 9-7

Finally, for some $M > 0$, $|f(\xi)| \leq M$ for ξ on C_ρ. Putting it all together, we have, for ξ on C_ρ and $|h| < \delta$:

$$\left|\frac{f(\xi)}{(\xi - z)^2(\xi - z - h)}\right| \leq \frac{M}{(2\delta)^2}\frac{1}{\delta} = \frac{M}{4\delta^3}.$$

Then

$$\left|\oint_{C_\rho} \frac{hf(\xi)\,d\xi}{(\xi - z)^2(\xi - z - h)}\right| \leq \frac{|h|M(\text{length of } C_\rho)}{4\delta^3}$$

$$= \frac{|h|M}{4\delta^3}2\pi\rho \to 0 \qquad \text{as } h \to 0.$$

Hence

$$\frac{f(z + h) - f(z)}{h} \to \frac{1}{2\pi i}\oint_{C_\rho} \frac{f(\xi)\,d\xi}{(\xi - z)^2}$$

as $h \to 0$, implying that

$$f'(z) = \frac{1}{2\pi i}\oint_{C_\rho} \frac{f(\xi)\,d\xi}{(\xi - z)^2}.$$

An inductive argument can now be used to show that the nth derivative of f at z is given by

$$f^{(n)}(z) = \frac{n!}{2\pi i}\oint_{C_\rho} \frac{f(\xi)\,d\xi}{(\xi - z)^{n+1}}.$$

Again, observe that this is the same as

$$\frac{1}{2\pi i}\oint_{C_\rho} \frac{d^n}{dz^n}\left(\frac{f(\xi)}{(\xi - z)}\right) d\xi,$$

making the formula easy to remember.

From this it follows that a function analytic in a domain D has derivatives of all orders in D. That is, existence of $f'(z)$ in D is sufficient to guarantee existence of $f''(z), f^{(3)}(z), \ldots$ in D. This amazing fact has no analogue for real-valued functions. By putting $z = z_0$ in the above, we have, in particular,

$$f^{(n)}(z_0) = \frac{n!}{2\pi i}\oint_{C_\rho} \frac{f(\xi)\,d\xi}{(\xi - z_0)^{n+1}}.$$

We now derive a bound for $f^{(n)}(z)$ in terms of a bound for $f(z)$. Suppose that

$|f(z)| \leq M$ for z in D. If ξ is on C_ρ, a circle within D about z and of radius ρ, then $|\xi - z| = \rho$, and

$$\left| \oint_{C_\rho} \frac{f(\xi)\, d\xi}{(\xi - z_0)^{n+1}} \right| \leq \frac{M(\text{length of } C_\rho)}{\rho^{n+1}}$$

$$= \frac{M 2\pi\rho}{\rho^{n+1}} = \frac{2\pi M}{\rho^n}.$$

Then

$$|f^{(n)}(z)| \leq \left| \frac{n!}{2\pi i} \right| \frac{M \cdot 2\pi}{\rho^n} = \frac{Mn!}{\rho^n}.$$

C. Morera's Theorem

We conclude this section with Morera's theorem, which is a converse of Cauchy's Theorem and will be useful in the next section.

Theorem

Let f be continuous in a domain D. Suppose that, for every closed path in D,

$$\oint_C f(z)\, dz = 0.$$

Then f is analytic in D.

We prove Morera's theorem as follows. Note that the hypothesis $\oint_C f(z)\, dz = 0$ for every closed path in D, implies that $\int_\gamma f(z)\, dz$ is independent of path in D. That is, $\int_\Gamma f(z)\, dz = \int_\gamma f(z)\, dz$ for any paths in D that have the same initial and terminal points. We now employ a device used before. Choose any z_0 in D, and, for any z in D, let $F(z) = \int_\gamma f(\xi)\, d\xi$, where γ is any path in D from z_0 to z. Then F is a complex function defined on D. We shall prove that $F'(z) = f(z)$ in D.

Since we are assuming nothing about the partials of $\text{Re}(f)$ and $\text{Im}(f)$, we cannot use any of the machinery of Section 9.3. So we proceed directly to an examination of $[F(z + h) - F(z)]/h$ as $h \to 0$. First, choose some r so that, if $|h| < r$, then $z + h$ is in D. Then the line segment from z to $z + h$ is in D whenever $|h| < r$. Then write

$$F(z) = \int_\gamma f(\xi)\, d\xi$$

and

$$F(z + h) = \int_\gamma f(\xi)\, d\xi + \int_L f(\xi)\, d\xi,$$

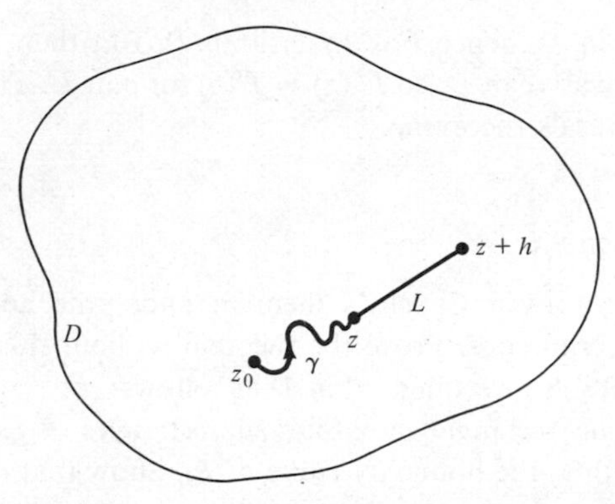

Fig. 9-8

where γ is any path in D from z_0 to z and L is the line segment from z to $z + h$ (Fig. 9-8). This is in D if $|h| < r$. Then

$$\begin{aligned}\frac{F(z+h) - F(z)}{h} &= \frac{1}{h}\int_L f(\xi)\, d\xi \\ &= \frac{1}{h}\int_L (f(z) - f(z) + f(\xi))\, d\xi \\ &= \frac{1}{h}\int_L f(z)\, d\xi + \frac{1}{h}\int_L (f(\xi) - f(z))\, d\xi.\end{aligned}$$

Now

$$\frac{1}{h}\int_L f(z)\, d\xi = \frac{f(z)}{h}(z + h - z) = f(z).$$

So we have

$$\frac{F(z+h) - F(z)}{h} = f(z) + \frac{1}{h}\int_L (f(\xi) - f(z))\, d\xi.$$

Since f is continuous at z, given $\varepsilon > 0$, there is some $\delta > 0$ such that $|f(\xi) - f(z)| < \varepsilon$ if $|\xi - z| < \delta$. If we choose $|h| < \delta$, then L has length $|h| < \delta$, so $|\xi - z| < \delta$ for ξ on L. Then $|f(\xi) - f(z)| < \varepsilon$ for ξ on L, so

$$\left|\frac{1}{h}\int_L (f(\xi) - f(z))\, d\xi\right| \le \frac{1}{|h|}\,\varepsilon(\text{length of } L) = \varepsilon.$$

Thus

$$\left|\frac{F(z+h) - F(z)}{h} - f(z)\right| = \left|\frac{1}{h}\int_L (f(\xi) - f(z))\, d\xi\right| < \varepsilon \qquad \text{if } |h| < \delta.$$

Then $F'(z) = f(z)$ in D, hence F is analytic in D. But then, from Section 9.6B, F has a second derivative in D, so $F''(z) = f'(z)$ for z in D. Then f is also analytic in D, proving Morera's theorem.

Exercises for Section 9.6

1. In the text, we proved Cauchy's theorem under the additional assumption that $f'(z)$ was continuous. Prove the theorem without this assumption for the special case that γ is a rectangle R in D as follows:

(a) Subdivide the rectangle into four subrectangles $R_1, \ldots, R_4$ as shown in Fig. 9-9, with γ_j the boundary curve of R_j. Show that

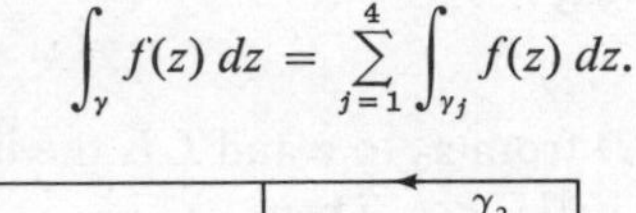

$$\int_{\gamma} f(z)\,dz = \sum_{j=1}^{4} \int_{\gamma_j} f(z)\,dz.$$

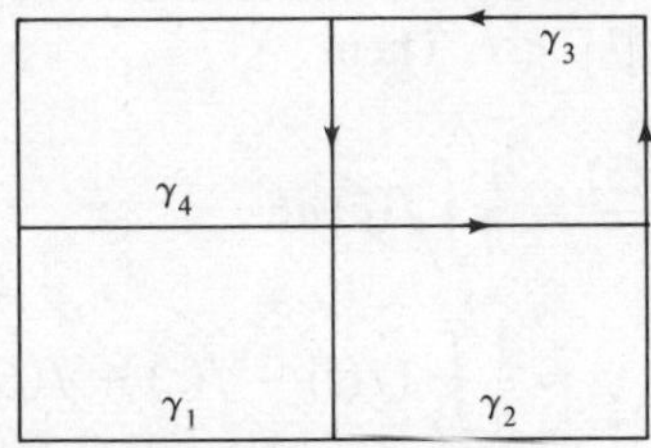

Fig. 9-9

(b) Show that $\left|\int_{\gamma_K} f(z)\,dz\right| \geq \frac{1}{4}\left|\oint_{\gamma} f(z)\,dz\right|$, for at least one value of K. Denote the rectangle R_K as $R^{(1)}$ and its boundary as $\gamma^{(1)}$. Subdivide $R^{(1)}$ as was just done with R to produce a subrectangle $R^{(2)}$ of $R^{(1)}$ with

$$\left|\int_{\gamma^{(2)}} f(z)\,dz\right| \geq \frac{1}{4}\left|\int_{\gamma^{(1)}} f(z)\,dz\right|,$$

with $\gamma^{(2)}$ the boundary of $R^{(2)}$. Thus

$$\left|\int_{\gamma^{(2)}} f(z)\,dz\right| \geq \frac{1}{4^2}\left|\oint_{\gamma} f(z)\,dz\right|.$$

(c) In this way, produce a sequence of rectangles $R^{(1)}, R^{(2)}, \ldots$, with $R^{(j+1)}$ a subrectangle of $R^{(j)}$, $R^{(1)}$ a subrectangle of R, $\gamma^{(j)}$ the boundary of $R^{(j)}$, and

$$\left|\oint_{\gamma^{(j)}} f(z)\,dz\right| \geq \frac{1}{4^j}\left|\oint_{\gamma} f(z)\,dz\right|.$$

(d) Show that there is a complex number z_0 in R with the property that, given $\tau > 0$, there is some N such that $R^{(n)}$ is within the disc $|z - z_0| < \tau$ if $n \geq N$.

(e) Let $\varepsilon > 0$. Show that, for some $\delta > 0$,

$$|f(z) - f(z_0) - (z - z_0)f'(z_0)| < \varepsilon|z - z_0| \qquad \text{if } |z - z_0| < \delta.$$

(f) Show that $\oint_{\gamma(n)} dz = \oint_{\gamma(n)} z\, dz = 0$.

(g) Show that

$$\left| \oint_{\gamma(n)} f(z)\, dz \right| \le \frac{\varepsilon}{4^n} dP,$$

where d is the length of the diagonal of R, P the perimeter of R, and n is sufficiently large that $R^{(n)}$ is within the disc $|z - z_0| < \delta$.

(h) Show that $\left|\oint_\gamma f(z)\, dz\right| \le dP\varepsilon$.

(i) Conclude that $\oint_\gamma f(z)\, dz = 0$.

2. Adapt the argument outlined in Exercise 1 to the case that γ is a triangle.

3. Compute the following integrals, using Cauchy's integral formula.

(a) $\displaystyle\int_\gamma \frac{e^{2z}}{z}$, γ the circle $|z| = 2$

(b) $\displaystyle\oint_\gamma \frac{(z^3 + 2z^2 + 2)\, dz}{z - 2i}$, γ the circle $|z - 2i| = \frac{1}{4}$

(c) $\displaystyle\int_\gamma \frac{\sin(2z^2 + 3z + 1)\, dz}{z - \pi}$, γ the circle $|z - \pi| = 1$

4. Carry out the inductive argument suggested in the text to derive the formula

$$f^{(n)}(z) = \frac{n!}{2\pi i} \oint_\gamma \frac{f(\xi)\, d\xi}{(\xi - z)^{n+1}},$$

which was based on Cauchy's integral formula.

5. Use the integral formula for the nth derivative of an analytic function to evaluate the following:

(a) $\displaystyle\oint_\gamma \frac{e^{z^2+2}\, dz}{(z - 2)^3}$, γ the circle $|z - 1| = 7$

(b) $\displaystyle\oint_\gamma \frac{\cos(3z^2 + 1)\, dz}{(z - \pi/8)^4}$, γ the circle $|z - \frac{1}{2}| = 4$

(c) $\displaystyle\oint_\gamma \frac{(3z^2 + 2z + \sin(z + 1))\, dz}{(z - 2)^2}$, γ the circle $|z - 2| = 1$

6. Let $P(z) = a_0 + a_1 z + \cdots + a_n z^n$, n a positive integer. Show that

$$\frac{1}{\pi} \iint_D |P(z)|^2\, dx\, dy = \sum_{j=1}^{n} \frac{|a_j|^2 R^{2j+2}}{j + 1},$$

where D is the domain bounded by the circle $|z| = R$.

7. In this example, we show how Cauchy's theorem can sometimes be used to evaluate real integrals. Let $f(z) = e^{-z^2}$. Then f is analytic in the whole plane. Let γ be the rectangle with vertices at $\pm a$ and $\pm a \pm ib$ $(a > 0, b > 0)$.

(a) Show that $\oint_\gamma f(z)\, dz = 0$.

(b) Write $\oint_\gamma f(z)\,dz$ as a sum of integrals over the four sides of γ, then show that, for given $b > 0$, the integrals over the two vertical sides go to zero as $a \to \infty$.

(c) Letting $a \to \infty$ and separating real and imaginary parts in the integrals over the horizontal sides of γ, show that

$$\int_{-\infty}^{\infty} e^{-x^2} \cos(2bx)\,dx = e^{-b^2} \int_{-\infty}^{\infty} e^{-x^2}\,dx.$$

Hence, using previous results, conclude that

$$\int_{-\infty}^{\infty} e^{-x^2} \cos(2bx)\,dx = \sqrt{\pi}e^{-b^2}.$$

8. Use the method outlined in Exercise 7 to show that

$$\int_0^{\infty} \cos(x^2)\,dx = \int_0^{\infty} \sin(x^2)\,dx = \frac{1}{2}\sqrt{\frac{\pi}{2}}.$$

Hint: Integrate e^{-z^2} over the pie-shaped sector formed by going along the real axis from $x = 0$ to $x = b$, then along the circle $|z| = R$ until an angle $\pi/4$ is reached with the positive real axis, then along a straight line back to the origin; show that the integral over the circular part goes to zero as $R \to \infty$, and consider the real and imaginary parts of the integrals along the straight-line segments.

9.7 Taylor Expansions of Analytic Functions

We want now to be able to develop complex functions in power series. First we shall say something about general series of complex functions.

Suppose that $f_1, f_2, \ldots$ are defined in a domain D. If $\sum_{n=1}^{\infty} f_n(z)$ converges at each z in D, say to $f(z)$, then we call f the *pointwise sum* of $\sum_{n=1}^{\infty} f_n$ on D. By direct analogy with the real case, we say that $\sum_{n=1}^{\infty} f_n$ converges uniformly to f on D if, given $\varepsilon > 0$, there is some N such that

$$\left| \sum_{j=1}^{n} f_j(z) - f(z) \right| < \varepsilon$$

whenever $n \geq N$ and z is in D.

Suppose now that $\sum_{n=1}^{\infty} f_n$ converges uniformly to f on D. We first note that f is continuous on D if each f_n is continuous on D. The proof of this is just like the proof of the corresponding theorem for real functions. Similarly, for any path γ in D, $\int_\gamma f(z)\,dz = \sum_{n=1}^{\infty} \int_\gamma f_n(z)\,dz$, the proof again being just like that of the theorem on term-by-term integration of real series.

The theorem on term-by-term differentiation is a little different. Since analyticity is really the property of interest, we shall show that *f is analytic on*

D if each f_n is analytic in D. The proof is easy. Let γ be any closed path in D. By Cauchy's theorem and the previous paragraph,

$$\oint_\gamma f(z)\,dz = \sum_{n=1}^{\infty} \oint_\gamma f_n(z)\,dz = 0.$$

Hence by Morera's theorem f is analytic on D.

We now particularize to power series. Let a be any complex number and $\{\alpha_n\}_{n=0}^{\infty}$ a sequence of complex numbers. A series $\sum_{n=0}^{\infty} \alpha_n(z-a)^n$ is called a *power series* about a with coefficient sequence $\{\alpha_n\}$. The argument used for real power series can be adapted to show the following:

(1) If the series converges at some $z_0 \neq a$, then it converges absolutely and uniformly in the disc $|z-a| < |z - z_0|$. Note here that we obtain convergence in a disc, as opposed to an interval in the real-variable case.
(2) Hence there are exactly three possibilities: (a) the series converges only for $z = a$; (b) the series converges for all z; (c) the series converges for some $z \neq a$ and diverges for some value of z.

In this event there is a positive number r, called the *radius of convergence*, such that the series converges for $|z-a| < r$ and diverges for $|z-a| > r$. For $|z-a| = r$, the series may converge or diverge. Convergence in the disc $|z-a| < r$ is absolute and uniform on any disc $|z-a| \leq r_0 < r$. Often we refer to the circle $|z-a| = r$ as the *circle of convergence*, since the series converges inside this circle. (Remember, however, that this term carries no implication of convergence at points *on* the circle itself.)

In case 2(c), r may be obtained as

$$\frac{1}{\lim_{n\to\infty} \left|\frac{\alpha_{n+1}}{\alpha_n}\right|} \qquad \text{if this limit exists,}$$

or as

$$\frac{1}{\lim_{n\to\infty} |\alpha_n|^{1/n}} \qquad \text{if this limit exists.}$$

When

$$\lim_{n\to\infty} \left|\frac{\alpha_{n+1}}{\alpha_n}\right| = 0,$$

then we are in case (b), and

$$\lim_{n\to\infty} \left|\frac{\alpha_{n+1}}{\alpha_n}\right| = \infty$$

corresponds to case (a).

Now suppose that $f(z) = \sum_{n=0}^{\infty} \alpha_n(z - a)^n$ for $|z - a| < r$, where $r > 0$. (We shall allow $r = \infty$ here, with the understanding that the set of z with $|z - a| < \infty$ is the whole complex plane.) Since each term of the series is analytic in $|z - a| < r$, then f is analytic there. Thus *a power series represents an analytic function inside its circle of convergence.*

It is a simple matter to show (as in the real case) that $\sum_{n=1}^{\infty} n\alpha_n(z - a)^{n-1}$ has the same circle of convergence as does $\sum_{n=0}^{\infty} \alpha_n(z - a)^n$, hence converges uniformly in any disc $|z - a| \leq r_0 < r$. This implies, as in the real case, that the original series can be differentiated term by term. That is,

$$f'(z) = \sum_{n=1}^{\infty} \alpha_n n(z - a)^{n-1}$$

for $|z - a| \leq r_0 < r$. A simple inductive argument now yields

$$\alpha_n = \frac{f^{(n)}(a)}{n!}.$$

Thus

$$f(z) = \sum_{n=0}^{\infty} \frac{f^{(n)}(a)(z - a)^n}{n!}.$$

Here f was defined by the power series and we derived an expression for the coefficient sequence in terms of derivatives of f evaluated at the center of the series. Conversely, if f is analytic in a disc $|z - a| < r$, we claim that f can be expanded in a power series about a. To see this, begin with the fact that $1/(1 - \xi) = \sum_{n=0}^{\infty} \xi^n$ for any complex number ξ with $|\xi| < 1$. The proof of this is simple. Write $S_n = \sum_{j=0}^{n} \xi^j$ and $\xi S_n = \sum_{j=0}^{n} \xi^{j+1}$. Then

$$S_n - \xi S_n = (1 - \xi)S_n, \qquad \text{so } S_n = \frac{1 - \xi^{n+1}}{1 - \xi} = \frac{1}{1 - \xi} - \frac{\xi^{n+1}}{1 - \xi}.$$

Now, as $n \to \infty$,

$$\left|\frac{\xi^{n+1}}{1 - \xi}\right| = \frac{|\xi^{n+1}|}{|1 - \xi|} \to 0 \qquad \text{if } |\xi| < 1.$$

Thus

$$S_n \to \frac{1}{1 - \xi} \qquad \text{as } n \to \infty.$$

Now write

$$\frac{1}{\xi - z} = \frac{1}{(\xi - a) - (z - a)} = \frac{1}{\xi - a} \frac{1}{1 - \left(\dfrac{z - a}{\xi - a}\right)}$$

$$= \frac{1}{\xi - a} \sum_{n=0}^{\infty} \left(\frac{z - a}{\xi - a}\right)^n = \sum_{n=0}^{\infty} \frac{(z - a)^n}{(\xi - a)^{n+1}} \qquad \text{for } \left|\frac{z - a}{\xi - a}\right| < 1.$$

If f is analytic in $|z - a| \leq R < r$, then

$$f(z) = \frac{1}{2\pi i} \oint_C \frac{f(\xi)\, d\xi}{(\xi - z)} \qquad \text{for each } z \text{ with } |z - a| \leq R,$$

where C is a circle $|\xi - a| = R_0$ and R_0 is chosen with $R < R_0 < r$. Now

$$\frac{f(\xi)}{\xi - z} = \sum_{n=0}^{\infty} \frac{f(\xi)(z - a)^n}{(\xi - a)^{n+1}}$$

and convergence is uniform on C. Hence

$$\frac{1}{2\pi i} \oint_C \frac{f(\xi)\, d\xi}{\xi - z} = f(z) = \sum_{n=0}^{\infty} \frac{1}{2\pi i} \oint_C \frac{f(\xi)(z - a)^n\, d\xi}{(\xi - a)^{n+1}}$$

$$= \sum_{n=0}^{\infty} \frac{1}{2\pi i} \oint_C \frac{f(\xi)\, d\xi (z - a)^n}{(\xi - a)^{n+1}}.$$

Now recall that

$$f^{(n)}(a) = \frac{n!}{2\pi i} \oint_C \frac{f(\xi)\, d\xi}{(\xi - a)^{n+1}}.$$

Then

$$f(z) = \sum_{n=0}^{\infty} \frac{f^{(n)}(a)(z - a)^n}{n!}.$$

This is the Taylor expansion of f about a and is valid inside any disc about a in which f is analytic. As might be expected, the series has the same form as the Taylor expansion of a real-valued function, except that our setting now is a disc, not just an interval on the real line.

In actual practice, we attempt to avoid evaluation of the coefficient series by either the integral formula,

$$\alpha_n = \frac{1}{2\pi i} \oint_C \frac{f(\xi)\, d\xi}{(\xi - a)^{n+1}}$$

or the equivalent derivative formula, $\alpha_n = f^{(n)}(a)/n!$, except in rare instances where all else fails. The derivative formula would give the Taylor expansions of for example, e^z and $\sin(z)$ about 0 easily enough:

$$e^z = \sum_{n=0}^{\infty} \frac{z^n}{n!}$$

and

$$\sin(z) = \sum_{n=0}^{\infty} \frac{(-1)^n z^{2n+1}}{(2n+1)!},$$

but in many instances we can utilize known series to derive the expansions we want, using a little algebra and an occasional clever trick.

EXAMPLE

Expand $1/(1 + z^2)$ in a series about 0. Write

$$\frac{1}{1+z^2} = \frac{1}{1-(-z^2)} = \sum_{n=0}^{\infty} (-z^2)^n = \sum_{n=0}^{\infty} (-1)^n z^{2n} \qquad \text{for } |z| < 1.$$

Another method:

$$\begin{aligned}\frac{1}{1+z^2} = \frac{1}{(1+zi)(1-zi)} &= \frac{\frac{1}{2}}{1+zi} + \frac{\frac{1}{2}}{1-zi} \\ &= \frac{1}{2}\frac{1}{1-(-zi)} + \frac{1}{2}\frac{1}{1-zi} \\ &= \frac{1}{2}\sum_{n=0}^{\infty} (-zi)^n + \frac{1}{2}\sum_{n=0}^{\infty} (zi)^n \\ &= \frac{1}{2}\sum_{n=0}^{\infty} (-i)^n z^n + \frac{1}{2}\sum_{n=0}^{\infty} i^n z^n \\ &= \sum_{n=0}^{\infty} \left(\frac{i^n + (-i)^n}{2}\right) z^n.\end{aligned}$$

Now

$$\frac{i^n + (-i)^n}{2} = \begin{cases} 1 & \text{for } n = 4m,\ m \text{ integer}, \\ 0 & \text{if } n = 4m+1 \text{ or } 4m+3, \\ -1 & \text{if } n = 4m+2. \end{cases}$$

Thus $1/(1 + z^2) = \sum_{m=0}^{\infty} (-1)^m z^{2m}$, as above.

Obviously, in this example, the first method was quicker. ∎

EXAMPLE

Expand $1/(3 + z^2)$ about 0. Write

$$\frac{1}{3+z^2} = \frac{1}{\sqrt{3}+zi}\,\frac{1}{\sqrt{3}-zi} = \frac{1/2\sqrt{3}}{3+zi} + \frac{1/2\sqrt{3}}{3-zi}$$

$$= \frac{1}{2\sqrt{3}}\,\frac{1}{\sqrt{3}(1+zi/\sqrt{3})} + \frac{1}{2\sqrt{3}}\,\frac{1}{\sqrt{3}(1-zi/\sqrt{3})}$$

$$= \frac{1}{6}\,\frac{1}{1-(-zi/\sqrt{3})} + \frac{1}{6}\,\frac{1}{1-(zi/\sqrt{3})}$$

$$= \frac{1}{6}\sum_{n=0}^{\infty}\frac{(-zi)^n}{3^{n/2}} + \frac{1}{6}\sum_{n=0}^{\infty}\frac{(zi)^n}{3^{n/2}}$$

$$= \frac{1}{6}\sum_{n=0}^{\infty}\left[\frac{(-i)^n + (i)^n}{3^{n/2}}\right]z^n.$$

We leave it to the reader to simplify the coefficient sequence, as in the previous example. ▮

EXAMPLE

Expand $1/(3 + z^2)$ about 1. Start out as in the last example:

$$\frac{1}{3+z^2} = \frac{1}{2\sqrt{3}}\,\frac{1}{\sqrt{3}+zi} + \frac{1}{2\sqrt{3}}\,\frac{1}{\sqrt{3}-zi}.$$

Now write

$$\frac{1}{\sqrt{3}+zi} = \frac{1}{\sqrt{3}+zi-i+i} = \frac{1}{(\sqrt{3}+i)+i(z-1)}$$

$$= \frac{1}{(\sqrt{3}+i)\left[1-\dfrac{(-i)(z-1)}{\sqrt{3}+i}\right]} = \frac{1}{\sqrt{3}+i}\sum_{n=0}^{\infty}\left[\frac{(-i)(z-1)}{\sqrt{3}+i}\right]^n$$

$$= \frac{1}{\sqrt{3}+i}\sum_{n=0}^{\infty}\left(\frac{-i}{\sqrt{3}+i}\right)^n(z-1)^n = \sum_{n=0}^{\infty}\frac{(-i)^n(z-1)^n}{(\sqrt{3}+i)^{n+1}}.$$

This is valid for

$$\left|\frac{i(z-1)}{\sqrt{3}+i}\right| = \frac{|z-1|}{|\sqrt{3}+i|} < 1, \qquad \text{or } |z-1| < 2.$$

Thus far we have

$$\frac{1}{2\sqrt{3}}\frac{1}{\sqrt{3}+zi} = \frac{1}{2\sqrt{3}}\sum_{n=0}^{\infty}\frac{(-i)^n}{(\sqrt{3}+i)^{n+1}}(z-1)^n.$$

Similarly,

$$\frac{1}{\sqrt{3}-zi} = \frac{1}{\sqrt{3}-i+i-zi} = \frac{1}{(\sqrt{3}-i)+i(z-1)}$$

$$= \frac{1}{(\sqrt{3}-i)\left[1-\dfrac{i(z-1)}{\sqrt{3}-i}\right]} = \frac{1}{\sqrt{3}-i}\sum_{n=0}^{\infty}\left(\frac{i(z-1)}{\sqrt{3}-i}\right)^n$$

$$= \sum_{n=0}^{\infty}\frac{i^n}{(\sqrt{3}-i)^{n+1}}(z-1)^n.$$

Then

$$\frac{1}{2\sqrt{3}}\frac{1}{\sqrt{3}-zi} = \frac{1}{2\sqrt{3}}\sum_{n=0}^{\infty}\frac{i^n}{(\sqrt{3}-i)^{n+1}}(z-1)^n.$$

This is valid in $|(z-1)/(\sqrt{3}-i)| < 1$, or $|z-1| < 2$. In summary, then, for $|z-1| < 2$, we have

$$\frac{1}{3+z^2} = \frac{1}{2\sqrt{3}}\sum_{n=0}^{\infty}\left[\frac{(-i)^n}{(\sqrt{3}+i)^{n+1}} + \frac{i^n}{(\sqrt{3}-i)^{n+1}}\right](z-1)^n.$$

It is clear that we cannot expand the disc $|z-1| < 2$ to obtain a larger circle of convergence. For, $z = 3i$ lies on the circle $|z-1| = 2$, and $3+z^2 = 0$ at $z = 3i$. There are, however, points on $|z-1| = 2$ at which the series converges. ∎

Exercises for Section 9.7

1. Find the Taylor expansion of each of the following about the given point z_0:

(a) $\dfrac{z+1}{z^2+2z+2}$, $z_0 = i$

(b) $\dfrac{z^2}{z^3+1}$, $z_0 = 0$

(c) $\sin(z)$, $z_0 = \dfrac{\pi}{2}$

(d) e^{-z^2}, $z_0 = 0$

(e) $\dfrac{1}{(1-z)^3}$, $z_0 = 0$ *Hint:* Differentiate $\dfrac{1}{1-z}$ twice.

2. Let f be analytic in a disc $|z| < R$, and let $\sum_{n=0}^{\infty} a_n z^n$ be the Taylor series for f about 0. Prove that

$$\frac{1}{2\pi}\int_0^{2\pi} |f(re^{i\theta})|^2 \, d\theta = \sum_{n=0}^{\infty} |a_n|^2 r^{2n}$$

for $0 \leq \theta \leq 2\pi$, $0 \leq r < R$. (This is called *Parseval's identity*.)

3. Prove that

$$\int_0^{2\pi} \left(\frac{\sin(n\theta/2)}{\sin(\theta/2)}\right)^2 d\theta = 2\pi n \qquad \text{for any positive integer } n.$$

Hint: Apply Parseval's identity (Exercise 2) to $(z^n - 1)/(z - 1)$, noting that $(z^n - 1)/(z - 1) = 1 + z + \cdots + z^{n-1}$.

4. Show that the radius of convergence r of $\sum_{n=0}^{\infty} a_n(z - a)^n$ is given by the Cauchy–Hadamard formula: $1/r = \lim\sup\{|a_n|^{1/n}\}$.

5. Prove a Weierstrass M-test for series of complex functions: If each f_n is defined on a set D of complex numbers, $|f_n(z)| \leq M_n$ for each z in D, and $\sum_{n=0}^{\infty} M_n$ converges, then $\sum_{n=0}^{\infty} f_n$ converges uniformly on D.

6. Show that $\sum_{n=0}^{\infty} z^n$ converges to $1/(1 - z)$ for $|z| < 1$ but that the convergence is not uniform.

7. Prove that $\sum_{n=0}^{\infty} a_n(z - a)^n$ and $\sum_{n=0}^{\infty} b_n(z - c)^n$ have the same radius of convergence if, for some N, $|a_n| = |b_n|$ for $n \geq N$. Is it necessary that $a_n = b_n$ for $n \geq N$? *Hint:* Note Exercise 4.

8. Let r be the radius of convergence of $\sum_{n=0}^{\infty} a_n(z - a)^n$. Prove the assertion made in the text that convergence is uniform on any disc $|z - a| \leq r^*$ for $0 < r^* < r$.

9. Prove the following summation-by-parts formula:

$$\sum_{n=0}^{K} a_n b_n = \sum_{n=0}^{K-1} (a_n - a_{n+1})B_n + a_K B_K, \qquad \text{where } B_n = \sum_{j=0}^{n} b_j.$$

Use this to prove that $\sum_{n=0}^{\infty} a_n f_n(z)$ converges uniformly on D if (1) $\lim_{n\to\infty} a_n = 0$, (2) $\sum_{n=0}^{\infty} |a_n - a_{n+1}|$ converges, and (3) for some M, $|f_n(z)| \leq M$ for all n and each z in D.

10. Suppose that $\sum_{n=0}^{\infty} a_n[\cos(n\theta) + i\sin(n\theta)]$ converges uniformly for $\alpha \leq \theta \leq \beta$. Prove that $\sum_{n=0}^{\infty} a_n z^n$ converges uniformly for $|z| \leq 1$ and $\alpha \leq \arg(z) \leq \beta$.

Hint: Make an estimate independent of θ for

$$\left| \sum_{n=0}^{\infty} a_n[\cos(n\theta) + i\sin(n\theta)] - \sum_{n=0}^{N} a_n[\cos(n\theta) + i\sin(n\theta)] \right|,$$

and show that this quantity goes to zero as $N \to \infty$.

9.8 Laurent Series

If f is analytic at a, we have just seen that f admits a power series expansion $f(z) = \sum_{n=0}^{\infty} \alpha_n(z - a)^n$ in some disc $|z - a| < R$ about a. Suppose, however, that f has some difficulty at a [e.g., $f'(a)$ does not exist] and is analytic only in a punctured disc $0 < |z - a| < R$. We shall now show that f can still be expanded in a series of powers of $(z - a)$, in some annulus $r < |z - a| < R$, if we admit negative powers in the expansion. That is, we now look for a representation $\sum_{n=-\infty}^{\infty} \alpha_n(z - a)^n$ for $f(z)$ that is valid in some circular ring $r < |z - a| < R$.

To begin, suppose that f is analytic in $0 < |z - a| < R^*$, and choose any r and R with $0 < r < R < R^*$. Let C_R be a circle of radius R about a, and C_r a circle of radius r about a. Let z be any complex number in the region between C_r and C_R (Fig. 9-10).

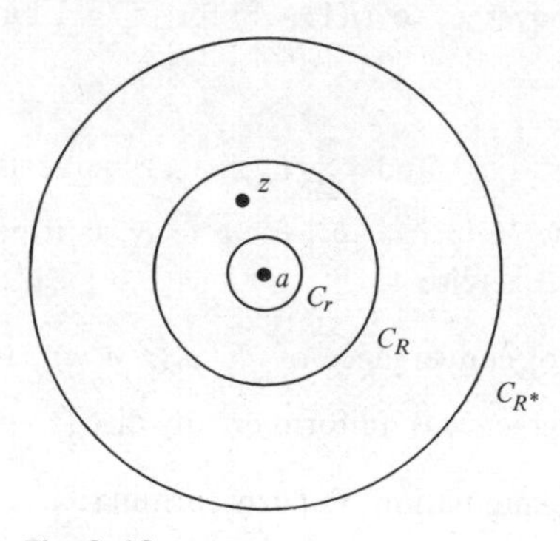

Fig. 9-10

By Cauchy's integral formula for non-simply connected regions.

$$f(z) = \frac{1}{2\pi i} \oint_{C_R} \frac{f(\xi)\,d\xi}{\xi - z} - \frac{1}{2\pi i} \oint_{C_r} \frac{f(\xi)\,d\xi}{\xi - z}.$$

Now, for ξ on C_R, $|(z - a)/(\xi - a)| < 1$, so

$$\frac{1}{\xi - z} = \frac{1}{(\xi - a) - (z - a)} = \sum_{n=0}^{\infty} \frac{(z - a)^n}{(\xi - a)^{n+1}}.$$

Then

$$\frac{1}{2\pi i}\oint_{C_R}\frac{f(\xi)\,d\xi}{\xi - z} = \sum_{n=0}^{\infty}\left[\frac{1}{2\pi i}\oint_{C_R}\frac{f(\xi)\,d\xi}{(\xi - a)^{n+1}}(z - a)^n\right] = \sum_{n=0}^{\infty}\alpha_n(z - a)^n,$$

where

$$\alpha_n = \frac{1}{2\pi i}\oint_{C_R}\frac{f(\xi)\,d\xi}{(\xi - a)^{n+1}} \qquad \text{for } n = 0, 1, 2, \ldots.$$

Next, for ξ on C_r, write

$$\frac{1}{\xi - z} = \frac{1}{(\xi - a) - (z - a)} = \frac{-1}{z - a}\,\frac{1}{1 - \left(\dfrac{\xi - a}{z - a}\right)}.$$

On C_r, $|(\xi - a)/(z - a)| < 1$, so that

$$\frac{1}{1 - \left(\dfrac{\xi - a}{z - a}\right)} = \sum_{n=0}^{\infty}\left(\frac{\xi - a}{z - a}\right)^n.$$

Then

$$\frac{1}{\xi - z} = -\sum_{n=0}^{\infty}\frac{(\xi - a)^n}{(z - a)^{n+1}} = -\sum_{n=1}^{\infty}\frac{(\xi - a)^{n-1}}{(z - a)^n}.$$

Then

$$\frac{f(\xi)}{\xi - z} = -\sum_{n=1}^{\infty}\frac{f(\xi)(\xi - a)^{n-1}}{(z - a)^n}.$$

The series on the right converges uniformly on C_r, so we have

$$\frac{-1}{2\pi i}\oint_{C_r}\frac{f(\xi)\,d\xi}{\xi - z} = \sum_{n=1}^{\infty}\left[\frac{1}{2\pi i}\oint_{C_r}f(\xi)(\xi - a)^{n-1}\,d\xi\right](z - a)^{-n}$$

$$= \sum_{n=1}^{\infty}\alpha_{-n}(z - a)^{-n},$$

where

$$\alpha_{-n} = \frac{1}{2\pi i}\oint_{C_r}f(\xi)(\xi - a)^{n-1}\,d\xi, \qquad n = 1, 2, \ldots.$$

Now, by previous results, we may replace C_R and C_r by any circle C about a between C_R and C_r without changing the integral expressions just derived for α_n, $n = 0, 1, 2, \ldots$, and α_{-n}, $n = 1, 2, \ldots$. Then

$$\alpha_n = \frac{1}{2\pi i} \oint_C \frac{f(\xi)\, d\xi}{(\xi - a)^{n+1}}, \qquad n = 0, 1, 2, \ldots$$

and

$$\alpha_{-n} = \frac{1}{2\pi i} \oint_C f(\xi)(\xi - a)^{n-1}\, d\xi, \qquad n = 1, 2, \ldots.$$

We may now combine these formulas into one:

$$\alpha_n = \frac{1}{2\pi i} \oint_C \frac{f(\xi)\, d\xi}{(\xi - a)^{n+1}}, \qquad n = 0, \pm 1, \pm 2, \ldots.$$

With this choice of the α_n's, we have

$$f(z) = \sum_{n=-\infty}^{\infty} \alpha_n (z - a)^n.$$

This is called a *Laurent expansion* of f about a in $r < |z - a| < R$. Note that, if f is analytic at a, then $\alpha_{-n} = 0$ for $n = 1, 2, \ldots$, by Cauchy's Theorem, and the Laurent expansion reduces to the Taylor expansion in the annulus.

Another point of view may provide some additional insight into what we are doing. In the above discussion, we have actually written f as a sum of two functions, g and h, where

$$h(z) = \sum_{n=0}^{-\infty} \alpha_n (z - a)^n$$

and

$$g(z) = \sum_{n=-\infty}^{-1} \alpha_n (z - a)^n = \sum_{n=1}^{\infty} \frac{\alpha_{-n}}{(z - a)^n}.$$

Note that $\sum_{n=0}^{\infty} \alpha_n (z - a)^n$ is a Taylor series, and so represents an analytic function. Thus h is analytic at least in $|z - a| < R$.

On the other hand, $\sum_{n=1}^{\infty} \alpha_{-n}/(z - a)^n$ is not a power series at all. However, put $\xi - a = 1/(z - a)$. Then

$$\sum_{n=1}^{\infty} \frac{\alpha_{-n}}{(z - a)^n} = \sum_{n=1}^{\infty} \alpha_{-n} (\xi - a)^n,$$

and this is a power series about a which must represent an analytic function $G(\xi)$ in some disc about a. Where does this power series converge? If the series $\sum_{n=1}^{\infty} \alpha_{-n}(z-a)^{-n}$ converges at, say, z_0, then $\sum_{n=1}^{\infty} \alpha_{-n}(\xi-a)^n$ converges at ξ_0, where $\xi_0 - a = 1/(z_0 - a)$. Being a power series, $\sum_{n=1}^{\infty} \alpha_{-n}(\xi-a)^n$ will converge at all ξ with $|\xi - a| < |\xi_0 - a|$. Then $\sum_{n=1}^{\infty} \alpha_{-n}(z-a)^{-n}$ will converge for $1/|z-a| < 1/|z_0 - a|$, or $|z-a| > |z_0 - a|$. Putting $r = |z_0 - a|$, $\sum_{n=1}^{\infty} \alpha_{-n}(z-a)^{-n}$ represents an analytic function in the region $|z-a| > r$.

In summary, we have written $f(z) = h(z) + g(z)$, where $h(z)$ is analytic in $|z-a| < R$ and $g(z)$ is analytic in $|z-a| > r$. The sum is analytic in the common region of analyticity, $r < |z-a| < R$.

We leave it as an exercise for the reader to show that the Laurent expansion converges uniformly in $\hat{r} \le |z-a| \le \hat{R}$, where $r < \hat{r} < \hat{R} < R$.

We conclude this section with several examples. As with Taylor series, we never compute the coefficients α_n directly from their integral formulas if we can circumvent this by utilizing known expansions.

EXAMPLE

Expand $1/(2+z)(1-z)$ in the annulus $1 < |z| < 2$. First, write

$$\frac{1}{(2+z)(1-z)} = \frac{1/3}{2+z} + \frac{1/3}{1-z}.$$

We now have $1/(2+z)(1-z)$ written as $h(z) + g(z)$, where $h(z) = \frac{1}{3}/(2+z)$ is analytic in $|z| < 2$, and $g(z) = \frac{1}{3}/(1-z)$ in $|z| > 1$. In $|z| < 2$,

$$\frac{\frac{1}{3}}{2+z} = \frac{1}{3}\,\frac{1}{2[1-(-z/2)]} = \frac{1}{6}\sum_{n=0}^{\infty}(-z/2)^n = \frac{1}{6}\sum_{n=0}^{\infty}\frac{(-1)^n}{2^n}z^n.$$

In order to expand $g(z)$ in powers of $1/z$ valid for $|z| < 1$, write

$$g(z) = \frac{-\frac{1}{3}}{z(1-1/z)} = \frac{-1}{3z}\sum_{n=0}^{\infty}\frac{1}{z^n}$$

$$= \sum_{n=0}^{\infty}\frac{-1}{3}\,\frac{1}{z^{n+1}} = \sum_{n=1}^{\infty}\frac{-1}{3}\,\frac{1}{z^n} = \sum_{n=-\infty}^{-1}\frac{-1}{3}z^n.$$

Thus, in the common region $1 < |z| < 2$,

$$\frac{1}{(2+z)(1-z)} = \sum_{n=0}^{\infty}\frac{(-1)^n}{6\cdot 2^n}z^n + \sum_{n=-\infty}^{-1}\frac{-1}{3}z^n = \sum_{n=-\infty}^{\infty}\alpha_n z^n,$$

where

$$\alpha_n = \begin{cases} \dfrac{(-1)^n}{6 \cdot 2^n} & \text{for } n = 0, 1, 2, \ldots, \\ \dfrac{-1}{3} & \text{for } n = -1, -2, -3, \ldots. \end{cases}$$

EXAMPLE

Expand $\sin(1/z)$ in $0 < |z| < \infty$. Here we can use the known expansion,

$$\sin(\xi) = \sum_{n=0}^{\infty} \frac{(-1)^n \xi^{2n+1}}{(2n+1)!}$$

to write

$$\sin\left(\frac{1}{z}\right) = \sum_{n=0}^{\infty} \frac{(-1)^n (1/z)^{2n+1}}{(2n+1)!}.$$

In this example, only negative powers of z appear.

Exercises for Section 9.8

1. Obtain the Laurent expansion for $f(z)$ about a for each of the following:

(a) $f(z) = \dfrac{1}{1+z^2}$, $a = i$

(b) $f(z) = \dfrac{\sin(z)}{z^3}$, $a = 0$

(c) $f(z) = e^{1/(z-1)}$, $a = 1$

(d) $f(z) = \dfrac{z^2 + 2z + 1}{z^2 - 1}$, $a = 1$

(e) $f(z) = \dfrac{\cos(z)}{z - \pi}$, $a = \pi$

2. Suppose that f is analytic in an annulus $r < |z - a| < R$. Show that the Laurent series for f about a converges uniformly in $\hat{r} \leq |z - a| \leq \hat{R}$, whenever $r < \hat{r} < \hat{R} < R$.

3. Expand $1/(z - i)(z + 2)$ in a Laurent series in the annulus $1 < |z| < 2$. *Hint:* Use partial fractions to write the function as a sum $g(z) + h(z)$, with g analytic in $|z| < 2$ and h analytic in $|z| > 1$. Expand these using algebraic manipulations and the geometric series.

9.9 Isolated Singularities

We shall often encounter the situation in which f is analytic in a punctured disc $0 < |z - a| < R$, but not at a itself. In many instances $f(a)$ is not even defined.

In such a case we say that f has an *isolated singularity* at a. We shall now see that isolated singularities can be separated into three categories.

First, it may happen that we can assign a value to $f(a)$ in such a way that the extended f [extended to have a value $f(a)$] is analytic in the whole disc $|z - a| < R$. The classic example of this is the function f given by $f(z) = \sin(z)/z$ for $z \neq 0$. At $z = 0$, f is undefined. However, observe that, for $z \neq 0$,

$$\frac{\sin(z)}{z} = \frac{1}{z}\sum_{n=0}^{\infty}\frac{(-1)^n z^{2n+1}}{(2n+1)!} = \sum_{n=0}^{\infty}\frac{(-1)^n z^{2n}}{(2n+1)!}$$

and the series on the right is analytic in the whole plane and has the value 1 at 0. We can therefore assign $f(0)$ the value 1, and the new f,

$$f(z) = \begin{cases} \dfrac{\sin(z)}{z} & z \neq 0, \\ 1, & z = 0, \end{cases}$$

is analytic at 0.

In such a case we say that f has a *removable singularity* at a. We can make f analytic in the whole disc $|z - a| < R$ simply by properly assigning a value to $f(a)$. Note that analyticity at a implies continuity at a. Hence a necessary condition that f have a removable singularity at a is that $\lim_{z\to a} f(z)$ exists. The only possible candidate for $f(a)$ is then the value of this limit. There remains to determine whether or not the resulting function is analytic at a.

To see what the other two possibilities alluded to above are, expand f in a Laurent series in $0 < |z - a| < R$:

$$f(z) = \sum_{n=-\infty}^{\infty} \alpha_n (z-a)^n.$$

Now, if all the coefficients of negative powers of $z - a$ vanish, the Laurent series becomes a Taylor series, and f is analytic at a or has a removable singularity at a. This suggests that we look at the other cases: (1) Thc Laurent series has only finitely many negative powers; and (2) the series has infinitely many negative powers.

In the first instance, we have, for some m,

$$f(z) = \frac{\alpha_{-m}}{(z-a)^m} + \frac{\alpha_{-m+1}}{(z-a)^{m-1}} + \cdots + \frac{\alpha_{-1}}{(z-a)} + \alpha_0 + \alpha_1(z-a) + \cdots,$$

where $\alpha_{-m} \neq 0$ but $\alpha_{-k} = 0$ for $k > m$. Here we say that f has a pole of order m at a. In this case, we can write

$$f(z) = \frac{g(z)}{(z-a)^m},$$

where g is analytic in $|z - a| < R$.

If $m = 1$ above, we say that f has a *simple pole* at a. As examples, $1/(z - a)^t$ has a pole of order t at a, if t is a positive integer, while $\sin(z)/z^2$ has a simple pole at 0. It is easy to show in general that $f(z) \cdot g(z)$ has a pole of order m at a whenever f is analytic at a (or has a removable singularity at a) and g has a pole of order m at a. For example, $e^z/(z - 3)^3$ has a pole of order 3 at 3.

In the remaining case, we have

$$f(z) = \sum_{n=-\infty}^{\infty} \alpha_n(z - a)^n$$

and $\alpha_n \neq 0$ for infinitely many negative integer values of n. Here we say that f has an essential singularity at a. As an example, of this

$$e^{1/z} = \sum_{n=0}^{\infty} \frac{1}{n!\, z^n}$$

for $0 < |z|$, hence $e^{1/z}$ has an essential singularity at 0. Similarly, $\sin[1/(z - 2)]$ has an essential singularity at 2.

As with poles, it is easy to show that $f(z) \cdot g(z)$ has an essential singularity at a if f is analytic (or has a removable singularity) at a and g has an essential singularity at a. For example,

$$\frac{\sin(z)}{e^{6z^2}}\, e^{1/(z-1)^3}$$

has an essential singularity at 1, because of the factor $e^{1/(z-1)^3}$, $\sin(z)/e^{6z^2}$ being analytic at 1.

Functions exhibit extremely remarkable behavior close to essential singularities. An illustration of this is the Casorati–Weierstrass Theorem: *Suppose that f has an essential singularity at a. Let α be any complex number. Then, given any $\varepsilon > 0$ and $\delta > 0$, there exists ξ such that $|\xi - a| < \delta$ and $|f(\xi) - \alpha| < \varepsilon$.* Put more simply: In *every* neighborhood of an essential singularity, $f(z)$ comes arbitrarily close to *every* complex number. A proof is outlined in the exercises.

It is often useful to have some way of characterizing the point at infinity in terms of singularities. The natural way to do this is to look at $f(1/z)$ at $z = 0$. To be specific, suppose that

$$f\left(\frac{1}{z}\right) = \sum_{n=-\infty}^{\infty} \alpha_n z^n \qquad \text{for } 0 < |z| < r.$$

If only finitely many negative powers of z appear, say

$$f\left(\frac{1}{z}\right) = \frac{\alpha_{-m}}{z^m} + \frac{\alpha_{-m+1}}{z^{m-1}} + \cdots + \alpha_0 + \alpha_1 z + \alpha_2 z^2 + \cdots,$$

where $\alpha_{-m} \neq 0$ but $\alpha_{-k} = 0$ for $k > m$, then we say that f has a *pole of order m* at ∞. Similarly, if there are infinitely many negative powers of z in the Laurent expansion of $f(1/z)$ about 0, we say that f has an *essential singularity* at ∞.

As examples, e^z has an essential singularity at ∞, since $e^{1/z}$ has an essential singularity at 0, while $(z^3 - z^2 + z)/(z - 1)$ has a pole of order 2 at ∞. To see the latter, put

$$f(z) = \frac{z^3 - z^2 + 2}{z - 1}.$$

Then

$$f\left(\frac{1}{z}\right) = \frac{1 - z + z^2}{z^2(1 - z)} = \frac{1}{z^2} + \frac{1}{1 - z},$$

which has a pole of order 2 at 0.

We conclude this section with some additional terminology which is frequently encountered in the literature. We say that f has a *zero* at a if $f(a) = 0$. The zero is of multiplicity m if $f(a) = f'(a) = \cdots = f^{(m-1)}(a) = 0$, but $f^{(m)}(a) \neq 0$. In terms of Taylor series, supposing f to be analytic at a, this means that $f(z) = \sum_{n=m}^{\infty} \alpha_n(z - a)^n$ and $\alpha_m \neq 0$. Note that, if f is analytic at a, then f has a zero of order m at a exactly when $f(1/z)$ has a pole of order m at a.

Exercises for Section 9.9

1. Classify the kind of singularity of f at the given point a for each of the following:

(a) e^{-1/z^2}, $a = 0$

(b) $\dfrac{\sin(z)}{(z - \pi)^2}$, $a = \pi$

(c) $e^{1/(z-1)}$, $a = 1$

(d) $\dfrac{\cos(z^2 + 1)}{\sin(z)}$, $a = 0$

(e) $\dfrac{\sin(z)}{z - \pi}$, $a = \pi$

(f) $\dfrac{z^2 + 2z + 1}{z + 1}$, $a = -1$

2. In the text we treated only isolated singularities. Show that $1/\sin(1/z)$ has a nonisolated singularity at 0 in the sense that the function fails to be analytic in any punctured disc $0 < |z| < r$, regardless of how small we choose r.

3. For each of the functions in Exercise 1, determine whether or not there is a pole or essential singularity at ∞; if a pole, give the order of the pole.

4. Let f be analytic in the whole complex plane, and have a nonessential singularity at ∞. Prove that f is a polynomial.

5. Establish the following characterizations of isolated singularities. Suppose that f is analytic in $0 < |z - a| < \delta$.
 (a) f has a removable singularity at a if and only if $\lim_{z \to a} (z - a)f(z) = 0$.
 (b) Suppose that for some integer $k > 0$, $\lim_{z \to a} |z - a|^{\alpha}|f(z)| = 0$ for $\alpha > k$, and $\lim_{z \to a} |z - a|^{\alpha}|f(z)| = \infty$ for $\alpha < k$. Then f has a pole of order k at a.
 (c) Suppose that neither $\lim_{z \to a} |z - a|^{\alpha}|f(z)| = 0$ nor $\lim_{z \to a} |z - a|^{\alpha}|f(z)| = \infty$ holds for any real α. Then f has an essential singularity at a.

6. Fill in the details of the following proof of the Casorati–Weierstrass theorem. Suppose that the theorem is false. Produce a complex function f, having an essential singularity at a and a complex number ξ and some $\delta > 0$ such that $|f(z) - \xi| > \delta$ for $0 < |z - a| < \delta$. Show that, for some $r > 0$,

$$\lim_{z \to a} |z - a|^{r}|f(z) - \xi| = 0.$$

Next, show that $\lim_{z \to a} |z - a|^{r}|\xi| = 0$, hence conclude that $\lim_{z \to a} |z - a|^{r}|f(z)| = 0$. This contradicts Exercise 5(c).

9.10 Residue Theorem

We shall now use Cauchy's theorem together with the results of Section 9.9 on singularities to develop the extremely powerful residue theorem. Suppose that f is analytic in a domain D, except at a, where f has an isolated singularity. We can then expand

$$f(z) = \sum_{n=-\infty}^{\infty} \alpha_n (z - a)^n$$

for some ring $0 < |z - a| < R$ about a. Here

$$\alpha_n = \frac{1}{2\pi i} \oint_{\gamma} \frac{f(\xi)\, d\xi}{(\xi - a)^{n+1}}$$

and γ is any closed path around a in $0 < |z - a| < R$. In particular,

$$\alpha_{-1} = \frac{1}{2\pi i} \oint_{\gamma} f(\xi)\, d\xi,$$

which is $\oint_{\gamma} f(z)\, dz$ except for the constant factor $1/2\pi i$. The number α_{-1}, which is the coefficient of $1/(z - a)$ in the Laurent expansion of f about a, is called the *residue* of f at a, denoted $\text{Res}(f, a)$. The above formula is then

$$\oint_{\gamma} f(z)\, dz = 2\pi i \,\text{Res}(f, a).$$

Note that so far γ encloses only one singularity of f. However, this extends immediately to the case in which γ encloses more than one isolated singularity of f. For suppose that $a_1, \ldots, a_n$ are isolated singularities of f inside γ.

Enclose each a_j by a circle C_j of sufficiently small radius that no two of γ, $C_1, \ldots, C_n$ intersect, each C_i is within γ, and f is analytic inside C_j except at a_j. Then by Cauchy's integral formula,

$$\oint_\gamma f(z)\,dz = \sum_{j=1}^{\infty} \oint_{C_j} f(z)\,dz.$$

But by the above discussion,

$$\oint_{C_j} f(z)\,dz = 2\pi i \operatorname{Res}(f, a_j) \qquad \text{for each } j = 1, \ldots, n.$$

Then

$$\oint_\gamma f(z)\,dz = 2\pi i \sum_{j=1}^{\infty} \operatorname{Res}(f, a_j).$$

This is the residue theorem.

On the face of it, the residue theorem may appear almost tautological, in view of the way we have defined the numbers $\operatorname{Res}(f, a_i)$. Its tremendous importance lies in the fact that the residues $\operatorname{Res}(f, a_i)$ can often be calculated by some trick that does not involve explicit integration of anything, giving us in such instances a means of evaluating $\oint_\gamma f(z)\,dz$ without actually carrying out any integration explicitly. In particular, whenever we can write the Laurent expansion of f about a, we automatically have the coefficient of $1/(z - a)$, hence $\operatorname{Res}(f, a)$. Here is a nontrivial illustration to demonstrate the power of this method.

EXAMPLE

Compute $\oint_C e^{1/z}\,dz$, where C is any closed path having 0 in its interior. The Laurent series for $e^{1/z}$ about 0 is

$$\sum_{n=0}^{\infty} \frac{1}{n!\,z^n}.$$

The coefficient of $1/z$ in this expansion is 1, so

$$\oint_C e^{1/z}\,dz = 2\pi i \operatorname{Res}(e^{1/z}, 0) = 2\pi i.$$

Similarly, $\oint_C e^{1/z^2}\,dz = 0$, since

$$e^{1/z^2} = \sum_{n=0}^{\infty} \frac{1}{n!}\frac{1}{z^{2n}},$$

so $\operatorname{Res}(e^{1/z^2}, 0) = 0$. ∎

One instance in which we can usually compute the residue of f at a without integrating is when f has a pole at a. To see how, suppose that f has a pole of order m at a. Then

$$f(z) = \frac{\alpha_{-m}}{(z-a)^m} + \frac{\alpha_{-m+1}}{(z-a)^{m-1}} + \cdots + \frac{\alpha_{-1}}{z-a} + \alpha_0 + \alpha_1(z-a) + \cdots$$

in some annulus $0 < |z-a| < R$. Put

$$\begin{aligned} g(z) = (z-a)^m f(z) &= \alpha_{-m} + \alpha_{-m+1}(z-a) + \alpha_{-m+2}(z-a)^2 + \cdots \\ &\quad + \alpha_{-1}(z-a)^{m-1} + \alpha_0(z-a)^m + \alpha_1(z-a)^{m+1} + \cdots \\ &= \alpha_{-m} + \alpha_{-m+1}(z-a) + \cdots + \alpha_{-1}(z-a)^{m-1} + (z-a)^m h(z), \end{aligned}$$

where $h(z) = \sum_{n=0}^{\infty} \alpha_n (z-a)^n$. Both h and g are analytic at a, being represented by Taylor series there. We can therefore differentiate both g and h as many times as we like around a. Start differentiating:

$$\begin{aligned} g'(z) &= \alpha_{-m+1} + 2\alpha_{-m+2}(z-a) + \cdots + (m-1)\alpha_{-1}(z-a)^{m-2} \\ &\quad + \frac{d[(z-a)^m h(z)]}{dz}, \\ g''(z) &= 2\alpha_{-m+2} + \cdots + (m-1)(m-2)\alpha_{-1}(z-a)^{m-2} \\ &\quad + \frac{d^2[(z-a)^m h(z)]}{dz^2}, \end{aligned}$$

and so on.

Finally, we get, at the $(m-1)$st step,

$$g^{(m-1)}(z) = \alpha_{-1}(m-1)(m-2)\cdots(2)(1) + \frac{d^{m-1}[(z-a)^m h(z)]}{dz^{m-1}}.$$

Now take the limit as $z \to a$. We get

$$\lim_{z\to a} g^{(m-1)}(z) = \alpha_{-1}(m-1)! + \lim_{z\to a} \frac{d^{m-1}[(z-a)^m h(z)[}{dz^{m-1}}.$$

We claim that the limit on the right is zero. It is easy to prove this by induction on m, or simply by observing that the $(m-1)$st derivative of $(z-a)^m h(z)$ will involve a sum of terms, all of which have a positive power of $(z - a)$ as a factor and hence vanish as $z \to a$. Recalling that $g(z) = (z - a)^m f(z)$, we now have

$$\operatorname{Res}(f, a) = \frac{1}{(m-1)!} \lim_{z \to a} \frac{d^{m-1}[(z-a)^m f(z)]}{dz^{m-1}}.$$

This provides an effective method for computing residues at poles. In particular, if f has a simple pole at a, then $m = 1$, and, using the conventions that $0! = 1$ and the zero derivative is the function itself, we have

$$\operatorname{Res}(f, a) = \lim_{z \to a} [(z - a)f(z)].$$

There is a special case in which the above formula for residues at simple poles yields a handy formula. Suppose that $f(z) = g(z)/h(z)$, where g is analytic at a and h has a simple zero at a. Then f has a simple pole at a. Since $h(a) = 0$ but $h'(a) \neq 0$, we have

$$\begin{aligned}\operatorname{Res}(f, a) &= \lim_{z \to a} (z - a)f(z) = \lim_{z \to a} \frac{(z-a)g(z)}{h(z)}\\ &= \lim_{z \to a} g(z) \frac{z - a}{h(z) - h(a)}\\ &= \frac{g(a)}{h'(a)}.\end{aligned}$$

EXAMPLE

Evaluate

$$\oint_C \frac{\sin(3z)\, dz}{(z-4)^3},$$

where C is any closed path not passing through 4. Consider two cases:

(1) 4 is not interior to C. Then

$$\oint_C \frac{\sin(3z)\, dz}{(z-4)^3} = 0,$$

by Cauchy's theorem.

(2) 4 is interior to C. Now

$$\oint_C \frac{\sin(3z)\, dz}{(z-4)^3} = 2\pi i \operatorname{Res}\left(\frac{\sin(3z)}{(z-4)^3}, 4\right).$$

To calculate this residue, note that $\sin(3z)/(z - 4)^3$ has a pole of order 3 at 4. For $z \neq 4$, we find

$$\frac{d^2}{dz^2} (z - 4)^3 \left[\frac{\sin(3z)}{(z-4)^3}\right] = -9 \sin(3z).$$

Then

$$\operatorname{Res}(\sin(3z)/(z-4)^3, 4) = \lim_{z\to 4}(-9\sin(3z)) = -9\sin(12).$$

Then

$$\oint_C \frac{\sin(3z)\,dz}{(z-4)^3} = -18\pi i \sin(12). \qquad \blacksquare$$

EXAMPLE

Evaluate

$$\oint_C \frac{dz}{(z-\alpha)(z-\beta)},$$

where α and β are distinct complex numbers and C is any closed path not passing through either α or β. We have three essentially different cases to consider.

(1) Neither α nor β is interior to C. Then

$$\oint_C \frac{dz}{(z-\alpha)(z-\beta)} = 0,$$

by Cauchy's Theorem.

(2) α is interior to C, β exterior. Now $1/(z-\alpha)(z-\beta)$ has one singularity inside C, a simple pole at $z=\alpha$. Then

$$\oint_C \frac{dz}{(z-\alpha)(z-\beta)} = 2\pi i \operatorname{Res}\left(\frac{1}{(z-\alpha)(z-\beta)}, \alpha\right)$$

$$= 2\pi i \lim_{z\to\alpha}(z-\alpha)\frac{1}{(z-\alpha)(z-\beta)} = 2\pi i(\alpha-\beta).$$

If β is interior to C and α exterior, we obtain $2\pi i(\beta-\alpha)$.

(3) α and β are both interior to C. Now

$$\oint_C \frac{dz}{(z-\alpha)(z-\beta)} = 2\pi i\left[\operatorname{Res}\left(\frac{1}{(z-\alpha)(z-\beta)}, \alpha\right) + \operatorname{Res}\left(\frac{1}{(z-\alpha)(z-\beta)}, \beta\right)\right]$$

$$= 2\pi i\left(\frac{1}{\alpha-\beta} + \frac{1}{\beta-\alpha}\right) = 0. \qquad \blacksquare$$

EXAMPLE

Evaluate

$$\oint_C \frac{e^{2z^2}\,dz}{\sin(z)},$$

where C is the path shown in Fig. 9-11. The only singularity of $e^{2z^2}/\sin(z)$ inside

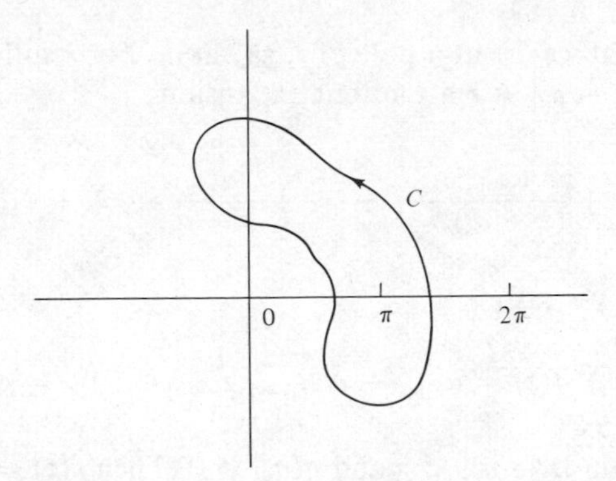

Fig. 9-11

C is a simple pole at π, Putting $g(z) = e^{2z^2}$ and $h(z) = \sin(z)$ in the formula derived above, we have

$$\operatorname{Res}\left(\frac{e^{2z^2}}{\sin(z)}, \pi\right) = \frac{g(\pi)}{h'(\pi)} = \frac{e^{2\pi^2}}{\cos(\pi)} = -e^{2\pi^2}.$$

Then

$$\oint_C \frac{e^{2z^2}\,dz}{\sin(z)} = -2\pi i e^{2\pi^2}. \qquad \blacksquare$$

We shall return to the use of the residue theorem in evaluating integrals in the next section. For now, we shall give an application of a different kind.

Suppose that f is analytic in a domain D, except for poles of order m_j at $a_j, j = 1, \ldots, M$. Let γ be a closed path in D enclosing $a_1, \ldots, a_n$ in its interior. Suppose also that f has zeros of order n_j at b_j interior to $\gamma, j = 1, \ldots, N$. We claim then that

$$\frac{1}{2\pi i} \oint_\gamma \frac{f'(z)\,dz}{f(z)} = \sum_{j=1}^{N} n_j - \sum_{j=1}^{M} m_j.$$

Put another way,

$$\frac{1}{2\pi i} \oint_\gamma \frac{f'(z)\,dz}{f(z)}$$

equals the number of zeros minus the number of poles of f within γ, each zero or pole counted as many times as its order calls for. So, for example, a pole of order 2 is counted twice in the above sum, and so on.

The proof is a straightforward exercise in using the residue theorem. Look at $f'(z)/f(z)$ for z inside γ. The only possible singularities are at zeros or poles of f.

Suppose first that we are at a pole of f, say at a_s. For r sufficiently small, we can write in $0 < |z - a_s| < r$ a Laurent expansion:

$$f(z) = \frac{\alpha_{-m_s}}{(z - a_s)^{m_s}} + \frac{\alpha_{-m_s+1}}{(z - a_s)^{m_s-1}} + \cdots + \frac{\alpha_{-1}}{z - a_s} + \alpha_0 + \alpha_1(z - a_s) + \cdots.$$

Then

$$(z - a_s)^{m_s} f(z) = \alpha_{-m_s} + \alpha_{-m_s+1}(z - a) + \cdots = g(z),$$

where g is analytic in $|z - a_s| < r$ and $g(a_s) \neq 0$. Then $f(z) = (z - a_s)^{-m_s} g(z)$, and

$$\frac{f'(z)}{f(z)} = \frac{-m_s(z - a_s)^{-m_s-1}g(z) + (z - a_s)^{-m_s}g'(z)}{(z - a_s)^{-m_s}g(z)}$$

$$= \frac{-m_s}{z - a_s} + \frac{g'(z)}{g(z)}.$$

Since $g'(z)/g(z)$ is analytic at a_s, then $f'(z)/f(z)$ has at a_s a simple pole, and $\operatorname{Res}(f'(z)/f(z), a_s) = -m_s$. [Actually, $g'(z)/g(z)$ has a Taylor expansion $\sum_{j=0}^{\infty} \beta_j(z - a_s)^j$ about a_s, and

$$\frac{-m_s}{z - a_s} + \sum_{j=0}^{\infty} \beta_j(z - a_s)^j$$

becomes the Laurent expansion of $f'(z)/f(z)$ in $0 < |z - a_s| < r$. The residue is then the coefficient of $1/(z - a_s)$.]

Now look for the residue of $f'(z)/f(z)$ at a zero of f, say at b_t. Since b_t is a zero of multiplicity n_t, then, in some neighborhood $|z - b_t| < R$, we have a Taylor expansion

$$f(z) = \alpha_{n_t}(z - b_t)^{n_t} + \alpha_{n_t+1}(z - b_t)^{n_t+1} + \cdots$$

$$= (z - b_t)^{n_t} h(z),$$

where h is analytic in $|z - b_t| < R$ and $h(b_t) \neq 0$. Then

$$\frac{f'(z)}{f(z)} = \frac{n_t(z - b_t)^{n_t-1}h(z) + (z - b_t)^{n_t}h'(z)}{(z - b_t)^{n_t}h(z)}$$

$$= \frac{n_t}{z - b_t} + \frac{h'(z)}{h(z)}.$$

Since $h'(z)/h(z)$ is analytic at b_t, then $\operatorname{Res}(f'(z)/f(z), b_t) = n_t$.

Totaling everything up gives us, by the residue theorem,

$$\frac{1}{2\pi i}\oint_\gamma \frac{f'(z)\,dz}{f(z)} = \text{sum of residues of } \frac{f'(z)}{f(z)} \text{ inside } \gamma$$

$$= \sum_{j=1}^{N} n_j - \sum_{j=1}^{M} m_j.$$

This is called the *argument principle*, and it is often of use in estimating the number of zeros or poles of a function in a given domain. Sometimes it is of use in evaluating integrals as well.

EXAMPLE

Compute $\int_\gamma \tan(z)\,dz$, where γ is the square shown below in Fig. 9-12. Write

$$\oint_\gamma \tan(z)\,dz = \oint_\gamma \frac{-d(\cos(z))\,dz}{\cos(z)}$$

$$= -2\pi i[\text{number of zeros of } \cos(z) \text{ inside } \gamma\text{-number of poles of } \cos(z) \text{ inside } \gamma].$$

Now $\cos(z)$ has no poles in the finite plane, and its zeros within γ are at $-\pi/2$ and $\pi/2$, both of multiplicity 1. Thus

$$\oint_\gamma \tan(z)\,dz = -2\pi i(2 - 0) = -4\pi i.$$

As a check, we can use the residue theorem to write

$$\oint_\gamma \tan(z)\,dz = 2\pi i[\text{sum of residues of } \tan(z) \text{ at poles within } \gamma].$$

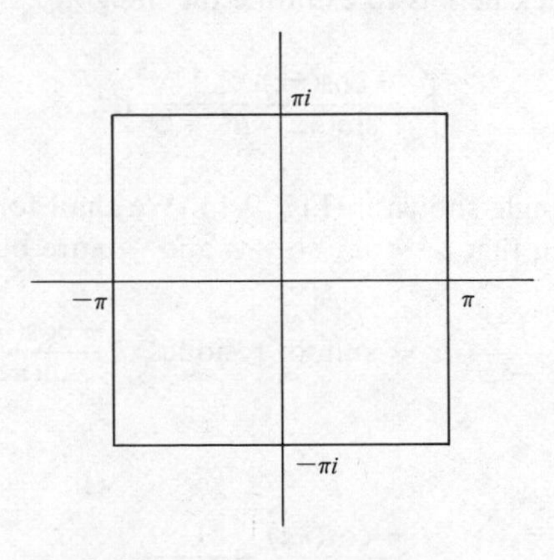

Fig. 9-12

Inside γ, $\tan(z)$ has only simple poles at $\pi/2$ and $-\pi/2$. Calculate

$$\operatorname{Res}\left(\tan(z), \frac{\pi}{2}\right) = \operatorname{Res}\left(\frac{\sin(z)}{\cos(z)}, \frac{\pi}{2}\right) = \frac{\sin(\pi/2)}{\left.\dfrac{d(\cos(z))}{dz}\right|_{\pi/2}} = -1$$

and

$$\operatorname{Res}\left(\tan(z), \frac{-\pi}{2}\right) = \operatorname{Res}\left(\frac{\sin(z)}{\cos(z)}, \frac{-\pi}{2}\right) = \frac{\sin(-\pi/2)}{\left.\dfrac{d(\cos(z))}{dz}\right|_{-\pi/2}} = -1.$$

Thus, by the residue theorem

$$\oint_{\gamma} \tan(z)\, dz = 2\pi i(-1 \ -1) = -4\pi i.$$

in agreement with the result derived by the argument principle. ∎

We conclude this section with a surprising application. Under some circumstances, a very clever use of the residue theorem can result in rather general results on the summation of series. We shall give one fully detailed example, and leave some others to the exercises.

EXAMPLE

Evaluate $\sum_{n=-\infty}^{\infty} 1/(a^2 - n^2)$, where a is any real number that is not an integer. The (nonobvious) trick here is to examine the integral

$$\oint_{\gamma_n} \frac{\pi \cos(\pi z)}{\sin(\pi z)} \frac{1}{a^2 - z^2}\, dz,$$

where γ_n is the rectangle shown in Fig. 9-13. We shall for convenience assume that n is large enough that $a^2 < n^2$, so $+a$ and $-a$ are both inside γ_n. Now

$$\frac{1}{2\pi i} \oint_{\gamma_n} \frac{\pi \cos(\pi z)}{\sin(\pi z)} \frac{1}{a^2 - z^2}\, dz = \text{sum of residues of } \frac{\pi \cos(\pi z)}{\sin(\pi z)} \frac{1}{a^2 - z^2} \text{ inside } \gamma_n.$$

By choice of n,

$$\frac{\pi \cos(\pi z)}{\sin(\pi z)} \frac{1}{a^2 - z^2}$$

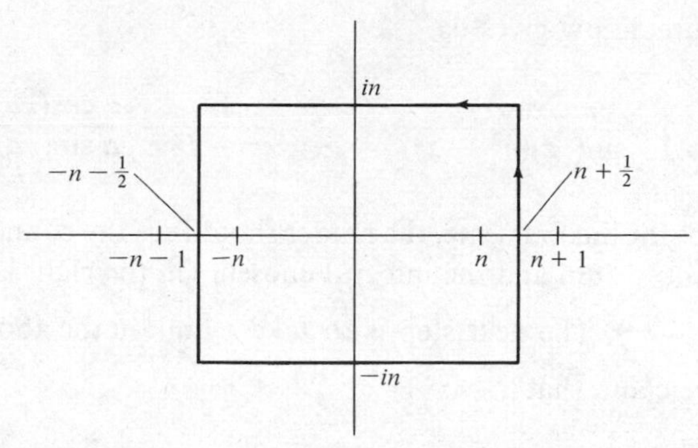

Fig. 9-13

has, within γ_n, simple poles at $0, \pm 1, \ldots, \pm n, \pm a$. We compute the residues at these points. For the residue at an integer pole, say at k, write

$$\frac{\pi\cos(\pi z)}{\sin(\pi z)}\frac{1}{a^2 - z^2} = \frac{\pi\cos(\pi z)}{a^2 - z^2}\frac{1}{\sin(\pi z)}.$$

Since $\cos(\pi z)/(a^2 - z^2)$ is analytic at k, then

$$\operatorname{Res}\left(\frac{\pi\cos(\pi z)}{a^2 - z^2}\frac{1}{\sin(\pi z)}, k\right) = \frac{\pi\cos(\pi k)}{a^2 - k^2}\frac{1}{\left.\dfrac{d(\sin(\pi z))}{dz}\right|_{z=k}}$$

$$= \frac{1}{a^2 - k^2}.$$

To find the residue at a, write

$$\frac{\pi\cos(\pi z)}{\sin(\pi z)}\frac{1}{a^2 - z^2} = \frac{\pi\cos(\pi z)}{(a + z)\sin(\pi z)}\frac{1}{a - z},$$

so that

$$\operatorname{Res}\left(\frac{\pi\cos(\pi z)}{\sin(\pi z)(a^2 - z^2)}, a\right) = \frac{\pi\cos(\pi a)}{\sin(\pi a)(2a)}\frac{1}{\left.\dfrac{d(a - z)}{dz}\right|_{z=a}} = \frac{-\pi\cos(\pi a)}{2a\sin(\pi a)}.$$

Similarly, at $-a$, the residue is

$$\operatorname{Res}\left(\frac{\pi\cos(\pi z)}{\sin(\pi z)(a - z)}\frac{1}{a + z}, -a\right) = \frac{\pi\cos(-\pi a)}{\sin(-\pi a)(-2a)}\frac{1}{\left.\dfrac{d(a + z)}{dz}\right|_{z=-a}}$$

$$= \frac{-\pi\cos(\pi a)}{2a\sin(\pi a)}.$$

The residue theorem now gives us

$$\frac{1}{2\pi i}\oint_{\gamma_n}\frac{\pi\cos(\pi z)\,dz}{\sin(\pi z)(a^2-z^2)}=\sum_{j=-n}^{n}\frac{1}{a^2-j^2}-\frac{\pi\cos(\pi a)}{a\sin(\pi a)}.$$

Here, perhaps for the first time, the reader should see the connection between the series we want to sum and the integral chosen. On the right is a partial sum for $\sum_{n=-\infty}^{\infty} 1/(a^2-n^2)$. The next step is to take a limit in the above equation as $n\to\infty$. First, we claim that

$$\left|\oint_{\gamma_n}\frac{\pi\cos(\pi z)\,dz}{\sin(\pi z)(a^2-z^2)}\right|\to 0$$

as $n\to\infty$. The proof of this requires some estimates. Note that the length of γ_n is $4n+1$. Further,

$$|a^2-z^2|=|z^2-a^2|\geq |z|^2-|a|^2\geq n^2-a^2$$

for z on γ_n. Then

$$\frac{1}{|a^2-z^2|}\leq\frac{1}{n^2-a^2}$$

for z on γ_n, so

$$\left|\oint_{\gamma_n}\frac{\pi\cos(\pi z)}{\sin(\pi z)(a^2-z^2)}\,dz\right|\leq\frac{4n+1}{n^2-a^2}M$$

if we can produce M such that

$$\left|\frac{\pi\cos(\pi z)}{\sin(\pi z)}\right|\leq M\qquad\text{for } z \text{ on } \gamma_n.$$

We now produce such an M. Note that, on the right side of γ_n, $z=n+\frac{1}{2}+iy$, $-n\leq y\leq n$. Thus

$$\begin{aligned}\frac{\pi\cos(\pi z)}{\sin(\pi z)}&=\frac{\pi\cos[\pi(n+\frac{1}{2}+iy)]}{\sin[\pi(n+\frac{1}{2}+iy)]}\\&=\pi\left|\frac{\cos[\pi(n+\frac{1}{2})]\cos(\pi iy)-\sin[\pi(n+\frac{1}{2})]\sin(\pi iy)}{\sin[\pi(n+\frac{1}{2})]\cos(\pi iy)+\cos[\pi(n+\frac{1}{2})]\sin(\pi iy)}\right|\\&=\pi\left|\frac{\sin[\pi(n+\frac{1}{2})]\sin(\pi iy)}{\sin[\pi(n+\frac{1}{2})]\cos(\pi iy)}\right|=\pi\left|\frac{-i\sinh(\pi y)}{\cosh(\pi y)}\right|=\pi|\tanh(\pi y)|.\end{aligned}$$

Here we have used the facts that $\cos[\pi(n+\frac{1}{2})]=0$, $-i\sin(iz)=\sinh(z)$, and

$\cos(iz) = \cosh(z)$. Now, $|\tanh(\pi y)| \leq 1$ for all real y. Thus, on the right side of γ_n,

$$\left|\frac{\pi \cos(\pi z)}{\sin(\pi z)}\right| \leq \pi.$$

The same estimate works for the left side of γ_n, where $z = -(n + \frac{1}{2}) + iy$, $-n \leq y \leq n$.

Next, on top of γ_n, $z = x + in$, $-(n + \frac{1}{2}) \leq x \leq n + \frac{1}{2}$. We then have

$$\begin{aligned}
\left|\frac{\cos(\pi z)}{\sin(\pi z)}\right|^2 &= \left|\frac{\cos[\pi(x + in)]}{\sin[\pi(x + in)]}\right|^2 \\
&= \left|\frac{\cos(\pi x)\cos(\pi in) - \sin(\pi x)\sin(\pi in)}{\cos(\pi x)\sin(\pi in) + \cos(\pi in)\sin(\pi x)}\right|^2 \\
&= \left|\frac{\cos(\pi x)\cosh(\pi n) + i\sin(\pi x)\sinh(\pi n)}{-i\cos(\pi x)\sinh(\pi n) + \cosh(\pi n)\sin(\pi x)}\right|^2 \\
&= \frac{\cos^2(\pi x)\cosh^2(\pi n) + \sin^2(\pi x)\sinh^2(\pi n)}{\cos^2(\pi x)\sinh^2(\pi n) + \cosh^2(\pi n)\sin^2(\pi x)} \\
&= \frac{\frac{1}{4}\cos^2(\pi x)(e^{2\pi n} + e^{-2\pi n} + 2) + \frac{1}{4}\sin^2(\pi x)(e^{2\pi n} + e^{-2\pi n} - 2)}{\frac{1}{4}\cos^2(\pi x)(e^{2\pi n} + e^{-2\pi n} - 2) + \frac{1}{4}\sin^2(\pi x)(e^{2\pi n} + e^{-2\pi n} + 2)} \\
&= \frac{\frac{1}{2}(e^{2\pi n} + e^{-2\pi n}) + [\cos^2(\pi x) - \sin^2(\pi x)]}{\frac{1}{2}(e^{2\pi n} + e^{-2\pi n}) - [\cos^2(\pi x) - \sin^2(\pi x)]} \\
&= \frac{\cos(2\pi n) + \cos(2\pi x)}{\cosh(2\pi n) - \cos(2\pi x)} \leq \frac{\cosh(2\pi n) + 1}{\cosh(2\pi n) - 1} = \frac{1 + \dfrac{1}{\cosh(2\pi n)}}{1 - \dfrac{1}{\cosh(2\pi n)}}.
\end{aligned}$$

Now, $\cosh(2\pi n) \to \infty$ as $n \to \infty$. Choose N sufficiently large that $1/\cosh(2\pi n) \leq \frac{1}{2}$ for $n \geq N$. Then, for $n \geq N$,

$$\frac{1 + 1/\cosh(2\pi n)}{1 - 1/\cosh(2\pi n)} \leq \frac{1 + \frac{1}{2}}{1 - \frac{1}{2}} = 3.$$

Then

$$\left|\frac{\pi \cos(\pi z)}{\sin(\pi z)}\right| \leq 3\pi$$

on the top side of γ_n, if $n \geq N$. A similar derivation gives the same result for the bottom side of γ_n. In summary, then, for $n \geq N$, and z on γ_n,

$$\left|\frac{\pi \cos(\pi z)}{\sin(\pi z)}\right| \leq 3\pi.$$

Then, for $n \geq N$,

$$\left| \oint_{\gamma_n} \frac{\pi \cos(\pi z)}{\sin(\pi z)(a^2 - z^2)} \, dz \right| \leq \frac{4n + 1}{n^2 - a^2} 3\pi \to 0$$

as $n \to \infty$. Hence

$$\frac{1}{2\pi i} \oint_{\gamma_n} \frac{\pi \cos(\pi z)}{\sin(\pi z)(a^2 - z^2)} \, dz = \sum_{j=-n}^{n} \frac{1}{a^2 - j^2} - \frac{\pi \cos(\pi a)}{a \sin(\pi a)} \to 0$$

as $n \to \infty$. Looking at the right side, we now conclude that

$$\sum_{n=-\infty}^{\infty} \frac{1}{a^2 - n^2} = \frac{\pi \cos(\pi a)}{a \sin(\pi a)}.$$ ∎

The last example illustrates two facts. First, complex-variable methods provide a powerful tool for handling problems that seem inaccessible by real methods. The series to be summed was real, but the method involved complex function theory. Second, the appropriate way to wring the result out of complex function theory is usually not obvious, and requires considerable skill and experience on the part of the user.

In the next section we shall see complex integration methods applied to the evaluation of real integrals.

Exercises for Section 9.10

1. Let g be analytic in a domain D, and let f be analytic in D except at a finite number of simple poles. Let γ be a closed path in D enclosing poles $a_1, \ldots, a_n$ of f, simple zeros $b_1, \ldots, b_k$ of f, and not passing through any poles of f. Show that

$$\frac{1}{2\pi i} \oint_{\gamma} g(z) \frac{f'(z) \, dz}{f(z)} = \sum_{j=1}^{k} g(b_j) - \sum_{j=1}^{n} g(a_j).$$

2. Use the residue theorem to do these problems of Section 9.5: Exercise 3(a), (b), (c); and Exercise 5(a), (b), (c).

3. Use the residue theorem to compute:

(a) $\displaystyle\oint_{\gamma} \frac{e^{z^2} \, dz}{(z - i)(z + 2)^2(z - 4)}$, γ the circle $|z| = 3$

(b) $\displaystyle\oint_{\gamma} \frac{z^2 + 2z + 1}{(z - 3)^3}$, γ the circle $|z| = 8$

(c) $\displaystyle\oint_{\gamma} \frac{\sin(3z) \, dz}{\cos(z) - \sin(z)}$, γ the circle $|z| = \pi/2$

(d) $\displaystyle\oint_{\gamma} \frac{[z^{1/2}]_P \, dz}{z + i}$, γ the circle $|z - i| = \frac{1}{2}$

4. Consider the problem of evaluating

$$\oint_\gamma \frac{[e^{2z} + \sin(z)]\,dz}{(z-4)^3(z-3)^2(z-1)^2(z+i)(z-i)(z+8)},$$

where γ is the circle $|z| = 24$.

(a) Work the problem using the residue theorem.

(b) Redo the problem by splitting the integrand into partial fractions and using Cauchy's integral formula or Cauchy's integral formulas for derivatives on each part separately.

5. Evaluate

$$\oint_\gamma \frac{dz}{(z-1)(z+8)} \text{ on } \gamma\colon |z| = 5,$$

(a) By the residue theorem.

(b) By splitting the integrand into partial fractions and integrating each part by parametrization of γ (or perhaps of a suitable, convenient replacement for γ).

6. Evaluate the following by the residue theorem.

(a) $\displaystyle\oint_\gamma \frac{\sin(3z)\,dz}{e^z(z-4)^2}, \gamma\colon |z| = 5$

(b) $\displaystyle\oint_\gamma \frac{e^{1/z}\,dz}{z^2}, \gamma\colon |z| = 2$

(c) $\displaystyle\oint_\gamma \frac{2z^2\,dz}{(z+2)^{1/3}}, \gamma\colon |z| = 7$

(d) $\displaystyle\oint_\gamma \frac{dz}{z^6+1}, \gamma\colon |z| = \tfrac{3}{2}$

7. Compute the following, using the argument principle, and then by using the residue theorem.

(a) $\displaystyle\oint_\gamma \frac{(3z^2+2)\,dz}{z^3+2z+8}, \gamma\colon |z| = 8$

(b) $\displaystyle\oint_\gamma \frac{dz}{(1+z)\log(z+1)}, \gamma\colon |z| = \tfrac{1}{2}$

8. Prove *Rouche's theorem: If f and g are analytic in a domain D, γ is a closed path in D, and $|f(z)| > |g(z)|$ for z on γ, then $f(z)$ and $f(z) + g(z)$ have the same number of zeros within γ.*

Hint: Apply the argument principle to

$$\frac{f(z)+g(z)}{f(z)}.$$

9. Use the method of the last example of this chapter to show that, for a real,

(a) $\displaystyle\sum_{n=-\infty}^{\infty} \frac{a}{n^2+a^2} = \frac{\pi}{\tanh(\pi a)}, a \neq 0.$

(b) $\displaystyle\sum_{n=-\infty}^{\infty} \frac{1}{(n+a)^2} = \left[\frac{\pi}{\sin(\pi a)}\right]^2, a \neq \text{integer}.$

Hint: Consider

$$\oint_{\gamma_n} \frac{f(z)\pi \cos(\pi z)\, dz}{\sin(\pi z)}$$

for appropriate γ_n and $f(z)$, and then let $n \to \infty$.

10. Show that

(a) $\displaystyle\sum_{n=-\infty}^{\infty} \frac{(-1)^n \sin(na)n}{b^2 - n^2} = \frac{\pi \sin(ab)}{2 \sin(b\pi)}$, b real, noninteger, $-\pi < a < \pi$

(b) $\displaystyle\sum_{n=-\infty}^{\infty} \frac{(-1)^n}{(a+n)^2} = \frac{\pi^2 \cot(\pi a)}{\sin(\pi a)}$, a real, noninteger

Hint: Consider $\oint_{\gamma_n} f(z)\pi \csc(\pi z)\, dz$ for appropriate $f(z)$ and γ_n, and let $n \to \infty$.

11. Show that limits of the form

$$\lim_{n\to\infty} \sum_{j=-n}^{n} \frac{1}{P(j)},$$

with $P(z)$ a polynomial of degree ≥ 2 with real coefficients and no integer zeros, can be evaluated by

(a) Applying the residue theorem to

$$\oint_{\gamma_n} \frac{\pi \cos(\pi z)\, dz}{\sin(\pi z)P(z)}$$

with γ_n the rectangle through $\pm n + 1/2$, and $\pm(n + 1/2)i$, for n sufficiently large that all zeros of $P(z)$ are inside γ_n;

(b) Then showing that

$$\oint_{\gamma_n} \frac{\pi \cos(\pi z)\, dz}{\sin(\pi z)P(z)} \to 0 \qquad \text{as } n \to \infty.$$

12. Show that limits of the form

$$\lim_{n\to\infty} \sum_{j=-n}^{n} \frac{(-1)^j}{P(j)}$$

with $P(z)$ a polynomial of degree ≥ 2 with real coefficients and no integer zeros, can be evaluated by

(a) Applying the residue theorem to

$$\oint_{\gamma_n} \frac{\pi \csc(\pi z)\, dz}{P(z)},$$

with γ_n the path of Exercise 11(a).

(b) Then showing that

$$\oint_{\gamma_n} \frac{\pi \csc(\pi z)\, dz}{P(z)} \to 0 \qquad \text{as } n \to \infty.$$

13. Sum the following series by residue methods:

(a) $\sum_{n=1}^{\infty} \frac{1}{n^2}$

(b) $\sum_{n=1}^{\infty} \frac{1}{n^4}$

(c) $\sum_{n=1}^{\infty} \frac{(-1)^{n-1}}{n^2}$

14. Let $P(z) = (z - a_1)\cdots(z - a_n)$, with $a_1, \ldots, a_n$ complex numbers (not necessarily distinct). Let γ be a closed path with $a_1, \ldots, a_n$ interior to γ. Show that

$$\oint_\gamma \frac{P'(z)\,dz}{P(z)} = 2\pi i n$$

(a) By using the argument principle.
(b) By computing $P'(z)/P(z)$ explicitly and using the residue theorem.

9.11 Application of the Residue Theorem to Evaluation of Real Integrals

The residue theorem turns out to be an extremely powerful tool in the evaluation of certain kinds of real integrals. We shall use this section to give some idea of the techniques involved.

Let us first consider real integrals of the form

$$\int_0^{2\pi} f(\cos(\theta), \sin(\theta))\,d\theta,$$

where $f(x, y)$ is a rational function (i.e., quotient of polynomials) in x and y. The idea is to transform this to a complex integral that can then be evaluated by the residue theorem or some other means.

Recall that

$$\cos(\theta) = \tfrac{1}{2}(e^{i\theta} + e^{-i\theta}) \quad \text{and} \quad \sin(\theta) = \frac{1}{2i}(e^{i\theta} - e^{-i\theta}).$$

Put $z = e^{i\theta}$ (so that we are on the unit circle) to write

$$\cos(\theta) = \frac{1}{2}\left(z + \frac{1}{z}\right) \quad \text{and} \quad \sin(\theta) = \frac{1}{2i}\left(z - \frac{1}{z}\right).$$

Then

$$\int_0^{2\pi} f(\cos(\theta), \sin(\theta))\,d\theta = \oint_\gamma f\left(\frac{1}{2}\left(z + \frac{1}{z}\right), \frac{1}{2i}\left(z - \frac{1}{z}\right)\right)\frac{1}{iz}\,dz,$$

where $\gamma(\theta) = e^{i\theta}$, $\theta\colon 0 \to 2\pi$.

Assuming that

$$\frac{1}{iz} f\left(\frac{1}{2}\left(z + \frac{1}{z}\right), \frac{1}{2i}\left(z - \frac{1}{z}\right)\right)$$

has no singularities on the unit circle, we can use the residue theorem on

$$\frac{1}{i} \oint_\gamma \frac{1}{z} f\left(\frac{1}{2}\left(z + \frac{1}{z}\right), \frac{1}{2i}\left(z - \frac{1}{z}\right)\right) dz$$

to evaluate

$$\int_0^{2\pi} f(\cos(\theta), \sin(\theta))\, d\theta.$$

EXAMPLE

Evaluate

$$\int_0^{2\pi} \frac{d\theta}{[2 + \cos(\theta)]^2}.$$

Write

$$\frac{1}{[2 + \cos(\theta)]^2} = \frac{1}{[2 + \frac{1}{2}(z + 1/z)]^2} = \frac{4z^2}{(z^2 + 4z + 1)^2}.$$

Then

$$\int_0^{2\pi} \frac{d\theta}{[2 + \cos(\theta)]^2} = \frac{1}{i} \oint_\gamma \frac{4z^2\, dz}{(z^2 + 4z + 1)^2} \frac{dz}{z} = \frac{1}{i} \oint_\gamma \frac{4z\, dz}{(z^2 + 4z + 1)^2}$$

$$= \frac{4}{i} \oint_\gamma \frac{z\, dz}{[z - (-2 + \sqrt{3})]^2 [z - (-2 - \sqrt{3})]^2},$$

where γ is the unit circle $|z| = 1$. The integrand has double poles at $-2 + \sqrt{3}$ and at $-2 - \sqrt{3}$. Only $-2 + \sqrt{3}$ lies within the unit circle, so we need only compute

$$\operatorname{Res}\left(\frac{z}{(z^2 + 4z + 1)^2}, -2 + \sqrt{3}\right).$$

This is

$$\lim_{z \to -2+\sqrt{3}} \left(\frac{d}{dz}\left([z - (-2 + \sqrt{3})]^2 \frac{z}{(z^2 + 4z + 1)^2}\right)\right)$$

$$= \lim_{z \to -2+\sqrt{3}} \left[\frac{d}{dz}\left(\frac{z}{[z - (-2 - \sqrt{3})]^2}\right)\right]$$

$$= \lim_{z \to -2+\sqrt{3}} \left(\frac{(z - (-2 - \sqrt{3}))^2 - 2z[z - (-2 - \sqrt{3})]}{[z - (-2 - \sqrt{3})]^4}\right)$$

$$= \frac{\sqrt{3}}{18}.$$

Then

$$\oint_\gamma \frac{4z\,dz}{(z^2+4z+1)^2} = 4(2\pi i)\frac{\sqrt{3}}{18} = \frac{4\pi i\sqrt{3}}{9}.$$

Then

$$\int_0^{2\pi} \frac{d\theta}{[2+\cos(\theta)]^2} = \frac{1}{i}\oint_\gamma \frac{4z\,dz}{(z^2+4z+1)^2} = \frac{4\pi\sqrt{3}}{9}. \quad ∎$$

EXAMPLE

As a second example, we shall compute $\int_0^{2\pi} \cos^6(\theta)\,d\theta$. Write

$$\cos^6(\theta) = \frac{1}{2}\left(z + \frac{1}{z}\right)^6 = \frac{1}{2^6}\frac{(z^2+1)^6}{z^6}.$$

Then

$$\int_0^{2\pi} \cos^6(\theta)\,d\theta = \frac{1}{i}\oint_\gamma \frac{1}{2^6}\frac{(z^2+1)^6}{z^6}\frac{dz}{z} = \frac{1}{2^6 i}\oint_\gamma \frac{(z^2+1)\,dz}{z^7}.$$

Here the integrand has a pole of order 7 at 0, which is interior to γ. There are two ways we can find the residue at 0. First,

$$\operatorname{Res}\left(\frac{(z^2+1)^6}{z^7}, 0\right) = \lim_{z\to 0}\left(\frac{1}{6!}\frac{d^6}{dz^6}\frac{z^7(z^2+1)^6}{z^7}\right)$$
$$= \frac{1}{6!}\lim_{z\to 0}\frac{d^6}{dz^6}[(z^2+1)^6].$$

A straightforward but tedious calculation will now certainly give the residue. However, a simpler way is to write, using the binomial theorem,

$$\frac{(1+z^2)^6}{z^7} = \frac{1}{z^7}\sum_{j=0}^{6}\binom{6}{j}(z^2)^j = \sum_{j=0}^{6}\binom{6}{j}z^{2j-7}.$$

This must be the Laurent expansion of $(1+z^2)^6/z^7$ about 0. The coefficient of $\frac{1}{z}$ (obtained when $j=3$) is $\binom{6}{3}$, so the residue is $\binom{6}{3}$, or $6!/3!\,3!$, which works out to 20. Hence

$$\int_0^{2\pi} \cos^6(\theta)\,d\theta = \frac{1}{2^6 i}\oint_\gamma \frac{(z^2+1)^6\,dz}{z^7} = \frac{1}{2^6 i}(2\pi i)(20) = \frac{5\pi}{8}. \quad ∎$$

We shall now consider evaluation of improper integrals of the form $\int_{-\infty}^{\infty} R(x)\,dx$. We shall suppose that R is a rational function (quotient of polynomials) and that R has no poles on the real axis. Finally, we shall assume that

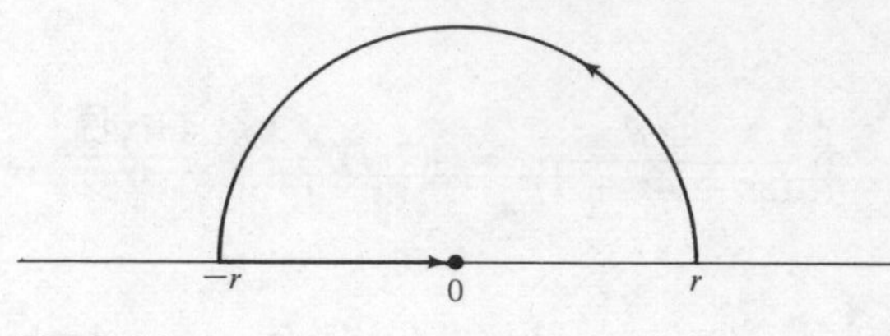

Fig. 9-14

$\int_{-\infty}^{\infty} |R(x)|\, dx$ converges. To ensure this, we suppose that the degree of the denominator exceeds that of the numerator by at least 2.

The trick now is to note that the only singularities a rational function can have are at zeros of the denominator. If the denominator is a polynomial of degree n, then $R(z)$ has at most n singularities, and these are all poles. We choose a number r such that each pole in the upper half-plane is interior to the closed path γ formed by taking the semicircle $z = re^{i\theta}$, $\theta: 0 \to \pi$, together with the segment $-r \le x \le r$ on the real axis (Fig. 9-14). Then we have

$$\oint_\gamma R(z)\, dz = 2\pi i(\text{sum of residues of } R \text{ at poles in the upper half-plane})$$

$$= \int_L R(z)\, dz + \int_C R(z)\, dz,$$

where L is the line segment and C the semicircle part of γ. The first integral is $\int_{-r}^{r} R(x)\, dx$, since we can parametrize L as $L(x) = x$, $x: -r \to r$.

By the above assumption on degrees, the degree of the denominator of $z^2R(z)$ is at least as large as that of the numerator, so, for some M, $|z^2R(z)| \le M$ for $|z|$ sufficiently large, say $|z| \ge K$. Then $|R(z)| \le M/|z|^2$ for $|z| \ge K$. Choose $r \ge K$. Then for z on C, $|z| = r \ge K$, so $|R(z)| \le M/r^2$, and we have

$$\left| \int_C R(z)\, dz \right| \le \frac{M}{r^2}(\text{length of } C) = \frac{M(\pi r)}{r^2} = \frac{\pi M}{r} \to 0$$

as $r \to \infty$. Hence, as $r \to \infty$ we have

$$\int_{-\infty}^{\infty} R(z)\, dz = 2\pi i(\text{sum of residues of } R \text{ at poles in upper half-plane}).$$

There is a fine point that we have glossed over so far. Note that the method gives us $\lim_{r \to \infty} \int_{-r}^{r} R(x)\, dx$, which is the Cauchy principal value of $\int_{-\infty}^{\infty} R(x)\, dx$ If $\lim_{\substack{\alpha \to -\infty \\ \beta \to \infty}} \int_\alpha^\beta R(x)\, dx$ exists, then it certainly equals the Cauchy principal value. However, it is possible for the Cauchy principal value to exist and for the more general limit $\lim_{\substack{\alpha \to -\infty \\ \beta \to \infty}} \int_\alpha^\beta R(x)\, dx$ to not exist. Keep in mind, then, that we are working here with the Cauchy principal value.

EXAMPLE

Compute

$$\int_{-\infty}^{\infty} \frac{x^2 dx}{x^4 + 16}.$$

Here we choose $R(z) = z^2/(z^4 + 16)$. The first thing we must do is to find the poles of R, which are the zeros of $z^4 + 16$. Thus we need the fourth roots of -16.

To find these, write $-16 = 16e^{\pi i}$. The fourth roots of -16 are then

$$\xi_1 = 2e^{\pi i/4} = \sqrt{2} + i\sqrt{2},$$
$$\xi_2 = 2e^{3\pi i/4} = -\sqrt{2} + i\sqrt{2},$$
$$\xi_3 = 2e^{5\pi i/4} = -\sqrt{2} - i\sqrt{2},$$

and

$$\xi_4 = 2e^{7\pi i/4} = \sqrt{2} - i\sqrt{2}.$$

Of these, only ξ_1 and ξ_2 lie in the upper half-plane. Thus

$$\int_{-\infty}^{\infty} \frac{x^2\, dx}{x^4 + 16} = 2\pi i\left[\operatorname{Res}\left(\frac{z^2}{z^4 + 16}, \xi_1\right) + \operatorname{Res}\left(\frac{z^2}{z^4 + 16}, \xi_2\right)\right].$$

We have only to calculate these residues. Write

$$z^4 + 16 = (z - \xi_1)(z - \xi_2)(z - \xi_3)(z - \xi_4).$$

Then

$$\begin{aligned}\operatorname{Res}\left(\frac{z^2}{z^4 + 16}, \xi_1\right) &= \lim_{z \to \xi_1} \left(\frac{(z - \xi_1)z^2}{z^4 + 16}\right) \\ &= \frac{\xi_1^2}{(\xi_1 - \xi_2)(\xi_1 - \xi_3)(\xi_1 - \xi_4)} \\ &= \frac{(\sqrt{2} + i\sqrt{2})^2}{2\sqrt{2}(2\sqrt{2} + 2i\sqrt{2})(2i\sqrt{2})} = \frac{1 + i}{8i}\end{aligned}$$

and

$$\begin{aligned}\operatorname{Res}\left(\frac{z^2}{z^4 + 16}, \xi_2\right) &= \lim_{z \to \xi_2} \left(\frac{(z - \xi_2)z^2}{z^4 + 16}\right) \\ &= \frac{\xi_2^2}{(\xi_2 - \xi_1)(\xi_2 - \xi_3)(\xi_2 - \xi_4)} \\ &= \frac{(-\sqrt{2} + i\sqrt{2})^2}{-2\sqrt{2}(2i\sqrt{2})(-2\sqrt{2} + 2i\sqrt{2})} = \frac{1 - i}{8i}.\end{aligned}$$

Then

$$\int_{-\infty}^{\infty} \frac{x^2\,dx}{x^4 + 16} = 2\pi i\left[\frac{1+i}{8i} + \frac{1-i}{8i}\right] = \frac{\pi}{2}. \qquad \blacksquare$$

We should note before going on to something else that we could utilize the poles in the lower half-plane just as easily as we have those in the upper half-plane. Since we are assuming that none of the poles [which are zeros of the denominator of $R(z)$] are real, then the poles appear in conjugate pairs. Then for every one in the upper half-plane, there is one in the lower half-plane, and we could take γ as the segment from $-r$ to r, together with the semicircle $re^{i\theta}$, $\theta: \pi \to 2\pi$.

The method we have just discussed is often useful in evaluating integrals of the form $\int_{-\infty}^{\infty} R(x)\cos(ax)\,dx$ and $\int_{-\infty}^{\infty} R(x)\sin(ax)\,dx$, with $a > 0$ (constant) and $R(x)$ as above (a rational function with denominator of degree at least 2 greater than the degree of the numerator, and no poles on the real axis). Using the same γ as above we have,

$$\oint_{\gamma} R(z)e^{iaz}\,dz = 2\pi i\,[\text{sum of the residues of } R(z)e^{iaz} \text{ in the upper half-plane.}]$$

The connection between $\oint_{\gamma} R(z)e^{iaz}\,dz$ and the real integrals we want is this. Note that

$$\oint_{\gamma} R(z)e^{iaz}\,dz = \int_{-r}^{r} R(x)\cos(ax)\,dx + i\int_{-r}^{r} R(x)\sin(ax)\,dx + \int_{C} R(z)e^{iaz}\,dz,$$

where C is the semicircle part of γ. Now

$$\begin{aligned} |R(z)e^{iaz}| &= |R(z)|\,|e^{ia(x+iy)}| \\ &= |R(z)|\,|e^{iax}|\,|e^{-ay}| = |R(z)|\,|e^{-ay}| \le |R(z)|, \end{aligned}$$

since $a > 0$ and $y \ge 0$ imply that $e^{-ay} \le 1$. If z is on C, then

$$|R(z)| \le \frac{\pi M}{r},$$

as before, for some constant M, and so also

$$|R(z)e^{iaz}| \le \frac{\pi M}{r}.$$

Then, as $r \to \infty$, $\left|\int_C R(z)e^{iaz}\,dz\right| \to 0$, and we have

$$\begin{aligned} &\int_{-\infty}^{\infty} R(x)\cos(ax)\,dx + i\int_{-\infty}^{\infty} R(x)\sin(ax)\,dx \\ &\qquad = 2\pi i\,[\text{sum of residues of } R(z)e^{iaz} \text{ in the upper half-plane}]. \end{aligned}$$

The real part of the right side then yields $\int_{-\infty}^{\infty} R(x)\cos(ax)\,dx$, and the imaginary part, $\int_{-\infty}^{\infty} R(x)\sin(ax)\,dx$.

EXAMPLE

We shall use the above method to evaluate

$$\int_{-\infty}^{\infty} \frac{\cos(ax)\,dx}{x^4 + 16}, \qquad \text{where } a > 0.$$

Look at $e^{iaz}/(z^4 + 16)$, which has simple poles at the fourth roots of 16. These were computed in the last example. There we found that the fourth roots of 16 in the upper half-plane are $\xi_1 = \sqrt{2}(1 + i)$ and $\xi_2 = \sqrt{2}(-1 + i)$. We therefore compute, using some of the work of the last example,

$$\begin{aligned}\operatorname{Res}\left(\frac{e^{iaz}}{z^4 + 16}, \sqrt{2}(1 + i)\right) &= \lim_{z \to \sqrt{2}(1+i)} \frac{e^{iaz}}{(\xi_1 - \xi_2)(\xi_1 - \xi_3)(\xi_1 - \xi_4)} \\ &= \frac{e^{ia\sqrt{2}(1+i)}}{16i\sqrt{2}(1 + i)}\end{aligned}$$

and

$$\begin{aligned}\operatorname{Res}\left(\frac{e^{iaz}}{z^4 + 16}, \sqrt{2}(-1 + i)\right) &= \lim_{z \to \sqrt{2}(-1+i)} \frac{e^{iaz}}{(\xi_2 - \xi_1)(\xi_2 - \xi_3)(\xi_2 - \xi_4)} \\ &= \frac{e^{ia\sqrt{2}(-1+i)}}{16i\sqrt{2}(1 - i)}.\end{aligned}$$

To simplify these expressions for the residues, write

$$e^{ia\sqrt{2}(1+i)} = e^{-a\sqrt{2}}[\cos(a\sqrt{2}) + i\sin(a\sqrt{2})]$$

and

$$e^{ia\sqrt{2}(-1+i)} = e^{-a\sqrt{2}}[\cos(a\sqrt{2}) - i\sin(a\sqrt{2})].$$

Then

$$\begin{aligned}&\int_{-\infty}^{\infty} \frac{\cos(ax)\,dx}{x^4 + 16} + i\int_{-\infty}^{\infty} \frac{\sin(ax)\,dx}{x^4 + 16} \\ &\qquad = \frac{2\pi i}{16i\sqrt{2}}\, e^{-a\sqrt{2}}\left[\frac{\cos(a\sqrt{2}) + i\sin(a\sqrt{2})}{1 + i} + \frac{\cos(a\sqrt{2}) - i\sin(a\sqrt{2})}{1 - i}\right] \\ &\qquad = \frac{\pi e^{-a\sqrt{2}}}{8\sqrt{2}}\,[\cos(a\sqrt{2}) + \sin(a\sqrt{2})].\end{aligned}$$

Then

$$\int_{-\infty}^{\infty} \frac{\cos(ax)\,dx}{x^4 + 16} = \frac{\pi e^{-a\sqrt{2}}}{8\sqrt{2}} [\cos(a\sqrt{2}) + \sin(a\sqrt{2})]$$

and

$$\int_{-\infty}^{\infty} \frac{\sin(ax)\,dx}{x^4 + 16} = 0.$$

The latter could of course have been seen immediately from the fact that $\sin(ax)/(x^4 + 16)$ is an odd function on the real line. ∎

It is possible to go on at great length discussing complex techniques for evaluation of real integrals. In many instances considerable ingenuity is required in choosing a contour for the complex integral that yields the desired result. We shall give two fairly simple examples of this.

EXAMPLE

Suppose that we want to evaluate

$$\int_0^{\infty} \frac{x^{a-1}\,dx}{1 + x}, \qquad \text{where } 0 < a < 1.$$

Here the integrand has singularities at 0 and -1. We do two things that may surprise the reader who is new at this game. First, we consider the complex function $f(z) = z^{a-1}/(1 - z)$, *not* $z^{a-1}/(1 + z)$. This has singularities at 0 and $+1$. Second, we take as γ the contour shown in Fig. 9-15, with inner circular part C_δ of radius $\delta < 1$, and outer circular part C_Δ of radius $\Delta > 1$. Now we must

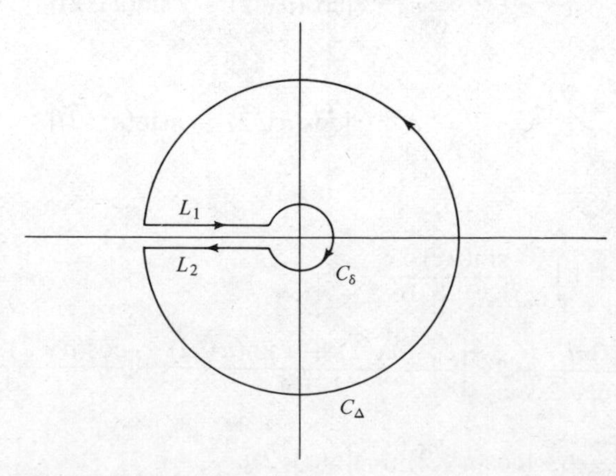

Fig. 9-15

agree on a meaning for z^{a-1}. We take $-\pi < \text{Arg}(z) \leq \pi$ and put $z^{a-1} = e^{(a-1)\,\text{Log}(z)}$. Then f is analytic on and within γ except at $+1$. By the residue theorem,

$$\oint_\gamma \frac{z^{a-1}\,dz}{1-z} = 2\pi i \,\text{Res}\left(\frac{z^{a-1}}{1-z}, 1\right)$$

$$= 2\pi i \lim_{z \to 1}\left((z-1)\frac{z^{a-1}}{1-z}\right) = 2\pi i \lim_{z \to 1}(-z^{a-1}) = -2\pi i.$$

We now compute $\oint_\gamma z^{a-1}\,dz/(1-z)$ over the individual pieces of γ. On C_δ, $z = \delta e^{i\theta}$, say $\theta\colon \pi - \varepsilon \to -\pi + \varepsilon$, so

$$\int_{C_\delta} \frac{z^{a-1}\,dz}{1-z} = \int_{\pi-\varepsilon}^{-\pi+\varepsilon} \frac{(\delta e^{i\theta})^{a-1} i\delta e^{i\theta}\,d\theta}{1-\delta e^{i\theta}} = \int_{\pi-\varepsilon}^{-\pi+\varepsilon} \frac{\delta^a e^{ia\theta}\,d\theta}{1-\delta e^{i\theta}}.$$

On C_Δ, $z = e^{i\theta}$, say $\theta\colon -\pi + \varepsilon \to \pi - \varepsilon$. So

$$\int_{C_\Delta} \frac{z^{a-1}\,dz}{1-z} = \int_{-\pi+\varepsilon}^{\pi-\varepsilon} \frac{\Delta^a e^{ia\theta}\,d\theta}{1-\Delta e^{i\theta}}.$$

On L_1, we write $z = re^{i\pi}$, $r\colon \Delta \to \delta$, to obtain

$$\int_{L_1} \frac{z^{a-1}\,dz}{1-z} = \int_\Delta^\delta \frac{(re^{i\pi})^{a-1} e^{i\pi}\,dr}{1-re^{i\pi}} = \int_\Delta^\delta \frac{r^{a-1} e^{i\pi a}\,dr}{1+r}.$$

On L_2, $z = re^{-i\pi}$, $r\colon \delta \to \Delta$, so

$$\int_{L_2} \frac{z^{a-1}\,dz}{1-z} = \int_\delta^\Delta \frac{(re^{-i\pi})^{a-1} e^{-i\pi}\,dr}{1-re^{-i\pi}} = \int_\delta^\Delta \frac{r^{a-1} e^{-i\pi a}\,dr}{1+r}.$$

Then

$$\int_{L_1} \frac{z^{a-1}\,dz}{1-z} + \int_{L_2} \frac{z^{a-1}\,dz}{1-z} = \int_\delta^\Delta \frac{r^{a-1}}{1+r}(e^{-i\pi a} - e^{i\pi a})\,dr.$$

We now estimate

$$\int_{C_\Delta} \frac{z^{a-1}\,dz}{1-z} \quad \text{and} \quad \int_{C_\delta} \frac{z^{a-1}\,dz}{1-z}$$

with a view toward letting $\delta \to 0$ and $\Delta \to \infty$. On C_δ,

$$\left|\frac{\delta^a e^{ia}}{1-\delta e^{i\theta}}\right| = \frac{|\delta^a|}{|1-\delta e^{i\theta}|} \leq \frac{\delta^a}{1-\delta}.$$

Here we have used $|1 - \delta e^{i\theta}| \geq |1| - |\delta e^{i\theta}| = 1 - \delta$. On C_Δ,

$$\left|\frac{\Delta^a e^{ia}}{1 - \Delta e^{i\theta}}\right| = \frac{|\Delta^a|}{|1 - \Delta e^{i\theta}|} \leq \frac{\Delta^a}{\Delta - 1},$$

since $|1 - \Delta e^{i\theta}| = |\Delta e^{i\theta} - 1| \geq |\Delta e^{i\theta}| - |1| = \Delta - 1$. As $\Delta \to \infty$ and $\delta \to 0$, $\Delta^a/(\Delta - 1) \to 0$ and $\delta^a/(1 - \delta) \to 0$, since $0 < a < 1$.

Thus, in the limit as $\Delta \to \infty$ and $\delta \to 0$, we obtain

$$\int_0^\infty \frac{-r^{a-1}(e^{\pi ia} - e^{-\pi ia})\, dr}{1 + r} = -2\pi i.$$

That is,

$$\left(\int_0^\infty \frac{x^{a-1}\, dx}{1 + x}\right) 2i \sin(\pi a) = 2\pi i,$$

or, equivalently,

$$\int_0^\infty \frac{x^{a-1}\, dx}{1 + x} = \frac{\pi}{\sin(\pi a)}. \quad \blacksquare$$

EXAMPLE

Suppose that we want to evaluate $\int_0^\infty \sin(x)\, dx/x$. We have already done this in Section 7.3, but here is another way, using contour integration. Consider $\oint_C (e^{iz}/z)\, dz$, where C is the contour shown below. Since e^{iz}/z is analytic on and inside C, then $\oint_C (e^{iz}/z)\, dz = 0$. Consider C as broken up into semicircles C_R, C_r, and lines L_1 and L_2 as shown in Fig. 9-16.

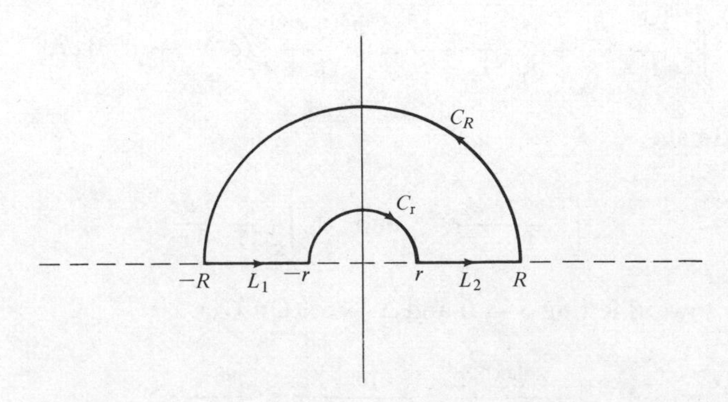

Fig. 9-16

We evaluate the integral of e^{iz}/z on each piece of C separately. First, on C_R, $|z| = R$ and we can write $z = R\cos(\theta) + iR\sin(\theta)$, $\theta\colon 0 \to \pi$. Then

$$\frac{e^{iz}}{z} = \frac{e^{i[R\cos(\theta) + iR\sin(\theta)]}}{R\cos(\theta) + iR\sin(\theta)}$$

$$= \frac{e^{-R\sin(\theta)}e^{iR\cos(\theta)}}{R[\cos(\theta) + i\sin(\theta)]}, \qquad \text{and so } \left|\frac{e^{iz}}{z}\right| = \frac{e^{-R\sin(\theta)}}{R} \text{ on } C_R.$$

Then

$$\left|\oint_{C_R} \frac{e^{iz}}{z}\,dz\right| = \left|\int_0^{\pi} \frac{e^{-R\sin(\theta)}e^{iR\cos(\theta)}iRe^{i\theta}\,d\theta}{R\cos(\theta) + iR\sin(\theta)}\right|$$

$$\leq \int_0^{\pi} |e^{-R\sin(\theta)}e^{iR\cos(\theta)}i|\,d\theta \leq \int_0^{\pi} e^{-R\sin(\theta)}\,d\theta \leq \int_0^{\pi/2} e^{-R\sin(\theta)}\,d\theta.$$

Now, for $0 \leq \theta \leq \pi/2$, $\sin(\theta) \geq \theta$, since $\sin(\theta) = \theta - \theta^3/3! + \cdots$. Then $e^{-R\sin(\theta)} \leq e^{-R\theta}$, so

$$\int_0^{\pi/2} e^{-R\sin(\theta)}\,d\theta \leq \int_0^{\pi/2} e^{-R\theta}\,d\theta = \frac{-1}{R}(e^{-R\pi/2} - 1) = \frac{1}{R}(1 - e^{-R\pi/2}).$$

Then

$$\left|\int_{C_R} \frac{e^{iz}}{z}\,dz\right| \leq \frac{1}{R}(1 - e^{-R\pi/2}) \to 0 \qquad \text{as } R \to \infty.$$

Next we take the integral over the smaller semicircle:

$$\int_{C_r} \frac{e^{iz}\,dz}{z} = \int_{C_r} \frac{dz}{z} + \int_{C_r} \frac{(e^{iz} - 1)\,dz}{z}.$$

But

$$\int_{C_r} \frac{dz}{z} = \int_{\pi}^{0} \frac{1}{re^{i\theta}}(ire^{i\theta})\,d\theta = -\pi i$$

and

$$\left|\int_{C_r} \frac{(e^{iz} - 1)\,dz}{z}\right| = \left|\int_{\pi}^{0} \frac{e^{i[r\cos(\theta) + ir\sin(\theta)]} - 1}{re^{i\theta}}\,ire^{i\theta}\,d\theta\right|$$

$$= \left|\int_0^{\pi} [e^{-r\sin(\theta)}e^{ir\cos(\theta)} - 1]\,d\theta\right|$$

$$\leq \int_0^{\pi} |e^{r\sin(\theta)}e^{ir\cos(\theta)} - 1|\,d\theta \to 0 \qquad \text{as } r \to 0,$$

since $e^{-r\sin(\theta)}e^{ir\cos(\theta)} \to 1$ as $r \to 0$. Hence we finally have

$$\int_{C_r} \frac{e^{iz}\,dz}{z} \to -\pi i \qquad \text{as } r \to 0.$$

Finally,

$$\int_{L_1} \frac{e^{iz}\,dz}{z} + \int_{L_2} \frac{e^{iz}\,dz}{z} = \int_{-R}^{-r} \frac{e^{ix}\,dx}{x} + \int_r^R \frac{e^{ix}\,dx}{x}$$

$$= \int_R^r \frac{e^{-it}(-dt)}{-t} + \int_r^R \frac{e^{ix}\,dx}{x}$$

$$= -\int_r^R \frac{e^{-it}\,dt}{t} + \int_r^R \frac{e^{ix}\,dx}{x} = 2i\int_r^R \frac{\sin(x)\,dx}{x}$$

$$\to 2i\int_0^\infty \frac{\sin(x)\,dx}{x} \qquad \text{as } r \to 0 \text{ and } R \to \infty.$$

Then as $r \to 0$ and $R \to \infty$, we have

$$0 = 2i\int_0^\infty \frac{\sin(x)\,dx}{x} - \pi i.$$

Then

$$\int_0^\infty \frac{\sin(x)\,dx}{x} = \frac{\pi}{2}.$$

∎

Exercises for Section 9.11

1. Derive each of the following by residue methods:

(a) $\displaystyle\int_0^{2\pi} \frac{d\theta}{1 - 2a\cos(\theta) + a^2} = \frac{2\pi}{1 - a^2}$ for $0 < a < 1$

(b) $\displaystyle\int_0^{2\pi} \frac{\cos^2(3\theta)\,d\theta}{1 - 2a\cos(\theta) + a^2} = \frac{\pi(1 - a + a^3)}{1 - a}$ for $0 < a < 1$

(c) $\displaystyle\int_0^{2\pi} \frac{d\theta}{[a + b\cos(\theta)]^2} = \frac{2\pi a}{(a^2 - b^2)^{3/2}}$ for $0 < b < a$

(d) $\displaystyle\int_0^{2\pi} \frac{d\theta}{a + \sin(\theta)} = \frac{2\pi}{a^2 - 1}$ for $a > 1$

(e) $\displaystyle\int_0^{\pi} \cos^{2n}(\theta)\,d\theta = \frac{\pi(2n)!}{2^{2n}(n!)^2}$, n any positive integer (Can your derive this result using gamma functions?)

2. Derive each of the following by residue methods:

(a) $\displaystyle\int_0^\infty \frac{x^2\,dx}{x^6+1} = \frac{\pi}{6}$ *Hint:* $\displaystyle\int_0^\infty \frac{x^2\,dx}{x^6+1} = \frac{1}{2}\int_{-\infty}^\infty \frac{x^2\,dx}{x^6+1}$

(b) $\displaystyle\int_{-\infty}^\infty \frac{dx}{(x^2+1)^2} = \frac{3\pi}{8}$

(c) $\displaystyle\int_{-\infty}^\infty \frac{x^4\,dx}{(a+bx^2)^4} = \frac{\pi}{16a^{3/2}b^{5/2}}$ for $a > 0,\ b > 0$

(d) $\displaystyle\int_{-\infty}^\infty \frac{dx}{1+x^4} = \frac{\pi}{\sqrt{2}}$

(e) $\displaystyle\int_{-\infty}^\infty \frac{x^2\,dx}{(1+x^2)^2} = \frac{\pi}{2}$

(f) $\displaystyle\int_0^\infty \frac{x^2\,dx}{(a^2+x^2)^3} = \frac{\pi}{16a^3}$ for $a > 0$

3. Derive each of the following using residue methods.

(a) $\displaystyle\int_{-\infty}^\infty \frac{\cos(x)\,dx}{a^2+x^2} = \frac{\pi e^{-a}}{a},\ a > 0$

(b) $\displaystyle\int_{-\infty}^\infty \frac{\cos(x)\,dx}{(1+x^2)^2} = \frac{\pi}{e}$

(c) $\displaystyle\int_{-\infty}^\infty \frac{\cos(x)\,dx}{(a^2+x^2)(b^2+x^2)} = \frac{\pi}{a^2-b^2}\left(\frac{e^{-b}}{b} - \frac{e^{-a}}{a}\right),\quad a > 0,\ b > 0,$ and $a \neq b$

4. Show that

$$\int_{-\infty}^\infty \frac{e^{-x}\,dx}{e^x+1} = \frac{\pi}{\sin(\pi a)},\quad 0 < a < 1.$$

Hint: Consider $\oint_\gamma e^{az}\,dz/(1+e^z)$ with γ the rectangle with vertices at $\pm r$ and $\pm r + 2\pi i$. The integrals along the vertical sides may be shown to go to zero as $r \to \infty$.

5. Show that

$$\int_{-\infty}^\infty \frac{\cos(x)\,dx}{a^2-x^2} = \frac{\pi\sin(a)}{a}\quad \text{for } a > 0.$$

Hint: Consider $\oint_\gamma e^{iz}\,dz/(a^2-z^2)$ with γ the path shown in Fig. 9-17.

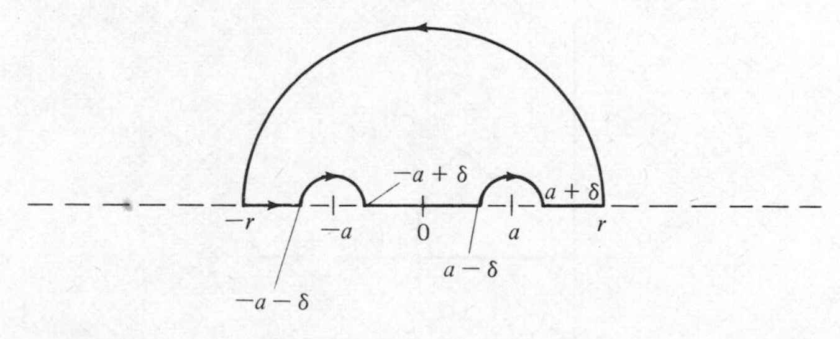

Fig. 9-17

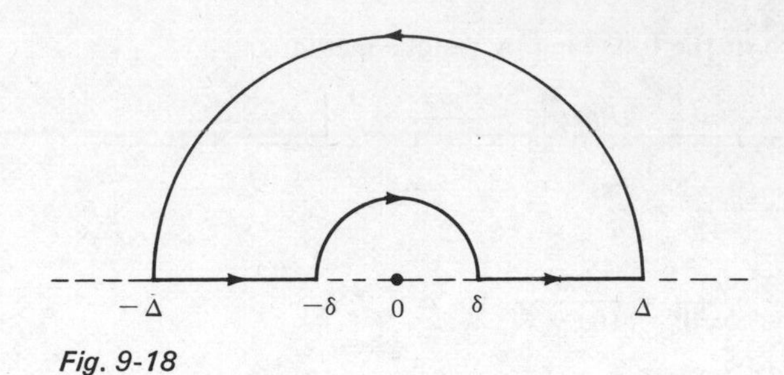

Fig. 9-18

6. Show that

$$\int_0^\infty \frac{[\ln(x)]^2\,dx}{1+x^2} = \frac{\pi^3}{8}.$$

Hint: Consider

$$\oint_\gamma \frac{[\mathrm{Log}(z)]^2\,dz}{1+z^2},$$

with γ the path shown in Fig. 9-18.

7. Evaluate the following by residue methods:

(a) $\displaystyle\int_0^\infty \frac{x\sin(x)\,dx}{a^2+x^2}$, a real

(b) $\displaystyle\int_0^{\pi/2} \frac{dx}{a+\sin^2(x)}$, $a > 0$

(c) $\displaystyle\int_{-\infty}^\infty \frac{dx}{(1+x^2)^n}$, $n = 1, 2, 3, \ldots$ $\left(\textit{Answer: } \dfrac{\pi(2n-2)!}{2^{2n-2}[(n-1)!]^2}.\right)$

Can you do this one with gamma functions?

(d) $\displaystyle\int_{-\infty}^\infty \frac{x\sin(bx)\,dx}{a^4+x^4}$, for a, b real

(e) $\displaystyle\int_0^{2\pi} \frac{d\theta}{a^2\cos^2(\theta)+b^2\sin^2(\theta)}$, for a, b real

(f) $\displaystyle\int_0^{2\pi} e^{\cos(\theta)}\,d\theta$ [Here the best you can do for an answer is an infinite series $2\pi\sum_{n=0}^{\infty} 1/(n!)^2$.]

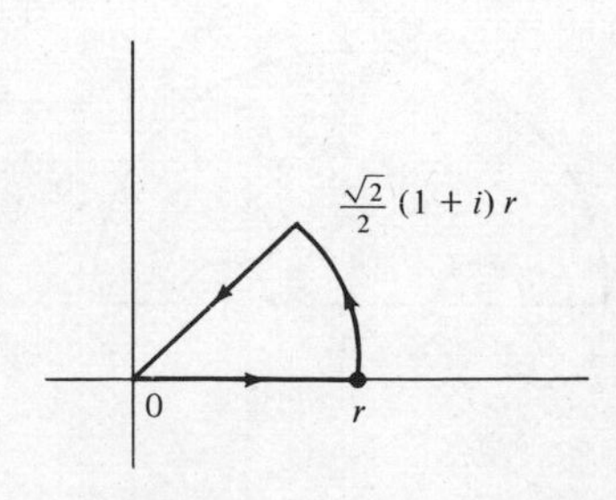

Fig. 9-19

8. Evaluate Fresnel's integrals:

$$\int_0^\infty \cos(x^2)\,dx = \int_0^\infty \sin(x^2)\,dx = \frac{1}{2}\sqrt{\frac{\pi}{2}}$$

by using the residue theorem on $\oint_\gamma e^{-z^2}\,dz$, where γ goes from 0 to r along the real axis, then along $|z| = r$ to the point $re^{i\pi/4}$, then along the straight line from $re^{i\pi/4}$ back to the origin; and then letting $r \to \infty$ (see Fig. 9-19).

9.12 Conformal Mapping

Suppose we have a function f analytic in a domain D. It is often useful to write $w = f(z)$ for each z in D and to think of f as specifying a correspondence between the z-plane and the w-plane. In this context f is often called a *transformation* or *mapping*.

The mapping is one-to-one (1-1) if different points have different images; that is, $f(z_1) \neq f(z_2)$ if $z_1 \neq z_2$. The mapping is onto a set D^* of the w-plane if every w in D^* is the image of some z in D. That is, if w is in D^*, there is some z in D with $f(z) = w$.

As an example, $f(z) = z^2$ for $|z| \leq 1$ maps the unit disc $|z| \leq 1$ onto the unit disc $|w| \leq 1$ but is not 1-1, as, for example, $f(\frac{1}{2}) = f(-\frac{1}{2})$. But $g(z) = z$ maps $|z| \leq 1$ into (but not onto) the disc $|w| \leq 2$ in 1-1 fashion. Thus the notions of 1-1 and onto are independent.

Functions that preserve angles are particularly important in complex analysis. To see what we mean by "preserves angles," let f be a mapping from a domain D in the z-plane to a region D^* in the w-plane.

Look at any z_0 in D, with $w_0 = f(z_0)$ in D^*. Let γ_1 and γ_2 be smooth curves in D passing through z_0. Then f maps γ_1 and γ_2 to curves γ_1^* and γ_2^* in D^* passing through w_0. Suppose that γ_1^* and γ_2^* are also smooth. The angle between γ_1 and γ_2 at z_0 is the angle between their tangents at z_0. Similarly, we can compute the angle between γ_1^* and γ_2^* at w_0. If these angles are the same, for any smooth curves γ_1 and γ_2 through z_0, then we say that f *preserves angles* at z_0. If f preserves angles at each point in D, then we say that f preserves angles in D. The general idea of this paragraph is depicted in Fig. 9-20.

We can also speak of a mapping as being sense-preserving in addition to being angle-preserving. It may happen that, as we rotate *from* the tangent $\gamma_1'(z_0)$ *to* the tangent $\gamma_2'(z_0)$, the rotation in the w-plane is reversed, from $\gamma_2^{*\prime}(z_0)$ to $\gamma_1^{*\prime}(z_0)$. We then say that f is *sense-reversing*. If orientation is preserved, we say that f is *sense-preserving*.

As an example, let $f(z) = \bar{z}$ for $|z| \leq 1$. Then f is one-to-one on $|z| \leq 1$ onto $|w| \leq 1$. To be specific, take $z_0 = 0$, $\gamma_1(t) = t$, $-\frac{1}{2} \leq t \leq \frac{1}{2}$, and $\gamma_2(t) = it$, $-\frac{1}{2} \leq t \leq \frac{1}{2}$. Obviously, γ_1 intersects γ_2 at 0 at an angle of $\pi/2$. Now $\gamma_1^*(t) = f(\gamma_1(t)) = \bar{t} = t$ for $-\frac{1}{2} \leq t \leq \frac{1}{2}$, and $\gamma_2^*(t) = f(\gamma_2(t)) = \overline{it} = -it$, $-\frac{1}{2} \leq t \leq \frac{1}{2}$. Now γ_1^* and γ_2^* also intersect at an angle of $\pi/2$ at 0. However, as we rotate from $\gamma_1'(0) = 1$ to $\gamma_2'(0) = i$ about the origin in the z-plane, a counterclockwise

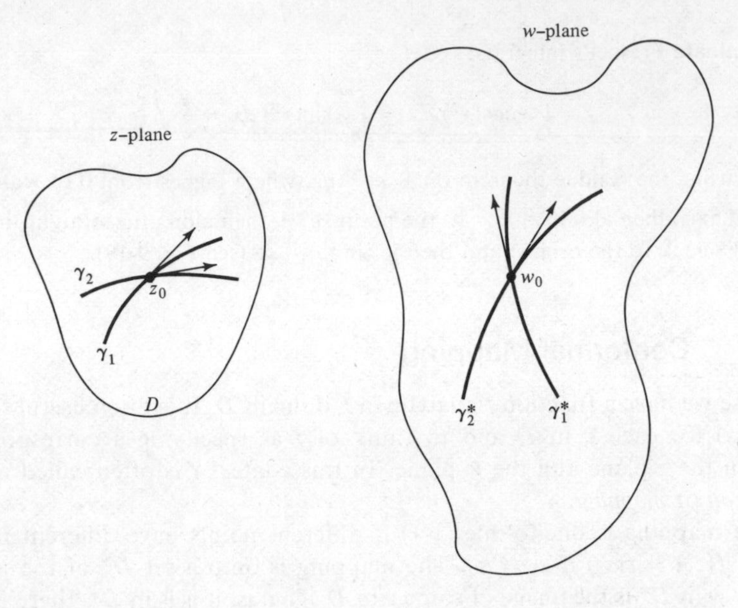

Fig. 9-20

rotation, the corresponding rotation from $(\gamma_1^*)'(0) = 1$ to $(\gamma_2^*)'(0) = -i$ about the origin in the w-plane is in a reversed (clockwise) sense (Fig. 9-21). Here f is sense-reversing but angle-preserving.

In general, a function that is both angle- and sense-preserving on a domain D is said to be *conformal* on D. Such a function is called a *conformal map.*

We could give examples of conformal maps right away, but it is easier to prove the following, from which we can obtain all the examples we want rather easily.

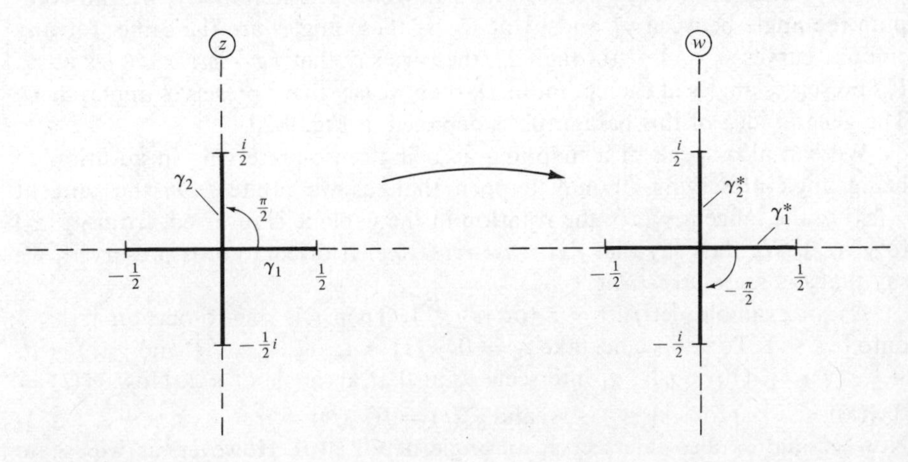

Fig. 9-21

Theorem

Suppose that f maps a domain D onto a domain D^* in a one-to-one fashion. Then

(1) *If f is analytic on D and $f'(z) \neq 0$ for each z in D, then f is conformal on D.*
(2) *If $f = u + iv$ is conformal on D, and u and v have continuous first partials on D satisfying*

$$u_x v_y - u_y v_x \neq 0,$$

then f is analytic on D.

Note that the derivative condition in (2) is really a Jacobian condition, and may be written

$$\frac{\partial(u, v)}{\partial(x, y)} \neq 0.$$

In proving the theorem, it is easy to see why (1) is true. Suppose that γ is a smooth curve through z_0 in D. The image curve γ^* in D^* is given by $\gamma^*(t) = f(\gamma(t))$, and has tangent $\gamma^{*\prime}(t_0) = f'(\gamma(t_0)) \cdot \gamma'(t_0)$ at $z_0{}^* = \gamma^*(t_0)$. Now

$$\operatorname{Arg}(f'(\gamma(t_0)) \cdot \gamma'(t_0)) = \operatorname{Arg}(f'(\gamma(t_0)) + \operatorname{Arg}(\gamma'(t_0)).$$

Then

$$\operatorname{Arg}(\gamma^{*\prime}(t_0)) - \operatorname{Arg}(\gamma'(t_0)) = \operatorname{Arg}(f'(z_0)).$$

This means that

$$\operatorname{Arg}((\gamma_1^*)'(t_0)) - \operatorname{Arg}((\gamma_2^*)'(t_0)) = \operatorname{Arg}(\gamma_1'(t_0)) - \operatorname{Arg}(\gamma_2'(t_0))$$

for any smooth curves γ_1, γ_2 with $\gamma_1(t_0) = \gamma_2(t_0) = z_0$. The left side is the angle from γ_2^* to γ_1^* at $f(z_0)$, and the right side is the angle from γ_2 to γ_1 at z_0. Thus f preserves angles and sense of rotation and so is conformal.

Statement (2) is not as easy to see. What we must do is show that u and v satisfy the Cauchy–Riemann equations in D, assuming the hypothesis of (2).

Fix $z_0 = x_0 + iy_0$ in D. The lines $x = x_0$, $y = y_0$ intersect at right angles in D. Hence the image curves $f(x_0 + iy)$ and $f(x + iy_0)$ intersect at right angles in D^*. The tangent vector to $f(x_0 + iy)$ at $f(z_0)$ is $(u_y, v_y)_{z_0}$ and that to $f(x + iy_0)$ at $f(z_0)$ is $(u_x, v_x)_{z_0}$ (Fig. 9-22). Since these vectors must be perpendicular, their dot product is 0:

$$(u_x u_y + v_x v_y)_{z_0} = 0.$$

To save writing, we shall now agree that all derivatives in this discussion are to be evaluated at (x_0, y_0).

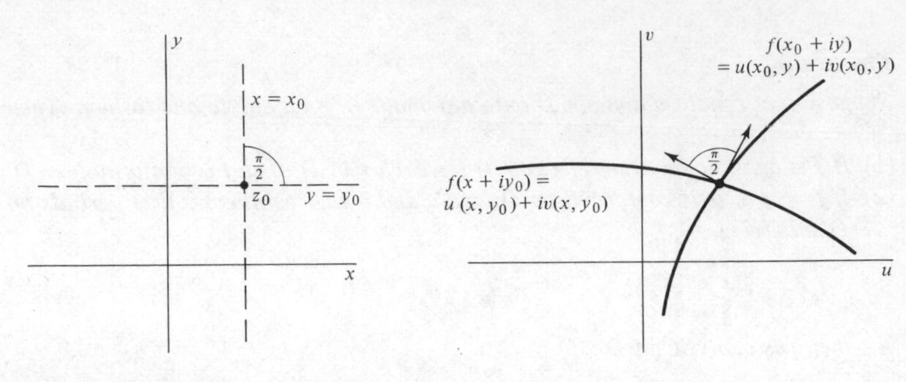

Fig. 9-22

Now look at the lines $y = y_0$, $y = y_0 + (x - x_0)$ in the z-plane. These intersect at an angle of $\pi/4$ radians. Then their images, $f(x, y_0)$ and $f(x, y_0 + x - x_0)$, must also intersect at $f(z_0)$ at an angle of $\pi/4$. Now the tangent to $f(x, y_0)$ at $f(z_0)$ has slope given by $\tan(\alpha) = v_x/u_x$ and the tangent to $f(x, y_0 + x - x_0)$ at $f(z_0)$ has slope given by

$$\tan(\beta) = \frac{\dfrac{d(v(x, y_0 + x - x_0))}{dx}}{\dfrac{d(u(x, y_0 + x - x_0))}{dx}}.$$

Since f is conformal and sense-preserving then $\beta - \alpha = \pi/4$, as shown in Fig. 9.23. Then

$$1 = \tan(\beta - \alpha) = \frac{\tan(\beta) - \tan(\alpha)}{1 + \tan(\beta)\tan(\alpha)}$$

$$= \left(\frac{\dfrac{d(v(x, y_0 + x - x_0))}{dx}}{\dfrac{d(u(x, y_0 + x - x_0))}{dx}} - \frac{\dfrac{\partial v}{\partial x}}{\dfrac{\partial u}{\partial x}}\right) \Bigg/ \left(1 + \frac{v_x \dfrac{d(v(x, y_0 + x - x_0))}{dx}}{u_x \dfrac{d(u(x, y_0 + x - x_0))}{dx}}\right).$$

After some manipulation, this becomes

$$u_x v_y - u_y v_x = u_x^2 + v_x^2 + u_x u_y + v_x v_y.$$

Since

$$u_x u_y + v_x v_y = 0,$$

we have

$$u_x v_y - u_y v_x = u_x^2 + v_x^2.$$

At least one of these partials is nonzero, say $u_x \neq 0$. Then

$$\frac{-1}{u_x}(v_x v_y) = u_y.$$

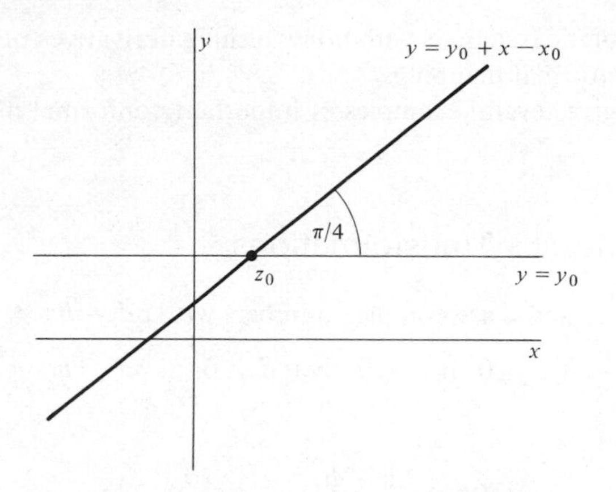

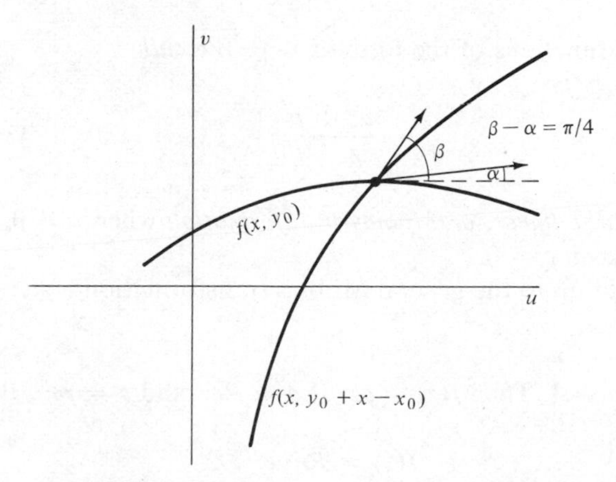

Fig. 9-23

Hence

$$u_x v_y = \left[\frac{-1}{u_x}(v_x v_y) v_x\right] = u_x^2 + v_x^2.$$

Then

$$u_x^2 v_y + v_x^2 v_y = u_x(u_x^2 + v_x^2),$$

or, equivalently,

$$v_y(u_x^2 + v_x^2) = u_x(u_x^2 + v_x^2).$$

Since $u_x^2 + v_x^2 \neq 0$, we conclude that $v_y = u_x$. Then also

$$\frac{-1}{u_x}(v_x v_y) = -v_x = u_y,$$

and we have the Cauchy–Riemann equations.

To sum up, analytic functions with nonvanishing derivatives provide us with a large class of conformal mappings.

We shall now give several examples of important conformal mappings.

A. Möbius Transformations

Suppose that a, b, c, and d are complex numbers with $ad - bc \neq 0$. Put $f(z) = \dfrac{az + b}{cz + d}$ for $z \neq -\dfrac{d}{c}$, if $c \neq 0$. If $c = 0$, then $d \neq 0$ (as $ad - bc \neq 0$), so we can write

$$f(z) = \frac{az + b}{d} = \frac{a}{d}z + \frac{b}{d},$$

a special case of functions of the form $\alpha z + \beta$. *We call*

$$\frac{az + b}{cz + d}$$

a *linear fractional, bilinear, or Möbius transformation* when $c \neq 0$, and a *linear transformation* when $c = 0$.

We shall build up to the general Möbius transformation a step at a time by considering cases.

(1) $c = b = 0$, $d = 1$. Then $f(z) = az$. If $a = Re^{i\xi}$ and $z = re^{i\theta}$, then

$$f(z) = Rr \cdot e^{i(\xi + \theta)}.$$

Then $|f(z)| = |a|\,|z|$, so the distance from the origin to z is changed by a factor of $|a|$, and $\arg(f(z)) = \arg(z) + \arg(a)$ (to within arbitrary additional terms of $2n\pi$), which means that the vector from 0 to z is rotated by an angle ξ in the positive (counterclockwise) sense. If a is real, then f is a straight stretching ($a > 1$) or shrinking ($0 < a < 1$), or a stretching or shrinking together with a reversal of direction (if $a < -1$ or $-1 < a < 0$). If $f(z) = -z$, then f simply flips z around in the opposite direction, with no stretching or shrinking.

(2) $a = d = 1$, $c = 0$. Then $f(z) = z + b$. If we write $z = (x, y)$ and $b = (b_1, b_2)$, then $f(z) = (x + b_1, y + b_2) = (x + b_1) + i(y + b_2)$, which is a translation in the plane.

(3) $c = 0$, $d = 1$. Then $f(z) = az + b$, the general linear transformation. This is a rotation–stretching [as in (1)], accompanied by a translation (2). Often it is useful to decompose a mapping into a sequence of mappings that can then be examined a piece at a time, and the effects taken into account cumulatively. Here we can think of $z \to az + b$ as $z \to \xi \to w$, where $\xi = az$

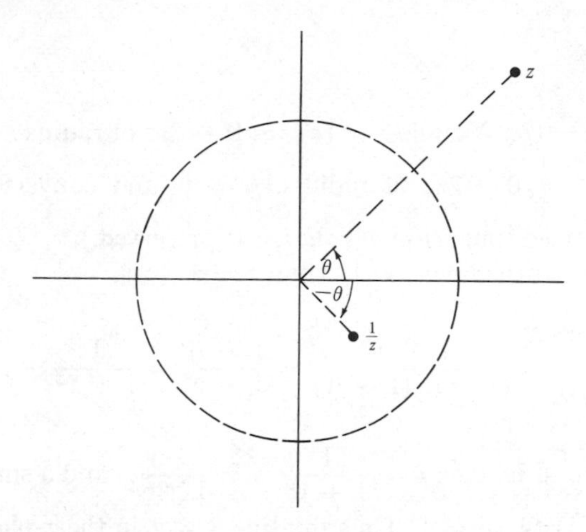

Fig. 9-24

and $w = \xi + b$. Thus the end result is a rotation–stretching ($\xi = az$) coupled with a translation ($w = \xi + b$).

(4) $a = d = 0$, $b = c = 1$. Then $f(z) = 1/z$. If z has distance r from 0, then $w = 1/z$ has distance $1/r$ from 0. Points outside $x^2 + y^2 = 1$ are mapped inside, and conversely. Further, if $\arg z = \theta$, then $\arg(1/z) = -\theta$. Geometrically, then, we obtain $1/z$ by stretching (or shrinking) the vector of length r from 0 to z, to one of length $1/r$, then reflecting the resulting vector across the real axis. We call this transformation inversion (Fig. 9-24).

(5) We now consider $f(z) = (az + b)/(cz + d)$, in general, with $c \neq 0$ [if $c = 0$ and $d \neq 0$, we are back in (3)]. In this case we can always visualize the end result as a succession of rotation–stretchings, inversions, and translations:

$$\underset{\text{rotation–stretching}}{z \to cz} \underset{\text{translation}}{\to cz + d} \underset{\text{inversion}}{\to} \frac{1}{cz + d}$$

$$\underset{\text{rotation–stretching}}{\to} \frac{bc - ad}{c} \frac{1}{cz + d}$$

$$\underset{\text{translation}}{\to} \frac{bc - ad}{c} \frac{1}{cz + d} + \frac{a}{c} = \frac{az + b}{cz + d}.$$

It is easy to show analytically (and easy to visualize geometrically in terms of rotation, stretchings, and translations) that Möbius transformations always map (circles and straight lines) to (circles and straight lines). The parentheses are inserted because a circle may map to a circle or straight line, and a straight line to a circle or straight line. The proof is left to the exercises, but we give an illustration now.

EXAMPLE

Let $w = f(z) = 1/z$. A circle $z = re^{i\theta}$, $\theta: 0 \to 2\pi$, of radius $r > 1$, maps to a circle $z = \frac{1}{r} e^{-i\theta}$, $\theta: 0 \to 2\pi$, of radius $1/r < 1$, and conversely. (Note that *orientation* is reversed, but *sense of rotation* is preserved.)

However, look at the line $x = 1$. Then $z = 1 + iy$, $-\infty < y < \infty$, so

$$w = \frac{1}{x + iy} = \frac{1 - iy}{(1 + iy)(1 - iy)} = \frac{1 - iy}{1 + y^2} = \frac{1}{1 + y^2} - i\frac{y}{1 + y^2}.$$

Writing $w = u + iv$, then $u = \frac{1}{1 + y^2}$, $v = \frac{-y}{1 + y^2}$, and a simple calculation shows that $(u - \frac{1}{2})^2 + v^2 = \frac{1}{4}$. Thus the line $x = 1$ in the z-plane maps to the circle $(u - \frac{1}{2})^2 + v^2 = \frac{1}{4}$ in the $w = u + iv$ plane. ∎

One convenient fact about Möbius transformations is that three points and their images are sufficient to completely determine a Möbius transformation. Suppose that we want a Möbius transformation mapping $z_1 \to w_1$, $z_2 \to w_2$, $z_3 \to w_3$, where $z_1, \ldots, w_3$ are given. The reader can check the algebra of the statement that solution of

$$\frac{w_1 - w}{w_1 - w_2}\frac{w_3 - w_2}{w_3 - w} = \frac{z_1 - z}{z_1 - z_2}\frac{z_3 - z_2}{z_3 - z}$$

for w in terms of z gives such a mapping.

As an example, suppose that we want $1 \to i$, $2 \to -i$, and $i \to 3$. We then solve for w in

$$\frac{(i - w)(3 + i)}{2i(3 - w)} = \frac{(1 - z)(i - 2)}{(-1)(i - z)}.$$

This eventually yields

$$w = \frac{(7 + 9i)z + (-9 - 13i)}{(-1 + 3i) + (-3 - i)}.$$

Having specified where we want three given points to go, the mapping is completely and uniquely determined, and we have no control over the images of the other points.

If we want to map some point, say z_3, to ∞, we simply omit all terms involving w_3 on the left above. For example, if we want to map $0 \to 0$, $i \to i$, and $-i \to \infty$, solve for w in

$$\frac{-w}{-i} = \frac{-z(-2i)}{(-i)(-i - z)}$$

to obtain

$$w = \frac{2iz}{z + i}.$$

Often, in constructing maps for specific purposes, it is convenient to map one point to ∞ because of the savings in time spent on algebraic manipulation. If, say, we want $z_3 \to \infty$, then we solve for w in

$$\frac{w_1 - w}{w_1 - w_2} = \frac{(z_1 - z)(z_3 - z_2)}{(z_1 - z_2)(z_3 - z)}.$$

Perhaps surprisingly, we now have a means of constructing, in many instances, a conformal mapping of one given domain or region onto another. The method is based upon the facts that Möbius transformations are conformal and that we can construct Möbius transformations mapping certain specified points to certain specified images. We shall illustrate the method with a concrete example.

EXAMPLE

Construct a conformal mapping of the right half-plane onto the unit disc. Here we want to map R onto D, where $R = \{z \mid \text{Re}(z) \geq 0\}$ and $D = \{z \mid |z| \leq 1\}$ (see Fig. 9-25).

First, by continuity, we should expect that boundary points of R will map to boundary points of D, and conversely. Further, straight lines $\to$ straight lines or circles. So we begin by trying to map the y-axis, which is the boundary of R, onto the unit circle, which is the boundary of D.

Pick three convenient boundary points of R, say 0, i, and $-i$, and three of D, say 1, i, and -1. Map $0 \to 1$, $i \to i$, $-i \to -1$. We then must solve for w in

$$\frac{(1 - w)(-1 + i)}{(1 - i)(-1 - w)} = \frac{(-z)(-2i)}{(-i)(-i - z)}.$$

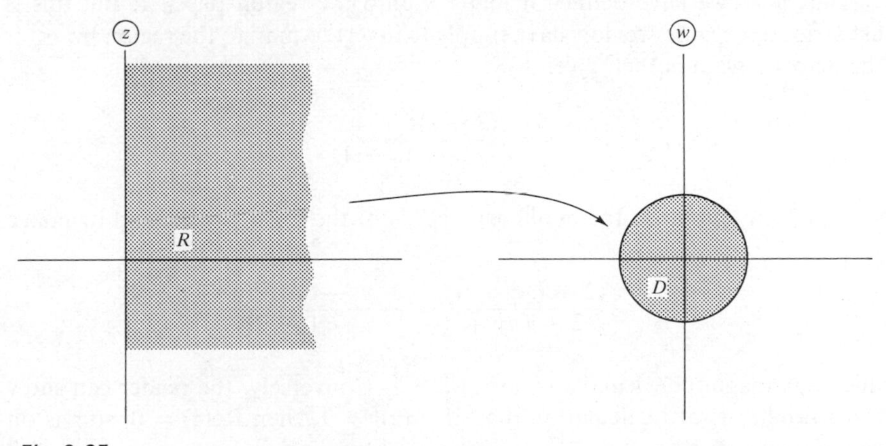

Fig. 9-25

We obtain

$$w = \frac{(-2 + i)z - 1}{(2 + i)z - 1}.$$

Now check to see what this map does. Pick any point interior to R, say 1. Then $w(1) = (-3 + i)/(1 + i)$, which lies *outside* the unit circle $|w| = 1$. But if one interior point of R maps to $|w| > 1$, they all do. Otherwise, we would have, say, z_1 interior to R mapping to D, and z_2 interior to R mapping to $|w| > 1$. In this case, draw a curve from z_1 to z_2 not hitting the boundary of R (Fig. 9-26). By continuity of the mapping, this curve must map to a curve connecting w_1 to w_2, and this curve would have to cross the boundary of D, say at ξ. But then some ζ on the curve from z_1 to z_2 would map to ξ. This would give us a non-boundary-point mapping to a boundary point, a contradiction.

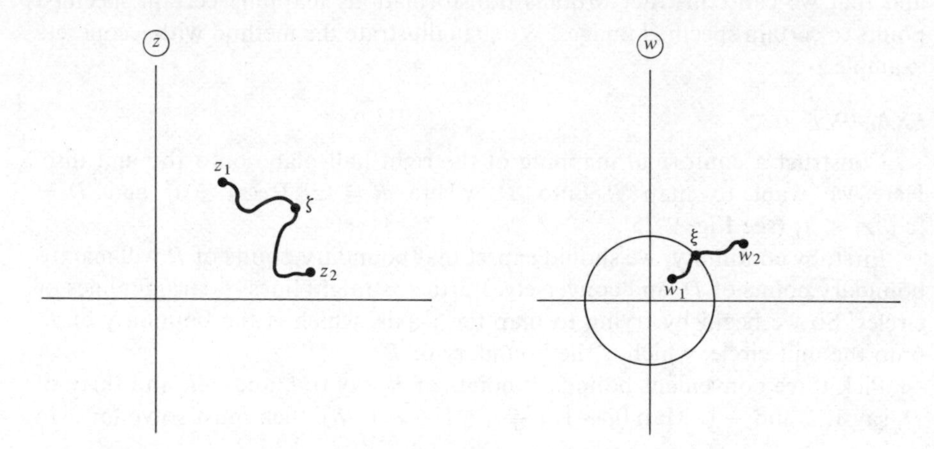

Fig. 9-26

Thus w as we have defined it maps R onto the region $|w| \geq 1$. But this is just as good, because we now have simply to invert to map to the region $|w| \leq 1$. The map we want is then given by

$$w = \frac{(2 + i)z - 1}{(-2 + i)z - 1}.$$

Note that any point on the boundary of R is of the form $z = iy$, and its image is

$$w = \frac{(2 + i)iy - 1}{(-2 + i)iy - 1} = \frac{-y - 1 - 2iy}{-y - 1 + 2iy},$$

which has magnitude 1 and so is on $|w| = 1$. Conversely, the reader can show by a straightforward calculation that, if $|w(z)| = 1$, then $\text{Re}(z) = 0$, so z is on the boundary of R. ∎

B. Joukowski Transformation

The Joukowski transformation, or mapping, given by

$$w = J(z) = z + \frac{a^2}{z}$$

for $z \neq 0$ (a is a positive constant), is important in fluid dynamics, as we shall see in the next section. Obviously J is not one-to-one. In fact, solving for z gives

$$z = \frac{w \pm \sqrt{w^2 - 4a^2}}{2},$$

so there are two values of z that map to each value of w except 0. However, by suitably restricting the domain under consideration, J can be considered one-to-one from one domain to another. In most applications, one chooses the inverse,

$$z = \frac{w + \sqrt{w^2 - 4a^2}}{2}.$$

If we write $z = x + iy$ and $w = u + iv$, then

$$\begin{aligned} w = u + iv = x + iy + \frac{a^2}{x + iy} &= x + iy + \frac{a^2(x - iy)}{(x + iy)(x - iy)} \\ &= x + iy + \frac{a^2 x}{x^2 + y^2} - \frac{iya^2}{x^2 + y^2} \\ &= x\left(1 + \frac{a^2}{x^2 + y^2}\right) + iy\left(1 - \frac{a^2}{x^2 + y^2}\right), \end{aligned}$$

so

$$u = x\left(1 + \frac{a^2}{x^2 + y^2}\right)$$

and

$$v = y\left(1 - \frac{a^2}{x^2 + y^2}\right).$$

With one exception, circles about the origin in the z-plane map to ellipses with foci on the real axis in the w-plane. For suppose $|z| = r \neq a$. It is easy to check that

$$\frac{u^2}{(1 + a^2/r^2)^2} + \frac{v^2}{(1 - a^2/r^2)^2} = r^2,$$

and this is an ellipse with foci at $(2a, 0)$ and $(-2a, 0)$. The exception occurs when $r = a$. Then $u = 2x$ and $v = 0$, so the circle $|z| = a$ maps to a line segment $w = 2x$, $-a \leq x \leq a$. That is, $|z| = a$ maps to $[-2a, 2a]$.

Conformal mappings are of sufficient interest and difficulty that extensive lists have been compiled, recording various conformal maps between different regions. We have done just enough here to show the idea without becoming lost in the details. The fundamental theorem on conformal maps is due to Riemann, and says that, given a (simply connected) domain D in the z-plane, then there is an analytic function f mapping D one-to-one onto the disc $|w| < 1$. Given D, however, the task of actually producing such a function may involve insurmountable technical difficulties. We shall defer further consideration of these matters to the exercises and to the next section, where we shall look at some applications which necessitate the concoction of specific conformal mappings.

Exercises for Section 9.12

1. Prove that a Möbius transformation always maps circles and straight lines to circles and straight lines (i.e., each circle maps to a circle or straight line, and each straight line maps to a circle or straight line).

2. Show that the map $w(z) = \bar{z}$ is not a Möbius transformation.

3. Let $w = f(z)$ be a Mobius transformation. We call z_0 a fixed point if $f(z_0) = z_0$.
   (a) Does every Möbius transformation have at least one fixed point?
   (b) How many fixed points can a Möbius transformation have before it reduces to the identity mapping $w(z) = z$?

4. Find Möbius transformations that make the following correspondences:
   (a) $0 \to i$, $2 \to 3i$, $1 \to 2$
   (b) $6 \to 6$, $7 \to 7$, $8 \to i$
   (c) $2i \to i$, $3i \to 4$, $i \to 0$
   (d) $|z| \leq 1$ maps to $|w| \leq 2$.
   (e) $|z| \leq 1$ maps to $|w| \geq 4$.

5. Let

$$w(z) = \frac{i(z - a)}{(z + \bar{a})},$$

   where a is any complex number with $\text{Im}(a) > 0$. Show that w maps the upper half-plane $\text{Im}(z) > 0$ onto the unit disc $|w| < 1$. Is w conformal?

6. Let $w(z) = 1/(1 - z)$ for $z \neq 1$. Show that w maps $|z| < 1$ onto the half-plane $\text{Re}(z) \geq -\frac{1}{2}$ and is conformal.

7. Does there exist a Möbius transformation besides $w(z) = z$ with the property that $w(w(z)) = z$?

8. Let $w(z) = az + bz^2$, with a and b complex constants. Find the image of $|z| < 1$ under this map, and verify that the mapping is conformal.

9. Verify that the semicircular region $|z| < 1$, $\mathrm{Im}(z) > 0$ is mapped by

$$w = \left(\frac{z+1}{z-1}\right)^2$$

conformally onto the upper half-plane $\mathrm{Im}(w) > 0$.

10. Verify that the circular wedge $|z| < 1$, $0 < \mathrm{Arg}(z) < \pi/n$ $(n = 1, 2, 3, \ldots)$ is mapped conformally onto the upper half-plane $\mathrm{Im}(w) > 0$ by

$$w = \left(\frac{z^n+1}{z^n-1}\right)^2.$$

9.13 Additional Applications

Thus far we have built up a great deal of machinery with only relatively small return. Certainly such things as evaluation of improper integrals and summation of series are interesting and important, but we now want to concentrate on applications in physical problems.

A. Application to Fluid Dynamics

Complex variable techniques have proved to be extremely effective in the mathematical treatment of fluid flow. We shall give some illustrations of this now.

Let us consider a two-dimensional fluid flow, that is, a flow in which every variable under consideration is a function of just two space coordinates, say x and y (and possibly a time coordinate t). At each point and time, the fluid has a pressure $p(x, y, t)$, a temperature $T(x, y, t)$, a density $\rho(x, y, t)$, a viscosity $\mu(x, y, t)$, and a velocity $v(x, y, t)$. If ρ = constant, we call the fluid *incompressible* (e.g., water or, at low velocities, air). If $\mu = 0$, we say that the flow is nonviscous. This is often a good approximation to the way many fluids actually behave under certain conditions. If curl $v = (0, 0, 0)$ we call the flow irrotational (see Section 6.2 for the interpretation of curl as a measure of fluid rotation about a point). In a simply connected region, irrotationality is equivalent to the existence of a potential function, i.e., a real-valued function φ with $v = -\nabla\varphi$. (The minus sign is just a convention.)

We now assume that our flow is not only two-dimensional, but irrotational and incompressible. In addition to the resulting potential function φ, we introduce a stream function ψ as follows. Fix a point 0 for reference, and let $P(x, y)$ be any other point. Joint 0 to P by two distinct paths, γ_1 and γ_2, enclosing a region D (Fig. 9-27). At any time t, conservation of mass dictates that the mass per unit time flowing into D across γ_1 must equal that flowing out of D across γ_2. Then the mass per unit time crossing γ_1 counterclockwise is the same as that crossing γ_2. This mass is then a function only of (x, y). We denote it $\psi(x, y)$, and call ψ the stream function of the flow.

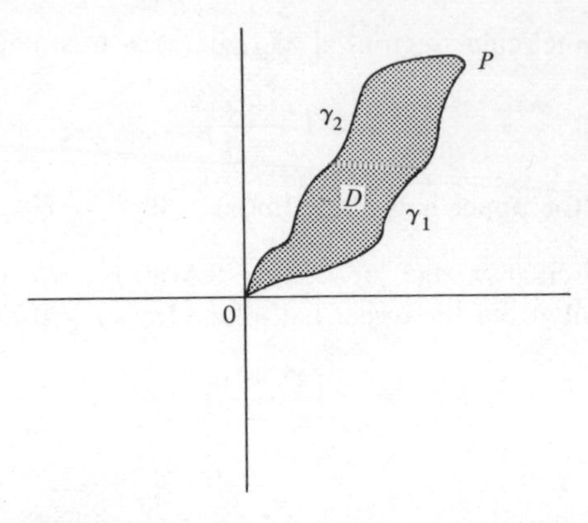

Fig. 9-27

If we displace P to $Q(x + \Delta x, y + \Delta y)$, with γ a path from P to Q, then the mass crossing γ per unit time counterclockwise is $\psi(x + \Delta x, y + \Delta y) - \psi(x, y)$, or $\Delta\psi$, which we can approximate by the differential

$$\frac{\partial\psi}{\partial x}\Delta x + \frac{\partial\psi}{\partial y}\Delta y.$$

But also $\Delta\psi \sim \rho(x, y)v_2(x, y)\,\Delta x - \rho(x, y)v_1(x, y)\,\Delta y$. Here $(v_1, v_2) = v$, the velocity, with the positive direction of v_1 to the right, and that of v_2 upward (Fig. 9-28). The quantity $\rho(x, y)\,\Delta x$ is the mass across the line from P to $(x + \Delta x, y)$, and $\rho(x, y)\,\Delta y$, that across the line from $(x + \Delta x, y)$ to Q, per unit time. Then, adjusting units so that $\rho = 1$,

$$\frac{\partial\psi}{\partial x} = \rho v_2 = v_2 \qquad \text{and} \qquad \frac{\partial\psi}{\partial y} = -\rho v_1 = -v_1.$$

But

$$v_1 = \frac{-\partial\varphi}{\partial x} \qquad \text{and} \qquad v_2 = \frac{-\partial\varphi}{\partial y}.$$

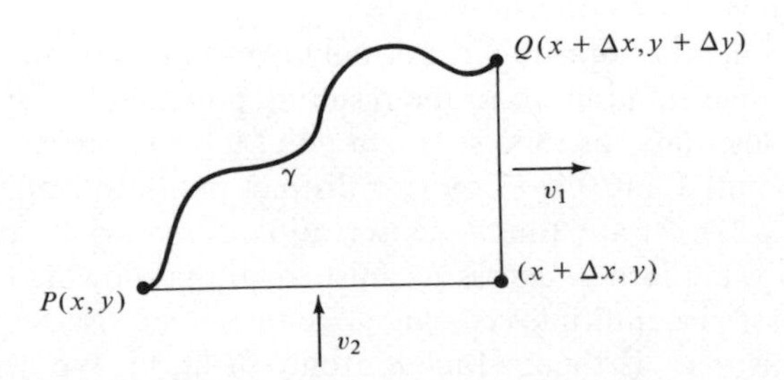

Fig. 9-28

Then

$$\frac{\partial \psi}{\partial x} = \frac{-\partial \varphi}{\partial y} \quad \text{and} \quad \frac{\partial \psi}{\partial y} = \frac{\partial \varphi}{\partial x}.$$

Thus ψ and φ satisfy the Cauchy–Riemann equations. Assuming continuity of the partials, we can then define $f(x + iy) = \varphi(x, y) + i\psi(x, y)$ and obtain an analytic function f with real part the potential function and imaginary part the stream line of the flow.

Note also that φ and ψ are harmonic. This is true in general for the real and imaginary parts of any analytic function, but follows from other considerations as well. Since our flow is irrotational, curl v is the zero vector. Hence

$$\begin{aligned} 0 = \frac{\partial v_2}{\partial x} - \frac{\partial v_1}{\partial y} &= \frac{\partial}{\partial x}\left(\frac{-\partial \varphi}{\partial y}\right) - \frac{\partial}{\partial y}\left(\frac{-\partial \varphi}{\partial x}\right) \\ &= \frac{\partial}{\partial x}\left(\frac{\partial \psi}{\partial x}\right) - \frac{\partial}{\partial y}\left(\frac{-\partial \psi}{\partial y}\right) = \frac{\partial^2 \psi}{\partial x^2} + \frac{\partial^2 \psi}{\partial y^2}, \end{aligned}$$

so ψ is harmonic. Then also φ is harmonic, as by the Cauchy–Riemann equations,

$$\begin{aligned} \varphi_{xx} + \varphi_{yy} &= \frac{\partial}{\partial x}\left(\frac{\partial \varphi}{\partial x}\right) + \frac{\partial}{\partial y}\left(\frac{\partial \varphi}{\partial y}\right) \\ &= \frac{\partial}{\partial x}(\psi_y) + \frac{\partial}{\partial y}(\psi_x) = \frac{\partial^2 \psi}{\partial x\, \partial y} - \frac{\partial^2 \psi}{\partial y\, \partial x} = 0. \end{aligned}$$

Note the velocity has magnitude

$$\begin{aligned} \|v\| &= \|(-\varphi_x, -\varphi_y)\| = \sqrt{\varphi_x^2 + \varphi_y^2} \\ &= \sqrt{\psi_x^2 + \psi_y^2}. \end{aligned}$$

Stream lines of the flow are defined to be solutions to $dy/dx = v_2/v_1$. Thus stream lines are solutions to $dy/dx = -\psi_x/\psi_y$, or $d\psi = 0$ (hence $\psi =$ constant), and have the property that the tangent to a stream line at any point is in the direction of the velocity of the flow at that point.

We now consider some specific examples of flows.

EXAMPLE—UNIFORM STREAM

Let $f(z) = -ae^{i\theta} \cdot z$, where θ and a are positive constants. Then

$$\begin{aligned} f(z) &= -a[\cos(\theta) - i\sin(\theta)](x + iy) \\ &= -a[\cos(\theta)x + \sin(\theta)y] - ai[\cos(\theta)y - \sin(\theta)x]. \end{aligned}$$

Here

$$\varphi(x, y) = -a[x\cos(\theta) + y\sin(\theta)]$$

and

$$\psi(x, y) = -a[y\cos(\theta) - x\sin(\theta)].$$

The velocity vector is given by

$$v = (-\varphi_x, -\varphi_y) = (a\cos(\theta), a\sin(\theta)) = ae^{i\theta},$$

with magnitude a and direction that of a vector making an angle θ with the positive real axis. We therefore have a uniform flow as shown in Fig. 9-29.

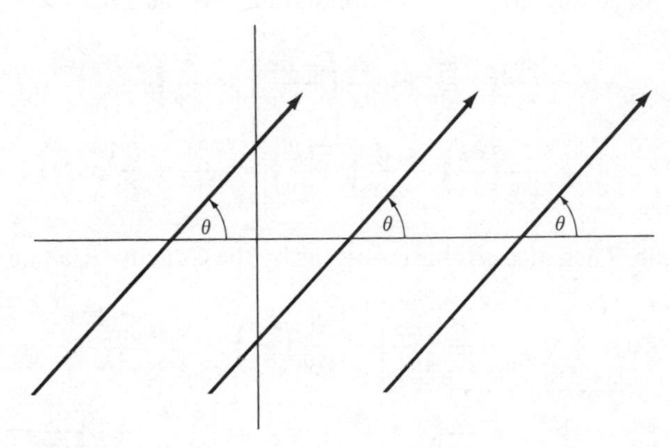

Fig. 9-29

Here the stream lines are given by $dy/dx = v_2/v_1 = \tan(\theta)$. The stream lines are therefore the straight lines $y = \tan(\theta) \cdot x +$ constant, which are along the direction of the flow at any point. ∎

EXAMPLE—VORTEX AT THE ORIGIN

Let a be real and constant and put

$$f(z) = \frac{ia}{2\pi}\,\mathrm{Log}(z).$$

Then

$$f(z) = \frac{ia}{2\pi}[\ln(r) + i\theta] = \frac{-a\theta}{2\pi} + \frac{ia}{2\pi}\ln(r).$$

Then

$$\varphi(r, \theta) = -\theta \qquad \text{and} \qquad \psi(r, \theta) = \frac{a}{2\pi}\ln(r).$$

The stream lines are $\psi(r, \theta) =$ constant, or, equivalently, $r =$ constant; thus the stream lines are circles about the origin. Since stream lines have at each point the direction of the velocity at that point, we can think of the flow as a whirlpool about the origin. The origin is then thought of as a vortex.

The number a has a physical significance. The circulation of the flow about a closed curve γ is defined as $\oint_\gamma v_1\,dx + v_2\,dy$, where $(v_1, v_2) =$ velocity. Now

$$\psi(r, \theta) = \frac{a}{2\pi}\ln(\sqrt{x^2 + y^2}) = \frac{a}{4\pi}\ln(x^2 + y^2).$$

Then

$$v_1 = -\varphi_x = -\psi_y = \frac{a}{2\pi}\frac{-y}{x^2 + y^2}$$

and

$$v_2 = -\varphi_y = \psi_x = \frac{a}{2\pi}\frac{x}{x^2 + y^2}.$$

If γ is the circle $|z| = r$ about the origin, then the circulation about γ is

$$\frac{a}{2\pi}\oint_\gamma \frac{-y\,dx}{x^2 + y^2} + \frac{x\,dy}{x^2 + y^2} = \frac{a}{2\pi}\int_0^{2\pi} \frac{-r\sin(\theta)[-r\sin(\theta)] + r\cos(\theta)r\cos(\theta)}{r^2}\,d\theta$$

$$= \frac{a}{2\pi}\int_0^{2\pi} d\theta = a.$$

If we introduce a solid boundary by putting a circular wall at, say, $r = A$, then the flow outside the cylinder is irrotational with no singularities. Thus $f(z) = \dfrac{ia}{2\pi}\operatorname{Log}(z)$, $|z| \geq A$, gives irrotational flow about a cylinder with circulation a about the cylinder. ∎

EXAMPLE—FLOWS AROUND PLATES AND ELLIPTICAL BARRIERS

As an application of conformal mapping, recall that the Joukowski transformation takes most circles about the origin to ellipses. Put

$$w = z + \frac{A^2}{z} \quad \text{and} \quad z = \frac{w + \sqrt{w^2 - 4A^2}}{2}.$$

The circle $x^2 + y^2 = r^2$ maps to the ellipse

$$\frac{u^2}{(1 + A^2/r^2)^2} + \frac{v^2}{(1 - A^2/r^2)^2} = r^2$$

in the $w = u + iv$ plane, if $r > A$. Thus

$$f(w) = \frac{ia}{2\pi} \operatorname{Log}\left(\frac{w + \sqrt{w^2 - 4A^2}}{2}\right)$$

may be regarded as describing the flow in the (u, v)-plane past an elliptical barrier, with a vortex at the origin. In the degenerate case $r = A$, the circle $x^2 + y^2 = r^2$ maps to the line segment $[-2A, 2A]$ on the real axis, and f describes flow around a flat plate.

If we combine a rotation with the Joukowski transformation, by putting $w = e^{i\xi}(z + A^2/z)$, then we obtain flow about an elliptical barrier with foci on a line making an angle ξ with the positive real axis, and having a vortex at the origin.

In general, the Joukowski transformation $w = z + a^2/z$ maps a circle $|z - z_0| = r$ to a curve known as a Joukowski aerofoil. Given a function describing a particular kind of flow about a circular barrier at $|z - z_0| = r$, we can then determine flow about a Joukowski aerofoil, which can be made to assume a great variety of shapes by choosing z_0 and r in various ways (see Fig. 9-30). ∎

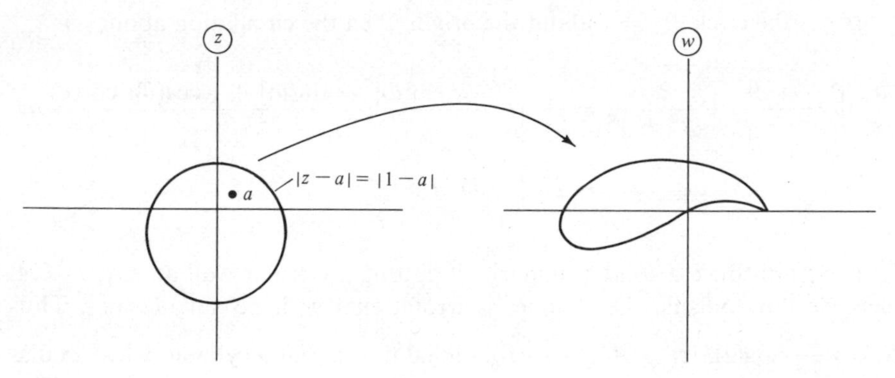

Fig. 9-30

As a brief glimpse of how residues enter into the complex analysis of flows, we shall state without proof a theorem of Blasius. Suppose that f represents flow around a barrier with cross section curve γ. Then Blasius's theorem says that

$$\text{(horizontal component of thrust)} - i\text{(vertical component of thrust)} = \tfrac{1}{2}i\rho \oint_\gamma (f'(z))^2\, dz$$

and

$$\text{moment of the thrust about the origin} = \operatorname{Re}\left(-\tfrac{1}{2}\rho \oint_\gamma z(f'(z))^2\, dz\right).$$

In practice, these integrals are usually evaluated by the residue theorem. This gives

$$\oint_\gamma (f'(z))^2\, dz = 2\pi i[\text{sum of the residues of } (f'(z))^2 \text{ at singularities inside } \gamma]$$

and

$$\oint_\gamma z(f'(z))^2\, dz = 2\pi i[\text{sum of the residues of } z(f'(z))^2 \text{ inside } \gamma].$$

B. Application to the Dirichlet Problem

The Dirichlet problem for a given plane region R consists of solving Laplace's equation

$$\frac{\partial^2 \varphi}{\partial x^2} + \frac{\partial^2 \varphi}{\partial y^2} = 0,$$

subject to the condition that values $\varphi(x, y)$ are specified at the boundary points of R. This is a reasonable problem in view of our previous results on harmonic functions, in particular, the fact that the value of such a function at a point is the arithmetic mean of its values on the boundary of a disc centered at the point (see Sections 6.5C and 6.16D).

We shall illustrate here the use of conformal mappings in solving the Dirichlet problem for a variety of regions. The basic strategy is to first consider the problem on a region sufficiently simple that an explicit solution can be obtained. We then automatically have a solution for any other region which is the image of this one under a conformal mapping. In order to implement this plan, we first must manufacture a connection between real-valued harmonic functions of two variables and analytic functions. In one direction, the connection is clear. If f is analytic, then $\text{Re}(f)$ and $\text{Im}(f)$ are harmonic. This is immediate from the Cauchy–Riemann equations. We now show a converse of this. Let φ be harmonic in a domain D. We claim that there is a harmonic function ψ on D such that $\varphi + i\psi$ is analytic in D.

To produce ψ, define

$$g(x + iy) = \frac{\partial \varphi(x, y)}{\partial x} - i\frac{\partial \varphi(x, y)}{\partial y}.$$

Assume that φ has continuous partials; then

$$\frac{\partial(\text{Re}(g))}{\partial x} = \frac{\partial^2 \varphi}{\partial x^2} = -\frac{\partial^2 \varphi}{\partial y^2} = \frac{\partial(\text{Im}(g))}{\partial y}$$

and

$$\frac{\partial(\mathrm{Re}(g))}{\partial y} = \frac{\partial^2 \varphi}{\partial y\, \partial x} = \frac{\partial^2 \varphi}{\partial x\, \partial y} = \frac{-\partial(\mathrm{Im}(g))}{\partial x}.$$

Thus $\mathrm{Re}(g)$ and $\mathrm{Im}(g)$ satisfy the Cauchy–Riemann equations in D. From the continuity of φ_x and φ_y in D, it follows then that g is analytic in D. Since D is simply connected, there is then some G analytic on D such that $G'(z) = g(z)$ for z in D. Now

$$G'(z) = \frac{\partial(\mathrm{Re}(g))}{\partial x} - i\frac{\partial(\mathrm{Re}(G))}{\partial y} = g(z) = \frac{\partial \varphi}{\partial x} - i\frac{\partial \varphi}{\partial y}.$$

Hence

$$\frac{\partial(\mathrm{Re}(G))}{\partial x} = \frac{\partial \varphi}{\partial x} \quad \text{and} \quad \frac{\partial(\mathrm{Re}(G))}{\partial y} = \frac{\partial \varphi}{\partial y} \text{ on } D.$$

Then, for some real constant C, $\mathrm{Re}(G) = \varphi + C$ on D. Put $f(z) = G(z) - C$. Then f is analytic on D. Now put $\psi = \mathrm{Im}(f)$. Then ψ is harmonic in D, and $\varphi + i\psi = \mathrm{Re}(G) - C + i\,\mathrm{Im}(G) = \mathrm{Re}(f) + i\,\mathrm{Im}(f) = f$.

We now look for some reasonable region for which the Dirichlet problem can be solved. It turns out that such a region is the unit disc.

Suppose then that we seek φ harmonic in $x^2 + y^2 < 1$, with $\varphi(x, y) = u(x, y)$, for $x^2 + y^2 = 1$. Here u is a given function and specifies the values φ is to take on the boundary of the unit disc. We want to be able to obtain values of φ inside the disc from values of u on the boundary.

For purposes of the derivation, suppose that φ is harmonic on a slightly larger disc $|z| < 1 + \varepsilon$. We can then produce analytic f on $|z| < 1 + \varepsilon$ such that $\varphi = \mathrm{Re}(f)$. By adding a suitable constant, adjust f so that $\varphi = \mathrm{Re}(f)$ and also $f(0) = \varphi(0, 0)$.

Expand f in a Taylor series:

$$f(z) = \sum_{n=0}^{\infty} \alpha_n z^n, \qquad |z| < 1 + \varepsilon.$$

Now

$$\begin{aligned}
\varphi(x, y) &= \mathrm{Re}(f(x + iy)) = \tfrac{1}{2}[f(x + iy) + \overline{f(x + iy)}] \\
&= \tfrac{1}{2}\left(\sum_{n=0}^{\infty} \alpha_n z^n + \sum_{n=0}^{\infty} \bar{\alpha}_n \bar{z}^n\right) \\
&= \tfrac{1}{2}(\alpha_0 + \bar{\alpha}_0) + \tfrac{1}{2}\sum_{n=1}^{\infty} (\alpha_n z^n + \bar{\alpha}_n \bar{z}^n).
\end{aligned}$$

But

$$\alpha_0 = f(0) = \varphi(0, 0) = \bar{\alpha}_0.$$

So

$$\varphi(x, y) = \varphi(0, 0) + \tfrac{1}{2} \sum_{n=1}^{\infty} (\alpha_n z^n + \bar{\alpha}_n \bar{z}^n).$$

Now, if $|z| = 1$, then $z\bar{z} = 1$, so $\bar{z} = 1/z$. Then

$$\varphi(x, y) = u(x, y) = \varphi(0, 0) + \tfrac{1}{2} \sum_{n=1}^{\infty} \alpha_n z^n + \frac{\bar{\alpha}_n}{z^n}.$$

Let $\gamma(\theta) = e^{i\theta}$, $\theta: 0 \to 2\pi$. Then, for any integer m,

$$\begin{aligned} \frac{1}{2\pi i} \oint_\gamma \varphi(x, y) z^m \, dz &= \frac{\varphi(0, 0)}{2\pi i} \oint_\gamma z^m \, dz + \frac{1}{2} \frac{1}{2\pi i} \oint_\gamma \sum_{n=1}^{\infty} \left(\alpha_n z^n + \frac{\bar{\alpha}_n}{z^n}\right) z^m \, dz \\ &= \frac{\varphi(0, 0)}{2\pi i} \oint_\gamma z^m \, dz + \frac{1}{4\pi i} \sum_{n=1}^{\infty} \oint_\gamma (\alpha_n z^{n+m} + \bar{\alpha}_n z^{n-m}) \, dz. \end{aligned}$$

Now, by a simple calculation, for any integer t,

$$\oint_\gamma z^t \, dz = \begin{cases} 2\pi i & \text{for } t = -1, \\ 0 & \text{for } t \neq -1. \end{cases}$$

Then, by choosing $m = -1$, we have

$$\frac{1}{2\pi i} \oint_\gamma \varphi(x, y) z^{-1} \, dz = \frac{\varphi(0, 0)}{2\pi i} (2\pi i) = \varphi(0, 0) = \alpha_0,$$

and, by choosing $m = -k - 1$ $(k = 1, 2, \ldots)$,

$$\frac{1}{2\pi i} \oint_\gamma \varphi(x, y) z^{-k-1} \, dz = \frac{1}{4\pi i} \alpha_k (2\pi i) = \frac{\alpha_k}{2}.$$

Then

$$\alpha_n = \frac{1}{\pi i} \oint_\gamma \varphi(x, y) z^{-n-1} \, dz, \qquad n = 1, 2, \ldots.$$

We have now determined the coefficient sequence α_n of the power series for

f, in terms of the data function u, whose values are given at points on γ. Substituting back gives, for $|z| < 1$,

$$\begin{aligned} f(z) &= \sum_{n=0}^{\infty} \alpha_n z^n = \alpha_0 + \sum_{n=1}^{\infty} \alpha_n z^n \\ &= \frac{1}{2\pi i} \oint_\gamma u(\xi) \frac{1}{\xi}\, d\xi + \sum_{n=1}^{\infty} \frac{2}{2\pi i} \oint_\gamma u(\xi) \frac{1}{\xi^{n+1}}\, d\xi\, z^n \\ &= \frac{1}{2\pi i} \oint_\gamma u(\xi) \left(\frac{1}{\xi} + 2 \sum_{n=1}^{\infty} \frac{z^n}{\xi^{n+1}} \right) d\xi \\ &= \frac{1}{2\pi i} \oint_\gamma u(\xi) \left[1 + 2 \sum_{n=1}^{\infty} \left(\frac{z}{\xi} \right)^n \right] \frac{1}{\xi}\, d\xi. \end{aligned}$$

Now

$$\sum_{n=1}^{\infty} \left(\frac{z}{\xi} \right)^n = \frac{1}{1 - z/\xi} - 1 = \frac{z}{\xi - z},$$

since $|z| < 1$ and $|\xi| = 1$ imply that $|z/\xi| < 1$. Then

$$\begin{aligned} f(z) &= \frac{1}{2\pi i} \oint_\gamma u(\xi) \left(1 + \frac{2z}{\xi - z} \right) \frac{1}{\xi}\, d\xi \\ &= \frac{1}{2\pi i} \oint_\gamma \frac{u(\xi)}{\xi} \frac{\xi + z}{\xi - z}\, d\xi. \end{aligned}$$

Our solution is then

$$\varphi(x, y) = \operatorname{Re}(f(z)) = \operatorname{Re}\left[\frac{1}{2\pi i} \oint_\gamma \frac{u(\xi)}{\xi} \frac{\xi + z}{\xi - z}\, d\xi \right],$$

which gives φ at (x, y) in $|z| < 1$ in terms of an integral involving the data function u on $|z| = 1$.

This formula is the Poisson integral formula and solves the Dirichlet problem for the unit disc. If we put $\xi = e^{i\theta}$ on γ, we obtain

$$\begin{aligned} \frac{1}{2\pi i} \oint_\gamma \frac{u(\xi)}{\xi} \frac{\xi + z}{\xi - z}\, d\xi &= \frac{1}{2\pi i} \int_0^{2\pi} \frac{u(\cos(\theta), \sin(\theta))}{e^{i\theta}} \frac{e^{i\theta} + z}{e^{i\theta} - z}\, i e^{i\theta}\, d\theta \\ &= \frac{1}{2\pi} \int_0^{2\pi} u(\cos(\theta), \sin(\theta)) \frac{e^{i\theta} + z}{e^{i\theta} - z}\, d\theta. \end{aligned}$$

Now put $z = re^{i\zeta}$ and write

$$\frac{e^{i\theta} + z}{e^{i\theta} - z} = \frac{e^{i\theta} + re^{i\zeta}}{e^{i\theta} - re^{i\zeta}} = \frac{1 + re^{i(\zeta-\theta)}}{1 - re^{i(\zeta-\theta)}}$$

$$= \frac{[1 + r\cos(\zeta - \theta)] + ir\sin(\zeta - \theta)}{[1 - r\cos(\zeta - \theta)] - ir\sin(\zeta - \theta)} \frac{[1 - r\cos(\zeta - \theta)] + ir\sin(\zeta - \theta)}{[1 - r\cos(\zeta - \theta)] + ir\sin(\zeta - \theta)}$$

$$= \frac{[1 + r\cos(\zeta - \theta)][1 - r\cos(\zeta - \theta)] - r^2\sin^2(\zeta - \theta)}{[1 - r\cos(\zeta - \theta)]^2 + r^2\sin^2(\zeta - \theta)}$$

$$+ \frac{ir\sin(\zeta - \theta)[1 + r\cos(\zeta - \theta)] + ir\sin(\zeta - \theta)[1 - r\cos(\zeta - \theta)]}{[1 - r\cos(\zeta - \theta)]^2 + r^2\sin^2(\zeta - \theta)}$$

$$= \frac{[1 - r^2\cos^2(\zeta - \theta)] - r^2\sin^2(\zeta - \theta) + 2ir\sin(\zeta - \theta)}{[1 + r^2\cos^2(\zeta - \theta)] - 2r\cos(\zeta - \theta) + r^2\sin^2(\zeta - \theta)}$$

$$= \frac{1 - r^2 + 2ir\sin(\zeta - \theta)}{1 + r^2 - 2r\cos(\zeta - \theta)}.$$

Then

$$f(z) = \frac{1}{2\pi}\int_0^{2\pi} \frac{u(\cos(\theta), \sin(\theta))}{1 + r^2 - 2r\cos(\zeta - \theta)} [1 - r^2 + 2ir\sin(\zeta - \theta)]\, d\theta$$

and our solution φ is given by

$$\varphi(x, y) = \text{Re}(f(x + iy)) = \frac{1 - r^2}{2\pi}\int_0^{2\pi} \frac{u(\cos(\theta), \sin(\theta))\, d\theta}{1 + r^2 - 2r\cos(\zeta - \theta)},$$

where $x + iy = re^{i\zeta}$, $0 \le r < 1$, $0 \le \zeta < 2\pi$. This gives us a somewhat more explicit formulation of Poisson's integral solution to the Dirichlet problem for the unit disc.

Having solved the Dirichlet problem for the unit disc about the origin, we shall now consider solutions for some other regions. The method may be outlined briefly as follows:

PROBLEM

Find a function φ harmonic in a region D, and satisfying $\varphi(x, y) = u(x, y)$ for (x, y) on the boundary of D, where u is a given function.

SOLUTION

Let $\mathscr{D}$ denote the region $|w| \le 1$ in the w-plane. Assume that we can produce a conformal mapping T from D onto $\mathscr{D}$. Then the boundary ∂D of D maps onto the boundary $\partial\mathscr{D}$ of $\mathscr{D}$. Denote the inverse map by T^{-1}, so $T^{-1}(w) = z$ if $w = T(z)$. The situation is depicted in Fig. 9-31.

Now, $u(x, y)$ is specified for (x, y) on ∂D. Define, for $w = \alpha + i\beta$ on $\partial\mathscr{D}$, $U(\alpha, \beta) = u(T^{-1}(w))$. We can now consider the Dirichlet problem for $\mathscr{D}$: $\nabla^2\Phi = 0$ on $\mathscr{D}$, $\Phi(\alpha, \beta) = U(\alpha, \beta)$ on $\partial\mathscr{D}$.

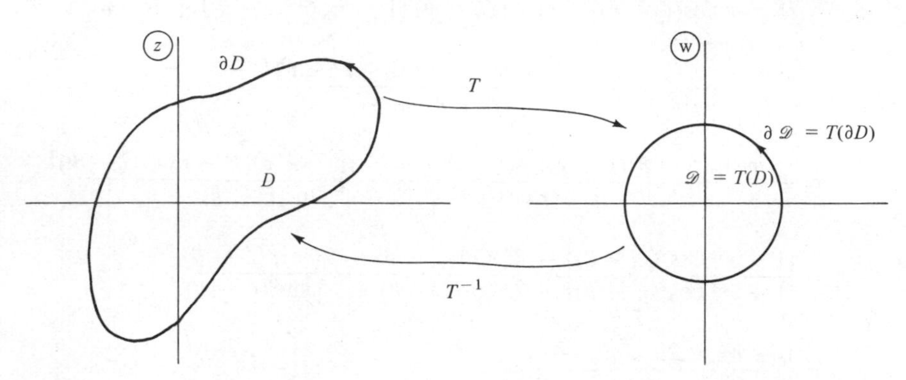

Fig. 9-31

The connection between this new problem and the original one is this. Suppose that Φ is a solution to the Dirichlet problem for $\mathscr{D}$. Put $\varphi(x, y) = \Phi(T(x + iy))$ for $(x, y) = x + iy$ in D. We claim also that $\nabla^2\varphi = 0$ on D. This is pretty much a straightforward calculation, which we leave to the reader. (Put $\varphi(x, y) = \Phi(T(x + iy)) = \Phi(\alpha(x, y), \beta(x, y))$, use the chain rule to compute $\nabla^2\varphi$, and use the facts that $\dfrac{\partial^2\Phi}{\partial\alpha^2} + \dfrac{\partial^2\Phi}{\partial\beta^2} = 0$ and α and β satisfy the Cauchy-Riemann equations).

We now derive an integral formula for φ. We have one for Φ:

$$\Phi(\alpha, \beta) = \mathrm{Re}\left(\frac{1}{2\pi i}\oint_{\partial\mathscr{D}} \frac{U(\zeta)}{\zeta}\left(\frac{\zeta + \alpha + i\beta}{\zeta - \alpha - i\beta}\right) d\zeta\right).$$

The theorem for change of variables in Riemann integrals goes over in the same form to complex integrals. Putting $\alpha + i\beta = T(x + iy)$ for $x + iy$ in D, and $\zeta = T(\xi)$, we have

$$\varphi(x, y) = \Phi(T(x + iy)) = \mathrm{Re}\left(\frac{1}{2\pi i}\int_{\partial D} \frac{U(T(\xi))}{T(\xi)}\left(\frac{T(\xi) + T(x + iy)}{T(\xi) - T(x + iy)}\right) T'(\xi)\, d\xi\right)$$

$$= \mathrm{Re}\left\{\frac{1}{2\pi i}\int_{\partial D} \frac{u(\xi)}{T(\xi)}\left(\frac{T(\xi) + T(x + iy)}{T(\xi) - T(x + iy)}\right) T'(\xi)\, d\xi.\right.$$

We now compute some specific examples.

(1) *Dirichlet problem for a disc* $|z| \leq \rho$, *where* $\rho > 0$. Here we can put $T(z) = (1/\rho)z$. This transforms our region $D: |z| \leq \rho$, conformally onto the region $\mathscr{D}: |w| \leq 1$. Then the solution is, for $x^2 + y^2 < \rho^2$,

$$\varphi(x, y) = \text{Re}\left\{\frac{1}{2\pi i}\oint_{|z|=\rho}\left[\frac{u(\xi)}{\xi/\rho}\frac{\xi/\rho + (x+iy)/\rho}{\xi/\rho - (x+iy)/\rho}\frac{d\xi}{\rho}\right]\right\}$$

$$= \text{Re}\left[\frac{1}{2\pi i}\oint_{|z|=\rho}\frac{u(\xi)}{\xi}\frac{\xi + (x+iy)}{\xi - (x+iy)}\,d\xi\right].$$

Parametrize $|\xi| = \rho$ as $\xi = \rho e^{i\theta}$, $\theta: 0 \to 2\pi$. Then

$$\varphi(x, y) = \text{Re}\left[\frac{1}{2\pi i}\int_0^{2\pi}\frac{u(\rho\cos(\theta), \rho\sin(\theta))}{\rho e^{i\theta}}\frac{\rho e^{i\theta} + x + iy}{\rho e^{i\theta} - x - iy}\,i\rho e^{i\theta}\,d\theta\right]$$

$$= \text{Re}\left[\frac{1}{2\pi i}\int_0^{2\pi} u(\rho\cos(\theta), \rho\sin(\theta))\frac{e^{i\theta} + x + iy}{e^{i\theta} - x - iy}\,d\theta\right].$$

Putting $x + iy = re^{i\zeta}$, a calculation like that done before with the Poisson formula yields

$$\varphi(x, y) = \frac{\rho^2 - r^2}{2\pi}\int_0^{2\pi}\frac{u(\rho\cos(\theta), \rho\sin(\theta))\,d\theta}{\rho^2 - 2r\rho\cos(\zeta - \theta) + r^2}.$$

When $\rho = 1$, this of course reduces to the Poisson formula for the unit disc.

(2) *Dirichlet problem for a disc* $|z - z_0| \leq \rho$. Here we can use the map $w = T(z) = (1/\rho)(z - z_0)$. This takes $|z - z_0| \leq \rho$ onto the disc $|w| \leq 1$. The solution at $x + iy$ in $|z - z_0| \leq \rho$ is then

$$\varphi(x, y) = \text{Re}\left[\frac{1}{2\pi i}\oint_{|z-z_0|=\rho}\frac{u(\xi)}{(\xi - z_0)/\rho}\frac{(\xi - z_0)/\rho + (x + iy - z_0)/\rho}{(\xi - z_0)/\rho - (x + iy - z_0)/\rho}\frac{d\xi}{\rho}\right]$$

$$= \text{Re}\left[\frac{1}{2\pi i}\oint_{|z-z_0|=\rho}\frac{u(\xi)}{\xi - z_0}\frac{\xi + x + iy - 2z_0}{\xi - x - iy}\,d\xi\right].$$

Parametrize $|z - z_0| = \rho$: $\xi = z_0 + \rho e^{i\theta}$, $\theta: 0 \to 2\pi$. Then

$$\varphi(x, y) = \text{Re}\left[\frac{1}{2\pi i}\int_0^{2\pi}\frac{u(z_0 + \rho e^{i\theta})}{\rho e^{i\theta}}\frac{\rho e^{i\theta} + x + iy - 2z_0}{\rho e^{i\theta} - x - iy}\,i\rho e^{i\theta}\,d\theta\right]$$

$$= \text{Re}\left[\frac{1}{2\pi}\int_0^{2\pi} u(z_0 + \rho e^{i\theta})\frac{\rho e^{i\theta} + x + iy - 2z_0}{\rho e^{i\theta} - x - iy}\,d\theta\right].$$

When $z_0 = 0$, this reduces to (1) above.

(3) *Dirichlet problem for the exterior of the unit disc.* Here D is the region $|z| > 1$. We map D onto the unit disc by a simple inversion: $w = T(z) = 1/z$. The boundary of $|w| < 1$ is the curve γ, where $\gamma(\theta) = e^{i\theta}$, $\theta: 0 \to 2\pi$. Note here

that the image of γ under T^{-1} is $T^{-1}(\gamma(t)) = 1/e^{i\theta} = e^{-i\theta}$, $\theta: 0 \to 2\pi$, which traverses the unit circle clockwise. We then have, for $x^2 + y^2 > 1$,

$$\begin{aligned}
\varphi(x, y) &= \mathrm{Re}\left[\frac{1}{2\pi i}\oint_{T^{-1}(\gamma)} \frac{u(\xi)}{T(\xi)}\frac{T(\xi) + T(z)}{T(\xi) - T(z)} T'(\xi)\, d\xi\right] \\
&= \mathrm{Re}\left[\frac{1}{2\pi i}\oint_{T^{-1}(\gamma)} \frac{u(\xi)}{1/\xi}\frac{1/\xi + 1/z}{1/\xi - 1/z}(-1/\xi^2)\, d\xi\right] \\
&= \mathrm{Re}\left[\frac{1}{2\pi i}\int_0^{2\pi} \frac{u(e^{-i\theta})}{1/e^{-i\theta}}\frac{1/e^{-i\theta} + 1/z}{1/e^{-i\theta} - 1/z}(-1/e^{-2i\theta})(-ie^{-i\theta})\, d\theta\right] \\
&= \mathrm{Re}\left[\frac{1}{2\pi i}\int_0^{2\pi} \frac{u(e^{-i\theta})}{e^{i\theta}}\frac{z + e^{-i\theta}}{z - e^{-i\theta}}(i/e^{-i\theta})\, d\theta\right] \\
&= \mathrm{Re}\left[\frac{1}{2\pi}\int_0^{2\pi} u(e^{-i\theta})\frac{z + e^{-i\theta}}{z - e^{-i\theta}}\, d\theta\right].
\end{aligned}$$

Put $z = re^{i\zeta}$ and simplify to obtain

$$\begin{aligned}
\varphi(x, y) &= \mathrm{Re}\left[\frac{1}{2\pi} u(e^{-i\theta})\frac{-1 + r^2 - 2i\sin(\zeta + \theta)}{1 - 2r\cos(\zeta + \theta) + r^2}\, d\theta\right] \\
&= \frac{r^2 - 1}{2\pi}\int_0^{2\pi} \frac{u(\cos(\theta), -\sin(\theta))\, d\theta}{1 - 2r\cos(\zeta + \theta) + r^2},
\end{aligned}$$

where $x + iy = re^{i\zeta}$, $r > 1$, $0 \le \zeta < 2\pi$. Note that $r^2 - 1 > 0$, as $r > 1$ for z to lie outside the unit circle.

(4) *Dirichlet problem for the upper half-plane.* Here D is the region $\mathrm{Im}(z) > 0$, with the real axis as boundary. Suppose that u is given on the boundary, say $u(t, 0) = f(t)$ for t real. First, we must produce a mapping of D onto the region $|w| < 1$. Using the method discussed in Section 9.12, we obtain as one such mapping

$$w = T(z) = \frac{i(z - i)}{z + i}.$$

Then

$$z = T^{-1}(w) = \frac{-i(w + i)}{w - i}.$$

Writing, as usual, the boundary of $\mathscr{D}$ as $\gamma(\theta) = e^{i\theta}$, $\theta: 0 \to 2\pi$, we have

$$z = T^{-1}(\gamma(\theta)) = T^{-1}(e^{i\theta}) = -i\left(\frac{e^{i\theta} + i}{e^{i\theta} - i}\right) = \frac{\cos(\theta)}{1 - \sin(\theta)}.$$

Note what happens if we let $\theta: 0 \to 2\pi$ as usual. As $\theta: 0 \to \pi/2$, $z: 1 \to +\infty$. When $\theta: \pi/2 \to \pi$, $z: -\infty \to -1$. As $\theta: \pi \to 3\pi/2$, $z: -1 \to 0$. And, as

$\theta: 3\pi/2 \to 2\pi$, $z: 0 \to 1$. Thus, in letting $\theta: 0 \to 2\pi$, we have $e^{i\theta}$ traversing $|w| = 1$ once counterclockwise from $+1$ to -1 to $+1$, and $z = T^{-1}(e^{i\theta})$ traversing the real axis in a rather unusual way: $1 \to \infty$, then $-\infty \to -1 \to 0 \to 1$. We therefore rearrange the interval for θ so that, as $\theta: a \to b$, we still have $e^{i\theta}$ going once counterclockwise around $|w| = 1$, but at the same time $T^{-1}(e^{i\theta})$ traverses the real axis from $-\infty$ to ∞. The student can check that this is achieved by taking $\theta: -3\pi/2 \to \pi/2$. Now parametrize the real axis $T^{-1}(\gamma)$ by $T^{-1}(\gamma)(t) = t$, $-\infty < t < \infty$. Then

$$\varphi(x, y) = \operatorname{Re}\left\{\frac{1}{2\pi i}\int_{T^{-1}(\gamma)}\left[\frac{u(\xi)}{T(\xi)}\frac{T(\xi) + T(z)}{T(\xi) - T(z)}T'(\xi)\right]d\xi\right\}$$

$$= \operatorname{Re}\left[\frac{1}{2\pi i}\int_{-\infty}^{\infty}\frac{u(t, 0)}{i\left(\frac{t - i}{t + i}\right)}\frac{i\left(\frac{t - i}{t + i}\right) + i\left(\frac{z - i}{z + i}\right)}{i\left(\frac{t - i}{t + i}\right) - i\left(\frac{z - i}{z + i}\right)}\frac{-2}{(t + i)^2}\,dt\right]$$

$$= \operatorname{Re}\left[\frac{1}{\pi}\int_{-\infty}^{\infty}\frac{u(t, 0)}{(t - i)(t + i)}\frac{\frac{t - i}{t + i} + \frac{z - i}{z + i}}{\frac{t - i}{t + i} - \frac{z - i}{z + i}}\,dt\right]$$

$$= \operatorname{Re}\left[\frac{1}{\pi}\int_{-\infty}^{\infty}\frac{u(t, 0)}{t^2 + 1}\frac{(t - i)(z + i) + (z - i)(t + i)}{(t - i)(z + i) - (t + i)(z - i)}\,dt\right]$$

$$= \operatorname{Re}\left[\frac{1}{\pi}\int_{-\infty}^{\infty}\frac{u(t, 0)}{t^2 + 1}\frac{zt + 1}{i(t - z)}\,dt\right].$$

Put $z = x + iy$ and do some manipulations to write

$$\frac{zt + 1}{t - z} = \frac{xt^2 + t - x^2t - x - y^2t + i(yt^2 + y)}{(t - x)^2 + y^2}.$$

Then

$$\operatorname{Re}\left(\frac{u(t, 0)}{t^2 + 1}\frac{1}{i}\frac{zt + 1}{t - z}\right) = \frac{u(t, 0)}{t^2 + 1}\frac{yt^2 + y}{(t - x)^2 + y^2} = \frac{yu(t, 0)}{(t - x)^2 + y^2}.$$

The solution is then

$$\varphi(x, y) = \frac{1}{\pi}\int_{-\infty}^{\infty}\frac{yu(t, 0)\,dt}{(t - x)^2 + y^2} = \frac{y}{\pi}\int_{-\infty}^{\infty}\frac{f(t)\,dt}{(t - x)^2 + y^2} \qquad \text{for } x \text{ real and } y > 0.$$

In all of Examples (1) through (4), it must be understood that in general it is impossible to actually integrate the resulting expression for $\varphi(x, y)$ in closed form. Nevertheless, the integral provides a means of computing numerical values of the solution at given points, or of making estimates, for example, on the magnitude of the solution. As an exercise in differentiation, the reader should check that

$$\frac{\partial^2 \varphi}{\partial x^2} + \frac{\partial^2 \varphi}{\partial y^2} = 0$$

when

$$\varphi(x, y) = \frac{y}{\pi} \int_{-\infty}^{\infty} \frac{f(t)\, dt}{(t - x)^2 + y^2}, \qquad y > 0.$$

We shall compute one more example of a solution to a Dirichlet problem in an unbounded region, this one somewhat different from the last.

(5) *Dirichlet problem for a quarter-plane* $x > 0, y > 0$. Suppose that we want $\nabla^2 \varphi = 0$, for $x > 0$ and $y > 0$, and

$$\begin{aligned} \varphi(0, y) &= 0, \qquad & y > 0, \\ \varphi(x, 0) &= f(x), \qquad & x > 0. \end{aligned}$$

As usual, we begin by mapping the region D, consisting of points (x, y) with $x > 0, y > 0$, to the interior of the unit circle. However, we need not start from scratch to find such a mapping, as we already have a mapping from the upper half-plane to the unit circle. All we need do is find a mapping from the present region D to the upper half-plane, then compose the maps. The idea of this is shown in Fig. 9-32.

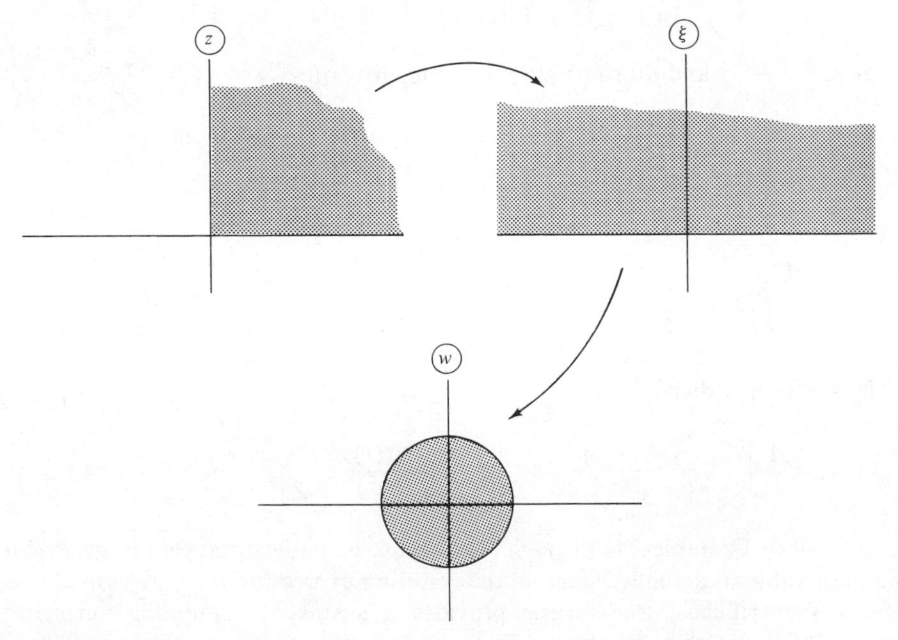

Fig. 9-32

A map from the quarter plane $x > 0$, $y > 0$ to the half-plane $\text{Im}(z) > 0$ is easy. Note, if $z = re^{i\theta}$ and $0 < \theta < \pi/2$, then $z^2 = r^2e^{i2\theta}$, and $0 < 2\theta < \pi$. Thus the map $\xi = z^2$ maps $x > 0$, $y > 0$ in the z-plane to the upper half-plane $\text{Im}\,\xi > 0$ in the ξ-plane. From example (4)

$$w = \frac{i(\xi - i)}{(\xi + i)}$$

maps $\text{Im}\,\xi > 0$ to $|w| < 1$ in the w-plane. Thus

$$w = T(z) = \frac{i(z^2 - i)}{z^2 + i}$$

takes $x > 0$, $y > 0$ to $|w| < 1$. Further, as each step used in constructing the final mapping is conformal, so is the final mapping itself. We now have, for $x > 0$ and $y > 0$,

$$\varphi(x, y) = \text{Re}\left[\frac{1}{2\pi i}\int_{\partial D} \frac{u(\tau)}{T(\tau)}\frac{T(\tau) + T(z)}{T(\tau) - T(z)}\,T'(\tau)\,d\tau\right].$$

Consider ∂D for this region D. Clearly ∂D consists of the nonnegative x- and y-axes. We can thus parametrize ∂D as

$$\gamma_1(\alpha) = \alpha, \qquad \alpha\colon 0 \to \infty,$$

and

$$\gamma_2(\beta) = i\beta, \qquad \beta\colon \infty \to 0.$$

The reason for taking $\beta\colon \infty \to 0$ is orientation. We want ∂D oriented counterclockwise with respect to D, so we think of the positive sense coming down the y-axis to the origin, then out the x-axis. Splitting $\int_{\partial D}$ into $\int_{\gamma_1}$ and $\int_{\gamma_2}$, we then have

$$\varphi(x, y) = \text{Re}\left[\frac{1}{2\pi i}\int_0^\infty \frac{u(\alpha, 0)}{T(\alpha)}\frac{T(\alpha) + T(z)}{T(\alpha) - T(z)}\,T'(\alpha)\,d\alpha \right.$$
$$\left. - \frac{1}{2\pi i}\int_0^\infty \frac{u(0, \beta)}{T(\beta)}\frac{T(\beta) + T(z)}{T(\beta) - T(z)}\,T'(\beta)\,d\beta\right].$$

(Here we wrote $\int_\infty^0 \cdots d\beta$ as $-\int_0^\infty \cdots d\beta$.)

Now, according to our initial conditions,

$$u(0, \beta) = 0 \qquad \text{for } \beta > 0,$$

and

$$u(\alpha, 0) = f(\alpha) \qquad \text{for } \alpha > 0.$$

Then

$$\varphi(x, y) = \operatorname{Re}\left[\frac{1}{2\pi i}\int_0^\infty \frac{f(\alpha)}{\dfrac{i(\alpha^2 - i)}{(\alpha^2 + i)}} \frac{\dfrac{i(\alpha^2 - i)}{(\alpha^2 + i)} + \dfrac{i(z^2 - i)}{(z^2 + i)}}{\dfrac{i(\alpha^2 - i)}{(\alpha^2 + i)} - \dfrac{(z^2 - i)}{(z^2 + i)}} \frac{-4\alpha\, d\alpha}{(\alpha^2 + i)^2}\right].$$

A long but straightforward calculation yields

$$\operatorname{Re}\left[\frac{1}{2\pi i} \frac{f(\alpha)}{\dfrac{i(\alpha^2 - i)}{(\alpha^2 + i)}} \frac{\dfrac{(\alpha^2 - i)}{(\alpha^2 + i)} + \dfrac{(z^2 - i)}{(z^2 + i)}}{\dfrac{(\alpha^2 - i)}{(\alpha^2 + i)} - \dfrac{(z^2 - i)}{(z^2 + i)}} \frac{-4\alpha}{\alpha^2 + 1}\right]$$

$$= \frac{yf(\alpha)}{\pi}\left[\frac{1}{y^2 + (\alpha - x)^2} - \frac{1}{y^2 + (\alpha + x)^2}\right].$$

Thus our solution is

$$\varphi(x, y) = \frac{y}{\pi}\int_0^\infty f(\alpha)\left[\frac{1}{y^2 + (\alpha - x)^2} - \frac{1}{y^2 + (\alpha + x)^2}\right] d\alpha.$$

We shall, by way of illustration, derive the same result in Chapter 10 by Fourier integral methods.

C. Liouville's Theorem and the Fundamental Theorem of Algebra

Here is a mathematical application of some of the results of this chapter. Before we saw some functions that were entire (analytic in the whole plane), such as e^z, $\sin(z)$, and $\cosh(z)$. None of these were bounded. We now show why.

Liouville's Theorem

Let f be a nonconstant, entire function. Then f is not bounded.

PROOF

Suppose that f is bounded, say $|f(z)| \le M$ for all z. Let γ_r be a circle of radius r about 0: $\gamma_r(\theta) = re^{i\theta}$, $\theta: 0 \to 2\pi$. By Cauchy's formula for higher derivatives,

$$\begin{aligned}
|f^{(n)}(0)| &= \left| \frac{n!}{2\pi i} \oint_{\gamma_r} \frac{f(\xi)\, d\xi}{\xi^{n+1}} \right| \\
&= \frac{n!}{2\pi} \left| \int_0^{2\pi} \frac{f(re^{i\theta}) ire^{i\theta}\, d\theta}{r^{n+1} e^{i(n+1)\theta}} \right| \\
&= \frac{n!}{2\pi} \left| \int_0^{2\pi} \frac{f(re^{i\theta})\, d\theta}{r^n e^{in\theta}} \right| \\
&\leq \frac{n!}{2\pi} \int_0^{2\pi} \left| \frac{f(re^{i\theta})}{r^n e^{in\theta}} \right| d\theta \\
&\leq \frac{n!}{2\pi} \frac{M}{r^n} 2\pi = \frac{M}{r^n} n!.
\end{aligned}$$

If $n > 0$, we can make $Mn!/r^n$ as small as we like by taking r larger. We conclude that $f^{(n)}(0) = 0$ for $n = 1, 2, 3, \ldots$. The Taylor series for f about 0 is then

$$f(z) = \sum_{n=0}^{\infty} \frac{f^{(n)}(0) z^n}{n!} = f(0).$$

Then f is constant, a contradiction. Hence f is not bounded. ∎

As an application of Liouville's theorem, we give a short proof of the fundamental theorem of algebra. Let $p(z)$ be a nonconstant polynomial with complex coefficients. The fundamental theorem of algebra says that $p(z)$ has at least one complex root. That is, for some z, $p(z) = 0$.

To prove this, suppose that $p(z) \neq 0$ for all z. Let $f(z) = 1/p(z)$. Then f is entire. It is easy to check that $|p(z)| \to \infty$ as $|z| \to \infty$. Hence, for some $R > 0$, $|p(z)| \geq 1$ for $|z| \geq R$. Then $|f(z)| \leq 1$ for $|z| \geq R$. But f is continuous, hence bounded, on the compact set $|z| \leq R$. Thus f is bounded in the entire plane, contradicting Liouville's theorem. Thus f cannot be entire, hence p has a zero.

D. More on Analytic Functions

We conclude this section by applying a good deal of the theory we have accumulated to the derivation of some additional general properties of analytic functions.

(1) *The zeros of a nonzero analytic function in a domain D are isolated.* Suppose that f is analytic in a domain D, but that $f(z) \neq 0$ for at least one z in D. Suppose that z_0 is in D and $f(z_0) = 0$. We claim that there is a disc about z_0 in D in which z_0 is the only zero of f. This means that the zeros of f in D are isolated in the sense that each is contained in a disc not containing any other zero.

To prove this, expand f in a Taylor series about z_0:

$$f(z) = \sum_{n=0}^{\infty} a_n (z - z_0)^n.$$

Now $a_0 = f(z_0) = 0$. If z_0 is a zero of order m, then $a_0 = a_1 = \cdots = a_{m-1} = 0$, but $a_m \neq 0$. Then, in some neighborhood $|z - z_0| < \varepsilon$ in D:

$$f(z) = \sum_{n=m}^{\infty} a_n(z - z_0)^n = \sum_{n=0}^{\infty} a_{n+m}(z - z_0)^{n+m}$$

$$= (z - z_0)^m \sum_{n=0}^{\infty} a_{n+m}(z - z_0)^n.$$

Put $g(z) = \sum_{n=0}^{\infty} a_{n+m}(z - z_0)^n$. Then g is analytic at z_0, and $g(z_0) = a_m \neq 0$. Since g is continuous and nonzero at z_0, there is a neighborhood $|z - z_0| < \delta$ in which $g(z) \neq 0$. Choose r as the smaller of the numbers ε and δ. Then, in $|z - z_0| < r$, $(z - z_0)^m g(z)$ vanishes only at z_0, hence f has no zero other than z_0 in $|z - z_0| < r$. This proves our assertion.

(2) *Equality of analytic functions in a domain.* Let f and g be analytic in a domain D. We would like to know when $f = g$ on D.

Obviously this is the case if $f(z) = g(z)$ for each z in D. But is it necessary to check each point of D? We now show that it is not.

Theorem

Suppose that $\{z_n\}_{n=1}^{\infty}$ converges to a point of D, and that each z_n is in D. Suppose that $f(z_n) = g(z_n)$ for each n. Then $f(z) = g(z)$ for each z in D.

This is really quite an amazing result, as there is no apparent connection between values of f and g at other points of D. The proof is surprisingly simple.

PROOF

Let $h(z) = f(z) - g(z)$, and suppose that $\lim_{n\to\infty} z_n = z_0$ in D. Note that $h(z_n) = 0$ for all n, and, by continuity of h,

$$h(z_0) = \lim_{n\to\infty} h(z_n) = 0.$$

Then z_0 is a nonisolated zero of h in D. This violates (1) unless $h(z) = 0$ for all z in D. ∎

(3) *If $|f|$ is constant on a domain D, then so is f.* At first, this result seems reasonable, but it is not entirely obvious. Put $f(z) = u(x, y) + iv(x, y)$. Then, $|f(z)| = [u(x, y)^2 + v(x, y)^2]^{1/2}$. Conceivably this could be constant, but f nonconstant, if it is possible for u and v to vary in such a way that $u^2 + v^2$ remains constant. We now show that this cannot happen, so that in fact f itself is constant if $|f|$ is.

Suppose that $[u(x, y)^2 + v(x, y)^2]^{1/2}$ is constant for $x + iy$ in D, say $u^2 + v^2 = C$ in D. Then

$$2u\frac{\partial u}{\partial x} + 2v\frac{\partial v}{\partial x} = 0$$

and

$$2u \frac{\partial u}{\partial y} + 2v \frac{\partial v}{\partial y} = 0.$$

But, by the Cauchy–Riemann equations,

$$\frac{\partial u}{\partial x} = \frac{\partial v}{\partial y} \quad \text{and} \quad \frac{\partial u}{\partial y} = \frac{-\partial v}{\partial x}.$$

Then

$$u \frac{\partial u}{\partial x} - v \frac{\partial u}{\partial y} = 0$$

and

$$v \frac{\partial u}{\partial x} + u \frac{\partial u}{\partial y} = 0.$$

Then

$$uv \frac{\partial u}{\partial x} - v^2 \frac{\partial u}{\partial y} = 0$$

and

$$vu \frac{\partial u}{\partial x} + u^2 \frac{\partial u}{\partial y} = 0.$$

Then

$$(u^2 + v^2) \frac{\partial u}{\partial y} = 0 = C \frac{\partial u}{\partial y}.$$

If $C = 0$, then $u = v = 0$ in D, so $f(z) = 0$ for all z in D. If $C \neq 0$, then $\partial u/\partial y = 0$ in D. Then also $\partial v/\partial x = 0$ in D. Similarly, $\partial u/\partial x = \partial v/\partial y = 0$ in D. Then $u(x, y) =$ constant and $v(x, y) =$ constant for (x, y) in D, and hence f is constant in D.

(4) *The maximum modulus principle.* Suppose that f is analytic and non-constant in a domain D and on its boundary curve γ. What can we say about maximum values of $|f|$ on the compact set R consisting of D and γ? Clearly $|f|$ achieves a maximum somewhere on R. Simply put $f = u + iv$, so $|f| = (u^2 + v^2)^{1/2}$, and treat $|f|$ as a continuous function of two variables on a compact set R.

Next, where does the maximum occur? From our experience with real-valued functions, we would not expect to be able to make any general statement about location. We shall now show, however, in the present case of f analytic on R that $|f|$ takes on its maximum value on γ. This is called the *maximum modulus principle.*

To prove this, suppose that z_0 is in D and that $|f(z_0)| \geq |f(z)|$ for all z in some disc $|z - z_0| \leq \varepsilon$ in D. Let C be the circle $|z - z_0| = \varepsilon$, and use Cauchy's integral formula to write

$$f(z_0) = \frac{1}{2\pi i} \oint_C \frac{f(\xi)\, d\xi}{\xi - z_0}.$$

Parametrize $C(t) = z_0 + \varepsilon e^{it}$, $t: 0 \to 2\pi$. Then

$$f(z_0) = \frac{1}{2\pi i} \int_0^{2\pi} \frac{f(z_0 + \varepsilon e^{it})}{\varepsilon e^{it}} \varepsilon i e^{it}\, dt$$

$$= \frac{1}{2\pi} \int_0^{2\pi} f(z_0 + \varepsilon e^{it})\, dt.$$

Then

$$|f(z_0)| \leq \frac{1}{2\pi} \int_0^{2\pi} |f(z_0 + \varepsilon e^{it})|\, dt.$$

Now

$$|f(z_0 + \varepsilon e^{it})| \leq |f(z_0)| \qquad \text{for } t: 0 \to 2\pi.$$

Then

$$\frac{1}{2\pi} \int_0^{2\pi} |f(z_0 + \varepsilon e^{it})|\, dt \leq \frac{1}{2\pi} |f(z_0)| 2\pi = |f(z_0)|.$$

Hence

$$|f(z_0)| = \frac{1}{2\pi} \int_0^{2\pi} |f(z_0 + e^{it})|\, dt.$$

Then

$$\int_0^{2\pi} [|f(z_0)| - |f(z_0 + e^{it})|]\, dt = 0.$$

Since $|f(z_0)| - |f(z_0 + \varepsilon e^{it})| \geq 0$ for $0 \leq t \leq 2\pi$, then we conclude that $|f(z_0)| = |f(z_0 + re^{it})|$ for $0 \leq t \leq 2\pi$ and $0 \leq r \leq \varepsilon$. Then $|f(z)| = |f(z_0)|$ for $|z - z_0| \leq \varepsilon$. Hence $|f|$ is constant on $|z - z_0| \leq \varepsilon$. By (3), then, f is constant on $|z - z_0| \leq \varepsilon$. Then, by (2), f is constant in D, a contradiction. Hence f cannot achieve a local maximum at any point in D, and so must achieve its maximum on the boundary γ.

(5) *An application of the argument principle*. Suppose that f is analytic and nonconstant in a domain D and on its boundary curve γ. We can think of f as a mapping from a z-plane to a w-plane by putting $w = f(z)$. Let $\Gamma = \{f(z) \mid z$ is on $\gamma\}$. Then Γ is the image under the mapping of γ, and is also a closed (but not necessarily simple) curve in the w-plane. Note that D is mapped by f to the region interior to Γ (shaded in Fig. 9-33), not to the exterior of Γ. This is immediate, as $|f|$ must remain bounded by the maximum modulus principle.

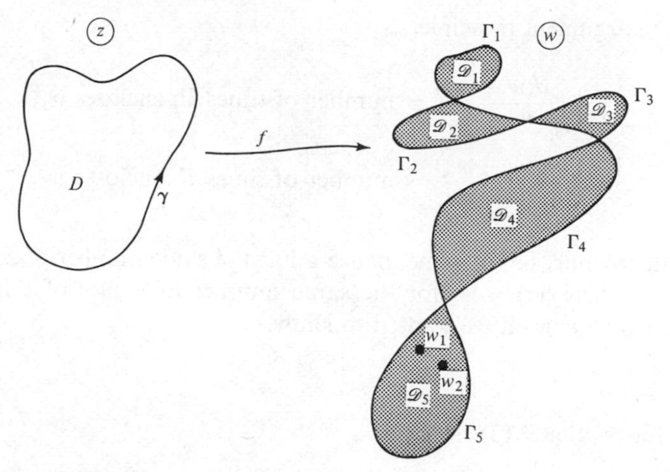

Fig. 9-33

The region $\mathscr{D}$ interior to Γ is not necessarily a domain, but may be thought of as being made up of domains $\mathscr{D}_1, \mathscr{D}_2, \ldots$, with boundary curves $\Gamma_1, \Gamma_2, \ldots$ formed from pieces of Γ (Fig. 9-33).

Choose two points w_1 and w_2 in some $\mathscr{D}_j$. We now claim that the number of points in D mapped to w_1 by f is the same as the number of points mapped to w_2 by f. (We do *not*, however, claim that this is the same as the number of points to, say, v in some other $\mathscr{D}_k$.)

To prove this, we use the argument principle. Let

$$g(z) = f(z) - w_1$$
$$h(z) = f(z) - w_2$$

for z in D and on γ. Then g and h are analytic in D and on γ, hence have no poles there. Then

(number of zeros of g within γ-number of poles of g within γ)

$$\begin{aligned} &= \text{number of zeros of } g \text{ within } \gamma \\ &= \frac{1}{2\pi i}\oint_\gamma \frac{g'(z)\,dz}{g(z)} \\ &= \frac{1}{2\pi i}\oint_\gamma \frac{f'(z)\,dz}{f(z) - w_1} \\ &= \frac{1}{2\pi i}\oint_{\Gamma_j} \frac{dw}{w - w_1} = \frac{1}{2\pi i}\oint_{\Gamma_j} \frac{d(w - w_1)}{w - w_1}. \end{aligned}$$

Similarly,

$$\text{number of zeros of } h \text{ within } \gamma = \frac{1}{2\pi i}\oint_\gamma \frac{f'(z)\,dz}{f(z) - w_2} = \frac{1}{2\pi i}\oint_{\Gamma_j} \frac{d(w - w_2)}{w - w_2}.$$

Now, by the argument principle,

$$\frac{1}{2\pi i}\oint_{\Gamma_j}\frac{d(w-w_1)}{w-w_1} = \text{number of times } \Gamma_j \text{ encloses } w_1,$$

$$\frac{1}{2\pi i}\oint_{\Gamma_j}\frac{d(w-w_2)}{w-w_2} = \text{number of times } \Gamma_j \text{ encloses } w_2.$$

These numbers must be the same, hence g has the same number of zeros inside γ as does h. Then $f(z) = w_1$ for the same number of values of z inside γ as $f(z) = w_2$, which is what we wanted to show.

Exercises for Section 9.13

1. Let a and b be real constants, and $a > 0$. Describe the flow given by

$$\varphi + i\psi = \frac{a}{r}\, e^{i(b-\theta)}.$$

2. Describe the flow given by $f(z) = -az - b\,\text{Log}(z)$, where a and b are positive constants. *Hint:* Consider f as a superposition of two flows: $f(z) = f_1(z) + f_2(z)$, where $f_1(z) = -az$ and $f_2(z) = -b\,\text{Log}(z)$.

3. Solve the Dirichlet problem for the lower half-plane:

$$\nabla^2\varphi = 0, \qquad -\infty < x < \infty,\ y < 0,$$
$$\varphi(x, 0) = u(x), \qquad -\infty < x < \infty.$$

4. Solve the Dirichlet problem for the right quarter-plane more generally than in Section 9.13B(5) by considering the problem

$$\nabla^2\varphi = 0, \qquad x > 0,\ y > 0,$$
$$\varphi(x, 0) = f(x), \qquad x > 0,$$
$$\varphi(0, y) = g(y), \qquad y > 0.$$

[we have $g(y) = 0$ in the text]. *Hint:* Consider two problems:

(1)		(2)	
$\nabla^2\varphi_1 = 0,$	$x > 0,\ y > 0,$	$\nabla^2\varphi_2 = 0,$	$x > 0,\ y > 0,$
$\varphi_1(x, 0) = f(x),$	$x > 0,$	$\varphi_2(x, 0) = 0,$	$x > 0,$
$\varphi_1(0, y) = 0,$	$y > 0.$	$\varphi_2(0, y) = g(y),$	$y > 0.$

Solve these separately and then put $\varphi = \varphi_1 + \varphi_2$.

5. Solve the Dirichlet problem for the half-plane $\text{Im}(z) > 1$:

$$\nabla^2\varphi = 0, \qquad -\infty < x < \infty,\ y > 1,$$
$$\varphi(x, 1) = f(x), \qquad -\infty < x < \infty.$$

6. Solve the Dirichlet problem for the quarter-plane $\mathrm{Re}(z) < 0$, $\mathrm{Im}(z) < 0$:

$$\begin{aligned} \nabla^2\varphi &= 0, && x < 0,\ y < 0,\\ \varphi(x, 0) &= f(x), && x < 0,\\ \varphi(0, y) &= g(y), && y < 0. \end{aligned}$$

Hint: Look at Exercise 4.

7. Solve the Dirichlet problem for the half-plane $\mathrm{Re}(z) > 2$:

$$\begin{aligned} \nabla^2\varphi &= 0, && x > 2,\ -\infty < y < \infty,\\ \varphi(2, y) &= g(y), && -\infty < y < \infty. \end{aligned}$$

Hint: Use Exercise 5, and map the half-plane $\mathrm{Im}(z) > 1$ conformally onto the half-plane $\mathrm{Re}(z) > 2$ by a translation and rotation.

8. Solve the Dirichlet problem for the elliptical region $(x^2/a^2) + (y^2/a^2) < 1$ (a, b positive constants):

$$\begin{aligned} \nabla^2\varphi &= 0, && \frac{x^2}{a^2} + \frac{y^2}{b^2} < 1,\\ \varphi(x, y) &= u(x, y), && \frac{x^2}{a^2} + \frac{y^2}{b^2} = 1. \end{aligned}$$

9. Let f be analytic in a domain D and on its boundary γ. Suppose that $f(z) \neq 0$ for z in D. Show that $|f|$ achieves its minimum on γ. Use this to give another proof of the fundamental theorem of algebra.

10. Prove the following complex form of L'Hospital's Rule. Suppose that f and g are analytic and nonconstant in a neighborhood of z_0, and that f and g both have a simple zero at z_0. Suppose that

$$\lim_{z \to z_0} \frac{f'(z)}{g'(z)} = L.$$

Then

$$\lim_{z \to z_0} \frac{f(z)}{g(z)} = L.$$

Hint: Imitate part of the proof in Section 9.13D(2).

11. Let f be an entire function and suppose, for some positive integer n, that $|f(z)| \leq |z|^n$ for all z. Prove that f is a polynomial.

12. Let f be an entire function, and suppose that $|\mathrm{Re}(f(z))| \leq K$ for all z and some constant K. Prove that f is constant. *Hint:* Apply Liouville's theorem to $e^{f(z)}$.

10

Fourier Series, Integrals, and Transforms

10.1 Some Introductory Remarks

In 1807, Joseph Fourier presented to the French Academy a paper on the mathematical theory of heat, in the course of which he asserted that arbitrary functions could be represented by series of sines and cosines. Such a statement so startled the Academy that one of its leading members, Lagrange, immediately denied the possibility of its validity.

In the very strictest sense, Lagrange was right. However, history has vindicated Fourier as the founder of a theory that enables one to write functions satisfying fairly unrestrictive conditions as sums of sines and cosines. Although later mathematicians filled in many of the details and placed Fourier's ideas on firm mathematical footing, nevertheless the original insight was Fourier's.

Several questions come to mind at this point. Why, first, should anyone want to represent a function as a series of sines and cosines? What was Fourier working on which led him to think that this might be an interesting thing to be able to do?

We shall attempt to answer both questions with an example. Consider the problem of determining the temperature distribution in a thin bar of length L and uniform cross section, made of some homogeneous material, in which the ends are kept at zero temperature and the initial temperature at each interior point is known. Put a coordinate system so that the length of the bar extends from 0 to L on the x-axis, and suppose the temperature u at any time t and in the cross section of bar at point x depends only on x and t. The differential equation for $u(x, t)$ is then

$$\frac{\partial u}{\partial t} = k\frac{\partial^2 u}{\partial x^2}, \qquad 0 < x < L, t > 0,$$

where k is a constant depending on the material of the bar. This is the equation we derived in Section 5.2 and again in Section 6.16G. The condition of zero temperature at the ends gives us boundary conditions:

$$u(0, t) = u(L, t) = 0, \qquad t \geq 0;$$

and the given initial temperature gives us an initial condition

$$u(x, 0) = f(x),$$

where $f(x)$ is given for $0 \leq x \leq L$.

In order to solve the problem [i.e., find $u(x, t)$], we try a method called *separation of variables*. We shall look for a solution of the form $u(x, t) = A(x)B(t)$. Substitute into

$$\frac{\partial u}{\partial t} = k\frac{\partial^2 u}{\partial x^2}$$

to obtain

$$AB' = kA''B,$$

or

$$\frac{B'}{kB} = \frac{A''}{A}.$$

Having separated the variables x and t to opposite sides of the equation, we now argue as follows. The left side depends only on t, the right side only on x. Since x and t are independent, both sides must be constant. For some λ, then,

$$\frac{B'}{kB} = \frac{A''}{A} = \lambda.$$

We now have two ordinary differential equations:

$$B' - \lambda kB = 0, \qquad \text{and} \qquad A'' - \lambda A = 0.$$

We also have a new quantity to determine, λ.

Note that the boundary conditions tell us something about A. Since $u(0, t) = A(0)B(t) = 0$ for all $t > 0$, then $A(0) = 0$ or $B(t) = 0$ for all t. The latter gives us a trivial solution, so we conclude that $A(0) = 0$. Similarly, $u(L, t) = A(L)B(t) = 0$ implies that $A(L) = 0$.

Now consider the problem for A:

$$A''(x) - \lambda A(x) = 0,$$
$$A(0) = A(L) = 0.$$

Since we do not know λ, we consider the possibilities. On physical grounds, we expect λ to be real, as $A(x)$ should be real-valued. This leaves three cases.

(1) $\lambda = 0$. Then $A'' = 0$, so $A(x) = \alpha x + \beta$, for constants α and β. But $A(0) = 0 = \beta$; and $A(L) = \alpha L = 0$ implies that $\alpha = 0$. This gives us a trivial solution for A, and so we reject this case.
(2) $\lambda > 0$. Say $\lambda = \varphi^2$, where $\varphi > 0$. Then $A'' - \varphi^2 A = 0$, with general solution $A(x) = \alpha e^{\varphi x} + \beta e^{-\varphi x}$, with α and β constants (see Appendix I of this chapter). Now $A(0) = 0$ implies that $\alpha = -\beta$. Then $A(x) = 2\alpha \cosh(\varphi x)$, and $A(L) = 2\alpha \cosh(\varphi L) = 0$ implies that $\alpha = 0$ or $\varphi L = 0$. Since $L \neq 0$, then $\alpha = 0$, and we again reach a trivial solution. Case (2) is out.
(3) $\lambda < 0$, say $\lambda = -\varphi^2$, where φ is positive. Then $A'' + \varphi^2 A = 0$, with general solution $A(x) = \alpha \cos(\varphi x) + \beta \sin(\varphi x)$. Since $A(0) = \alpha = 0$, then $A(x) = \beta \sin(\varphi x)$. Finally, $A(L) = \beta \sin(\varphi L) = 0$. We do not want $\beta = 0$, as then we have the trivial solution again. So we choose φL to make $\sin(\varphi L) = 0$. We can do this by making φL an integer multiple of $\pi: \varphi L = n\pi$. Since $\varphi > 0$, then $n = 1, 2, 3, \ldots$, and $\varphi = n\pi/L$. In this case, then, for each positive integer n, we obtain a value for λ, which we shall index by n:

$$\lambda_n = \frac{-n^2\pi^2}{L^2}$$

and a solution to $A'' - \lambda_n A = 0$, $A(0) = A(L) = 0$:

$$A_n(x) = \beta_n \sin\left(\frac{n\pi x}{L}\right).$$

Here β_n is a still undetermined constant. The numbers λ_n are called eigenvalues of the problem $A'' - \lambda A = 0$, $A(0) = A(L) = 0$, and the corresponding solutions A_n, eigenfunctions.

Now, go back to the equation for B: $B' - \lambda k B = 0$. For each n, we have an equation

$$B' - \lambda_n k B = 0 = B' + \frac{n^2\pi^2 k B}{L^2}.$$

This has general solution $B(t) = \xi e^{-n^2\pi^2 kt/L^2}$, where ξ is a constant. To indicate that this solution for B is tied to the choice of n, we write

$$B_n(t) = \xi_n e^{-n^2\pi^2 kt/L^2}.$$

Finally go back to the beginning of our solution. Recall that we are attempting to find $u(x, t)$ as a product of $A(x)$ and $B(t)$. Since we now have solutions

$A_n(x)$ and $B_n(t)$ for each positive integer n, we put $u_n(x, t) = A_n(x)B_n(t) = \zeta_n \sin(n\pi x/L)e^{-n^2\pi^2 kt/L^2}$. (Here we renamed $\beta_n \xi_n = \zeta_n$.)

Note that, for each positive integer n, $u_n(x, t)$ satisfies

(1) The differential equation

$$\frac{\partial u}{\partial t} = k\frac{\partial^2 u}{\partial x^2}, \qquad 0 < x < L, t > 0.$$

(2) The boundary conditions

$$u(0, t) = u(L, t) = 0, \qquad t > 0.$$

We have remaining the condition $u(x, 0) = f(x)$, $0 \leq x \leq L$. Now $u_n(x, 0) = \zeta_n \sin(n\pi x/L)$, and we can make this equal $f(x)$ only if $f(x)$ is a multiple of $\sin(n\pi x/L)$, which certainly need not be the case. How, then, can we satisfy the initial condition? It was at this point that a brilliant stroke by Fourier provided the final key. Observe that, if F and G satisfy

$$\frac{\partial u}{\partial t} = k\frac{\partial^2 u}{\partial x^2}, \qquad u(0, t) = u(L, t) = 0,$$

then so does the superposition or sum $F + G$ (just substitute and check it out). This extends immediately to finite sums: Any finite sum of the functions $u_n(x, t)$ will also satisfy the differential equation and the boundary conditions. But such a finite sum still may not satisfy the initial condition $u(x, 0) = f(x)$. We are therefore led to a highly nonobvious step—an infinite superposition. That is, we try for a solution

$$u(x, t) = \sum_{n=1}^{\infty} u_n(x, t) = \sum_{n=1}^{\infty} \zeta_n \sin\left(\frac{n\pi x}{L}\right)e^{-kn^2\pi^2 t/L^2}.$$

This raises all sorts of delicate questions, some of which are still the object of considerable mathematical investigation. But, putting aside the fine points for the moment, let us continue and see what happens.

Assume that the series converges for $0 \leq x \leq L$, $t > 0$, thus defining $u(x, t)$. Assume that we can differentiate the series term by term; it is then easy to check that

$$\frac{\partial u}{\partial t} = k\frac{\partial^2 u}{\partial x^2},$$

because

$$\frac{\partial u_n}{\partial t} = k\frac{\partial^2 u_n}{\partial x^2} \qquad \text{for each term } u_n.$$

Further, $u(0, t) = u(L, t) = 0$, as each $u_n(0, t) = u_n(L, t) = 0$. Finally, we want $f(x) = u(x, 0)$. This means that we need

$$f(x) = \sum_{n=1}^{\infty} u_n(x, 0) = \sum_{n=1}^{\infty} \zeta_n \sin\left(\frac{n\pi x}{L}\right).$$

The problem, then, is to choose the ζ_n's so that the trigonometric series on the right converges to the given function. At this point we have answered one question posed at the beginning of this discussion. Here is a physically interesting problem whose solution calls for us to write a given function f as an infinite series of sines. Rather than stop here, however, we shall indicate a clever trick for finding the constants ζ_n and bring the problem to a final solution. Write

$$f(x) \sin\left(\frac{m\pi x}{L}\right) = \sum_{n=1}^{\infty} \zeta_n \sin\left(\frac{n\pi x}{L}\right) \sin\left(\frac{m\pi x}{L}\right),$$

where m is any positive integer. Integrate from 0 to L, assuming that $\sum_{n=1}^{\infty} \int_0^L = \int_0^L \sum_{n=1}^{\infty}$. Then

$$\int_0^L f(x) \sin\left(\frac{m\pi x}{L}\right) dx = \sum_{n=1}^{\infty} \zeta_n \int_0^L \sin\left(\frac{n\pi x}{L}\right) \sin\left(\frac{m\pi x}{L}\right) dx.$$

The point to this calculation is that

$$\int_0^L \sin\left(\frac{n\pi x}{L}\right) \sin\left(\frac{m\pi x}{L}\right) = \begin{cases} \dfrac{L}{2} & \text{if } m = n. \\ 0 & \text{if } m \neq n. \end{cases}$$

In the series on the right, then, all terms vanish except $\zeta_m \int_0^L \sin^2(m\pi x/L)\, dx$, which is $(L/2)\zeta_m$. Solving for ζ_m then gives us

$$\zeta_m = \frac{2}{L} \int_0^L f(x) \sin\left(\frac{m\pi x}{L}\right) dx,$$

and we can compute ζ_m, knowing $f(x)$.

The proposed solution to the heat-conduction problem, at least formally, is then

$$u(x, t) = \frac{2}{L} \sum_{n=1}^{\infty} \left[\int_0^L f(\xi) \sin\left(\frac{n\pi \xi}{L}\right) d\xi\right] \sin\left(\frac{n\pi x}{L}\right) e^{-kn^2\pi^2 t/L^2},$$

with ξ used in place of x as the dummy-integration variable to avoid confusion.

Of course, we have not justified any of the last steps, and a little reflection will suggest several serious difficulties. First, in solving for the ζ_n's, we interchanged $\sum_{n=1}^{\infty}$ and $\int_0^L$, which may not work in any given case. Second, we have said nothing about convergence of any of the series involved. Finally, even if the series $\sum_{n=1}^{\infty} \zeta_n \sin(n\pi x/L)$ does converge, perhaps it does not converge to $f(x)$, as is required to satisfy $u(x, 0) = f(x)$. The reason for this is in the solution for the constants:

$$\zeta_n = \frac{2}{L} \int_0^L f(\xi) \sin\left(\frac{n\pi\xi}{L}\right) d\xi.$$

If we define a new function $g(x)$ by changing the value of $f(x)$ at a single point y, then the integral determining ζ_n is unchanged. But obviously $\sum_{n=1}^{\infty} \zeta_n \sin(n\pi x/L)$ cannot converge to both $f(y)$ and $g(y)$.

Let us pause to sum up the discussion thus far. In attempting to solve a problem in heat flow, we were led by the separation-of-variables method to attempt a trigonometric Fourier series for the data function. At least formally, the method produced a series representation of the solution. In order to be sure that this series does in fact converge to the solution to the problem, we must investigate more carefully the individual calculations made above. In particular, we must concern ourselves with the question of whether the Fourier series for a function on an interval converges to the function on that interval.

All this will be done next section. For now, let us clarify Fourier's role in the development of the series that bear his name. First, the idea of representing certain functions as series of sines and/or cosines was not due to Fourier, nor was the trick we used in the example to determine the coefficients in the proposed series. These were known at least to the Swiss mathematician Euler as early as 1777, and were probably known to Euler and d'Alembert around 1750. But these great mathematicians missed the point seized upon by Fourier. To Euler, d'Alembert, and others working in this area, Fourier's method of determining the coefficients was highly suspect, and was to be trusted only when the resulting series could be shown by other means to converge to the function being represented. Fourier's contribution lay in understanding, if only partly rigorously, the generality and power of the method, and in applying it to a wide variety of problems, particularly in the mathematical development of the theory of heat flow.

10.2 Pointwise Convergence of Fourier Series

Let $f(x)$ be defined for $-L \le x \le L$. We define the Fourier series for $f(x)$ to be the series

$$\frac{a_0}{2} + \sum_{n=1}^{\infty} \left[a_n \cos\left(\frac{n\pi x}{L}\right) + b_n \sin\left(\frac{n\pi x}{L}\right)\right],$$

where

$$a_n = \frac{1}{L}\int_{-L}^{L} f(\xi)\cos\left(\frac{n\pi\xi}{L}\right) d\xi, \qquad n = 0, 1, 2, \ldots$$

and

$$b_n = \frac{1}{L}\int_{-L}^{L} f(\xi)\sin\left(\frac{n\pi\xi}{L}\right) d\xi, \qquad n = 1, 2, \ldots.$$

In making this definition, we are assuming only that f is sufficiently well behaved that these integrals exist. The questions we want to answer are:

(1) For $-L \leq x \leq L$, to what (if any) real number does the Fourier series for $f(x)$ converge?
(2) What conditions on f will guarantee that the series converges to $f(x)$?

First, a word about the numbers $a_0, a_1, a_2, \ldots, b_1, b_2, \ldots$. These are called the *Fourier coefficients* of $f(x)$ on $[-L, L]$. To see why they are chosen according to the above formulas, suppose that

$$f(x) = \frac{a_0}{2} + \sum_{n=1}^{\infty}\left[a_n\cos\left(\frac{n\pi x}{L}\right) + b_n\sin\left(\frac{n\pi x}{L}\right)\right] \qquad \text{for } -L \leq x \leq L.$$

For any positive integer m then

$$f(x)\cos\left(\frac{m\pi x}{L}\right) = \left(\frac{a_0}{2}\right)\cos\left(\frac{m\pi x}{L}\right) + \sum_{n=1}^{\infty}\left[a_n\cos\left(\frac{m\pi x}{L}\right)\cos\left(\frac{n\pi x}{L}\right) + b_n\sin\left(\frac{n\pi x}{L}\right)\cos\left(\frac{m\pi x}{L}\right)\right].$$

Integrate from $-L$ to L. Assume that $\sum_{n=1}^{\infty}$ and $\int_{-L}^{L}$ can be interchanged, and use the formulas

$$\int_{-L}^{L}\cos\left(\frac{n\pi x}{L}\right)\cos\left(\frac{m\pi x}{L}\right) dx = \begin{cases} L & \text{if } m = n, \\ 0 & \text{if } m \neq n, \end{cases}$$

$$\int_{-L}^{L}\cos\left(\frac{m\pi x}{L}\right)\sin\left(\frac{n\pi x}{L}\right) dx = 0$$

to conclude that

$$a_m = \frac{1}{L}\int_{-L}^{L} f(\xi)\cos\left(\frac{m\pi\xi}{L}\right) d\xi.$$

Similarly, if we multiply $f(x)$ and the series by $\sin(m\pi x/L)$, interchange $\sum_{n=1}^{\infty}$ and $\int_{-L}^{L}$, and use the formula

$$\int_{-L}^{L} \sin\left(\frac{n\pi\xi}{L}\right) \sin\left(\frac{m\pi\xi}{L}\right) d\xi = \begin{cases} L & \text{if } m = n, \\ 0 & \text{if } m \neq n, \end{cases}$$

we obtain

$$b_m = \frac{1}{L}\int_{-L}^{L} f(\xi) \sin\left(\frac{m\pi\xi}{L}\right) d\xi.$$

Finally, simply integrating from $-L$ to L and interchanging $\sum_{n=1}^{\infty}$ and $\int_{-L}^{L}$ yields

$$\int_{-L}^{L} f(\xi)\, d\xi = \frac{a_0}{2}\int_{-L}^{L} dx = a_0 L,$$

hence

$$a_0 = \frac{1}{L}\int_{-L}^{L} f(\xi)\, d\xi.$$

This is exactly

$$\frac{1}{L}\int_{-L}^{L} f(\xi) \cos\left(\frac{m\pi\xi}{L}\right) d\xi \qquad \text{when } m = 0.$$

These considerations prove nothing but suggest that the choice of coefficients made in the last definition is the "right" one if a useful theory is to be worked out.

Note that the constant term in the Fourier series is called $a_0/2$ instead of just a_0. This is pretty much a matter of taste but allows us to write a single formula for the a_n's, $n = 0, 1, 2, \ldots$, instead of one for a_0 and another for a_n when $n > 0$. Note also that the Fourier series we are examining are on an interval $[-L, L]$, not $[0, L]$, as arose in the heat-flow problem of Section 10.1. We shall see later that Fourier expansions in either sines or cosines on $[0, L]$ are easily derivable from the type of Fourier series we are considering on $[-L, L]$.

We shall now prepare the way for the statement and proof of a convergence theorem for Fourier series. Some terminology and notation are needed first.

Recall that f is said to be monotone on $[a, b]$ if $f(x) \leq f(y)$ whenever $a \leq x \leq y \leq b$ (then f is monotone nondecreasing) or $f(x) \leq f(y)$ whenever $a \leq y \leq x \leq b$ (f is monotone nonincreasing). We shall say that f satisfies Dirichlet's condition on $[a, b]$ if f is bounded and if we can insert finitely many points $a = t_0 < t_1 < \cdots < t_n = b$ in such a way that f is monotone on each open subinterval (t_{i-1}, t_i), $i = 1, \ldots, n$. This amounts to saying that f has only a finite number of relative maxima and minima on $[a, b]$.

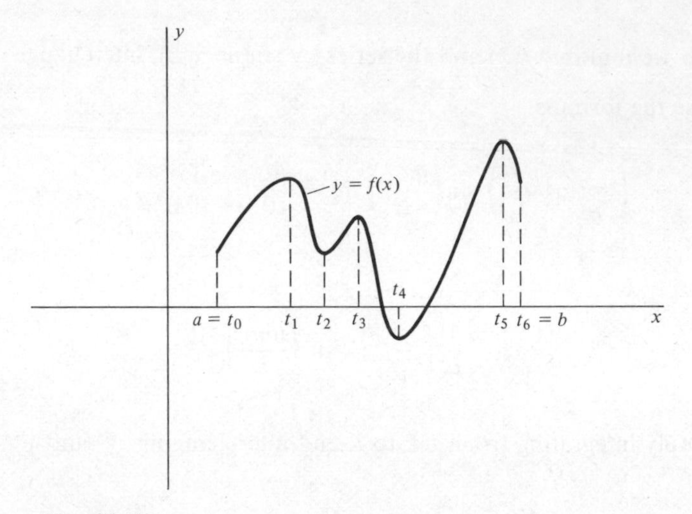

Fig. 10-1

For example, let f be the function graphed below (Fig. 10-1). With $t_1, \ldots, t_5$ chosen as indicated, f is monotone on $(t_0, t_1), (t_1, t_2), \ldots, (t_4, t_5)$, and (t_5, t_6). The renaming of a as t_0 and b as t_n is purely a notational convenience.

Not all functions satisfy Dirichlet's condition. For example, let

$$f(x) = \begin{cases} \sin\left(\dfrac{1}{x}\right), & 0 < x \leq 1, \\ 0, & x = 0. \end{cases}$$

Then f has a maximum or minimum at each point $x = 2/(2n + 1)\pi$ for $n = 0, 1, 2, \ldots$, and there are too many such points to allow f to satisfy Dirichlet's condition on $[0, 1]$.

Note also that f need not be continuous to satisfy Dirichlet's condition (see, for example, f as graphed in Fig. 10-2). In fact, f may have infinitely many discontinuities, as with

$$f(x) = x + \frac{1}{n}, \qquad \frac{1}{n+1} < x \leq \frac{1}{n}, \qquad n = 1, 2, \ldots.$$

Here f is monotone, nondecreasing on $(0, 1]$, but has a discontinuity at $1/n$ for $n = 2, 3, \ldots$.

One important characteristic of functions satisfying a Dirichlet condition is the existence of right and left limits. That is, if f satisfies a Dirichlet condition on $[a, b]$, then, for any x with $a < x < b$,

$$\lim_{\varepsilon \to 0+} f(x + \varepsilon) \quad \text{and} \quad \lim_{\varepsilon \to 0+} f(x - \varepsilon)$$

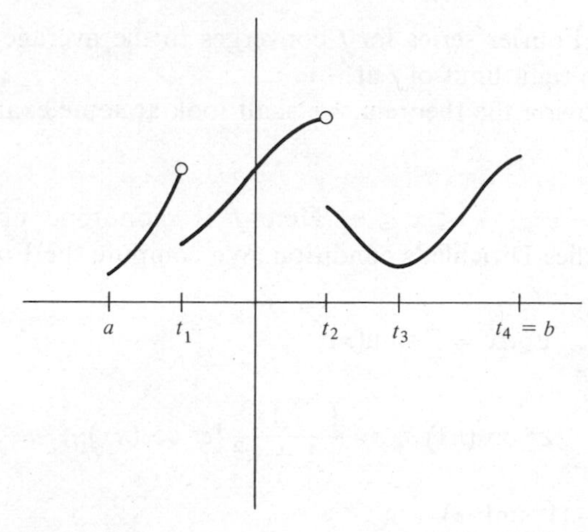

Fig. 10-2

both exist. Here $\lim_{\varepsilon \to 0+}$ denotes a limit as $\varepsilon \to 0$ through positive values. In addition,

$$\lim_{\varepsilon \to 0+} f(a + \varepsilon) \quad \text{and} \quad \lim_{\varepsilon \to 0+} f(b - \varepsilon)$$

exist. These facts are easy to show and are left to the reader. It will be useful to have notation for these limits. We shall put, for any function f,

$$\lim_{\varepsilon \to 0+} f(x + \varepsilon) = f(x + 0) \quad \text{and} \quad \lim_{\varepsilon \to 0+} f(x - \varepsilon) = f(x - 0)$$

whenever these limits exist. When $x = 0$, we shorten $f(x + 0)$ and $f(x - 0)$ to $f(0+)$ and $f(0-)$, respectively.

We can now state our convergence theorem. As a convenience, we use the interval $[-\pi, \pi]$. A simple change of variables will then give us a corresponding theorem for any interval $[-L, L]$.

Theorem on Pointwise Convergence of Fourier Series

Suppose that f satisfies Dirichlet's condition on $[-\pi, \pi]$. Then

(1) *For any x with $-\pi < x < \pi$, the Fourier series for f converges at x to* $\frac{1}{2}[f(x + 0) + f(x - 0)]$.
(2) *The Fourier series for f converges at π and at $-\pi$ to* $\frac{1}{2}[f(\pi - 0) + f(-\pi + 0)]$.

Geometrically, this says that, at each point x between π and $-\pi$, the Fourier series for f converges to the average of the right and left limits of f at x. At π

and $-\pi$, the Fourier series for f converges to the average of the left limit of f at π and the right limit of f at $-\pi$.

Before proving the theorem, we shall look at some examples.

EXAMPLE

Let $f(x) = e^x$, $-\pi \leq x \leq \pi$. Here f is monotone nondecreasing, hence certainly satisfies Dirichlet's condition. We compute the Fourier coefficients:

$$a_0 = \frac{1}{\pi}\int_{-\pi}^{\pi} e^x\,dx = \frac{2}{\pi}\sinh(\pi);$$

$$a_n = \frac{1}{\pi}\int_{-\pi}^{\pi} e^x \cos(nx)\,dx = \frac{1}{\pi}\frac{1}{1+n^2}\left[e^x\cos(nx) + ne^x\sin(nx)\right]\Big|_{-\pi}^{\pi}$$

$$= \frac{2(-1)^n \sinh(\pi)}{\pi(1+n^2)}$$

$$b_n = \frac{1}{\pi}\int_{-\pi}^{\pi} e^x \sin(nx)\,dx = \frac{1}{\pi}\frac{1}{1+n^2}\left[e^x\sin(nx) - ne^x\cos(nx)\right]\Big|_{-\pi}^{\pi}$$

$$= \frac{n}{\pi(1+n^2)}\cos(n\pi)(e^{-\pi} - e^{\pi}) = \frac{2n}{\pi(1+n^2)}(-1)^{n+1}\sinh(\pi).$$

Thus the Fourier series for e^x on $[-\pi, \pi]$ is

$$\frac{1}{\pi}\sinh(\pi) + \frac{2}{\pi}\sinh(\pi)\sum_{n=1}^{\infty}\left[\frac{(-1)^n\cos(nx)}{1+n^2} + \frac{n(-1)^{n+1}\sin(nx)}{1+n^2}\right]$$

or, equivalently,

$$\frac{1}{\pi}\sinh(\pi) + \frac{2}{\pi}\sinh(\pi)\sum_{n=1}^{\infty}\left[\frac{(-1)^n}{1+n^2}\left[\cos(nx) - n\sin(nx)\right]\right].$$

Since f is continuous on $[-\pi, \pi]$, then $f(x+0) = f(x-0) = f(x)$ for $-\pi < x < \pi$, and the series converges to e^x for such x. At $-\pi$ and at π, the series converges to

$$\tfrac{1}{2}\left(\lim_{\varepsilon\to 0+} e^{\pi-\varepsilon} + \lim_{\varepsilon\to 0+} e^{-\pi+\varepsilon}\right) = \tfrac{1}{2}(e^{\pi} + e^{-\pi}) = \cosh(\pi),$$

which is the average of the values of e^x at π and at $-\pi$ (see Fig. 10-3). As a side remark, sometimes a result like this can be used to sum a series. Putting $x = \pi$ into the series, we have

$$\cosh(\pi) = \frac{1}{\pi}\sinh(\pi) + \frac{2}{\pi}\sinh(\pi)\sum_{n=1}^{\infty}\frac{(-1)^n\cos(n\pi)}{1+n^2}$$

$$= \frac{1}{\pi}\sinh(\pi) + \frac{2}{\pi}\sinh(\pi)\sum_{n=1}^{\infty}\frac{1}{1+n^2}.$$

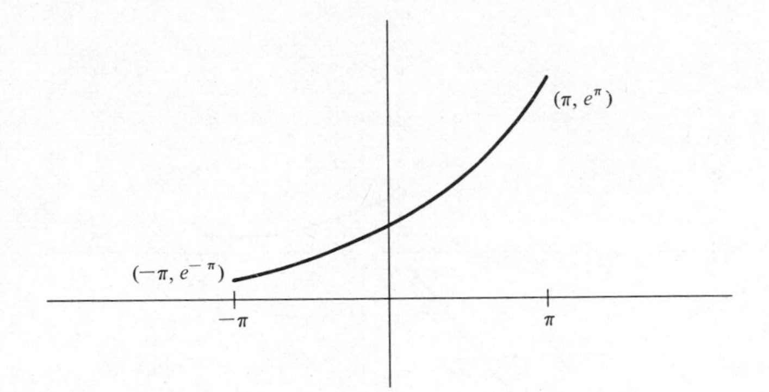

Fig. 10-3

Thus, dividing by $(1/\pi)\sinh(\pi)$ gives us

$$1 + 2\sum_{n=1}^{\infty} \frac{1}{1+n^2} = \pi \coth(\pi).$$

[Compare this with Exercise 9(a), Section 9.10.] This technique for summing series is usually effective only by hindsight or fantastic cleverness. For example, given the problem of evaluating $\sum_{n=1}^{\infty} 1/(1+n^2)$, who would think to start with the Fourier series for e^x? ∎

EXAMPLE

Let

$$f(x) = \begin{cases} x, & -\pi \le x < 0, \\ 3, & x = 0, \\ e^{-x}, & 0 < x \le \pi. \end{cases}$$

Then f satisfies Dirichlet's condition, as f is monotone nondecreasing on $[-\pi, 0)$ and monotone nonincreasing on $[0, \pi]$, as shown in Fig. 10-4.

The Fourier coefficients, after a tedious computation, are given by

$$a_0 = \frac{1}{\pi}\int_{-\pi}^{0} x\,dx + \frac{1}{\pi}\int_{0}^{\pi} e^{-x}\,dx = -\frac{\pi^2}{2}\frac{1}{\pi} + (1 - e^{-\pi})\frac{1}{\pi}$$

$$= \frac{1}{\pi}(1 - e^{-\pi}) - \frac{\pi}{2},$$

$$a_n = \frac{1}{\pi}\int_{-\pi}^{0} x\cos(nx)\,dx + \frac{1}{\pi}\int_{0}^{\pi} e^{-x}\cos(nx)\,dx$$

$$= \frac{1}{\pi n^2}[1 - (-1)^n] + \frac{1}{\pi(1+n^2)}[1 - (-1)^n e^{-\pi}],$$

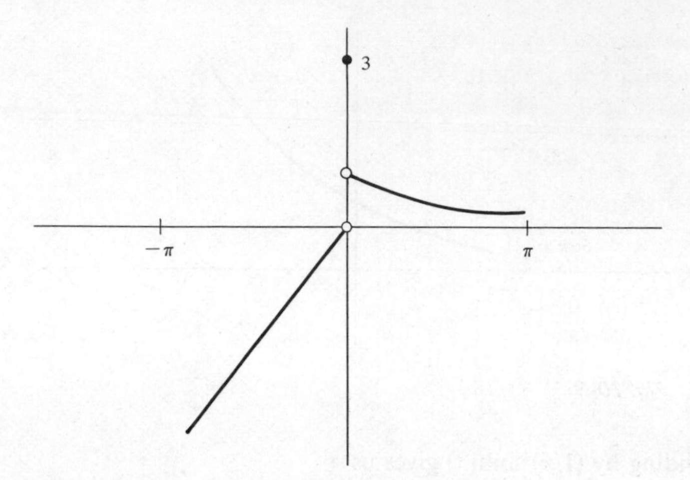

Fig. 10-4

and

$$b_n = \frac{1}{\pi}\int_{-\pi}^{0} x\sin(nx)\,dx + \frac{1}{\pi}\int_{0}^{\pi} e^{-x}\sin(nx)\,dx$$

$$= \frac{(-1)^{n+1}}{n} + \frac{n}{\pi(1+n^2)}[1-(-1)^n e^{-\pi}],$$

for $n = 1, 2, \ldots$.

The Fourier series for f is then

$$-\frac{\pi}{4} + \frac{1}{2\pi}(1-e^{-\pi}) + \sum_{n=1}^{\infty}\left\{\frac{1}{n^2}(1-(-1)^n) + \frac{1}{\pi(1+n^2)}[1-(-1)^n e^{-\pi}]\right\}\cos(nx)$$

$$+ \sum_{n=1}^{\infty}\left\{\frac{(-1)^{n+1}}{n} + \frac{n[1-(-1)^n e^{-\pi}]}{\pi(1+n^2)}\right\}\sin(nx).$$

According to the convergence theorem, we conclude the following: At $-\pi < x < 0$ and $0 < x < \pi$, the series converges to $f(x)$; at π and $-\pi$, the series converges to

$$\tfrac{1}{2}[f(\pi-0)+f(-\pi+0)] = \tfrac{1}{2}[e^{-\pi}+(-\pi)] = \frac{-\pi+e^{-\pi}}{2};$$

at 0, the series converges to $\frac{1}{2}[f(0+)+f(0-)] = \frac{1}{2}(3+0) = \frac{3}{2}$. ▮

We now turn to a proof of the theorem. Some preliminary lemmas are needed, but for the sake of motivation we shall begin by sketching the framework of the proof and then go back to fill in details as their need becomes apparent.

For the time being, let x be any point in the interval $[-\pi, \pi]$. In order to investigate convergence of the Fourier series for f at x, we must investigate the partial sums $S_n(x)$ of the series, where

$$S_n(x) = \frac{a_0}{2} + \sum_{j=1}^{n} [a_j \cos(jx) + b_j \sin(jx)], \qquad n = 1, 2, \ldots.$$

Put in the values of a_j and b_j and write

$$\begin{aligned} S_n(x) &= \frac{1}{2\pi}\int_{-\pi}^{\pi} f(\xi)\, d\xi \\ &\quad + \sum_{j=1}^{n}\left[\frac{1}{\pi}\int_{-\pi}^{\pi} f(\xi)\cos(j\xi)\, d\xi \cos(jx) + \frac{1}{\pi}\int_{-\pi}^{\pi} f(\xi)\sin(j\xi)\, d\xi \sin(jx)\right] \\ &= \frac{1}{2\pi}\int_{-\pi}^{\pi} f(\xi)\left\{1 + 2\sum_{j=1}^{n} [\cos(j\xi)\cos(jx) + \sin(j\xi)\sin(jx)]\right\} d\xi \\ &= \frac{1}{2\pi}\int_{-\pi}^{\pi} f(\xi)\left\{1 + 2\sum_{j=1}^{n} \cos[j(\xi - x)]\right\} d\xi. \end{aligned}$$

At this point we invoke an identity due to Lagrange:

$$1 + 2\sum_{j=1}^{n} \cos(j\varphi) = \frac{\sin[(n + \frac{1}{2})\varphi]}{\sin(\varphi/2)} \qquad [\text{when } \sin(\varphi/2) \neq 0],$$

to be proved later [or perhaps already proved by the reader in response to Exercise 7(a), Section 9.4]. Substituting this yields

$$\begin{aligned} S_n(x) &= \frac{1}{2\pi}\int_{-\pi}^{\pi} f(\xi)\frac{\sin[(n + \frac{1}{2})(\xi - x)]\, d\xi}{\sin[(\xi - x)/2]} \\ &= \frac{1}{2\pi}\int_{-\pi}^{x} f(\xi)\frac{\sin[(n + \frac{1}{2})(\xi - x)]\, d\xi}{\sin[(\xi - x)/2]} + \frac{1}{2\pi}\int_{x}^{\pi} f(\xi)\frac{\sin[(n + \frac{1}{2})(\xi - x)]\, d\xi}{\sin[(\xi - x)/2]}. \end{aligned}$$

We now have to be a little more specific about x. First suppose that $-\pi < x < \pi$. In this case put $\xi - x = -2\varphi$ in the first integral and $\xi - x = 2\varphi$ in the second to obtain

$$\begin{aligned} S_n(x) &= \frac{1}{2\pi}\int_{(x+\pi)/2}^{0} f(x - 2\varphi)\frac{\sin[(n + \frac{1}{2})(-2\varphi)](-2)\, d\varphi}{\sin(-\varphi)} \\ &\quad + \frac{1}{2\pi}\int_{0}^{(\pi - x)/2} f(x + 2\varphi)\frac{\sin[(n + \frac{1}{2})(2\varphi)]2\, d\varphi}{\sin(\varphi)} \\ &= \frac{1}{\pi}\int_{0}^{(x+\pi)/2} f(x - 2\varphi)\frac{\sin[(2n + 1)\varphi]\, d\varphi}{\sin(\varphi)} \\ &\quad + \frac{1}{\pi}\int_{0}^{(\pi - x)/2} f(x + 2\varphi)\frac{\sin[(2n + 1)\varphi]\, d\varphi}{\sin(\varphi)}. \end{aligned}$$

Leave this for the moment, and consider the cases $x = \pi$ or $x = -\pi$, say $x = \pi$ first. Then

$$\int_x^\pi f(\xi)\,\frac{\sin[(n+\frac{1}{2})(\xi - x)]\,d\xi}{\sin[(\xi - x)/2]} = 0,$$

so

$$S_n(\pi) = \frac{1}{2\pi}\int_{-\pi}^{\pi} f(\xi)\,\frac{\sin[(n+\frac{1}{2})(\xi - \pi)]\,d\xi}{\sin[(\xi - \pi)/2]}.$$

Make the change of variables $\xi = \pi - 2\varphi$ to obtain

$$S_n(\pi) = \frac{1}{\pi}\int_0^\pi f(\pi - 2\varphi)\,\frac{\sin[(2n+1)\varphi]\,d\varphi}{\sin(\varphi)}.$$

Choosing τ as any number between 0 and π, we then have

$$\begin{aligned}
S_n(\pi) &= \frac{1}{\pi}\int_0^{\pi-\tau} f(\pi - 2\varphi)\,\frac{\sin[(2n+1)\varphi]\,d\varphi}{\sin(\varphi)} \\
&\quad + \frac{1}{\pi}\int_{\pi-\tau}^{\pi} f(\pi - 2\varphi)\,\frac{\sin[(2n+1)\varphi]\,d\varphi}{\sin(\varphi)} \\
&= \frac{1}{\pi}\int_0^{\pi-\tau} f(\pi - 2\varphi)\,\frac{\sin[(2n+1)\varphi]\,d\varphi}{\sin(\varphi)} \\
&\quad + \frac{1}{\pi}\int_0^{\tau} f(-\pi + 2\varphi)\,\frac{\sin[(2n+1)\varphi]\,d\varphi}{\sin(\varphi)}.
\end{aligned}$$

[Here we put $\zeta = \pi - \varphi$ to transform

$$\begin{aligned}
\int_{\pi-\tau}^{\pi} f(\pi - 2\varphi)\,\frac{\sin[(2n+1)\varphi]\,d\varphi}{\sin(\varphi)} &= \int_\tau^0 f(-\pi + 2\zeta)\,\frac{\sin[(2n+1)(\pi - \zeta)](-d\zeta)}{\sin(\pi - \zeta)} \\
&= \int_0^\tau f(-\pi + 2\zeta)\,\frac{\sin[(2n+1)\zeta]\,d\zeta}{\sin(\zeta))},
\end{aligned}$$

and then rewrote the integration variable as φ again for uniformity of notation.]

Similarly, at $x = -\pi$, we have

$$\int_{-\pi}^{x} f(\xi)\,\frac{\sin[(n+\frac{1}{2})(\xi - x)]\,d\xi}{\sin[(\xi - x)/2]} = 0.$$

Then

$$\begin{aligned}
S_n(-\pi) &= \frac{1}{2\pi}\int_{-\pi}^{\pi} f(\xi)\,\frac{\sin[(n+\frac{1}{2})(\xi + \pi)]\,d\xi}{\sin[(\xi + \pi)/2]} \\
&= \frac{1}{\pi}\int_0^\pi f(-\pi + 2\varphi)\,\frac{\sin[(2n+1)\varphi]\,d\varphi}{\sin(\varphi)}
\end{aligned}$$

with the change $\xi = -\pi + 2\varphi$.

Again, choosing τ as any number between 0 and π, we have

$$\begin{aligned} S_n(-\pi) &= \frac{1}{\pi}\int_0^{\pi-\tau} f(-\pi + 2\varphi)\frac{\sin[(2n+1)\varphi]\,d\varphi}{\sin(\varphi)} \\ &\quad + \frac{1}{\pi}\int_{\pi-\tau}^{\pi} f(-\pi + 2\varphi)\frac{\sin[(2n+1)\varphi]\,d\varphi}{\sin(\varphi)} \\ &= \frac{1}{\pi}\int_0^{\pi-\tau} f(-\pi + 2\varphi)\frac{\sin[(2n+1)\varphi]\,d\varphi}{\sin(\varphi)} \\ &\quad + \frac{1}{\pi}\int_0^{\tau} f(\pi - 2\varphi)\frac{\sin[(2n+1)\varphi]\,d\varphi}{\sin(\varphi)}. \end{aligned}$$

We now have integral expressions for $S_n(x)$ in all the cases: $-\pi < x < \pi$, $x = \pi$, and $x = -\pi$. We are, of course, interested in $\lim_{n\to\infty} S_n(x)$. The key now is to note that in all cases $S_n(x)$ has the same general form:

$$\frac{1}{\pi}\int_0^{\cdots} f(\cdots)\frac{\sin((2n+1)\varphi)\,d\varphi}{\sin(\varphi)} + \frac{1}{\pi}\int_0^{\cdots} f(\cdots)\frac{\sin[(2n+1)\varphi]\,d\varphi}{\sin(\varphi)}.$$

We have left blank the places where things change depending upon case. The point is: If we want to evaluate $\lim_{n\to\infty} S_n(x)$, we must be able to handle a limit of the form

$$\lim_{n\to\infty}\int_0^{a} f(g(\varphi))\frac{\sin[(2n+1)\varphi]\,d\varphi}{\sin(\varphi)}.$$

We shall now carry out a sequence of steps culminating in the determination of this kind of limit.

First, we observe that any function f that satisfies Dirichlet's condition can be written as a difference $f(x) = \varphi(x) - \psi(x)$, with φ and ψ both monotone nondecreasing on $[a, b]$. The details of proof of this are very messy, but an example should suffice to clarify the idea. Suppose that f is given by

$$f(x) = \begin{cases} -x, & -1 \le x \le 0, \\ \frac{1}{2}, & 0 < x < 1, \\ 2 + x, & 1 \le x < 2, \\ \dfrac{1}{e^x}, & 2 \le x \le 3. \end{cases}$$

We want to write $f(x) = \varphi(x) - \psi(x)$, where φ and ψ are bounded and monotone

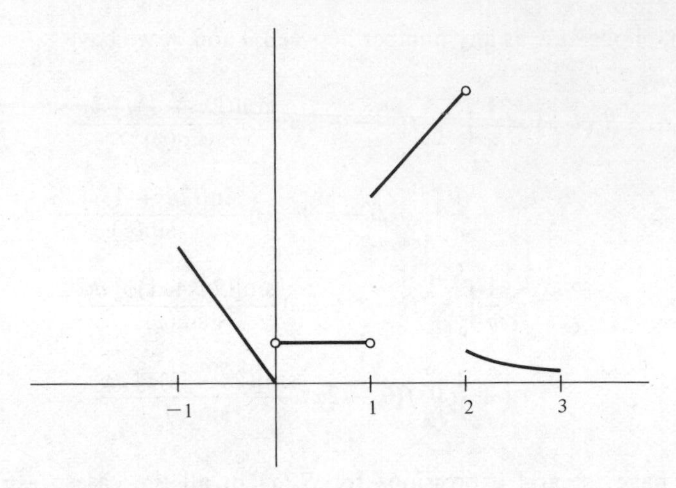

Fig. 10-5

nondecreasing on $[-1, 3]$. From the graph of f (Fig. 10-5), we are led to put

$$\varphi(x) = \begin{cases} 0, & -1 \le x \le 0, \\ \frac{1}{2}, & 0 < x < 1, \\ 2 + x, & 1 \le x < 2, \\ 4, & 2 \le x \le 3 \end{cases}$$

and

$$\psi(x) = \begin{cases} x, & -1 \le x \le 0, \\ 0, & 0 < x < 2, \\ 4 - e^{-x}, & 2 \le x \le 3. \end{cases}$$

The rest of the steps we need to determine the kind of limit that we are interested in will be given as a sequence of lemmas.

Lemma 1

Let $0 < a < b$, and suppose that f satisfies Dirichlet's condition on $[a, b]$. Then

$$\lim_{n \to \infty} \int_a^b f(\xi) \frac{\sin(n\xi)\, d\xi}{\xi} = 0.$$

PROOF

Suppose first that f is monotone on $[a, b]$. By a mean value theorem for integrals (see Appendix II of this chapter), there is some α, $a \le \alpha \le b$, such that

$$\int_a^b f(\xi) \frac{\sin(n\xi)\, d\xi}{\xi} = f(a + 0) \int_a^\alpha \frac{\sin(n\xi)\, d\xi}{\xi} + f(b - 0) \int_\alpha^b \frac{\sin(n\xi)\, d\xi}{\xi}.$$

Put $\zeta = n\xi$. Then

$$\int_a^\alpha \frac{\sin(n\xi)\,d\xi}{\xi} = \int_{na}^{n\alpha} \frac{\sin(\zeta)\,d\zeta}{(\zeta/n)\,n} = \int_{na}^{n\alpha} \frac{\sin(\zeta)\,d\zeta}{\zeta}.$$

As $n \to \infty$, for $\alpha \neq a$, we have

$$\left| \int_{na}^{n\alpha} \frac{\sin(\zeta)\,d\zeta}{\zeta} \right| \le \frac{1}{n(\alpha - a)}\frac{1}{na} \to 0.$$

Then

$$\int_a^\alpha \frac{\sin(n\xi)\,d\xi}{\xi} \to 0 \qquad \text{as } n \to \infty.$$

Similarly,

$$\int_\alpha^b \frac{\sin(n\xi)\,d\xi}{\xi} \to 0 \qquad \text{as } n \to \infty.$$

Now drop the restriction that f is monotone. Since f satisfies Dirichlet's condition, then $f(x) = \varphi(x) - \psi(x)$ for some monotone φ and ψ on $[a, b]$. Then, by what we have proved so far,

$$\int_a^b \varphi(\xi)\frac{\sin(n\xi)\,d\xi}{\xi} \to 0 \qquad \text{and} \qquad \int_a^b \psi(\xi)\frac{\sin(n\xi)\,d\xi}{\xi} \to 0,$$

as $n \to \infty$, hence

$$\int_a^b f(\xi)\frac{\sin(n\xi)\,d\xi}{\xi} \to 0 \qquad \text{as } n \to \infty. \qquad \blacksquare$$

Lemma 2

$$\int_0^K \frac{\sin(\xi)\,d\xi}{\xi} \le \int_0^\pi \frac{\sin(\xi)\,d\xi}{\xi} \qquad \textit{for any } K > 0.$$

PROOF

If $0 < K < \pi$, the conclusion is immediate, as $\sin(\xi)/\xi > 0$ for $0 < x < \pi$. If $\pi < K < 2\pi$, write

$$\int_0^K \frac{\sin(\xi)\,d\xi}{\xi} = \int_0^\pi \frac{\sin(\xi)\,d\xi}{\xi} + \int_\pi^K \frac{\sin(\xi)\,d\xi}{\xi}.$$

Now $\sin(\xi)/\xi < 0$ for $\pi < \xi < K$, so $\int_\pi^K \sin(\xi)\,d\xi/\xi < 0$, hence

$$\int_0^K \frac{\sin(\xi)\,d\xi}{\xi} < \int_0^\pi \frac{\sin(\xi)\,d\xi}{\xi}.$$

Now suppose that $K > 2\pi$. Then, for some integer $m \geq 2$,

$$m\pi \leq K \leq (m+1)\pi.$$

Put

$$\sigma_j = \int_{j\pi}^{(j+1)\pi} \frac{\sin(\xi)\,d\xi}{\xi} \qquad \text{for } j = 0, 1, 2, \ldots$$

Then $\sigma_{2n} > 0$ and $\sigma_{2n+1} < 0$ for $n = 0, 1, 2, \ldots$ Further,

$$|\sigma_j| = \left| \int_{j\pi}^{(j+1)\pi} \frac{\sin(\xi)\,d\xi}{\xi} \right| \leq \int_{j\pi}^{(j+1)\pi} \left| \frac{\sin(\xi)}{\xi} \right| d\xi$$
$$\leq \pi\left(\frac{1}{j\pi}\right) = \frac{1}{j},$$

so $|\sigma_j|$ decreases monotonically to zero as $j \to \infty$. Now write

$$\int_0^K \frac{\sin(\xi)\,d\xi}{\xi} = \sigma_0 + \sigma_1 + \sigma_2 + \cdots + \sigma_{m-1} + \int_{m\pi}^K \frac{\sin(\xi)\,d\xi}{\xi}.$$

Consider two cases:

(1) m is even. Then $\sin(\xi) \geq 0$ for $m\pi \leq \xi \leq K$, so

$$\int_{m\pi}^K \frac{\sin(\xi)\,d\xi}{\xi} > 0.$$

Further, $\sigma_{m-1} < 0$, so

$$\int_0^K \frac{\sin(\xi)\,d\xi}{\xi} \leq \sigma_0 + \sigma_1 + \cdots + \sigma_{m-1} < \sigma_0 = \int_0^\pi \frac{\sin(\xi)\,d\xi}{\xi}.$$

(2) m is odd. Then $\sin(\xi) \leq 0$ for $m\pi \leq \xi \leq K$, so

$$\int_{m\pi}^K \frac{\sin(\xi)\,d\xi}{\xi} \geq \int_{m\pi}^{(m+1)\pi} \frac{\sin(\xi)\,d\xi}{\xi} = \sigma_m,$$

and $\sigma_m < 0$. Then

$$\int_0^K \frac{\sin(\xi)\,d\xi}{\xi} \leq \sigma_0 + \cdots + \sigma_{m-1} + \sigma_m < \sigma_0,$$

and Lemma 2 is proved. ∎

Lemma 3

Let $a > 0$ and suppose that f satisfies Dirichlet's condition on $[0, a]$. *Then*

$$\lim_{n\to\infty}\int_0^a f(\xi)\,\frac{\sin(n\xi)\,d\xi}{\xi} = \frac{\pi}{2}f(0+).$$

Suppose first that f is monotone on $[0, a]$. Put

$$\int_0^a f(\xi)\,\frac{\sin(n\xi)\,d\xi}{\xi} = f(0+)\int_0^a \frac{\sin(n\xi)\,d\xi}{\xi} + \int_0^a (f(\xi) - f(0+))\,\frac{\sin(n\xi)\,d\xi}{\xi}.$$

Put $\varphi = n\xi$ to obtain

$$\lim_{n\to\infty} f(0+)\int_0^a \frac{\sin(n\xi)\,d\xi}{\xi} = f(0+)\int_0^\infty \frac{\sin(\varphi)\,d\varphi}{\varphi} = \frac{\pi}{2}f(0+),$$

from the last example of Section 9.7, or from Section 7.3.

We must now show that

$$\int_0^a (f(\xi) - f(0+))\,\frac{\sin(n\xi)\,d\xi}{\xi} \to 0 \text{ as } n \to \infty.$$

Let $\varepsilon > 0$. We shall produce N such that

$$\left|\int_0^a (f(\xi) - f(0+))\,\frac{\sin(n\xi)\,d\xi}{\xi}\right| < \varepsilon \qquad \text{if } n \geq N.$$

To do this, let

$$M = \int_0^\pi \frac{\sin(n\xi)\,d\xi}{\xi}$$

and note that $M > 0$. Since $\lim_{\varepsilon\to 0+} f(\xi) = f(0+)$, then, for some δ, $0 < \delta < a$ and $|f(\xi) - f(0+)| < \varepsilon/4M$ if $0 < \xi < \delta$. Now write

$$\int_0^a [f(\xi) - f(0+)]\,\frac{\sin(n\xi)\,d\xi}{\xi} = \int_0^\delta [f(\xi) - f(0+)]\,\frac{\sin(n\xi)\,d\xi}{\xi}$$
$$+ \int_\delta^a [f(\xi) - f(0+)]\,\frac{\sin(n\xi)\,d\xi}{\xi}.$$

Consider the integrals on the right one at a time. By a mean value theorem for integrals (again), there is some ζ, $0 < \zeta < \delta$, such that

$$\int_0^\delta [f(\xi) - f(0+)] \frac{\sin(n\xi)\, d\xi}{\xi} = [f(0+) - f(0+)] \int_0^\zeta \frac{\sin(n\xi)\, d\xi}{\xi}$$

$$+ [f(\delta - 0) - f(0+)] \int_\zeta^\delta \frac{\sin(n\xi)\, d\xi}{\xi}$$

$$= [f(\delta - 0) - f(0+)] \int_\zeta^\delta \frac{\sin(n\xi)\, d\xi}{\xi}.$$

Now put $n\xi = \tau$. Then, using Lemma 2 at the last step, we obtain

$$\left| \int_\zeta^\delta \frac{\sin(n\xi)\, d\xi}{\xi} \right| = \left| \int_{n\zeta}^{n\delta} \frac{\sin(\tau)}{\tau/n} \frac{d\tau}{n} \right| = \left| \int_{n\zeta}^{n\delta} \frac{\sin(\tau)\, d\tau}{\tau} \right|$$

$$= \left| \int_0^{n\delta} \frac{\sin(\tau)\, d\tau}{\tau} - \int_0^{n\zeta} \frac{\sin(\tau)\, d\tau}{\tau} \right| \le 2 \int_0^\pi \frac{\sin(\tau)\, d\tau}{\tau} = 2M.$$

Hence

$$\left| \int_0^\delta [f(\xi) - f(0+)] \frac{\sin(n\xi)\, d\xi}{\xi} \right|$$

$$= |f(\delta - 0) - f(0+)| \left| \int_\zeta^\delta \frac{\sin(n\xi)\, d\xi}{\xi} \right| \le \frac{\varepsilon}{2M} \left| \int_\zeta^\delta \frac{\sin(n\xi)\, d\xi}{\xi} \right|$$

$$\le \frac{\varepsilon}{4M}(2M) = \frac{\varepsilon}{2}.$$

There remains to consider

$$\int_\delta^a [f(\xi) - f(0+)] \frac{\sin(n\xi)\, d\xi}{\xi}.$$

But, by Lemma 1,

$$\lim_{n \to \infty} \int_\delta^a [f(\xi) - f(0+)] \frac{\sin(n\xi)\, d\xi}{\xi} = 0.$$

Hence, for some N,

$$\left| \int_\delta^a [f(\xi) - f(0+)] \frac{\sin(n\xi)\, d\xi}{\xi} \right| < \frac{\varepsilon}{2} \qquad \text{for } n \ge N.$$

Then, for $n \geq N$,

$$\left| \int_0^a [f(\xi) - f(0+)] \frac{\sin(n\xi)\, d\xi}{\xi} \right| \leq \left| \int_0^\delta [f(\xi) - f(0+)] \frac{\sin(n\xi)\, d\xi}{\xi} \right| + \left| \int_\delta^a [f(\xi) - f(0+)] \frac{\sin(n\xi)\, d\xi}{\xi} \right| < \frac{\varepsilon}{2} + \frac{\varepsilon}{2} = \varepsilon.$$

This proves Lemma 3 when f is monotone.

In general, when f satisfies Dirichlet's condition on $[0, a]$, write $f(x) = \varphi(x) - \psi(x)$, where φ and ψ are monotone nondecreasing on $[0, a]$, and apply the above result to φ and ψ separately. ∎

Lemma 4

Suppose that f satisfies Dirichlet's condition on $[0, a]$ *and on* $[a, b]$, *where* $0 < a < b < \pi$. Then

$$\lim_{n\to\infty} \int_0^a f(\xi) \frac{\sin(n\xi)\, d\xi}{\sin(\xi)} = \frac{\pi}{2} f(0+),$$

and

$$\lim_{n\to\infty} \int_a^b f(\xi) \frac{\sin(n\xi)\, d\xi}{\sin(\xi)} = 0.$$

PROOF

Write

$$f(\xi) \frac{\sin(n\xi)}{\sin(\xi)} = \frac{f(\xi)\xi}{\sin(\xi)} \frac{\sin(n\xi)}{\xi}.$$

By Lemma 3,

$$\lim_{n\to\infty} \int_0^a \frac{f(\xi)\, \xi}{\sin(\xi)} \frac{\sin(n\xi)\, d\xi}{\xi} = \frac{\pi}{2} \lim_{\varepsilon\to 0+} \frac{f(\varepsilon)\varepsilon}{\sin(\varepsilon)} = \frac{\pi}{2} f(0+), \qquad \text{as } \frac{\varepsilon}{\sin(\varepsilon)} \to 1 \qquad \text{as } \varepsilon \to 0.$$

By Lemma 1,

$$\lim_{n\to\infty} \int_a^b \frac{f(\xi)\xi}{\sin(\xi)} \frac{\sin(n\xi)\, d\xi}{\xi} = 0,$$

completing the proof of Lemma 4. ∎

We can finally complete the limit part of the proof of the convergence theorem. Consider three cases:

(1) $-\pi < x < \pi$. Then we have

$$S_n(x) = \frac{1}{\pi}\int_0^{(x+\pi)/2} f(x-2\varphi)\frac{\sin((2n+1)\varphi)\,d\varphi}{\sin(\varphi)} + \frac{1}{\pi}\int_0^{(\pi-x)/2} f(x+2\varphi)\frac{\sin((2n+1)\varphi)\,d\varphi}{\sin(\varphi)}.$$

Now, for $0 \leq \varphi \leq (x+\pi)/2$, we have $-\pi \leq x - 2\varphi \leq x$. Then $f(x-2\varphi)$ satisfies Dirichlet's condition on $0 \leq \varphi \leq (x+\pi)/2$. Further, the right limit of $f(x-2\varphi)$ at $\varphi = 0$ is the left limit of f at x, since as $\varphi \to 0+$, $x - 2\varphi \to x - 0$. Note that $2n+1 \to \infty$ as $n \to \infty$; then we have by Lemma 4,

$$\lim_{n\to\infty} \frac{1}{\pi}\int_0^{(x+\pi)/2} f(x-2\varphi)\frac{\sin[2(n+1)\varphi]\,d\varphi}{\sin(\varphi)} = \frac{1}{\pi}\frac{\pi}{2}f(x-0) = \frac{1}{2}f(x-0).$$

Similarly, as $0 \leq \varphi \leq (\pi - x)/2$, then $x \leq x + 2\varphi \leq \pi$, and $f(x+2\varphi)$ satisfies Dirichlet's condition on $0 \leq \varphi \leq (\pi - x)/2$. Further, as $\varphi \to 0+$, $x + 2\varphi \to x + 0$, so the right limit of $f(x+2\varphi)$ at $\varphi = 0$ is $f(x+0)$. By Lemma 4,

$$\lim_{n\to\infty} \frac{1}{\pi}\int_0^{(\pi-x)/2} f(x+2\varphi)\frac{\sin[(2n+1)\varphi]\,d\varphi}{\sin(\varphi)} = \frac{1}{\pi}\frac{\pi}{2}f(x+0) = \frac{1}{2}f(x+0).$$

Then

$$S_n(x) \to \frac{1}{2}(f(x+0) + f(x-0)) \qquad \text{as } n \to \infty.$$

(2) $x = \pi$. Then we have

$$S_n(\pi) = \frac{1}{\pi}\int_0^{\pi-\tau} f(\pi-2\varphi)\frac{\sin[(2n+1)\varphi]\,d\varphi}{\sin(\varphi)} + \frac{1}{\pi}\int_0^{\tau} f(-\pi+2\varphi)\frac{\sin[2(n+1)\varphi]\,d\varphi}{\sin(\varphi)}.$$

Arguing as above, and using Lemma 4, we note that $f(\pi - 2\varphi)$ has the right limit $f(\pi - 0)$ as $\varphi \to 0+$, $f(-\pi + 2\varphi)$ has the right limit $f(-\pi + 0)$ as $\varphi \to 0+$, and so, as $n \to \infty$,

$$S_n(\pi) \to \frac{1}{\pi}\frac{\pi}{2}f(\pi-0) + \frac{1}{\pi}\frac{\pi}{2}f(-\pi+0) = \frac{1}{2}[f(\pi-0) + f(-\pi+0)].$$

(3) $x = -\pi$. Then we have

$$S_n(-\pi) = \frac{1}{\pi}\int_0^{\pi-\tau} f(-\pi + 2\varphi)\frac{\sin[(2n+1)\varphi]\,d\varphi}{\sin(\varphi)}$$
$$+ \frac{1}{\pi}\int_0^{\tau} f(\pi - 2\varphi)\frac{\sin[(2n+1)\varphi]\,d\varphi}{\sin(\varphi)}$$
$$\to \frac{1}{2}[f(-\pi+0) + f(\pi-0)] \qquad \text{as } n \to \infty.$$

There is one loose end to tie up: Lagrange's identity.

Lemma 5 (Lagrange's identity)

$$1 + 2\sum_{j=1}^{n}\cos(j\varphi) = \frac{\sin[(n+1/2)\varphi]}{\sin(\varphi/2)} \qquad \textit{for } \sin\frac{\varphi}{2} \neq 0.$$

PROOF

We use the fact that

$$\sum_{j=1}^{n} a^j = \frac{a - a^{n+1}}{1-a} \qquad \text{if } a \neq 1.$$

Then

$$2\sum_{j=1}^{n}\cos(j\varphi) = 2\sum_{j=1}^{n}\frac{e^{ij\varphi} + e^{-ij\varphi}}{2} = \sum_{j=1}^{n}(e^{i\varphi})^j + \sum_{j=1}^{n}(e^{-i\varphi})^j$$
$$= \frac{e^{i\varphi} - e^{i(n+1)\varphi}}{1 - e^{i\varphi}} + \frac{e^{-i\varphi} - e^{-i(n+1)\varphi}}{1 - e^{-i\varphi}}$$
$$= \frac{e^{i\varphi/2} - e^{i(n+1/2)\varphi}}{e^{-i\varphi/2} - e^{i\varphi/2}} - \frac{e^{-i\varphi/2} - e^{-i(n+1/2)\varphi}}{e^{-i\varphi/2} - e^{i\varphi/2}}$$
$$= \frac{e^{-i\varphi/2} - e^{i\varphi/2} + e^{i(n+1/2)\varphi} - e^{-i(n+1/2)\varphi}}{e^{i\varphi/2} - e^{-i\varphi/2}}$$
$$= -1 + \frac{\sin[(n+1/2)\varphi]}{\sin(\varphi/2)},$$

provided that $\sin(\varphi/2) \neq 0$. ∎

This takes care of all the details, and the convergence theorem is established.

For the record, we shall state the theorem for Fourier series on an interval $[-L, L]$.

Theorem

Suppose that f satisfies Dirichlet's condition on $[-L, L]$.

(1) *If* $-L < x < L$, *then the Fourier series for f on* $[-L, L]$ *converges at x to* $\frac{1}{2}(f(x+0) + f(x-0))$.

(2) *At L and at $-L$, the Fourier series for f on $[-L, L]$ converges to*

$$\tfrac{1}{2}[f(L - 0) + f(-L + 0)].$$

PROOF

Apply the previous theorem and the change of variables $y = \pi x/L$. ▮

The convergence theorem that we have just proved is the culmination of a long effort by a number of mathematicians, among them Dirichlet, Fejer, and Riemann. Theorems such as these put Fourier expansions on a firm theoretical footing for the first time, and provided scientists using the expansions in their work fairly simple ways of determining the exact relationship between a function and its Fourier expansion.

It is obvious that uniform convergence is in general too much to ask for in a Fourier series. Each function in the series is continuous, being a constant, a sine or a cosine, while the function being expanded need not be continuous everywhere. In the exercises we shall pursue the problem of determining conditions on f to ensure that the Fourier series for f converges uniformly to f on the given interval. We shall also indicate in the exercises some interesting properties of the Fourier coefficients.

Exercises for Section 10.2

1. For each of the following, write the Fourier series for f on the given interval, and determine the sum of the Fourier series at each point in the interval.

(a) $f(x) = x^2$, on $[-1, 1]$

(b) $f(x) = |x|$, on $[-2, 2]$

(c) $f(x) = \cos^2(x) - \sin^2(x)$, on $[-\pi, \pi]$ *Hint:* $\cos^2(x) - \sin^2(x) = \cos(2x)$.

(d) $f(x) = e^{3x}$, on $[-1, 1]$

(e) $f(x) = \cosh(x)$, on $[-\pi, \pi]$

(f) $f(x) = \ln(2 + x)$ on $[-1, 1]$

(g) $f(x) = \begin{cases} |x|, & -1 \le x \le \frac{1}{2} \\ 3, & x = \frac{1}{2} \\ 2 + x, & \frac{1}{2} < x \le 1 \end{cases}$

(h) $f(x) = \begin{cases} e^x, & -\pi \le x < 0 \\ 1, & x = 0 \\ 2x, & 0 < x \le \pi \end{cases}$

(i) $f(x) = \begin{cases} 3x^2, & -2 \le x < 1 \\ 2, & x = 1 \\ 3, & 1 < x \le \frac{3}{2} \\ 2x, & \frac{3}{2} < x \le 2 \end{cases}$

2. In this problem we indicate some properties of Fourier coefficients.

(a) Best mean approximation: Let

$$\varphi_0(x) = \frac{1}{L}, \qquad \varphi_{2j}(x) = \frac{1}{L}\cos\left(\frac{j\pi x}{L}\right), \qquad \varphi_{2j-1}(x) = \frac{1}{L}\sin\left(\frac{j\pi x}{L}\right),$$

$$j = 1, 2, \ldots, m, \; -L \le x \le L.$$

Let f be continuous on $[-L, L]$, except possibly at finitely many points, where f is still bounded. Determine constants $\alpha_0, \alpha_1, \ldots, \alpha_{2m}$ to minimize

$$\int_{-L}^{L} \left[f(x) - \sum_{j=0}^{2m} \alpha_j\varphi_j(x)\right]^2 dx.$$

Hint: Compute

$$\left[f(x) - \sum_{j=0}^{2m} \alpha_j\varphi_j(x)\right]^2 = f(x)^2 + \sum_{j=0}^{2m} \alpha_j^2\varphi_j(x)^2 - 2\sum_{j=0}^{2m} \alpha_j f(x)\varphi_j(x),$$

then show that

$$\int_{-L}^{L}\left[f(x) - \sum_{j=0}^{2m} \alpha_j\varphi_j(x)\right]^2 dx = \int_{-L}^{L} f(x)^2\,dx + \sum_{j=0}^{2m} \alpha_j^2 - 2\sum_{j=0}^{2m} \alpha_j c_j,$$

where the c_j's are the Fourier coefficients: $c_{2j} = a_j$ for $j = 0, 1, \ldots, m$, and $c_{2j-1} = b_j$ for $j = 1, \ldots, m$. Show that the last equation can be written

$$\int_{-L}^{L}\left[f(x) - \sum_{j=0}^{2m} \alpha_j\varphi_j(x)\right]^2 dx + \int_{-L}^{L} f(x)^2\,dx - \sum_{j=0}^{2m} c_j^2 + \sum_{j=0}^{2m} (\alpha_j - c_j)^2.$$

Finally, show that we must choose $\alpha_j = c_j$ for $j = 0, \ldots, 2m$ in order to minimize $\int_{-L}^{L}\left[f(x) - \sum_{j=0}^{2m} \alpha_j\varphi_j(x)\right]^2 dx$. [With this choice of the α_j's, we call $\sum_{j=0}^{2m} \alpha_j\varphi_j(x)$ the best mean-square approximation to $f(x)$ with respect to the functions $\varphi_0, \varphi_1, \ldots$.]

(b) Prove Bessel's inequality:

$$\sum_{j=0}^{2m} c_j^2 \le \frac{1}{L}\int_{-L}^{L} f(x)^2\,dx,$$

where the c_j's are the Fourier coefficients, as in (a).

(c) Using (b), show that $\sum_{j=0}^{\infty} c_j^2$ converges.

(d) Using (c), show that $\lim_{n\to\infty} c_n = 0$.

3. Let f and f' be continuous on $[-\pi, \pi]$ and suppose that $f(\pi) = f(-\pi)$. Suppose that f and f' satisfy Dirichlet's condition on $[-\pi, \pi]$. Show that the Fourier series for f converges absolutely and uniformly to f on $[-\pi, \pi]$. *Hint:* Consider the Fourier series for $f'(x)$ on $[-\pi, \pi]$:

$$\frac{\alpha_0}{2} + \sum_{n=1}^{\infty} [\alpha_n\cos(nx) + \beta_n\sin(nx)],$$

where

$$\alpha_n = \frac{1}{\pi} \int_{-\pi}^{\pi} f'(\xi) \cos(n\xi)\, d\xi$$

and

$$\beta_n = \frac{1}{\pi} \int_{-\pi}^{\pi} f'(\xi) \sin(n\xi)\, d\xi.$$

Show that $\alpha_0 = 0$ and that $\alpha_n = nb_n$ and $\beta_n = -na_n$ for $n = 1, 2, \ldots$. Hence show that

$$\sum_{n=1}^{K} \sqrt{a_n^2 + b_n^2} = \sum_{n=1}^{K} \frac{1}{n} \sqrt{\alpha_n^2 + \beta_n^2}, \qquad K = 1, 2, 3, \ldots.$$

Now use Cauchy's inequality:

$$\left(\sum_{j=1}^{K} A_j B_j\right)^2 \le \left(\sum_{j=1}^{K} A_j^2\right)\left(\sum_{j=1}^{K} B_j^2\right)$$

(see Exercise 3, Section 9.1) to show that

$$\sum_{n=1}^{K} \sqrt{a_n^2 + b_n^2} \le \sqrt{\left(\sum_{n=1}^{K} \frac{1}{n^2}\right)\left[\sum_{n=1}^{K} (\alpha_n^2 + \beta_n^2)\right]}.$$

Now use Bessel's inequality [Exercise 2(b)] to show that

$$\lim_{K \to \infty} \sqrt{\left(\sum_{n=1}^{K} \frac{1}{n^2}\right)\left[\sum_{n=1}^{K} (\alpha_n^2 + \beta_n^2)\right]}$$

exists finite. Hence conclude that $\sum_{n=1}^{\infty} \sqrt{a_n^2 + b_n^2}$ converges. From this, show that $\sum_{n=1}^{\infty} (|a_n| + |b_n|)$ converges. Now use the Weierstrass M-test and the convergence theorem of this section to show that $\sum_{n=1}^{\infty} [a_n \cos(nx) + b_n \sin(nx)]$ converges uniformly on $[-\pi, \pi]$ to $f(x) - a_0/2$. Finally, conclude that

$$\frac{a_0}{2} + \sum_{n=1}^{\infty} [a_n \cos(nx) + b_n \sin(nx)]$$

converges uniformly and absolutely to $f(x)$ on $[-\pi, \pi]$.

4. Let f be continuous on $[-\pi, \pi]$, and $f(\pi) = f(-\pi)$. Let f' be continuous on $[-\pi, \pi]$, and suppose that f, f', and f'' satisfy Dirichlet's condition on $[-\pi, \pi]$. Then show that

$$f'(x) = \sum_{n=1}^{\infty} [-na_n \sin(nx) + nb_n \cos(nx)], \qquad -\pi < x < \pi$$

[i.e., show that the Fourier series for $f'(x)$ is the series obtained by term-by-term differentiation of the Fourier series for f].

5. Let f be continuous on $[-\pi, \pi]$ and $f(\pi) = f(-\pi)$. Suppose also that f satisfies Dirichlet's condition on $[-\pi, \pi]$. Show that

$$\int_{-\pi}^{x} f(\xi)\, d\xi + \frac{a_0(x+\pi)}{2} + \sum_{n=1}^{\infty} \left\{\frac{a_n \sin(nx)}{n} - \frac{b_n}{n}[\cos(nx) - (-1)^n]\right\}.$$

(Note that the result on the right is what we obtain by integrating the Fourier series for f on $[-\pi, \pi]$ term by term. Further, the equality holds whether or not the Fourier series for f converges to $f(x)$ on $[-\pi, \pi]$). *Hint:* Let

$$g(x) = \int_{-\pi}^{x} f(\xi)\, d\xi - \frac{a_0 x}{2}.$$

Show that the Fourier series for $g(x)$,

$$\frac{\alpha_0}{2} + \sum_{n=1}^{\infty} [\alpha_n \cos(nx) + \beta_n \sin(nx)],$$

converges to $g(x)$ on $[-\pi, \pi]$, and that

$$\alpha_n = \frac{-b_n}{n} \quad \text{and} \quad \beta_n = \frac{a_n}{n}, \qquad n = 1, 2, \ldots.$$

Further, show that

$$\frac{\alpha_0}{2} = \frac{a_0\pi}{2} + \sum_{n=1}^{\infty} \frac{b_n(-1)^n}{n}.$$

Hence conclude the desired result.

6. Expand x^2 in a Fourier series on the interval $[-\pi, \pi]$ to sum the series

$$\sum_{n=1}^{\infty} \frac{(-1)^{n+1}}{n^2} \quad \text{and} \quad \sum_{n=1}^{\infty} \frac{1}{n^2}.$$

10.3 Fourier Sine and Cosine Series

We now have conditions on f which enable us in many cases to determine the sum of the Fourier series for f on $[-L, L]$. In many applications, however, we want a pure sine or cosine series on $[0, L]$, as with the heat-flow problem of Section 10.1. We now show how to obtain such expansions.

First, suppose that we want to expand $f(x)$ in a cosine series on $[0, L]$; $f(x)$ need not even be defined for $-L \le x < 0$. Define

$$g(x) = \begin{cases} f(x), & -L \le x \le 0, \\ f(-x), & 0 < x \le L \end{cases}$$

(see Fig. 10-6). Then g is defined on $[-L, L]$, and $g(-x) = g(x)$, so g is an even extension of f to $[-L, L]$. Whatever f looks like on $[0, L]$, g looks like its mirror

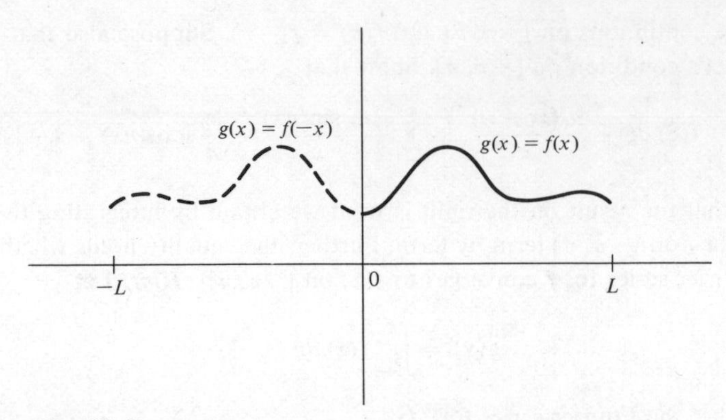

Fig. 10-6

image across the y-axis in $[-L, 0]$. Now expand g in a Fourier series on $[-L, L]$. The series is

$$\frac{a_0}{2} + \sum_{n=1}^{\infty} \left[a_n \cos\left(\frac{n\pi x}{L}\right) + b_n \sin\left(\frac{n\pi x}{L}\right) \right],$$

where

$$a_n = \frac{1}{L} \int_{-L}^{L} g(\xi) \cos\left(\frac{n\pi\xi}{L}\right) d\xi, \qquad n = 0, 1, 2, \ldots$$

and

$$b_n = \frac{1}{L} \int_{-L}^{L} g(\xi) \sin\left(\frac{n\pi\xi}{L}\right) d\xi, \qquad n = 1, 2, \ldots.$$

Now

$$a_n = \frac{1}{L} \int_{-L}^{0} g(\xi) \cos\left(\frac{n\pi\xi}{L}\right) d\xi + \frac{1}{L} \int_{0}^{L} g(\xi) \cos\left(\frac{n\pi\xi}{L}\right) d\xi$$
$$= \frac{1}{L} \int_{-L}^{0} f(-\xi) \cos\left(\frac{n\pi\xi}{L}\right) d\xi + \frac{1}{L} \int_{0}^{L} f(\xi) \cos\left(\frac{n\pi\xi}{L}\right) d\xi.$$

Now,

$$\int_{-L}^{0} f(-\xi) \cos\left(\frac{n\pi\xi}{L}\right) d\xi = \int_{0}^{L} f(\tau) \cos\left(\frac{n\pi\tau}{L}\right) d\tau.$$

So

$$a_n = \frac{2}{L} \int_{0}^{L} f(\xi) \cos\left(\frac{n\pi\xi}{L}\right) d\xi.$$

Next,

$$\begin{aligned} b_n &= \frac{1}{L}\int_{-L}^{0} g(\xi)\sin\left(\frac{n\pi\xi}{L}\right) d\xi + \frac{1}{L}\int_{0}^{L} g(\xi)\sin\left(\frac{n\pi\xi}{L}\right) d\xi \\ &= \frac{1}{L}\int_{-L}^{0} f(-\xi)\sin\left(\frac{n\pi\xi}{L}\right) d\xi + \frac{1}{L}\int_{0}^{L} f(\xi)\sin\left(\frac{n\pi\xi}{L}\right) d\xi \\ &= -\frac{1}{L}\int_{0}^{L} f(\tau)\sin\left(\frac{n\pi\tau}{L}\right) d\tau + \frac{1}{L}\int_{0}^{L} f(\xi)\sin\left(\frac{n\pi\xi}{L}\right) d\xi = 0. \end{aligned}$$

(Here the change of variable $\xi = -\tau$ was used in the first integral on the next-to-last line.) It therefore turns out that the Fourier series for $g(x)$ has only cosines, and we obtain a Fourier cosine series for g on $[-L, L]$. But $g(x) = f(x)$ on $[0, L]$, so we have a Fourier cosine series for f on $[0, L]$:

$$\frac{a_0}{2} + \sum_{n=1}^{\infty} a_n \cos\left(\frac{n\pi x}{L}\right),$$

where

$$a_n = \frac{2}{L}\int_{0}^{L} f(\xi)\cos\left(\frac{n\pi\xi}{L}\right) d\xi, \qquad n = 0, 1, 2, \ldots.$$

One must then use convergence theorems to determine the value of the cosine series at any x, $0 \le x \le L$. As an example, let $f(x) = e^x$, $0 \le x \le 3$. The Fourier coefficients for a cosine expansion are

$$a_0 = \tfrac{2}{3}\int_{0}^{3} e^{\xi}\, d\xi = \tfrac{2}{3}(e^3 - 1)$$

and, for $n = 1, 2, \ldots,$

$$\begin{aligned} a_n &= \tfrac{2}{3}\int_{0}^{3} e^{\xi}\cos\left(\frac{n\pi\xi}{L}\right) d\xi = \tfrac{2}{3}e^{\xi}\left.\frac{\cos(n\pi\xi/3) + \dfrac{n\pi}{3}\sin(n\pi\xi/3)}{1 + n^2\pi^2/9}\right|_{0}^{3} \\ &= \frac{6}{9 + n^2\pi^2}\,[e^3(-1)^n - 1]. \end{aligned}$$

The Fourier cosine series for e^x on $[0, 3]$ is then

$$\tfrac{1}{3}(e^3 - 1) + \sum_{n=1}^{\infty} \frac{6}{9 + n^2\pi^2}\,[(-1)^n e^3 - 1]\cos\left(\frac{n\pi x}{3}\right).$$

To see what this series converges to for $0 \le x \le 3$, note that we have really expanded $g(x)$ on $[-3, 3]$ in a Fourier series, with

$$g(x) = \begin{cases} e^x, & 0 \le x \le 3, \\ e^{-x}, & -3 \le x < 0 \end{cases}$$

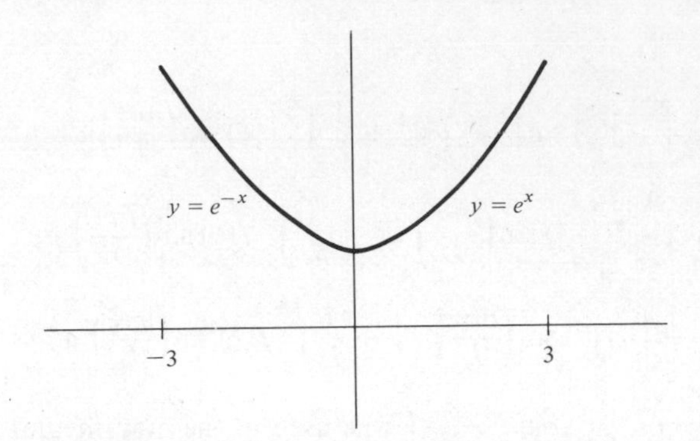

Fig. 10-7

(see Fig. 10-7). At $0 < x < 3$, the Fourier series for g converges to $\frac{1}{2}[g(x+0) + g(x-0)]$, which is e^x. At 0, the Fourier series for g converges to $\frac{1}{2}[g(0+) + g(0-)] = \frac{1}{2}(1+1) = 1$. At 3, the series for g converges to $\frac{1}{2}[g(3-0) + g(-3+0)] = \frac{1}{2}(e^3 + e^3) = e^3$. Thus the cosine series for e^x converges to e^x on $0 \le x \le 3$.

The same kind of trick can be used to obtain a Fourier sine series for f on $[0, L]$. This time, extend f to an odd function g on $[-L, L]$ by putting

$$g(x) = \begin{cases} f(x), & 0 \le x \le L, \\ -f(-x), & -L \le x < 0. \end{cases}$$

For $-L \le x \le 0$, $g(x)$ looks like $f(x)$ reflected across both the x- and y-axes, and we have $g(-x) = -g(x)$. Expand g in a Fourier series on $[-L, L]$ and examine the coefficients. A simple computation yields

$$\begin{aligned} a_0 &= \frac{1}{L}\int_{-L}^{L} g(\xi)\, d\xi = \frac{1}{L}\int_{-L}^{0} g(\xi)\, d\xi + \frac{1}{L}\int_{0}^{L} g(\xi)\, d\xi \\ &= \frac{1}{L}\int_{-L}^{0} -f(-\xi)\, d\xi + \frac{1}{L}\int_{0}^{L} f(\xi)\, d\xi \\ &= -\frac{1}{L}\int_{0}^{L} f(\xi)\, d\xi + \frac{1}{L}\int_{0}^{L} f(\xi)\, d\xi = 0. \end{aligned}$$

For $n = 1, 2, 3, \ldots,$

$$\begin{aligned} a_n &= \frac{1}{L}\int_{-L}^{0} g(\xi)\cos\left(\frac{n\pi\xi}{L}\right) d\xi + \frac{1}{L}\int_{0}^{L} g(\xi)\cos\left(\frac{n\pi\xi}{L}\right) d\xi \\ &= \frac{1}{L}\int_{-L}^{0} -f(-\xi)\cos\left(\frac{n\pi\xi}{L}\right) d\xi + \frac{1}{L}\int_{0}^{L} f(\xi)\cos\left(\frac{n\pi\xi}{L}\right) d\xi \\ &= -\frac{1}{L}\int_{0}^{L} f(\xi)\cos\left(\frac{n\pi\xi}{L}\right) d\xi + \frac{1}{L}\int_{0}^{L} f(\xi)\cos\left(\frac{n\pi\xi}{L}\right) d\xi \\ &= 0. \end{aligned}$$

Finally, the same kind of calculation yields

$$b_n = \frac{2}{L}\int_0^L f(\xi)\sin\left(\frac{n\pi\xi}{L}\right)d\xi.$$

Thus, g has a Fourier sine series (i.e., no cosines appear) on $[-L, L]$:

$$\sum_{n=1}^{\infty} b_n \sin\left(\frac{n\pi x}{L}\right) \qquad b_n = \frac{2}{L}\int_0^L f(\xi)\sin\left(\frac{n\pi\xi}{L}\right)d\xi.$$

Since $g(x) = f(x)$ for $0 \leq x \leq L$, this is also a sine series for f on $[0, L]$. Again, one must appeal to convergence theorems to determine the sum of the series at any x. As an example, we shall expand e^x in a sine series on $[0, 3]$. Compute

$$\begin{aligned} b_n &= \tfrac{2}{3}\int_0^3 f(\xi)\sin\left(\frac{n\pi\xi}{3}\right)d\xi \\ &= \tfrac{2}{3}e^{\xi}\,\frac{\sin(n\pi\xi/3) - \dfrac{n\pi}{3}\cos(n\pi\xi/3)}{1 + n^2\pi^2/9}\Bigg|_0^3 = \frac{2n}{9 + n^2\pi^2}\,[e^3(-1)^{n+1} + 1]. \end{aligned}$$

The Fourier sine series for e^x on $[0, 3]$ is then

$$\sum_{n=1}^{\infty} \frac{2n}{9 + n^2\pi^2}\,[e^3(-1)^{n+1} + 1]\sin\left(\frac{n\pi x}{3}\right).$$

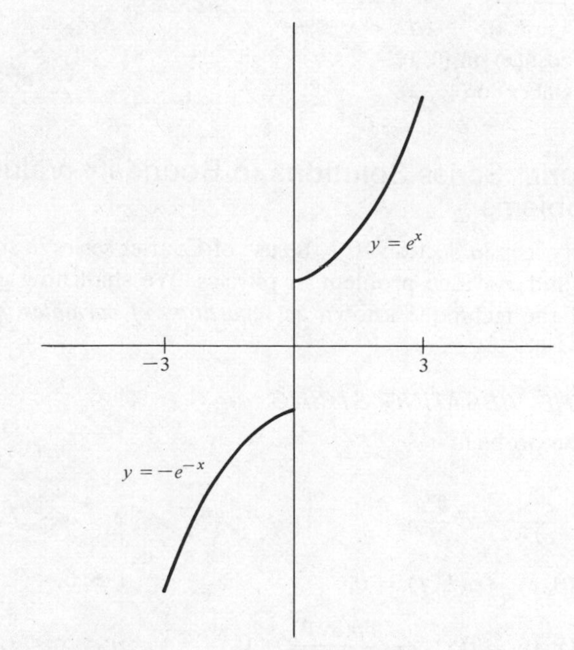

Fig. 10-8

We leave it to the reader to determine the sum of this series for each x in $[0, 3]$.

Just by way of illustration, the function $g(x)$ is given by

$$g(x) = \begin{cases} e^x, & 0 \le x \le 3, \\ -e^{-x}, & -3 \le x < 0, \end{cases}$$

as shown in Fig. 10-8.

Exercises for Section 10.3

1. For each of the following, expand f in a Fourier sine and a Fourier cosine series on the given interval. For each such expansion, determine the sum of the series at each point in the given interval.

(a) $f(x) = 2x^3$ on $[0, 2]$

(b) $f(x) = \cos^2(x)$ on $[0, 3]$

(c) $f(x) = xe^x$ on $[0, \pi]$

(d) $f(x) = x^2$ on $[0, 4\pi]$

(e) $f(x) = \begin{cases} x^2, & 0 \le x \le 1 \\ 2x + 1, & 1 < x < 3 \\ e^{2x}, & 3 \le x \le 4 \end{cases}$

(f) $f(x) = \begin{cases} 8, & 0 \le x < 1 \\ 10, & 1 \le x \le 2 \\ 12, & 2 < x \le 3 \end{cases}$

(g) $f(x) = \begin{cases} \cos^2(x), & 0 \le x \le \pi/2 \\ \sin^2(x), & \pi/2 < x \le \pi \end{cases}$

(h) $f(x) = \cosh(x)$ on $[0, 1]$

(i) $f(x) = \sinh(x)$ on $[0, 2]$.

10.4 Fourier Series Solutions to Boundary-Value Problems

We have already seen in Section 10.1 the use of Fourier series in solving, at least formally, a boundary-value problem in physics. We shall now give two other illustrations of the technique known as *separation of variables*, or the *Fourier method.*

EXAMPLE—THE VIBRATING STRING

Consider the problem

$$\frac{\partial^2 u}{\partial t^2} = a^2 \frac{\partial^2 u}{\partial x^2}, \qquad 0 < x < L,$$

$$u(0, t) = u(L, t) = 0, \qquad t > 0,$$

$$u(x, 0) = f(x), \quad \frac{\partial u(x, 0)}{\partial t} = 0, \qquad 0 < x < L.$$

This is the mathematical formulation of the vibrating-string problem for a homogeneous string of length L with fixed ends, initial position given by f, and zero initial velocity. The number a^2 is a constant that depends upon the material in the string (see Sections 5.2 and 8.6A).

Try a solution $u(x, t) = A(x)B(t)$. Substituting in the differential equation gives us

$$AB'' = a^2 A'' B,$$

or

$$\frac{A''}{A} = \frac{B''}{a^2 B}.$$

Since the left side depends only on x and the right only on t, and x and t are independent, both sides equal a constant, say λ. For such λ, then,

$$A'' - \lambda A = 0 \quad \text{and} \quad B'' - \lambda a^2 B = 0.$$

Now, $u(0, t) = A(0)B(t) = 0$ implies that $A(0) = 0$. Similarly, $u(L, t) = 0$ implies that $A(L) = 0$. We first concentrate on the problem

$$A'' - \lambda A = 0,$$
$$A(0) = A(L) = 0.$$

Consider the possibilities for real λ.

(1) $\lambda = 0$. Then $A(x) = \alpha x + \beta$. But $A(0) = 0$ implies that $\beta = 0$, and $A(L) = 0$ yields $\alpha = 0$, so we obtain a trivial solution, which we discard.

(2) $\lambda > 0$, say $\lambda = \varphi^2$, where $\varphi > 0$. Then $A'' - \varphi^2 A = 0$, with general solution $A(x) = \alpha e^{\varphi x} + \beta e^{-\varphi x}$. Since $A(0) = 0$, then $\alpha + \beta = 0$, so $\alpha = -\beta$. Then $A(x) = 2\alpha \sinh(\varphi x)$. But $A(L) = 0 = 2\alpha \sinh(\varphi L)$ implies that $\alpha = 0$ or $\varphi L = 0$. Since $L \neq 0$, then $\alpha = 0$, and we get a trivial solution again.

(3) $\lambda < 0$, say $\lambda = -\varphi^2$, $\varphi > 0$. Then $A'' + \varphi^2 A = 0$, with general solution

$$A(x) = \alpha \cos(\varphi x) + \beta \sin(\varphi x).$$

Now $A(0) = 0$, so $\alpha = 0$; then $A(x) = \beta \sin(\varphi x)$. Next, $A(L) = \beta \sin(\varphi L) = 0$. We do not want $\beta = 0$, so we choose φL as any integer multiple of π, $\varphi L = n\pi$. Then $\varphi = n\pi/L$. Since we need $\varphi > 0$, admissible values of n are $1, 2, 3, \ldots$ Corresponding to each such n, we obtain a solution for A:

$$A_n(x) = \beta_n \sin\left(\frac{n\pi x}{L}\right).$$

Now look at $B'' - \lambda a^2 B = 0$. For $\lambda = -\varphi^2 = -n^2\pi^2/L^2$, we have

$$B'' + \frac{n^2\pi^2 a^2}{L^2} B = 0.$$

This has general solution $B(t) = \gamma \cos(n\pi at/L) + \delta \sin(n\pi at/L)$. Now recall that $\frac{\partial u}{\partial t}(x, 0) = 0$. Then $A(x)B'(0) = 0$, hence $B'(0) = 0$. Then $\delta = 0$, so for each $n = 1, 2, 3, \ldots$, we have a solution

$$B_n(t) = \gamma_n \cos\left(\frac{n\pi at}{L}\right).$$

For $n = 1, 2, \ldots$, then,

$$A_n(x)B_n(t) = c_n \sin\left(\frac{n\pi x}{L}\right) \cos\left(\frac{n\pi at}{L}\right)$$

satisfies

$$\frac{\partial^2 u}{\partial t^2} = a^2 \frac{\partial^2 u}{\partial x^2},$$

$$u(0, t) = u(L, t) = 0,$$

and

$$\frac{\partial u}{\partial t}(x, 0) = 0.$$

To satisfy the initial condition $u(x, 0) = f(x)$, we attempt a superposition

$$u(x, t) = \sum_{n=1}^{\infty} A_n(x)B_n(t) = \sum_{n=1}^{\infty} c_n \sin\left(\frac{n\pi x}{L}\right) \cos\left(\frac{n\pi at}{L}\right).$$

We then need

$$u(x, 0) = f(x) = \sum_{n=1}^{\infty} c_n \sin\left(\frac{n\pi x}{L}\right).$$

We therefore need a Fourier sine expansion of f on $[0, L]$, and choose accordingly

$$c_n = \frac{2}{L} \int_0^L f(\xi) \sin\left(\frac{n\pi \xi}{L}\right) d\xi.$$

We therefore propose

$$u(x, t) = \frac{2}{L} \sum_{n=1}^{\infty} \left[\int_0^L f(\xi) \sin\left(\frac{n\pi\xi}{L}\right) d\xi\right] \sin\left(\frac{n\pi x}{L}\right) \cos\left(\frac{n\pi at}{L}\right)$$

as our solution.

Note that, for any t, $u(x, t)$ is a sum of multiples of sines with different periods. Physically this is what we might have expected, since it is reasonable to think of the motion of a vibrating string as being composed of a sum of fundamental modes of vibration, each such mode being represented by a sine term. ▮

EXAMPLE

Solve the following Dirichlet problem for a rectangle:

$$\begin{aligned} &u_{xx} + u_{yy} = 0 && 0 < x < a,\ 0 < y < b, \\ &u(0, y) = 0 = u(a, y), && 0 < y < b, \\ &u(x, 0) = f(x), && 0 < x < a, \\ &u(x, b) = 0, && 0 < x < a. \end{aligned}$$

Put $u(x, t) = A(x)B(t)$, to obtain

$$\frac{A''}{A} = \frac{-B''}{B}.$$

Since x and t are independent, both sides must equal some constant, say λ. Then $A'' - \lambda A = 0$, $B'' + \lambda B = 0$. We also have from the boundary conditions

$$A(0) = A(a) = 0 \quad \text{and} \quad B(b) = 0.$$

Take cases on λ:

(1) $\lambda = 0$. Then $A(x) = \alpha x + \beta$. But $A(0) = A(a) = 0$ then forces $\alpha = \beta = 0$, so we discard this case.

(2) $\lambda > 0$, say $\lambda = \varphi^2$, $\varphi > 0$. Then $A(x) = \alpha e^{\varphi x} + \beta e^{-\varphi x}$. Since $A(0) = 0$, then $\alpha = -\beta$. Then $A(x) = 2\alpha \sinh(\varphi x)$. Then $A(a) = 2\alpha \sinh(\varphi a) = 0$ implies that $\alpha = 0$, again a trivial solution.

(3) $\lambda < 0$, say $\lambda = -\varphi^2$, $\varphi > 0$. Then $A(x) = \alpha \cos(\varphi x) + \beta \sin(\varphi x)$. Since $A(0) = 0$, then $\alpha = 0$. Since $A(a) = \beta \sin(\varphi a) = 0$, φa must be a positive integer multiple of π, $\varphi a = n\pi$, $n = 1, 2, \ldots$ For each such n, we have a solution

$$A_n(x) = \alpha_n \sin\left(\frac{n\pi x}{a}\right).$$

Now, for $n = 1, 2, \ldots$, we have admissible values of λ, say $\lambda_n = -n^2\pi^2/a^2$. Then the equation for B is $B''(y) - \dfrac{n^2\pi^2}{a^2} B(y) = 0$, with general solution

$B(y) = \gamma e^{n\pi y/a} + \delta e^{-n\pi y/a}$. Since $B(b) = 0$, then $\delta = -\gamma e^{2n\pi b/a}$. Then, for $n = 1, 2, \ldots$, we have solutions for B:

$$\begin{aligned}
B_n(y) &= \gamma_n(e^{n\pi y/a} - e^{-n\pi y/a}e^{2n\pi b/a}) \\
&= \gamma_n e^{n\pi b/a}(e^{n\pi y/a}e^{-n\pi b/a} - e^{-n\pi y/a}e^{n\pi b/a}) \\
&= \gamma_n e^{n\pi b/a}(e^{n\pi(y-b)/a} - e^{-n\pi(y-b)/a}) \\
&= 2\gamma_n e^{n\pi b/a} \sinh\left[\frac{n\pi(y-b)}{a}\right] \\
&= \xi_n e^{n\pi b/a} \sinh\left[\frac{n\pi(b-y)}{a}\right], \qquad \text{with } \xi_n = -2\gamma_n.
\end{aligned}$$

For each $n = 1, 2, \ldots$, we now have

$$A_n(x)B_n(y) = \zeta_n \sin\left(\frac{n\pi x}{a}\right) e^{n\pi b/a} \sinh\left[\frac{n\pi(b-y)}{a}\right].$$

In order to satisfy $u(x, 0) = f(x)$, try a superposition,

$$u(x, t) = \sum_{n=1}^{\infty} \zeta_n \sin\left(\frac{n\pi x}{a}\right) e^{n\pi b/a} \sinh\left(\frac{n\pi(b-y)}{a}\right).$$

Then

$$u(x, 0) = f(x) = \sum_{n=1}^{\infty} \zeta_n \sin\left(\frac{n\pi x}{a}\right) e^{n\pi b/a} \sinh\left(\frac{n\pi b}{a}\right).$$

We then have

$$\zeta_n e^{n\pi b/a} \sinh\left(\frac{n\pi b}{a}\right) = \frac{2}{a}\int_0^a f(\xi) \sin\left(\frac{n\pi\xi}{a}\right) d\xi,$$

leading us to choose

$$\zeta_n = \frac{2}{ae^{n\pi b/a}\sinh(n\pi b/a)}\int_0^a f(\xi)\sin\left(\frac{n\pi\xi}{a}\right) d\xi,$$

and try as a solution

$$\begin{aligned}
u(x, t) &= \sum_{n=1}^{\infty} \frac{2e^{n\pi b/a}}{ae^{n\pi b/a}\sinh(n\pi b/a)}\int_0^a f(\xi)\sin\left(\frac{n\pi\xi}{a}\right) d\xi \cdot \sin\left(\frac{n\pi x}{a}\right) \sinh\left(\frac{n\pi(b-y)}{a}\right) \\
&= \sum_{n=1}^{\infty} \frac{2}{a}\frac{1}{\sinh(n\pi b/a)}\int_0^a f(\xi)\sin\left(\frac{n\pi\xi}{a}\right) d\xi \cdot \sin\left(\frac{n\pi x}{a}\right) \sinh\left[\frac{n\pi(b-y)}{a}\right].
\end{aligned}$$

These examples should suffice to indicate the method. Each time the method is used, extreme care must be taken in two places:

(1) The prescribed function f in the initial or boundary conditions which gives rise to the Fourier series should be sufficiently nice that the Fourier series converges to it on the given interval. Otherwise we fail to satisfy that initial or boundary condition. Here convergence theorems on Fourier series can help.
(2) It is not always obvious that the final series proposed as a solution actually satisfies the differential equation, and in many instances the series does not justify term-by-term differentiation. There is no set method for handling this difficulty, and each problem may require considerable ingenuity to verify that the series "solution" is, in fact, a solution.

We shall illustrate the handling of these problems for the wave and heat equations already solved.

In Section 10.1 we obtained

$$u(x, t) = \sum_{n=1}^{\infty} \zeta_n \sin\left(\frac{n\pi x}{L}\right) e^{-kn^2\pi^2t/L^2},$$

where

$$\zeta_n = \frac{2}{L}\int_0^L f(\xi) \sin\left(\frac{n\pi\xi}{L}\right) d\xi,$$

as a "solution" to the problem

$$\frac{\partial u}{\partial t} = k\frac{\partial^2 u}{\partial x^2}$$

$$u(0, t) = u(L, t) = 0,$$

$$u(x, 0) = f(x).$$

It is obvious that our function $u(x, t)$ satisfies $u(0, t) = u(L, t) = 0$. Further, if f is reasonable, then the series $\sum_{n=1}^{\infty} \zeta_n \sin(n\pi x/L)$ will converge to $f(x)$ on $[0, L]$, giving us the initial condition. Finally, does

$$\frac{\partial u}{\partial t} = k\frac{\partial^2 u}{\partial x^2}?$$

In this case, we can justify term-by-term differentiation of the series to verify directly that u satisfies the differential equation. Note first that the series

$$\sum_{n=1}^{\infty} \zeta_n \sin\left(\frac{n\pi x}{L}\right)\left(\frac{-kn^2\pi^2}{L^2}\right) e^{-kn^2\pi^2t/L^2}$$

obtained by differentiating each term with respect to t, converges uniformly in t in any interval $t \geq t_0 > 0$. This follows from the Weierstrass M-test. For, using the bound $|\zeta_n| \leq (2/L)\int_{-L}^{L} |f(\xi)|\,d\xi$, which is independent of n, we have, for $t \geq t_0$,

$$\left|\zeta_n \sin\left(\frac{n\pi x}{L}\right) e^{-kn^2\pi^2 t/L^2}\left(\frac{-kn^2\pi^2}{L^2}\right)\right| \leq \frac{2}{L^3}\left[\int_0^L |f(\xi)|\,d\xi\right] k\pi^2 n^2 e^{-kn^2\pi^2 t_0/L^2},$$

and the series

$$\sum_{n=1}^{\infty} n^2 e^{-kn^2\pi^2 t_0/L^2}$$

converges. Thus,

$$\frac{\partial u}{\partial t} = \sum_{n=1}^{\infty} \zeta_n\left(\frac{-kn^2\pi^2}{L^2}\right)\sin\left(\frac{n\pi x}{L}\right) e^{-kn^2\pi^2 t/L^2}.$$

Now differentiate the original series term by term with respect to x:

$$\sum_{n=1}^{\infty} \zeta_n\left(\frac{n\pi}{L}\right)\cos\left(\frac{n\pi x}{L}\right) e^{-kn^2\pi^2 t/L^2}.$$

For any $t > 0$ and $0 < x < L$,

$$\left|\zeta_n\left(\frac{n\pi}{L}\right)\cos\left(\frac{n\pi x}{L}\right) e^{-kn^2\pi^2 t/L^2}\right| \leq |\zeta_n|\left(\frac{n\pi}{L}\right) e^{-kn^2\pi^2 t/L^2}$$

$$\leq \frac{2}{L}\left[\int_0^L |f(\xi)|\,d\xi\right]\frac{n\pi}{L} e^{-kn^2\pi^2 t/L^2}.$$

Since $\sum_{n=1}^{\infty} ne^{-n^2\pi^2 t/L^2}$ converges, then

$$\sum_{n=1}^{\infty} \zeta_n(n\pi/L)\cos(n\pi x/L)\, e^{-kn^2\pi^2 t/L^2}$$

converges uniformly on $(0, L)$ for any $t > 0$. Hence

$$\frac{\partial u}{\partial x} = \sum_{n=1}^{\infty} \zeta_n\left(\frac{n\pi}{L}\right)\cos\left(\frac{n\pi x}{L}\right) e^{-kn^2\pi^2 t/L^2}.$$

The same kind of argument yields

$$\frac{\partial^2 u}{\partial x^2} = \sum_{n=1}^{\infty} -\zeta_n\left(\frac{n^2\pi^2}{L^2}\right)\sin\left(\frac{n\pi x}{L}\right) e^{-kn^2\pi^2 t/L^2}.$$

By direct comparison, then, we have

$$\frac{\partial u}{\partial t} = k\frac{\partial^2 u}{\partial x^2} \qquad \text{for } 0 < x < L \text{ and } t > 0$$

and we have a solution.

This kind of argument will not work in verifying our solution to the wave equation. There the problem was

$$a^2\frac{\partial^2 u}{\partial x^2} = \frac{\partial^2 u}{\partial t^2}, \qquad 0 < x < L,\ t > 0,$$

$$u(0, t) = u(L, t) = 0,$$

$$u(x, 0) = f(x), \qquad \frac{\partial u(x, 0)}{\partial t} = 0$$

and our proposed solution was

$$u(x, t) = \sum_{n=1}^{\infty} \zeta_n \sin\left(\frac{n\pi x}{L}\right)\cos\left(\frac{n\pi at}{L}\right),$$

where

$$\zeta_n = \frac{2}{L}\int_0^L f(\xi)\sin\left(\frac{n\pi\xi}{L}\right)d\xi.$$

Watch what happens if you try a term-by-term differentiation here. Differentiating with respect to x gives us

$$\sum_{n=1}^{\infty} \zeta_n\left(\frac{n\pi}{L}\right)\cos\left(\frac{n\pi x}{L}\right)\cos\left(\frac{n\pi at}{L}\right).$$

Once more gives us

$$\sum_{n=1}^{\infty} -\zeta_n\left(\frac{n^2\pi^2}{L^2}\right)\sin\left(\frac{n\pi x}{L}\right)\cos\left(\frac{n\pi at}{L}\right).$$

At, say, $x = L/\pi$ and $t = L/a$, we get the series

$$\sum_{n=1}^{\infty} -\zeta_n\left(\frac{n^2\pi^2}{L^2}\right)\sin(n)\cos(n\pi) = \sum_{n=1}^{\infty} -\zeta_n\left(\frac{n^2\pi^2}{L^2}\right)\sin(n)(-1)^n$$

$$= \sum_{n=1}^{\infty} (-1)^{n+1}\zeta_n\left(\frac{n^2\pi^2}{L^2}\right),$$

which in general diverges.

We conclude that there is no hope of differentiating term by term and substituting into the differential equation. Another kind of trick comes to the rescue here.

Observe that

$$\sin\left(\frac{n\pi x}{L}\right)\cos\left(\frac{n\pi at}{L}\right) = \tfrac{1}{2}\sin\left(\frac{n\pi x}{L} - \frac{n\pi at}{L}\right) + \tfrac{1}{2}\sin\left(\frac{n\pi x}{L} + \frac{n\pi at}{L}\right)$$

$$= \tfrac{1}{2}\sin\left[\frac{n\pi(x - at)}{L}\right] + \tfrac{1}{2}\sin\left[\frac{n\pi(x + at)}{L}\right].$$

Then our series becomes

$$\sum_{n=1}^{\infty} \tfrac{1}{2}\zeta_n \sin\left[\frac{n\pi(x - at)}{L}\right] + \sum_{n=1}^{\infty} \tfrac{1}{2}\zeta_n \sin\left[\frac{n\pi(x + at)}{L}\right].$$

Define a function g by

$$g(\xi) = \sum_{n=1}^{\infty} \zeta_n \sin\left(\frac{n\pi\xi}{L}\right).$$

Then, our $u(x, t)$ becomes $\frac{1}{2}g(x - at) + \frac{1}{2}g(x + at)$.

Further, note that $g(x) = f(x)$ for $0 \le x \le L$ and $t = 0$, assuming that the Fourier sine series for f converges to $f(x)$ at each x in $[0, L]$. Then, in terms of our data function f,

$$u(x, t) = \tfrac{1}{2}(f(x + at) + f(x - at)).$$

If f is twice-differentiable, then we can now verify directly that

$$a^2 \frac{\partial^2 u}{\partial x^2} = \frac{\partial^2 u}{\partial t^2}$$

by chain-rule differentiation.

The solution, $u(x, t) = \frac{1}{2}[f(x + at) + f(x - at)]$, is due to d'Alembert (refer to Section 5.4) and was known before Fourier methods came into practice. In view of the uniqueness of solutions to such problems, however, it is not surprising that the solution derived by Fourier analysis should agree with that obtained by d'Alembert by other means. ∎

These two examples were given to illustrate two different techniques by which solutions can be verified. In general different problems call for different techniques, and we conclude this section on the pessimistic note that, should the boundary-value problem be a difficult one, verification of its solution by Fourier methods will in general be difficult as well.

Exercises for Section 10.4

1. Use separation of variables to derive Fourier series solutions for each of the following:

(a) $4u_{xx} = u_{tt}, \quad 0 < x < 2, t > 0$

$u(0, t) = u(2, t) = 0, t > 0$

$$u(x, 0) = \begin{cases} x, & 0 \le x \le 1 \\ 2 - x, & 1 \le x \le 2 \end{cases}$$

$u_t(x, 0) = 0, 0 < x < 2$

(b) $u_t = 8u_{xx}, 0 < x < 4, t > 0$

$u(0, t) = u(4, t) = 0, t > 0$

$u(x, 0) = 3x^2, 0 \le x \le 4$

(c) $4u_{xx} = u_{tt}, 0 < x < 2, t > 0$

$u(0, t) = u(2, t) = 0, t > 0$

$$u(x, 0) = \begin{cases} x, & 0 \le x \le 1 \\ 2 - x, & 1 \le x \le 2 \end{cases}$$

$u_t(x, 0) = 3, \quad 0 < x < 2$

Hint: Split this into two problems: one with initial velocity zero and the given initial displacement, one with no initial displacement and the given initial velocity. Then add the solutions to these problems.

(d) $4u_{xx} = u_{tt}, 0 < x < \pi, t > 0$

$u(0, t) = u(L, t) = 0, t > 0$

$u(x, 0) = 0, 0 < x < \pi$

$u_t(x, 0) = e^x, 0 \le x \le \pi$

(e) $u_t = u_{xx} - u, 0 < x < 4, t > 0$

$u(0, t) = 0, u(4, t) = 1, t > 0$

$u(x, 0) = 0, 0 \le x \le 4$

(f) $u_{xx} + u_{yy} = 0, 0 < x < a, 0 < y < b$

$$\left.\begin{aligned} u(0, y) &= 2y, \\ u(a, y) &= 2y^2 \end{aligned}\right\} 0 \le y \le b$$

$$\left.\begin{aligned} u(x, 0) &= x^2 \\ u(x, b) &= 3x \end{aligned}\right\} 0 \le x \le a$$

Hint: This is a Dirichlet problem. Split it into four problems, in each of which u is to be zero on three sides of the rectangle, and take the given values on the fourth side. Then add the solutions to these problems.

(g) $u_t = 3u_{xx}, 0 < x < 2, t > 0$

$u(0, t) = 0, u_x(2, t) = -u(2, t), t > 0$

$u(x, 0) = 2x, 0 < x < 2$

Hint: Here the eigenvalues are solutions to a difficult equation. Do not attempt to determine the eigenvalues numerically, but show from their determining equation that there is one for each positive integer.

(h) $u_{tt} + 2au_t + bu = c^2u_{xx}$, $0 < x < 1$, $t > 0$

$u(0, t) = u(1, t) = 0$, $t > 0$

$u(x, 0) = 0$, $0 < x < 1$

$u_t(x, 0) = f(x)$, $0 < x < 1$

The differential equation here is called the *telegrapher's equation*, and arises in the transmission of electrical impulses along a wire.

(i) $\dfrac{\partial^2 u}{\partial r^2} + \dfrac{1}{r}\dfrac{\partial u}{\partial r} + \dfrac{1}{r^2}\dfrac{\partial^2 u}{\partial \theta^2} = 0$, $0 < r < 1$, $u(1, \theta) = f(\theta)$, $0 \leq \theta \leq 2\pi$

[This is a Dirichlet problem for the unit disc, with the differential equation Laplace's equation in polar coordinates and $u(1, \theta) = f(\theta)$ giving the values of u on the boundary of the unit disc.] Obtain a Fourier series solution, then show that this solution coincides with that obtained by using Poisson's integral formula (Section 9.13B).

(j) $u_{xx} + u_{yy} + u_x = 0$, $0 < x < 2$, $0 < y < 2$

$u(x, 0) = u(x, 2) = 0$, $0 < x < 2$

$u(0, y) = 0$, $0 < y < 2$

$u(2, y) = f(y)$, $0 < y < 2$

(k) $u_t + au = ku_{xx}$, $0 < x < 1$, $t > 0$

$u(0, t) = u_x(1, t) = 0$, $t > 0$

$u(x, 0) = f(x)$, $0 < x < 1$

2. In this problem we give a heuristic approach to Fourier series in two variables. Suppose that f is a function of two variables, say continuous in some region D containing a rectangle $-a \leq x \leq a$, $-b \leq y \leq b$. For a given y (think of y as fixed for the moment), $-b \leq y \leq b$, expand $f(x, y)$ in a Fourier series in x:

$$f(x, y) = \frac{a_0(y)}{2} + \sum_{n=1}^{\infty}\left[a_n(y)\cos\left(\frac{n\pi x}{a}\right) + b_n(y)\sin\left(\frac{n\pi x}{a}\right)\right].$$

Next, for $-b < y < b$, expand $a_0(y)$, $a_n(y)$, $b_n(y)$, for $n = 1, 2, \ldots$, in Fourier series:

$$a_0(y) = \frac{a_{00}}{2} + \sum_{m=1}^{\infty}\left[a_{0m}\cos\left(\frac{m\pi y}{b}\right) + b_{0m}\sin\left(\frac{m\pi y}{b}\right)\right],$$

$$a_n(y) = \frac{a_{n0}}{2} + \sum_{m=1}^{\infty}\left[a_{nm}\cos\left(\frac{m\pi y}{b}\right) + b_{nm}\sin\left(\frac{m\pi y}{b}\right)\right],$$

and

$$b_n(y) = \frac{c_{n0}}{2} + \sum_{m=1}^{\infty}\left[c_{nm}\cos\left(\frac{m\pi y}{b}\right) + d_{nm}\sin\left(\frac{m\pi y}{b}\right)\right], \qquad n = 1, 2, 3, \ldots.$$

Show by a heuristic argument (as we used in the beginning of Section 10.2 to show how the Fourier coefficients should be chosen) that

$$a_{nm} = \frac{1}{ab}\int_{-b}^{b}\int_{-a}^{a} f(x, y)\cos\left(\frac{n\pi x}{a}\right)\cos\left(\frac{m\pi y}{b}\right)dx\,dy,$$

$$b_{nm} = \frac{1}{ab}\int_{-b}^{b}\int_{-a}^{a} f(x, y)\cos\left(\frac{n\pi x}{a}\right)\sin\left(\frac{m\pi y}{b}\right)dx\,dy,$$

$$c_{nm} = \frac{1}{ab}\int_{-b}^{b}\int_{-a}^{a} f(x, y)\sin\left(\frac{n\pi x}{a}\right)\cos\left(\frac{m\pi y}{b}\right)dx\,dy,$$

and

$$d_{nm} = \frac{1}{ab}\int_{-b}^{b}\int_{-a}^{a} f(x, y)\sin\left(\frac{n\pi x}{a}\right)\sin\left(\frac{m\pi y}{b}\right)dx\,dy,$$

for $n, m = 0, 1, 2, \ldots.$

The resulting series,

$$\begin{aligned}\tfrac{1}{4}a_{00} &+ \tfrac{1}{2}\sum_{m=1}^{\infty}\left[a_{0m}\cos\left(\frac{m\pi y}{b}\right) + b_{0m}\sin\left(\frac{m\pi y}{b}\right)\right] \\ &+ \tfrac{1}{2}\sum_{n=1}^{\infty}\left[a_{n0}\cos\left(\frac{n\pi x}{a}\right) + c_{n0}\sin\left(\frac{n\pi x}{a}\right)\right] \\ &+ \sum_{n=1}^{\infty}\sum_{m=1}^{\infty}\left[a_{nm}\cos\left(\frac{n\pi x}{a}\right)\cos\left(\frac{m\pi y}{b}\right) + b_{nm}\cos\left(\frac{n\pi x}{a}\right)\sin\left(\frac{m\pi y}{b}\right)\right. \\ &\qquad \left. + c_{nm}\sin\left(\frac{n\pi x}{a}\right)\cos\left(\frac{m\pi y}{b}\right) + d_{nm}\sin\left(\frac{n\pi x}{a}\right)\sin\left(\frac{m\pi y}{b}\right)\right],\end{aligned}$$

is called the *double Fourier series* for f on $-a \le x \le a$, $-b \le y \le b$. Under fairly unrestrictive conditions on f, this series will converge to $f(x, y)$ for $-a \le x \le a$, $-b \le y \le b$.

3. Use the results of Exercise 2 to solve the following Dirichlet problem for a cube:

$$\begin{aligned} \nabla^2 u &= 0, & 0 < x < 1, 0 < y < 1, 0 < z < 1, \\ u(0, y, z) = u(1, y, z) &= 0, & 0 < y < 1, 0 < z < 1, \\ u(x, 0, z) = u(x, 1, z) &= 0, & 0 < x < 1, 0 < z < 1, \\ \left.\begin{aligned} u(x, y, 0) &= 0 \\ u(x, y, 1) &= f(x, y) \end{aligned}\right\} & & 0 < x < 1, 0 < y < 1. \end{aligned}$$

4. Solve the following two-dimensional heat-flow problem:

$$\begin{aligned} \frac{\partial u}{\partial t} &= \nabla^2 u, & 0 < x < \pi, 0 < y < \pi, t > 0, \\ \frac{\partial u(0, y, t)}{\partial x} = \frac{\partial u(\pi, y, t)}{\partial x} &= 0, & 0 < y < \pi, t > 0, \\ \frac{\partial u(x, 0, t)}{\partial y} = \frac{\partial u(x, \pi, t)}{\partial y} &= 0, & 0 < x < \pi, t > 0, \\ u(x, y, 0) &= f(x, y), & 0 < x < \pi, 0 < y < \pi. \end{aligned}$$

5. Solve the following two-dimensional wave equation:

$$\frac{\partial^2 u}{\partial t^2} = \nabla^2 u, \qquad 0 < x < \pi, 0 < y < \pi, t > 0,$$

$$u(0, y, t) = u(\pi, y, t) = 0, \qquad 0 < y < \pi, t > 0,$$

$$u(x, 0, t) = u(x, \pi, t) = 0, \qquad 0 < x < \pi, t > 0,$$

$$\frac{\partial u(x, y, 0)}{\partial t} = f(x, y), \qquad 0 < x < \pi, 0 < y < \pi.$$

10.5 Fourier Integral Expansions

Consider the following boundary-value problem:

$$\frac{\partial u}{\partial t} = k\frac{\partial^2 u}{\partial x^2}, \qquad -\infty < x < \infty, t > 0.$$

This represents heat flow in an infinitely long medium, with internal temperature given by $u(x, t)$. Motivated by physical considerations, we add on the additional condition that our solution must be bounded, say

$$|u(x, t)| \leq M \qquad \text{for all } x \text{ and for } t > 0.$$

Try solving this by the Fourier method and watch what happens. Put $u(x, t) = A(x)B(t)$ as usual and obtain

$$A'' - \lambda A = 0, \qquad B' - \lambda k B = 0,$$

where λ is the separation constant.

Consider three cases on λ:

(1) $\lambda = 0$. Then $A(x) = \alpha x + \beta$. For A to be bounded, we need $\alpha = 0$. Thus $\lambda = 0$ is acceptable with $A(x) =$ constant.

(2) $\lambda > 0$, say $\lambda = \varphi^2$ for $\varphi > 0$. Then $A(x) = \alpha e^{\varphi x} + \beta e^{-\varphi x}$. On the interval $(-\infty, 0)$, $e^{-\varphi x}$ is unbounded unless $\beta = 0$; and on $(0, \infty)$, $e^{\varphi x}$ is unbounded unless $\alpha = 0$, so we obtain a trivial solution in this case.

(3) $\lambda < 0$, say $\lambda = -\varphi^2$, $\varphi > 0$. Then $A(x) = \alpha\cos(\varphi x) + \beta\sin(\varphi x)$. This gives us a bounded solution for any $\varphi > 0$.

At this point note the difference between this problem and the heat-flow problem of Section 10.1. In the absence of any further boundary conditions, we must admit all values of $\varphi \geq 0$ as possibilities, and, corresponding to any $\varphi \geq 0$, a solution for A of the form

$$A_\varphi(x) = \alpha_\varphi \cos(\varphi x) + \beta_\varphi \sin(\varphi x)$$

(the case $\lambda = 0$ is included in this expression).

Turning to B, we have, for $\varphi = 0$, $B'(t) = 0$, so $B(t) =$ constant. And, for $\varphi > 0$, $B' - \varphi^2 kB = 0$ has solution $B(t) = \xi e^{-k\varphi^2 t}$, with ξ constant. This is bounded in $0 \leq t < \infty$. Thus, for any $\varphi \geq 0$, we have as a possibility for $u(x, t)$, the product

$$A_\varphi(x)B_\varphi(t) = (\zeta_\varphi \cos(\varphi x) + \delta_\varphi \sin(\varphi x))\, e^{-\varphi^2 kt}.$$

The effect of having to admit any $\varphi \geq 0$ is that we cannot form a superposition $\sum_\varphi A_\varphi(x)B_\varphi(t)$, as there are "too many" values for φ. There is a way out, however, suggested by the analogy between $\sum_{n=0}^{\infty}$ and $\int_0^\infty$. We try a "superposition"

$$u(x, t) = \int_0^\infty [\zeta_\varphi \cos(\varphi x) + \delta_\varphi \sin(\varphi x)]\, e^{-k\varphi^2 t}\, d\varphi.$$

This is a Fourier integral (as opposed to a Fourier series) solution. The problem is to choose the ζ_φ's and δ_φ's so as to satisfy the initial condition:

$$u(x, 0) = f(x) = \int_0^\infty [\zeta_\varphi \cos(\varphi x) + \delta_\varphi \sin(\varphi x)]\, d\varphi.$$

We now leave the example, which was designed to justify the work we are about to put in, and concentrate on developing the theory of Fourier integral expansions suitable for handling problems such as the above posed on unbounded intervals.

In order to develop some feeling for what we want, start with an ordinary Fourier expansion on $[-L, L]$:

$$\frac{1}{2L}\int_{-L}^{L} f(\xi)\, d\xi + \frac{1}{L}\sum_{n=1}^{\infty}\left[\int_{-L}^{L} f(\xi)\cos\left(\frac{n\pi\xi}{L}\right) d\xi \cos\left(\frac{n\pi x}{L}\right) + \int_{-L}^{L} f(\xi)\sin\left(\frac{n\pi\xi}{L}\right) d\xi \sin\left(\frac{n\pi x}{L}\right)\right].$$

This reduces, as in the proof of the convergence theorem in Section 10.2, to

$$\frac{1}{2L}\int_{-L}^{L} f(\xi)\, d\xi + \frac{1}{L}\sum_{n=1}^{\infty}\int_{-L}^{L} f(\xi)\cos\left[\frac{n\pi(\xi - x)}{L}\right] d\xi.$$

We envision eventually letting $L \to \infty$. Assume that $\int_{-\infty}^{\infty} f(\xi)\, d\xi$ converges; then $(1/2L)\int_{-\infty}^{\infty} f(\xi)\, d\xi \to 0$, so we shall omit this term. For the rest of the series, put $\Delta\alpha = \pi/L$ to obtain

$$\sum_{n=1}^{\infty}\frac{1}{\pi}\Delta\alpha\int_{-L}^{L} f(\xi)\cos[n\,\Delta\alpha(\xi - x)]\, d\xi,$$

or, slightly rewritten,

$$\frac{1}{\pi}\sum_{n=1}^{\infty}\left\{\int_{-L}^{L} f(\xi)\cos[n\,\Delta\alpha(\xi - x)]\,d\xi\,\Delta\alpha\right\}.$$

Now let $L \to \infty$. On the face of it, this sum would appear to approach

$$\frac{1}{\pi}\int_{0}^{\infty}\int_{-\infty}^{\infty} f(\xi)\cos[\alpha(\xi - x)]\,d\xi\,d\alpha.$$

Since, under certain conditions, the Fourier series converges to

$$\tfrac{1}{2}[f(x+0) + f(x-0)] \qquad \text{for } -L < x < L,$$

we might also expect that under similar conditions,

$$\frac{1}{\pi}\int_{0}^{\infty}\int_{-\infty}^{\infty} f(\xi)\cos[\alpha(\xi - x)]\,d\xi\,d\alpha = \tfrac{1}{2}[f(x+0) + f(x-0)].$$

Now rewrite

$$\cos[\alpha(\xi - x)] = \cos(\alpha\xi)\cos(\alpha x) + \sin(\alpha\xi)\sin(\alpha x),$$

and we obtain

$$\tfrac{1}{2}[f(x+0) + f(x-0)]$$
$$= \frac{1}{\pi}\int_{0}^{\infty}\int_{-\infty}^{\infty} f(\xi)\cos(\alpha\xi)\,d\xi\cos(\alpha x)\,d\alpha + \frac{1}{\pi}\int_{0}^{\infty}\int_{-\infty}^{\infty} f(\xi)\sin(\alpha\xi)\,d\xi\sin(\alpha x)\,d\alpha.$$

This is an expansion of the form

$$\int_{0}^{\infty}[\zeta_\alpha\cos(\alpha x) + \delta_\alpha\sin(\alpha x)]\,d\alpha,$$

where

$$\zeta_\alpha = \frac{1}{\pi}\int_{-\infty}^{\infty} f(\xi)\cos(\alpha\xi)\,d\alpha$$

and

$$\delta_\alpha = \frac{1}{\pi}\int_{-\infty}^{\infty} f(\xi)\sin(\alpha\xi)\,d\alpha,$$

which is what we wanted.

The above is of course not a rigorous treatment but is intended only to suggest the feasibility of the kind of expansion we are looking for. We shall now state a theorem such as the one on convergence of Fourier series, but adapted to the present circumstances.

Theorem

Suppose that f satisfies Dirichlet's condition on every interval $[a, b]$, *and that* $\int_{-\infty}^{\infty} |f(\xi)|\, d\xi$ *converges. Then*

$$\frac{1}{\pi}\int_0^\infty \int_{-\infty}^\infty f(\xi)\cos[\alpha(\xi - x)]\, d\xi\, d\alpha = \tfrac{1}{2}[f(x+0) + f(x-0)]$$

for every real x.

PROOF

We shall sketch a proof, leaving some details to the reader. Choose any real x, and observe that $f(x+0)$ and $f(x-0)$ both exist. Note that

$$\int_a^\infty f(\xi)\cos[\alpha(\xi - x)]\, d\xi$$

converges uniformly on $-\infty < x < \infty$, for every real a, since

$$|f(\xi)\cos[\alpha(\xi - x)]| \le |f(\xi)|, \text{ and } \int_{-\infty}^\infty |f(\xi)|\, d\xi \text{ converges.}$$

Then, for any a and b,

$$\begin{aligned}\int_0^b \int_a^\infty f(\xi)\cos[\alpha(\xi - x)]\, d\xi\, d\alpha &= \int_a^\infty \int_0^b f(\xi)\cos[\alpha(\xi - x)]\, d\alpha\, d\xi \\ &= \int_a^\infty f(\xi)\, \frac{\sin[\alpha(\xi - x)]}{\xi - x}\Bigg|_0^b d\xi \\ &= \int_a^\infty f(\xi)\, \frac{\sin[b(\xi - x)]}{\xi - x}\, d\xi.\end{aligned}$$

It is easy to check that this integral converges uniformly on $-\infty < b < \infty$. Now choose any $a > x$. Given $\varepsilon > 0$, there is some $N > a$ such that

$$\left|\int_A^\infty f(\xi)\, \frac{\sin[b(\xi - x)]}{\xi - x}\, d\xi\right| < \frac{\varepsilon}{2} \qquad \text{if } A \ge N.$$

Now

$$\lim_{b\to\infty} \int_a^A f(\xi)\, \frac{\sin[b(\xi - x)]}{\xi - x}\, d\xi = 0,$$

since we can change variables by $\xi - x = \zeta$ to write this limit as

$$\lim_{b \to \infty} \int_{a-x}^{A-x} f(\zeta + x) \frac{\sin(b\zeta)}{\zeta} \, d\zeta$$

and use Lemma 1 of Section 10.2. Then, for some $M > 0$,

$$\left| \int_a^A f(\xi) \frac{\sin[b(\xi - x)]}{\xi - x} \, d\xi \right| < \frac{\varepsilon}{2} \qquad \text{when } b \geq M.$$

Then

$$\begin{aligned} &\left| \int_a^\infty f(\xi) \frac{\sin[b(\xi - x)]}{\xi - x} \, d\xi \right| \\ &\qquad \leq \left| \int_a^A f(\xi) \frac{\sin[b(\xi - x)]}{\xi - x} \, d\xi \right| + \left| \int_A^\infty f(\xi) \frac{\sin[b(\xi - x)]}{\xi - x} \, d\xi \right| \\ &\qquad < \frac{\varepsilon}{2} + \frac{\varepsilon}{2} = \varepsilon \text{ when } b \geq \max(M, N). \end{aligned}$$

Hence

$$\lim_{b \to \infty} \int_a^\infty f(\xi) \frac{\sin[b(\xi - x)]}{\xi - x} \, d\xi = 0$$

for any real $a > x$, so

$$\int_0^\infty \int_a^\infty f(\xi) \cos[\alpha(\xi - x)] \, d\xi \, d\alpha = 0.$$

Now, write

$$\begin{aligned} &\int_0^b \int_x^a f(\xi) \cos[\alpha(\xi - x)] \, d\xi \, d\alpha \\ &\qquad = \int_x^a \int_0^b f(\xi) \cos[\alpha(\xi - x)] \, d\alpha \, d\xi = \int_x^a f(\xi) \frac{\sin[b(\xi - x)]}{\xi - x} \, d\xi \\ &\qquad = \int_0^{a-x} f(\zeta + x) \frac{\sin(b\zeta)}{\zeta} \, d\zeta, \end{aligned}$$

by putting $\zeta = \xi - x$. Letting $b \to \infty$, we get

$$\begin{aligned} \int_0^\infty \int_x^a f(\xi) \cos[\alpha(\xi - x)] \, d\xi \, d\alpha &= \lim_{b \to \infty} \int_0^{a-x} f(\zeta + x) \frac{\sin(b\zeta)}{\zeta} \, d\zeta \\ &= \frac{\pi}{2} f(x + 0), \end{aligned}$$

by Lemma 3 of Section 10.2.

Combining our results now yields

$$\int_0^\infty \int_x^a f(\xi)\cos[\alpha(\xi - x)]\,d\xi\,d\alpha + \int_0^\infty \int_a^\infty f(\xi)\cos[\alpha(\xi - x)]\,d\xi\,d\alpha$$
$$= \int_0^\infty \int_x^\infty f(\xi)\cos[\alpha(\xi - x)]\,d\xi\,d\alpha = \frac{\pi}{2} f(x + 0).$$

A similar argument yields

$$\int_0^\infty \int_{-\infty}^x f(\xi)\cos[\alpha(\xi - x)]\,d\xi\,d\alpha = \frac{\pi}{2} f(x - 0).$$

Hence

$$\frac{1}{\pi}\int_0^\infty \int_{-\infty}^\infty f(\xi)\cos[\alpha(\xi - x)]\,d\xi\,d\alpha = \tfrac{1}{2}[f(x + 0) + f(x - 0)],$$

completing the proof. ∎

Thus the solution to the heat problem considered as an introduction to this section is, under suitable conditions on f,

$$u(x, t) = \frac{1}{\pi}\int_0^\infty \int_{-\infty}^\infty f(\xi)[\cos(\alpha\xi)\cos(\alpha x) + \sin(\alpha\xi)\sin(\alpha x)]e^{-k\alpha^2 t}\,d\xi\,d\alpha.$$

EXAMPLE

Let

$$f(x) = \begin{cases} e^x, & -\infty < x < 0, \\ e^{-x}, & 0 < x < \infty, \\ 0, & x = 0. \end{cases}$$

Then

$$\int_{-\infty}^\infty |f(x)|\,dx = \int_{-\infty}^0 e^x\,dx + \int_0^\infty e^{-x}\,dx,$$

and these converge. Further, f satisfies Dirichlet's condition on any interval $[a, b]$. We shall compute the Fourier integral for f.

We need first to compute $\int_{-\infty}^\infty f(\xi)\cos[\alpha(\xi - x)]\,d\xi$. A simple integration gives us

$$\int_{-\infty}^0 e^\xi \cos[\alpha(\xi - x)]\,d\xi + \int_0^\infty e^{-\xi}\cos[\alpha(\xi - x)]\,d\xi = \frac{\cos(\alpha x)}{\alpha^2 + 1} \qquad \text{for } \alpha > 0.$$

Then the Fourier integral expansion of $f(x)$ is

$$\frac{1}{\pi}\int_0^\infty \frac{\cos(\alpha x)}{\alpha^2+1}\,d\alpha.$$

Note that this expansion converges to $f(x)$ on $-\infty < x < \infty$. Note also that, for $\alpha = 0$,

$$\int_{-\infty}^\infty f(\xi)\cos[\alpha(\xi - x)]\,d\xi = \int_{-\infty}^\infty f(\xi)\,d\xi = 2 \neq \left.\frac{\cos(\alpha x)}{\alpha^2+1}\right|_{\alpha=0}.$$

However, in forming $\int_0^\infty f(\xi)\cos[\alpha(\xi - x)]\,d\xi\,d\alpha$, the value of

$$\int_0^\infty f(\xi)\cos[\alpha(\xi - x)]\,d\xi$$

at a single point is irrelevant, so we did not bother to compute

$$\int_{-\infty}^\infty f(\xi)\cos[\alpha(\xi - x)]\,d\xi \qquad \text{for } \alpha = 0$$

in forming the integral expansion. ∎

Fourier sine and cosine integral expansions can be developed in a manner entirely analogous to Fourier sine and cosine series. To be specific, suppose that $\int_0^\infty |f(x)|\,dx$ converges and that f satisfies Dirichlet's condition on each interval $[a, b]$ with $0 \le a < b$. Define

$$g(x) = \begin{cases} f(x), & x \ge 0, \\ f(-x), & x < 0. \end{cases}$$

At each x we then have

$$\tfrac{1}{2}[g(x+0) + g(x-0)] = \int_0^\infty [\zeta_\alpha \cos(\alpha x) + \delta_\alpha \sin(\alpha x)]\,d\alpha,$$

where

$$\begin{aligned}\zeta_\alpha &= \frac{1}{\pi}\int_{-\infty}^\infty g(\xi)\cos(\alpha\xi)\,d\xi = \frac{1}{\pi}\left[\int_{-\infty}^0 g(\xi)\cos(\alpha\xi)\,d\xi + \int_0^\infty g(\xi)\cos(\alpha\xi)\,d\xi\right] \\ &= \frac{1}{\pi}\left[\int_{-\infty}^0 f(-\xi)\cos(\alpha\xi)\,d\xi + \int_0^\infty f(\xi)\cos(\alpha\xi)\,d\xi\right] = \frac{2}{\pi}\int_0^\infty f(\xi)\cos(\alpha\xi)\,d\xi\end{aligned}$$

and

$$\delta_\alpha = \frac{1}{\pi}\int_{-\infty}^{\infty} g(\xi)\sin(\alpha\xi)\,d\xi$$

$$= \frac{1}{\pi}\left[\int_{-\infty}^{0} f(-\xi)\sin(\alpha\xi)\,d\xi + \int_{0}^{\infty} f(\xi)\sin(\alpha\xi)\,d\xi\right] = 0.$$

Thus, for $x > 0, f(x) = g(x)$ and we have a Fourier cosine integral expansion:

$$\tfrac{1}{2}[f(x+0) + f(x-0)] = \frac{2}{\pi}\int_0^\infty \int_0^\infty f(\xi)\cos(\alpha\xi)\cos(\alpha x)\,d\xi\,d\alpha.$$

Similarly, if we put

$$g(x) = \begin{cases} f(x), & x \geq 0, \\ -f(-x), & x < 0, \end{cases}$$

and expand this g in a Fourier integral, we obtain, for each $x > 0$ such that $f(x + 0)$ and $f(x - 0)$ exist:

$$\tfrac{1}{2}[f(x+0) + f(x-0)] = \frac{2}{\pi}\int_0^\infty \int_0^\infty f(\xi)\sin(\alpha\xi)\sin(\alpha x)\,d\xi\,d\alpha,$$

a Fourier sine integral expansion.

We shall illustrate the use of Fourier integrals with a Dirichlet problem for an unbounded region.

EXAMPLE

Solve

$$\frac{\partial^2 u}{\partial x^2} + \frac{\partial^2 u}{\partial y^2} = 0, \qquad x > 0, y > 0,$$

$$u(0, y) = 0, \qquad y > 0,$$

$$u(x, 0) = f(x), \qquad x > 0.$$

We also impose the usual condition that the solution be bounded. Putting $u(x, y) = A(x)B(y)$, we obtain in the usual way $(A''/A) = (-B''/B) = \lambda$, where λ is the separation constant. Then $A'' - \lambda A = 0$ and $B'' + \lambda B = 0$. Since $u(0, y) = 0$ for $y > 0$, then $A(0) = 0$. Consider three cases on λ:

(1) $\lambda = 0$. Then $A(x) = \alpha x + \beta$. Since A is to be bounded, then $\alpha = 0$. Since $A(0) = 0$, then $\beta = 0$ also, and this case gives us a trivial solution.

(2) $\lambda > 0$, say $\lambda = \varphi^2$ for some $\varphi > 0$. Then we have as general solution $A(x) = \alpha e^{\varphi x} + \beta e^{-\varphi x}$. In the region $x > 0$, $e^{\varphi x}$ is unbounded, so we must

have $\alpha = 0$. Since $A(0) = 0$, then we must have $\beta = 0$, and we again have a trivial solution.

(3) $\lambda < 0$, say $\lambda = -\varphi^2$ for $\varphi > 0$. The solution for A in this case is of the form $A(x) = \gamma\cos(\varphi x) + \delta\sin(\varphi x)$. Both cosine and sine are bounded on the whole real line. Since $A(0) = \gamma = 0$, then for each $\varphi > 0$, we obtain a bounded solution $A_\varphi(x) = \delta_\varphi\sin(\varphi x)$ to the differential equation $A'' + \varphi^2 A = 0$.

With $\lambda = -\varphi^2$, we have $B'' - \varphi^2 B = 0$, so $B(y) = ae^{\varphi y} + be^{-\varphi y}$. Since $e^{\varphi y}$ is unbounded for $y > 0$, we need $a = 0$. Then $B_\varphi(y) = be^{-\varphi y}$ is a bounded solution to $B'' - \varphi^2 B = 0$ on $y > 0$. We then have, for each $y > 0$, a bounded solution

$$u_\varphi(x, y) = \zeta_\varphi\sin(\varphi x)e^{-\varphi y}$$

to the equation $\nabla^2 u = 0$ in $x > 0$, $y > 0$. Here ζ_φ is an as-yet-undetermined constant. For each $\varphi > 0$, we also have $u_\varphi(0, y) = 0$ for $y > 0$.

To satisfy $u(x, 0) = f(x)$ for $x > 0$, we attempt a "superposition"

$$u(x, y) = \int_0^\infty u_\varphi(x, y)\,d\varphi = \int_0^\infty \zeta_\varphi\sin(\varphi x)e^{-\varphi y}\,d\varphi.$$

We must now choose the constants ζ_φ so that

$$u(x, 0) = \int_0^\infty \zeta_\varphi\sin(\varphi x)\,d\varphi = f(x).$$

That is, we must expand f in a Fourier sine integral on $x > 0$. To do this, choose

$$\zeta_\varphi = \frac{2}{\pi}\int_0^\infty f(\xi)\sin(\varphi\xi)\,d\xi.$$

Assume that f is sufficiently nice that this integral sine expansion of f converges to $f(x)$ for $x > 0$; we then have as solution

$$u(x, y) = \frac{2}{\pi}\int_0^\infty\int_0^\infty f(\xi)\sin(\varphi\xi)\sin(\varphi x)e^{-\varphi y}\,d\xi\,d\varphi.$$

This can be put into a somewhat more recognizable form. Note that

$$\begin{aligned}\sin(\varphi\xi)\sin(\varphi x) &= \tfrac{1}{2}[\cos(\varphi\xi - \varphi x) - \cos(\varphi\xi + \varphi x)]\\ &= \tfrac{1}{2}\{\cos[\varphi(\xi - x)] - \cos[\varphi(\xi + x)]\}.\end{aligned}$$

Then

$$u(x, y) = \frac{1}{\pi}\int_0^\infty \int_0^\infty f(\xi)\{\cos[\varphi(\xi - x)] - \cos[\varphi(\xi + x)]\}e^{-\varphi y}\, d\xi\, d\varphi.$$

Interchange the order of integration to write

$$\begin{aligned} u(x, y) &= \frac{1}{\pi}\int_0^\infty \int_0^\infty f(\xi)\{\cos[\varphi(\xi - x)] - \cos[\varphi(\xi + x)]\}e^{-\varphi y}\, d\varphi\, d\xi \\ &= \frac{1}{\pi}\int_0^\infty f(\xi)\left(\int_0^\infty \{\cos[\varphi(\xi - x)] - \cos[\varphi(\xi + x)]\}e^{-\varphi y}\, d\varphi\right) d\xi. \end{aligned}$$

Now, for any $b > 0$ and $y > 0$, two integrations by parts gives us

$$\begin{aligned} \int_0^b \cos[\varphi(\xi - x)]e^{-\varphi y}\, d\varphi = \frac{-1}{y}\, e^{-by}\cos[b(\xi - x)] + \frac{1}{y} \\ + \frac{1}{y^2}(\xi - x)e^{-by}\sin[b(\xi - x)] \\ - \frac{1}{y^2}(\xi - x)^2\int_0^b e^{-\varphi y}\cos[\varphi(\xi - x)]\, d\varphi. \end{aligned}$$

Let $b \to \infty$. Then we obtain

$$\int_0^\infty \cos[\varphi(\xi - x)]e^{-\varphi y}\, d\varphi = \frac{1}{y} - \frac{1}{y^2}(\xi - x)^2\int_0^\infty e^{-\varphi y}\cos[\varphi(\xi - x)]\, d\varphi.$$

Hence

$$\int_0^\infty e^{-\varphi y}\cos[\varphi(\xi - x)]\, d\varphi = \frac{1}{y}\,\frac{1}{1 + \dfrac{(\xi - x)^2}{y^2}} = \frac{y}{y^2 + (\xi - x)^2}.$$

A very similar calculation yields

$$\int_0^\infty e^{-\varphi y}\cos[\varphi(\xi + x)]\, d\varphi = \frac{y}{y^2 + (\xi + x)^2}.$$

Then

$$\begin{aligned} u(x, y) &= \frac{1}{\pi}\int_0^\infty f(\xi)\left[\frac{y}{y^2 + (\xi - x)^2} - \frac{y}{y^2 + (\xi + x)^2}\right] d\xi \\ &= \frac{y}{\pi}\int_0^\infty f(\xi)\left[\frac{1}{y^2 + (\xi - x)^2} - \frac{1}{y^2 + (\xi + x)^2}\right] d\xi, \end{aligned}$$

for $x > 0$ and $y > 0$.

This agrees with the solution to this problem derived by conformal-mapping methods in Section 9.13B(5). ∎

Exercises for Section 10.5

1. Find the Fourier integral expansion for each of the following, and determine, for each point, the value to which the integral converges:

(a) $f(x) = \dfrac{x}{1 + x^2}$

(b) $f(x) = \begin{cases} 0, & x \le 0 \\ e^{-x}, & x > 0 \end{cases}$

(c) $f(x) = \begin{cases} 0, & x \le 1 \\ \dfrac{1}{x^2}, & x > 1 \end{cases}$

(d) $f(x) = \begin{cases} c, & -b \le x \le b \\ 0 & x > b \text{ or } x < -b \end{cases}$ (c constant, $b > 0$)

2. Find the Fourier sine and cosine integral expansions for each of the following, and determine, for each point, the value to which the integrals converge:

(a) $f(x) = \dfrac{x}{1 + x^4}, \quad x \ge 0$

(b) $f(x) = e^{-x}, \quad x \ge 0$

(c) $f(x) = \begin{cases} \cos(x), & 0 \le x \le \pi \\ 0, & x > \pi \end{cases}$

(d) $f(x) = \begin{cases} 2, & 0 < x \le 3 \\ 3, & 3 < x \le 4 \\ 0, & x > 4 \end{cases}$

3. Use Fourier integrals to solve the following (in all cases we are looking for bounded solutions):

(a) $\dfrac{\partial u}{\partial t} = k \dfrac{\partial^2 u}{\partial x^2}, \quad x > 0, t > 0$

$u(0, t) = 0, \quad t > 0$

$u(x, 0) = f(x), \quad x > 0$

(b) $u_{tt} = k^2 u_{xx}, \quad -\infty < x < \infty, t > 0$

$\left.\begin{aligned} u(x, 0) &= f(x) \\ u_t(x, 0) &= 0 \end{aligned}\right\} -\infty < x < \infty$

(c) $u_{tt} = k^2 u_{xx}, \quad -\infty < x < \infty, t > 0$

$\left.\begin{aligned} u(x, 0) &= 0 \\ u_t(x, 0) &= f(x) \end{aligned}\right\} -\infty < x < \infty$

(d) $u_{xx} + u_{yy} = 0, \quad -\infty < x < \infty, y > 0$

$u(x, 0) = f(x), \quad -\infty < x < \infty$

(e) $u_{xx} + u_{yy} = 0, \quad -\infty < x < \infty, 0 < y < 1$

$$\left.\begin{aligned} u(x, 0) &= f(x) \\ u(x, 1) &= g(x) \end{aligned}\right\} -\infty < x < \infty$$

(f) $u_{xx} + u_{yy} = 0, \quad -1 < x < 1, -\infty < y < \infty$

$$\left.\begin{aligned} u(-1, y) &= 0 \\ u(1, y) &= f(y) \end{aligned}\right\} -\infty < y < \infty$$

4. Use Fourier integrals to show that

$$\int_0^\infty \frac{y \sin(xy)\, dy}{1 + y^2} = \begin{cases} \dfrac{\pi}{2} e^{-x}, & x > 0, \\ -\dfrac{\pi}{2} e^{x}, & x < 0. \end{cases}$$

Show also that

$$\int_0^\infty \frac{\cos(xy)\, dy}{1 + y^2} = \frac{\pi}{2} e^{-|x|} \text{ for } -\infty < x < \infty.$$

10.6 Fourier Transforms

In this section we shall give a brief, nonrigorous introduction to the use of Fourier transforms in solving boundary-value problems. As a starting point, begin with a function f defined on the whole real line, and consider the Fourier integral

$$\frac{1}{\pi} \int_0^\infty \int_{-\infty}^\infty f(\xi) \cos[\alpha(\xi - x)]\, d\xi\, d\alpha.$$

As a matter of convenience, we shall suppose that f is such that this integral converges to $f(x)$ for all real x. Now, observe that $\cos[\alpha(\xi - x)]$ is an even function of α on $-\infty < \alpha < \infty$, hence

$$\begin{aligned} f(x) &= \frac{1}{\pi} \int_0^\infty \int_{-\infty}^\infty f(\xi) \cos[\alpha(\xi - x)]\, d\xi\, d\alpha \\ &= \frac{1}{2\pi} \int_{-\infty}^\infty \int_{-\infty}^\infty f(\xi) \cos[\alpha(\xi - x)]\, d\xi\, d\alpha. \end{aligned}$$

Further, $\sin[\alpha(\xi - x)]$ is an odd function of α on $-\infty < \alpha < \infty$, so

$$\int_{-\infty}^\infty \int_{-\infty}^\infty f(\xi) \sin[\alpha(\xi - x)]\, d\xi\, d\alpha = 0.$$

Thus

$$f(x) = \frac{1}{2\pi}\int_{-\infty}^{\infty}\int_{-\infty}^{\infty} f(\xi)\{\cos[\alpha(\xi - x)] + i\sin[\alpha(\xi - x)]\}\,d\xi\,d\alpha$$

$$= \frac{1}{2\pi}\int_{-\infty}^{\infty}\int_{-\infty}^{\infty} f(\xi)e^{i\alpha(\xi - x)}\,d\xi\,d\alpha.$$

If we put

$$\hat{f}(\alpha) = \frac{1}{\sqrt{2\pi}}\int_{-\infty}^{\infty} f(\xi)e^{i\alpha\xi}\,d\xi,$$

for $-\infty < \alpha < \infty$, then

$$f(x) = \frac{1}{\sqrt{2\pi}}\int_{-\infty}^{\infty} \hat{f}(\alpha)e^{-i\alpha x}\,d\alpha.$$

We call $\hat{f}$ the *Fourier transform* of f, and the last expression, from which we retrieve f from $\hat{f}$, is called the *inversion formula.*

If $f(x)$ is defined only on $[0, \infty]$, it is possible to produce cosine and sine Fourier transforms in exactly the same way that Fourier cosine and sine series and integrals were obtained. For a cosine transform, extend f to an even function g:

$$g(x) = \begin{cases} f(x), & x \geq 0, \\ f(-x), & x < 0. \end{cases}$$

Then

$$g(\alpha) = \frac{1}{\sqrt{2\pi}}\int_{-\infty}^{\infty} g(\xi)e^{i\alpha\xi}\,d\xi$$

$$= \frac{1}{\sqrt{2\pi}}\left[\int_{-\infty}^{0} f(-\xi)e^{i\alpha\xi}\,d\xi + \int_{0}^{\infty} f(\xi)e^{i\alpha\xi}\,d\xi\right]$$

$$= \frac{1}{\sqrt{2\pi}}\left[\int_{0}^{\infty} f(\xi)e^{-i\alpha\xi}\,d\xi + \int_{0}^{\infty} f(\xi)e^{i\alpha\xi}\,d\xi\right]$$

$$= \frac{1}{\sqrt{2\pi}}\int_{0}^{\infty} f(\xi)(e^{i\alpha\xi} + e^{-i\alpha\xi})\,d\xi$$

$$= \sqrt{\frac{2}{\pi}}\int_{0}^{\infty} f(\xi)\cos(\alpha\xi)\,d\xi.$$

Since $g(\alpha) = f(\alpha)$ for $\alpha > 0$, then we are led to define

$$\hat{f}_C(\alpha) = \sqrt{\frac{2}{\pi}}\int_{0}^{\infty} f(\xi)\cos(\alpha\xi)\,d\xi$$

as the *Fourier cosine transform* of f on $[0, \infty)$.

Similarly, by extending f to an odd function on $(-\infty, \infty)$, we obtain a motivation for defining the *Fourier sine transform* of f on $[0, \infty)$ by

$$\hat{f}_S(\alpha) = \sqrt{\frac{2}{\pi}} \int_0^\infty f(\xi) \sin(\alpha\xi)\, d\xi.$$

If the reader actually carries out the calculation leading to $\hat{f}_S$, as we did for that leading to $\hat{f}_C$, he will find a factor of i on the right side. We omit this in the definition of $\hat{f}_S(\alpha)$ in order to make $\hat{f}_S(\alpha)$ real-valued.

The inversion formulas for $\hat{f}_C$ and $\hat{f}_S$ are, as we might expect,

$$f(x) = \sqrt{\frac{2}{\pi}} \int_0^\infty \hat{f}_C(\alpha) \cos(\alpha x)\, d\alpha$$

for the cosine transform, and

$$f(x) = \sqrt{\frac{2}{\pi}} \int_0^\infty \hat{f}_S(\alpha) \sin(\alpha x)\, d\xi\, d\alpha$$

for the sine transform. These follow again by looking at Fourier integral expansions. For example, if f has a Fourier cosine integral representation, then

$$\begin{aligned} f(x) &= \frac{2}{\pi} \int_0^\infty \int_0^\infty f(\xi) \cos(\alpha\xi) \cos(\alpha x)\, d\xi\, d\alpha \\ &= \sqrt{\frac{2}{\pi}} \int_0^\infty \left(\sqrt{\frac{2}{\pi}} \int_0^\infty f(\xi) \cos(\alpha\xi)\, d\xi \right) \cos(\alpha x)\, d\alpha = \sqrt{\frac{2}{\pi}} \int_0^\infty \hat{f}_C(\alpha) \cos(\alpha x)\, d\alpha. \end{aligned}$$

All this is fine so far, but what is the point? We shall now show that Fourier transforms are a powerful tool for solving boundary-value problems. But first we must observe several properties of Fourier transforms.

(1) $$\widehat{(f + g)}(\alpha) = \hat{f}(\alpha) + \hat{g}(\alpha).$$

That is, the Fourier transform of a sum is the sum of the Fourier transforms (assuming, of course, that $\hat{f}$ and $\hat{g}$ exist).

(2) For any real number a,

$$\widehat{(af)}(\alpha) = a\hat{f}(\alpha).$$

Both (1) and (2) are immediate from the definition of the transform and the linearity of the integral.

(3) $$\widehat{(f')}(\alpha) = -i\alpha\hat{f}(\alpha) \text{ if } \lim_{\alpha \to \infty} f(\alpha) = \lim_{\alpha \to -\infty} f(\alpha) = 0.$$

This relates the transform of a derivative to the transform of the function itself.

To prove (3), which is not so obvious, we write

$$
\begin{aligned}
\widehat{(f')}(\alpha) &= \frac{1}{\sqrt{2\pi}} \int_{-\infty}^{\infty} f'(\xi) e^{i\alpha\xi}\, d\xi \\
&= \frac{1}{\sqrt{2\pi}} \lim_{b\to\infty} \int_{-b}^{b} f'(\xi) e^{i\alpha\xi}\, d\xi \\
&= \frac{1}{\sqrt{2\pi}} \lim_{b\to\infty} \left[(f(\xi)e^{i\alpha\xi}) \Big|_{-b}^{b} - \int_{-b}^{b} f(\xi) i\alpha e^{i\alpha\xi}\, d\xi \right] \\
&= -i\alpha \hat{f}(\alpha).
\end{aligned}
$$

(4) By an easy induction on (3), we have

$$\widehat{(f^{(n)})}(\alpha) = (-i\alpha)^n \hat{f}(\alpha),$$

whenever

$$\lim_{\alpha\to\pm\infty} f(\alpha) = \lim_{\alpha\to\pm\infty} f'(\alpha) = \cdots = \lim_{\alpha\to\pm\infty} f^{(n-1)}(\alpha) = 0.$$

To see how these properties enable us to solve certain boundary and initial value problems, consider the following examples.

EXAMPLE

Consider

$$
\begin{aligned}
\frac{\partial u}{\partial t} &= k \frac{\partial^2 u}{\partial x^2}, && -\infty < x < \infty,\ t > 0, \\
u(x, 0) &= f(x), && -\infty < x < \infty.
\end{aligned}
$$

This is a problem of heat flow in an infinite rod. We seek a bounded solution.

For any $t > 0$, take the Fourier transform with respect to x (thinking of t as fixed) of both sides of the partial differential equation. This gives us:

$$\widehat{\frac{\partial u}{\partial t}} = k \widehat{\frac{\partial^2 u}{\partial x^2}}.$$

Now, $\widehat{\dfrac{\partial u}{\partial t}} = \dfrac{1}{\sqrt{2\pi}} \int_{-\infty}^{\infty} \dfrac{\partial u}{\partial t}(\xi, t) e^{i\alpha t}\, d\xi$. Assuming that $\dfrac{\partial}{\partial t}$ and $\int_{-\infty}^{\infty} \cdots d\xi$ can be interchanged, we then have that $\widehat{\dfrac{\partial u}{\partial t}} = \dfrac{\partial \hat{u}}{\partial t}$. On the other side of the equation, the transform and derivatives are taken with respect to the same variable, so we must use (2) and (4) above. These give us

$$k \widehat{\frac{\partial^2 u}{\partial x^2}} = k \widehat{\frac{\partial^2 u}{\partial x^2}} = k(-i\alpha)^2 \hat{u}(\alpha, t) = -k\alpha^2 \hat{u}(\alpha, t).$$

The differential equation then transforms to

$$\frac{\partial \hat{u}}{\partial t}(\alpha, t) = -k\alpha^2 \hat{u}(\alpha, t).$$

This can be solved by inspection, giving us $\hat{u}(\alpha, t) = A(\alpha)e^{-k\alpha^2 t}$, with $A(\alpha)$ to be determined.

Now, $u(\alpha, 0) = f(\alpha)$, so $\hat{u}(\alpha, 0) = \hat{f}(\alpha) = A(\alpha)$. Thus

$$\hat{u}(\alpha, t) = \hat{f}(\alpha)e^{-k\alpha^2 t}.$$

Stop here for a moment to observe what has happened. The Fourier transform has enabled us to substitute the problem

$$\frac{\partial \hat{u}}{\partial t} = -k\alpha^2 \hat{u}, \qquad \hat{u}(\alpha, 0) = \hat{f}(\alpha),$$

whose solution is immediately obtainable, for the original problem, whose solution is far from obvious.

We now have, not u, but its Fourier transform, $\hat{u}$. There remains to apply the inversion formula to produce u. For each $t > 0$ we have

$$u(x, t) = \frac{1}{\sqrt{2\pi}} \int_{-\infty}^{\infty} \hat{u}(\alpha, t)e^{-i\alpha x}\, d\alpha.$$

Thus

$$\begin{aligned} u(x, t) &= \frac{1}{\sqrt{2\pi}} \int_{-\infty}^{\infty} \hat{f}(\alpha)e^{-k\alpha^2 t}e^{-i\alpha x}\, d\alpha \\ &= \frac{1}{\sqrt{2\pi}} \int_{-\infty}^{\infty} \left[\frac{1}{\sqrt{2\pi}} \int_{-\infty}^{\infty} f(\xi)e^{i\alpha\xi}\, d\xi\right] e^{-k\alpha^2 t}e^{-i\alpha x}\, d\alpha \\ &= \frac{1}{2\pi} \int_{-\infty}^{\infty} \int_{-\infty}^{\infty} f(\xi)e^{i\alpha(\xi - x)}e^{-k\alpha^2 t}\, d\xi\, d\alpha. \end{aligned}$$

At this point recall that we have previously solved the current problem by Fourier integral methods in Section 10.5. In order to show that the solution just obtained by transform agrees with that derived before by Fourier integral, proceed as follows. Write the solution above as

$$\begin{aligned} u(x, t) &= \frac{1}{2\pi} \int_{-\infty}^{\infty} \int_{-\infty}^{\infty} f(\xi)\{\cos[\alpha(\xi - x)] + i \sin[\alpha(\xi - x)]\}e^{-k\alpha^2 t}\, d\xi\, d\alpha \\ &= \frac{1}{2\pi} \int_{-\infty}^{\infty} \int_{-\infty}^{\infty} f(\xi) \cos[\alpha(\xi - x)]e^{-k\alpha^2 t}\, d\xi\, d\alpha \\ &\quad + \frac{i}{2\pi} \int_{-\infty}^{\infty} \int_{-\infty}^{\infty} f(\xi) \sin[\alpha(\xi - x)]e^{-k\alpha^2 t}\, d\xi\, d\alpha. \end{aligned}$$

Since $u(x, t)$ must be real for $-\infty < x < \infty$, and $t > 0$, we must have

$$\int_{-\infty}^{\infty} \int_{-\infty}^{\infty} f(\xi) \sin[\alpha(\xi - x)]e^{-k\alpha^2 t}\, d\xi\, d\alpha = 0.$$

Finally, $\cos[\alpha(\xi - x)]e^{-k\alpha^2 t}$ is an even function of α on $-\infty < \alpha < \infty$, so for any $t > 0$,

$$\int_{-\infty}^{\infty} \cos[\alpha(\xi - x)]e^{-k\alpha^2 t}\, d\alpha = 2\int_{0}^{\infty} \cos[\alpha(\xi - x)]e^{-k\alpha^2 t}\, d\alpha.$$

Thus we obtain, by transform methods,

$$u(x, t) = \frac{1}{\pi}\int_{0}^{\infty} \int_{-\infty}^{\infty} f(\xi) \cos[\alpha(\xi - x)]e^{-k\alpha^2 t}\, d\xi\, d\alpha,$$

in agreement with the previous solution. ∎

EXAMPLE

As another example of a problem solved before by other means (in this case, by conformal mapping), consider the Dirichlet problem for the upper half-plane:

$$\frac{\partial^2 u}{\partial x^2} + \frac{\partial^2 u}{\partial y^2} = 0, \qquad -\infty < x < \infty, y > 0,$$

$$u(x, 0) = f(x), \qquad -\infty < x < \infty.$$

For any given $y > 0$, we can take a Fourier transform in x:

$$\widehat{\frac{\partial^2 u(\alpha, y)}{\partial x^2}} + \widehat{\frac{\partial^2 u(\alpha, y)}{\partial y^2}} = \hat{0}(\alpha) = 0 = (-i\alpha)^2\hat{u}(\alpha, y) + \widehat{\frac{\partial^2 u}{\partial y^2}}.$$

Assuming that

$$\int_{-\infty}^{\infty} \frac{\partial^2 u}{\partial y^2}(\xi, y)e^{i\alpha\xi}\, d\xi = \frac{\partial^2}{\partial y^2}\int_{-\infty}^{\infty} u(\xi, y)e^{i\alpha\xi}\, d\xi,$$

we have $\widehat{(u_{yy})} = (\hat{u})_{yy}$. Then, the equation $\nabla^2 u = 0$ is transformed into

$$\frac{\partial^2 \hat{u}}{\partial y^2}(\alpha, y) = \alpha^2\hat{u}(\alpha, y),$$

which again is a simple equation to solve. By inspection, $\hat{u}(\alpha, y)$ must be of the form $\hat{u}(\alpha, y) = A_\alpha e^{\alpha y} + B_\alpha e^{-\alpha y}$, where A_α and B_α are constants to be determined.

It is usually understood that we want a bounded solution. We therefore want $\hat{u}$ to be bounded as well. Since $y > 0$, we must put

$$A_\alpha = 0 \qquad \text{for } \alpha > 0,$$
$$B_\alpha = 0 \qquad \text{for } \alpha < 0.$$

Then

$$\hat{u}(\alpha, y) = C_\alpha e^{-|\alpha| y},$$

where

$$C_\alpha = \begin{cases} B_\alpha & \text{for } \alpha > 0, \\ A_\alpha & \text{for } \alpha < 0. \end{cases}$$

Since $u(x, 0) = f(x)$, then $\hat{u}(\alpha, 0) = \hat{f}(\alpha) = C_\alpha$. Thus

$$\hat{u}(\alpha, y) = \hat{f}(\alpha)e^{-|\alpha| y}.$$

There remains to invert this to determine u. For any $y > 0$, we have

$$\begin{aligned} u(x, y) &= \frac{1}{\sqrt{2\pi}} \int_{-\infty}^{\infty} \hat{u}(\alpha, y)e^{-i\alpha x}\, d\alpha \\ &= \frac{1}{\sqrt{2\pi}} \int_{-\infty}^{\infty} \hat{f}(\alpha)e^{-|\alpha| y}e^{-i\alpha x}\, d\alpha \\ &= \frac{1}{2\pi} \int_{-\infty}^{\infty} \int_{-\infty}^{\infty} f(\xi)e^{i\alpha\xi}e^{-|\alpha| y}e^{-i\alpha x}\, d\xi\, d\alpha. \end{aligned}$$

Thus our solution is

$$u(x, y) = \frac{1}{2\pi} \int_{-\infty}^{\infty} \int_{-\infty}^{\infty} f(\xi)e^{i\alpha(\xi - x)}e^{-|\alpha| y}\, d\xi\, d\alpha.$$

Again, this does not look like the solution derived in Section 9.13B(4) by conformal mapping. However, we can rewrite the present solution as

$$\begin{aligned} u(x, y) &= \frac{1}{2\pi} \int_{-\infty}^{0} \int_{-\infty}^{\infty} f(\xi)e^{i\alpha(\xi - x)}e^{-|\alpha| y}\, d\xi\, d\alpha \\ &\quad + \frac{1}{2\pi} \int_{0}^{\infty} \int_{-\infty}^{\infty} f(\xi)e^{i\alpha(\xi - x)}e^{-|\alpha| y}\, d\xi\, d\alpha \\ &= \frac{1}{2\pi} \int_{-\infty}^{0} \int_{-\infty}^{\infty} f(\xi)e^{i\alpha(\xi - x)}e^{\alpha y}\, d\xi\, d\alpha + \frac{1}{2\pi} \int_{0}^{\infty} \int_{-\infty}^{\infty} f(\xi)e^{i\alpha(\xi - x)}e^{-\alpha y}\, d\xi\, d\alpha \\ &= \frac{1}{2\pi} \int_{0}^{\infty} \int_{-\infty}^{\infty} f(\xi)e^{-i\tau(\xi - x)}e^{-\tau y}\, d\xi\, d\tau + \frac{1}{2\pi} \int_{0}^{\infty} \int_{-\infty}^{\infty} f(\xi)e^{i\alpha(\xi - x)}e^{-\alpha y}\, d\xi\, d\alpha \\ &= \frac{2}{2\pi} \int_{0}^{\infty} \int_{-\infty}^{\infty} f(\xi)\, \frac{e^{i\alpha(\xi - x)} + e^{-i\alpha(\xi - x)}}{2}\, e^{-\alpha y}\, d\xi\, d\alpha \\ &= \frac{1}{\pi} \int_{0}^{\infty} \int_{-\infty}^{\infty} f(\xi) \cos[\alpha(\xi - x)]e^{-\alpha y}\, d\xi\, d\alpha \\ &= \frac{1}{\pi} \int_{-\infty}^{\infty} f(\xi)\left(\int_{0}^{\infty} \cos[\alpha(\xi - x)]e^{-\alpha y}\, d\alpha \right) d\xi. \end{aligned}$$

The last step was a switch in order of integration. Now compute

$$\int_0^\infty \cos[\alpha(\xi - x)]e^{-\alpha y}\, d\alpha = e^{-\alpha y} \frac{\{-y\cos[\alpha(\xi - x)] + (\xi - x)\sin[\alpha(\xi - x)]\}}{y^2 + (\xi - x)^2}\Bigg|_0^\infty$$

$$= \frac{y}{y^2 + (\xi - x)^2}.$$

Then

$$u(x, y) = \frac{1}{\pi}\int_{-\infty}^{\infty} f(\xi)\frac{y}{y^2 + (\xi - x)^2}\, d\xi$$

$$= \frac{y}{\pi}\int_{-\infty}^{\infty} \frac{f(\xi)\, d\xi}{y^2 + (\xi - x)^2},$$

in agreement with the previously derived solution. ∎

We conclude this section with an example that involves use of Fourier sine and cosine transforms. Of the four properties of Fourier transforms that we have utilized in our applications, (1) and (2) go over directly to Fourier sine and cosine transforms:

$$\widehat{(f + g)}_C(\alpha) = \hat{f}_C(\alpha) + \hat{g}_C(\alpha),$$

$$\widehat{(f + g)}_s(\alpha) = \hat{f}_s(\alpha) + \hat{g}_s(\alpha),$$

$$\widehat{(af)}_C(\alpha) = a \cdot \hat{f}_C(\alpha),$$

and

$$\widehat{(af)}_s(\alpha) = a \cdot \hat{f}_s(\alpha).$$

However, (3) and (4) come out quite differently. Suppose that f is defined on $[0, \infty)$, and that $\lim_{\alpha \to \infty} f(\alpha) = 0$. Then

$$\widehat{(f'_C)}(\alpha) = \sqrt{\frac{2}{\pi}}\int_0^\infty f'(\xi)\cos(\alpha\xi)\, d\xi$$

$$= \sqrt{\frac{2}{\pi}}\left\{[f(\xi)\cos(\alpha\xi)]\Big|_0^\infty - \int_0^\infty f(\xi)[-\alpha\sin(\alpha\xi)]\, d\xi\right\}$$

$$= -\sqrt{\frac{2}{\pi}}f(0) + \sqrt{\frac{2}{\pi}}\,\alpha\int_0^\infty f(\xi)\sin(\alpha\xi)\, d\xi$$

$$= -\sqrt{\frac{2}{\pi}}f(0) + \alpha\hat{f}_S(\alpha).$$

By a similar calculation, which the reader should carry out,

$$\widehat{(f_S')}(\alpha) = -\alpha \hat{f}_C(\alpha).$$

If, in addition, $\lim_{\alpha \to \infty} f'(\alpha) = 0$, it is easy to check that

$$\widehat{(f_C'')}(\alpha) = -\sqrt{\frac{2}{\pi}} f'(0) - \alpha^2 \hat{f}_C(\alpha),$$

and

$$\widehat{(f_S'')}(\alpha) = \sqrt{\frac{2}{\pi}}\, \alpha f(0) - \alpha^2 \hat{f}_S(\alpha).$$

Similar formulas can be derived for higher derivatives, but these are all we need for the following.

EXAMPLE

Consider (again) the Dirichlet problem for the upper half plane:

$$u_{xx} + u_{yy} = 0, \qquad -\infty < x < \infty, y > 0,$$
$$u(x, 0) = f(x), \qquad -\infty < x < \infty.$$

Again, we are looking for a bounded solution.

For each x, $-\infty < x < \infty$, take a Fourier sine transform in y:

$$\widehat{(u_{xx})}_S(x, \alpha) + \widehat{(u_{yy})}_S(x, \alpha) = \hat{0}_S(\alpha) = 0.$$

Assume that $\widehat{(u_{xx})}_S(x, \alpha) = (\hat{u}_S)_{xx}(x, \alpha)$. Then we have

$$(\hat{u}_S)_{xx}(x, \alpha) + \sqrt{\frac{2}{\pi}}\, \alpha u(x, 0) - \alpha^2 \hat{u}_S(x, \alpha) = 0.$$

Then

$$(\hat{u}_S)_{xx}(x, \alpha) + \sqrt{\frac{2}{\pi}}\, \alpha f(x) - \alpha^2 \hat{u}_S(x, \alpha) = 0.$$

This equation is not nearly as pleasant to solve as the one obtained from $\nabla^2 u = 0$ by Fourier transform. However, a solution is

$$\hat{u}_S(x, \alpha) = A_\alpha e^{\alpha x} + B_\alpha e^{-\alpha x} - \frac{1}{\sqrt{2\pi}} \int_0^x f(\xi) e^{-\alpha\xi} e^{\alpha x}\, d\xi$$
$$+ \frac{1}{\sqrt{2\pi}} \int_0^x f(\xi) e^{\alpha\xi} e^{-\alpha x}\, d\xi,$$

where A_α and B_α are constants to be determined. In order to do this, recall that we want $u(x, y)$ to be bounded for $-\infty < x < \infty$, $y > 0$. For a given $\alpha > 0$, the integral $\int_0^x f(\xi)e^{-\alpha\xi}e^{\alpha x}\,d\xi$ grows exponentially as $x \to \infty$. So we choose

$$A_\alpha = \frac{1}{\sqrt{2\pi}} \int_0^\infty f(\xi)e^{-\alpha\xi}e^{\alpha x}\,d\xi.$$

This gives us, for terms in $\hat{u}_S(x, \alpha)$ involving $e^{\alpha x}$,

$$A_\alpha e^{\alpha x} - \frac{1}{\sqrt{2\pi}} \int_0^x f(\xi)e^{-\alpha\xi}e^{\alpha x}\,d\xi = \frac{1}{\sqrt{2\pi}} \int_x^\infty f(\xi)e^{-\alpha\xi}e^{\alpha x}\,d\xi.$$

Similarly, $e^{-\alpha x} \to \infty$ as $x \to -\infty$ for any $\alpha > 0$, so we choose

$$B_\alpha = \int_0^{-\infty} f(\xi)e^{\alpha\xi}e^{-\alpha x}\,d\xi.$$

Our solution for $\hat{u}_S(x, \alpha)$ is then

$$\begin{aligned} \hat{u}_S(x, \alpha) &= \frac{1}{\sqrt{2\pi}} \left[\int_x^\infty f(\xi)e^{-\alpha\xi}e^{\alpha x}\,d\xi + \int_{-\infty}^x f(\xi)e^{\alpha\xi}e^{-\alpha x}\,d\xi \right] \\ &= \frac{1}{\sqrt{2\pi}} \left[\int_x^\infty f(\xi)e^{\alpha(x-\xi)}\,d\xi + \int_{-\infty}^x f(\xi)e^{\alpha(\xi-x)}\,d\xi \right]. \end{aligned}$$

To find $u(x, y)$, we must now invert this. According to the inversion formula for sine transforms, we have

$$u(x, y) = \sqrt{\frac{2}{\pi}} \int_0^\infty \hat{u}_S(x, \alpha)\sin(\alpha y)\,d\alpha.$$

Then

$$\begin{aligned} u(x, y) &= \frac{1}{\pi} \int_0^\infty \int_x^\infty f(\xi)e^{-\alpha(\xi-x)}\sin(\alpha y)\,d\xi\,d\alpha \\ &\quad + \frac{1}{\pi} \int_0^\infty \int_{-\infty}^x f(\xi)e^{\alpha(\xi-x)}\sin(\alpha y)\,d\xi\,d\alpha. \end{aligned}$$

Now,

$$\begin{aligned} \int_0^\infty e^{-\alpha(\xi-x)}\sin(\alpha y)\,d\alpha &= e^{-\alpha(\xi-x)} \frac{(\xi - x)\sin(\alpha y) - y\cos(\alpha y)}{(\xi - x)^2 + y^2} \Bigg|_0^\infty \\ &= \frac{y}{y^2 + (\xi - x)^2}, \qquad x < \xi < \infty, \end{aligned}$$

since $e^{-\alpha(\xi - x)} \to 0$ as $\alpha \to \infty$ for $x < \xi < \infty$. Similarly,

$$\int_0^\infty e^{\alpha(\xi - x)} \sin(\alpha y)\, d\alpha = e^{\alpha(\xi - x)} \frac{(\xi - x)\sin(\alpha y) - y\cos(\alpha y)}{y^2 + (\xi - x)^2}\Bigg|_0^\infty$$

$$= \frac{y}{y^2 + (\xi - x)^2}, \qquad -\infty < x < \xi,$$

since $e^{\alpha(\xi - x)} \to 0$ as $\alpha \to \infty$ for $-\infty < x < \xi$. Then

$$u(x, y) = \frac{1}{\pi}\left[\int_x^\infty f(\xi)\frac{y}{y^2 + (\xi - x)^2}\, d\xi + \int_{-\infty}^x f(\xi)\frac{y}{y^2 + (\xi - x)^2}\, d\xi\right]$$

$$= \frac{y}{\pi}\int_0^\infty f(\xi)\frac{1}{y^2 + (\xi - x)^2}\, d\xi,$$

as we have obtained twice before. ∎

Note that, for the Dirichlet problem in the upper half-plane, Fourier transforms were easier than conformal mappings, which in turn were easier than Fourier sine transforms. This is not a general rule. However, it is generally true that different methods applied to a problem can yield different-looking solutions, although for the Dirichlet problem in the upper half-plane we showed explicitly that the solutions derived by conformal mapping, Fourier transform, and Fourier sine transform were identical.

One final note on the use of Fourier sine transforms in the last example. We took the transform in y, not in x. This was because x was to range between $-\infty$ and ∞, whereas y varied from 0 to ∞. The latter range is suitable for a sine or cosine transform, the former is not.

Exercises for Section 10.6

1. Suppose that $\lim_{\alpha \to \infty} f(\alpha) = \lim_{\alpha \to \infty} f'(\alpha) = \lim_{\alpha \to \infty} f''(\alpha) = 0$. Derive formulas for $(\widehat{f_c^{(3)}})(\alpha)$ and $(\widehat{f_s^{(3)}})(\alpha)$ similar to those given in the text for $(\widehat{f_c'})(\alpha)$, $(\widehat{f_c''})(\alpha)$, $(\widehat{f_s'})(\alpha)$, and $(\widehat{f_s''})(\alpha)$.

2. Find the Fourier transforms of the following functions:

(a) $f(x) = \begin{cases} e^x, & -\infty < x \le 0 \\ e^{-x}, & 0 < x < \infty \end{cases}$

(b) $f(x) = \begin{cases} \cos(x), & 0 < x < \pi \\ 0, & x \ge 0 \text{ or } x \le 0 \end{cases}$

(c) $f(x) = \begin{cases} c, & -a \le x \le a \\ 0, & x > 0 \text{ or } x < -a \end{cases}$ (c is constant, a is a positive constant)

(d) $f(x) = e^{-x^2}, \quad -\infty < x < \infty$

3. Find the Fourier sine and cosine transforms of each of the following:

(a) $f(x) = \begin{cases} c, & 0 \le x < b \\ 0, & x > b \end{cases}$ (c constant, b a positive constant)

(b) $f(x) = e^{-x^2}, \quad x \ge 0$

(c) $f(x) = \begin{cases} 0, & 0 \le x < \pi \\ \sin(x), & \pi \le x \le 2\pi \\ 0, & x > 2\pi \end{cases}$

(d) $f(x) = \begin{cases} 0, & 0 \le x \le \pi \\ \dfrac{1}{x}, & x > \pi \end{cases}$

4. Show that

$$\int_0^\infty \hat{f}_S(\alpha)\hat{g}_S(\alpha)\, d\alpha = \int_0^\infty f(x)g(x)\, dx$$

and

$$\int_0^\infty \hat{f}_C(\alpha)\hat{g}_C(\alpha)\, d\alpha = \int_0^\infty f(x)g(x)\, dx.$$

(Assume whatever conditions you need on f and g.)

5. Show that

$$\int_{-\infty}^\infty \hat{f}(\alpha)\hat{g}(\alpha)e^{-i\alpha x}\, d\alpha = \int_{-\infty}^\infty f(t)g(x-t)\, dt.$$

(Assume any reasonable conditions you need on f and g.)

6. Solve the following using transform methods (i.e., Fourier transform or Fourier sine or cosine transforms):

(a) $u_t = ku_{xx}, \quad 0 < x < \infty, t > 0$

$u(0, t) = 0, \quad t > 0$

$u(x, 0) = f(x), \quad 0 < x < \infty$

(b) Do (a) when

$$f(x) = \begin{cases} c, & 0 \le x \le b, \\ 0, & x > b, \end{cases}$$

where c and b are constants and $b > 0$.

(c) A Dirichlet problem for the right quarter-plane:

$$u_{xx} + u_{yy} = 0, \qquad x > 0 \text{ and } y > 0,$$
$$u(x, 0) = f(x), \qquad x > 0,$$
$$u(0, y) = 0, \qquad y > 0.$$

(d) A Dirichlet problem for the right quarter-plane:

$$u_{xx} + u_{yy} = 0, \qquad x > 0, y > 0;$$
$$u(x, 0) = f(x), \qquad x > 0;$$
$$u(0, y) = g(y), \qquad y > 0.$$

[In (a), (c), and (d), assume any reasonable conditions you need for $f(x)$ and $g(x)$.]

7. For each of the following, use the inversion formula to find the function whose Fourier transform is given:

(a) $\dfrac{1}{x^2 + a^2}$, a constant and nonzero

(b) $\dfrac{\sin(ax)}{x}$, $a > 0$

(c) $f(x - a)$, a constant

(c) $\sqrt{\dfrac{a}{\pi}}\sin(ax^2)$, a constant

Appendix I: General Solutions to Elementary Ordinary Differential Equations

We are not assuming a knowledge of differential equations as a prerequisite to this book. There are, however, two differential equations which arise very frequently and which we would like to be able to solve in order to illustrate certain methods and techniques. These differential equations are

(1) $y'' - a^2y = 0$,
(2) $y'' + a^2y = 0$,

with a any positive constant.

There is a general method for handling such equations. Attempt a solution of the form $y(x) = e^{rx}$, with r to be determined. First, try this in (1). Substituting $y = e^{rx}$ into (1), gives us

$$r^2e^{rx} - a^2e^{rx} = 0,$$

or

$$r^2 - a^2 = 0.$$

Then $r = a$ or $r = -a$, and we have two solutions, e^{ax} and e^{-ax}. Obviously αe^{ax} and βe^{-ax} are still solutions for any constants α and β, and so is the sum $\alpha e^{ax} + \beta e^{-ax}$. It can be shown, and is usually done in differential equations courses, that *any* solution to $y'' - a^2y = 0$ can be written in the form $\alpha e^{ax} + \beta e^{-ax}$ by appropriately choosing α and β. For this reason, $\alpha e^{ax} + \beta e^{-ax}$ is called the *general solution* to $y'' - a^2y = 0$.

Now for (2). Again we try $y(x) = e^{rx}$, and substitute, but this time we obtain

$$r^2e^{rx} + a^2e^{rx} = 0,$$

or $r^2 + a^2 = 0$, for which we obtain (recalling that a is real), $r = \pm ai$. Then a^{aix} and e^{-aix} are solutions, as is $\alpha e^{aix} + \beta e^{-aix}$ for any constants α and β.

If we put $\alpha = \beta = \frac{1}{2}$, then we have $\frac{1}{2}(e^{aix} + e^{-aix}) = \cos(ax)$ as a solution. Similarly, putting $\alpha = 1/2i$ and $\beta = -1/2i$ gives us $(1/2i)(e^{aix} - e^{-aix}) = \sin(ax)$ as a solution. Then, for any constants A and B, $A\cos(ax) + B\sin(ax)$ is also a solution. Again, it can be shown that any solution to $y'' - a^2y = 0$ is of this form, for appropriate choice of A and B. Thus $A\cos(ax) + B\sin(ax)$ is the general solution to $y'' - a^2y = 0$.

Appendix II: Mean Value Theorem for Integrals

In the course of proving a convergence theorem for Fourier series, we made use of a mean-value-type theorem for Riemann integrals which the reader has probably not seen before. We shall state and prove the theorem here.

Theorem

Suppose that f is bounded and monotonic on $[a, b]$, and that g is bounded and $\int_a^b g(x)\,dx$ exists. Suppose also that g has only a finite number of sign changes in $[a, b]$. Then, for some ξ in $[a, b]$,

$$\int_a^b f(x)g(x)\,dx = f(a+0)\int_a^{\xi} g(x)\,dx + f(b-0)\int_{\xi}^b g(x)\,dx.$$

To prove the theorem, we begin with several special cases and work our way toward the above conclusion.

Step 1 Suppose that $g(x) \geq 0$ on $[a, b]$. Let m be the greatest lower bound, M the least upper bound, of $\{f(x) \mid a \leq x \leq b\}$. We claim that, for some θ, $m \leq \theta \leq M$ and

$$\int_a^b f(x)g(x)\,dx = \theta\int_a^b g(x)\,dx.$$

To prove this, note first that

$$m \leq f(x) \leq M \qquad \text{for } a \leq x \leq b.$$

Since $g(x) \geq 0$ on $[a, b]$, then

$$mg(x) \leq f(x)g(x) \leq Mg(x) \qquad \text{for } a \leq x \leq b.$$

Then

$$\int_a^b mg(x)\,dx = m\int_a^b g(x)\,dx \le \int_a^b f(x)g(x)\,dx$$

$$\le \int_a^b Mg(x)\,dx = M\int_a^b g(x)\,dx.$$

Hence, for some θ, $m \le \theta \le M$ and

$$\int_a^b f(x)g(x)\,dx = \theta\int_a^b g(x)\,dx.$$

A similar argument yields the same conclusion if $g(x) \le 0$ on $[a, b]$.

Finally, if f is continuous on $[a, b]$, then $f(\alpha) = m$ and $f(\beta) = M$ for some α, β in $[a, b]$, and by the intermediate value theorem, $f(\xi) = \theta$ for some ξ in $[a, b]$. Then

$$\int_a^b f(x)g(x)\,dx = f(\xi)\int_a^b g(x)\,dx.$$

Step 2 Suppose that f is bounded and monotonic on $[a, b]$, and $f(x) \ge 0$ on $a \le x \le b$. Let $g(x)$ be bounded and change signs only finitely many times on $[a, b]$, and let $\int_a^b g(x)\,dx$ exist. Then, for some ξ in $[a, b]$,

$$\int_a^b f(x)g(x)\,dx = f(a)\int_a^\xi g(x)\,dx.$$

To prove this, partition $a = x_0 < x_1 < \cdots < x_n = b$ so $g(x)$ has the same sign throughout (x_{i-1}, x_i) for $i = 1, \ldots, n$. Write

$$\int_a^b f(x)g(x)\,dx = \sum_{i=1}^n \int_{x_{i-1}}^{x_i} f(x)g(x)\,dx.$$

Now suppose that f is monotone nonincreasing on $[a, b]$. By step 1 there are numbers θ_i, $f(x_{i-1}) \le \theta_i \le f(x_i)$, such that

$$\int_{x_{i-1}}^{x_i} f(x)g(x)\,dx = \theta_i\int_{x_{i-1}}^{x_i} g(x)\,dx.$$

Write $G(t) = \int_a^t g(x)\,dx$, so that

$$\int_{x_{i-1}}^{x_i} g(x)\,dx = \int_a^{x_i} g(x)\,dx - \int_a^{x_{i-1}} g(x)\,dx = G(x_i) - G(x_{i-1}).$$

Then

$$\int_{x_{i-1}}^{x_i} f(x)g(x)\,dx = \theta_i(G(x_i) - G(x_{i-1})).$$

Then

$$\int_a^b f(x)g(x)\,dx = \sum_{i=1}^{n} \theta_i(G(x_i) - G(x_{i-1})).$$

Now $G(a) = 0$, so by rearranging terms, we have

$$\int_a^b f(x)g(x)\,dx = [f(a) - \theta_1]G(a) + \sum_{i=2}^{n} (\theta_{i-1} - \theta_i)G(x_{i-1}) + \theta_n G(x_n).$$

Now $\theta_{i-1} - \theta_i \geq 0$ for $i = 1, 2, \ldots, n$, as f is monotone nonincreasing. Thus for some K between $\max(G(x_0), \ldots, G(x_n))$ and $\min(G(x_0), \ldots, G(x_n))$, we can write

$$\begin{aligned}[f(a) - \theta_1]G(a) &+ \sum_{i=2}^{n} (\theta_{i-1} - \theta_i)G(x_{i-1}) + \theta_n G(x_n) \\ &= K\{[f(a) - \theta_1] + \sum_{i=2}^{n} (\theta_{i-1} - \theta_i) + \theta_n\} \\ &= Kf(a).\end{aligned}$$

Now G is continuous on $[a, b]$, so, for some ξ, $a \leq \xi \leq b$, and $K = G(\xi)$. But $G(\xi) = \int_a^\xi g(x)\,dx$, so we have

$$f(a)\int_a^\xi g(x)\,dx = \int_a^b f(x)g(x)\,dx.$$

A similar argument establishes a similar result if f is monotone nondecreasing on $[a, b]$. Here we obtain

$$\int_a^b f(x)g(x)\,dx = f(b)\int_\xi^b g(x)\,dx$$

for some ξ in $[a, b]$.

Step 3 If f is bounded and monotonic on $[a, b]$, and g is bounded, changes sign only finitely many times on $[a, b]$, and $\int_a^b g(x)\,dx$ exists, then

$$\int_a^b f(x)g(x)\,dx = f(a)\int_a^\xi g(x)\,dx + f(b)\int_\xi^b g(x)\,dx$$

for some ξ in $[a, b]$.

To prove this, suppose first that f is monotone nonincreasing. Let $h(x) = f(x) - f(b)$. Then h is monotone nonincreasing on $[a, b]$, and $h(x) \geq 0$ on $[a, b]$. By step 2,

$$\int_a^b h(x)g(x)\,dx = h(a)\int_a^\xi g(x)\,dx$$

for some ξ in $[a, b]$. Then,

$$\begin{aligned}\int_a^b (f(x) - f(b))g(x)\,dx &= \int_a^b f(x)g(x)\,dx - f(b)\int_a^b g(x)\,dx \\ &= f(a)\int_a^\xi g(x)\,dx - f(b)\int_a^\xi g(x)\,dx.\end{aligned}$$

That is,

$$\begin{aligned}\int_a^b f(x)g(x)\,dx &= f(a)\int_a^\xi g(x)\,dx + f(b)\left[\int_a^b g(x)\,dx - \int_a^\xi g(x)\,dx\right] \\ &= f(a)\int_a^\xi g(x)\,dx + f(b)\int_\xi^b g(x)\,dx.\end{aligned}$$

A similar argument applies when f is monotone nondecreasing, with $h(x) = f(x) - f(a)$ in place of the above choice of h.

Step 4 We can now prove the mean value theorem cited in the beginning of this section. Simply redefine f to have the value $f(a + 0)$ at a and the value $f(b - 0)$ at b. This does not change the monotone character of f or the values of the integrals in step 3.

Appendix: Cylindrical and Spherical Coordinates

Depending upon the setting of the problem, rectangular coordinates may not be the most convenient. Generally speaking, any set of numbers that uniquely determine the position of a point in space may prove useful. In particular, the following two schemes arise very frequently.

Cylindrical Coordinates

Any point (x, y, z) in R^3 may be located by specifying the polar coordinates r and θ of the projection $(x, y, 0)$ onto the xy-plane, and then giving the vertical

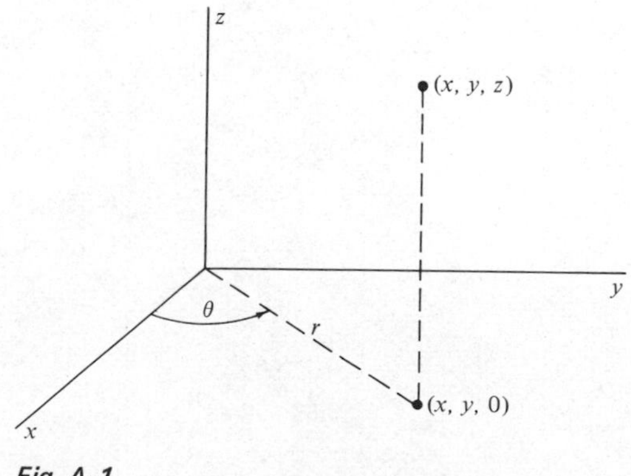

Fig. A-1

coordinate z. Thus cylindrical coordinates of (x, y, z) are (r, θ, z), where

$$x = r\cos(\theta) \qquad \text{and} \qquad y = r\sin(\theta),$$

$r \geq 0$ and $0 \leq \theta \leq 2\pi$.

Spherical Coordinates

We can also specify the point (x, y, z) in spherical coordinates (ρ, θ, φ), where ρ is the length of the vector from the origin to (x, y, z), φ is the angle of declination from the positive z axis of this vector, and θ is the polar angle of rotation from the positive x-axis to the projection $(x, y, 0)$ in the xy-plane. Rectangular and spherical coordinates are related by

$$x = \rho\cos(\theta)\sin(\varphi),$$
$$y = \rho\sin(\theta)\sin(\varphi),$$

and

$$z = \rho\cos(\varphi),$$

where $\rho \geq 0$, $0 \leq \theta \leq 2\pi$, and $0 \leq \varphi \leq \pi$.

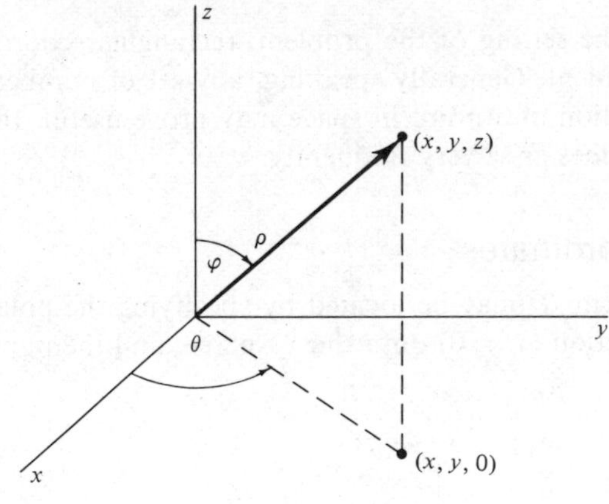

Fig. A-2

Solutions to Selected Exercises

This section contains solutions and hints for solution of a limited number of exercises. In many instances only the basic idea of the solution is given, and the reader is expected to fill in the missing details.

Section 1.1

3. (a) Write

$$u_n = \frac{\sqrt{n}(\sqrt{n+1} - \sqrt{n})}{\sqrt{n+1} + \sqrt{n}}(\sqrt{n+1} + \sqrt{n}) = \frac{\sqrt{n}}{\sqrt{n+1} + \sqrt{n}}$$

$$= \frac{1}{\sqrt{1 + \dfrac{1}{n}} + 1} \to \frac{1}{2} \qquad \text{as } n \to \infty.$$

(f) Examine the graph of $\ln(x)/x^b$.

(g) Write $[\ln(n)^a/n^b] = [\ln(n)/n^{b/a}]^a$ and use (f).

5. Write $a^{1/n} = e^{\ln(a)/n}$ and note that $\ln(a)/n \to 0$ as $n \to \infty$.

6. Write $n^{1/n} = e^{\ln(n)/n}$ and examine $\lim_{n\to\infty} \ln(n)/n$.

25. First, show that $u_{n+1} \geq u_n$. Next, show that $\{u_n\}$ is bounded above. To do this, write $\ln(n) = \int_1^n dt/t$ and, comparing $\sum_{j=1}^{n} 1/j$ and $\int_1^n dt/t$ as areas under $y = 1/x$,

$1 \le x \le n$, show that $\sum_{j=2}^{n} (1/j) - \ln(n) \le 0$, hence that $u_n - 1 \le 0$, $n = 1, 2, \ldots$.

Section 1.2

1. (c) For $x \neq 0$,

$$\frac{nx}{1 + nx} = \frac{x}{x + 1/n} \to 1 \qquad \text{as } n \to \infty.$$

Clearly, $f_n(0) \to 0$ as $n \to \infty$. Convergence is not uniform, as the limit function is not continuous.

(e) $f_n(x) \to 0$ as $n \to \infty$ for $x \ge 0$. Convergence is uniform. To show this, first show that f_n achieves its maximum on $[0, \infty)$ at $x = 1/n$, and note that $f_n(1/n) = 1/en^2 \to 0$ as $n \to \infty$. Thus, given $\varepsilon > 0$, it suffices to produce N such that $1/en^2 < \varepsilon$ for $n \ge N$.

Section 1.3

1. (h) Show that $\ln(n) \le \sqrt{n}$ for n sufficiently large, and compare $\sum_{n=2}^{\infty} \ln(n)/(2n^2 + 6)$ with $\sum_{n=2}^{\infty} \sqrt{n}/(2n^2 + 6)$. For this latter series, note that $\sqrt{n}/(2n^2 + 6) \le \sqrt{n}/n^2 = 1/n^{3/2}$, and show that $\sum_{n=2}^{\infty} 1/n^{3/2}$ converges.

(k) Investigate $\lim_{n \to \infty} \arctan(n)/n$. In doing this, pay particular attention to what happens when n passes through odd multiples of $\pi/2$.

(l) Show first that $n^{1/n} \to 1$ as $n \to \infty$. [*Hint:* $n^{1/n} = e^{\ln(n)/n}$, and $\ln(n)/n \to 0$ as $n \to \infty$.] For some K, $1 \le n^{1/n} \le 2$ for $n \ge K$. Then, for $n \ge K$, $1/2 \le 1/n^{1/n}$. Then $1/2n \le 1/nn^{1/n}$ for $n \ge K$. Now use the fact that $\sum_{n=1}^{\infty} 1/2n$ diverges.

Section 1.5

9. Write

$$\left(\sum_{n=0}^{\infty} \frac{x^n}{n!}\right)\left(\sum_{n=0}^{\infty} \frac{x^n}{n!}\right) = \sum_{n=0}^{\infty} c_n,$$

where

$$c_n = \sum_{j=0}^{n} \frac{x^j}{j!} \frac{x^{n-j}}{(n-j)!} = x^n \sum_{j=0}^{n} \frac{1}{j!\,(n-j)!}.$$

Now observe that

$$2^n = (1 + 1)^n = \sum_{j=0}^{n} \binom{n}{j} = \sum_{j=0}^{n} \frac{n!}{j!\,(n-j)!},$$

so

$$\sum_{j=0}^{n} \frac{1}{j!\,(n-j)!} = \frac{2^n}{n!}.$$

36. Use

$$R_k(x) = \frac{f^{(k+1)}(\theta)(x-a)^{k+1}}{(k+1)!}.$$

Compute

$$\frac{d^{k+1}}{dx^{k+1}}[(1+x)^m] = m(m-1)\cdots(m-k)(1-k)^{m-k-1}.$$

Now show that

$$|R_k(x)| \leq \frac{m(m-1)\cdots(m-k)}{(k+1)!} \to 0 \qquad \text{as } k \to \infty.$$

Section 2.2

3. $A \cdot B = \|A\|\,\|B\| \cos(\theta) = \|A\|\,\|B\|$ if (1) A or B is the zero vector, or (2) neither A nor B is the zero vector, and $\cos(\theta) = 1$. In (2), A and B are parallel.

4. $\|A + B\| = \|A\| + \|B\|$ if and only if $\|A + B\|^2 = (\|A\| + \|B\|)^2$ if and only if $(A + B) \cdot (A + B) = \|A\|^2 + \|B\|^2 + 2\|A\|\,\|B\|$. This reduces to $A \cdot B = \|A\|\,\|B\|$. Now use 3.

Section 2.4

4. **(b)** Note that $(0, -4, 0)$ and $(3/2, 0, 0)$ are points on the given plane. Then $(\frac{3}{2}, 4, 0)$ is a vector in the plane. Similarly, $(0, 0, \frac{3}{4})$ is a point in the plane, so $(0, 4, \frac{3}{4})$ is a vector in the plane. Now compute $(\frac{3}{2}, 4, 0) \times (0, 4, \frac{3}{4})$.

Section 3.1

2. **(c)** h is continuous at $(0, 0)$. Observe that, if $x \neq 0$, then

$$\left|\frac{x^2y}{x^2 + y^4}\right| \leq \left|\frac{x^2y}{x^2}\right| = |y|,$$

and, if $x = 0$, $h(x, y) = 0$.

8. **(a)** Show that $(D^C)^C$ and D consist of exactly the same points.

(b) Suppose first that D is open. Let P be in ∂D. Then every neighborhood of P has a point of D and a point not of D in it. Then P is in $\partial(D^C)$. Since D is open, D^C is closed, so P is in D^C. Then P is not in D.

Conversely, suppose that D and ∂D have no points in common. Let Q be in $\partial(D^C)$ but not in D^C. Then Q is in ∂D and in D, a contradiction. Hence every point of $\partial(D^C)$ is in D^C, so D^C is closed.

13. Let $\varepsilon > 0$. We seek $\delta > 0$ so that $|z^2 + x^2 + y^2 + 2xyz - 17| < \varepsilon$ if $\|(x, y, z) - (1, 1, 3)\| < \delta$. First, write

$$\begin{aligned}
&|z^2 + x^2 + y^2 + 2xyz - 17| \\
&\qquad = |z^2 - 9 + x^2 - 1 + y^2 - 1 + 2xyz - 6| \\
&\qquad \leq |z - 3||z + 3| + |x - 1||x + 1| + |y - 1||y + 1| + 2|xyz - 3|.
\end{aligned}$$

First restrict $|z - 3|$, $|x - 1|$, and $|y - 1|$ to be less than 1, giving us a first estimate $|z^2 + x^2 + y^2 + 2xyz - 17| \leq 7|z - 3| + 2|x - 1| + |y - 1| + 2|xyz - 3|$. Next, write

$$\begin{aligned}
&2|xyz - 3| \\
&\qquad = 2|(x - 1)(y - 1)(z - 3) + 3y(x - 1) + x(z - 3) + z(y - 1)| \\
&\qquad \leq 2|x - 1||y - 1||z - 3| + 6|y||x - 1| + 2|x||z - 3| + 2|z||y - 1| \\
&\qquad \leq 2|x - 1||y - 1||z - 3| + 12|x - 1| + 4|z - 3| + 8|y - 1|.
\end{aligned}$$

Then, so far,

$$\begin{aligned}
&|z^2 + x^2 + y^2 + 2xyz - 17| \\
&\qquad \leq 2|x - 1||y - 1||z - 3| + 11|z - 3| + 10|y - 1| + 14|x - 1|.
\end{aligned}$$

Now pick δ so that the right side is less than ε if $\|(x, y, z) - (1, 1, 3)\| < \delta$.

Section 3.2

1. (a) $\lim_{(x,y)\to(0,0)} x^4y/(y^2 + x^4) = 0$. To prove this, use the estimate

$$\left|\frac{x^4y}{x^4 + y^2}\right| \leq \left|\frac{x^4y}{x^4}\right| = |y| \qquad \text{if } x \neq 0.$$

Section 4.2

7. Let f be monotone nondecreasing on $[a, b]$. (The argument is similar if f is monotone nonincreasing.) For any partition $P: a = x_0 < x_1 \cdots < x_n = b$, $\underline{S}(P) = \sum_{i=1}^{n} f(x_{i-1})(x_i - x_{i-1})$ and $\overline{S}(P) = \sum_{i=1}^{n} f(x_i)(x_i - x_{i-1})$. Then, $\overline{S}(P) - \underline{S}(P) = \sum_{i=1}^{n} [f(x_i) - f(x_{i-1})](x_i - x_{i-1})$. Let $\varepsilon > 0$. If $f(a) = f(b)$, then $\overline{S}(P) - \underline{S}(P) = 0 < \varepsilon$. If $f(b) \neq f(a)$, choose P so that $x_i - x_{i-1} \leq \varepsilon/(f(b) - f(a))$, and show that $\overline{S}(P) - \underline{S}(P) < \varepsilon$. Then use Exercise 4.

Section 4.3

1. Let $P_1, \ldots, P_n$ be points in the plane. Let $\varepsilon > 0$. Produce a grid G such that each P_i is in exactly one rectangle of G, and the sum of the areas of these rectangles is less than ε. (For example, make the rectangles of dimension $\sqrt{\varepsilon/2n}$ by $\sqrt{\varepsilon/2n}$.)

Section 4.7

1. (b)

$$\iint_D y\sin(x) = \int_0^{\pi/4}\int_0^{2y} y\sin(x)\,dx\,dy = \int_0^{\pi/4} [-y\cos(x)]\Big|_0^{2y}\,dy$$

$$= \int_0^{\pi/4} [-y\cos(2y) + y]\,dy = \frac{\pi^2}{32} - \frac{\pi}{8} + \frac{1}{4}.$$

In the other order,

$$\iint_D y\sin(x) = \int_0^{\pi/2}\int_{x/2}^{\pi/4} y\sin(x)\,dy\,dx$$

$$= \frac{\pi^2}{32}\int_0^{\pi/2}\sin(x)\,dx - \frac{1}{8}\int_0^{\pi/2} x^2\sin(x)\,dx.$$

7. (d)

$$\text{Volume of } D = \iiint_D 1 = \int_0^1\int_{-z}^{z}\int_{-\sqrt{z^2-x^2}}^{\sqrt{z^2-x^2}} dy\,dx\,dz$$

$$= \int_0^1\int_{-z}^{z} 2\sqrt{z^2-x^2}\,dx\,dz$$

$$= \int_0^1 [x\sqrt{z^2-x^2} + z^2\arcsin(x/z)]\Big|_{-z}^{z}\,dz$$

$$= \int_0^1 z^2\pi\,dz = \frac{\pi}{3}.$$

Section 4.9

7. (b) $\iiint_D (x+y+z) = \int_0^1\int_0^{1-x}\int_0^{1-x-y} (x+y+z)\,dz\,dy\,dx$, reducing the problem to three single Riemann integrations.

Section 4.10

1. (d) $\iint_V \cos(x)e^{-xy} = \lim_{r\to\infty}\iint_{T_r}\cos(x)e^{-xy}$, where T_r consists of all points (x, y) with $0 \le x \le 2\pi$ and $0 \le y \le r$. Now

$$\iint_{T_r}\cos(x)e^{-xy} = \int_0^r\int_0^{2\pi}\cos(x)e^{-xy}\,dx\,dy = \int_0^r \frac{y}{1+y^2}(1 - e^{-2\pi y})\,dy.$$

Integrate this and take the limit as $r \to \infty$.

Section 5.1

3. (a)

$$f_x(0, 0) = \lim_{\Delta x\to 0}\frac{f(\Delta x, 0) - f(0, 0)}{\Delta x} = \lim_{\Delta x\to 0}\frac{0}{\Delta x} = 0.$$

Similarly, $f_y(0, 0) = 0$.

(b)

$$\frac{\partial^2 f}{\partial x\,\partial y}(0,0) = \frac{\partial}{\partial x}\left(\frac{\partial f}{\partial y}\right)(0,0) = \lim_{\Delta x \to 0} \frac{f_y(\Delta x, 0) - f_y(0,0)}{\Delta x}.$$

Now $f_y(\Delta x, 0) = (\Delta x)^3/[(\Delta x)^2]^2 = 1/\Delta x$ for $\Delta x \neq 0$. Then

$$\frac{\partial^2 f}{\partial x\,\partial y}(0,0) = \lim_{\Delta x \to 0} \frac{1/\Delta x - 0}{\Delta x} = \lim_{\Delta x \to 0} \frac{1}{(\Delta x)^2}$$

does not exist finite.

5. **(a)** The desired line L has slope $f_x(1, 3) = 7$. Since L passes through $(1, 3, 25)$, then L has equation $z - 25 = 7(x - 1)$, $y = 3$.

(b) The desired line has slope $f_y(1, 3) = 13$, so has equation $z - 25 = 13(y - 3)$, $x = 1$.

(c) Pick two points on L, say $(1, 3, 25)$ and $(2, 3, 32)$. The vector $(1, 0, 7)$ is then along L. Similarly, using (b), we get a vector, say $(0, 1, 13)$, along the line of (b). Then $(1, 0, 7) \times (0, 1, 13) = (-7, -13, 1)$ is normal to the tangent plane, so the tangent plane is $-7(x - 1) - 13(y - 3) + z - 25 = 0$.

Section 5.3

3. **(c)** It must be shown that

$$\lim_{(h,k)\to(0,0)} \frac{e^h(1 + k^2) - 1 - h^2 - 2k^2}{\sqrt{h^2 + k^2}} = 0.$$

To do this, show that

$$\lim_{(h,k)\to(0,0)} \frac{e^h(1 + k^2) - 1}{\sqrt{h^2 + k^2}} = 0$$

and

$$\lim_{(h,k)\to(0,0)} \frac{h^2 + 2k^2}{\sqrt{h^2 + k^2}} = 0.$$

10. Show that $z_x(0, 0)$ does not exist by arguing directly from the definition of partial derivative.

Section 5.4

2. Write $w = F(f(u, v, p), g(u, v, p), h(u, v, p), a(u, v, p))$. Then $w_u = F_x f_u + F_y g_u + F_z h_u + F_t a_u$, with similar expressions for w_v and w_p. Next,

$$\begin{aligned} w_{uu} = {} & F_x f_{uu} + F_y g_{uu} + F_z h_{uu} + F_t a_{uu} \\ & + f_u(F_{xx} f_u + F_{xy} g_u + F_{xz} h_u + F_{xt} a_u) \\ & + g_u(F_{yx} f_u + F_{yy} g_u + F_{yz} h_u + F_{yt} a_u) \\ & + a_u(F_{tx} f_u + F_{ty} g_u + F_{tz} h_u + F_{tt} a_u). \end{aligned}$$

Similar formulas hold for the other partials.

3. Compute $w_r = u_x x_r + u_y y_r = u_x \cos(\theta) + u_y \sin(\theta)$,

$$\begin{aligned} w_{rr} &= (u_{xx}\cos(\theta) + u_{xy}\sin(\theta))\cos(\theta) + (u_{xy}\cos(\theta) + u_{yy}\sin(\theta))\sin(\theta) \\ &= u_{xx} + u_{yy} + 2u_{xy}\sin(\theta)\cos(\theta), \end{aligned}$$

assuming that $u_{xy} = u_{yx}$. Next,

$$w_\theta = u_x(-r\sin(\theta)) + u_y r\cos(\theta)$$

and

$$w_{\theta\theta} = u_{xx}r^2 + u_{yy}r^2 - 2u_{xy}r^2\sin(\theta)\cos(\theta) - ru_x\cos(\theta) - ru_y\sin(\theta).$$

Now compute $w_{rr} + w_{\theta\theta}/r^2 + w_r/r$.

Section 5.6

3. (c) Compute $f_x = f_y = 0$ to be the equation

$$2x[1 + \ln(x^2 + y^2)] = 2y[1 + \ln(x^2 + y^2)].$$

This has no solution on $2 < x^2 + y^2 < 4$. On the boundary points with $x^2 + y^2 = 2$, we have $f(x, y) = 2\ln(2)$, a minimum. On each boundary point with $x^2 + y^2 = 4$, the function has a maximum value of $f(x, y) = 4\ln(4)$. [This is also the answer to 8(c).]

4. Minimize $g(x, y) = (x - 1)^2 + (y - 2)^2 + (x^2 + y^2 - 3)^2$. Note that there does not exist a maximum.

Section 5.7

4. Write

$$\begin{aligned} F(f(y), y, g(y), h(y)) &= 0, \\ G(f(y), y, g(y), h(y)) &= 0, \\ H(f(y), y, g(y), h(y)) &= 0, \end{aligned}$$

for all y in some appropriate interval. For such y,

$$F_1 f'(y) + F_2 + F_3 g'(y) + F_4 h'(y) = 0,$$
$$G_1 f'(y) + G_2 + G_3 g'(y) + G_4 h'(y) = 0,$$

and

$$H_1 f'(y) + H_2 + H_3 g'(y) + H_4 h'(y) = 0.$$

Solve these for $f'(y)$, $g'(y)$, and $h'(y)$. From these, compute the second derivatives.

Section 5.8

2. Write $F(x, f(x), g(x)) = 0 = G(x, f(x), g(x))$. Then

$$dF = F_x\,dx + F_y\,dy + F_z\,dz = 0 \quad \text{and} \quad dG = G_x\,dx + G_y\,dy + G_z\,dz = 0.$$

Substitute $dy = f'(x)\,dx$ and $dz = g'(x)\,dx$, then solve for $f'(x)$ and $g'(x)$.

Section 5.9

5. Let $g(x, y, z) = ayz + bzx + cxy - abc$ and $f(x, y, z) = xyz$, and work with $f - \lambda g = h$. Solve $h_x = h_y = h_z = 0$, to obtain

$$\lambda = \frac{yz}{cy + bz} = \frac{xz}{az + cx} = \frac{xy}{cy + bx}.$$

Use these to obtain $y = bx/a$ and $z = cx/a$. Put these into the constraint equation $g(x, y, z) = 0$ to obtain $x = a$, hence conclude that $y = b$ and $z = c$. Now show that these values produce a maximum, not a minimum.

Section 6.1

1. (a)

$$\int_C (x, y, z) = \int_2^3 (t^2, 2t, 0) \cdot (2t, 2, 0)\, dt = \int_2^3 (2t^3 + 4t)\, dt.$$

2. (c) The line is $x = 6t$, $y = 1 + 7t$, $z = 4t$, $t: 0 \to 1$. Then

$$\int_C a\, dx + b\, dy + c\, dz = \int_0^1 a(6)\, dt + \int_0^1 b(7)\, dt + \int_0^1 c(4)\, dt$$
$$= 6a + 7b + 4c.$$

(e) Write C as $C(x) = (x, x^2, 0)$, $x: 0 \to 2$. Then

$$\int_C z\, dx + x\, dy + y\, dz = \int_0^2 0\, dx + x(2x)\, dx + x^2\, d(0) = \int_0^2 2x^2\, dx = \tfrac{16}{3}.$$

Section 6.3

1. (c)

$$\int_0 p\, dx = \int_0^{\pi/2} [\cos^2(t) + \sin^2(t)]^{1/2}[-\sin(t)]\, dt,$$

and

$$\int_C p\, ds = \int_0^{\pi/2} [\cos^2(t) + \sin^2(t)]^{1/2}\{[-\sin(t)]^2 + [\cos(t)]^2 + (2t)^2\}^{1/2}\, dt$$
$$= \int_0^{\pi/2} \sqrt{1 + 4t^2}\, dt.$$

Section 6.5

1. (a) For $C(t) = (\cos(t), \sin(t))$, $t: 0 \to 2\pi$, we have $\eta(t) = (\cos(t), \sin(t))$. Then $dG/d\eta = \nabla G \cdot \eta = 2x\cos(t) + 2y\sin(t)$. Then

$$\int_C F \frac{dG}{d\eta}\, ds = \int_0^{2\pi} \cos(t)\sin(t)[2\cos^2(t) + 2\sin^2(t)]\sqrt{\sin^2(t) + \cos^2(t)}\, dt = 0.$$

Next, $\iint_D F\,\nabla^2 G = \iint_D xy(2+2) = 4\int_0^{2\pi}\int_0^1 r\cos(\theta)\, r\sin(\theta)\, r\,dr\,d\theta = 0$, and $\iint_D \nabla F \cdot \nabla G = \iint_D (y, x)\cdot(2x, 2y) = 0$.

7. Write $C_r(t) = (a + r\cos(t), b + r\sin(t))$, $t: 0 \to 2\pi$. Then

$$\frac{1}{2\pi r}\int_{C_r} u\,ds = \frac{1}{2\pi r}\int_0^{2\pi} 2(a + r\cos(t))(b + r\sin(t))\sqrt{r^2}\,dt$$

$$= \frac{1}{2\pi r}\int_0^{2\pi} [2ab + 2ar\sin(t) + 2br\cos(t) + 2r^2\sin(t)\cos(t)]\,r\,dt$$

$$= \frac{1}{2\pi}(2ab)(2\pi).$$

10. (c)

$$\iint_D \frac{x}{y+10} = \int_2^4\int_0^{2\pi} \frac{r\cos(\theta) r\,dr\,d\theta}{10 + r\sin(\theta)}.$$

Now,

$$\int_0^{2\pi} \frac{\cos(\theta)\,d\theta}{10 + r\sin(\theta)} = \frac{1}{r}\ln(10 + r\sin(\theta))\Big|_0^{2\pi} = 0,$$

so

$$\iint_D \frac{x}{y+10} = 0.$$

12. Put $u = x - y$, $v = x + y$. The region D in the xy plane is mapped to the region D^* in the uv-plane bounded by $v = 2$, $v = 4$, $u = -v$, $u = v$. Now the Jacobian is computed directly to be $\frac{1}{2}$, so

$$\iint_D \cos\left(\frac{x-y}{x+y}\right) = \iint_{D^*} \tfrac{1}{2}\cos\left(\frac{u}{v}\right) = \int_3^4\int_{-v}^{v} \tfrac{1}{2}\cos\left(\frac{u}{v}\right) du\,dv$$

$$= \sin(1)\int_2^4 v\,dv = 6\sin(1).$$

18. (e) By direct computation, $Q_x = P_y$ in (simply connected) D, so $\int_C P\,dx + Q\,dy$ is independent of path in D. To find F with $\nabla F = (P, Q)$, write $F_x = P$ to obtain $F(x, y) = \sin(x)\cos(y) + (x^3/3)y + xe^y + \varphi(y)$. Put $F_y = Q$ to obtain $\varphi'(y) = 0$, so we may choose $\varphi(y) = 0$.

Section 6.7

1. (a)

$$\eta(u, v) = \left(\begin{vmatrix} 0 & 2 \\ 2u & 6v\end{vmatrix}, \begin{vmatrix} 2u & 6v \\ 1 & 0\end{vmatrix}, \begin{vmatrix} 1 & 0 \\ 0 & 2\end{vmatrix}\right) = (-4u, -6v, 2).$$

The tangent plane at any point $(x_0, y_0, z_0) = \Sigma(u_0, v_0)$ is then $-4u_0(x - x_0) - 6v_0(y - y_0) + 2(z - z_0) = 0$. The surface in rectangular coordinates is $z = x^2 + \frac{3}{4}y^2$, for $x^2 + (y^2/4) \le 2$.

Look at a typical point, say $\Sigma(\frac{1}{2}, \frac{1}{2}) = (\frac{1}{2}, 1, 1)$. The normal at this point is $\eta(\frac{1}{2}, \frac{1}{2}) = (-2, -3, 2)$, which points away from the inside of Σ. Thus the positive side of Σ is the inside (the graph of Σ is shaped like a bowl). To find the area of Σ, compute $\iint_D 2\sqrt{4u^2 + 9v^2 + 1}$, where D is the region $u^2 + v^2 \leq 2$. Use polar coordinates to obtain the answer $(2\pi/9)(2^{3/2} - 1)$.

Section 6.8

1. (c)

$$\iint_\Sigma F = \iint_D [e^u, \sin(v), v + u - 3] \cdot (-1, -1, 1)$$

$$= \iint_D [-e^u - \sin(v) + v + u - 3],$$

where D is the region $0 \leq u \leq 1$, $0 \leq v \leq 1$.

2. (b) $\iint_\Sigma f\,dx\,dy = \iint_\Sigma (0, 0, f) = \iint_D [0, 0, \sin(x)y(x^2 + y^2)] \cdot (-2x, -2y, 1)$, where D is the region $x^2 + y^2 \leq 2$.

Section 6.10

1. (a) $\eta(u, v) = (-2u, -2v, 1)$, so

$$\iint_\Sigma y^2x^2z^2 = \iint_D v^2u^2(u^2 + v^2)^2\sqrt{4u^2 + 4v^2 + 1}\,du\,dv,$$

where D is the region $u^2 + v^2 \leq 1$. Use polar coordinates to evaluate this integral.

Section 6.12

1. Consider $F(x, y, z) = (x + y + z, xz, y^2)$ and $\Sigma(u, v) = (u, v, u^2 + v^2)$, $0 \leq u \leq 1$, $0 \leq v \leq 1$. Now, curl $F = (2y - x, 1, z - 1)$ and $\eta(u, v) = (-2u, -2v, 1)$. Then

$$\iint_\Sigma \text{curl } F = (2v - u, 1, u^2 + v^2) \cdot (-2u, -2v, 1)\,du\,dv = -\tfrac{5}{3}.$$

Next, write the boundary C of D in four pieces:

$$C_1(u) = (u, 0), \quad u: 0 \to 1;$$
$$C_2(v) = (1, v), \quad v: 0 \to 1;$$
$$C_3(u) = (u, 1), \quad u: 1 \to 0;$$
$$C_4(v) = (0, v), \quad v: 1 \to 0.$$

Then $\partial\Sigma$ is in four pieces:

$$\partial\Sigma_1(u) = \Sigma(C_1(u)) = \Sigma(u, 0) = (u, 0, u^2), \quad u: 0 \to 1;$$
$$\partial\Sigma_2(v) = \Sigma(1, v) = (1, v, 1 + v^2), \quad v: 0 \to 1;$$
$$\partial\Sigma_3(u) = \Sigma(u, 1) = (u, 1, 1 + u^2), \quad u: 1 \to 0;$$
$$\partial\Sigma_4(v) = \Sigma(0, v) = (0, v, v^2), \quad v: 1 \to 0.$$

Compute

$$\int_{\partial\Sigma_1} F = \int_0^1 (u + u^2, u^3, 0) \cdot (1, 0, 2u)\, du = \tfrac{5}{6},$$

$$\int_{\partial\Sigma_2} F = \int_0^1 (2 + v + v^2, 1 + v^2, v^2) \cdot (0, 1, 2v)\, dv = \tfrac{11}{6},$$

$$\int_{\partial\Sigma_3} F = \int_1^0 (2 + u + u^2, u + u^3, 1) \cdot (1, 0, 2u)\, du = -\tfrac{23}{6},$$

$$\int_{\partial\Sigma_4} F = \int_1^0 (v + v^2, 0, v^2) \cdot (0, 1, 2v)\, dv = -\tfrac{1}{2}.$$

Add these to obtain

$$\int_{\partial\Sigma} F = -\tfrac{5}{3}.$$

Section 6.13

4. **(b)** To have

$$\frac{\partial\Phi}{\partial x} = yz + \frac{1}{x + y + z},$$

we need $\Phi(x, y, z) = xyz + \ln(x + y + z) + \varphi(y, z)$. Next we need

$$\frac{\partial\Phi}{\partial y} = xz + \frac{1}{x + y + z} + \frac{\partial\varphi}{\partial y} = e^x z + \frac{1}{x + y + z},$$

so $\partial\varphi/\partial y = xz - e^x z$, impossible if φ is a function of y and z only. Thus $\int_C F$ is not independent of path. Alternatively, show that curl $F \neq (0, 0, 0)$, noting that D is simply connected.

Section 6.14

1. Put $F(x, y, z) = (x, y, z^2)$ and $\Sigma(\theta, \varphi) = [2\cos(\theta)\sin(\varphi), 2\sin(\theta)\sin(\varphi), 2\cos(\varphi)]$, for $0 \leq \theta \leq 2\pi$ and $0 \leq \varphi \leq \pi$. Then div $F = 2 + 2z$ and

$$\iiint_{V(\Sigma)} \text{div } F = \int_{-\sqrt{2}}^{\sqrt{2}} \int_{-\sqrt{2-x^2}}^{\sqrt{2-x^2}} \int_{-\sqrt{2-x^2-y^2}}^{\sqrt{2-x^2-y^2}} (2 + 2z)\, dz\, dy\, dz = \frac{64\pi}{3}.$$

Next,

$$\iint_{\Sigma} F = \int_0^{\pi} \int_0^{2\pi} [2\cos(\theta)\sin(\varphi), 2\sin(\theta)\sin(\varphi), 4\cos^2(\varphi)]$$

$$\cdot [4\cos(\theta)\sin^2(\varphi), 4\sin(\theta)\sin^2(\varphi), 4\sin(\varphi)\cos(\varphi)]\, d\theta\, d\varphi$$

$$= \int_0^{\pi} \int_0^{2\pi} [8\sin^3(\varphi) + 16\sin(\varphi)\cos^3(\varphi)]\, d\theta\, d\varphi = \frac{64\pi}{3}.$$

(Note that the outer normal was used throughout the calculation.)

Section 6.15

4. The cone $z^2 = x^2 + y^2$ intersects the sphere $x^2 + y^2 + z^2 = 1$ (in the region $z \geq 0$) in the circle $x^2 + y^2 = \frac{1}{2}$, $z = \frac{1}{2}$. This projects in the xy-plane onto the circle $x^2 + y^2 = \frac{1}{2}$. The region V inside the sphere and outside the cone, in $z \geq 0$, consists of two regions, which can be described as follows:

$$V_1: \quad x^2 + y^2 \leq \tfrac{1}{2}, \qquad 0 \leq z \leq \sqrt{x^2 + y^2},$$

$$V_2: \quad x^2 + y^2 \geq \tfrac{1}{2}, \qquad 0 \leq z \leq \sqrt{1 - x^2 - y^2}.$$

Then $\iiint_V z = \iiint_{V_1} z + \iiint_{V_2} z$. Use cylindrical coordinates:

$$\iiint_V z = \int_0^{2\pi} \int_0^{1/\sqrt{2}} \int_0^r zr\, dz\, dr\, d\theta = \frac{\pi}{16},$$

$$\iiint_{V_2} z = \int_0^{2\pi} \int_{1/\sqrt{2}}^1 \int_0^{\sqrt{1-r^2}} zr\, dz\, dr\, d\theta = \frac{\pi}{16}.$$

Section 7.2

2. (b)

$$F'(x) = \int_{x^2}^{x^4} \frac{1}{x} \ln(xt)\, dt + 4x^2 \ln(x^5) - 2 \ln(x^3).$$

Section 7.3

1. (d) Since

$$\left| \frac{t \tanh(tx)}{x^4 + t^3} \right| \leq \frac{t}{t^3} = \frac{1}{t^2} \qquad \text{for } -\infty < x < \infty \text{ and } t \geq 1,$$

and $\int_1^\infty (1/t^2)\, dt$ converges, then the integral converges uniformly on $(-\infty, \infty)$.

4. Put

$$F(x) = \int_0^\infty \frac{dt}{1 + x^2 t^2}.$$

Convergence is uniform on any interval $[a, b]$ with $a > 0$, as

$$\frac{1}{1 + x^2 t^2} \leq \frac{1}{1 + a^2 t^2} \qquad \text{and} \qquad \int_0^\infty \frac{dt}{1 + a^2 t^2}$$

converges. Then

$$\int_a^b F(x)\, dx = \int_a^b \int_0^\infty \frac{dt\, dx}{1 + x^2 t^2} = \int_0^\infty \int_a^b \frac{dx\, dt}{1 + x^2 t^2}$$

$$= \int_0^\infty \frac{\arctan(bt) - \arctan(at)}{t}\, dt.$$

Now find $F(x)$ in closed form. We have

$$F(x) = \int_0^\infty \frac{dt}{1 + x^2t^2} = \lim_{r\to\infty} \int_0^r \frac{dt}{1 + x^2t^2}$$

$$= \lim_{r\to\infty} \frac{1}{x}\,[\arctan(rx) - \arctan(0)] = \frac{\pi}{2x}.$$

Then

$$\int_a^b F(x)\,dx = \int_a^b \frac{\pi}{2x}\,dx = \frac{\pi}{2}\ln\left(\frac{b}{a}\right).$$

Section 7.4

1. (a) First show that

$$B(x, 1 - x) = \int_0^1 \frac{t^{x-1} + t^{-x}}{1 + t}\,dt.$$

Here it is useful to show by change of variables that

$$\int_0^\infty \frac{t^{x-1}}{1 + t}\,dt = \int_0^1 \frac{t^{-x}}{1 + t}\,dt \qquad \text{for } 0 < x < 1.$$

Now conclude that

$$\Gamma(x)\Gamma(1 - x) = \int_0^1 \frac{t^{x-1} + t^{-x}}{1 + t}\,dt.$$

Expand $1/(1 + t)$ in a MacLauren series and carry out the above integration to obtain

$$\Gamma(x)\Gamma(1 - x) = \frac{1}{x} + 2x \sum_{n=1}^{\infty} (-1)^{n+1} \frac{1}{n^2 - x^2}.$$

Now use the method of Exercise 12, Section 9.10, to show that

$$\frac{1}{x} + 2x \sum_{n=1}^{\infty} (-1)^{n+1} \frac{1}{n^2 - x^2} = \pi \csc(\pi x).$$

Alternatively, take this result on faith, or use a Fourier series expansion (Chapter 10) to write

$$\cos(xt) = \frac{2x\sin(x\pi)}{\pi}\left[\frac{1}{2x^2} + \sum_{n=1}^{\infty} \frac{(-1)^n \cos(nt)}{x^2 - n^2}\right],$$

and put $t = 0$ to obtain

$$1 = \frac{2x\sin(x\pi)}{\pi}\left[\frac{1}{2x^2} + \sum_{n=1}^{\infty} (-1)^n \frac{1}{x^2 - n^2}\right],$$

then solve for $\csc(\pi x)$.

(h) Put $t = \sin^2(\theta)$ into the definition $B(x, y) = \int_0^1 t^{x-1}(1 - t)^{y-1}\, dy$ to obtain $B(x, y) = 2\int_0^{\pi/2} \sin^{2x-1}(\theta)\cos^{2y-1}(\theta)\, d\theta$. First, we want $2x - 1 = 2n$ and $2y - 1 = 0$, so choose $x = (2n + 1)/2$ and $y = 1/2$. Then

$$\tfrac{1}{2}B(n + \tfrac{1}{2}, \tfrac{1}{2}) = \int_0^{\pi/2} \sin^{2n}(\theta)\, d\theta = \frac{\tfrac{1}{2}\Gamma(n + \tfrac{1}{2})\Gamma(\tfrac{1}{2})}{\Gamma(n + 1)}$$

$$= \frac{\sqrt{\pi}}{2}\,\frac{\Gamma(n + \tfrac{1}{2})}{n!}.$$

Now express $\Gamma(n + \tfrac{1}{2})$ in terms of factorials. Similarly, choose $y = \tfrac{1}{2}$ and $x = n + 1$ to evaluate $\int_0^{\pi/2} \sin^{2n+1}(\theta)\, d\theta$.

4. Put $u = (x/a)^\alpha$, $v = (y/b)^\beta$, and $w = (z/c)^\gamma$. Then V is transformed to the region V^* bounded by $u = 0$, $v = 0$, $w = 0$, and $u + v + w = 1$. Compute directly

$$\frac{\partial(x, y, z)}{\partial(u, v, w)} = \frac{abc}{\alpha\beta\gamma}\, u^{(1/\alpha)-1} v^{(1/\beta)-1} w^{(1/\gamma)-1}.$$

Then

$$\iiint_V x^{P-1}y^{Q-1}z^{R-1} = \frac{a^P b^Q c^R}{\alpha\beta\gamma} \iiint_{V^*} u^{(P/\alpha)-1} v^{(Q/\beta)-1} w^{(R/\gamma)-1}$$

$$= \frac{a^P b^Q c^R}{\alpha\beta\gamma} \int_0^1 \int_0^{1-u} \int_0^{1-u-v} u^{(P/\alpha)-1} v^{(Q/\beta)-1} w^{(R/\gamma)-1}\, dw\, dv\, du$$

$$= \frac{a^P b^Q c^R}{\alpha\beta\gamma} \int_0^1 \int_0^{1-u} u^{(P/\alpha)-1} v^{(Q/\beta)-1} (1 - u - v)^{R/\gamma} \left(\frac{\gamma}{R}\right) dv\, du.$$

For the next integration, put $v = (1 - u)t$ to write

$$\int_0^{1-u} v^{(Q/\beta)-1}(1 - u - v)^{R/\gamma}\, dv$$

$$= \int_0^1 (1 - u)^{(Q/\beta)-1} t^{(Q/\beta)-1} (1 - u)^{R/\gamma} (1 - t)^{R/\gamma} (1 - u)\, dt$$

$$= (1 - u)^{Q/\beta + R/\gamma} B\left(\frac{Q}{\beta}, 1 + \frac{R}{\gamma}\right).$$

Then our integral is, finally,

$$\frac{a^P b^Q c^R}{\alpha\beta\gamma} \int_0^1 B\left(\frac{Q}{\beta}, 1 + \frac{R}{\gamma}\right)\left(\frac{\gamma}{R}\right) u^{(P/\alpha)-1}(1 - u)^{Q/\beta + R/\gamma}\, du$$

$$= \frac{a^P b^Q c^R}{\alpha\beta\gamma}\, B\left(\frac{Q}{\beta}, 1 + \frac{R}{\gamma}\right)\left(\frac{\gamma}{R}\right) B\left(\frac{P}{\alpha}, 1 + \frac{Q}{\beta} + \frac{R}{\gamma}\right),$$

and this reduces to the desired result upon using the formulas

$$\frac{\Gamma(x)\Gamma(y)}{\Gamma(x + y)} = B(x, y) \qquad \text{and} \qquad x\Gamma(x) = \Gamma(1 + x).$$

Section 8.2

2. Put $y = y_m + \varepsilon\eta$, where y_m is the solution function and η is twice differentiable on $[a, b]$ and $\eta(a) = \eta(b) = \eta'(a) = \eta'(b) = 0$. Put $I(\varepsilon) = \int_a^b f(x, y_m + \varepsilon\eta, y_m' + \varepsilon\eta', y_m'' + \varepsilon\eta'')\,dx$, for any such η. Then, putting $I'(0) = 0$, we have

$$\int_a^b \left(\frac{\partial f}{\partial y_m}\eta + \frac{\partial f}{\partial y_m'}\eta' + \frac{\partial f}{\partial y_m''}\eta''\right) dx = 0.$$

As before,

$$\int_a^b \frac{\partial f}{\partial y_m}\eta'\,dx = -\int_a^b \eta\frac{\partial}{\partial x}\left(\frac{\partial f}{\partial y_m'}\right) dx.$$

A similar integration by parts gives us

$$\int_a^b \frac{\partial f}{\partial y_m''}\eta''\,dx = \int_a^b \frac{\partial^2}{\partial x^2}\left(\frac{\partial f}{\partial y_m''}\right)\eta\,dx.$$

Then

$$\int_a^b \left[\frac{\partial f}{\partial y_m} - \frac{\partial}{\partial x}\left(\frac{\partial f}{\partial y_m'}\right) + \frac{\partial^2}{\partial x^2}\left(\frac{\partial f}{\partial y_m''}\right)\right]\eta\,dx = 0.$$

Now show that the conclusion of the lemma used in the section still holds if we add on the assumption that η'' exists and $\eta'(a) = \eta'(b) = 0$. Hence conclude (dropping the subscript) that the solution function to the problem must satisfy

$$\frac{\partial f}{\partial y} - \frac{\partial}{\partial x}\left(\frac{\partial f}{\partial y'}\right) + \frac{\partial^2}{\partial x^2}\left(\frac{\partial f}{\partial y''}\right) = 0, \qquad y(a) = \alpha,\ y(b) = \beta.$$

Section 8.3

1. (c) Write a path C between P_0 and P_1 as $C(t) = (\rho\cos(\theta)\sin(\varphi), \rho\sin(\theta)\sin(\varphi), \rho\cos(\varphi))$, $t: a \to b$, with $\rho = 1$ and $\theta = \theta(t)$, $\varphi = \varphi(t)$. Then the length of C is

$$\int_a^b \left[\left(\frac{d}{dt}[\cos(\theta)\sin(\varphi)]\right)^2 + \left(\frac{d}{dt}[\sin(\theta)\sin(\varphi)]\right)^2 + \left(\frac{d}{dt}[\cos(\varphi)]\right)^2\right]^{1/2} dt$$

$$= \int_a^b [\dot\theta^2\sin^2(\varphi) + \dot\varphi^2]^{1/2}\,dt = \int_a^b f(\theta, \varphi, \dot\theta, \dot\varphi, t)\,dt.$$

It is now easy to write out the two Euler–Lagrange equations for the problem.

Section 8.5

1. Putting $f(x, y, z, \dot x, \dot y, \dot z, t) = \dot x^2 + \dot y^2 + \dot z^2$, the Euler-Lagrange equations are

$$\frac{\partial}{\partial x}(f + \lambda g) - \frac{d}{dt}\frac{\partial}{\partial \dot x}(f + \lambda g) = \lambda\frac{\partial g}{\partial x} - 2\ddot x = 0,$$

$$\lambda\frac{\partial g}{\partial y} - 2\ddot y = 0, \qquad \lambda\frac{\partial g}{\partial z} - 2\ddot z = 0.$$

These can be rewritten in the form requested in the problem.

Section 9.1

4. Putting $z = a + ib$, compute

$$\frac{z}{1+z^2} = \frac{a(1-a^2-b^2) + 2ab^2 + i(1+a^2-b^2)b - 2a^2b}{(1+a^2-b^2)^2 + 4a^2b^2}.$$

Then

$$\operatorname{Im}\left(\frac{z}{1+z^2}\right) = \frac{b(1-|z|^2)}{(1+a^2-b^2)^2 + 4a^2b^2}.$$

If $|z| = 1$, then $\operatorname{Im}[z/(1+z^2)] = 0$. If $\operatorname{Im}[z/(1+z^2)] = 0$, then $b = 0$ or $1 - |z|^2 = 0$. But $b \neq 0$ by assumption, so $|z| = 1$ in this case.

Section 9.2

1. (c)

$$f(z) = \frac{1}{1-x-iy} = \frac{1-x}{(1-x)^2+y^2} + i\left(\frac{y}{(1-x)^2+y^2}\right)$$
$$= u(x, y) + iv(x, y).$$

Next, $\bar{f}(x+iy) = u(x, y) - iv(x, y)$, so $\alpha(x, y) = u(x, y)$ and $\beta(x, y) = -v(x, y)$. Finally,

$$|f(x+iy)| = \left\{\frac{(1-x)^2}{[(1+x)^2+y^2]^2} + \frac{y^2}{[(1-x)^2+y^2]^2}\right\}^{1/2}.$$

Section 9.3

1. (c) Suppose first that $z \neq 0$. Then

$$f'(z) = \lim_{h\to 0} \frac{|z+h| - |z|}{h} = \lim_{h\to 0} \frac{h\bar{h} + \bar{z}h + \bar{h}z}{h(|z+h| + |z|)}.$$

First take h real approaching zero to obtain the value $(z + \bar{z})/2|z|$ for this limit. Next take h imaginary and approaching zero to obtain $(\bar{z} - z)/2|z|$. These can be equal only if $z = -z$, or $z = 0$, a contradiction to the present case. Thus $f'(z)$ does not exist for $z \neq 0$. If $z = 0$, $f'(0) = \lim_{h\to 0} (|h|/h)$, which also does not exist. Next compute $z = u + iv = \sqrt{x^2+y^2} + i \cdot 0$. It is now easy to verify that the Cauchy–Riemann equations are satisfied nowhere.

Section 9.4

1. (b) $\cos(i) = \frac{1}{2}(e^{i^2} + e^{-i^2}) = \frac{1}{2}(e + 1/e)$, so $\operatorname{Re}(\cos(i)) = \frac{1}{2}(e + 1/e)$ and $\operatorname{Im}(\cos(i)) = 0$.

2. (d) $\sin(i^{2i}) = (1/2i)(e^{i^{2i+1}} - e^{-i^{2i+1}})$. Now

$$i^{2i+1} = e^{(2i+1)\log(i)} = e^{(2i+1)[\ln(|i|)+i\arg(i)]} = e^{-\pi-4n\pi}e^{i\pi/2}e^{2n\pi i}$$

$$= ie^{-(4n+1)\pi} \qquad \text{for integer } n.$$

Then,

$$\sin(i^{2i}) = \frac{1}{2i}(e^{ie^{-(4n+1)\pi}} - e^{-ie^{-(4n+1)\pi}}) = \sin(e^{-(4n+1)\pi}), \qquad n \text{ any integer.}$$

Section 9.5

1. (d) First, compute

$$\cosh((1+i)z) = \cos(x+y)\cosh(x-y) + i\sin(x+y)\sinh(x-y).$$

Then

$$\int_\gamma \cosh[(1+i)z]\,dz$$
$$= \int_\gamma \cos(x+y)\cosh(x-y)\,dx - \sin(x+y)\sinh(x-y)\,dy$$
$$+ i\int_\gamma \sin(x+y)\sinh(x-y)\,dx + \cos(x+y)\cosh(x-y)\,dy.$$

Now

$$\gamma(t) = \begin{cases} it, & t: 0 \to 1, \\ t-1+i, & t: 1 \to 2. \end{cases}$$

Alternatively, we can write

$$\gamma(t) = \begin{cases} (0, t), & t: 0 \to 1, \\ (t, 1), & t: 0 \to 1. \end{cases}$$

It is now easy to compute the four real line integrals above and evaluate $\int_\gamma \cosh[(1+i)z]\,dz$.

Another (simpler) method is this. Put

$$F(z) = \frac{\sinh[(1+i)z]}{1+i}.$$

Then $F' = f$, and F' is continuous in the whole complex plane. Thus

$$\int_\gamma f(z)\,dz = \frac{1}{1+i}\sinh[(1+i)z]\Big|_0^{1+i} = \frac{1}{1+i}\sinh[(1+i)^2].$$

Section 9.6

3. (b)

$$\oint_\gamma \frac{(z^3+2z^2+2)}{z-2i}\,dz = \oint_\gamma \frac{f(z)\,dz}{z-2i}$$

$$= 2\pi i f(2i), \qquad \text{where } f(z) = z^3 + 2z^2 + 2.$$

5. (a)

$$\oint_\gamma \frac{e^{z^2+2}}{(z-2)^3}\,dz = \frac{2\pi i}{2!}\frac{2!}{2\pi i}\oint_\gamma \frac{f(\xi)\,d\xi}{(\xi-2)^3} = \pi i f^{(2)}(2), \qquad \text{where } f(z) = e^{z^2+2}.$$

6. Write $P(z) = \sum_{j=0}^{\infty} a_j z^j$. Put $z = re^{i\theta}$, $0 \le r \le R$ and $0 \le \theta \le 2\pi$. Then, $\bar{z} = re^{-i\theta}$, and

$$|P(z)|^2 = P(z)\overline{P(z)} = \left(\sum_{j=0}^{n} a_j r^j e^{ij\theta}\right)\left(\sum_{j=0}^{n} \bar{a}_j r^j e^{-ij\theta}\right)$$
$$= \sum_{j=0}^{n} a_j \bar{a}_j r^{2j} + p(r,\theta),$$

where $p(r, \theta)$ contains the terms with factors $e^{ik\theta}$, $k \ne 0$. Then

$$\iint_D |P(z)|^2\,dx\,dy = \sum_{j=0}^{n} |a_j|^2 \int_0^{2\pi}\int_0^R r^{2j+1}\,dr\,d\theta + \int_0^{2\pi}\int_0^R p(r,\theta) r\,dr\,d\theta$$
$$= \sum_{j=0}^{n} |a_j|^2 2\pi\left(\frac{R^{2j+2}}{2j+2}\right),$$

since

$$\int_0^{2\pi} e^{ij\theta}\,d\theta = \begin{cases} 0 & \text{for } j \ne 0, \\ 2\pi & \text{for } j = 0. \end{cases}$$

Section 9.7

2.

$$\frac{1}{2\pi}\int_0^{2\pi} |f(re^{i\theta})|^2\,d\theta = \frac{1}{2\pi}\int_0^{2\pi}\left|\sum_{n=0}^{\infty} a_n r^n e^{in\theta}\right|^2 d\theta$$
$$= \frac{1}{2\pi}\sum_{n=0}^{\infty}\sum_{k=0}^{n} a_k \bar{a}_{n-k} r^n \int_0^{2\pi} e^{i(2k-n)\theta}\,d\theta,$$

using a Cauchy product series. See the remark in the solution to Exercise 6, Section 9.6, about $\int_0^{2\pi} e^{ij\theta}\,d\theta$ to evaluate the above series as

$$\frac{1}{2\pi}\sum_{n=0}^{\infty} a_n \bar{a}_n r^{2n}(2\pi).$$

Section 9.8

1. (e) First expand $\cos(z)$ in a Taylor series about π:

$$\cos(z) = 1 + \sum_{n=1}^{\infty} \frac{(-1)^{n+1}}{(2n)!}(z-\pi)^{2n}.$$

Now divide by $z - \pi$.

Section 9.9

1. (e) $\lim_{z\to\pi} \sin(z)/(z-\pi) = -1$. [Expand $\sin(z)$ about π to see this.] Thus $\sin(z)/(z-\pi)$ has a removable singularity at π.

2. Note that $1/\sin(1/z)$ has a singularity at $1/n\pi$ for $n = 1, 2, \ldots$, and that $1/n\pi \to 0$ as $n \to \infty$.

4. Expand f in a Taylor series: $f(z) = \sum_{n=0}^{\infty} a_n z^n$. Then $f(1/z) = \sum_{n=0}^{\infty} a_n/z^n$. Since f has a nonessential singularity at ∞, then $f(1/z)$ has a nonessential singularity at 0. This requires that, for some K, $a_n = 0$ for $n \geq K$, so f is a polynomial.

Section 9.10

3. (c) The only singularity of $\sin(3z)/(\cos(z) - \sin(z))$ inside $|z| = \pi/2$ is at $\pi/4$, and this is a simple pole [note Exercise 5(b), Section 9.9]. The residue at $\pi/4$ is

$$\left.\frac{\sin(3z)}{-\sin(z) - \cos(z)}\right|_{z=\pi/4} = -\tfrac{1}{2}.$$

The integral is then $-\pi i$.

Section 9.11

3. (b) Consider

$$\oint_{\gamma} \frac{e^{iz}\,dz}{(1+z^2)^2},$$

with γ the closed curve consisting of the real axis from $-r$ to r, and the semicircle in the upper half-plane of radius r about the origin. The integrand has poles of order 2 at i and at $-i$ (choose $r > 1$). The residue at i is

$$\lim_{z\to i} \frac{d[e^{iz}/(z+i)^2]}{dz} = \frac{-i}{2e}.$$

The residue at $-i$ is 0. Then

$$\oint_{\gamma} \frac{e^{iz}\,dz}{(1+z^2)^2} = \frac{\pi}{e}.$$

Hence we conclude that

$$\int_{-\infty}^{\infty} \frac{\cos(x)\,dx}{(1+x^2)^2} = \frac{\pi}{e}.$$

Section 9.13

5. We first need a conformal map from the region $\mathrm{Im}(z) > 1$ to the unit disc. We may construct such a map as a composition: the first step is a translation, the second uses a mapping already obtained.

$$\mathrm{Im}(z) > 1 \longrightarrow \mathrm{Im}(p) > 0 \longrightarrow |w| < 1.$$

$$p = z - i \qquad\qquad w = \frac{i(p-i)}{p+i}$$

Thus we have

$$w = \frac{i(z-i-i)}{z-i+i} = \frac{i(z-2i)}{z}.$$

Parametrize ∂D by $\partial D(t) = t + i$, $-\infty < t < \infty$. For $-\infty < x < \infty$, and $y > 1$, we then have as our solution:

$$\varphi(x, y) = \mathrm{Re}\left\{\frac{1}{2\pi i}\int_{-\infty}^{\infty} \frac{f(t)}{\dfrac{i(t+i-2i)}{t+i}} \left(\frac{\dfrac{i(t+i-2i)}{t+i} + \dfrac{i(x+iy-2i)}{x+iy}}{\dfrac{i(t+i-2i)}{t+i} - \dfrac{i(x+iy-2i)}{x+iy}}\right) \times \left(\frac{-2}{(t+i)^2}\right) dt\right\}$$

$$= \mathrm{Re}\left\{\frac{1}{2\pi i}\int_{-\infty}^{\infty} \frac{-2f(t)}{i(t-i)(t+i)} \left(\frac{\dfrac{t-i}{t+i} + \dfrac{x+iy-2i}{x+iy}}{\dfrac{t-i}{t+i} - \dfrac{x+iy-2i}{x+iy}}\right) dt\right\}.$$

After some algebra, we get:

$$\varphi(x, y) = \frac{1}{\pi}\int_{-\infty}^{\infty} f(t)\,\frac{\{(tx+1)(y-1) + (t-x)t(y-1)\}}{(t^2+1)\{(y-1)^2 + (t-x)^2\}}\,dt.$$

This simplifies further to

$$\varphi(x, y) = \frac{y-1}{\pi}\int_{-\infty}^{\infty} \frac{f(t)\,dt}{(y-1)^2 + (t-x)^2}.$$

Section 10.4

1. (a) Put $u(x, t) = A(x)B(t)$ to obtain, for some constant λ, $A'' - \lambda A = 0$ and $B'' - 4\lambda B = 0$. Show that $A(0) = A(2) = 0$ force the case $\lambda = -\alpha^2$, $\alpha > 0$, in order to obtain nontrivial solutions for A. Then $A(x) = a\cos(\alpha x) + b\sin(\alpha x)$. Since $A(0) = 0$, $a = 0$. Since $A(2) = 0$, then $b\sin(2\alpha) = 0$, so we choose $2\alpha = n\pi$, $n = 1, 2, \ldots$. Then $\alpha = n\pi/2$, so $A_n(x) = b_n\sin(n\pi x/2)$, $n = 1, 2, \ldots$. Then $B'' + n^2\pi^2 B = 0$ has solutions $B_n(t) = c_n\cos(n\pi t) + d_n\sin(n\pi t)$. Since $B'(0) = 0$, then $d_n = 0$, so $B_n(t) = c_n\cos(n\pi t)$, $n = 1, 2, \ldots$. Now try $u(x, t) = \sum_{n=1}^{\infty} h_n\sin(n\pi x/2)\cos(n\pi t)$. Then

$$u(x, 0) = \sum_{n=1}^{\infty} h_n \sin\left(\frac{n\pi x}{2}\right).$$

To find the coefficients, now expand $u(x, 0)$ in a Fourier sine series on [0, 2]. To do this, we must choose

$$h_n = \frac{2}{2}\int_0^2 f(\xi)\sin\left(\frac{n\pi\xi}{2}\right)d\xi$$
$$= \int_0^1 \xi\sin\left(\frac{n\pi\xi}{2}\right)d\xi + \int_1^2 (2-\xi)\sin\left(\frac{n\pi\xi}{2}\right)d\xi = \frac{8}{n^2\pi^2}\sin\left(\frac{n\pi}{2}\right).$$

Thus the solution is

$$u(x, t) = \frac{8}{\pi^2}\sum_{n=1}^{\infty}\frac{\sin(n\pi/2)}{n^2}\sin\left(\frac{n\pi x}{2}\right)\cos(n\pi t)$$
$$= \frac{8}{\pi^2}\sum_{n=1}^{\infty}\frac{\sin((2n-1)\pi/2)}{(2n-1)^2}\sin\left(\frac{(2n-1)\pi x}{2}\right)\cos((2n-1)\pi t).$$

(f) Consider four problems: $\nabla^2 u = 0,\ 0 < x < a,\ 0 < y < b$, with:

(1) $u(0, y) = 2y,\ u(a, y) = 0,\ 0 < y < b,\ u(x, 0) = u(x, b) = 0,\ 0 < x < a$;

(2) $u(a, y) = 2y^2,\ u(0, y) = 0,\ 0 < y < b,\ u(x, 0) = u(x, b) = 0,\ 0 < x < a$;

(3) $u(0, y) = u(a, y) = 0,\ 0 < y < b,\ u(x, 0) = x^2,\ u(x, b) = 0,\ 0 < x < a$;

(4) $u(0, y) = u(a, y) = 0,\ 0 < y < b,\ u(x, 0) = 0,\ u(x, b) = 3x,\ 0 < x < a$.

Solve these individually to obtain, respectively:

$$u_1(x, y) = \frac{4b}{\pi}\sum_{n=1}^{\infty}\frac{(-1)^{n+1}}{\sinh(n\pi a/b)}\sinh(n\pi(a-x)/b)\sin(n\pi y/b),$$
$$u_2(x, y) = \frac{4b^2}{\pi^3}\sum_{n=1}^{\infty}\frac{(-n^2\pi^2(-1)^n + 2(-1)^n - 2)}{n^3\sinh(n\pi a/b)}\sin(n\pi y/b)\sinh(n\pi x/b),$$
$$u_3(x, y) = \frac{2a^2}{\pi^3}\sum_{n=1}^{\infty}\frac{((n^2\pi^2 - 2)(-1)^{n+1} - 2)}{n^3\sinh(n\pi b/a)}\sin(n\pi x/a)\sinh(n\pi(b-y)/a),$$
$$u_4(x, y) = \frac{6a}{\pi}\sum_{n=1}^{\infty}\frac{(-1)^{n+1}}{\sinh(n\pi b/a)}\sin(n\pi x/a)\sinh(n\pi y/a).$$

The solution to the problem is then $u(x, y) = \sum_{j=1}^{4} u_j(x, y)$.

Section 10.5

1. (b) The Fourier integral expansion of $f(x)$ is

$$\frac{1}{\pi}\int_0^{\infty}\int_{-\infty}^{\infty} f(\xi)\cos(\alpha\xi)\,d\xi\cos(\alpha x)\,d\alpha + \frac{1}{\pi}\int_0^{\infty}\int_{-\infty}^{\infty} f(\xi)\sin(\alpha\xi)\,d\xi\sin(\alpha x)\,d\alpha.$$

Since $f(x) = 0$ for $x \leq 0$, and $f(x) = e^{-x}$ for $x > 0$, these become

$$\frac{1}{\pi}\int_0^{\infty}\int_0^{\infty} e^{-\xi}\cos(\alpha\xi)\,d\xi\cos(\alpha x)\,d\alpha + \frac{1}{\pi}\int_0^{\infty}\int_0^{\infty} e^{-\xi}\sin(\alpha\xi)\,d\xi\sin(\alpha x)\,d\alpha.$$

Now,

$$\int_0^\infty e^{-\xi}\cos(\alpha\xi)\,d\xi = \frac{1}{1+\alpha^2} \quad \text{and} \quad \int_0^\infty e^{-\xi}\sin(\alpha\xi)\,d\xi = \frac{\alpha}{1+\alpha^2}.$$

Then the integral expansion is

$$\frac{1}{\pi}\int_0^\infty \frac{\cos(\alpha x)\,d\alpha}{1+\alpha^2} + \frac{1}{\pi}\int_0^\infty \frac{\alpha\sin(\alpha x)\,d\alpha}{1+\alpha^2}.$$

The student can evaluate these using residue methods (see Exercises 3(a) and 7(a), Section 9.11, and Exercise 4, this section).

4. Note that these integrals arise in finding the Fourier integral expansion of $f(x)$ as given in 1(b).

Index

D

M